Geophysical Monograph 28

Magnetospheric Currents

Thomas A. Potemra
Editor

American Geophysical Union
Washington, D.C.
1984

Published under the aegis of the AGU Geophysical Monograph Board: Donald Eckhardt, Chairman; Elaine Oran, James Papike, John Schaake, and Sean Solomon, members.

Magnetospheric Currents

Based on papers presented at the Chapman Conference on "Magnetospheric Currents" held at Irvington, Virginia, April 5-8, 1983

Library of Congress Cataloging in Publication Data

Main entry under title:

Magnetospheric currents.

   (Geophysical monograph ; 28)
   Selection from papers presented at Chapman Conference on Magnetospheric Currents, held at the Tides Inn in Irvington, Va., Apr. 5-8, 1983.
     Bibliography: p.
   1. Magnetospheric currents--Congresses. 2. Plasma instabilities--Congresses. I. Potemra, Thomas A.
II. American Geophysical Union. III. Chapman Conference on Magnetospheric Currents (1983 : Irvington, Va.)
IV. Series: Geophysical monograph ; no. 28.
QC809.M35M324 1983   538'766   83-21427
ISBN 0-87590-055-0

Copyright 1984 by
American Geophysical Union
2000 Florida Avenue, N.W.
Washington, D.C. 20009

Figures, tables and short excerpts may be reprinted in scientific books and journals if the source is properly cited, all other rights reserved.

Printed in the United States of America.

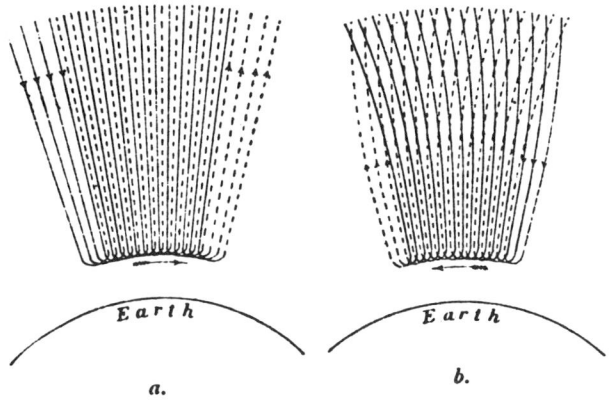

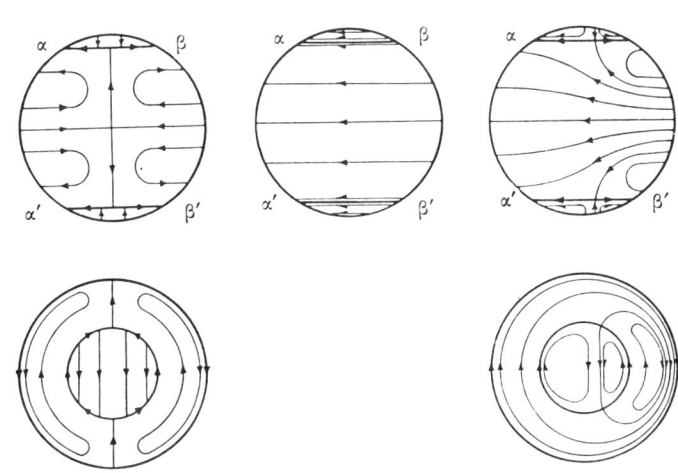

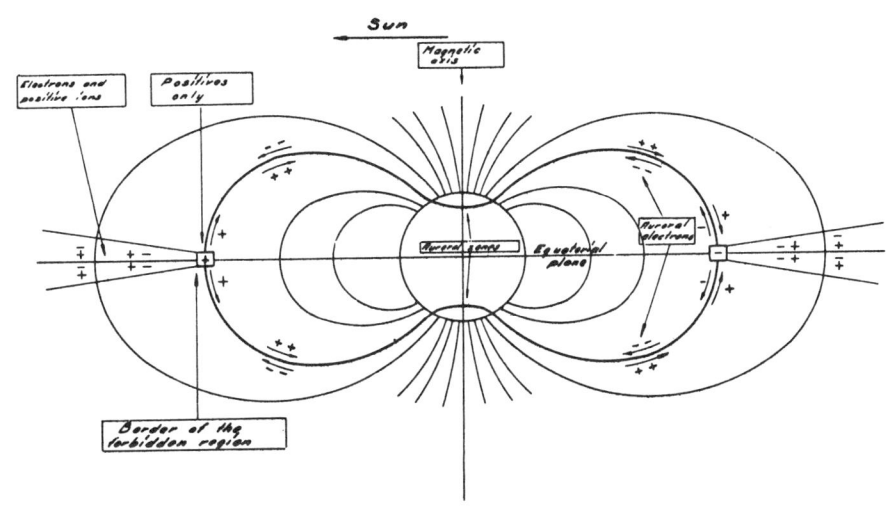

Historical Diagrams of Current Systems:

Top: The system of field-aligned currents suggested by Kristian Birkeland in 1908 (Figure 50 from "The Norwegian Aurora Polaris Expedition 1902-3, Vol. 1, On the Cause of Magnetic Storms and the Origin of Terrestrial Magnetism," by K. Birkeland, H. Aschehoug & Co., Christiana (Oslo), Norway, 1908; see the articles by Dessler and Egeland in this volume).

Center: Five figures showing the atmospheric current systems developed by Sydney Chapman in 1927 (Figures 7, 8, and 9 from "On Certain Average Characteristics of World Wide Magnetic Disturbance," by S. Chapman, *Proceedings of the Royal Society, A*, Vol. 115, pp. 242-267, 1927). The upper three figures in this panel represent the view of the terrestrial current systems as seen from the sun, and the lower two figures are the views over the north pole (with "noon" at bottom and "midnight" at the top of each) corresponding to Figures 7 and 9 (see the article by Sugiura in this volume).

ttom: The system of auroral currents proposed by Hannes Alfvén in 1939 (Figure 5 in "A Theory of Magnetic Storms and of the Aurorae," *Proceedings of the Royal Swedish Academy of Sciences*, Vol. 18, No. 3, 1939. Also published in *EOS*, Vol. 51, pp. 180-194, March 1970 with comments by A. J. Dessler and J. M. Wilcox).

CONTENTS

PREFACE ix

EARLY HISTORY

KRISTIAN BIRKELAND: THE MAN AND THE SCIENTIST 1
Alv Egeland

SYDNEY CHAPMAN AND HIS EARLY STUDIES OF MAGNETIC DISTURBANCES 17
Masahisa Sugiura

THE EVOLUTION OF ARGUMENTS REGARDING THE EXISTENCE OF FIELD-ALIGNED CURRENTS 22
A. J. Dessler

INTRODUCTION TO MAGNETOSPHERIC CURRENTS

THE MAGNETOSPHERIC CURRENTS: AN INTRODUCTION 29
S.-I. Akasofu

INTRODUCTION TO THE TOPOLOGY OF MAGNETOSPHERIC CURRENT SYSTEMS 49
W. P. Olson

FUNDAMENTALS OF CURRENT DESCRIPTION 63
Vytenis M. Vasyliunas

SURFACE OBSERVATIONS

ESTIMATION OF ELECTRIC FIELDS AND CURRENTS FROM GROUND-BASED MAGNETOMETER DATA 67
Y. Kamide and A. D. Richmond

ELECTRIC FIELDS AND CURRENTS ASSOCIATED WITH ACTIVE AURORA 77
W. Baumjohann and H. J. Opgenoorth

POLAR CAP CURRENT SYSTEMS 86
E. Friis-Christensen

NEAR-SPACE OBSERVATIONS

RELATIONSHIPS BETWEEN FIELD-ALIGNED CURRENTS, ELECTRIC FIELDS, AND PARTICLE PRECIPITATION AS OBSERVED BY DYNAMICS EXPLORER-2 96
M. Sugiura, T. Iyemori, R. A. Hoffman, N. C. Maynard,
J. L. Burch, and J. D. Winningham

A STUDY OF HIGH LATITUDE CURRENT SYSTEMS DURING QUIET GEOMAGNETIC CONDITIONS USING MAGSAT DATA 104
J. R. Burrows, T. J. Hughes, and Margaret D. Wilson

FIELD-ALIGNED CURRENTS DURING NORTHWARD IMF 115
Takesi Iijima

THREE-DIMENSIONAL BIRKELAND-IONOSPHERIC CURRENT SYSTEM,
DETERMINED FROM MAGSAT   123
    L. J. Zanetti, W. Baumjohann, T. A. Potemra, and P. F. Bythrow

VARIATION OF THE AURORAL BIRKELAND CURRENT PATTERN ASSOCIATED WITH
THE NORTH-SOUTH COMPONENT OF THE IMF   131
    P. F. Bythrow, T. A. Potemra, and L. J. Zanetti

DISTRIBUTION OF AURORA AND IONOSPHERIC CURRENTS OBSERVED
SIMULTANEOUSLY ON A GLOBAL SCALE   137
    J. D. Craven, Y. Kamide, L. A. Frank, S.-I. Akasofu,
    and M. Sugiura

DISTANT SPACE OBSERVATIONS

CURRENTS IN THE EARTH'S MAGNETOTAIL   147
    L. A. Frank, C. Y. Huang, and T. E. Eastman

CHARACTERISTICS OF THE CROSS-TAIL CURRENT IN THE EARTH'S MAGNETOTAIL 158
    A. T. Y. Lui

FIELD-ALIGNED CURRENTS NEAR THE MAGNETOSPHERE BOUNDARY   171
    Edward W. Hones, Jr.

IONOSPHERIC EFFECTS

MODELS OF AURORAL-ZONE CONDUCTANCES   180
    Patricia H. Reiff

COORDINATED GROUND AND SATELLITE OBSERVATIONS OF CONDUCTIVITIES,
ELECTRIC FIELDS, AND FIELD-ALIGNED CURRENTS   192
    R. M. Robinson

THEORY AND MODELS

MAGNETOSPHERIC DYNAMO PROCESSES   200
    David P. Stern

MAGNETOSPHERIC TOPOLOGY OF FIELDS AND CURRENTS   208
    Walter J. Heikkila

A NEW THEORY OF SOURCES OF BIRKELAND CURRENTS   223
    K. D. Cole

DIELECTRIC AND PERMEABILITY EFFECTS IN COLLISIONLESS PLASMAS   234
    K. D. Cole

FIELD-ALIGNED CURRENT SHEETS AS TANGENTIAL AND ROTATIONAL
DISCONTINUITIES   241
    G. Atkinson

ELECTRODYNAMICS OF CONVECTION IN THE INNER MAGNETOSPHERE   247
    R. W. Spiro and R. A. Wolf

COUPLING OF BIRKELAND CURRENT RINGS   260
    G. L. Siscoe and N. U. Crooker

REGION ONE BIRKELAND CURRENTS CONNECTING TO SUNWARD CONVECTING
FLUX TUBES   269
    J. L. Karty, R. A. Wolf, and R. W. Spiro

CORRECTED GEOMAGNETIC COORDINATES FOR EPOCH 1980   276
  Georg Gustafsson

HIGH-LATITUDE FIELD-ALIGNED CURRENTS PATTERNS IN CONNECTION
WITH MAGNETOSPHERIC STRUCTURE   284
  Y. I. Feldstein, R. G. Afonina, B. A. Belov, A. E. Levitin,
  D. S. Faermark, and V. Y. Gajdukov

PLASMA INSTABILITIES

ELECTRIC FIELDS AND CURRENTS OBSERVED BY S3-2 IN THE VICINITY
OF DISCRETE ARCS   294
  William J. Burke

ASSOCIATION OF FIELD-ALIGNED CURRENTS WITH SMALL-SCALE
AURORAL PHENOMENA   304
  Cynthia Cattell

THREE-DIMENSIONAL POTENTIAL STRUCTURE ASSOCIATED WITH BIRKELAND
CURRENTS   315
  Lars Block

THE ROLE OF CURRENTS IN PLASMA REDISTRIBUTION   325
  G. Atkinson

THE ROLE OF FIELD-ALIGNED CURRENT FILAMENTS IN GENERATING MORNING
SECTOR PI 1 PULSATIONS AT SUB-AURORAL LATITUDES   331
  M. J. Engebretson, S. J. Solberg, L. J. Cahill, Jr., and
  R. L. Arnoldy

CURRENT SYSTEMS IN OTHER MAGNETOSPHERES

ROTATIONALLY-INDUCED BIRKELAND CURRENT SYSTEMS   340
  T. W. Hill

DYNAMICS OF FIELD-ALIGNED CURRENT SOURCES AT EARTH AND JUPITER   350
  D. D. Barbosa

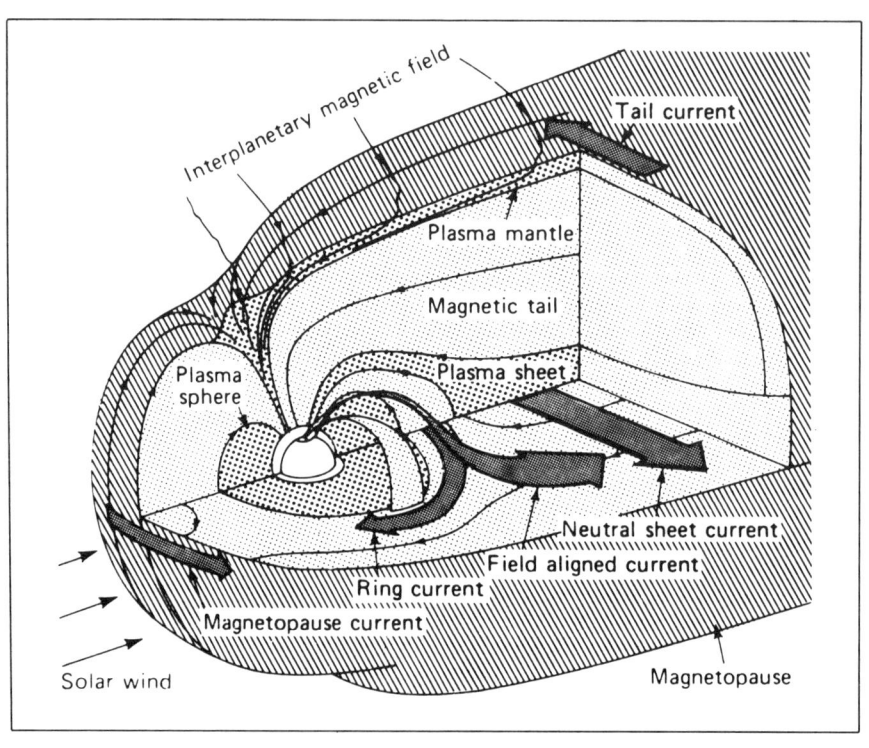

PREFACE

When viewed from outer space, the earth's magnetic field does not resemble a simple dipole but is severely distorted into a comet-shaped configuration by the continuous flow of solar wind plasma. A complicated system of currents flows within this distorted magnetic field configuration called the "magnetosphere." For example, the compression of the geomagnetic field by the solar wind plasma on the dayside of the earth is associated with a large-scale current flowing across the geomagnetic field lines, called the Chapman-Ferraro or magnetopause current. The magnetospheric system includes large-scale currents that flow in the "tail," "Birkeland" currents that flow along geomagnetic field lines into and away from the two auroral regions, the ring current that flows at high altitudes around the equator of the earth, and a complex system of currents that flows completely within the layers of the ionosphere, the earth's ionized atmosphere. The intensities of these various currents reach millions of amperes and are closely related to solar activity.

A Chapman Conference on Magnetospheric Currents was held at the Tides Inn in Irvington, Virginia, during April 5-8, 1983, for the purpose of bringing together scientists and students interested in electric currents in the earth's and other planets' magnetospheres. The knowledge in this area was reviewed and remaining questions were identified. Over 90 registrants from four continents (including countries such as Japan, the People's Republic of China, Australia, Germany, Sweden, Norway, Denmark, and Canada) participated in formal presentations, poster sessions, and informal discussions. Forty-two of the conference papers were submitted, and following a "peer review" of each paper, thirty-seven were accepted for publication in this volume.

It was very appropriate to have this conference as part of the series named in honor of Sydney Chapman because he contributed so much to our understanding of currents in the earth's ionosphere. This volume includes historical articles describing some of Chapman's work, the contributions that Kristian Birkeland and Hannes Alfvén made to an understanding of magnetospheric currents, and reflections on the scientific interactions between these three great scientists. These articles are included with the hope that some insight can be gained into the complex scientific and personal processes involved in the development of an understanding of our physical environment. It is hoped that the scientific articles included in this volume achieve the conference goals of consolidating and summarizing the present knowledge of magnetospheric currents and of stimulating new research areas.

I would like to express my gratitude to all the magnetospheric scientists who authored or reviewed the papers in this book and to the American Geophysical Union staff for their help in running the conference and publishing the book.

Financial support was provided by the National Science Foundation, National Aeronautics and Space Administration, Office of Naval Research, Air Force Geophysics Laboratory, The Johns Hopkins University Applied Physics Laboratory, Southwest Research Institute, TRW Defense and Space Systems Group, Ball Brothers/Aerospace Division, RCA Astro-Electronics, and Martin Marietta/Denver Aerospace.

T. A. Potemra

Kristian Birkeland
1867-1917
(Photo courtesy of A. Egeland, University of Oslo, Norway.)

Sydney Chapman
1888-1970
(Photo courtesy of the Geophysical Institute, University of Alaska).

Hannes Alfvén
1908-
Nobel Prize for Physics, 1970
(Photo courtesy of the Royal Institute of Technology, Stockholm, Sweden).

KRISTIAN BIRKELAND: THE MAN AND THE SCIENTIST

Alv Egeland

Laboratory for Extraterrestrial Physics, NASA/GSFC, Greenbelt, MD 20771

Institute of Physics, University of Oslo, Oslo, Norway

Background

Before reviewing Birkeland's (1867-1917) outstanding contributions to auroral theory and in particular to the foundation of modern magnetospheric physics, I will briefly summarize what was known about the physics of the northern lights and geomagnetic disturbances 100 years ago. The most important discoveries are listed in Figure 1. For further details see Brekke and Egeland, 1983.

In 1741 Hiorter and Celsius, in Uppsala, Sweden, made the important discovery that auroral occurrence and geomagnetic disturbances were closely related. Shortly afterward correspondence with Graham in London revealed that both auroras and magnetic variations occurred simultaneously over large areas. Little further progress was made in this field until Birkeland's work at the end of the last century.

While Schwabe recognized the sunspot cycle in the 1840's, it was Sabine who in the 1850's found a similar periodicity in geomagnetic disturbances. As the aurora was related to magnetic activity, the intensity and frequency of auroras were also expected to follow the same cycle. During the last part of the 19th century several physicists tried to confirm this relationship. Based on contemporary and older data (all obtained from visual observations only) on the occurrence of the northern lights, some evidence for an 11 year periodicity was indeed claimed. Furthermore, it was clearly documented by Fritz (1881) that the northern light has a maximum zone close to 23° from the magnetic poles.

The most disputed property of the aurora during the 19th century was the height, with proposed values ranging from the tree tops to well above 1000 km. The Finnish scientist Lemström thought that the northern lights were due to electric discharges between the earth and the sky. However, the two most reasonable auroral theories in the 1880's were, according to Tromholt (1885) who wrote the most penetrating works on aurora before 1890, the cosmic and the electrical theory. Several other auroral properties established by that time are listed in Figure 1, which is self-explanatory.

The latter part of this paper is devoted to glimpses of Birkeland as a human being. He was an interesting but unusual person, with several characteristics similar to those of Dr. Stockman in Ibsen's famous play "An Enemy of the People". Dr. Stockman, both intelligent and hard working, strongly advocated solutions to difficult environmental problems, in spite of public resentment which destroyed his closest family financially and politically. Only today, almost a century after Ibsen, would the public identify itself with Dr. Stockman's ideals, rather than with the views of his opponents. In several respects, the same thing has happened with Birkeland.

During his later years Birkeland started to study zodiacal light and focused on cosmological theories, but his contributions to these fields will not be discussed here. Neither will I go into any details of Birkeland's remarkable achievements in technology and applied physics, which made him both famous and rich.

Birkeland's First Years in Research

In 1885, when Birkeland was only 18 years old, he wrote a paper which was deposited in the Norwegian Academy. This paper, "Une Methode Enumerative de la Geometrie" was not published until 1914 (and then co-authored by a mathematician). Birkeland considered this work his most intellectual achievement. He also published a paper on mathematics in 1886 and another one in 1887.

Birkeland had originally planned to be a mathematician, but after a couple of years at the University he switched to theoretical physics. His mathematical training was an excellent foundation for his work on Maxwell's theory, which was his first field of endeavor in physics. He started to work with electromagnetic waves in conductors, and he later continued with wave propagation in the space.

However, his main interest and contribution was on energy transfer by means of electromagnetic waves. His general expression for the Poynting vector, derived in 1894, is still valid. In 1895 he published an important paper based on this research in which the first general solution of

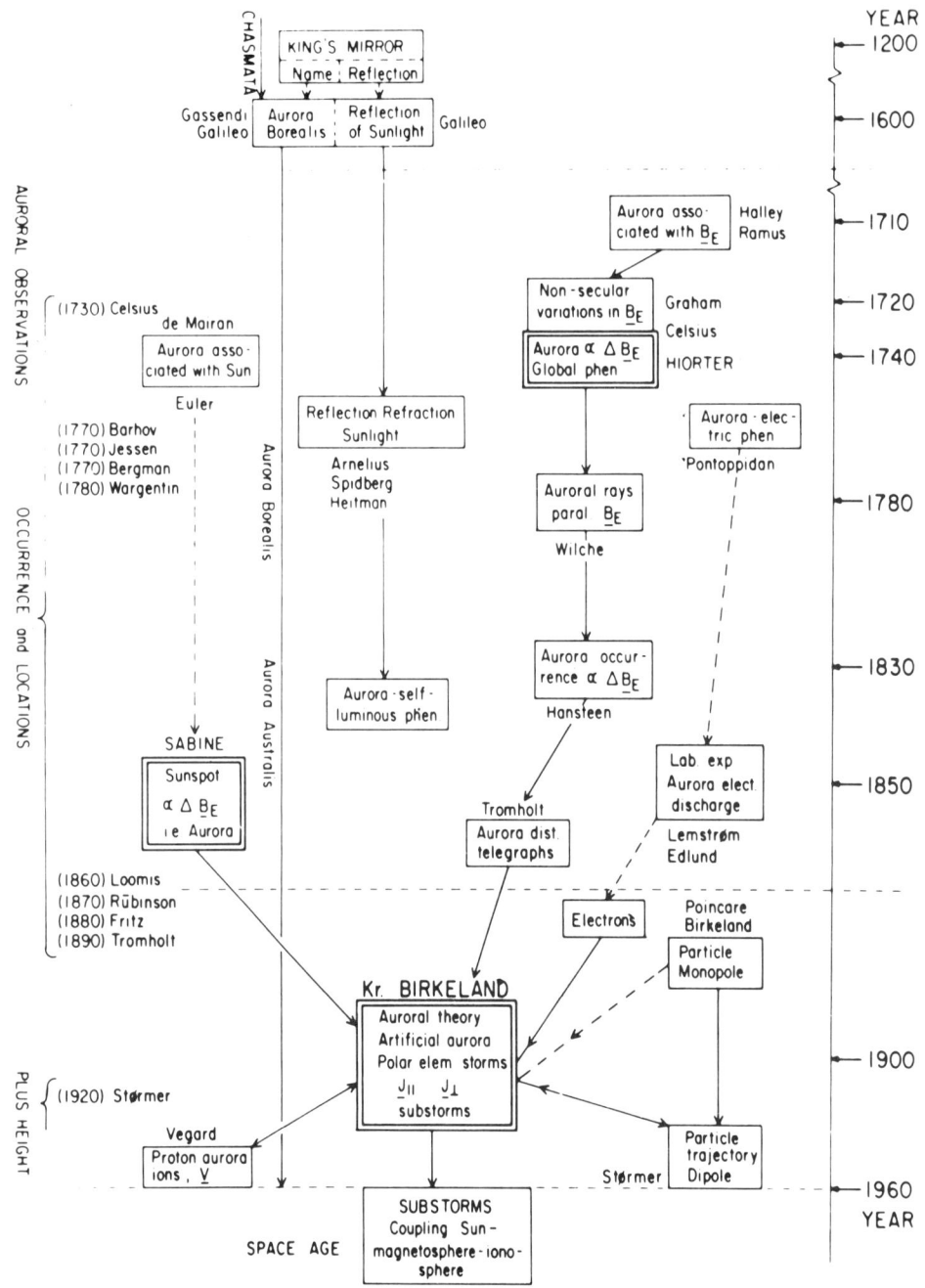

Fig. 1. Important events in the history of magnetic disturbances and aurora.

Maxwell's equations was given for a homogeneous, isotropic medium (Figure 2). This was one of the main problems in physics in the early 90's, and Birkeland's contribution was an important one. Unfortunately, this pioneer paper on wave propagation has been almost completely overlooked (Romer, 1982). His scientific interest in those early years was to a great extent concentrated on electromagnetic waves, and he continued to work on wave propagation both experimentally and theoretically long after he became internationally known as an auroral/geomagnetic expert.

Birkeland's Auroral Work

It was the discovery of x-rays and electrons in the 1890's that redirected young Birkeland's interests. These new fields of research received all his attention, and he started to experiment with them in the university course which he taught. Also, low-vacuum technology, with which he soon became acquainted, advanced significantly around this period of time.

In 1895, while experimenting with cathode rays near a magnet, Birkeland noticed that the electrons were guided towards the magnetic pole. "Maybe the aurora is produced in a similar way," he reasoned. In 1896 he extended this to an auroral theory, in a paper published by the French Academy. The main point of his theory was that energetic electrons were ejected from sunspots on the solar surface. These particles were later captured by the Earth's magnetic field and were guided to the nightside polar regions, where they produced the visible aurora. Birkeland's theory was the first detailed, realistic explanation of how an aurora was created.

In order to verify his theory, Birkeland performed extensive laboratory experiments in which the Sun-Earth system was for the first time simulated in a vacuum chamber. His model experiments were simple but at the same time ingenious. In a large (up to 1.5 m³) evacuated tank he attached a little sphere, a model of the Earth's globe which he called a terrella. Inside the terrella he mounted an electric coil which simulated the Earth's magnetic field, suitably scaled to produce similar particle orbits. He directed beams of electrons towards this simulated earth, and by coating the terrella with fluorescent paint, he was able to see where they impinged on it. Artificial aurora was produced by this terrella experiment for the first time in 1896 (Figure 3). After this initial experiment Birkeland carried out a long series of experiments, with increasing scale sizes and particle energies. Even before the turn of the century Birkeland became convinced that solar electrons were the direct cause of the aurora. In his experiments Birkeland showed that beams of electrons were always directed towards the polar regions, and that they reproduced many features of the different kinds of aurora. It is really amazing that he formulated and demonstrated his new, modern auroral theory in such a short time period.

As an extension of his terrella experiments he studied the motion of electrons in a magnetic dipole field. By interposing several coated screens in different meridian planes of the terrella, he obtained important information about the trajectories of charged particles in a dipole field. In several ways Birkeland's terrella experiments remind us of how a television picture is produced by fast electrons striking a phosphor screen. In the 1890's his experiments were among the most advanced in experimental physics. In later years, Birkeland's terrella experiments were repeated at several universities and research institutes.

Theoretically, the motion of a charged particle in a unipolar magnetic field was solved by Birkeland and Poincare before the turn of the century. Birkeland did not have time for further work in this direction and asked the young Norwegian mathematician Carl Størmer (1874-1957) to study the motions of charged particles in a magnetic dipole field. Størmer's results appeared in several papers from 1904 and onward. Birkeland was delighted with the agreement between experiment and theory (Størmer, 1955). In Figure 4 we have combined Birkeland's terrella experiments with Størmer's calculated trajectories.

Solution générale des équations de Maxwell pour un milieu absorbant homogène et isotrope;

Par M. BIRKELAND.

« Dans un Mémoire qui paraîtra prochainement dans les *Archives de Genève*, je viens d'examiner comment se développe un ébranlement électromagnétique quelconque dans un milieu homogène et isotrope ayant les coefficients d'induction électrostatique et magnétique ε et μ et la conductibilité spécifique λ.

» J'en donnerai ici les résultats en les discutant succinctement.

» Désignons par les fonctions X, Y, Z les composantes de la force électrique au point (x, y, z), à l'époque t; de même, par L, M, N les composantes de la force magnétique. Ces six fonctions, *qu'il s'agit de trouver*, dépendent alors des coordonnées x, y, z et du temps t.

» A l'époque $t = 0$, ces mêmes fonctions se réduisent à six fonctions X_0, Y_0, Z_0 et L_0, M_0, N_0, *que nous supposerons données* et qui ne dépendent donc que des trois variables x, y, z.

» Les variations que subissent les quantités X, Y, Z, L, M, N au cours du temps sont données par les équations de Maxwell qui, dans la notation de Hertz, ont la forme suivante (¹) :

$$(I) \begin{cases} A\varepsilon \frac{dX}{dt} = \frac{dM}{dz} - \frac{dN}{dy} - 4\pi\lambda AX, & A\mu \frac{dL}{dt} = \frac{dY}{dz} - \frac{dZ}{dy}, \\ A\varepsilon \frac{dY}{dt} = \frac{dN}{dx} - \frac{dL}{dz} - 4\pi\lambda AY, & A\mu \frac{dM}{dt} = \frac{dX}{dz} - \frac{dZ}{dx}, \\ A\varepsilon \frac{dZ}{dt} = \frac{dL}{dy} - \frac{dM}{dx} - 4\pi\lambda AZ, & A\mu \frac{dN}{dt} = \frac{dY}{dx} - \frac{dX}{dy}, \end{cases}$$

A étant la vitesse de la lumière dans le vide.

(¹) Hertz, *OEuvres complètes*, t. II, p. 218.
B.

Fig. 2. A short excerpt from Birkeland's 1895 paper where the first general solution of Maxwell's equations was given.

The Discovery of the Polar Elementary Storm

In order to test his auroral theory, Birkeland needed experimental data. He very efficiently organized several expeditions and established a network of high latitude magnetic stations.

In 1897 he made two trips to the auroral zone mainly to look for suitable station sites and to

Fig. 3. Birkeland in his terrella-laboratorium in 1905. His personal wealth allowed him to engage assistants and buy new equipment for his laboratory. His biggest current generator could deliver 0.3 Amp at 20 kV.

plan the logistics of longer-lasting expeditions. At that time, the best known auroral physicist in the Nordic countries was Professor Lemstrom, in Finland, who reported that auroras were generated by leakage of electrical charges from high, sharp mountain tops. This turned Birkeland's attention towards mountain-top observatories: he was also interested in atmospheric electricity and wished to investigate whether it was connected with aurora. Furthermore, he planned to determine the auroral heights by triangulation and therefore he needed two similar mountain tops separated by more than a few kilometers.

Due to these facts Birkeland and his assistants, in the winter of 1897, checked out the highest mountains in Finnmark (70° geographic north), in northern Norway. During the first trip they were trapped by terrible weather, and had to spend a day and a night in the open, in a snow storm at a temperature of −25°C. Birkeland nearly lost his life, and the trip was a failure. Later that year, however, he went back to the same area and selected two mountain tops, Haldde and Talvik, for further work (cf. Figure 5). Two small magnetic observatories were built of stone, at a height of more than 900 meters above sea level, during 1896-99. The observatories were 3.4 km apart and were connected by telephone lines. It was a great disappointment to Birkeland that the triangulation measurements of the aurora did not give reliable data.

Even though the winter storms, with wind speeds up to 40 m/s (90 miles per hour), were severe, the expedition during the winter 1899 to 1900 was still scientifically successful. The results of this appeared in Birkeland's first extensive publication on geomagnetic storms in 1901. Its subtitle was "On the Cause of Magnetic Storms and the Origin of Terrestrial Magnetism". Thus, this publication only dealt with the magnetic results.

It was during his stay at Haldde observatory in

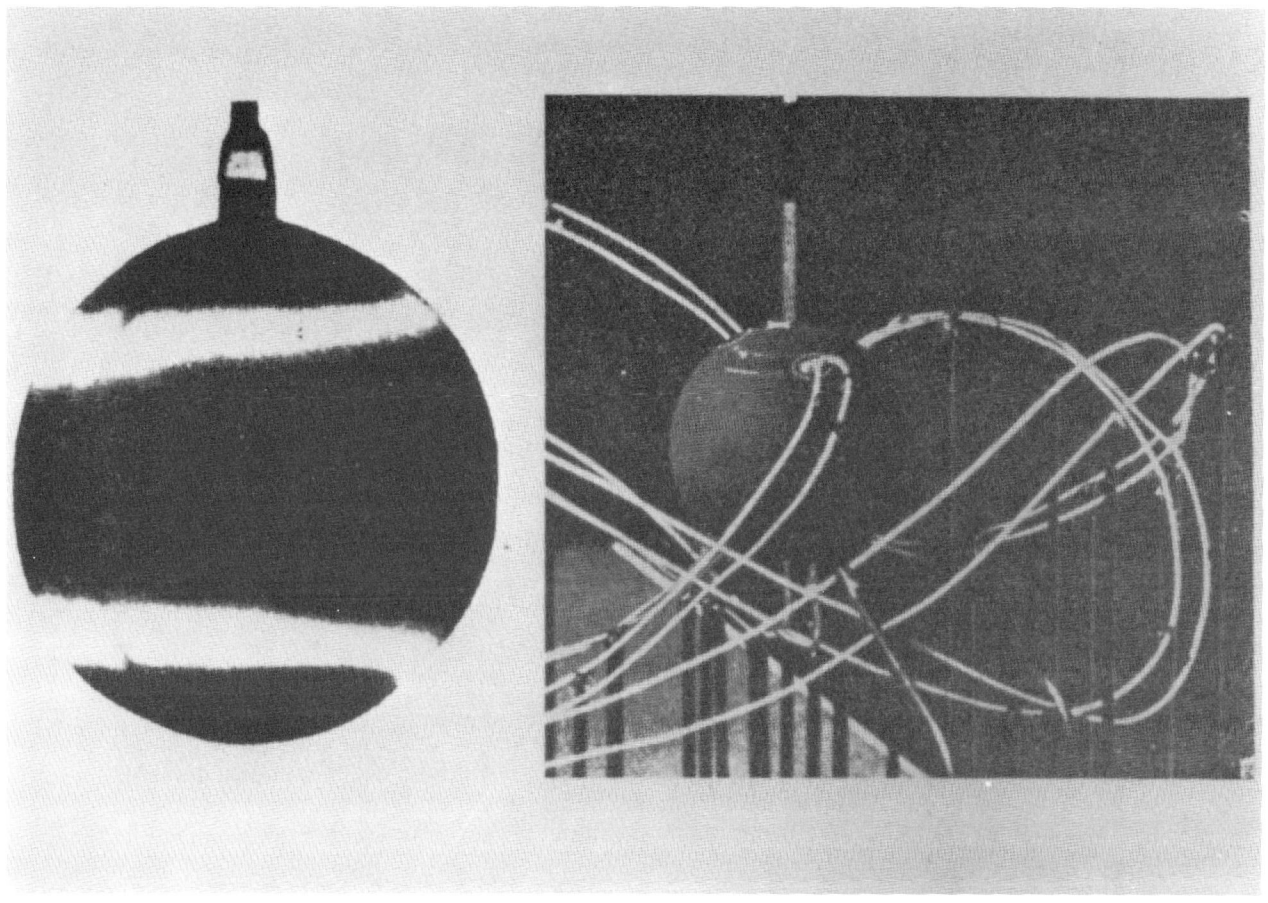

Fig. 4. A front view of the terrella showing electrons hitting the sphere in two zones in the northern and southern hemisphere similar to the auroral zones. The figure to the right shows one of Störmer's wire models displaying his calculated particle trajectories through the magnetosphere.

1899/1900 that the idea occurred to him that the same particles that produced auroras also caused geomagnetic perturbations, and that the main source for the magnetic disturbances was in the upper atmosphere. From this time on, Birkeland mainly concentrated on the study of disturbances in the Earth magnetic field. In his later work Birkeland also suggested that aurora and magnetic activity may have important meteorological connections.

Based on his first large expedition, Birkeland concluded that he needed simultaneous magnetic data from a network of high-latitude stations separated by about 1000 km from one another. It was really his second expedition, in 1902-1903, which produced comprehensive data for the studies of geomagnetic perturbations. Figure 6 shows the location of his magnetic stations.

"The Norwegian Aurora Polaris Expedition 1902-1903" was published in two large volumes (totalling 800 pages with 42 plates, not including preface and table of contents), by Birkeland. Both were subtitled "Volume I, On the Cause of Magnetic Storms and the Origin of Terrestrial Magnetism," and both the first section (in 1908) and second one (in 1913) were produced by the publisher H. Aschehoug and Company in Christiania (cf. Figure 7).

These two sections are devoted to investigations of magnetic storms by means of diurnal registrations from up to 25 observatories. Birkeland illustrated magnetic disturbances by "overhead current arrows" (Figure 8), and he reached the conclusion that world geomagnetic disturbances could be classified into the following "elementary perturbations":

a. Equatorial perturbations (positive and negative)
b. Polar elementary storms (positive and negative)
c. Cyclo-median storms

The first ones, "equatorial perturbations", are the same as the phenomena which we now call geomagnetic storms. His "cyclo-median storms" are

Fig. 5. The first observatory Birkeland built in 1899, at the 800 meter high mountain Haldde, in Finmark.

now attributed to hard solar flare photons, and they occur almost exclusively in the sunlit hemisphere. "Polar elementary storms" are now called "polar substorms": they represented an entirely new class of magnetic disturbances, associated with auroral currents in the ionosphere. They were not previously identified because they were relatively local to the auroral zone and were not at all distinct at middle latitudes, where most magnetic observatories were located. Birkeland showed that the magnetic perturbations for the simplest negative polar elementary storm could be represented by a line current above the observer.

It is really remarkable that he was able to deduce the general pattern of these geomagnetic perturbations from his early study. His explanation of polar magnetic perturbations was confirmed in his 1913 publication, by the analysis of data obtained during the First International Polar Year 1882-1883, when 10 high-latitude magnetic stations were set up around the auroral zone. For 50 years (i.e., until the IGY in 1957/58) Birkeland's collection of data remained unsurpassed in coverage and quality. His analysis of the earlier 1882-83 data supplemented his 1902-03 expedition results and helped in defining the pattern of electric current flow in high latitudes during polar elementary storms.

It must be remarked here, that his early results and ideas were completely, and unwisely, neglected for more than 40 years after his death (in 1917); that is until after the idea of "substorms" was revived in the 1960's. This may have been due to the fact that, to a large extent, polar geomagnetic studies concentrated on "average" geomagnetic disturbance morphology rather than on "individual events" as Birkeland had done. The discovery of the ionosphere in 1925 might also have influenced the geomagnetists' interest in horizontal electric currents in the ionosphere rather than in electric current streams connecting the Earth with its surrounding environment.

Birkeland was the first to succeed in obtaining a physical picture of the electric current responsible for worldwide geomagnetic perturbations (cf. next section). His two great volumes on magnetic storms (1908 and 1913) rank among the greatest work in this field ever published (cf. Figure 7).

Birkeland had planned to publish more material, particularly on earth currents and aurora as related to magnetic perturbations, as Volume II. Unfortunately, the second volume never appeared. Sadly, he died just before reaching 50 years of age.

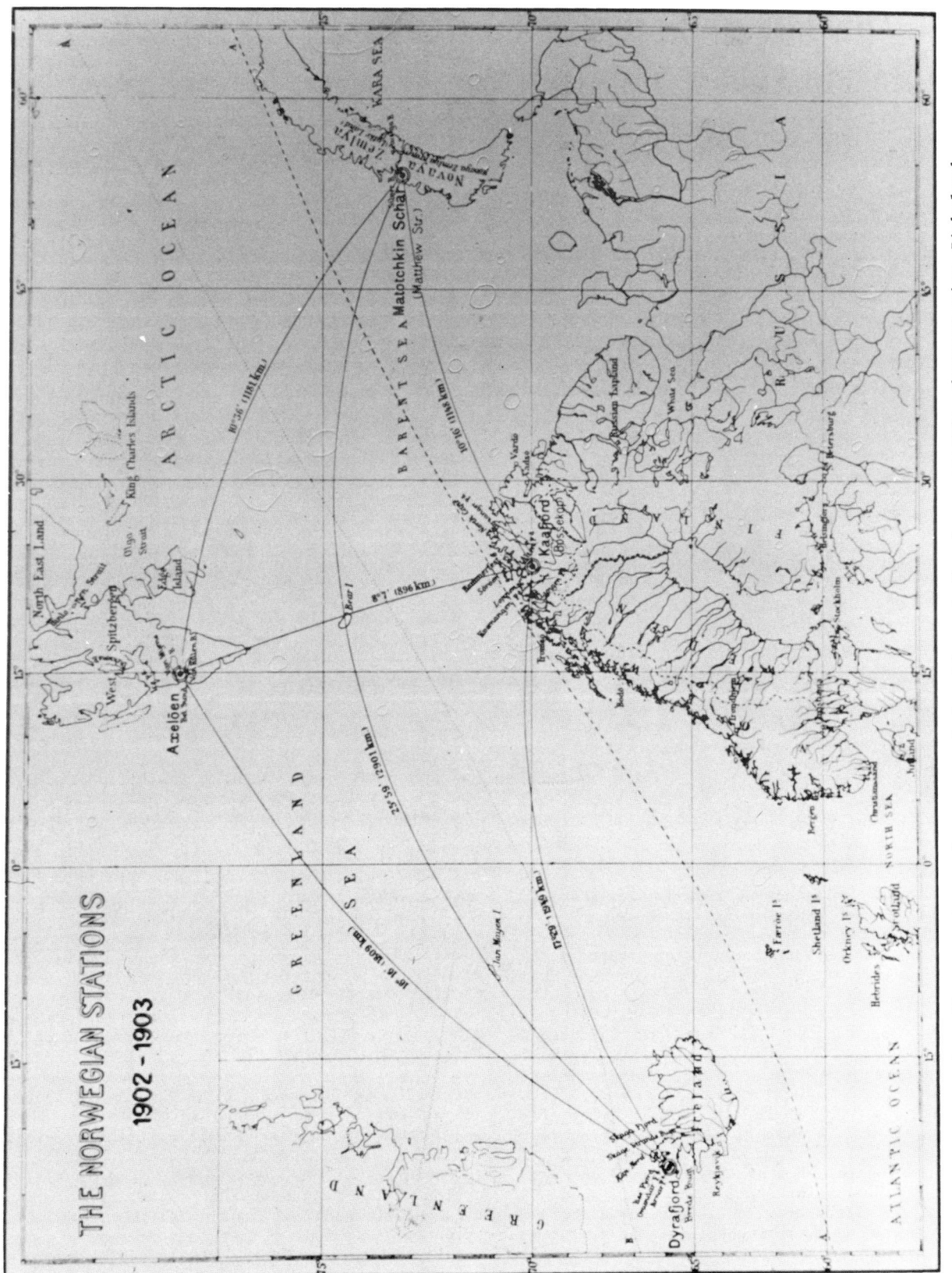

Fig. 6. The map shows the name and location of the four magnetic observatories Birkeland operated in the year 1902 - 1903.

THE NORWEGIAN AURORA POLARIS EXPEDITION 1902–1903

VOLUME I

ON THE CAUSE OF MAGNETIC STORMS AND
THE ORIGIN OF TERRESTRIAL MAGNETISM

BY

KR. BIRKELAND

FIRST SECTION

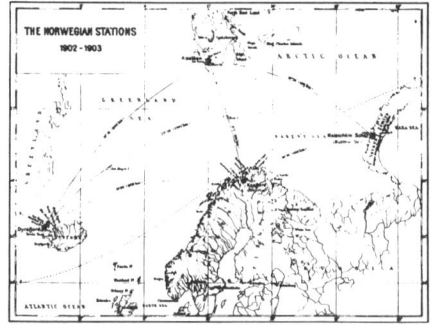

CHRISTIANIA
H. ASCHEHOUG & CO.

LEIPZIG LONDON, NEW YORK PARIS
JOHANN AMBROSIUS BARTH LONGMANS, GREEN & CO. C. KLINCKSIECK

Fig. 7. The front page of Birkeland's great work on magnetic storm studies. In the preface he wrote: "the origin of terrestrial magnetism, and the origin of the sun's heat, may be carried out upon a far wider basis than I have been able to employ, without making the expenses connected therewith too great a deterrent".

How Birkeland Explained Polar Elementary Storms
and Deduced Auroral Electric Currents

A sketch of the current system proposed by Birkeland is shown in Figure 9. The vertical currents (nowadays called Birkeland currents) follow the Earth's magnetic field lines, while the horizontal portion parallels the Earth's surface at an altitude of approximately 120 km (Birkeland estimated it at ~ 300 km). Because Birkeland currents have been a focal point of space physics during the last 10 years, we will only briefly review later developments concerning this subject (see A. Dessler's paper in these proceedings).

Birkeland concentrated on single events and used simultaneous magnetic data from more than 20 stations. To illustrate the latitudinal distribution of geomagnetic perturbations, he drew so-called current arrows on charts, as shown in Figure 8. These arrows give the direction of the horizontal current above each station, required to produce the magnetic disturbance observed there. He pointed out that such arrows did not give the real current system, but were simply and solely a geometrical representation of the perturbing pattern. Birkeland found that the same pattern was repeated almost exactly from one event to the next, and therefore viewed it as "a typical current pattern" for an elementary storm.

Based on his current model, Birkeland explained the observations shown in Figure 8 in the following way: "We should then, in this perturbation of the 15th December, have to consider the effect of a long vertical current, which, in the case of negative corpuscles, must come near to the earth at about Dyrafjord, or somewhat west of it, responding to an ascending galvanic current. A little above the surface of the earth it turns eastward, or rather the aggregate effect of the cosmic current relative to the earth is that of a galvanic current that is directed westward, or more accurately towards the south west. In this descent of electric corpuscles, some will occasionally come so near the earth that they will be partially absorbed by its atmosphere, and will then eventually give rise to aurora" (page 98).

Thus, Birkeland explained the polar elementary storm as being caused by two vertical currents, in opposite directions, connected by a horizontal current (see Figure 9). He assumed that both the rising current and the descending one were carried by particles spiralling around magnetic field lines. This assumption was based on his study of a charged particle's motion in a dipole field.

As he pointed out, this current system can explain many of the properties of the polar storm. For instance, it explains why the disturbance field in the auroral zone shows large and sudden spatial and temporal variations, while at lower latitudes the changes take place very gradually. At the auroral stations we are much closer to the disturbing current. The fact that the horizontal flow parallels the auroral zone explains why the storm intensity drops off much more rapidly transverse to the auroral zone than along the zone.

In the second part of his magnificent book, Birkeland (1913) calculated the magnetic field from such a current system and compared the results with observations.

When he proposed his current system, the ionosphere had not been discovered (Appleton detected the F-layer in 1925). It is therefore not surprising that some modifications of this original work are needed. The horizontal current is now known to be located in the E-layer and to be carried by thermal ions, not by precipitated particles as proposed by Birkeland. Furthermore, ionospheric Hall currents were unknown to him. He estimated the height of the horizontal current by triangulation without taking earth-currents into account. Thus, his height estimates were high by roughly a factor of two. It should, however, be pointed out that he concluded that "much of the

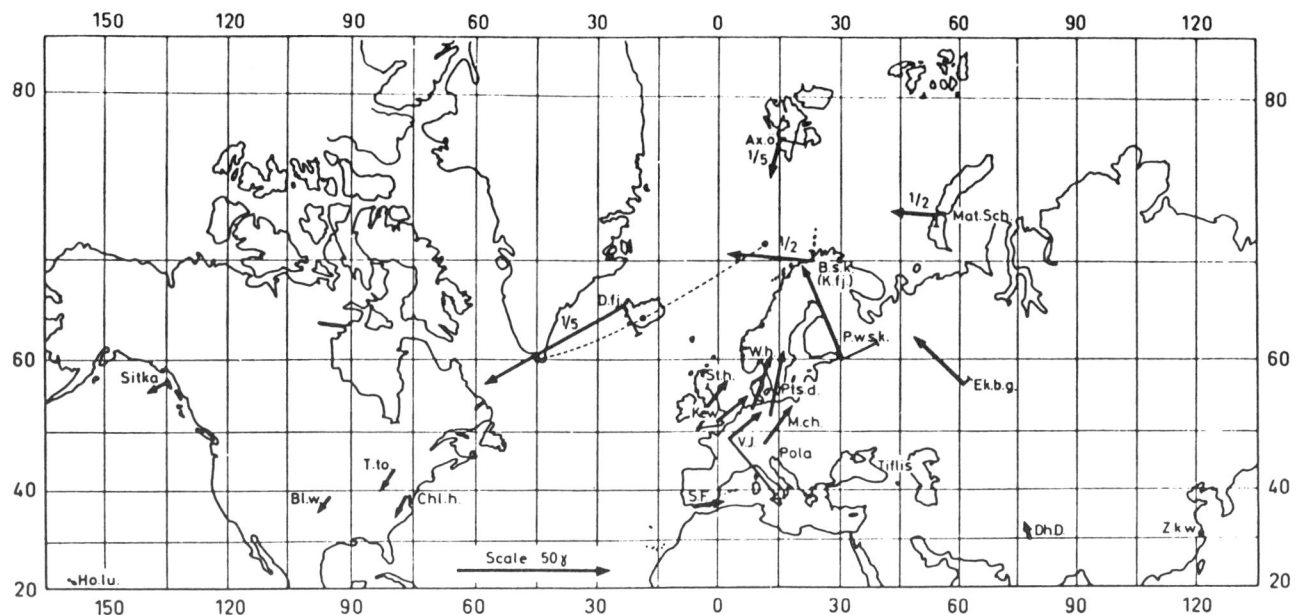

Fig. 8. Typical magnetic disturbances illustrated by overhead current arrows (cf. the text).

disagreement may be due to the earth-currents". Thus, this effect was well known to him.

In 1965, Alfvén wrote: "The current system proposed by Birkeland has been criticized, and other systems with all currents flowing horizontally in the ionosphere have been proposed. However, we will not repeat the arguments against Birkeland's model; we know that none of them are valid."

Here we will also mention that one of the first to propose the possibility of the auroral particles representing a source of heat in the upper atmosphere at high latitudes was Birkeland (1913). He concluded that the particles needed to

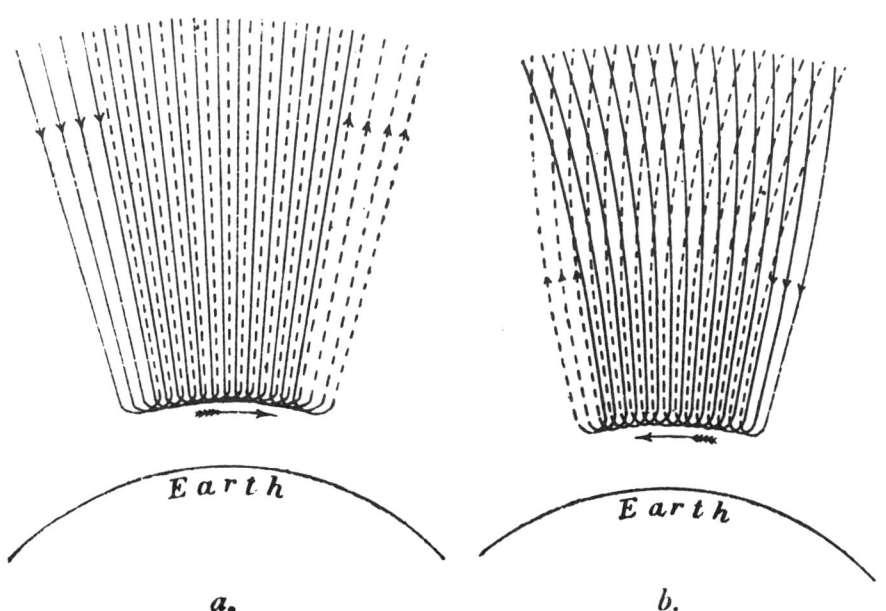

Fig. 9. Birkeland's original figure of the current system for polar elementary storms. The vertical currents which follow the Earth's magnetic field are now called Birkeland currents.

carry the currents of the order of 1 million Amperes – which he estimated to be presented in the upper atmosphere, during elementary storms as he called it, would represent a power of the order of 10^{18} ergs/sec. Compared with the modern view on this matter, Birkeland's result is rather remarkable (cf. Cole, 1966 and Akasofu, 1977) where many more references are given.

Glimpses of Birkeland as a Human Being

Birkeland's Background and Education

Birkeland's ancestors were small farmers, in a community called Birkeland, in the southernmost part of Norway. Some of them were interested in natural science but only as amateurs. Kristian's father moved to the capital of Norway, then known as Kristiania (now Oslo) and became a merchant, and Kristian was born there on December 13, 1867. At an early age Kristian showed a keen interest in physical science and mathematics, and as a young boy he carried out several experiments in both physics and chemistry. It is therefore not surprising to learn that the first thing he bought with money he had earned was a magnet.

In 1885 he started at the University of Kristiania, and during his first 2 years there he studied mathematics. Then he switched to theoretical physics and obtained his Candidatus Realium Degree (approximately equivalent to a masters' degree) in 1890, when he was only 23 years old. While at the University he taught mathematics at a local high school in order to finance his education, and he continued to work as a full-time teacher at the same school after he finished his degree. However, he also had a part-time job at the University from 1890.

In 1893 Birkeland became a full-time research assistant at the University and shortly after that he was offered a scholarship. He decided to go abroad for advanced training and study. Birkeland first visited Henri Poincaré's Institute in Paris for half a year, and there he worked mainly on the theory of propagation of electromagnetic waves in a conductor and with Maxwell's equations in general. This must have been a stimulating stay because he published four papers in the French Academy's journal "Comptes Rendus" and collaborated closely with Poincaré. During the following 15 years, Birkeland maintained his contact with the famous French scientist and through him published a long series of research notes in "Comptes Rendus." Thus, most of Birkeland's new ideas and experiments were first printed in French.

From Paris, Birkeland, in 1894, went to Switzerland, where he worked for a few months with Sarasin and de la Rive. De la Rive was interested in auroral physics and his theory was similar to Lemstörm's "electric discharge" hypothesis. Later in 1894 he went to Bonn where he studied under the well-known scientist Heinrich Hertz. There for the first time he worked on wave propagation in space. Birkeland and Hertz did not get along well, and Birkeland's stay in Bonn was cut to two months when Hertz died.

Birkeland then went back to Kirstiania and in a short time he expanded his research to a level never before seen at the University, which is probably why he was appointed a professor of physics in 1898, when only 31 years old (the average age for appointing new professors was about 50).

At this time, research grants at the Institute of Physics were limited and were entirely inadequate to finance such an activity as his. This was true even though he obtained generous additional support from the Government. In order to realize his research programs (see sections 3 and 4), he needed large financial resources, preferably ones that he personally could control.

Therefore Birkeland for a few years did some work in technology and applied physics, and he clearly demonstrated that his research had practical applications.

Birkeland made numerous contributions to practical and applied physics. He held many different patents, including those of the electromagnetic gun (see Figure 11), electric blankets, hardening of oil to produce solid margarine, hearing aids, plus others in which he applied the technology learned from his experiences with the construction of the electromagnetic furnace for the nitrogen industry.

Most Norwegians associate Birkeland's name with the Birkeland-Eyde method (Figure 10) for the production of potassium nitrate, which soon became one of Norway's most important industries. Although this was an impressive accomplishment of considerable commercial value, it was merely one episode in the bountiful life work of Birkeland. After Norsk Hydro in 1907 put this method into large-scale operation, Birkeland's financial situation greatly improved. He received roughly 10 times more money from Hydro than from his salary as a professor.

In 1960 a letter from Birkeland to the Swedish banker Marcus Wallenberg (a member of one of the richest families in Sweden) was published in Stockholm. In the letter, dated 1906, Birkeland applied for funds in order to be able to work out methods for utilizing atomic energy. Birkeland realized fully, he said, how difficult the problem was and that he might perhaps not manage it. "But I have never had such a mind to take up a thing, as I have with this problem" he says. Wallenberg considered Birkeland's idea "titanic" and alluring, however he would not invest money until the other inventions of Birkeland had yielded profit.

Birkeland strongly advocated that practically all problems could be solved by research, particularly in sciences, and he advised his co-workers to maintain close contact with applied sciences.

Birkeland's Personality

In a short, but very hectic period, Birkeland built an active research environment at the Uni-

Fig. 10. Kristian Birkeland and Sam Eyde (another man of ideas) who patented the electromagnetic furnace for producing nitrogen fertilizers. They became the founders of one of Norway's major industries.

versity. He was tireless, extremely energetic and enthusiastic, constantly involved in several major projects simultaneously. When he did not receive the financial support he asked from the Government, he used his own money both for salaries and to organize expeditions. He never spared himself. Birkeland's work was extensive and of a high caliber. He introduced innumerable ideas and theories - some fundamental and revolutionary, others utterly worthless. Birkeland followed several such ideas and tested them out in detail, and a surprising number of them turned out to be correct. In all his work, he had a creative and lively imagination and an enormous faculty for intelligently putting pieces of information together to obtain correct solutions to problems (Figure 11).

Whenever Birkeland started new projects (which he did often) he was extremely enthusiastic, and he worked hard both days and nights. To all his work, he brought a creative and lively imagination along with a good sense of humor and self-irony.

Birkeland was intensely interested in auroral and geomagnetic studies, and he devoted great efforts to increase our knowledge about these phenomena. He employed several young, enthusiastic scientists to work for him, among whom were Vegard, Krogness, Saeland, Skolem, and Devik. His direct and indirect influence was great. Through his many projects, Birkeland awakened an interest in auroral research in these young colleagues, and he inspired and stimulated them to do their very best.

In 1903 Birkeland showed some of his terrella experiments to his colleague, C. Störmer. Knowing that Poincaré had calculated the trajectories of

Fig. 11. Birkeland demonstrated his electromagnetic cannon in the festival hall at the University. Birkeland relates how the official demonstration ended when he used an extra-large model of the cannon: "It was at the University's old banquet hall on January 6, 1903. The cannon was placed in the hall and pointed toward the target which was a three-inch-thick plank wall of solid wood. I had closed off the space on both sides of the projectile's track, but except for this area the hall was full. In the first section of seats were representatives from Armstrong and Krupp--the large weapon forger in Europe. I went through the principles on which the cannon was based and 'Ladies and Gentlemen,' I said, 'you may calmly be seated. When I push the switch on, you will neither see nor hear anything except the bang of the projectile against the target. With this I pushed the switch. There was a flash, a deafening and hissing noise, a bright arc of light due to three thousand amperes being short circuited and a flame shot out of the mouth of the cannon. Some of the ladies shrieked and a moment later there was a panic. It was the most dramatic moment in my life--with this show, I shot my stock down from a value of 300 to zero. But the projectile hit the bull's eye." This unsuccessful demonstration illustrates his most distinctive characteristic. His creative imagination immediately saw the meaning of the unexpected result of this experiment. This enormous artificially created lightning served as the idea that created Norsk Hydro's nitrogen fertilizer industry.

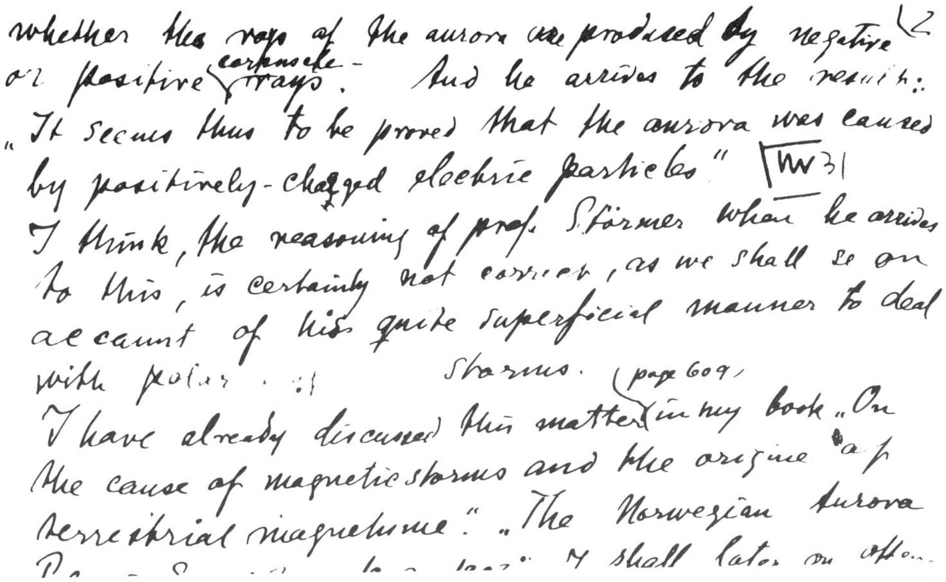

whether the rays of the aurora are produced by negative or positive corpuscule rays. And he arrives to the result:
"It seems thus to be proved that the aurora was caused by positively-charged electric particles" [MW 31]
I think, the reasoning of prof. Störmer when he arrives to this, is certainly not correct, as we shall see on account of his quite superficial manner to deal with polar ... Storms. (page 609)
I have already discussed this matter in my book "On the cause of magnetic storms and the origine of terrestrial magnetisme". "The Norwegian Aurora ...

Fig. 12. A short excerpt from one of Birkeland's hand-written manuscripts showing his style.

cathode rays in the case of a single pole, Störmer found it most interesting to try to find the trajectories in the case of a magnetic dipole. This started the famous research work of Störmer on the Polar Aurora (Störmer, 1955).

The idea of exercising as a means of staying in good physical shape was strange to Birkeland. He worked far too hard (both days and nights) and was careless concerning regular meals. Birkeland was in his 30's when he married, but he lived with his wife for only one year, and they had no children. His scientific experiments in the 1890's caused him serious hearing defects, and his general health deteriorated rapidly in his 40's.

Birkeland moved to Egypt in 1913 to study the zodiacal light (he had been down there on shorter trips earlier). However, another equally important reason for his move to Egypt was his health. Birkeland died June 15, 1917, at a time when a working committee was in the process of nominating him for the Nobel Prize in Physics.

At the University commemoration of Birkeland's death, Vice-Chancellor Saeland characterised Birkeland as a scientific explorer by the grace of God. In the eyes of many Norwegians he had an ideal life: he had become both rich and famous in his lifetime.

In fact Birkeland had many problems both in his scientific and private life. He had few close friends, though those he had really loved him. He had many antagonists and was often attacked in papers and by colleagues.

What Was He Really Like?

Unfortunately, none of his friends and co-workers have written in any detail about Birkeland as a human being. Thus, the following characteristics are based on what I have found in scattered notes and on talks with Dr. Harang and Dr. Devik, his last assistant. Devik would never say anything derogatory about anybody, so his information

Fig. 13. A cartoon in an Oslo newspaper from 1906. The text is "Birkeland shot the parrot", and is an illustration to his unsuccessful demonstration (see Figure 11). This clearly illustrates that Birkeland and his work were well-known, and that the papers were not gracious on this occasion. Birkeland took his misfortune with graceful humor and self-irony.

Fig. 14. A picture of the small hotel Seiyoken in Tokyo where Birkeland died June 15, 1917. The hotel burned down in the 1930's.

tends to present only one side of Birkeland.

Birkeland was absent-minded and very disorganized in his daily life. He never kept notebooks and to a large extent trusted his memory. When the University officials asked him to document some of his expenses he replied "I remember the sum." He left in many different places small notes, jotted on single sheets of paper, with new ideas or with the budget for the next expedition, everything noted down in a hurry. He wrote few letters, but sent many telegrams to assistants and colleagues. For several years I have tried to compile a complete list of his publications, but I would not be surprised if some items belonging there are still missing. He also wrote several articles, on different subjects, in public journals and newspapers, and those have still not been collected in any systematic way (Figure 12).

Birkeland's lectures as a university teacher were often surprising to the students. Olaf Devik appropriately describes this in his book "Amidst Fisherman, Physicists and other People": "Birkeland had little time for lectures, but when he occasionally lectured on a subject of which he was especially fond, he brought a breath of fresh air into the classroom which was not the usual staid classroom atmosphere. Thus, he operated scarce electrical lecture equipment far beyond its rated capacity and burned out 100 ampere fuses with dignified nonchalance. Then he would stop and in a royal manner untie the ruffles of his ermine jacket and dry his glasses in order to better see his last miscalculation on the blackboard." Thus, Birkeland was impressively dressed when he lectured to the students. However, he very often (for many years) neglected teaching at the University. Out of his own pocket he paid an assistant or colleague to take care of his duty as a teacher.

The correspondence between the Dean and/or the Vice-Chancellor and Birkeland clearly indicates that he ignored their authority. He once informed them "that I have taken half of the lecture hall for a new laboratory. By putting the students closer together there is enough space in the reduced hall."

He did not like the formal criteria (based only on the numbers of publications) by which the University appointed new professors, and during votes he was normally in the minority. His spirit was independent and did not accept authority. He was also not afraid to disagree with experts in

Fig. 15. Birkeland's commemorative plaque at the cemetery for famous Norwegians, placed there by the University. The following is a translation of the text on the stone: He combined atmospheric nitrogen in his electromagnetic furnace. He investigated the nature of the northern light, the sun's radiations, and the Earth's magnetic field.

fields where he himself was no expert.

In several respects he was childishly naive. Normally people don't publicise their latest idea or thought, before they had time to think it through, but Birkeland could not keep any secrets (not even his innermost thoughts). In this way he was in the "center" of many public controversies-- started either when he published a new idea, or when some journalist asked for his comments or points of view on controversial matters in a field in which he was not an expert. By doing this he laid himself open to arguments, quarrels and disagreements, even among nonscientists. On the other hand, Birkeland disliked to be attacked officially (Figure 13).

Several colleagues within his faculty were envious of Birkeland because he was rather generously supported by the government.

The last 2-3 years of Birkeland's life (while he stayed in Egypt) were particularly difficult. At the start of World War I in 1914, his Norwegian assistants were asked to return to Norway for military service, and after that he felt isolated and lonely. His health deteriorated rapidly. In addition he suddenly felt he was a suspicious foreigner in Egypt, constantly watched. He became unsure of himself and nervous, and he slept poorly. In 1917 he therefore decided that he would return to Norway and also would celebrate his 50th birthday in Oslo. Owing to the war, he had to travel a long way, via the Far East and the USSR, to get safely back. He died in Tokyo less than halfway home (cf. Figure 14).

In conclusion, it should be stated that Birkeland, during the period 1893-1913, contributed greatly to the study of the polar aurora and of the Earth's magnetosphere. He introduced many ideas which still remain central to that area; for instance, particle motion in a dipole field, magnetic substorms, and the idea of large electric currents flowing into and out of the upper atmosphere, now known in his honor as Birkeland currents. Although much of this work remained unrecognized for many years, it was truly the foundation of modern magnetospheric physics.

After the immediate space around the Earth was surveyed in detail by satellites and rockets, the attitude toward Birkeland's theory changed to one of great admiration and almost total acceptance. In retrospect it may be stated that Birkeland's work of 1896-1908 was far ahead of other auroral and geomagnetic research of that time. Because of Birkeland's unconventional personality, a great part of the scientific community ignored his

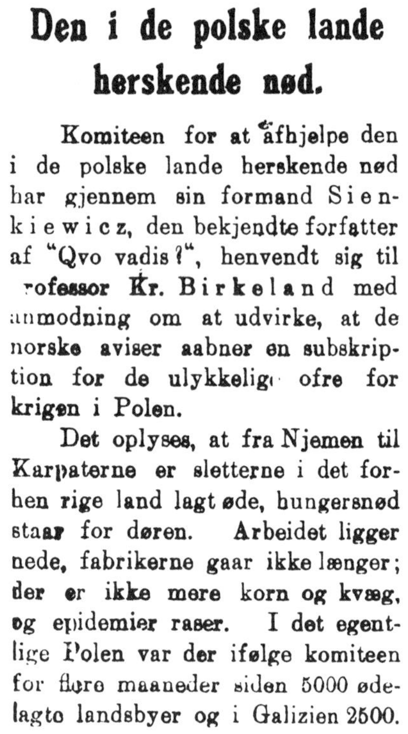

Den i de polske lande herskende nød.

Komiteen for at afhjelpe den i de polske lande herskende nød har gjennem sin formand Sienkiewicz, den bekjendte forfatter af "Qvo vadis?", henvendt sig til Professor Kr. Birkeland med anmodning om at udvirke, at de norske aviser aabner en subskription for de ulykkelige ofre for krigen i Polen.

Det oplyses, at fra Njemen til Karpaterne er sletterne i det forhen rige land lagt øde, hungersnød staar for døren. Arbeidet ligger nede, fabrikerne gaar ikke længer; der er ikke mere korn og kvæg, og epidemier raser. I det egentlige Polen var der ifølge komiteen for flere maaneder siden 5000 ødelagte landsbyer og i Galizien 2500.

Fig. 16. Birkeland often sent notes and longer articles to the newspapers. This is a copy of a short note from 1916 published together with Sienkiewicz the author of the famous book "Quo Vadis?", where they asked Norwegians to give money to starving people in Poland.

pioneering discoveries. Only now, in the age of space exploration, can we properly appreciate the prophetic nature of those discoveries.

Acknowledgments. Dr. D. Stern, of Goddard Space Flight Center, has read this paper and has provided numerous valuable corrections and suggestions. The main part of this paper was written while I was a senior postdoctorate fellow at the National Academy of Sciences (U.S.) at the Goddard Space Flight Center.

References

Akasofu, S.-I., Physics of mangetospheric substorms, D. Reidel Publ. Comp., Dordrecht, 1977.

Birkeland, Kr., Solution generale des equations de Maxwell pour un milieu homgene et isotrope, Comptes Rendus, CXXI, Paris, 1885.

Birkeland, Kr., Archives des Sciences Physiques et Naturelles, Vol. 1, p. 497, Geneva 4, 1896.

Birkeland, Kr., Expedition Norvegienne de 1899-1900 pour l'etude des aurores boreales, Resultats des recherches magnetiques, Videnskap. Skriftr, I (Math Naturv.) Christiania, 1901.

Birkeland, Kr., The Norwegian Aurora Polaris Expedition 1902-1903, Volume I, First Section, H. Aschehoug and Co., Christiania (pp. 1-315), 1908.

Birkeland, Kr., The Norwegian Aurora Polaris Expedition 1902-1903, Volume I, Second Section, H. Aschehoug and Co., Christiania, (pp. 319-801), 1913.

Brekke, A. and A. Egeland, The Northern Light; from mythology to space research, Springer-Verlag, Heidelberg, 1983.

Cole, K. D., Aust. J. Phys. 15, p. 223, 1961.

Devik, O., Kr. Birkeland as I knew him, in The Birkeland symposium on Aurora and Magnetic Storms, p. 13 (Eds. A. Egeland and J. Holtet) Centre National de la Rech. Sci., Paris, 1967.

Fritz, H., Das Polarlicht, Brockhaus, Leipzig, 1881.

Poincaré, H., Remarques sur une experience de M. Birkeland, Comptes Rendus, CXXIII, 930, Paris, 1896.

Ratchiffe, J. A., The formation of the ionosphere; Ideas of the early years: Geofysiske Publikasjoner, Geophysica Norvegica, Vol. 29, 13, 1972.

Romer, R. H., Alternatives to the Poynting vector for describing the flow of electromagnetic energy, Am. J. Phys., 50, 1166, 1982.

Störmer, C., The Polar Aurora, Clarendon Press, Oxford, 1955.

Tromholt, S., Under the Rays of the Aurora Borealis, Houghton-Mifflin, Boston, 1855.

SYDNEY CHAPMAN AND HIS EARLY STUDIES OF MAGNETIC DISTURBANCES

Masahisa Sugiura

Laboratory for Extraterrestrial Physics, Goddard Space Flight Center
Greenbelt, Maryland 20771

Abstract. Sydney Chapman's 1918 and 1927 papers on magnetic storms and disturbances of lesser intensity and the background in which these papers were prepared are described from a historical point of view. In spite of the greatly limited availability of observations, the average characteristics of the magnetic variations during these disturbances derived by Chapman are remarkably accurate. An attempt is made to portray Chapman's rigorous and conscientious method of data analysis.

Introduction

Although it is well known that contributions of Sydney Chapman (1888-1970) to the study of geomagnetic disturbances are extensive, it is often not fully appreciated that his papers published in 1918 and 1927 presented the morphology of magnetic storms and disturbances of lesser intensity quite accurately so far as their average features are concerned. The purpose of this paper is to show that some of the fundamental characteristics of the geomagnetic disturbances were already understood correctly in spite of the greatly limited extent of data available at the time. There are, of course, important features that were then observationally not known or conceptually misconceived and that had to be uncovered later by others. However, it is not the intent of the present paper to give a full account of the history of the study of geomagnetic disturbances or to discuss the controversies surrounding various developments.

The 1918 paper, entitled "An outline of a theory of magnetic storms" (Chapman, 1918b) was Chapman's first paper on magnetic storms, according to the list of publications by him and his collaborators. This list, which Chapman gave me in the early 1950's, divides the publications into the following categories (the range of years and the number of items under each category being given in parentheses): "Mathematics" (1910-1946; 11), "Astronomy" (1913-1944; 18), "Crystals" (1924-1927; 4), "Gases: Kinetic theory" (1911-1941; 23), "Terrestrial magnetism" (1913-1947; 75), "Lunar meteorological effects" (1918-1947; 28), "Upper atmosphere, aurora, ozone, ionosphere" (1920-1948; 39), and "Various" (1912-1948; 6). This list was prepared toward the end of 1948 or in 1949.

Beginning in 1910 and prior to 1918, Chapman wrote 8 papers (1911-1916) in "Mathematics", 7 papers (1913-1917) in "Astronomy", 8 papers (1911-1917) in "Gases: Kinetic theory", 6 papers (1913-1917) in "Terrestrial magnetism", 2 papers (both in 1912) in "Various"; among these, 7 had a coauthor, and the remaining 24 papers were by Chapman alone. His publications in 1918 seem to have been devoted to geomagnetism; he wrote one paper on lunar geomagnetic variation (Chapman, 1918a), 3 papers on magnetic storms (Chapman, 1918b, 1918c, 1918f), 2 papers of more general nature (Chapman, 1918d, 1918e) - a through f here being in the order given in Chapman's list, presumably in chronological order - and 2 papers on the lunar atmospheric tides (Chapman, 1918g, 1918h). Prior to 1918 he had written one paper on the solar and lunar geomagnetic variations (Chapman, 1913), and 5 papers on the lunar geomagnetic variations (Chapman, 1914a, 1914b, 1915, 1917a, 1917b). Thus he had a good understanding of the solar and lunar geomagnetic variations before preparing his first paper on magnetic storms (Chapman, 1918b). According to his own account (Chapman, 1952), he began his work on magnetic storms in 1917. In the following account, Chapman's original wording is used as much as possible to preserve the flavor of the classic work.

The 1918 Paper

Chapman's 1918 paper, although entitled "An outline of a theory of magnetic storms", consists of three parts: Part I. The Magnetic Data, pp.61-72; Part II. The Electric Current System and the Atmospheric Motions, pp.72-78; Part III. The Origin of the Atmospheric Motions, pp.78-83, and devotes more effort to the analysis of observational results than to theoretical considerations. In Part I, Chapman summarizes the results of his analysis of the hourly values of the three geomagnetic components observed at 12 observatories during 40 suddenly commencing magnetic storms of moderate intensity. These observatories are in low to middle magnetic latitudes (below 60°, and including one observatory in the southern hemisphere, i.e. Batavia). Data from Sitka (magnetic latitude 61°) were examined, but not included in the analysis in this paper. In the summary of the paper, Chapman states: "This paper briefly outlines some results of a study of magnetic storms, the object of which was to ascertain their broad general features, leaving aside individual details and cases." He announces his intention to write a follow-up paper in which he hoped "to deal more adequately with many points here lightly passed over, and especially with disturbance phenomena in polar regions, where magnetic conditions are scarcely ever quiet, and where the divergence between the magnetic and geographical axes of the earth introduces most complication." The follow-up paper is the 1927 paper to be described later.

Thus the 1918 paper discusses the average morphology of magnetic storms in low and middle latitudes. Chapman divided the storm variations into two parts: the "general storm variations" and the "additional diurnal variations" (or "local storm variations"). These correspond, in the present-day terminology, to the Dst and the SD components. The latter variations were determined for the first and second storm days, by subtracting the "average ordinary monthly mean diurnal variations" (i.e. Sq), and hence were SD. It was in Chapman's 1952 paper, which he always referred to as the "Annali paper" (Chapman, 1952), that he defined DS, the "disturbance local-time inequality" as a function of storm time; in his 1952 paper, "S" was used as a subscript, but later we (Sugiura and Chapman, 1960) adopted a notation DS for ease of typing, the same reason as Bartels adopted the notation Kp, keeping p on the same line as K.

It was already known that "great world-wide magnetic storms commence simultaneously to within a few seconds, over the earth, although small local magnetic fluctuations may sometimes mask the commencement at particular stations." Thus Chapman used "storm time", measured from the beginning of a storm, without reference to any individual locality. The term Dst, the storm time variation, was introduced in his 1952 paper; as in DS, "st" was then a subscript and later modified in the 1960 paper.

Referring to the general storm variation, i.e. Dst, Chapman states in the 1918 paper: "The principal average features of great magnetic storms, with which any theory of these phenomena must first concern itself, appear to be as follows:" "Over the whole earth, except, perhaps in the close vicinity of the poles, the net mean change in the horizontal force, during the first half-hour or so, is an increase. This is succeeded by a decrease of much greater amplitude, which lasts for several hours. A period of recovery then follows, and lasts for several days. Both the decrease and the recovery proceed most rapidly shortly after their initial stages, and gradually slow down." This description of the storm time variation at low and middle latitudes is remarkably accurate. He then describes, equally accurately, the storm time variations in the vertical force and declination.

As the second item of the principal features, he gives the local storm variations (i.e. SD).

He states: "The general storm variations are superposed on diurnal variations characteristic of each station. These depend on local time, but differ from the variations observed on ordinary days in a very definite manner. The residual diurnal variations at any station, obtained after subtracting the ordinary local-time changes corresponding to the place and season from the total diurnal variations occurring during a storm, will be termed the local storm variations. These are most intense during the first day of a storm, and gradually subside." Again the description of the local time variations is remarkably accurate.

The 1918 paper gives plots of the average Dst variations in the three magnetic elements for three groups of observatories for the first 48 hours of storm time. The description of the recovery phase is noteworthy. The paper says that the recovery after the [main phase] decrease has ceased lasts for many days, only about 10γ [nT] having been recovered by the end of the second day in the above storms. It is further stated: "This recovery shows itself in the non-cyclic variation on quiet days, though it begins, and proceeds most rapidly, at a time when the irregular storm variations may still be vigorously active." Chapman took the trouble of showing, by analysis, that the Dst variations in the horizontal force are strictly in this component and not a component of changes really relating to (geographic) north force.

Chapman recognized that the pre-storm level, that is, the level at the epoch half an hour before the storm commencement is several nT higher than the zero level which was defined by the monthly mean value for the months including the storms. He says that: "this is due to the depression of the monthly mean by the continued defect of force after the storm, and enables an estimate to be made of the duration of the period of recovery. The first value [i.e. the pre-storm value] represents the ordinary normal better than the mean monthly mean for these particular months." This shows the depth of his understanding of the data and the rigor with which he treated data.

The paper presents [equivalent] current systems for the two parts of the storm variations. For the general storm variations (i.e. Dst), the current is "approximately orthogonal to the magnetic meridian", that is, the current flow is round the parallel of magnetic latitude, symmetrically about the earth's magnetic axis. For the local storm variation (i.e. SD), a schematic current diagram is given, which is applicable only to low and middle latitudes. The current system is, however, drawn to the poles without the intense currents along the auroral zone which were to be introduced in the 1927 paper. In general pattern, the [equivalent] current system is essentially the same as those that later we became familiar with, so far as those latitudes to which the current system given was meant to apply were concerned.

After presenting the current systems, Chapman

says: "So far our concern has been either with facts drawn directly from observation, or, as in §8, the natural interpretation of these in terms of current systems. It is necessary now to proceed over slightly more hazardous ground, in the endeavour to explain the genesis of these current systems." Chapman attempted to explain the general storm variation by the currents induced by vertical motions of the atmosphere. To explain the local storm variation he combined the two current systems together, thus having stronger currents in the afternoon than in the morning, and proposed that the storm influences are more powerfully exercised over the P.M. than over the A.M. hemisphere. [In the paper, P.M. and A.M. are reversed here; obviously Chapman missed the error in reading proofs.] He initially considered a stronger vertical motion in the P.M. hemisphere, but he thought that a greater electric conductivity in this hemisphere would be a more likely cause for the stronger currents. Note that the storm effects in low and middle latitudes are dealt with in the 1918 paper; near the auroral zone the effects would be greater on the morning side.

Then Chapman writes: "One further and yet more hazardous step must now be taken, in order to account for the atmospheric movements of §9" [described above]. Taking into account the various magnetic storm features that he described in detail, he was led to suggest the following explanation. "A magnetic storm is generated by the entry into the earth's atmosphere of numbers of electric particles, mainly or entirely of the same sign of charge. They penetrate to a more or less definite level in the upper atmosphere, this level depending on the density and composition of the atmosphere, and upon the physical nature and velocity of the particles. Their velocity, upon entry, is considerable, and the communication of their momentum to the absorbing layers imparts to these layers a downward motion, which is to be identified with the initial downward movement during magnetic storms. The ionisation of the layers by the impact of the electric particles may also contribute to this downward motion; oxygen will be partly converted into ozone, with the double effect of diminishing the pressure and increasing the mean molecular weight of the layer."

"The electric particles being mainly or entirely of one sign, the absorbing layer will become charged, and the mutual repulsion of the entangled ions will produce an expansion of the layer, i.e. an upward motion - just as in the case of an electrified soap bubble. The entry of particles into the atmosphere may continue for some hours, but, in general, their downward momentum will be overborne by the upward expansion, which their absorption in the layer will only increase. Clearly, also, the more intense the injection, the more violent will be the initial depression and the subsequent expansion of the layer ....." No further descriptions would seem necessary. The explanation was, indeed, a hazardous venture.

In connection with particle precipitation, Chapman discusses the relevance of auroras to magnetic storms and refers to the studies by Störmer, Vegard and Krogness on the auroral heights.

The following quotation is interesting because it typifies Chapman's way of thinking. "The higher degree of magnetic disturbance over the P.M. hemisphere is not to be regarded as a mere consequence of the approach of the stream on that side of the earth, since the terrestrial magnetic field will modify the motion of the particles before they actually enter the atmosphere - many, indeed, will be deflected altogether away from the earth, according to Stormer's calculations. The latter have not yet been harmonised in detail with the observational facts regarding aurorae, nor with the indications of magnetic phenomena, as discussed in this paper. The analysis of the corpuscular paths is a very difficult mathematical task, however, and the success which Prof. Störmer has already obtained gives good hope that in time, as the initial postulates are brought into conformity with the physical evidence, so the mathermatical theory will accord with observation."

It is interesting to note that Chapman mentions in the 1918 paper "that the local storm variation current system, which is of world-wide extent during great disturbances, is present over a more restricted region round the poles at nearly all times."

The concluding remark reads as follows: "In conclusion, I would add that it is an attempt to survey a somewhat vague and complex subject, and to place certain definite features in theoretical relation to one another, that the discussion here presented should be judged. Some of the views expressed have been stated before in other context*, and in a more detailed and complete memoir I hope to refer to these and other theories on the subject. In this account I have mentioned, very briefly, only those papers which I have had actual occasion to use in the present discussion; but it is hardly necessary to state how much such an investigation must owe to the labours of others who have previously studied the many-sided phenomena dealt with. / *Since writing this paper, for example, I have noticed that the symmetry of the disturbance variation about the solar meridian plane had been remarked by van Bemmelen in 1903 ('Terrestrial Magnetism,' vol. 8, p.153), who also (ibid, vol.5, p.123) refer to a theory of "current-vortices" (cf. §11) by Schmidt ('Met. Zeitschrift,' 1889, p.385)."

The 1927 Paper

It took Chapman 9 years to write the promised follow-up of the 1918 paper. During the intervening years other subjects appear to have drawn Chapman's attention as well. He wrote one paper in astronomy ("Diffusion and viscosity in giant stars," 1922); 4 papers on crystals (1924-1927 in collaboration with other workers); 4 papers on gas kinetic theory" (1919-1924, 2 papers each in collaboration with a coauthor); 14 papers in geomagnetism (1919-1927, 2 papers written with T. T. Whitehead); 4 papers on tidal oscillations of the atmosphere (1919-1924, one paper with a co-author); 4 papers on aeronomy (1920-1926, one paper on "The composition, ionisation and viscosity of atmosphere at great heights" with E. A. Milne, who was Chapman's student; the other 3 papers were "On the changes of temperature in the lower atmosphere by eddy conditions or otherwise", "The electrical state of the upper atmosphere", and "Ionization of the upper atmosphere"); and a paper entitled "A note on the fluctuation of water-level in a tidal-power reservoir". He was beginning to be interested in the ionosphere, but his major work on the ionosphere was to come in 1931 (Chapman, 1931a, 1931b).

As the title "On certain average characteristics of world wide magnetic disturbance" indicates, the 1927 paper deals with a subject not limited to magnetic storms as the 1918 paper. In the 1927 paper Chapman intended to show the differences between the magnetic variations on magnetically ordinary and quiet days. He also intended to ascertain whether weak magnetic disturbance differs from intense disturbance not only in degree but also in type. Recognizing that mere inspection of magnetograph curves does not suffice to decide this question, he investigated this problem statistically, as he had determined the average features of magnetic storms in the 1918 paper.

First, he determined for middle and low latitudes (5 groups of observatories) the local time variations in the magnetic elements for quiet days (i.e. on the 5 quiet days per month selected internationally) for moderately strong disturbance and for relatively weak disturbance. He found that the variations for the two levels of disturbance had a considerable degree of similarity, while these variations were distinctly different from those for the quiet days. From the results of the analysis Chapman derived an important conclusion, namely, that "the average additional magnetic variations during magnetic disturbance, which are superposed on the quiet-day variations, do not differ much in type, as the intensity varies over a wide range."

For the disturbance variation corresponding to the storm time variation, Chapman used the differences between the daily mean values of the three elements for quiet days and the corresponding daily means for all days. He found that the difference was appreciable only in the horizontal force and that the latitude dependence of this difference was in general agreement with the expectation if one assumed that the average characteristics of this axially symmetric disturbance field are similar in type between slight and intense disturbances.

He carried out similar analyses for high latitude disturbance. The vector differences between all and quiet days were determined for 10 high latitude stations. The main feature found was that the variation in the horizontal force is a reduction, resembling the corresponding change in lower latitutdes, except that the magnitude of the reduction is larger in polar regions than in low latitudes.

Examining the local time variation (all days minus quiet days) in the vertical force, Chapman found that this variation preserved a constant phase from the equator to as far north as Sitka, the amplitude increasing greatly with latiitude. This increase persisted beyond the latitude of Sitka to that of Bossekop. Then at Nova Zembla, the amplitude suddenly became small, and on the polar side of this station the phase of this variation reversed and the amplitude became large again.

The paper states: "Within the auroral zone the normal quiet-day variation probably becomes insignificant compared with the disturbance diurnal variation, even in the reduced form of the latter on quiet days. If this is so, the similarity of the type between the whole diurnal variations, whether on quiet, average, or more than usually disturbed days (5 per month), which has been clearly demonstrated by Dr. Chree in reference to Antarctic magnetic records, constitutes a partial confirmation of the view that the general character of the disturbance field, in polar as well as in lower latitudes, does not vary greatly over a wide range of intensity."

The paper further suggests the possibility (supported by some observational evidence) that the auroral zone broadens and moves towards the equator during periods of intense disturbance, so that a station which ordinarily is on the equatorial side of the zone may be during magnetic storms under, or on the polar side of, the zone. The station Nova Zembla is shown to exemplify this situation.

The paper gives what is later termed "vectograms" describing the variation, during the course of the day, of the magnetic vector in the horizontal plane for five high latitude stations, one below the auroral zone, two on the auroral zone, and two in the polar cap. These vectograms were helpful in deducing the [equivalent] current system.

Based on the diurnal variations thus investigated in detail, Chapman derived a set of current systems, which are reproduced in the center panel of the frontispiece of this volume. The upper figures represent the views from the sun and the lower figures are views from above the north pole, with the sun towards the bottom of the figures. For the sake of clearness the angular distance of the auroral zone from the pole is shown as nearly 30° instead of about 23°.

Chapman seems to have considered these currents actually to flow in the upper atmosphere because he writes: "The most natural supposition is that the field is due to a system of electric currents, part being situated in the upper atmosphere, and part within the earth, the latter being secondary currents induced by the variations of the former. This view is supported by the known facts about aurorae, and by analogy with the theory, incomplete yet already to a large extent successful, of the ordinary solar and lunar diurnal magnetic variations. It has also been confirmed, in as direct a way as is ever likely to be possible, by Birkeland's study* of large perturbations of the horizontal and vertical components of magnetic force at adjacent stations in the Arctic region. He found a number of cases in which there was clear evidence of a linear current flowing overhead between two stations, the horizontal force perturbations being of the same sign, and the vertical force perturbations of opposite sign; the estimated heights of the centre-lines of such currents varied from 150 km. upwards. / *Birkeland, 'The Norwegian Aurora Polaris Expedition,' 1902-3, vol.i, §76, p.306. Currents of the order of 10^6 amperes were indicated during magnetic storms not of outstanding intensity."

Chapman considered the theory that he proposed in 1918 for magnetic storms to be "inadequate to explain the wider range of facts" presented in the 1927 paper, "apart from the valid objections made by Prof. Lindemann" (Lindemann, 1919). Chapman refers to the theory "advocated (particularly) by Birkeland, which attributes magnetic disturbance to the direct magnetic field of streams of charges moving freely outside, or at high levels in, the atmosphere." He comments that: "Schuster* has shown, however, that Birkeland's hypothesis is untenable owing, among other reasons, to the inability of streams, charged with sufficient intensity to produce the observed magnetic changes, to maintain themselves against the tendency to dispersion by the mutual electrostatic repulsion of the charges. / *A. Schuster, 'Roy. Soc. Proc.,' A, vol.lxxxv, p.44 (especially §6)(1911)."

Chapman abandoned the theory which he proposed in 1918, but he still thought that "the atmospheric motions, whatever their type, are always present, and with them the corresponding electromotive forces, but that the intensity of currents which they produce is dependent on the degree of ionization and conductivity of the air."

Concluding Remarks

I have attempted above to describe Chapman's 1918 and 1927 papers on the average morphology of magnetic storms and disturbance of weaker intensity and the background in which these papers were written. In so doing, I have tried to portray Chapman's methodical and meticulously detailed investigations giving results that apart from physical interpretations are to a large extent still valid.

References

Chapman, S., On the diurnal variations of the earth's magnetism produced by the moon and sun., Phil. Trans. Roy. Soc., A 213, 279-321, 1913.

Chapman, S., On the lunar variations of the earth's magnetism at Pavlovsk and Pola (1897-1903), Phil. Tran. Roy. Soc., A 214, 295-317, 1914a.

Chapman, S., The moon's influence on the earth's magnetism, Terr. Mag., 19, 39-44, 1914b.

Chapman, S., Lunar diurnal magnetic variation and its change with lunar distance, Phil. Trans. Roy. Soc., A 215, 161-176, 1915.

Chapman, S., On the influence of lunar declination of the lunar diurnal variation of magnetic declination at Zikawei, Terr. Mag., 22, 121-124, 1917a.

Chapman, S., Influence of solar activity on lunar diurnal magnetic variation, Terr. Mag., 22, 87-91, 1917b.

Chapman, S., The influence of changes in lunar distance upon the lunar diurnal magnetic variation, Terr. Mag., 23, 25-28, 1918a.

Chapman, S., An outline of a theory of magnetic storms, Proc. Roy. Soc., London, A 97, 61-83, 1918b.

Chapman, S., The energy of magnetic storms, Roy. Astr. Soc., Monthly Notices, 79, 70-83, 1918c.

Chapman, S., Diurnal changes of the earth's magnetism, Observatory, No. 522, 52-60, 1918d.

Chapman, S., Terrestrial magnetism, Trans. Victoria Inst., 16 pp., 1918e.

Chapman, S., On the times of sudden commencement of magnetic storms, Proc. Phys. Soc., 30., 205-212, 1918f.

Chapman, S., An example of the determination of a minute period variation as illustrative of the law of errors, Roy. Astr. Soc., Monthly Notices, 78, 635-638, 1918g.

Chapman, S., The lunar atmospheric tide at Greenwich, Quart. J. Roy. Met. Soc., 44, 271-280, 1918h.

Chapman, S., The absorption and dissociative or ionizing effect of monochromatic radiation in an atmosphere on a rotating earth, Part 1, Proc. Phys. Soc., 43, 26-45, 1931a.

Chapman, S., The absorption and dissociative or ionizing effect of monochromatic radiation in an atmosphere on a rotating earth, Part II. Grazing incidence, Proc. Phys. Soc., 43, 483-501. 1931b.

Chapman, S., The morphology of geomagnetic storms: an extension of the analysis of D_s, the disturbance local-time inequality, Annali di Geofisica, 5, 481-499, 1952.

Lindemann, F. A., Note on the theory of magnetic storms, Phil. Mag., 38, 669, 1919.

Sugiura, M., and S. Chapman, The average morphology of geomagnetic storms with sudden commencement, Abhandl. Akad. Wiss. Göttingen Math.-Phys. Kl., Sonderheft Nr. 4, 54 pp, Göttingen, 1960.

THE EVOLUTION OF ARGUMENTS REGARDING THE EXISTENCE OF FIELD-ALIGNED CURRENTS

A. J. Dessler

Space Science Laboratory, NASA/Marshall Space Flight Center, Huntsville, Alabama   35812

Abstract. We did not arrive at our present understanding of Birkeland (magnetically-field-aligned) currents by a direct, logical course. The story is rather more complex. Starting at the end of the 19th century, the Norwegian scientist Kristian Birkeland laid out a compelling case, supported by both theory and experiment, for the existence of field-aligned currents that cause both the aurora and polar geomagnetic disturbances. Sydney Chapman, the British geophysicist, became the acknowledged leader and opinion maker in the field in the decades following Birkeland's death. Chapman proposed, in contradistinction to Birkeland's ideas, equivalent currents that were restricted to flow in the ionosphere with no vertical or field-aligned components. Birkeland's ideas may have faded completely if it had not been for Hannes Alfvén, who became involved well after Chapman's ideas gained predominance. Alfvén kept insisting that Birkeland's current system made more sense because field-aligned currents were required to drive most of the ionospheric currents. I became personally involved when Zmuda et al. [1966] submitted to the Journal of Geophysical Research a paper reporting satellite data showing magnetic disturbances above the ionosphere that were consistent with field-aligned Birkeland currents, but which they did not interpret as being due to such currents.

Introduction

Most find browsing through old journals interesting, but sometimes the experience produces an ambivalent reaction. On one hand there is the undoubted good feeling that comes when you see that science is getting somewhere and that you know more on a particular topic than did a particular scientist in his day. However, there is sometimes a sinking feeling that follows from the realization that, on occasion, an entire scientific community had moved off in the wrong direction, worked hard, and accomplished essentially nothing, or even less. In this regard, I think the remarks of Samuel Pierpoint Langley [1889] in his address as retiring president of the American Association for the Advancement of Science are particularly appropriate. (This is the person who would have been the first to fly a heavier-than-air craft if the Wright brothers had not rushed into print with such undue haste; the NASA Langley Research Center is named after him in recognition of his efforts.) Langley preferred to think of the scientific community as "not wholly unlike a pack of hounds...where the louder-voiced bring many to follow them nearly as often in a wrong path as in a right one, where the entire pack even has been known to move off bodily on a false scent."

The story of how we arrived at our present understanding of Birkeland currents is a relatively modern example of Langley's analogy, which, he believed, "if a less dignified illustration, would be one which had the merit of having truth in it." Although Langley's pack of hounds "in the long-run perhaps catches its game," it is sobering to reflect that this inefficient, if not embarrassing, method of hunting for the truth is probably the way we work to this day.

Birkeland

The more I learn of Kristian Birkeland and his work, the more I realize how remarkable he was. He was energetic, he had lots of good ideas, and intellectually he towered above his contemporaries. He thought big, and he did big science. For example, he set up a chain of magnetic observatories in auroral latitudes that enabled him to deduce the concept of the polar elementary storm. (This name some find preferable to "magnetic substorm," which is our present term for this same phenomenon.) He did auroral-related laboratory experiments with what must have been the largest vacuum chambers to that time. He also was a superior theorist; he worked at the forefront of Maxwell electrodynamic theory, the most esoteric of theoretical subjects at that time. V. Vasyliunas (private communication, 1983) brought to my attention that one of Birkeland's papers from 1894 was recently quoted as relevant to the question of the uniqueness of the Poynting vector [Romer, 1982]. On top of all this, he was able to raise money for his research from an amazing variety of sources, which ranged from patenting and operating a nitrogen fixation process, to selling stock in an electromagnetic cannon corporation, to just talking local individuals and institutions into supporting his research (salesmanship at its finest). Details on these and other facets of Birkeland's life are covered in some detail in the article by Alv Egeland in this volume and more of Birkeland's scientific work in the review by Dessler [1967].

Mercury Poisoning. Before going into the development of Birkeland's ideas on field-aligned currents, I need to discuss a matter that will become a relevant factor when we consider possible reasons why Birkeland's ideas were neglected for so long. I will advance the hypothesis that Birkeland suffered from chronic mercury poisoning that affected both his health and research during the last few years of his life. This is not a novel thought regarding declines in the health of scientists of those times. For example, it has been suggested that Pascal, Faraday, Hertz, and Newton probably suffered from mercury poisoning [e.g., Johnson and Wolbarsht, 1979, and references therein].

Birkeland's early writing is marked by vigor, optimism, and a joy of both life and research. In the first section of his account of his 1902-1903 auroral expedition, Birkeland [1908] exhibits an incredibly strong will and an optimistic outlook when he wrote of establishing observatories on the highest mountaintops in northern Finland. In an earlier expedition to the auroral zone (February and March 1897), he and his small party nearly perished in a fearful blizzard. However, by autumn of that same year, he was back scaling mountains "to find a mountain that would do for my auroral investigations." He showed glowing pride when he wrote that he was able to build two observatories of stone and cement, completed in September 1899, "two of the best auroral observatories in the world."

Then it was back to the top of the mountain to spend the winter at the observatories (between September 1899 and April 1900), an experience "such as others have probably not passed through; for as far as is known, no one has ever before passed a winter upon the highest mountain-summits in Finland." The tone of his description of living conditions on these mountains is one of exhilaration. One reason he wanted to go to these mountains was because he had heard that, in northern Finland, the aurora came down so low that it touched the mountaintops. He wanted to see if this story was true, but if it were, he and his party would be engulfed in an aurora, the consequences of which could not be foretold, because the aurora was a phenomenon whose cause was unknown. His writing betrays no anxiety, beyond that of a scientist wanting some data.

Wind speeds of at least 100 miles per hour with temperatures of $-20°C$ (stronger winds that occurred destroyed their anemometers) made it uncomfortable inside the stone huts where, in strong wind, "it took three men with great effort to close our little door," and "water froze there a couple of yards from a glowing stove; and the lamp was blown out on the table in the middle of the room," but there is no sense of complaint. Rather, one feels he is reading the account of an explorer who has had a wonderful adventure. This mood is in contrast to the one of paranoia, despair, and ill health that marked his final expedition to Egypt [Egeland, this volume]. I believe this change in his mental state is discernible even between the first and second sections of his book on the 1902-1903 auroral expedition [Birkeland, 1908, 1913].

From descriptions of his laboratory, one could conclude that there was much mercury around--rotary mercury pumps, mercury rectifiers, and mercury manometers. In those days, there was essentially no awareness of health hazards from environmental conditions in the laboratory. Our present concerns are of suprisingly recent origin. The Merck Index lists symptoms of chronic mercury poisoning as including muscle tremors, personality changes, depression, irritability, and nervousness, while the Handbook of Poisoning lists tremor, erethism, nervousness, irritability, depression, insomnia, and drowsiness. Descriptions of Birkeland in his final years, plus the change in both the mood and quality of his last few research papers, seem to me to be totally consistent with symptoms of chronic mercury poisoning.

Field-Aligned Currents. Birkeland discovered polar magnetic substorms (which he called polar elementary storms). He noted that the magnetic disturbances were extremely localized in auroral regions, and he argued on the basis of analysis of numerous individual substorms that the observed magnetic perturbations were caused by horizontal currents that were maintained "by a constant supply of electricity from without." In other words, he deduced that the observations were to be accounted for by field-aligned currents that, when quite near the Earth, turned to flow for some distance approximately along the auroral zone, and then turned to flow back out into space. He further deduced that the horizontal portion of the current must flow at an altitude "rather low in relation to the Earth's dimensions." Comparison of the data with mathematical models led him to conclude that the horizontal portion of these currents flowed at an altitude of about 200 km. This conclusion is quite remarkable when one considers that the presence of a conducting ionosphere would not be recognized for another quarter century.

Birkeland argued that his magnetic observations could not be explained by currents that were entirely confined to a spherical shell about the Earth because (1) localized perturbations could not be produced (In his data, he could see horizontal currents that began and ended within a rather limited horizontal distance.), and (2) some powerful outside influence would be required both to maintain and to constrain the currents. With his detailed knowledge of Maxwell theory, he was able to present telling arguments to the effect that such currents were physically untenable. Furthermore, if the currents were really confined to a spherical shell, an explanation for the aurora would still be lacking. The economy of having the horizontal currents driven by vertical currents is that the particles that carry the vertical currents also account for the aurora.

Birkeland supported his thesis (that the polar magnetic disturbances are caused by horizontal currents that are maintained by vertical [essentially field-aligned] currents coming from the Sun in the form of electron beams) with ingenious and impressive laboratory experiments. With these experiments, he was able to present arguments by analogy that should have been persuasive. Not that there weren't problems with Birkeland's hypothesis. For example, he worried how the Sun could mantain a strong negative charge so it could function as a cathode. He pointed out that because the Sun is such a strong light source, one might suppose that there might be some way to convert light into electrical charge. But, he wrote, "As Maxwell's electromagnetic light theory at present stands, there is no opportunity..." And then the classic plea of the theorist, "It is thought by several that Maxwell's equations require a correcting term."

In summary, the idea that vertical currents flow into and out of the auroral zone to both drive horizontal currents near the Earth and cause the polar aurora was proposed and developed to a remarkable degree by Birkeland. He made it easy for the scientists that followed him to think and talk about corpuscular beams coming from the Sun and striking the Earth's magnetic field.

Chapman and Alfvén

Sydney Chapman entered British geomagnetic research at a time when it had not yet recovered from an almost casual comment by Lord Kelvin to the effect that the Sun could have no effect on geomagnetic activity, and "that the supposed connection between magnetic storms and sunspots is unreal" [Kelvin, 1892]. Kelvin was president of the Royal Society at the time; in terms of Langley's analogy, this position, plus his reputation, made him a hound with a loud voice, at least in British circles. His remark inspired a new field of research in which geomagnetic activity was hypothesized to be caused by atmospheric motion (the theory of free cyclonic electric-current-systems).

Although Chapman had written a few early papers on the production of a tidally-driven geomagnetic variation caused by the gravitational effect of the Moon on the Earth's atmosphere, his first major paper on a topic that could have been influenced by Birkeland's work did not appear until over a year following Birkeland's death [Chapman, 1918]. At that time, although he had referenced Birkeland [1908] in an earlier paper, this time Chapman does not reference him because, "I have mentioned only those papers which I have had actual occasion to use in the present discussion." However, he uses (in Part III) the idea that Birkeland made popular, namely, a corpuscular beam eminating from the Sun and interacting with the Earth's magnetic field and upper atmosphere. Thereafter his work diverged from Birkeland's, although there was occasional perfunctory reference to one or another of Birkeland's ideas. The references were never flattering, and there is no hint that he was ever knowingly influenced by any of Birkeland's papers.

Chapman's work was a significant departure in style from that of Birkeland. Chapman's background was that of a mathematician. He showed unusual skill in devising problems that were obviously relevant to some geomagnetic phenomenon, but for which the physical configuration was arranged so that the problem could be handled with complete mathematical rigor. For example, the magnetic effects of a plasma stream impacting the Earth's magnetic field were treated first as a perfectly-conducting plane surface and later as a cylinder surrounding the Earth. The advantage of these particular physical approximations is that the ensuing mathematics could be handled exactly, i.e., without any approximations. In contrast, Birkeland made mathematical approximations when necessary in order to maintain either the spacial configurations that he hypothesized for the solar-terrestrial system or one that he observed in his laboratory apparatus.

Another significant difference in research methodology between the two is that Chapman, following his remarkable successes in geomagnetic research on detecting lunar tidal effects in the Earth's ionosphere, was inclined to average large data sets in various ways to extract subtle effects. Birkeland treated each event separately to resolve fine spacial detail and obtained only general impressions of average geomagnetic behavior. For this reason, each was able to see things the other could not.

In order to put the geomagnetic data in a form that could be treated with mathematical rigor, Chapman made a simplifying physical assumption that the surface observations were to be accounted for by an "equivalent current system" that was restricted to flow in a spherical shell about the Earth, presumably the ionosphere. Then, spherical harmonic analysis could be applied, and geomagnetic data could be represented by a series of Legendre polynomials. With this approach, there was no place for Birkeland's field-aligned currents.

It must be acknowledged that, if one ignores the question of how auroral and geomagnetic activity are driven by solar activity, one cannot establish solely from surface magnetometer data whether Birkeland's or Chapman's current system is correct [Fukushima, 1969]. Thus, without unambiguous experimental evidence to decide between the two current systems, Chapman was naturally drawn to the more parsimonious of the two current system descriptions, plus he naturally favored the spherical system that was susceptible to exact mathematical treatment. (It is interesting to note that experimental evidence was published that, in retrospect, must have been misinterpreted in order to show Chapman's current system favored over Birkeland's [Vestine and Chapman, 1938].)

Some 20 years after Chapman's initial magnetic storm paper, Hannes Alfvén, 20 years Chapman's junior, became involved in the matter of where and how currents flowed in the magnetosphere to produce aurora and geomagnetic activity. His approach was similar in some respects to Birkeland's. Alfvén, a fellow Scandinavian, was trained as an electrical engineer. He, like Birkeland, was more interested in an approximate solution to a physically realistic model than an exact solution to an idealized model. Further, Alfvén held that useful insight could be gained from properly designed laboratory experiments. He agreed with Birkeland's [1913] statement that, "It will be immediately apparent what far-reaching consequences are here built upon our experimental analogies." In other words, Alfvén agreed with Birkeland in both methodology and spirit. Conflict with Chapman was inevitable.

One of Chapman's characteristics that made him so well liked was that he was not argumentative. As far as I am aware, he never engaged in either private or public debate on any scientific question. On the matter of field-aligned currents, he obviously could not have argued with Birkeland because Birkeland died before Chapman got interested in the subject. However, he never argued with Alfvén, although there were numerous opportunities. Chapman would, usually in some suitable review paper, make a brief statement as to why he disagreed, and then he was through with the matter, even though Alfvén was not.

Alfvén's first paper on geomagnetic theory was rooted, in part, in Chapman's previous work [Alfvén, 1939]. For an abridged, annotated version of this paper, see Dessler and Wilcox [1970]. Chapman dismissed Alfvén's efforts as "curious," which is undoubtedly exactly what Chapman thought they were. From this point on, Alfvén, who dearly loves a good argument, presisted in presenting the case for, among other things, field-aligned currents flowing into and out of the auroral zone. However, he did not have a good forum to present his case, and Chapman's opinions dominated. (See Dessler [1970] for additional discussion of this point.)

Alfvén made his views known most effectively through his now classic book, Cosmic Electrodynamics [Alfvén, 1950], and by his appearances at meetings where he would try to argue with Chapman. I was present at several of these meetings at which Chapman would present a paper on geomagnetic disturbances interpreted solely in terms of equivalent ionospheric current systems. Alfvén would argue that such currents were physically unrealistic because they left out the field-aligned currents described by Birkeland. During the last few years of making this argument, Alfvén would usually add a statement that went something like, "Now that satellites are orbiting the Earth, we shall see which current system is correct, the one described by Birkeland or the one described by Chapman and his students [principally Vestine]." If Alfvén presented a paper, Chapman would rise to state that Alfvén's views were not in keeping with those of Chapman and his colleagues and something to the effect that, "We are presently preparing a paper that will shortly be submitted that will clarify these issues." Alfvén protested, but Chapman sat down and would not debate.

Alfvén, as a young scientist, had a continuous uphill battle with Chapman, while, in contrast, Chapman never had to battle Birkeland. Alfvén had to struggle with Chapman's living presence for nearly 30 years, while Birkeland's 20-year career in auroral and geomagnetic physics ended just as Chapman's started. Chapman was a prolific author publishing enough papers to allow a meaningful correlation between his publication rate and the sunspot cycle [Campbell, 1968]. Alfvén's publication rate was significantly less. One cannot help but speculate how differently things may have developed if Birkeland had lived and remained vigorously active in the field for another 20 years (until Alfvén arrived). In such a scenario, Chapman would be the young scientist working to establish his ideas in the presence of the more senior Birkeland. But, largely because of Kelvin, Chapman moved into what was virtually a vacuum.

Eventual Outcome

Experiment, of course, is the final arbitrator. In early 1966, more than a half century after Birkeland had put forth his experimental and theoretical evidence regarding the existence of field-aligned currents, Zmuda, Martin, and Heuring submitted to the Journal of Geophysical Research a paper reporting the existence of "transverse magnetic disturbances" in the auroral zone as measured by a satellite-borne magnetometer. I was editor of the space physics portion of JGR at the time. I had been attuned to the possibility of the existence of field-aligned currents from listening to Alfvén at several meetings, reading some of his papers, and, perhaps most importantly, from talking with Carl-Gunne Fälthammar whose arguments influenced me greatly. In addition, Patel [1965, 1966] reported localized magnetic disturbances measured with the satellite Explorer 12. These too had been interpreted in terms of hydromagnetic waves. Although I was one of the early champions of the use of hydromagnetic theory to explain geomagnetic phenomena, it seemed clear that these localized, low-frequency magnetic disturbances did not fit the concept of a propagating wave. The observed disturbances must be caused by field-aligned currents.

In the case of the paper by Zmuda et al., the paper seemed straightforward enough that I did not feel I needed the advice of a referee, so I wrote the referee's report myself. (This practice is explained in a paper I wrote after retiring as editor [Dessler, 1972].) I recall writing (as referee) that the paper was fine except that the authors should acknowledge the possibility that the magnetic disturbances were caused by field-aligned currents. Zmuda did not wish to do so. My policy (as editor) was to accept such a paper immediately; as long as the paper was clear, I never felt an author was obliged to please or convince a referee [Dessler, 1972]. (My editorial files, complete with correspondence and referees reports, are on file in the archives of the Fondren Library of Rice University. The files will be fully open in 1992, should anyone wish to delve further into this or some other such story.)

I felt that field-aligned currents had finally been detected, and the claim was worthy of publication, so I tried to get a graduate student to work with me to write up the idea for publication. I had the good fortune to discuss the idea with W. David Cummings, who was a student at Rice University at the time, and the paper by Cummings and Dessler [1967] on field-aligned currents followed.

The final matter with which I was involved was naming field-aligned currents as Birkeland currents. As I more fully realized the importance of field-aligned currents to magnetospheric physics (for example, it was soon argued that these currents were the means of communication between the plasma sheet and the outer magnetosphere [e.g., Wolf and Dessler, 1969]), the more fitting it seemed that they should be named after Birkeland. Although the 1967 paper by Cummings and me used the term Birkeland currents, we did not do so in a deliberate way. It was later that I persuaded several of my colleagues to start using the term; among the first were Paul Cloutier and John Freeman [e.g., Schield et al., 1969; Cloutier, 1971]. The name caught on, and now it is even used for conference titles.

Analysis

It might be interesting to speculate on why the promising start made by Birkeland was so completely abandoned by the scientific community for such a long time. Why is Langley's analogy so apt when telling the story of Birkeland currents? There are, I believe, several factors that contributed to the course taken by the scientific community, the primary one being the dominant role of Chapman in determining the opinion of the scientific community.

Sydney Chapman is regarded by many as the most influential person in geomagnetic research during the past 100 years--and deservedly so. His ideas, consciously or unconsciously, dominated the thinking of the geomagnetic/magnetospheric research community from the mid-1920's almost until his death in 1970, a span of nearly half a century. The list of honors awarded him and the special offices and positions he has held is impressively long. His work guided the mainstream of geomagnetic research for nearly his entire professional life.

If Chapman had picked up the lead of Birkeland, how far might geomagnetic research have advanced? The geomagnetic community would have certainly followed him. However, I have already mentioned one reason why Chapman would not wish to adopt Birkeland's hypothesis of field-aligned currents—such models are not subject to being handled with complete mathematical rigor. In addition, there are two other factors that must be considered.

One is the opinion Chapman would have formed of Birkeland if, as seems logical, his introduction to Birkeland's work came from reading his most recent publications first. As I discussed earlier, Birkeland may have suffered from chronic mercury poisoning, and, as a result, the quality of his later writings was distinctly inferior. Chapman would have started with a poor impression of Birkeland that would have adversely affected his opinion of all of Birkeland's ideas.

The other factor I need to bring up is perhaps contentious, but it also is the most important for understanding why our thinking on Birkeland currents went astray. It appears that Chapman, for reasons to be discussed, was not willing to consider fairly the opinions of continental Europeans. In an age when the Sun never set on the British Empire, it is perhaps understandable that Chapman should be reluctant to accept the premise of a recently deceased Norwegian who spent a winter on a Finnish mountaintop where the Sun never rose.

Chapman was a scholar, a patriotic Englishman loyal to King/Queen and country, and, although of humble origins, a gentleman of Victorian moral principles. (He was 13 when Queen Victoria died.) Chapman was modest and naturally friendly. Although truth and moral principle were his shield, he avoided, as much as possible, beating people over the head with his shield.

Yet it appears he harbored deep-seated feelings that surfaced occasionally. Obituaries that Chapman wrote to mark the death of a distinguished colleague sometimes revealed these feelings. When he wrote of a fellow Englishman, the obituary was just what one would expect, the deceased was a fine person, a scholar, and an exemplary family man. Annoying habits or defects of character were gracefully left unmentioned. However, it was different when he wrote an obituary for a continental European.

The most startling such case is the obituary Chapman [1958] wrote for Störmer. It contains much standard biographical material and appreciation of Störmer's work, as one would expect in an obituary. But it also contains a section of criticism of Störmer's theoretical work. "To the writer [Chapman] it has long seemed that Störmer's theoretical work did not attain its main objective, and that auroral theory must be developed along radically different lines."

Near the end of the obituary, he wrote of matters that must have scandalized Chapman, the modest, generous man of Victorian moral principles. He wrote of how Störmer secretly took pictures of a girl without her knowing by using a buttonhole camera. On another occasion (different lady), "After she had gone he remarked to me that he had no idea she would be so 'tempting'." Then in a later conversation about a dinner party involving yet another lady, Chapman reported, "Rather naughtily I asked if she too was tempting. 'No,' he answered, 'not tempting, but cosy to talk to'." Chapman also wrote of Störmer's vanity and lack of generosity with his subordinates, "It is remarkable that none of his papers bears any name other than his own, though their text often refers to contributions by his assistants to their methods or ideas; some of those contributions were really substantial."

I am confident that the above remarks by Chapman are true, but I think most would agree that they are not appropriate in an obituary. Why would a man as sensitive to the feelings of others as Chapman do it? He must have disapproved of the described acts and behavior while Störmer was alive, but it was not until he was dead that Chapman felt it necessary to write of it. But what of the feelings of Störmer's friends? The truth was his shield, and, in a few such cases, he felt compelled to beat people over the head with his shield.

The 1967 Birkeland Symposium. A symposium was held in Sandefjord, Norway in September 1967 on the subject of aurora and magnetic storms. The date was picked to honor the 100th anniversary of the birth of one of Norway's greatest scientists. (Birkeland was actually born in mid-December, but the Organizing Committee mercifully moved the meeting date to a climatologically more pleasant season.) Chapman, the most eminent auroral and geomagnetic scientist of the time, was invited to give the opening scientific address. (This was 4 years before Alfvén would be awarded the Nobel Prize.) One would think that if he accepted the invitation, he would say only nice things about Birkeland. But this was not to be.

Chapman read his paper for the symposium verbatim. This paper [Chapman, 1968] and others from the symposium are in a special issue of Annales de Géophysique, Vol. 24, 1968. Although Chapman states that, "in science it is our good custom to appreciate the true discoveries and enlightening stimulus provided by great men, rather than to dwell on their misconceptions," he was, in this instance, unable to abide the custom. He even brought up Störmer, although this time he restricted his criticism to Störmer's failure to explain the aurora. But he wasn't even asked to talk about Störmer. Why then did he write, "The apparently unshakable hold, on Birkeland's mind, of his basic but invalid conception of intense electron beams, mingled error inextricably with truth in the presentation of his ideas and experiments on auroras and magnetic storms. As regards the aurora, somewhat the same may be said about Störmer's theoretical work. Birkeland's greater breadth of mind and wider interests led him astray further, in his theories of the main geomagnetic and solar magnetic fields and in his cosmogonic theories." This was hardly what the Norwegians had wanted or expected of Chapman.

We thus have the information we need to understand how geomagnetic research was misdirected for so long with regards to Birkeland currents. We must accept as an observed fact that the scientific community tends to stick together in their perception of what is right. For nearly half a century, Chapman's influence in geomagnetism was profound. Unless he endorsed an idea, it was seldom accepted by the community. Because Chapman felt safe only with ideas that were susceptible to exact mathematical solution, and because he apparently was, at least subconsciously, influenced by typical Victorian, British insular feelings regarding continental Europeans that kept him from giving fair consideration to their ideas, Birkeland's hypothesis of field-aligned currents was set aside until the weight of direct measurement made it impossible to ignore the concept any longer.

Stories such as this one lead me to appreciate how well science works despite occasional misdirection. No matter how complex the issue, the truth eventually burns through and is seen.

Acknowledgments. I wish to thank Jack Eddy for providing me with the reference to Barr [1960], which contained the Langley quotes, and A. Van Helden for the Johnson and Wolbarsht reference on mercury poisoning. I am also appreciative of help from John Isbell, D. Venkatesan, T. Moorehead, D. Dessler, and L. Dessler.

References

Alfvén, H., A theory of magnetic storms and the aurorae. I, Kungl. Sv. Vet.-Akademiens Handl., 18, 1-39, 1939.

Alfvén, H. Cosmical Electrodynamics, Oxford

University Press, London, 1950.

Barr, E. Scott, Historical survey of the early development of the infrared spectral region, Am. J. Phys., 28, 42-54, 1960.

Birkeland, Kr., The Norwegian Aurora Polaris Expedition, 1902-1903, Vol. 1, H. Aschehpug and Co., Christiania. First section, 1908; Second section, 1913.

Campbell, W. H., Correlation of sunspot numbers with the quantity of S. Chapman publications, Trans. Am. Geophys. Un., 49, 609-610, 1968.

Chapman, S., An outline of a theory of magnetic storms, Proc. Roy. Soc., 95, 61-83, 1918.

Chapman, S., Fredrik Carl Mulertz Störmer, 1874-1957, Biographical Memoirs of the Roy. Soc., 4, 257-279, 1958.

Chapman, S., Historical introduction to aurora and magnetic storms, Ann. de Geophys., 24, 497-505, 1968.

Cloutier, P. A., Ionospheric effects of Birkeland currents, Rev. Geophys. Space Phys., 9, 987-996, 1971.

Cummings, W. D., and A. J. Dessler, Field-aligned currents in the magnetosphere, J. Geophys. Res., 72, 1007-1013, 1967.

Dessler, A. J., Solar wind and interplanetary magnetic field, Rev. Geophys., 5, 1-41, 1967.

Dessler, A. J., 1970 Nobel Prize in physics, Science, 170, 604-606, 1970.

Dessler, A. J., Editing JGR--Space Physics, Eos Trans. AGU, 53, 4-13, 1972.

Dessler, A. J., and J. M. Wilcox, Magnetic storms and the aurorae: Comments and annotations on a paper by Hannes Alfvén, Eos Trans. AGU, 51, 180-194, 1970.

Fukushima, N., Equivalence in ground geomagnetic effect of Chapman-Vestine's and Birkeland-Alfvén's electric current-systems for polar magnetic storms, Rep. Ionos. Space Res. Jap., 23, 219-227, 1969.

Johnson, L. W., and M. L. Wolbarsht, Mercury poisoning: A probable cause of Isaac Newton's physical and mental ills, Notes and Records Roy. Soc. London, 34, 1-9, 1979.

Kelvin, William Thompson, Address to the Royal Society at their anniversary meeting, Nov. 30, 1892, Proc. Roy. Soc. London, A, 52, 302-310, 1892.

Langley, S. P., Address as retiring president of the AAAS, Am. J. Sci., Ser. 3, 37, 1-23, 1889.

Patel, V. L., Low frequency hydromagnetic waves in the magnetosphere: Explorer 12, Planetary Space Sci., 13, 485-506, 1965.

Patel, V. L., A note on the propagation of hydromagnetic waves in the magnetosphere, Earth Planetary Sci. Letters, 1, 282-283, 1966.

Romer, R. H., Alternatives to the Poynting vector for describing the flow of electromagnetic energy, Am. J. Phys., 50, 1166-1168, 1982.

Schield, M. A., J. W. Freeman, and A. J. Dessler, A source for field-aligned currents at auroral latitudes, J. Geophys. Res., 74, 247-256, 1969.

Vestine, E. H., and S. Chapman, The electric current-system of geomagnetic disturbances, Terr. Mag., 43, 351-382, 1938.

Wolf, R. A., and A. J. Dessler, Field-aligned currents in the magnetosphere, Comments on Astrophysics and Space Physics, 1, 117-121, 1969.

Zmuda, A. J., J. H. Martin, and F. T. Heuring, Transverse magnetic disturbances at 1100 km in the auroral region, J. Geophys. Res., 71, 5033-5045, 1966.

THE MAGNETOSPHERIC CURRENTS: AN INTRODUCTION

S.-I. Akasofu

Geophysical Institute, University of Alaska, Fairbanks, Alaska 99701

Abstract. One of the major goals of studying the magnetospheric currents is to gain understanding of the solar wind-magnetosphere-ionosphere system in terms of electric currents, more specifically their generation processes, their roles in coupling the magnetosphere and the ionosphere, their effects on the ionosphere, etc. This is quite natural because most magnetospheric phenomena are of electromagnetic nature and many magnetospheric processes are directly or indirectly related to electric currents. The basic physics associated with the magnetospheric currents is briefly described in terms of eight component current systems:

(1) the Sq current
(2) the Chapman-Ferraro current
(3) the ring current
(4) the solar wind-magnetosphere (SM) dynamo-generated currents
    (4.1) the cross-tail current
    (4.2) the substorm current
    (4.3) the polar cap current
    (4.4) the cusp current
    (4.5) the pulsation current

It is pointed out that our morphological knowledge on the magnetospheric currents has considerably improved during the last two decades, but that an intensive theoretical study of generation processes of the currents has just begun. In particular, the SMD dynamo-generated currents are vital in understanding the solar wind-magnetosphere coupling and energy transfer. There is a great opportunity for young theorists to make a significant contribution to this discipline.

Introduction

Our discipline on magnetospheric currents has grown out from *Geomagnetism*, in particular, geomagnetic storm studies. We owe S. Chapman for his contribution in establishing the present concept of geomagnetic storms (such as the initial phase and the main phase) and in identifying some of the responsible currents systems (such as the magnetopause (Chapman-Ferraro) current and the ring current). Both Birkeland (1908) and Chapman (1918) found intense eastward and westward currents in the polar region, the auroral electrojets (the term coined by Chapman). Both were keenly aware of the fact that magnetic field observations on the earth's surface $\Delta B (r = a, \theta, \phi)$ alone cannot determine uniquely the distribution of the responsible three-dimensional currents $J(r, \theta, \phi)$. Nevertheless, Birkeland was confident that they are fed from outside along the geomagnetic field lines, while Chapman was concerned with an accurate mathematical representation of the storm fields in terms of the so-called 'equivalent currents' on a spherical shell concentric to the earth. In Figure 1, their current systems are illustrated. Both current systems produce similar magnetic fields on the earth's surface.

In 1930-60, the Department of Terrestrial Magnetism (DTM), Carnegie Institution of Washington, Geophysikalisches Institut, Göttingen and Geophysical Institute, Tokyo University, became centers of geomagnetic storm studies. At the DTM and Tokyo University, magnetograms collected during the Second Polar Year were extensively analyzed in terms of the equivalent currents by A.G. McNish, E.H. Vestine, T. Nagata and others; see Vestine et al. (1947) and Fukunisha (1953). Chapman and J. Bartels (1940) worked together on the classic treatise, *Geomagnetism*, at the DTM until just before the World War II separated them. In the meantime, the importance of field-aligned currents in feeding the auroral electrojets was revived by H. Alfvén in his theory of magnetic storms (1950). Chapman-Ferraro's 'corpuscular' theory, Alfvén's 'electric field' theory, and E.O. Hulburt's 'ultraviolet radiation' theory were the three major rivalry theories in 1940-55.

It was also about that period when some scientists had thought that there was nothing more to be learned about geomagnetic storms and urged to close magnetic observatories. In 1950, a number of scientists led by Chapman, J.A. Van Allen and L.V. Berkner carefully began to plan the International Geophysical Year (IGY), the greatest international scientific enterprise on earth. The IGY became one of the most important milestones in our discipline of the magnetospheric currents. Our knowledge on geomagnetic storms, the ionosphere and the aurora was drastically improved by newly collected data during the IGY. The systematic and extensive observation of auroras and magnetic disturbance fields led us to the concept of the magnetospheric substorm which manifests itself as the auroral substorm, the polar magnetic substorm,

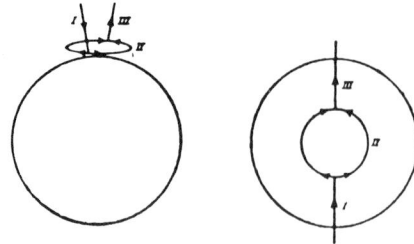
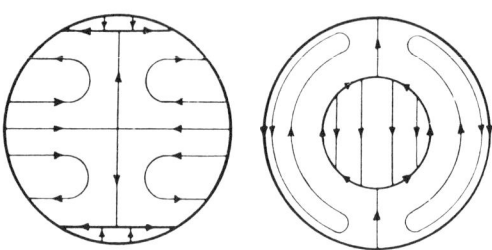

Fig. 1. Comparison between Birkeland's three-dimensional current system and Chapman's equivalent current system. Both current systems can reproduce fairly well the distribution of geomagnetic disturbance field on the earth's surface.

the ionospheric substorm, the formation of a mini-ring current belt, etc. All these substorm features are closely associated with a sudden growth of the auroral electrojets of the life time of the order of a few hours. It has also been found that the period of most major geomagnetic storms (namely, the magnetic manifestation of a magnetospheric storm) coincides with the period when intense substorms occur frequently. This finding led us to conclude that it is essential to understand magnetospheric substorms as basic elements of a magnetospheric storm.

The IGY opened also the new era in our discipline by introducing 'man-made satellites' as a means to explore space around the earth. This point will be discussed more quantitatively in "Current Systems Driven by the Solar Wind-Magnetosphere Dynamo (SMD)". Satellite-borne instruments enabled us to measure, for the first time, the geomagnetic field $B(r,\theta,\phi)$ well above the ionosphere. The presence of the ring current and the magnetopause (the Chapman-Ferraro boundary) was confirmed by some of the earliest satellites. The discovery of the magnetotail added a new dimension to our study of magnetospheric currents (Ness, 1965). We owe Zmuda and Armstrong (1974a,b) to confirm the presence of the field-aligned currents.

Our discipline has also been considerably enriched by a number of new methods of exploring the ionosphere and the magnetosphere, such as the direct measurements of the convective motion of plasmas and the associated electric fields by satellites (Bythrow et al., 1980, 1982; Hones and Schindler, 1979; Heelis and Hanson, 1980; Heelis et al., 1980), incoherent scatter radars (Foster et al., 1981, 1982; Evans et al., 1980; de la Beaujardiere et al., 1981; Robinson et al., 1981, 1982) and chemical releases (Föpple et al., 1967; Haerendel and Lüst, 1968; Heppner et al., 1971; Heppner et al., 1981) and the electric field measurements by satellites and rockets (Cauffman and Gurnett, 1971; Heppner, 1972a, 1977;) and balloons (Mozer and Serlin, 1969). Increasingly sophisticated particle detectors aboard rockets and satellites have identified current-carrying particles in different regions of the magnetosphere (Maier et al., 1980; Kaufman, 1981; Robinson et al., 1981; Frank et al., 1981; Frank and Huang, 1983; Burch et al., 1983). It is for these reasons that magnetospheric current studies have become increasingly interdisciplinary.

Meanwhile, the classical gound-based magnetic method has also been improved in modeling three-dimensional current systems (Kirkpatrick, 1952; Boström, 1964; Akasofu and Meng, 1969; Meng and Akasofu, 1969; Bonnevier et al., 1970; Baumjohann et al., 1980; Baumjohann and Opgenoorth, 1983; Walker et al., 1981). In particular, the International Magnetosphere Study (IMS) coincided with the development of advanced computer codes and also with the establishment of seven IMS meridian chains of magnetic observatories. Systematic data set from the meridian chains of magnetic observatories has been used as the input for the computer codes to determine the distribution of both ionosphere currents and field-aligned currents on an instantaneous basis (Kisabeth, 1979; Mishin et al., 1980; Kamide et al., 1981).

During the last decade, there has also been an increasing emphasis on a study of the roles of the solar wind and the interplanetary magnetic field (IMF) in generating magnetospheric current systems. It has been found that the growth and decay of the magnetospheric currents (expressed by various magnetic disturbance indices) during magnetospheric substorms depend critically on the IMF B_z component; for a list of the results, see Akasofu (1981). In particular, it was found that the total energy dissipation rate $U_T(t)$ of the magnetosphere is related to the power $\varepsilon(t)$ of the solar wind-magnetosphere (SM) dynamo or SMD (Perreault and Akasofu, 1978; Kan et al., 1980; Akasofu, 1981).

In spite of such a dramatic increase of our knowledge on the magnetospheric currents during the period between 1960-1975, there was and perhaps still is some reluctance among many theorists to study magnetospheric phenomena in terms of electric currents. On the basis of the MHD formalism, they replace the current term $\underline{J}$ by $\nabla \times \underline{B}$ and the Lorenz force $\underline{J} \times \underline{B}$ by pressure and

TABLE 1. Observational Methods Which Contribute to Studies of the Magnetic Currents

	Measured quantity	Inferred quantity
Ground-Based		
Magnetic obs.	ΔB	$J_\parallel$, $J_\perp$, E, Σ
Radar obs.	Ion drift speed Electron density	Σ, $J_\perp$ (ionosphere), $J_{\parallel i}$
Satellites and Rockets		
Magnetic obs.	ΔB	$J_\parallel$, $J_\perp$
Electric obs.	E	
Ion drift speed obs.	Ion drift speed	E
Particle flux obs.	Particle fluxes	$J_\parallel$, $J_\perp$
Balloon		
Electric field obs.	E	E (ionospheric)

stress. For example, they considered the ring current in terms of "inflation" and the solar wind-magnetosphere interaction magnetospheric substorm processes and auroral particle acceleration processes in terms of magnetic reconnection. It is obvious there are a number of problems which can most conveniently be studied in terms of $\nabla \times \underline{B}$ in magnetospheric physics, but that there are also a number of problems which can be understood best in terms of J, as pointed out by Alfvén (1977, 1981).

However, problems tend to arise when only one way of thinking becomes prevalent. Indeed, in considering auroral particle acceleration processes on the basis of the MHD formalism, an electric field $\underline{E}_\parallel$ along the geomagnetic field lines was ruled out at the outset and many satellites were sent to the magnetotail to search for reconnection-related acceleration processes. It was as late as 1977 when a significant potential drop along the geomagnetic field lines was discovered just above the aurora (Mozer et al., 1977). Further, it was found that the formation of the auroral potential structure can most easily be understood in terms of J; for a review of this subject, see Kan (1982).

The importance of considering electric currents in studying magnetospheric processes has indeed been revived after the discovery of the field-aligned currents and the auroral potential structure. The fact that the total energy dissipation rate $U_T(t)$ of the magnetosphere is approximately equal to the solar wind-magnetosphere (SM) dynamo power $\varepsilon(t)$ may encourage some theorists to study the solar wind-magnetosphere interaction explicitly in terms of dynamo processes.

In this paper, a brief description of the magnetospheric currents is given in terms of the following eight component current systems (Table 1).

1. Sq current system
2. Chapman-Ferraro system
3. Ring current (the symmetric component)
4. Solar wind-magnetosphere dynamo generated currents
    a. Cross-tail current system
    b. Substorm current system
    c. Polar cap current system
    d. Cusp current system
    e. Pulsation current system

For review articles on the magnetospheric currents, see Akasofu and Chapman (1972), Feldstein (1976), Alfvén (1977, 1981), Akasofu (1977), Nishida (1978), Potemra (1979), Greenwald (1979, 1982) Troshichev (1982), Kamide (1982) and Stern (1983).

Magnetospheric Currents

Sq Current

Motions of the ionospheric plasma, driven by solar thermal processes and the tidal force,

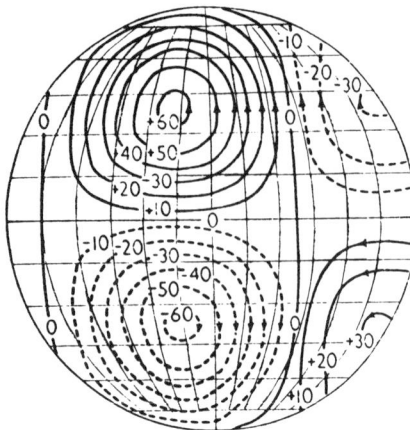

Fig. 2. The Sq current system at equinox seen from the direction of the sun.

across the earth's magnetic field produce two large-scale eddy current systems in the sunlit hemisphere (one in the northern hemisphere and the other in the southern hemisphere). Figure 2 shows the Sq current viewed from the sun. The two eddy currents are approximately fixed with respect to the sun. Thus, a point (magnetometer) on the earth experiences the <u>daily</u> magnetic variations called 'the solar quiet day magnetic variations (Sq)', as it moves under the eddy current once a day.

This ionospheric process has been explained by the dynamo theory in which the horizontal wind system is responsible for generating the required electromotive force.

The first comprehensive dynamo theory of the Sq currents was put forward by Chapman (1919). One of the interesting aspects of the Sq current is that it is very strongly concentrated along the magnetic dip equator. This concentrated portion is called the equatorial electrojet and arises from the fact that the effective conductivity is very high along the dip equator (Baker and Martyn, 1953). During the last decade, one of the main problems related to the Sq current was to identify wind systems which can provide the observed Sq magnetic variations (Matsushita, 1967; Richmond and Matsushita, 1976). It is also important to note that the Chapman-Ferraro current, which will be discussed in the next subsection, contributes significantly to the S_q variations (Olson, 1968).

Chapman-Ferraro Current

Chapman and Ferraro (1931) formulated their theory of the magnetosphere formation by assuming an advancing non-magnetic plasma flow towards a dipole field. Currents are induced in a thin layer of the advancing front of the plasma, shielding the earth's magnetic field in the main body of the plasma. As a result, the shielding current tends to confine the earth's magnetic field in the cavity formed around the earth. In this situation, the earth's magnetic field is said to be 'compressed'. The shielding current consists of two eddy currents, one in each hemisphere, centered around the so-called 'neutral point', where the magnetic field line is perpendicular to the plasma surface; for a non-magnetic plasma flow, the magnetic field is parallel to boundary surface, except at the neutral point. Microscopically, the shielding current arises from a partial gyration motion of both ions and electrons in the vicinity of the magnetopause as they advance from the non-magnetic plasma body and become 'exposed' to the earth's magnetic field.

Inside the magnetosphere the magnetic field produced by the shielding current can be approximated by the magnetic field of a dipole (of the same magnetic moment as that of the earth's dipole) which is located at a distance $2\ell_o$, where ℓ_o is the distance between the earth and the plasma front. This dipole is often called the image dipole. Mead and Beard (1964) gave an accurate expression for the magnetic potential V for the magnetic field produced by the shielding current.

Actually, Chapman and Ferraro (1931) formulated their theory to account for a step function-like change of the north-south component of the earth's magnetic field at storm onset, namely the so-called 'storm sudden commencement (ssc)'. This was because it had been thought then that the sun ejected the coronal gas only intermittently. Therefore, Chapman and Ferraro (1931) suggested that the magnetosphere formed only intermittently and thus that the Chapman-Ferraro current was generated also intermittently. It was after the discovery of the solar wind that we have found that the magnetosphere and the Chapman-Ferraro current are permanent features of the planet earth.

For a magnetized plasma such as the solar wind, the current distribution on the magnetopause is expected to be significantly different from the non-magnetized situation, depending on the nature of the field distribution across the boundary. In terms of the MHD discontinuity description, the boundary is expected to be either a tangential discontinuity or a rotational discontinuity. Some of the geomagnetic lines are expected to be interconnected with the interplanetary magnetic field (IMF) lines across the magnetopause. Such a magnetosphere is said to be 'open'. Therefore, the magnetopause is most likely to be a rotational type discontinuity. However, there is so far no study of the global current distribution on the magnetopause for a rotational discontinuity.

Ring Current (the Symmetric Component)

The ring current arises from motions of trapped particles in the inner magnetosphere. The three

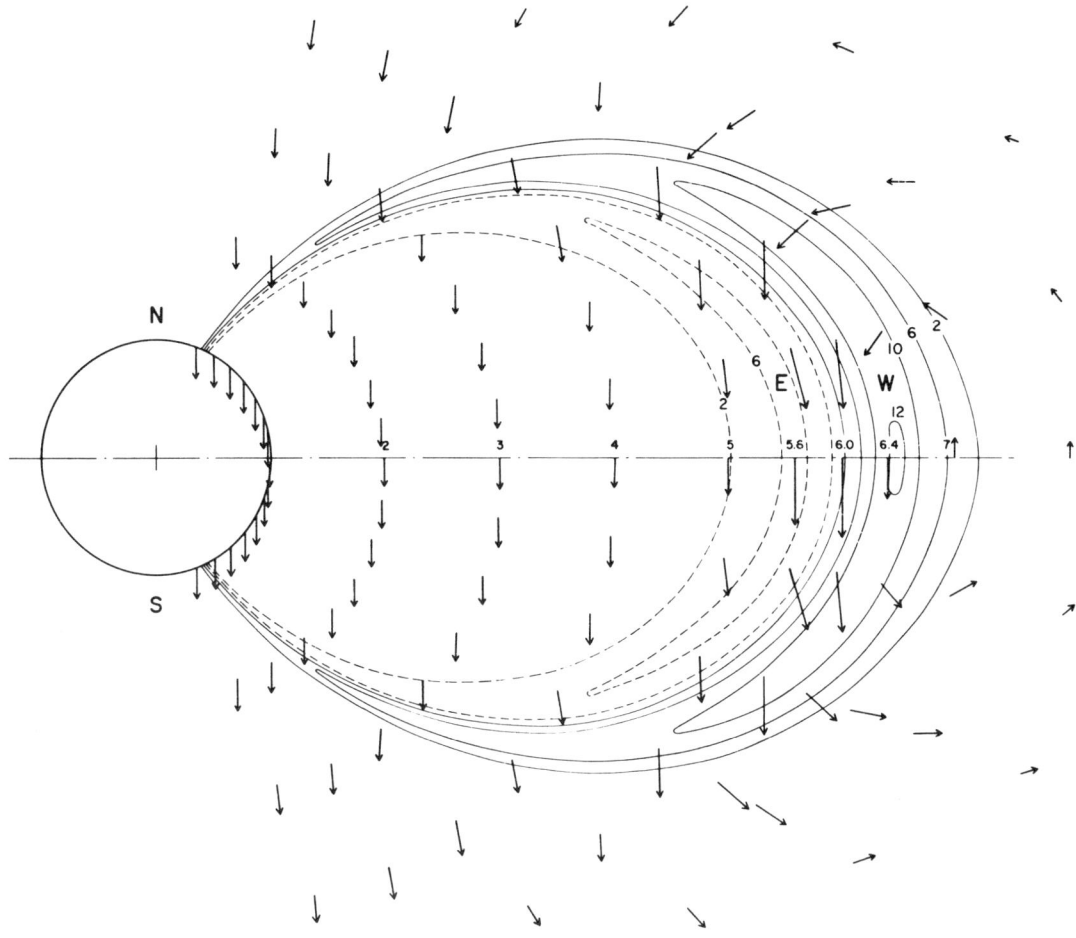

Fig. 3. The distribution of electric currents in the ring current and the magnetic fields (arrows) produced by the ring current; E and W indicates the eastward and westward currents, respectively (Akasofu and Chapman, 1961).

basic motions are the gradient drift motion, the curvature drift motion and the gyration. The perpendicular currents arising from them ($\underline{J}_G$, $\underline{J}_R$ and $\underline{J}_M$, respectively) are given by:

$$\underline{J}_G = -\frac{p_n}{B}\,\underline{b}\times\nabla\left(\frac{1}{B}\right) \qquad (1)$$

$$\underline{J}_R = \frac{p_s}{B^2}\,\underline{B}\times(\underline{b}\cdot\nabla)\,\underline{b} \qquad (2)$$

$$\underline{J}_M = p_n\,\underline{b}\times\nabla\left(\frac{1}{B}\right) - \left(\frac{1}{B}\right)\nabla\times(p_n\,\underline{b}) \qquad (3)$$

where p_n and p_s are the plasma pressure normal and parallel to $\underline{B}$ (of unit vector $\underline{b}$), respectively. The total current $\underline{J} = \underline{J}_G + \underline{J}_R + \underline{J}_M$ is given by

$$\underline{J} = \frac{1}{B}\,\underline{b}\times\nabla p_n + \frac{p_s - p_n}{B}(\underline{b}\times(\underline{b}\cdot\nabla)\,\underline{b}) \qquad (4)$$

It is important to note that the current at any point is always given by $\underline{J}$, not $\underline{J}_G$ or $\underline{J}_G + \underline{J}_R$. Equation (4) allows us to compute the current density $\underline{J}(r,\theta,\phi)$ at any given point for a given magnetic field configuration $\underline{B}\,(r,\,\theta,\,\phi)$ and the plasma pressure (p_n, p_s), provided that the plasma pressure p_n is much less than the magnetic pressure $B^2/8\pi$. Figure 3 gives an example of the current distribution in the ring current and the magnetic field produced by the current (Akasofu and Chapman, 1961). In general, an eastward current flows in the inner (earthward) side and a westward current in the outer side. The latter is slightly stronger than the former, so that the net current is directed westward. The resultant magnetic field in the vicinity of the earth is thus directed southward. Since a northward field is measured positive, the southward directed field is registered as depression of the north-south component. It is this southward field which is responsible for the symmetric part of the main phase depression.

The magnetic field $\Delta B_z(0)$ of the ring current at the earth's center can be given (Dessler and Parker, 1959) by

$$\Delta B_z(0) \sim -2.4 \times 10^{-21} \varepsilon_R \qquad (5)$$

where ε_R denotes the total kinetic energy of the ring current particles.

In addition to the current generated by the trapped particles in the ring current belt, there are westward-directed currents in the plasma sheet; they are generated by the guiding center motions and other processes (Frank and Huang, 1983). Their net effects are expected to make an appreciable contribution to the main phase decrease.

During the last decade, one of the important problems on the ring current has been to identify the particles which contribute most to the ring current. Terrestrial ions, such as He^+ and O^+, appear to contribute significantly to the ring current, as well as protons of solar wind origin and of terrestrial origin (Shelley et al., 1972; Lennartsson et al., 1981; and Balsiger et al., 1980). The life time of these ions is important to study the storm field during the recovery phase; for recent reviews on topics of the ring current, see Lyons and Williams (1980) and Williams (1983). Another important unsolved problem is 'injection' processes of the ring current particles from the plasma sheet to the inner magnetosphere (Ejiri et al., 1980; Moore et al., 1981).

Current Systems Driven by the Solar Wind-Magnetosphere Dynamo (SMD)

For a magnetized solar wind, the magnetosphere is expected to be open (Dungey, 1961), except for a very special situation in which the IMF is anti-parallel to the earth's dipole moment. Thus, some of the geomagnetic field lines are almost always interconnected with the IMF lines across the magnetopause. It is likely that a significant part of the electromotive force of the solar wind-magnetospheric (SM) dynamo is generated as the solar wind flows across the interconnected field lines near the magnetopause. Viewing the near-circular cross-section of the magnetotail from the earth, the current is generated along the magnetopause in the counterclockwise direction in the northern hemisphere and clockwise in the southern hemisphere. The induced voltage is expected to be approximately equal to the potential drop across the polar cap, namely ~ 100 kilovolts. Since the total magnetospheric current is the order of $\sim 5 \times 10^7$ amperes (Table 1), the total power generated by the dynamo is of the order of 5×10^{12} watt (5×10^{19} erg/sec).

The power of the SMD ε is given (Perreault and Akasofu, 1978) by

$$\varepsilon = VB^2 \sin^4\left(\frac{\theta}{2}\right) \ell_o^2 \qquad (6)$$

where V = the solar wind speed,
B = the magnitude of the IMF,
θ = the polar angle of the IMF, projected on the y-z plane,
ℓ_o = a constant (= $\sim 7 R_E$).

The power expressed by (6) is equal to the total power dissipation rate U_T in the inner magnetosphere and in the ionosphere. The power ε can vary considerably as the solar wind speed V, the IMF magntiude B and the orientation angle θ vary in time. The SM dynamo supplies much of the power for most magnetospheric phenomena. When the power ε is less than $\sim 10^{18}$ erg/sec (or $\sim 10^5$ M watts), the magnetosphere is said to be quiet or in the ground state. The magnetosphere is said to be disturbed when the power is greater than $\varepsilon \sim 10^{18}$ erg/sec. Akasofu and Kan (1982) suggested that when the power is between $\sim 10^{18}$ and $\sim 10^{19}$ erg/sec, the magnetosphere is in the substorm condition; if the power exceeds 10^{19} erg/sec, the magnetosphere is in the storm condition. As mentioned earlier in the Introduction, the period of most major magnetic storms coincides with the period when intense substorms occur frequently. More quantitatively, during a major storm, the intensity of substorms measured in terms of ε or U_T must exceed $\sim 10^{19}$ erg/sec and that a few such major substorms in a short period (~ 6 hrs) are required to build up an intense ring current belt.

Figure 4 shows the kinetic energy flux $K = \rho v^3 \ell_o^2$, the SM dynamo power ε, the total dissipation rate U_T of the magnetosphere and the geomagnetic indices AE and Dst for February 16-17, 1967. It can be seen that the SM dynamo power ε is correlated with the total dissipation rate U_T, but there is little relation between K and U_T. Further, it is important to note that the large main phase depression developed rapidly during the period when the power was high ($\varepsilon \gtrsim 10^{19}$ erg/sec).

The above correlation between $\varepsilon(t)$ and $U_T(t)$ suggests that a magnetospheric substorm is primarily a direct consequence of an enhanced power of the SM dynamo, namely of a directly driven process; see also McPherron (1983). Thus, the magnetosphere is primarily a driven system. However, there is no doubt that magnetic energy accumulated in the magnetotail contributes also to substorm processes (the unloading component). It is one of the future problems to determine the degree of contribution of both the directly driven and the unloading components.

Perhaps, one of the most important observational evidences for the directly driven nature of substorms is that the polar cap potential drop ϕ_{pc} is proportional to ε or more likely to $\sqrt{\varepsilon}$ (Reiff et al., 1981); their results are shown in Figure 5. This is because the polar cap potential drop ϕ_{pc} can be considered as the driving source of the auroral electrojets, as well as many other polar substorm features.

As mentioned earlier, the SM dynamo supplies

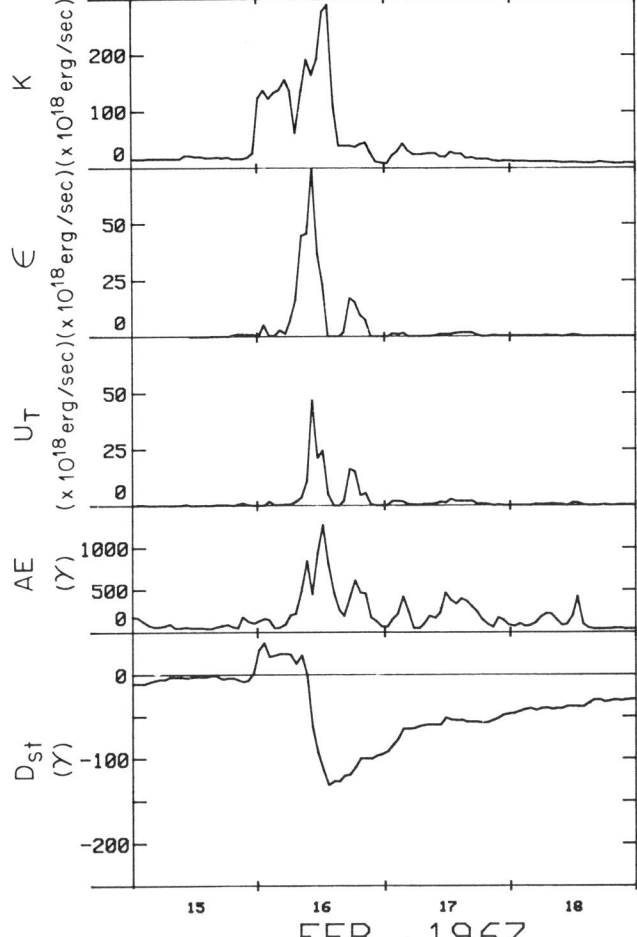

Fig. 4. From the top, the kinetic energy flux (K), the dynamo power (ϵ), the total energy dissipation rate (U_T) of the magnetosphere and the two geomagnetic indices AE and Dst for the February 16-17, 1967 storm (Akasofu, 1981).

much of the power for various 'disturbance' phenomena in the inner magnetosphere and ionosphere. It is convenient to describe the dynamo-generated currents in terms of following five systems:

(a) Cross-tail current system
(b) Substorm current system
(c) Current system when the IMF B_z is positive and large
(d) Cusp current system
(e) Pulsation current system

Cross-Tail Current System

Much of the currents generated by the SM dynamo flows across the magnetotail from the dawn side to the dusk side. As a result, the magnetotail consists of two 'solenoids'. It has been a long-standing problem to find out current carriers of the cross-tail current and the nature of its orbits (Sonnerup, 1971; Eastwood, 1975; Cowley, 1978; Wagner et al., 1979). It appears now that the main carrier is electrons (Frank and Huang, 1983). It is an interesting problem to find out why electrons, instead of protons, are responsible for carrying a significant amount of the cross-tail current.

Substorm Current System

Introduction. The substorm current system consists of ionospheric currents, field-aligned currents and all connected currents in the magnetosphere. The ionospheric currents consist mainly of the eastward and westward electrojets. The field-aligned currents include the so-called 'region 1 and 2 currents'. This current system is so named here, because it is greatly enhanced during magnetospheric substorms. The magnetic fields produced by this particular current system become measurable when the power ϵ exceeds $\sim 10^{18}$ erg/sec. In this situation, all typical features associated with the current become evident, and most magnetospheric 'disturbance' features have been described in terms of magnetospheric substorm features. The ionospheric current pattern undergoes significant changes during a magnetospheric substorm, but the two electrojets and the region 1 and 2 field-aligned currents remain as the basic features.

Average Ionospheric Current Pattern. In the following, we shall describe some details of the average substorm current system in the ionosphere. It was obtained on the basis of 50-day average horizontal disturbance vectors at the IMS Alaska meridian chain stations, using an advanced

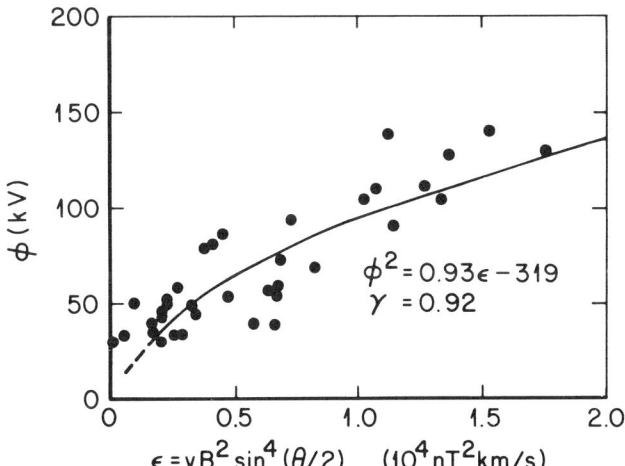

Fig. 5. The relationship between the dynamo power ϵ and the polar cap potential drop Φ (Reiff et al., 1981).

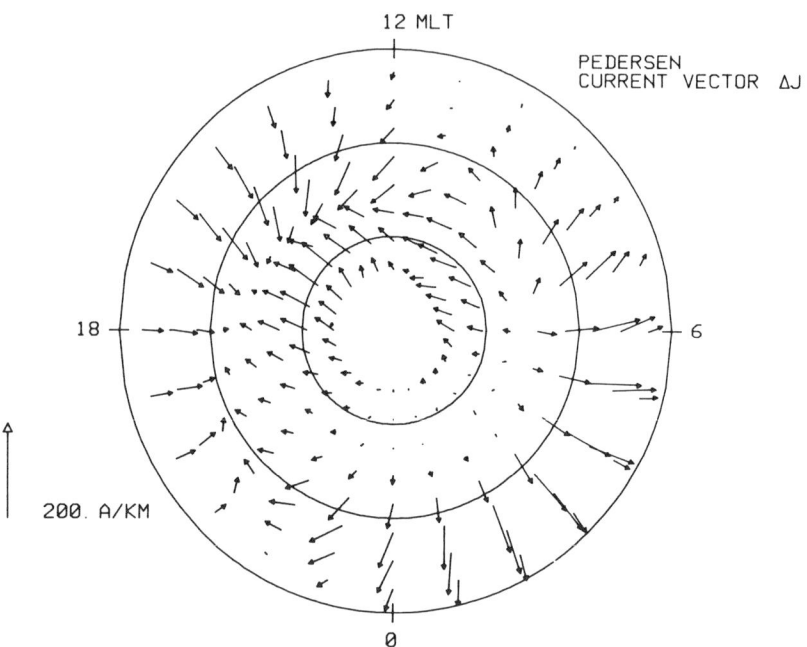

Fig. 6. The distribution of the Pedersen current in magnetic latitude-MLT coordinates which is obtained by the KRM method.

computer code developed by Kamide et al. (1981). In the ground-based magnetic method, the non-uniqueness of the results had been of serious concern to workers in the past. Thus, it is of great importance to compare the distributions of electric currents obtained by different methods. However, it was found that the deduced ionospheric current systems by different methods were in good agreement (Akasofu et al., 1981). Therefore, the non-uniqueness problem is, for practical purposes, removed as far as the distribution of the ionospheric currents is concerned.

In modeling ionospheric current systems on the basis of magnetic or electric field observations, by either ground-based or satellite-borne instruments, accurate knowledge of the conductivity of the polar ionosphere is essential. One solution to this problem is to have a large number of either ground-based or satellite-borne ionospheric sounders and/or incoherent scatter radars. Such a solution cannot be realized at this time. However, during the last decade, the conductivity in the auroral region has extensively been studied by incoherent scatter radars (Brekke et al., 1974; Vickery et al., 1981). Further, it has also become possible to construct conductivity models using electron data taken from polar orbiting satellites (Wallis and Budzinski, 1981; Spiro et al., 1982; Reiff, 1983). Such an improvement of our knowledge on the conductivity is very fortunate in modeling the current system.

Now, we shall examine the distribution of the Pedersen currents (Figure 6) and the Hall currents (Figure 7) by the KRM method using daily average data obtained by the Alaska meridian chain of magnetometers as the input; for details, see Kamide and Akasofu (1981). One of the main features of the distribution of the Pedersen currents is the southward directed component in the latitudinal belt between 62° and 72° in the morning sector and a northward directed component in the same latitudinal belt in the afternoon sector. As we shall see shortly, these north-south currents across the auroral oval are connected to the region 1 and 2 field-aligned currents. Another prominent feature of the Pedersen current is divergence of the current from about 76° in latitude in the midmorning sector and convergence toward the same latitude in the afternoon sector. Associated with this particular feature, there is a fairly intense eastward component in the latitudinal belt between 70° and 80° in latitude in the day sector.

The distribution of the Hall currents is characterized by an intense westward component in the latitudinal belt between 62° and 72° in the morning sector and an eastward component in the same latitudinal belt in the afternoon sector (Hughes and Rostoker, 1977, 1979; Rostoker and Hughes, 1979; and Rostoker, 1980a). Its diverging trend at about 75° in latitude in the day sector, together with two crescent flow patterns, are clearly seen.

In the past, many workers assumed that all Hall current lines close themselves in the ionosphere, forming two cells located in the morning-forenoon

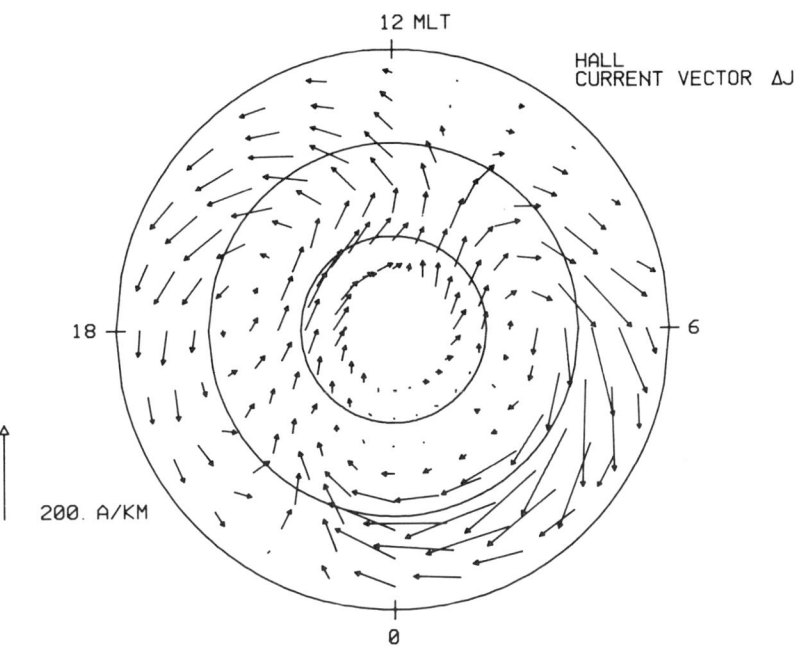

Fig. 7. The distribution of the Hall current.

and the afternoon-evening sectors, respectively. Such a practice is incorrect. At the "tips" of the two crescent patterns, the Hall current must be associated with field-aligned currents, a downward net current from about ~ 1100 MLT sector and an upward net current from about ~ 2200 MLT sector.

Example of an Instantaneous Current Pattern. It is also interesting to see how the ionospheric current system grows and decays during a single substorm. Here, we examine a typical example which is obtained using magnetic records from the six IMS meridian chains and others (the total number of stations being 71); the KRM method is used (Ahn, 1983). Figure 8 shows the total ionospheric current distributions at 1040, 1100, 1200, 1400, 1500 and 1700 UT on March 17. It is particularly interesting to note that the intrusion of the westward electrojet is the most prominent feature, in addition to the overall enhancement.

Field-Aligned Currents. The initial study of the field-aligned currents by Zmuda and Armstrong (1974a,b) was greatly refined by Iijima and Potemra (1978). Their result is reproduced here as Figure 9. On the basis of Figures 6 and 9, one can immediately infer that much of the Pedersen current is fed by the region 1 and 2 currents. The reason for this particular configuration of the field-aligned currents and the Pedersen currents can be understood by realizing that the Lorentz force generated by the current system is directed in both the dawn and dusk sides of the ionosphere, driving the plasma sunward.

Some of the important aspects of the region 1 and 2 currents are that the intensity of the region 1 current is about twice of the region 2 current and that both currents increase as magnetic activity increases (Sugiura and Potemra, 1976). The first point is very important in considering generation mechanisms of the field-aligned currents. For recent observations of the field-aligned currents, see Klumpar et al. (1976); McDiarmid et al. (1978); Saflekos et al. (1978); Shuman et al. (1981); Iijima (1983a); Burrow and Wilson (1983) and Sugiura et al. (1983).

We have so far only a few observations of the field-aligned currents in the distant magnetosphere (Harendel et al., 1971; Fairfield, 1973; McPherron and Barfield, 1980; Frank et al., 1981). It appears that the field-aligned currents are concentrated in the upper and lower boundaries of the plasma sheet in the magnetotail and that the directions are in general consistent with those of the region 1 currents.

Driving Mechanisms and Models. Generation mechanisms of the field-aligned currents are closely related to plasma flows, namely the solar wind flow near the magnetopause (including the low-latitude boundary layer) and the convection flow inside the magnetosphere. In a steady state, the equation of motion of such plasma flows can be rewritten as

$$J_\perp = \frac{B}{B^2} \times [\nabla p + \rho (\underline{v} \cdot \nabla) \underline{v} - \mu \nabla^2 v] \qquad (7)$$

where $\underline{J}_\perp$ denotes the perpendicular (height-

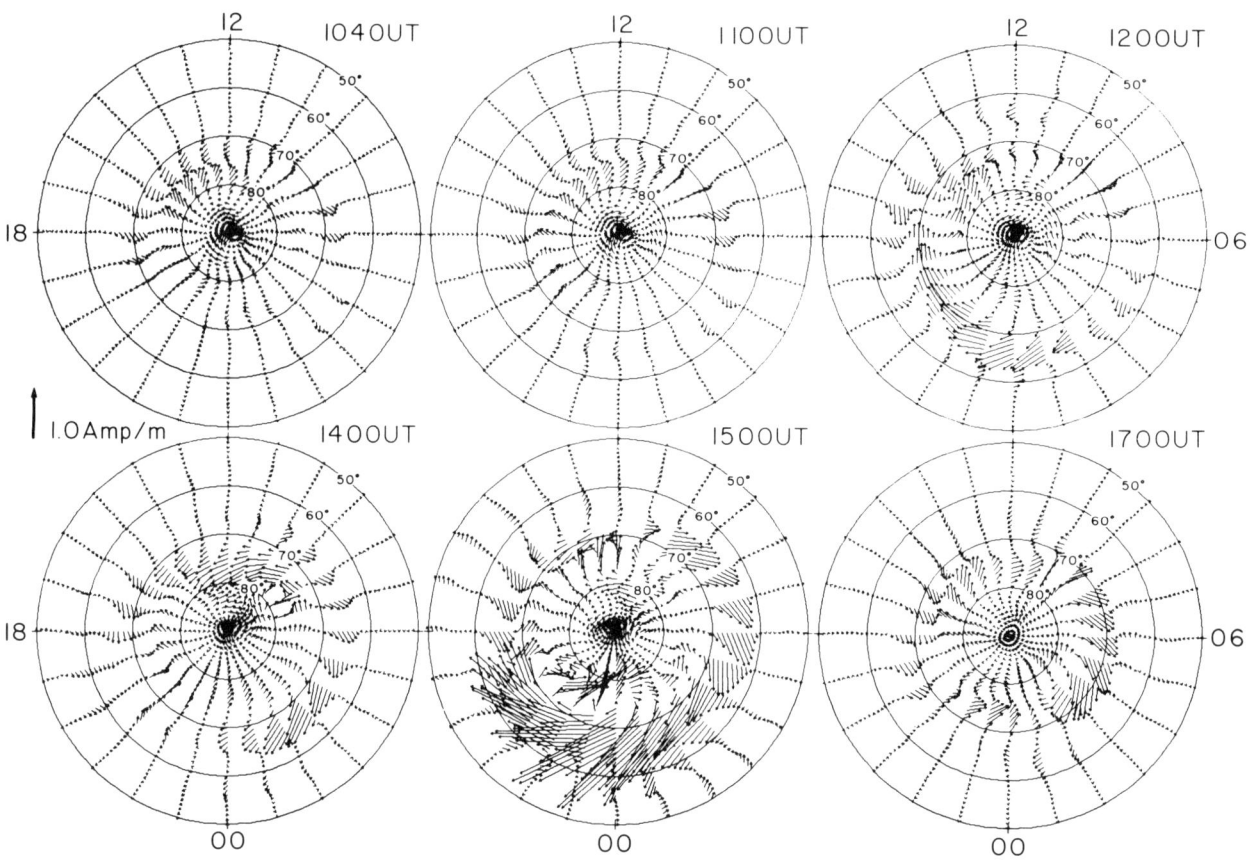

Fig. 8. An example of changes of the distribution of the total ionospheric current during one of the magnetospheric substorms on March 17, 1978; for the AE index, see Figure 14a.

integrated) current and

$$J_\parallel = - \nabla \cdot J_\perp \tag{8}$$

It is important to realize, first of all, that there will be no field-aligned current, unless there is the ionosphere, namely a conductive layer. Thus, $j_\parallel$ cannot be independent of the ionospheric conductivities (Σ_p, Σ_H). The higher the conductivities are, the higher the field-aligned currents are. This suggests that $\underline{v}$ and ∇p in (7) cannot be independent of $\overline{\Sigma}_p$ and $\overline{\Sigma}_H$ and thus that the plasma flow is somewhat modified by the presence of the ionosphere. A dynamo is meaningless unless a load is connected. As pointed out by Kan et al. (1983), most individual theoretical studies in the past dealt only with one of the terms in the brackets of (7).

(a) <u>Pressure Gradient Term</u>

The pressure gradient term (∇p) was considered by Vasyliunas (1970) who showed that the derivative of $J_\parallel$ along the magnetic field direction is related to ∇p by

$$\frac{\partial}{\partial \ell}\left(\frac{J_\parallel}{B}\right) = \frac{2(\underline{B} \times \nabla B) \cdot \nabla p}{B^4} = \frac{2}{B^4} \nabla B \cdot (\nabla p \times \underline{B}) \tag{9}$$

which can be integrated to give

$$J_\parallel = \frac{B_I}{2\ B_e} \nabla p \times \frac{B_e}{B_e} \cdot \nabla \frac{d\ell}{B} \tag{10}$$

where $\underline{B}_e$ and $\underline{B}_i$ denote the magnetic field at the equator and the ionosphere, respectively. Equation (9) indicates that field-aligned currents are present whenever the pressure gradient has a component along the direction of the ∇B drift. Wolf (1975), Harel et al. (1981a), Chen et al. (1982) and Karty et al. (1982) examined exten-

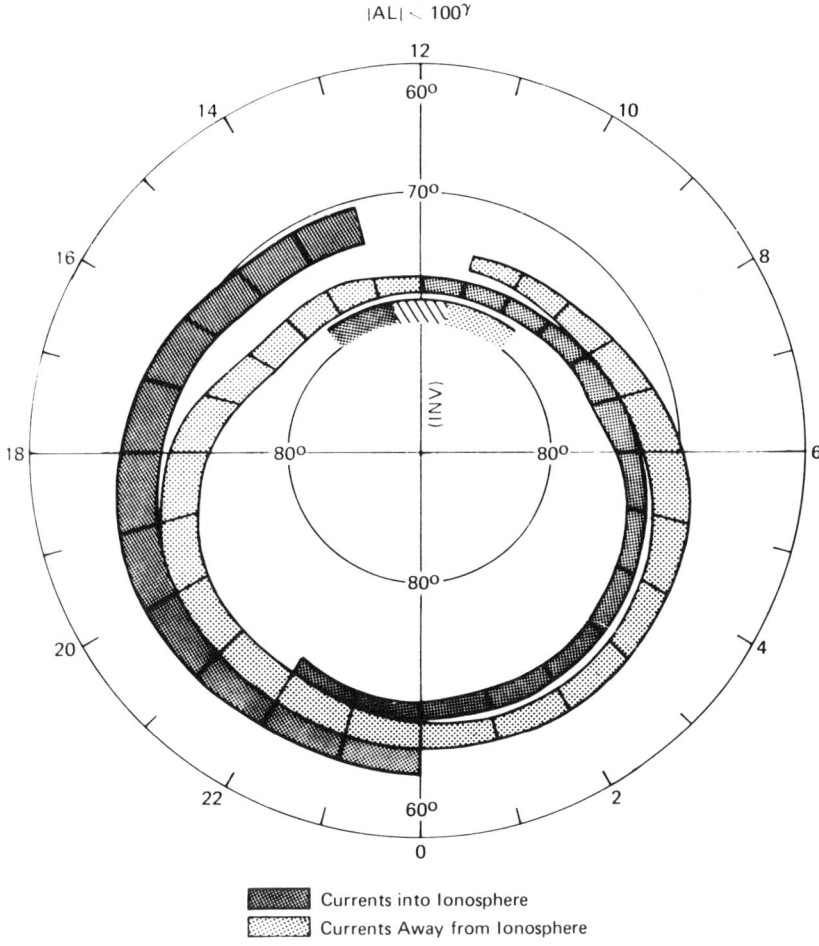

Fig. 9. The distribution of upward and downward field-aligned currents in invariant latitude-MLT coordinates (Iijima and Potemra, 1978).

sively the current systems associated with the ∇p term by their numerical model.

(b) <u>Inertia Term</u>

The inertia team was examined by Boström (1975), Rostoker and Boström (1976), Rostoker (1980b), Kan and Lee (1980) and Hasegawa and Sato (1980). Hasegawa and Sato (1980) expressed $J_\parallel$ by

$$J_\parallel = B \int_0^\ell \frac{en}{B} \frac{d}{dt} \frac{\Omega}{\omega} \, d\ell \qquad (11)$$

where Ω denotes the projection of the vorticity vector in the direction of the magnetic field, namely

$$\underline{\Omega} = (\nabla \times \underline{v}) \cdot \underline{b} \qquad (12)$$

(c) <u>Viscous Term</u>

Sonnerup (1980) examined the viscous term in (7) by considering the low latitude boundary layer (Eastman and Hones, 1977, Eastman et al., 1979) inside the dawn and dusk magnetopause.

(d) <u>Summary</u>

Our theoretical understanding of generation mechanisms of the field-aligned currents is very poor at the present time. The main problem is that most theoretical studies deal with only one of the terms in the brackets of (7). At present most models include quantitatively or qualitatively only one of the terms; for recent papers on theoretical models of the magnetospheric current system, see Rostoker (1980), Harel et al. (1981a,b), Crooker and Siscoe (1981), Stern (1983), Siscoe (1982) and Troshichev (1982).

<u>Joule Heat Production</u>. A part of the power generated by the solar wind-magnetosphere dynamo in the ionosphere. Since the KRM method provides

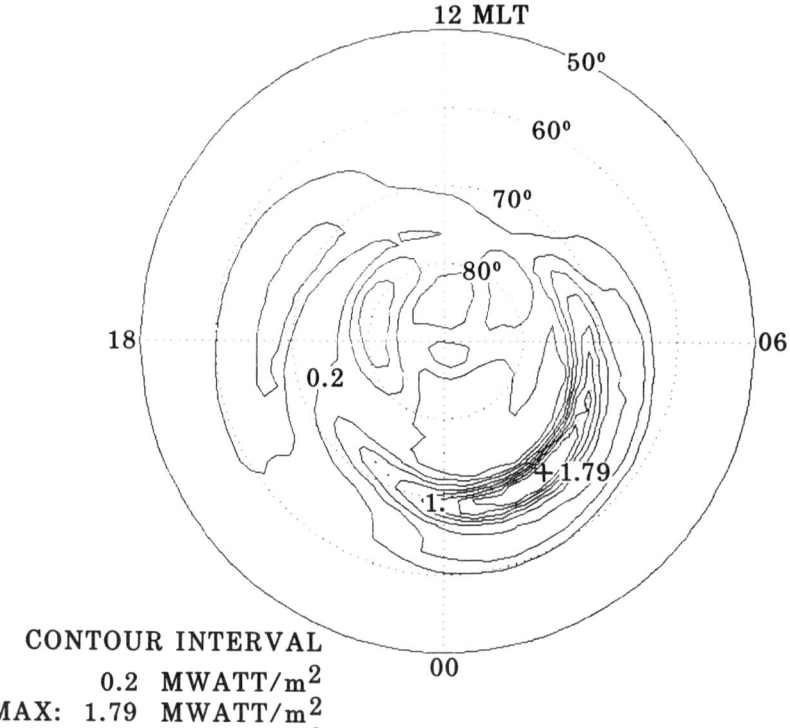

Fig. 10. The distribution of the Joule heat production rate in invariant latitude-MLT coordinates; see the caption for Figure 10 (Akasofu et al., 1983).

the electric potential distribution and the current distribution, it is possible to obtain the Joule heat production rate $u_J = (\underline{E} \cdot \underline{I})$ over the entire polar region. Figure 10 shows the result which corresponds to the situations examined earlier (Figures 6 and 7). One can see that the major Joule heat dissipation takes place in the morning sector along the auroral oval. There is also a secondary peak in the afternoon sector (Banks et al., 1981). It is expected that the heat thus produced will generate a global wind in the upper atmosphere. A global wind can also be generated by the fact that the ($\underline{E} \times \underline{B}$) convective motion of the plasma in the ionosphere tends to drag the neutral component (cf. Roble et al., 1982; Spencer et al., 1983; Foster et al., 1983). Such a wind may have a far-reaching consequence. Suppose as an extreme situation that the neutral component moves with the same speed as the ionospheric plasma. In such a situation, there will be no ionospheric current and no joule heat production. That is to say, the ionosphere has a limit in dissipating the power generated by the SM dynamo.

<u>Relations With Auroral Phenomena</u>. One of the most important reasons to study magnetospheric currents is that the upward field-aligned currents are closely related to the development of the auroral potential structure. This is because magnetospheric electrons cannot carry upward field-aligned currents of density of 10^{-6} A/m^2 or greater unless an electric potential drop of the order of 1 kV or greater is set up along the magnetic field lines (Knight, 1973; Fridman and Lemaire, 1980; Chiu and Cornwall, 1980; Lyons, 1981; Lyons et al., 1979; Kan, 1982). It has been suggested that it is this auroral potential structure which is responsible for accelerating auroral electrons and thus causing the variety of auroral phenomena.

The upward-directed field-aligned currents associated with the Hall currents may also be carried by downward moving electrons from the magnetosphere. However, as mentioned earlier, because of their mirroring tendency the downward moving electrons can carry only a limited amount of current density, $\sim 10^{-6}$ A/m^{-2}. Since the upward Hall-associated field-aligned current ($j_{\parallel H}$) occurs in the midnight and the late evening sectors (from both the westward and eastward electrojets), it is interesting to speculate that bright auroral arcs in the poleward expanding bulge and the westward traveling surge are partly produced by the electrons which carry the upward $j_{\|H}$.

Therefore, the ionosphere is only a passive part of the magnetosphere in terms of providing the closing circuit for the Pedersen current I_p, dissipating the power generated by the solar wind-magnetosphere dynamo ($\underline{E} \cdot \underline{I}_p > 0$). However, the ionosphere may play an active role by contributing to many substorm processes which were primarily thought to be due to magnetospheric processes in the past.

Current System During Periods When the IMF B_z Component is Positive and Large

When the IMF B_z component is positive and large (and thus $\varepsilon \lesssim 10^{17}$ erg/sec), the current system does not appear to be simply a weak substorm current system. This is, perhaps, partly because the geometry of the open field region deviates significantly from a contracted oval and because effects of the IMF B_y component becomes relatively more important than in the situation of the IMF $B_z < 0$ (Russell, 1972; Cowley, 1981a, b; Akasofu et al., 1983). The drastic changes of the open field line region are manifested by the presence of the so-called 'polar cap arcs' and the asymmetry of the auroral distribution with respect to the sun-earth line (Lassen and Danielsen, 1978; Gussenhoven, 1982; Akasofu and Roederer, 1983).

Maezawa (1976) showed that the polar cap current system during an extremely quiet period is different from that in an active period and thus the two-cell pattern degenerates into the four-cell pattern. There have been several studies of this interesting feature (Saflekos and Potemra, 1980; Horwitz and Akasofu, 1979; Burke et al., 1979, 1982; Smiddy et al., 1980; Reiff, 1982; Bythrow et al., 1983; Iijima, 1983b; Friis-Christensen, 1983). Figure 11 shows the four-cell convection pattern and the corresponding electric field observation (Burke et al., 1979). However, much work is still needed to understand the accurate current pattern during such extremely quiet periods.

Cusp Current System

The cusp region is a unique region of the polar upper atmosphere, where solar wind particles appear to have free access, because the magnetic field from the cusp is perpendicular to the magnetopause (Section 2.2). There seems to be a unique current system associated with the cusp region, and its circuit is greatly influenced by the IMF B_y component.

Since the effect of the B_y component of the IMF on the daily magnetic variations in the polar cap was found by Svalgaard (1968, 1972, 1973) and Mansurov (1969), a number of papers have been

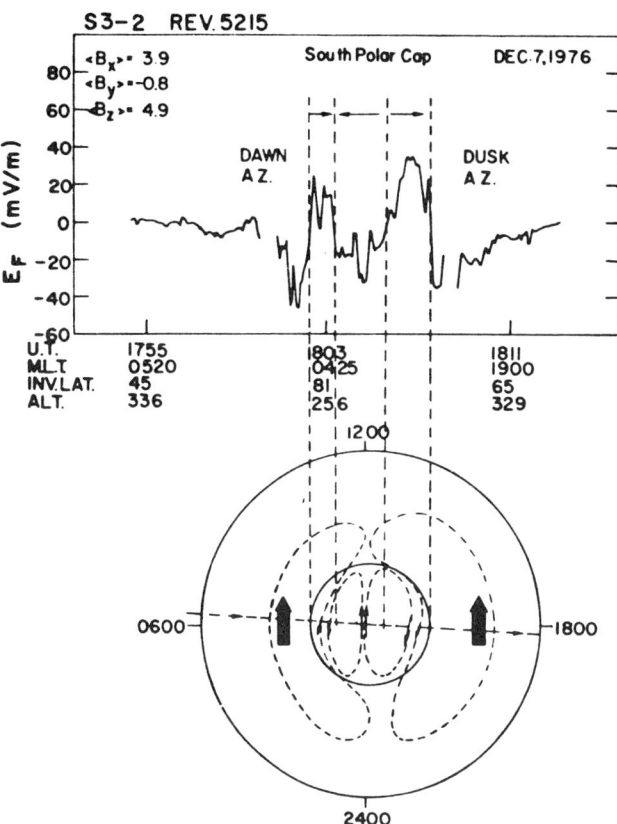

Fig. 11. The electric field pattern observed by the S3-2 satellite in the southern polar cap during a period of northward IMF. The resultant plasma flow is indicated in the lower part (Burke et al., 1979).

published on this subject (Jørgensen et al., 1972; Friis-Christensen et al., 1972; Heppner, 1972b; Sumaruk and Feldstein, 1973; Kawasaki and Akasofu, 1973; Berthelier et al., 1974; McDiarmid et al., 1978; Friis-Christensen and Wilhjelm, 1975; Feldstein, 1976; Wilhjelm et al., 1978; Iijima and Potemra, 1978; Saflekos and Potemra, 1980; Rostoker et al., 1982; Doyle et al., 1981; Bythrow et al., 1982, 1983).

One may consider the B_y effects as modulation effects of the SM dynamo by the IMF B_y component, causing an asymmetry of the current system with respect to the noon-midnight meridian. The dynamo becomes asymmetric with respect to the sun-earth line. Rostoker et al. (1982) showed that the IMF B_y component redistributes the currents in the cusp region and does not change significantly the total current and that the redistribution results in eastward ionospheric current being observed in the prenoon quadrant on the poleward side of the normally present west-

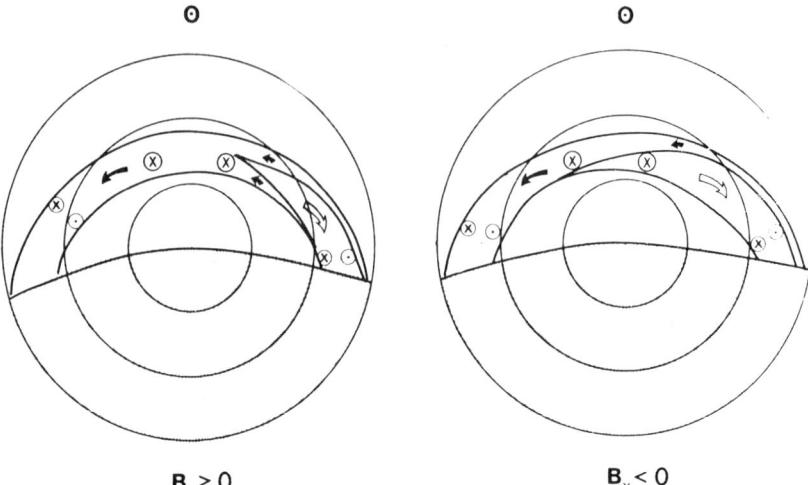

Fig. 12. Schematic diagram of the electrojets across local noon for conditions where the IMF By component has opposite polarities. In this model the balanced components of the field-aligned currents feature current out of the ionosphere at the poleward edge of the noon sector cleft for By > 0 and current into the ionosphere at the poleward edge of the noon sector cleft for By < 0 (Rostoker, 1982).

ward electrojet when $B_y > 0$, where as only westward current flows in the prenoon quadrant when $B_y > 0$. Figure 12 illustrates schematically the distribution of ionospheric currents for IMF $B_y > 0$ and $B_y < 0$, as well as the distribution of the field-aligned currents.

Pulsation Current System

Geomagnetic pulsations have mostly been studied in terms of Alfvén waves, so that there have been only a few papers in the past, which deal with pulsations in terms of magnetospheric currents. Holzer and Reid (1975) and Rostoker and Lam (1978) considered equivalent circuits for some of the pulsations. According to the latter authors, the circuit parameters for Pc-5 pulsations are as follows:

L_M (inductance) ~ 7H
C_M (capacitance of the magnetospheric currents) ~ 8×10^2 F
C_I (capacitance of the ionosphere) ~ 20 F
R_I (resistance of the ionosphere) ~ 0.5 Ω

In the circuit L_M and R_I are connected in series, but C_M and C_I are in parallel. The resulting resonance frequency is about 2 mHz which is within the Pc-5 frequency range.

During the last few years, there has been some trend to consider the field-aligned currents in terms of the Alfvén waves. A steady state may be represented by a standing Alfvén wave. A sudden injection of field-aligned currents to the ionosphere has been studied in terms of the propagating Alfvén waves. Such consideration appears to be successful in explaining the Pi-2 pulsations which tend to appear at substorm onset or intensification.

Concluding Remarks

Our morphological knowledge on the magnetospheric currents has been improved considerably during the last two decades by great efforts based on a variety of observational methods. However, an intensive theoretical study of generation processes of the currents has just begun. Both genius and diligent efforts are now needed to make a quantum progress in theoretical studies, and thus there is a great opportunity for young theorists to make a significant contribution.

In dealing with magnetospheric currents, it is important to consider the whole 'circuit' (Alfvén, 1981). An interesting question is then what are the basic elements in the magnetospheric circuit. Expecting considerable outcry, it is proposed here that they are an inductance, a resistor and a fluorescent lamp which are connected in series; Figure 13. The main part of the inductance arises from a large-scale circuit of the magnetosphere, particularly the magnetotail 'circuit'. The ionosphere can be considered as a resistor as a first approximation. The flourescent lamp represents the aurora. The proposed circuit emphasizes also that the magnetosphere is a directly driven system. In the past, most workers have considered that a Thyratron-like device (representing the magnetotail) is an essential element in the circuit, generating a series of saw-tooth-like signals (representing substorms) even for a constant supply of the power.

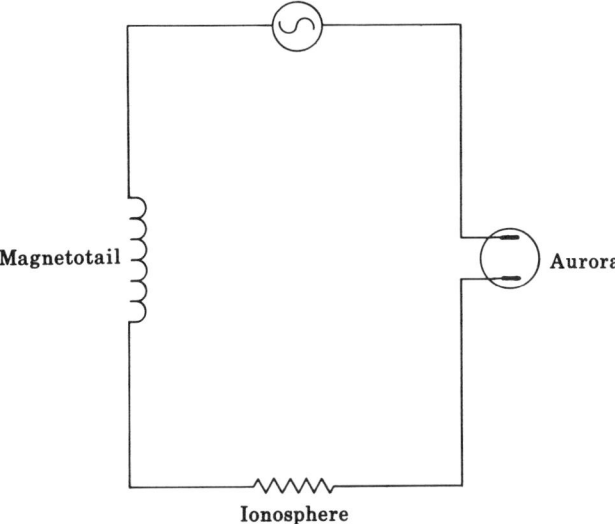

Fig. 13. An example of the basic magnetospheric current circuit.

If there is considerable disagreemnt on the basic circuit of the magnetosphere among magnetospheric physicists, it will indicate simply how little we share the common ground in studying magnetospheric currents and how little we know about the magnetospheric circuit as a whole. It is hoped that we shall have the common basic circuit in the next five years or so, upon which various complexities can be added to make the magnetospheric circuit as realistic as possible.

Acknowledgments. I would like to thank my colleagues who have collaborated with me in studying various aspects of the magnetospheric currents during the last two decades. This review work was supported in part by a grant from the National Science Foundation (ATM81-15321) and in part by a grant from the National Aeronautics and Space Administration (NSG-7447).

References

Ahn, B.-H., Electric conductivities, currents and energy dissipation in the polar ionosphere, Ph.D. Thesis, University of Alaska, May 1983.

Akasofu, S.-I., and S. Chapman, The ring current geomagnetic disturbances and the Van Allen radiation belt, J. Geophys. Res., 66, 1321, 1961.

Akasofu, S.-I., and C.-I. Meng, A study of polar magnetic substorms, J. Geophys. Res., 74, 293, 1969.

Akasofu, S.-I., and S. Chapman, Solar-Terrestrial Physics, Oxford University Press, 1972.

Akasofu, S.-I., Physics of Magnetospheric Sulstorms, D. Reidel Pub. Co., Dordrecht-Holland, 1977.

Akasofu, S.-I., Y. Kamide, and J. Kisabeth, Comparison of two modeling methods for three-dimensional current systems, J. Geophys. Res., 85, 3396, 1981.

Akasofu, S.-I., D.N. Covey, and C.-I. Meng, Dependence of the geometry of the region of open field lines on the interplanetary magnetic field, Planet. Space Sci., 29, 803, 1981.

Akasofu, S.-I., Energy coupling between the solar wind and the magnetosphere, Space Sci. Rev., 28, 121, 1981.

Akasofu, S.-I., Relationships between the AE and Dst indices during geomagnetic storms, J. Geophys. Res., 86, 4820, 1981.

Akasofu, S.-I., and J.R. Kan, Importance of initial ionospheric conductivity on substorm onset, Planet. Space Sci., 30, 1315, 1982.

Akasofu, S.-I., B.-H. Ahn, and G.J. Romick, A study of the polar current systems using IMS meridian chains of magnetometers, 1. Alaska meridian chain, Space Sci. Rev., (in press), 1983.

Akasofu, S.-I., and M. Roederer, Polar cap arcs and the open regions, Planet. Space Sci., (in press), 1983.

Alfvén, H., Electric currents in cosmic plasmas, Rev. Geophys. Space Phys., 15, 271, 1977.

Alfvén, H. Cosmic Plasma, D. Reidel Pub. Co., Dordrecht-Holland, 1981.

Baker, W.G., and D.F. Martyn, Electric currents in the ionosphere, 1. The conductivity, Phil. Trans. R. Soc., A246, 281, 1953.

Balsiger, H., P. Eberhardt, J. Geiss and D.T. Young, Magnetic storm injection of 0.9 - 16 keV solar and terrestrial ions into the high-altitude magnetosphere, J. Geophys. Res., 85, 1645, 1980.

Banks, P.M., J.C. Foster, and J.R. Doupnik, Chatanika radar observations relating to the latitudinal and local time variations of Joule heating, J. Geophys. Res., 86, 6869, 1981.

Baumjohann, W., J. Untied, and R.A. Greenwald, Joint two-dimensional observations of ground magnetic and ionospheric electric fields associated with auroral zone currents. 1. Three-dimensional current flows associated with a substorm-intensified eastward electrojet, J. Geophys. Res., 85, 1963, 1980.

Baumjohann, W., and H.J. Opgenoorth, Electric fields and currents associated with active aurorae, (this volume), 1983.

Berthelier, A., J.J. Berthelier, and C. Guerin, The effect of the east-west component of the interplanetary magnetic field on magnetic convection as deduced from magnetic perturbations at high latitudes, J. Geophys. Res., 79, 3187, 1974.

Bonnevier, B., R. Boström, and G. Rostoker, A three-dimensional model current system for polar magnetic substorms, J. Geophys. Res., 75, 107, 1970.

Boström, R., A model of the auroral electrojets, J. Geophys. Res., 69, 4983, 1964.

Boström, R., Mechanisms for driving Birkeland

currents, Physics of the Hot Plasma in the Magnetosphere, ed. by B. Hultqvist and L. Stenflo, p. 341, Plenum Press, New York, 1975.

Brekke, A., J.R. Doupnik, and P.M. Banks, Incoherent scatter measurement of E region conductivities and current in the auroral zone, J. Geophys. Res., 79, 3773, 1974.

Burch, J.L., P.H. Reiff, and M. Sugiura, Upward electron beams measured by DE-1: A preliminary source of dayside region 1 Birkeland currents, (this volume), 1983.

Burke, W.J., M.C. Kelley, R.C. Sagalin, M. Smiddy, and S.T. Lai, Polar cap electric field structures with a northward interplanetary magnetic field, Geophys. Res. Lett., 6, 21, 1979.

Burke, W.J., M.S. Gussenhoven, M.C. Kelley, D.A. Hardy, and F.J. Rich, Electric and magnetic field characteristics of discrete arcs in the polar cap, J. Geophys. Res., 87, 2431, 1982.

Burrow, J.R., and M.D. Wilson, A review of low altitude currents inferred from ISIS 2 and MAGSAT satellites, (this volume), 1983.

Bythrow, P.F., R.A. Heelis, W.B. Hanson, and R.A. Power, Simultaneous observations of field-aligned currents and plasma drift velocities by Atmospheric Explorer, C, J. Geophys. Res., 85, 151, 1980.

Bythrow, P.F., T.A. Potemra, and R.A. Hoffman, Observations of field-aligned currents, particles and plasma drift in the polar cusps near solstice, J. Geophys. Res., 87, 5131, 1982.

Bythrow, P.F., T.A. Potemra, and L.J. Zanetti, Variation of the auroral Birkeland current pattern associated with interplanetary and geomagnetic conditions, (this volume), 1983.

Cauffman, D.P., and D.A. Gurnett, Double-probe measurements of convection electric fields with the Injun-5 satellite, J. Geophys. Res., 76, 6014, 1971.

Chapman, S., and V.C.A. Ferraro, A new theory of magnetic storms. Part I. The initial phase, Terr. Mag. Atmosph. Elect., 36, 77, 1931.

Chapman, S., An outline of a theory of magnetic storms, Proc. Roy. Soc., A95, 61, 1918.

Chapman, S., The solar and lunar variations of terrestrial magnetism, Phil. Trans. R. Soc., A218, 1, 1919.

Chapman, S., and J. Bartels, Geomagnetism, Clarendon, Oxford, 1940.

Chen, C.-K., R.A. Wolf, M. Harel, and J.L. Karty, Theoretical magnetograms based on quantitative simulation of a magnetospheric substorm, J. Geophys. Res., 87, 6137, 1982.

Chiu, Y.T., and J.M. Cornwall, Electrostatic model of a quiet auroral arc, J. Geophys. Res., 85, 543, 1980.

Cowley, S.W.H., A note on the motion of charged particles in one-dimensional magnetic current sheet, Planet. Space Sci., 26, 539, 1978.

Cowley, S.W.H., Magnetospheric asymmetries associated with the Y-component of the IMF, Planet. Space Sci., 29, 79, 1981a.

Cowley, S.W.H., Magnetospheric and ionospheric flow and the interplanetary magnetic field, 4-1, The Physical Basis of the Ionosphere in the Solar-Terrestrial System, AGARD Conf. No. 295, Nenilly-sur-Seine, France 1981b.

Crooker, N.U., and G.L. Siscoe, Birkeland currents as the cause of the low-latitude asymmetric disturbance field, J. Geophys. Res., 86, 201, 1981.

de la Beujardiere, O., R. Vondrak, R. Heelis, W. Hanson, and R. Hoffman, Auroral arc electrodynamic parameters measured by AE-C and the Chatanika radar, J. Geophys. Res., 86, 4671, 1981.

Dessler, A.J., and E.N. Parker, Hydromagnetic theory of geomagnetic storms, J. Geophys. Res., 64, 2239, 1959.

Doyle, M.A., F.J. Rich, W.J. Burke, and M. Smiddy, Field-aligned currents and electric fields observed in the region of the dayside cusp, J. Geophys. Res., 86, 5656, 1981.

Dungey, J.W., Interplanetary magnetic field and the auroal zones, Phys. Rev. Lett., 6, 47, 1961.

Eastman, T.E., E.W. Hones, Jr., S.J. Bame, and J.R. Asbridge, The magnetospheric boundary layer: site of plasma, momentum and energy transfer from the magnetosheath into the magnetosphere, Geophys. Res. Lett., 3, 685, 1976.

Eastman, T.E., and E.W. Hones, Jr., Characteristics of the magnetospheric boundary layer and magnetopause layer as observed by Imp 6, J. Geophys. Res., 84, 2019, 1979.

Eastwood, J.W., Some properties of the current sheet in the geomagnetic tail, Planet. Space Sci., 23, 1, 1975.

Ejiri, M., R.A. Hoffman, and P.H. Smith, Energetic particle penetrations into the inner magnetosphere, J. Geophys. Res., 85, 653, 1980.

Evans, J.V., J.M. Holt, W.L. Oliver, and R.H. Ward, Millstone Hill incoherent scatter observation of auroral convection $60° \leq \Lambda \leq 70°$. 2. Initial results, J. Geophys. Res., 85, 41, 1980.

Fairfield, D.H., Magnetic field signatures of substorms on high-latitude field lines in the nightside magnetosphere, J. Geophys. Res., 78, 1553, 1973.

Feldstein, Ya. I., Magnetic field variations in the polar region during magnetically quiet periods and interplanetary magnetic fields, Space Sci. Rev., 18, 777, 1976.

Föppl, H., G. Haerendel, L. Haser, J. Loidl, P. Lütjens, R. Lüst, F. Melzner, B. Meyer, H. Neuss, and E. Rieger, Artificial strontium and barium clouds in the upper atmosphere, Planet. Space Sci., 15, 357, 1967.

Foster, J.C., J.R. Doupnik, and G.S. Stiles, Large scale patterns of auroral ionospheric convection observed with the Chatanika radar, J. Geophys. Res., 86, 11357, 1981.

Foster, J.C., P.M. Banks, and J.R. Doupnik, Electrostatic potential in the auroral ionosphere derived from Chatanika radar observations, J. Geophys. Res., 87, 7513, 1982.

Foster, J.C., J.-P. St. Maurice, and V.J. Abreu, Joule heating by auroral currents, (this volume), 1983.

Frank, L.A., and C.Y. Huang, Field-aligned and cross-field currents in the earth's magnetotail, (this volume).

Frank, L.A., R.L. McPherron, R.J. DeCoster, B.G. Burek, K.L. Ackerson and C.T. Russell, Field-aligned currents in the earth's magnetotail, J. Geophys. Res., 86, 687, 1981.

Fridman, M., and J. Lemaire, Relationship between auroral electron fluxes and field-aligned electric potential difference, J. Geophys. Res., 85, 664, 1980.

Friis-Christensen, E., K. Lassen, J. Wilhjelm, J.M. Wilcox, W. Gonzalez, and D.S. Colburn, Critical component of the interplanetary magnetic field responsible for large geomagnetic effects in the polar cap, J. Geophys. Res., 77, 3371, 1972.

Friis-Christensen, E., and J. Wilhjelm, Polar currents for different directions of the interplanetary magnetic field in the Y-Z plane, J. Geophys. Res., 80, 1248, 1975.

Friis-Christensen, E., Polar cap current systems, (this volume), 1983.

Fukushima, N., Polar magnetic storms and geomagnetic bays, J. Fac. Sci. Univ., Tokyo Sect. 2, 8, 293, 1953.

Greenwald, R.A., Studies of currents and electric fields in the auroral zone ionosphere using radar auroral backscatter, p. 213, Dynamics of the Magnetosphere, ed. by S.-I. Akasofu, D. Reidel Pub. Co., Dordrecht-Holland, 1979.

Greenwald, R.A., Recent advances in magnetosphere-ionosphere coupling, Rev. Geophys. Space Phys., 20, 577, 1982.

Gussenhoven, M.S., Extremely high latitude auroras, J. Geophys. Res., 87, 2401, 1982.

Haerendel, E., and R. Lüst, Electric fields in the upper atmosphere, Earth's Particles and Fields, p. 271, ed. by B.M. McCormac, Reinhold, New York, 1968.

Haerendel, G., P.C. Hedgecock and S.-I. Akasofu, Evidence of magnetic field-aligned currents during the substorms on March 18, 1969, J. Geophys. Res., 26, 2382, 1971.

Harel, M., R.A. Wolf, P.H. Reiff, R.W. Spiro, W.J. Burke, F.J. Rich and M. Smiddy, Quantitative simulation of a magnetospheric substorm, 1. Model logic and overview, J. Geophys. Res., 86, 2217, 1981a.

Harel, M., R.A. Wolf, R.W. Spiro, P.H. Reiff, C.-K. Chen, W.J. Burke, F.J. Rich, and M. Smiddy, Quantitative simulation of a magnetospheric substorm, 2. Comparison with observations, J. Geophys. Res. 86, 2242, 1981b.

Hasegawa, A., and T. Sato, Generation of field-aligned current during substorm, in Dynamics of the Magnetosphere, p. 529, ed. by S.-I. Akasofu, D. Reidel Pub. Co., Dordrecht-Holland, 1980.

Heelis, R.A., and W.B. Hanson, High-latitude ion convection in the night time F region, J. Geophys. Res., 85, 1995, 1980.

Heelis, R.A., J.D. Winningham, and W.B. Hanson, The relationships between high-latitude convection reversals and the energetic particle morphology observed by Atmospheric Explorer, J. Geophys. Res., 85, 3315, 1980.

Heppner, J.P., J.P. Stolarik, and E.M. Wescott, Electric-field measurements and the identification of currents causing magnetic disturbances in the polar cap, J. Geophys. Res., 76, 6028, 1971.

Heppner, J.P., Electric field variations during substorms, OGO-6 measurements, Planet. Space Sci., 20, 1475, 1972a.

Heppner, J.P. Polar cap electric field distributions related to their interplanetary magnetic field direction, J. Geophys. Res., 77, 4877, 1972b.

Heppner, J.P., Empirical models of high-latitude electric fields, J. Geophys. Res., 82, 1115, 1977.

Heppner, J.P., M.L. Miller, M.B. Pongratz, G.M. Smith, L.L. Smith, S.B. Mende, and N.R. Nath, The Cameo barium releases: $E_\parallel$ fields over the polar cap, J. Geophys. Res., 86, 3519, 1981.

Holzer, T.E., and G.C. Reid, The response of the dayside magnetosphere-ionosphere system to time-varying field line reconnection at the magnetopause, 1. Theoretical model, J. Geophys. Res., 80, 2041, 1975.

Hones, E.W., Jr., and K. Schindler, Magnetotail plasma flow during substorms: A survey with Imp 6 and Imp 8, J. Geophys. Res., 84, 7155, 1979.

Horwitz, J.L., and S.-I. Akasofu, On the relationship of the polar cap current system to the north-south component of the interplanetary magnetic field, J. Geophys. Res., 84, 2517, 1979.

Hughes, T.J., and G. Rostoker, A comprehensive model current system for high-latitude magnetic activity, I, The steady state system, Geophys. J. R. Astron. Soc., 58, 525, 1979.

Iijima, T., and T.A. Potemra, Large-scale characteristics of field-aligned currents associated with substorms, J. Geophys. Res., 83, 599, 1978.

Iijima, T., Field-aligned currents, (this volume), 1983a.

Iijima, T., Large-scale field-aligned currents over the polar cap with reversed polarities and reversed S_q^p associated with prolonged northward IMF, (this volume), 1983b.

Jörgensen, T.S., E. Friis-Christensen, and J. Wilhjelm, Interplanetary magnetic-field direction and high-latitude ionospheric currents, J. Geophys. Res., 77, 1976, 1972.

Kamide, Y., and S.-I. Akasofu, Global distribution of the Pedersen and Hall currents and the electric potential pattern during a moderately disturbed period, J. Geophys. Res., 86, 3665, 1981.

Kamide, Y., A.D. Richmond and S. Matsushita,

Estimation of ionospheric electric fields, ionospheric currents and field-aligned currents, 1, Quiet periods, J. Geophys. Res., 86, 801, 1981.

Kamide, Y., The relationship between field-aligned currents and the auroral electrojets: A review, Space Sci. Rev., 31, 127, 1982.

Kan, J.R. and L.C. Lee, Theory of imperfect magnetosphere-ionosphere coupling, Geophys. Res. Lett., I, 633, 1980.

Kan, J.R., L.C. Lee and S.-I. Akasofu, The energy coupling function and the power generated by the solar wind-magnetosphere dynamo, Planet. Space Sci., 28, 823, 1980.

Kan, J.R., Towards a unified theory of discrete auroras, Space Sci. Rev., 31, 71, 1982.

Kan, J.R., L.C. Lee, V.M. Vasyliunas and S.-I. Akasofu, Generation of region I and II field-aligned currents, J. Geophys. Res., (submitted), 1983.

Karty, J.L., C.-K. Chen, R.A. Wolf, M. Harel and R.W. Spiro, Modeling of high-latitude currents in a substorm, J. Geophys. Res., 87, 777, 1982.

Kaufmann, R.L., Auroral electron beams: Electric currents and energy source, J. Geophys. Res., 86, 7586, 1981.

Kawasaki, K. and S.-I. Akasofu, A possible current system associated with the S_q^p variation, Planet. Space Sci., 21, 329, 1973.

Kirkpatrick, C.B., On current systems proposed for SD in the theory of magnetic storms, J. Geophys. Res., 57, 511, 1952.

Kisabeth, J.L., On calculating magnetic and vector potential fields due to large-scale magnetospheric current systems and induced currents in an infinitely conducting earth, in Quantitative Modeling of Magnetospheric Processes, Geophys. Monog. Ser., 21, 473, ed. by W.P. Olson, Washington, D.C., 1979.

Klumpar, D.M., J.R. Burrows and M.D. Wilson, Simultaneous observations of field-aligned currents and particle fluxes in the post-midnight sectors, Geophys. Res. Lett., 3, 395, 1976.

Knight, S., Parallel electric fields, Planet. Space Sci., 21, 741, 1973.

Lassen, K. and C. Danielsen, Quiet time pattern of auroral arcs for different direction of the interplanetary magnetic field in the y-z plane, J. Geophys. Res., 83, 5277, 1978.

Lennartsson, W., R.D. Sharp, E.G. Shelley, and R.G. Johnson, J. Geophys. Res., 86, 4628, 1981.

Lyons, L.R., D.S. Evans and R. Lundin, An observed relation between magnetic field-aligned electric fields and downward electric energy fluxes in the vicinity of the auroral forms, J. Geophys. Res., 84, 457, 1979.

Lyons, L.R., and D.J. Williams, A source for the geomagnetic storm main phase ring current, J. Geophys. Res., 85, 523, 1980.

Lyons, L.R., Discrete aurora as the direct result of an inferred high-altitude generating potential distribution, J. Geophys. Res., 86, 1, 1981.

McDiarmid, I.B., J.R. Burrows and M.D. Wilson, Magnetic field perturbations in the dayside cleft and their relationship to the IMF, J. Geophys. Res., 83, 5753, 1978.

McPherron, R.L. and J.N. Barfield, A seasonal change in the effect of field-aligned currents at synchronous orbit, J. Geophys. Res., 85, 6743, 1980.

McPherron, R.L., Relation of the three-dimensional current systems to the solar wind, (this volume), 1983.

Maezawa, K., Magnetic convection induced by the positive and negative z components of the interplanetary magnetic field: Quantitative analysis using polar cap magnetic records, J. Geophys. Res., 81, 2289, 1976.

Maier, E.J., S.E. Kayser, J.R. Burrows and D.M. Klumpar, The suprathermal electron contributions to high-latitude Birkeland currents, J. Geophys. Res., 85, 2003, 1980.

Mansurov, S.M., New evidence of a relationship between magnetic fields in space and on earth, Geomagn. Aeron., 9, 622, 1969.

Matsushita, S., Solar quiet and lunar daily variation fields, Physics of Geomagnetic Phenomena, Vol. 1, ed. by S. Matsushita and W.H. Campbell, p. 301, Academic Press, New York, 1967.

Mead, G.D. and D.B. Beard, Shape of the geomagnetic field by the solar wind, J. Geophys. Res., 69, 1181, 1964.

Meng, C.-I. and S.-I. Akasofu, A study of polar magnetic substorms, 2, Three-dimensional current system, J. Geophys. Res., 78, 4035, 1969.

Mishin, V.M., A.D. Bazarzhapov and G.B. Shpynev, Electric fields and currents in the earth's magnetosphere, Dynamics of the Magnetosphere, p. 249, ed. by S.-I. Akasofu, D. Reidel Pub. Co., Dordrecht-Holland, 1980.

Moore, T.E., R.L. Arnoldy, J. Feynman and D.A. Hardy, Propagating substorm injection front, J. Geophys. Res., 86, 6713, 1981.

Mozer, F.S. and R. Serlin, Magnetospheric electric field measurements with balloons, J. Geophys. Res., 74, 4739, 1969.

Mozer, F.S., C.W. Carlson, M.K. Hudson, R.B. Torbert, B. Parady, J. Yatteau and M.C. Kelley, Observations of paired electrostatic shocks in the polar magnetosphere, Phys. Rev. Lett., 38, 292, 1977.

Ness, N.F., The earth's magnetic tail, J. Geophys. Res., 70, 2989, 1965.

Nishida, A., Geomagnetic Diagnosis of the Magnetosphere, Springer-Verlag, New York, 1978.

Olson, W., Magnetopause shape associated fields for various inclinations of the earth's dipole axis to the solar wind, Ph.D. Thesis, Univ. of California, Los Angeles, 1968.

Perreault, P. and S.-I. Akasofu, A study of

geomagnetic storms, Geophys. J. Roy. Astronom. Soc., 54, 547, 1978.

Potemra, T.A., Current systems in the earth's magnetosphere, Rev. Geophys. Space. Phys., 17, 640, 1979.

Reiff, P.H., R.W. Spiro and T.W. Hill, Dependence of polar cap potential drop on interplanetary parameters, J. Geophys. Res., 86, 7639, 1981.

Reiff, P.H., Sunward convection in both polar caps, J. Geophys. Res., 87, 5970, 1982.

Reiff, P.H., Models of auroral conductances, (this volume), 1983.

Richmond, A.D. and S. Matsushita, On the production mechanism of electric currents and fields in the ionosphere, J. Geophys. Res., 81, 547, 1976.

Robinson, R.M., E.A. Bering, R.R. Vondrak, H.R. Anderson and P.A. Cloutier, Simultaneous rocket and radar measurements of currents in an auroral arc, J. Geophys. Res., 86, 7703, 1981.

Robinson, R.M., R.R. Vondrak and T.A. Potemra, Electrodynamic properties of the evening sector ionosphere within the region 2 field-aligned current sheet, J. Geophys. Res., 87, 731, 1982.

Roble, R.G., R.E. Dickinson and E.C. Ridley, Global circulation and temperature structure of thermosphere with high-latitude plasma convection, J. Geophys. Res., 87, 1599, 1982.

Rostoker, G. and R. Boström, A mechanism for driving the gross Birkeland current configuration in the auroral oval, J. Geophys. Res., 81, 235, 1976.

Rostoker, G. and H.-L. Lau, A generation mechanism for Pc5 micropulsations in the morning sector, Planet. Space Sci., 26, 493, 1978.

Rostoker, G. and T.J. Hughes, A comprehensive model current system from high-latitude magnetic activity, II, The substorm component, Geophys. J. Roy. Astron. Soc., 58, 571, 1979.

Rostoker, G., The auroral electrojets, in Dynamics of the Magnetosphere, ed. by S.-I. Akasofu, p. 201, D. Reidel Pub. Co., Dordrecht-Holland, 1980a.

Rostoker, G., Magnetospheric and ionospheric currents in the polar cusp and their dependence on the By component of the interplanetary magnetic field, J. Geophys. Res., 85, 4167, 1980b.

Rostoker, G., M. Mareschal and J.C. Samson, Response of dayside net downward field-aligned current to changes in the interplanetary magnetic field and to substorm perturbations, J. Geophys. Res., 87, 3489, 1982.

Russell, C.T., The configuration of the magnetosphere, in Critical Problems of Magnetospheric Physics, ed. by E.R. Dyer, Jr., p. 1, Nat. Acad. Sci., Washington, D.C., 1972.

Saflekos, N.A., T.A. Potemra and T. Iijima, Small-scale transverse magnetic disturbances in the polar regions observed by Triad, J. Geophys. Res., 83, 1493, 1978.

Saflekos, N.A. and T.A. Potemra, The orientation of Birkeland current sheets in the dayside polar region and its relationship to the IMF, J. Geophys. Res., 85, 1987, 1980.

Shelley, E.G., R.G. Johnson, and R.D. Sharp, Satellite observations of energetic heavy ions during a geomagnetic storm, J. Geophys. Res., 77, 6104, 1972.

Shuman, B.M., R.P. Vancour, M. Smiddy, N.A. Saflekos and F.J. Rich, Field-aligned current, convection electric field and auroral particle measurements during a major magnetic storm, J. Geophys. Res., 86, 5561, 1981.

Siscoe, G.L., Energy coupling between regions 1 and 2 Birkeland current systems, J. Geophys. Res., 87, 5124, 1982.

Smiddy, M., W.J. Burke, M.C. Kelley, N.A. Saflekos, M.S. Gussenhoven, D.A. Hardy and F.J. Rich, Effects of high-latitude conductivity on observed convection electric fields and Birkeland currents, J. Geophys. Res., 85, 6811, 1980.

Sonnerup, B.U.O., Theory of the low-latitude boundary layer, J. Geophys. Res., 85, 2017, 1980.

Sonnerup, B.U.O., Adiabatic particle orbits in a magnetic null sheet, J. Geophys. Res., 76, 8211, 1971.

Spencer, N.W., L.E. Wharton, P.B. Hays and T.L. Killeen, Effects of magnetospheric currents on the heating and dynamics of the polar regions upper thermosphere, (this volume), 1983.

Spiro, R.W., P.H. Reiff and L.J. Maher, Jr., Precipitating electron energy flux and auroral zone conductances - an empirical model, J. Geophys. Res., 87, 8215, 1982.

Stern, D.P., The origin of Birkeland currents, Rev. Geophys. Space Phys., 21, 125, 1983.

Sugiura, M. and T.A. Potemra, Net field-aligned currents observed by Triad, J. Geophys. Res., 81, 2155, 1976.

Sugiura, M., N.C. Maynard, J.L. Burch and J.D. Winningham, Initial results of DE-1 and -2 observations of field-aligned currents, (this volume), 1983.

Sumaruk, P.V. and Y.-I. Feldstein, Seasonal changes of variations of the Z component in the polar region in connection with the sign of the YSE component of the interplanetary magnetic field, Geomag. Aeron., 13, 469, 1973.

Svalgaard, L., Sector structure of the interplanetary magnetic field and daily variation of the geomagnetic field at high latitudes, Pap. 6, Dansk. Meteorol. Inst. Geophys., Charlottenlund, Denmark, 1968.

Svalgaard, L., Interplanetary magnetic-sector structure 1926-1971, J. Geophys. Res., 77, 4027, 1972.

Svalgaard, L., Polar cap magnetic variations and their relationship with the interplanetary magnetic sector structure, J. Geophys. Res., 78, 2064, 1973.

Troshichev, O.A., Polar magnetic disturbances and

field-aligned currents, Space Sci. Rev., 32, 275, 1982.

Vasyliunas, V.M., Mathematical models of magnetospheric convection and its coupling to the ionosphere, in Particle and Fields in the Magnetosphere, ed. by B.M. McCormac, p. 60, D. Reidel Pub. Co., Hingham, Mass., 1970.

Vestine, E.H., I. Lange, L. Laporte and W.E. Scott, The Geomagnetic Field, its Description and Analysis, Carnegie Institution of Washington, publication 580, 1947.

Vickrey, J.F., R.R. Vondrak, S.J. Matthews, The diurnal and latitudinal variation of auroral zone ionospheric conductivity, J. Geophys. Res., 86, 65, 1981.

Wagner, J.S., J.R. Kan and S.-I. Akasofu, Particle dynamics in the plasma sheet, J. Geophys. Res., 84, 891, 1979.

Walker, J.K., J.A. Koehler, F. Creutzberg, A.G. McNamara, A. Vallance Jones and B.A. Whalen, Post substorm convection and auroral arc currents determined from multiple ionospheric, magnetic and electric field observations, J. Geophys. Res., 86, 9975, 1981.

Wallis, D.D. and E.E. Budzinski, Empirical models of height integrated conductivities, J. Geophys. Res., 86, 125, 1981.

Wilhjelm, J., E. Friis-Christensen and T.A. Potemra, The relationship between ionosphere and field-aligned currents in the dayside cusp, J. Geophys. Res., 83, 5586, 1978.

Williams, D.J., The earth's ring current: causes, generation and decay, Space Sci. Rev., 34, 223, 1983.

Wolf, R.A., Ionosphere-magnetosphere coupling, Space Sci. Rev., 17, 537, 1975.

Zmuda, A.J. and J.C. Armstrong, The diurnal flow pattern of field-aligned currents, J. Geophys. Res., 79, 4611, 1974a.

Zmuda, A.J. and J.C. Armstrong, The diurnal variation of the region with vector magnetic field changes associated with field-aligned currents, J. Geophys. Res., 79, 2501, 1974b.

INTRODUCTION TO THE TOPOLOGY OF MAGNETOSPHERIC CURRENT SYSTEMS

W. P. OLSON

McDonnell Douglas Astronautics Company, 5301 Bolsa Avenue, Huntington Beach, CA 92647

Abstract. A brief history of the observations and study of the major magnetospheric current systems is presented. These currents are usually inferred from the observation of their associated magnetic fields but are formed by magnetospheric plasmas. Although the energy sources for the magnetopause and ring currents are well understood, the particle and energy sources for the tail and Birkeland currents are still being sought. It is suggested that the direct entry of magnetosheath plasma into the magnetosphere is responsible for the tail and Region 1 Birkeland currents.

Introduction

There are four major magnetospheric current systems; the magnetopause, ring, tail and field aligned (Birkeland) currents. A description of their topology and a brief discussion of the history of their study are presented.

The Magnetopause Currents

In 1882, Balfour Stewart suggested that the upper atmosphere is the most probable location for the solar influence responsible for the observed solar day to night variations in the earth's surface magnetic field at mid latitude. For this idea he is credited with being the first person to suggest the presence of the ionosphere. However, since it is now known that the magnetospheric current systems also contribute to the earth's surface magnetic field, we might more accurately credit Balfour Stewart as the first person to suggest the existence of the region around the earth where ionized particles and electric currents persisted. Early workers were careful not to state explicitly what the source of these magnetic variations was, but instead to represent them as being produced by an "equivalent overhead current system". The form of this current system is shown in Figure 1. Its major feature is the existence of two counter flowing cells. We note that in the region between the cells (above the equator), the currents flow from west to east. In order to explain the sudden commencement portion of magnetic storms, Chapman and Ferraro [1930, 1932] suggested that streams of charged particles are intermittently ejected from the sun. As they approach the earth's magnetic field, a current system is formed at the boundary between the charged particle stream and the geomagnetic field. These currents produce a magnetic field that increases the horizontal component of the earth's surface magnetic field at mid-latitudes at all local times. In order to study this problem analytically, they examined a plasma approaching the earth whose boundary with the geomagnetic field was initially represented as a plane. Currents formed by the interaction between the streaming plasma and the geomagnetic field form at this

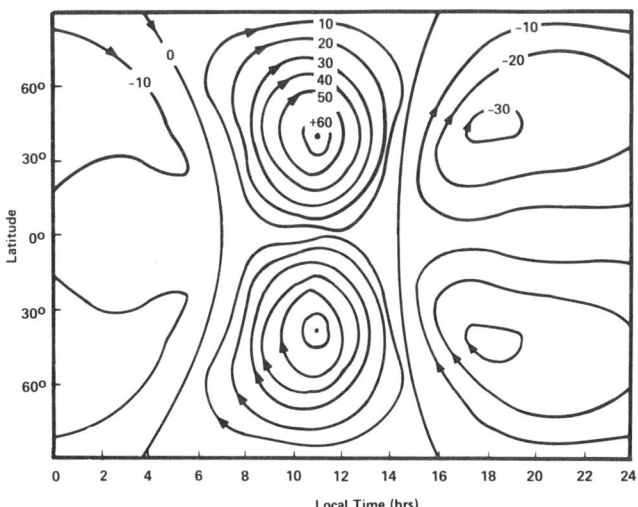

Fig. 1. An equivalent overhead current system responsible for the day to night variations in the earth's surface magnetic field as observed during quiet solar conditions. Its major features are two counter flowing cells of current separated by an equatorial current system flowing from west to east (from Chapman and Ferraro, 1932).

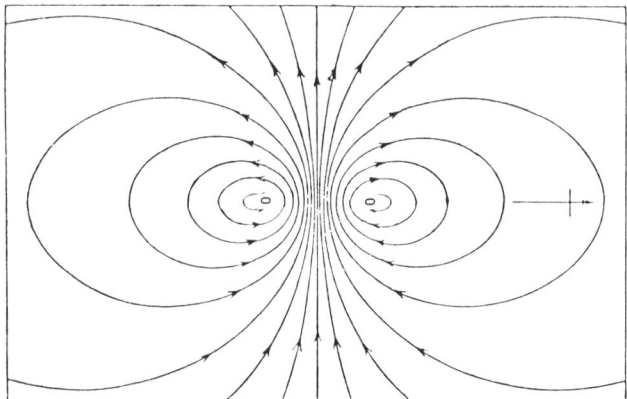

Fig. 2. Currents flowing on the leading edge of a plasma cloud as it interacts with the geomagnetic field (from Chapman and Ferraro, 1932). View is from the earth looking toward the sun so the equatorial currents again flow from west to east. The pattern is very similar in form to that shown in Figure 1.

intersection plane (see Figure 2). We note the similarity between this pattern of currents and that of the equivalent overhead current system depicted in Figure 1. Chapman and Ferarro realized that if these currents (and the associated plasma) flowed past the earth, they would not remain in a plane, but rather carve out a region for a time around the earth. An equatorial cross-section of this region is given in Chapman and Bartels' book and shown in Figure 3. They even determined the plasma density and velocity required to explain the magnetic storm sudden commencement phase (a density of approximately 0.5 hydrogen atoms per cubic centimeter and a velocity of approximately 1000 kilometers/second).

In the early fifties, Bierman [1951] noted that the direction of cometary tails (always pointing away from the sun) could not be explained in terms of solar electromagnetic radiation pressure. He therefore was led to suggest that there is a continuous flow of plasma which moves radially away from the sun. Later, Parker in a series of papers beginning in 1957 [see Parker, 1963] examined theoretically the continuous expansion of the solar corona, and by the late 1950's the concept of a continuous solar wind, a low energy neutral plasma, was well established. By then it was well recognized that the extension of the geomagnetic field into space would be limited at all times by the flow of the solar wind.

In the early 1960's, many people worked on determining the shape of the magnetosphere (that region of space to which the geomagnetic field is confined), and as a by-product determined also the strength and form of the currents flowing on the boundary between the magnetosphere and the solar wind (the magnetopause) [Beard, 1960; Midgley and Davis, 1963; Slutz & Winkelman, 1964; Mead and Beard, 1964]. The magnetopause was to be found where the kinetic pressure of the solar wind was balanced by the energy density (or pressure) of the geomagnetic field. For nominal solar wind velocities and densities, along the earth-sun line the magnetopause is found approximately 10-1/2 earth radii (R_E) from the earth's center. The pressure balance condition has been used successfully to predict the size and shape of the magnetosphere over the observed range of solar wind densities and velocities. The geocentric distance to the subsolar magnetopause is predicted to range from

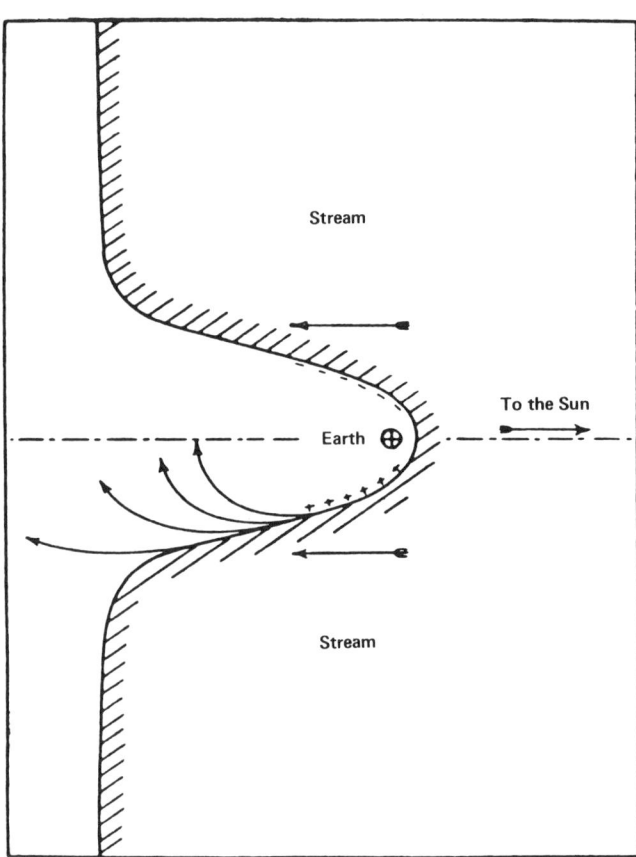

Fig. 3. Plasma field interaction shown in equatorial cross section (Chapman and Ferraro, 1932). As the plasma cloud flows past the earth it loses its planar front edge and carves out a hollow around the earth. Note also that Chapman and Ferraro attempted to explain the formation of the ring current by allowing a positive charge to enter the region around the earth on the dawn side and negative charge to enter on the dusk side.

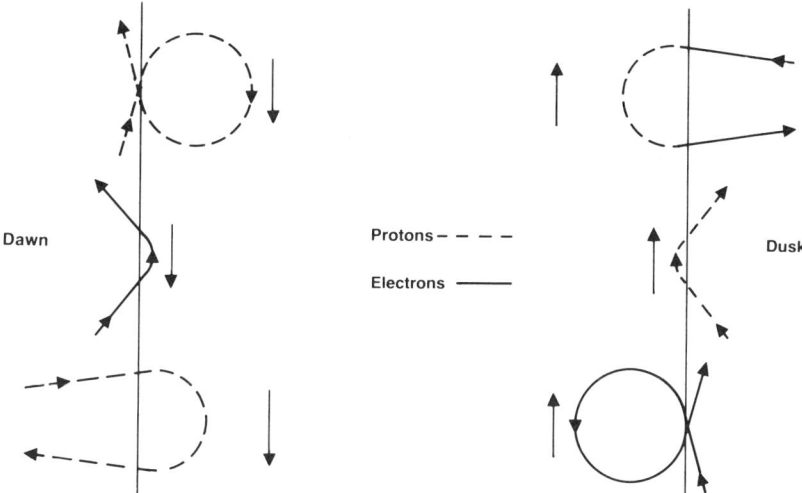

Fig. 4. Specular reflection of electrons and positive ions. Particle trajectories are shown for particles incident from a region of no magnetic field and then interacting with a uniform magnetic field (producing circular orbits), and finally being reflected specularly from the magnetosphere region.

over 15 R_E to about 6 R_E. The subsolar magnetopause has been observed many times inside of geosynchronous orbit. The magnetopause currents and their associated magnetic field also vary with solar wind pressure. Recently, time varying models of B have been developed for specific events in conjunction with the magnetospheric Coordinated Data Analysis Workshops [Olson and Pfitzer, 1982]. Models and observations agree well near local noon where the magnetopause currents predominate.

The generation of a current system at the magnetopause can be understood in terms of the trajectories of individual solar wind particles as they interact with the geomagnetic field. As shown in Figure 4, electrons and protons are deflected in opposite directions as they interact with the geomagnetic field. Also, since the scale length of the geomagnetic field is much larger than the gyroradius of the solar wind protons and electrons at the magnetopause, it has been assumed to good approximation that the magnetic field in its region of interaction with a particular solar wind particle can be considered as a constant. Thus all of the solar wind plasma is specularly reflected off of the geomagnetic field.

A complication to this simple picture arises because of the difference in proton and electron gyroradii. Although this sets up an electrostatic field in the magnetopause, several authors [e.g., see Ferraro, 1952; Dungey, 1958] have shown that while this polarization field influences individual electron and ion trajectories, it does not appreciably change the basic determination of magnetopause shape and the calculation of the magnetopause current system as determined using the pressure balance condition [e.g., Beard, 1966; Willis, 1971]. Also, Parker [1967 a, b] argued that the polarization field might be short-circuited by flow of charge along magnetic field lines connected to the ionosphere. (Note that this is an early suggestion of the need for current flow along magnetic field lines.)

The magnetopause current system, which persists at all times (although its strength may vary), is shown in Figure 5. Its close resemb-

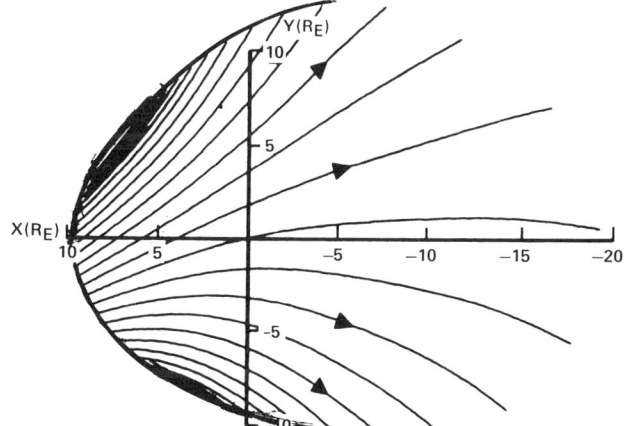

Fig. 5. Computer model of the magnetopause current system. The currents are viewed from outside of the dusk side of the magnetosphere. Note the similarity in form of the magnetopause current system to the Chapman and Ferraro intermittent currents and the equivalent overhead current system.

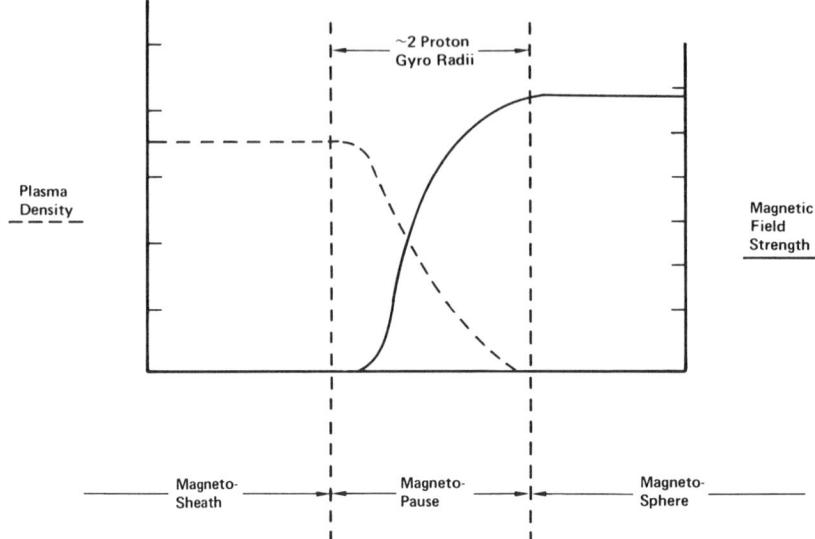

Fig. 6. Classical magnetopause structure. Over the magnetopause region, the magnetospheric magnetic field decreases from its nominal value to zero just beyond the magnetopause. Likewise, the magnetosheath plasma density decreases from its nominal value just outside the magnetosphere to zero just inside the magnetopause. In such models, the magnetosphere is closed, no charged particles may enter, and the geomagnetic field does not penetrate beyond the magnetopause.

lance to the intermittent Chapman-Ferraro currents and the equivalent overhead current system is noted. Thus it is not surprising that the magnetopause currents contribute to the quiet day to night variations in the surface magnetic field [Mead, 1964; Olson, 1970]. The structure of the magnetopause as given in these early models is depicted in Figure 6. The thickness of the magnetopause is on the order of the proton gyroradius, which in turn depends locally on the strength of the geomagnetic field. In the subsolar magnetosphere, therefore, the thickness of the magnetopause is on the order of 100 kilometers, while on the flanks (or sides) of the anti-sunward extent of the magnetosphere (the tail region), the magnetopause increases in thickness to about 1500 kilometers.

Because of the specular reflection of the solar wind particles, none of them is permitted to enter the magnetosphere. As shown in Figure 6, these early models suggested that over a region called the magnetopause, the geomagnetic field will diminish from its nominal field to zero outside of the magnetosphere, while the solar wind particle density will decrease from its ambient value to zero within the magnetosphere. Thus the magnetosphere, as given by these models, is said to be closed. That is, it is assumed no solar wind particles can enter it and the geomagnetic field is everywhere parallel to the magnetopause. Such a statement, of course, has many implications for the solar wind-geomagnetic field interaction.

The magnetospheric magnetic field associated with these currents was described by a scalar potential much the same way that the earth's main field had been described for decades. The use of a scalar potential representation of a magnetic field is restricted to those regions outside of the source currents. Thus in the first models of the magnetospheric magnetic field it was possible to simply add the scalar potential representations of the magnetopause and main fields.

In these pressure balance models of the magnetopause currents, the existence of two magnetically neutral points is a key feature. They correspond to the two foci of the currents shown in Figure 5. All magnetic field lines on the magnetopause are connected to these two neutral points, where the total magnetic field is zero and its direction undefined. It has been suggested that because of this topology, the entire magnetopause is an electric equipotential surface (in the approximation that a given field line, owing to its high electrical conductivity, is an electric equipotential). In 1969, Willis suggested that the magnetopause actually exhibits a cusped geometry as opposed to neutral points. Later, Heikkila and Winningham [1971] and Frank [1971] observed the presence of this cusp region on the dayside magnetosphere where magnetosheath particle entry extends deep into the magnetosphere along magnetic field lines. It is now clear that currents flowing through the dayside cusps contribute to the Birkeland current system observed at ionospheric heights [Iijima and Potemra, 1976]. At present there is still no adequate widely accepted theory explaining the cusp geometry and the currents that flow on and in the dayside cusp regions.

The observed magnetopause structure departs from early theory in one other important way. It is now apparent from much satellite data that instead of the classical magnetopause structure, there is a more complicated structure that includes a "boundary layer" of plasma just inside the magnetopause [Paschman et al., 1976; Palmer and Hones, 1978; Harendel et al., 1978; Eastman and Hones, 1979]. The boundary layer has been given several names depending on its location, but appears to exist over almost the entire magnetopause. The boundary layer thickness is generally found to be 10 to 25 times as large as the subsolar magnetopause thickness [Eastman and Hones, 1979]. Comparison of the particle distribution functions for the magnetosheath region and the boundary layer suggest that penetration of some magnetosheath plasma into the magnetosphere occurs at or through the boundary layer. More is said on the entry of magnetosheath plasma into the magnetosphere as it pertains to the tail and Birkeland current systems in the section on the Sources of Magnetospheric Currents.

Although Chapman and Ferraro only examined the formation of the region around the earth during the intermittent passage of a cloud of charged particles from the sun, their work exhibits many insights into the structure of the real magnetosphere. As shown in Figure 3, they attempted to get charged particles through the "magnetopause region" in order to supply current to the trapping region as is required to explain the main phase decrease of a magnetic storm. They also exhibited remarkable insight on the subject of the dayside cusp regions, as is evidenced by the following quotation from Chapman and Ferraro [1931], "...there will be two foci of the current-circulations, to north and south of the plane of the magnetic equator, where the lines of force of the field within the hollow are normal to the current-sheet. At these points the mechanical reaction of the field upon the current-sheet vanishes, so that the retarding force on the stream-surface becomes small near these points. Thus the surface will continue to advance in these two regions, while being retarded elsewhere; it would seem that two 'horns' would thus protrude outwards from the surface, along which the stream-gas can advance with little or no retardation. As the retarded hollow surface itself advances, the position of these horns may change; presumably they will become closer together. The matter passing along these horns seems likely to find its way towards the polar regions."

The Ring Current

In addition to their work on the sudden commencement phase of a magnetic storm, Chapman and Ferraro also attempted to explain the main phase decrease in the horizontal component of the earth's surface magnetic field which typically persists for several days and is the principal feature of the magnetic storm at mid latitudes at all local times. They suggested that this was caused by the presence of a ring of current that flowed at great altitudes that was somehow formed by the cloud of plasma flowing from the sun as it moved past the earth. Although there were problems with the early Chapman-Ferraro theory and others, modern theory of the ring current simply states that it is caused by pressure gradients in the particles trapped in the earth's inner magnetosphere. The density of the trapped particles increases with altitude out to perhaps a geocentric distance of 2.5 earth radii (R_E) beyond which the density decreases. The ring current flows eastward out to 2.5 R_E where ∇p changes sign and the current flows westward. The ring current signature is dominated by the westward currents and thus depresses the total field in the inner magnetosphere and at the earth's surface.

Sugiura et al. [1971] subtracted a model representation of the earth's main magnetic field from the total field observed on several satellites operating in the magnetosphere. They showed that this scalar difference, ΔB, persisted even during periods of magnetic quiet. These ΔB contours also demonstrated that the ring current was not confined to a flat region near the equator, but rather was distributed over a large portion of the inner magnetosphere (see Figure 7). Thus to represent the magnetic field produced by the ring current, it was necessary to use a vector potential so that $\nabla \times \vec{B} = 0$ in the region of the source currents.

The persistence of the ring current at quiet times had a profound effect on the modeling of the magnetospheric magnetic field. The early models of the magnetospheric magnetic field containing only the main field and the magnetopause field depicted the inner magnetosphere as having an inflated field strength (stronger than the field as given by the main field only). This led to a disagreement between observed solar cosmic ray cutoffs and predictions made using the magnetospheric field model to determine the Lorentz force trajectories theoretically. Several "diffusion-like" processes were invented to explain this discrepancy. Actually, all that was necessary was a better magnetic field model, one that included the quiet time ring current which tended to depress the total magnetic field in the inner magnetosphere. Differences in the northward component of the geomagnetic field along the sun-earth line are shown for several models in Figure 8. In order to represent the magnetic field in the region of these distributed currents, it was necessary to represent the magnetic field with a vector

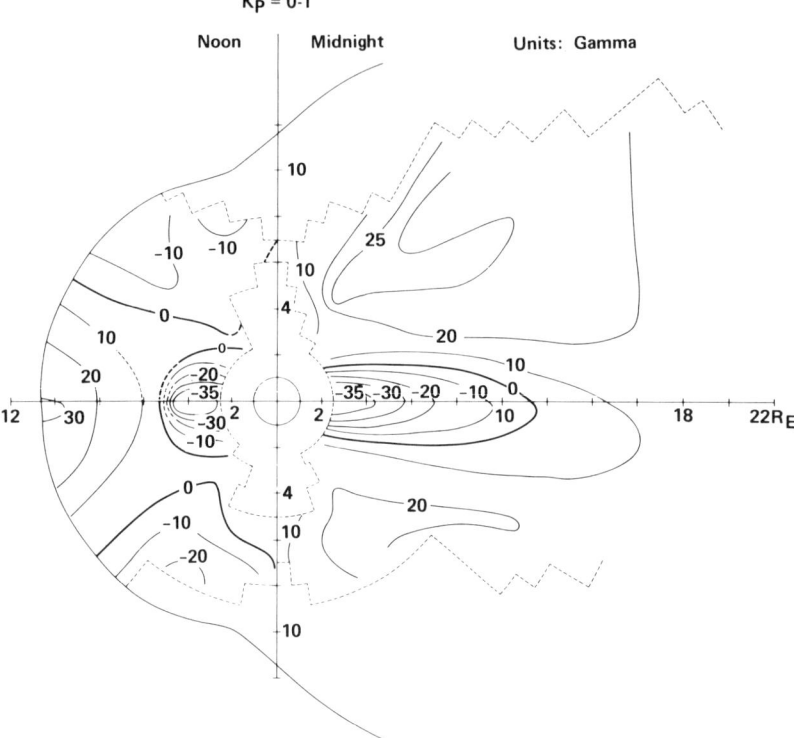

Fig. 7. ΔB contours (from Sugiura et al., 1971). The scalar quantity ΔB is determined by subtracting the strength of the main field (as determined from a quantitative model) from the magnitude of the observed field. These ΔB contours shown for very low magnetic activity levels demonstrate that the ring current persists during quiet times, and also that it is not confined to a narrow region near the magnetic equator, but rather flows throughout a large portion of the inner magnetosphere.

potential. Thus Olson and Pfitzer [1974a] developed a procedure for finding the magnetic field from such a distributed current system and representing it in the region of the source currents.

We note in passing that the study of solar cosmic rays indicates clearly that energetic particles have access to the magnetosphere, i.e., that the magnetosphere is not "closed" to them. Presumably this is because when they impact the magnetospheric magnetic field, unlike the less energetic solar wind particles, their gyroradii are so large that the magnetic field the particles "sample" is nonuniform thereby permitting some of them to enter. This leads naturally to the following question: What is the lowest energy charged particle at the magnetopause that will sample enough of a nonuniform magnetic field to have access to the magnetosphere? More is said about this question in the section on the sources of magnetospheric currents.

As mentioned previously, the ring current, during both quiet and disturbed times, is populated by charged particles trapped in the earth's magnetic field (in the inner magnetosphere). Of the four major magnetospheric currents, the ring current is the only one that is not directly linked to the solar wind (or magnetosheath plasma). In order to populate the ring current, it is necessary for the total magnetospheric magnetic field to change with time. (To move particles into the magnetic trapping region where the ring current persists, an electric field must be present. Electrostatic fields in the tail may play a minor role but are conservative and thus cannot act to permanently place plasma into the ring current. It is generally agreed that the non-conservative electric field associated with $\partial B/\partial t$ is the primary source for injection of charged particles into the trapping region and thus for the buildup of the ring current.) This happens during both magnetic storms and the magnetospheric substorm process. In both cases some particles are injected into the inner magnetosphere (into the "trapping region"), where they remain for a period of time forming the ring current. The decay of the ring current appears to proceed by many factors which contribute to placing some of the ring current particles into the pitch angle loss cone. Such

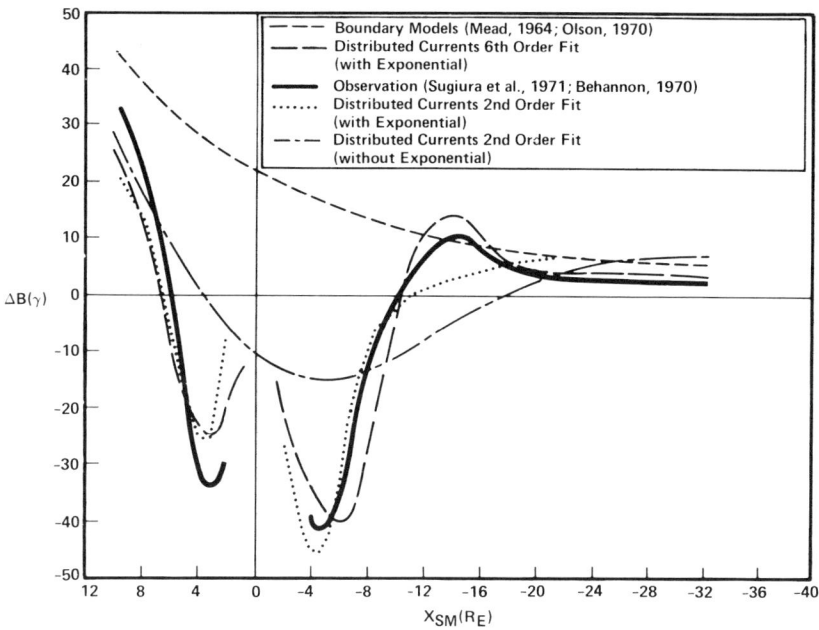

Fig. 8. ΔB along the earth's sun line. The difference in the total field and the main field (the magnetospheric contribution to the observed field) is shown for several model fields. Note that in the inner magnetosphere a magnetopause only field acts to strengthen the total field. Models containing the magnetic field associated with the quiet time ring current instead show that the inner magnetosphere has a field strength which is depressed from the field produced only by the main field.

particles are removed from the ring curent along magnetic field lines where they are lost to the ionosphere.

We note that the ΔB data set of Sugiura et al. [1971] cuts off at a geocentric distance of 2.5 earth radii. To date there have been only very limited measurements of the earthward extent of the ring current to altitudes less than 2.5 R_E. In order to accurately determine the earth's main magnetic field and the contribution of the ring current to the earth's surface magnetic field from satellite data it is necessary that the magnetic field component of the ring current earthward of 2.5 R_E be observed and studied.

The Tail Currents

Early observations of the magnetic field in the anti-solar region of the magnetosphere showed that it is split into two lobe regions - one with the field lines directed away from the earth, the other (in the northern hemisphere) with the field lines directed toward the earth. These two lobe regions are seen now to be separated by the plasma sheet region in the equatorial portion of the tail where the plasma density is relatively high. Also, there is a region in the center of the tail near the equator where the total magnetic field is northward and quite small. This structure could not be explained simply in terms of the magnetopause curents, which produced a tail field that is too "dipolar". The first model of the tail current system was provided by Williams and Mead [1965]. They used an infinitesimally thin sheet of current that flowed across the tail out through the magnetopause from plus to minus infinity in a direction perpendicular to the magnetic dipole axis and to the earth-sun line (from minus to plus infinity along the solar magnetospheric Y axis). There were, of course, problems with this representation. Any calculation of the resulting magnetic field that required the field strength near the front edge of the current sheet gave abnormally large fields which were, of course, an artifact of the simple current system.

In the late sixties and early seventies, several more realistic models of the tail currents were developed [Beard et al., 1970; Bird and Beard, 1972; Olson, 1974]. Olson and Pfitzer [1974a] found that in order to quantitatively represent the magnetic field structure in the tail (in both the lobe regions and near the equator), it was necessary to permit the currents to flow throughout the plasma sheet and not be constrained to a thin region at the equator. This result is not surprising. Currents flow in the magnetosphere wherever plasma is present, owing to the drifts of the charged particles in the geomagnetic

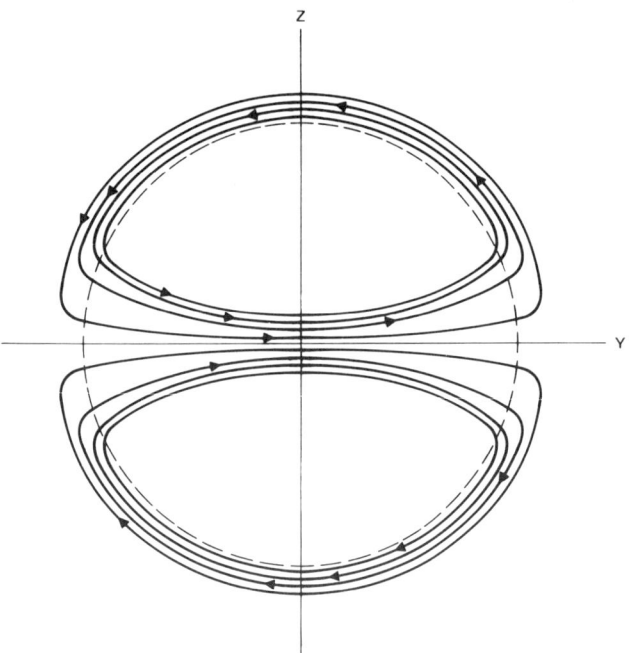

Fig. 9. Planar cross section of the tail currents. The tail curents flow from dawn to dusk across the center of tail throughout most of the plasma sheet region. In order to explain the lobe structure of the magnetic field, it is also necesary that a return path for these currents exists over the lobes just outside of the magnetosphere. Only in the distant tail ($X_{sm} < -20\ R_E$) can the tail currents be approximated as flowing in a plane.

field and the resulting gradients in plasma pressure. In order to represent the pronounced lobe structure, it was necessary to allow the cross-tail currents to return on or just beyond the magnetopause. A cross-section of the tail illustrating both the topology of the current system and the magnetic field structure is given in Figure 9. We note that this representation of the tail current system is an approximation since the currents flowing across the plasma sheet are not confined to a plane. Their path is best understood in terms of equatorial particles in a symmetric magnetosphere (one in which the solar wind is incident perpendicular to the earth's geomagnetic dipole axis). Particles on the magnetic equatorial plane move in a magnetic field that is everywhere directed northward. In order to conserve their first adiabatic invariant, 90° pitch angle particles move along paths of constant magnetic field. Contours of constant magnetic field are shown in the equatorial plane in Figure 10. It is seen that these contours wrap well around to the front side of the magnetosphere. If a fraction of the solar wind particles in the magnetosheath have access to the magnetosphere, it is seen that the cross-tail currents will extend almost to the nose of the magnetosphere.

The tail currents are tied intimately to the substorm process. As such, they must be considered simultaneously with the tail plasma and magnetic field. The tail of the magnetosphere (and thus the tail currents) responds dramatically to changes in the magnetosheath plasma and to the interplanetary magnetic field direction and strength. The topology of the tail currents is well enough understood to integrate over them to obtain an expression for the tail magnetic field. The dynamics of the tail, although now well described, are still not well understood, nor is the control of the tail by the magnetosheath plasma and interplanetary magnetic field.

The Birkeland (or Field Aligned) Currents

Birkeland [1908], in his observations of auroral activity, was the first to suggest that charged particles and electric currents may flow along field lines at high latitudes. Later, Van Sabben [1966] in his work on hemispheric asymmetries in the earth's surface magnetic variation field also suggested that it would be easy for currents to flow along magnetic field lines between the ionosphere in the northern and southern hemispheres. Also, Parker [1967 a, b], as mentioned previously, suggested that currents flow along field lines connecting the ionosphere to the magnetopause in order to short out the polarization charge formed by the difference in gyroradii of magnetosheath ions and electrons as they are deflected by the geomagnetic field. In the early seventies, magnetometer data from polar orbiting satellites were used to verify the existence of field aligned currents [Zmuda and Armstrong, 1974; Iijima and Potemra, 1976 a, b]. The magnetic signature of the field aligned currents just above the ionosphere is now well documented for both quiet and disturbed magnetic conditions. The field aligned currents are characterized by their limited extent in latitude and by the close proximity of toward and away currents near their intersection with the ionosphere. The location of the Region 1 and Region 2 currents is shown in Figure 11. Note that near local noon there is a small area of field aligned currents attributed to the flow of charged particles in the dayside cusp region. Since the Region 1 currents are always stronger than the Region 2 currents, it is likely that the Region 1 currents are connected to the source that drives the Birkeland current system. There is some evidence [see Review by Stern, 1983] that the Region 1 currents connect across the polar cap during hemispheric summer,

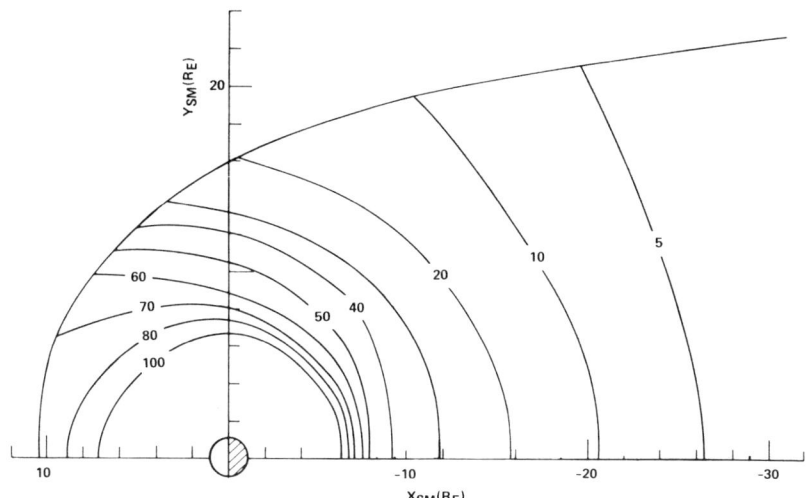

Fig. 10. Contours of constant magnetic field strength shown in the magnetic equatorial plane. Adiabatic particles with 90° pitch angles will flow along these contours. The cross tail currents exist only where these contours intersect the magnetopause (the ring current flows in the inner magnetosphere on closed constant field contours).

thus acting to diminish the Region 2 currents. The continuation of the Birkeland currents through the ionosphere has been studied in some detail [for example, see Wolf et al., 1982; Reiff, 1983]. The closure of the Birkeland currents in the magnetosphere, however, is not at this time well understood. There is some agreement that the Region 1 currents connect to the magnetopause, or boundary layer. The question of where the Region 2 currents close in the magnetosphere remains, although some authors suggest that it is connected to the inner edge of the plasma sheet [see Reviews by Potemra, 1979 and Stern, 1983].

The study of the Birkeland currents may provide important information on magnetospheric magnetic field topology. For example, their signature at ionospheric heights (see Figure 11) suggests limits in latitude and longitude for the dayside cusp. The location of the Region 1 currents also suggests that field lines near the magnetopause and inner edge of the plasma sheet map to an oval extending to almost all local times in the polar region.

The study of the closure of the Birkeland currents in the magnetosphere should also provide information required for the proper modeling of $\vec{B}$ near midnight. As shown in Figure 12, there is good agreement between the dynamic model of $\vec{B}$ and observations throughout the dayside magnetosphere [Olson and Pfitzer, 1982]. This is because the magnetopause and ring currents that contribute the most to $\vec{B}$ there are well understood and modeled. On the nightside, however, the agreement vanishes, probably because the models do not at present properly take into account the closure of the

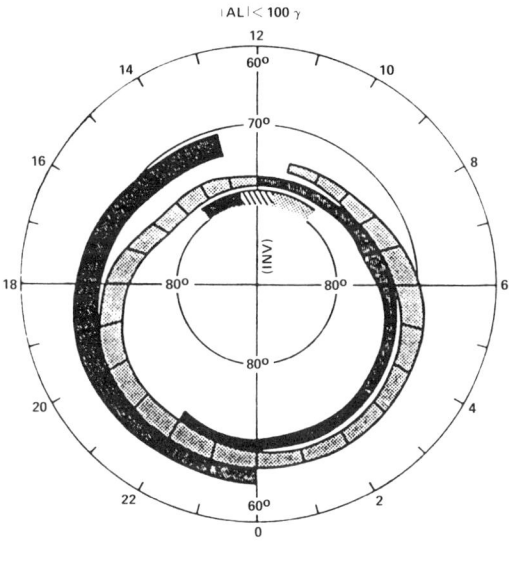

Fig. 11. The intersection of the Birkeland Region 1 and Region 2 currents with the ionosphere. According to Iijima and Potemra (1976), field aligned currents believed to flow in the dayside cusp region are also shown forward of the Region 1 currents near noon. The Region 1 currents flow toward the ionosphere on the dawn side of the earth, while the Region 2 currents flow toward the ionosphere on the dusk side of the earth.

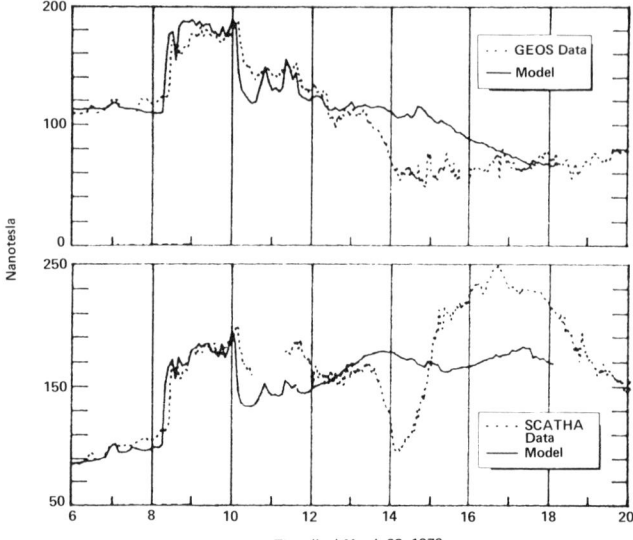

Fig. 12. The magnitude of the total observed magnetic field as shown at the location of the GEOS and SCATHA satellites for an event studied in detail at the Coordinated Data Analysis Workshop-6. The agreement between observations and dynamic model prediction is good until the satellites approach the midnight sector of approximately 2000 local time. It is believed that the discrepancy between model and observations near midnight is caused by the closure of the Birkeland Region 2 currents in the magnetosphere.

Region 2 Birkeland currents on or near the inner edge of the plasma sheet.

Sources of the Magnetospheric Currents

Although we now know that the magnetopause region is more structured than suggested by the early pressure balance models, the pressure balance formalism has done an adequate job in determining magnetopause shape, extent, and the location and form of the magnetopause current system. Likewise, the formation of the ring current is now well understood, although details of particle injection and decay remain under study.

The sources of the tail current system and of the Birkeland currents, however, have not been unambiguously defined. Several authors [Bird and Beard, 1972; Olson, 1974] have suggested that the cross tail currents flow throughout the plasma sheet and are ultimately supplied by the magnetosheath. Likewise, several authors [Eastman et al., 1976; Iijima and Potemra, 1976a] have suggested that the Region 1 Birkeland currents are connected to the magnetopause or the boundary layer. In addition, Heikkila [1974, 1978] has suggested that a current generator exists at or just outside of the subsolar magnetopause that drives several of the magnetospheric current systems. However, none of these papers have suggested explicitly the mechanism for driving either the generator or the current systems. It is suggested here that physically the mechanism driving all these current systems is simply the entry of low energy plasma (the magnetosheath plasma) into the magnetosphere, and that this entry occurs over almost the entire magnetopause, but is largest along the flanks of the tail and on the tailward side of the dayside cusp region [Olson and Pfitzer, 1983]. Because this entry causes a charge separation across the dawn and dusk sides of the magnetosphere, it drives both electric field and current systems in the magnetosphere. The details of this entry mechanism are briefly discussed. [For a more in-depth description of the entry mechanism and its consequences, see Olson and Pfitzer, 1983].

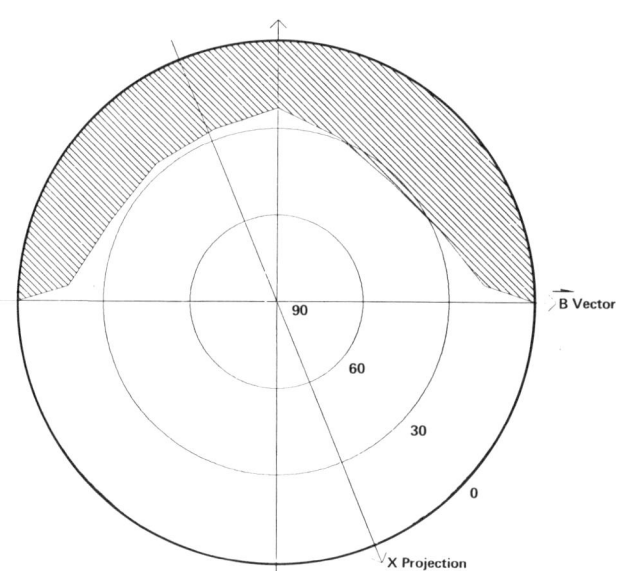

Fig. 13. Charged particle entry efficiency. A plane is defined tangent to the magnetopause at the point where entry efficiency is to be determined. The direction of particle impact is given in terms of its angular elevation above the plane. Location on the plane is defined by the direction of the magnetic field through the point and also the direction of the projection of the X axis. Entry is shown for 1 kilovolt protons incident at a point just above the magnetic equator on the dawn side of the magnetosphere. At a distance of 20 R_E down the tail the entry fraction is just under 4 percent for an isotropic plasma.

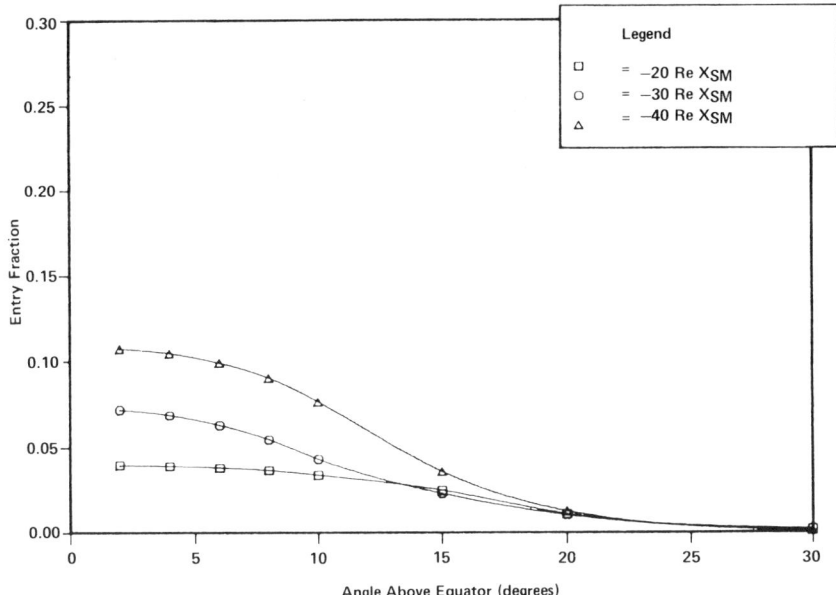

Fig. 14. Entry fraction as a function of location on the dawn flank of the tail for 1 kilovolt protons. Although a significant fraction of incident 1 kilovolt protons have access to the magnetosphere at the magnetic equator. This fraction falls off appreciably at an angle 15 or 20 degrees off of the equator.

Entry of solar cosmic ray particles into the polar regions of the magnetosphere has been observed and studied for almost two decades. As mentioned earlier, the agreement between predicted and observed solar cosmic ray cutoffs was not good until the contribution of the quiet time ring current was included in models of the magnetospheric magnetic field. These studies showed clearly that the magnetosphere is open to solar cosmic ray particles. However, since the magnetosphere was first studied it has been assumed that the solar wind (and magnetosheath) particles are specularly reflected off of the geomagnetic field. This assumption was made because the geomagnetic field is quite uniform over the region where it interacts with an impacting particle. However, the field is not precisely uniform and, in fact, possesses a gradient parallel to the magnetopause. Thus even low energy charged particles do not move in precisely circular orbits as they interact with the geomagnetic field.

Several authors have therefore suggested that even in the keV energy range, particles have access to the magnetosphere [Vestine, 1963; Fejer, 1965; Wentworth, 1965; Stevenson and Comstock, 1968; Bird and Beard, 1972; Cole, 1974; Olson and Pfitzer, 1974b, 1983]. Olson and Pfitzer have used a realistic magnetospheric model to define the magnetic field structure over its region of interaction with incident particles. This was done at several locations near the equator in the tail of the magnetosphere and at each point particles were allowed to impact from several directions. An example of particle entry at a point near the equatorial region of the tail is shown in Figure 13. Thus, for an isotropic 1 kev plasma, about 4 percent of the incident particles can enter the magnetosphere at a point 20 R_E down the tail and about 3 R_E above the equator. Note that since the magnetosheath plasma flows in the anti-solar direction, the entry fraction for the magnetosheath plasma in the same magnetic field topology may be larger than for an isotropic plasma (that is, for a plasma described simply in terms of its electron and ion temperatures and which does not possess a bulk velocity component). The dependence of the entry fraction on location at several points along the equator on the dawn flank of the tail is shown in Figure 14. This entry mechanism also exhibits a strong dependence on particle energy. Generally, charged particle entry at a given location at the magnetopause depends on the magnetic field structure (in particular, the component of ΔB parallel to the magnetopause and perpendicular to $\vec{B}$), the direction of the incident particle and its charge, mass and speed.

Another important aspect of charged particle entry permitted by gradients in the magnetospheric magnetic field is that only positively charged particles may enter on the dawn side of the magnetosphere, and only electrons may enter on the dusk side. This is shown schematically in Figure 15. Once a particle enters the magnetosphere, where it goes depends on the magnetic field topology and the particle's pitch

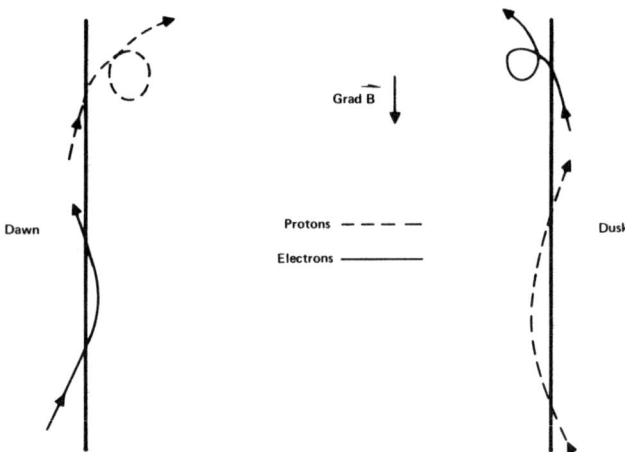

Fig. 15. Non-specular deflection and entry of charged particles. The small gradient in the strength of the magnetic field parallel to the magnetopause is directed toward the earth in the magnetotail region. This magnetic field topology permits only positive ions to enter on the dawn side of the tail of the magnetosphere, and only electrons on the dusk side.

angle. As shown schematically in Figure 16, a charged particle gaining access to a magnetic field line near the magnetopause on the dawn flank of the tail will drift across the tail and helps form the plasma sheet if its pitch angle is near 90°. However, if the particle's initial velocity vector makes an angle close to the direction of the magnetic field, that particle will move along the magnetic field line toward the ionosphere. Positive charge will flow toward the dawn side of the auroral ionosphere in both hemispheres (or alternatively the accumulation of net charge along the dawn flank of the magnetosphere will draw electrons away from the ionosphere along the field lines linking the regions). On the dusk of the magnetosphere electrons entering at or near the equatorial magnetopause also flow along field lines toward the ionosphere. Thus we believe that particles entering the magnetosphere (because of the nonuniform magnetic field at the magnetopause) will contribute to both the cross tail current system and to the Region 1 Birkeland currents. In general, this particle entry mechanism sets up a charge separation across the magnetosphere: plus charge on the dawn side, negative charge on the dusk side. The response of the magnetosphere in an attempt to short out this charge separation causes many things to happen including, we believe, the formation of the cross tail currents and at least a portion of the Region 1 Birkeland currents. A more detailed discussion of this entry mechanism and its consequences is found in Olson and Pfitzer [1983].

Summary

The first suggestions that currents flowed at high altitudes above the earth's surface were made near the turn of the century. The magnetopause current system was the first of the four major magnetospheric current systems to be studied and observed quantitatively. It must be considered the principal magnetospheric current system because it accounts for a major portion of the magnetic field topology of the magnetosphere. The source of the magnetopause current system is understood to be the interaction of the shocked solar wind plasma with the geomagnetic field. There has been controversy over this interaction and whether or not it acts to

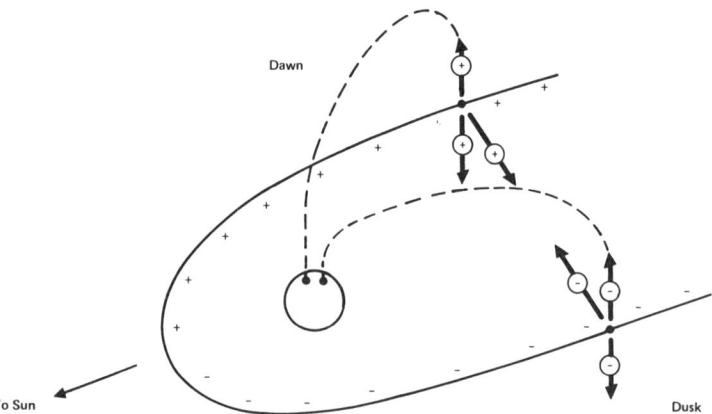

Fig. 16. Motion of Entering Particles. Once a particle has entered the magnetosphere, where it goes depends on its pitch angle. Particles with pitch angles near 90° will move across the tail through the plasma sheet. Particles with pitch angles directed along or anti-parallel to the magnetic field direction will move along magnetic field lines toward the ionosphere.

exclude all of the solar wind plasma from the magnetosphere. The ring current has been studied for over half a century and is known to be the cause of the main phase of magnetic storms. The particles forming the ring current are trapped in the geomagnetic field in the inner magnetosphere. The ring current particle population is supplied primarily by the plasma sheet during periods of magnetic disturbance. The decay of the ring current particles when the magnetosphere is in a quiescent state is still being intensely investigated. The cross tail current system is now believed to persist throughout the plasma sheet as is required to explain the pronounced lobe structure in the tail. It is necessary also to postulate that return currents flow through the magnetosheath region just beyond the magnetopause over the lobe regions of the tail. We believe that both the tail currents and the Region 1 Birkeland currents are driven by charge separation across the magnetosphere produced by the entry of a portion of the low energy charged particles incident on the magnetopause. Time variations in the sources of these current systems also produce induced electric fields which are now being studied. Models of the magnetospheric magnetic field are in good agreement with observations (even during large variations in the solar wind) on the dayside magnetosphere. However, near local midnight model field and observations are in disagreement that cannot be explained by the tail currents. It is probable that this discrepancy is caused by the present neglect in the models of the closure of the Birkeland Region 2 currents in the magnetosphere.

References

Beard, D. B., Radiation trapped in the earth's magnetic field, Billy McCormac, (Ed.), D. Reidel Publishing Co., Dordrecht, Holland, 1966.

Beard, D. B., The interaction of the terrestrial magnetic field with the solar Corpuscular Radiation, J. Geophys. Res., 65, 3559, 1960.

Beard, D. B., M. Bird, and Y. H. Huang, Self-consistent theory of the magnetotail, Planet. Space Sci., 18, 1349, 1970.

Bierman, L., Kometenschweife Und Solare Korpuskular Strahlung, Z. Astrophys., 29, 274, 1951.

Bird, M. K., and D. B. Beard, The Self-consistent geomagnetic tail under static conditions, Planet. Space Sci., 20, 2057, 1972.

Birkeland, C., [See Chapman, S., and J. Bartels, Geomagnetism, Oxford University Press, New York, 1940.].

Chapman, S., and V.C.A. Ferraro, A new theory of magnetic storms, Nature, 126, 129, 1930.

Chapman, S., and V.C.A. Ferraro, A new theory of magnetic storms, Terr. Magn., 37, 147, 1932.

Chapman, S., and J. Bartels, Geomagnetism, published by Oxford at the Clarendon Press, London, 1940.

Cole, K. D., Outline of a theory solar wind interaction with the magnetosphere, Planet. Space Sci., 22, 1075, 1974.

Dungey, J. W., Cosmic Electrodynamics, Cambridge University Press, London, 1958.

Eastman, T. E., E. W. Hones, Jr., S. J. Bame, and J. R. Asbridge, The magnetospheric boundary layer: Site of plasma, momentum and energy transfer from the magnetosheath into the magnetosphere, Geophys. Res. Lett., 3, 685, 1976.

Eastman, T. E., and E. W. Hones, Jr., Characteristics of the magnetospheric boundary layer and magnetopause layer as observed by IMP 6, J. Geophys. Res., 84, 2019, 1979.

Ferraro, V.C.A., On the theory of the first phase of a geomagnetic storm, J. Geophys. Res., 57, 15, 1952.

Frank, L. A., Plasma in the earth's polar magnetosphere, J. Geophys. Res., 76, 5202, 1971.

Fejer, J. A. Geometry of the magnetospheric tail and auroral current system, J. Geophys. Res., 70, 4972, 1965.

Haerendel, G., G. Paschmann, N. Scopke, H. Rosenbauer, and P. C. Hedgecock, The frontside boundary layer of the magnetosphere and the problem of reconnection. J. Geophys. Res., 83, 3195, 1978.

Heikkila, W. J., and J. D. Winningham, Penetration of magnetosheath plasma to low altitudes through the dayside magnetospheric cusps, J. Geophys. Res., 76, 883, 1971.

Heikkila, W. J., Outline of a magnetospheric theory, J. Geophys. Res., 79, 2496, 1974.

Heikkila, W. J., Electric field topology near the dayside magnetopause, J. Geophys. Res., 83, 1071, 1978.

Iijima, T., and T. A. Potemra, The amplitude distribution of field aligned currents at northern high latitudes observed by TRIAD, J. Geophys. Res., 81, 2165, 1976a.

Iijima, T., and T. A. Potemra, Field aligned currents in the dayside cusp observed by TRIAD, J. Geophys. Res., 81, 5971, 1976b.

Mead, C. D., and D. B. Beard, Shape of the geomagnetic field solar wind boundary, J. Geophys. Res., 69, 1169, 1964.

Mead, C. D., Reformation of the geomagnetic field by the solar wind, J. Geophys. Res., 69, 1181, 1964.

Midgley, J. E., and L. Davis, Calculation by a moment technique of the perturbation of the geomagnetic field by the solar wind, J. Geophys. Res., 68, 5111, 1963.

Olson, W. P., A Model of the distributed magnetospheric currents, J. Geophys. Res., 79, 3731, 1974.

Olson, W. P., and K. A. Pfitzer, A quantitative model of the magnetospheric magnetic field, J. Geophys. Res., 79, 3739, 1974a.

Olson, W. P., and K. A. Pfitzer, Magnetospheric boundaries and fields in correlated interplanetary and magnetospheric observations, D. E. Page (Ed.), D. Reidel Publishing Co., Dordrecht, Holland, 1974b.

Olson, W. P., and K. A. Pfitzer, Introduction to the Topology of Magnetospheric Current Systems, 1983.

Olson, W. P., Variations in the earth's surface magnetic field from the magnetopause current system, Planet. Space Sci., 18, 1471, 1970.

Olson, W. P. and K. A. Pfitzer, A dynamic model of the magnetospheric magnetic and electric fields for July 29, 1977. J. Geophys. Res., Vol 87, No. A8, pp. 5943-5948, August 1, 1982.

Palmer, I. D., and E. W. Hones, Jr., Characteristics of energetic electrons in the vicinity of the magnetospheric boundary layer at Vela orbit, J. Geophys Res., 83, 2584, 1978.

Parker, E. N., Interplanetary Dynamical Processes, Interscience, John Wiley, New York, 1963.

Parker, E. N., Confinement of a magnetic field by a beam of ions, J. Geophys. Res., 72, 2315, 1967a.

Parker, E. N., Small-scale nonequilibrium of the magnetopause and its consequences, J. Geophys. Res., 72, 4365, 1967b.

Paschmann, G., G. Haerendel, N. Schopke, H. Rosenbauer, and P. D. Hedgecock, Plasma and magnetic field characteristics of the distant polar cusp near local noon: The Entry Layer, J. Geophys. Res., 81, 2883, 1976.

Potemra, T. A., Current Systems in the earth's magnetosphere, Rev. Geophys. and Space Sci., 17, 640, 1979.

Reiff, P. H., Polar and auroral phenomena: A review of U.S. progress during 1979-1982, Rev. of Geophys and Sp. Phys., Vol. 21, No. 2, 418, March 1983, U.S. Report to IUGG 1979-1982.

Slutz, R. J., and J. R. Winkelman, Shape of the magnetospheric boundary under solar wind pressure, J. Geophys. Res., 69, 4933, 1964.

Stern, D. P., The origins of Birkeland currents, Rev. Geophys. and Space Sci., 21, 125, 1983.

Stevenson, T. E., and C. Comstock, Particles incident on magnetic field gradients, J. Geophys. Res., 73, 175, 1968.

Stewart, B., Terrestrial Magnetism, Encyl. Brit., 9th Ed., 1882.

Sugiura, M., B. G. Ledley, T. L. Skillman, and J. P. Heppner, Magnetospheric field distortions observed by OGO 3 and 5, J. Geophys. Res., 76, 7552, 1971.

Van Sabben, Magnetospheric currents associated with the N-S asymmetry of S_q, J. Atmos. and Terrestrial Res., 28, 965, 1966.

Vestine, E. H., Some comments on the ionosphere and geomagnetism (Paper presented at the XIV General Assembly, International Scientific Radio Union, Tokyo, Japan, September 1963.

Wentworth, R. C., Diamagnetic ring current theory of the neutral sheet and its affects on the topology of the anti-solar magnetosphere, Phys. Rev. Letters, 14, 1008, 1965.

Williams, D. J., and G. D. Mead, Nightside magnetosphere configuration as obtained from trapped electrons at 100 kilometers, J. Geophys. Res., 70, 3017, 1965.

Willis, D. M., The influx of charged particles at the magnetic cusps on the boundary of the magnetosphere, Planet. Space Sci., 17, 339, 1969.

Willis, D. M., Structure of the magnetopause, Rev. of Geophys. and Space Sci., 9, 953, 1971.

Wolf, R. A., M. Harel, R. W. Spiro, G.-H. Voigt, P. H. Reiff, and C.-K. Chen, Computer simulation of inner magnetospheric dynamics for the magnetic storm of July 29, 1977, J. Geophys. Res., 87, 5959-5962, 1982.

Zmuda, A. J., and J. C. Armstrong, The diurnal variation of the region with vector magnetic field changes associated with field-aligned currents, J. Geophys. Res., 79, 2501, 1974.

FUNDAMENTALS OF CURRENT DESCRIPTION

Vytenis M. Vasyliunas

Max-Planck-Institut für Aeronomie, D-3411 Katlenburg-Lindau 3, Federal Republic of Germany

<u>Abstract</u>. Electrical currents, defined in terms of the net flux of charge transported by particles, deduced from the curl of the magnetic field in most of the observational work to date as well as in many theoretical contexts, are related to plasma dynamics both by the momentum equation and by the MHD connection between plasma flow and the magnetic field configuration (relations not easily representable by circuit analogies). The momentum equation provides directly an expression for the current density perpendicular to the magnetic field and, in combination with the current continuity equation, also provides an expression for the field-aligned gradient of the parallel (Birkeland) current density; these expressions are here given with all inertial and pressure anisotropy effects taken into account.

Introduction

This paper reviews some basic principles applicable to the treatment of the large-scale configuration of electrical currents in the magnetosphere, with attention to distinguishing aspects approached from the point of view of physics, of empirical modeling, and of theoretical calculation. No claim of superiority for any one approach is intended; all three are important for progress in understanding, but nevertheless they are different (in ways not easily stated in abstract terms but apparent in concrete instances).

Definitions and determinations

The electrical current density $\underset{\sim}{J}$ at any space-time point $\underset{\sim}{r}$, t can be defined in terms of the particle velocity distribution functions:

$$\underset{\sim}{J}(\underset{\sim}{r}, t) = \sum_a q_a \int d^3v f_a(\underset{\sim}{r}, \underset{\sim}{v}, t) \qquad (1)$$

where f_a is the distribution function and q_a the particle charge for the ath species, and the summation is carried out over all the species present in the plasma. The same current density $\underset{\sim}{J}$ appears in one of Maxwell's equations:

$$\nabla \times \underset{\sim}{B} = (4\pi/c)\underset{\sim}{J} + (1/c)\partial \underset{\sim}{E}/\partial t \qquad (2)$$

Equations (1) and (2) are completely general and provide a description that is always correct and appropriate from a physical point of view. The decomposition of the current into various types (such as magnetization current, drift current, polarization or dielectric current, etc., discussed later in relation to stress balance) is essentially a calculational device that may occasionally be useful but has no direct physical significance in a plasma.

A consequence of (1) is that a non-zero $\underset{\sim}{J}$ may exist anywhere where there is a plasma. Thus, unlike the case of circuits where the current is confined to a multiply connected volume and its path fixed by the pre-existing wires, the magnetosphere together with its environs constitutes a simply connected volume where the current paths are not fixed a priori and in general $\underset{\sim}{J} \neq 0$ throughout the volume (although, of course, the magnitude of $\underset{\sim}{J}$ may be very small in some regions). This <u>is</u> one reason why circuit analogies are not appropriate for a physical discussion of the magnetosphere, whatever use they might have for modeling or calculation. As a striking illustration in a somewhat different context, to describe the generation of a planetary magnetic field remains a difficult and essentially unsolved problem in spite of the fact that it is a readily doable task in electrical engineering to design a self-exciting dynamo; the reason for the difference, according to Ian Lerche (private communication), is that the currents of the planetary dynamo are required to flow in a simply connected conducting volume.

There is no general reason to assume that $J_\parallel$, the component of $\underset{\sim}{J}$ parallel to $\underset{\sim}{B}$, either is zero or is constant or proportional to B along a field line. (This is in contrast to the electric field $\underset{\sim}{E}$, where $E_\parallel \approx 0$ is a valid first approximation in many cases.) Only in the ionosphere can one assume that, on horizontal length scales much larger than the vertical scale height, $J_{vertical} \ll J_{horizontal}$ which in the polar regions is nearly equivalent to $J_\parallel \ll J_\perp$.

The direct determination of $\underset{\sim}{J}$ from charged-particle observations via (1) is an exceedingly difficult task which has only recently been attempted with anything like a chance of success

(Frank et al., 1983 - this volume). Besides the technical/instrumental problem of measuring the particle intensities with sufficient accuracy and adequate three-dimensional velocity space coverage, there is a fundamental difficulty associated with spacecraft charging: regardless of the current density in the plasma, the distribution function of all the particles at the spacecraft surface (including secondary and photoelectrons from the spacecraft as well as particles from the plasma) must be such as to insure zero net current to the spacecraft. Only when most of the particles in the plasma have energies that are high in comparison to the spacecraft potential and the energies of secondary and photoelectrons may one expect a spacecraft-borne instrument to measure a plasma distribution function that is not seriously modified, the condition of zero net current being achieved solely by an appropriate adjustment of the secondary and photoelectron flux.

At present, our observational knowledge of the electric current systems in the magnetosphere is derived almost exclusively from measurements of the magnetic field, from which $\underset{\sim}{J}$ has been inferred by an application of (2); more precisely, the inferred quantity is $\underset{\sim}{J} + (1/4\pi)\partial \underset{\sim}{E}/\partial t$, but the second term is in most cases negligible compared to the first. Many theoretical treatments of current systems likewise first calculate $\underset{\sim}{B}$ and then take its curl to obtain $\underset{\sim}{J}$; examples include currents on the magnetopause (Mead and Beard, 1964) ($\underset{\sim}{B}$ calculated by solving a boundary value problem), in the magnetosheath (Spreiter et al., 1968) and in the solar wind (Parker, 1963) ($\underset{\sim}{B}$ calculated from the "draping" of field lines and from the Parker spiral, respectively, both obtained as hydromagnetic deformations of the field lines by a given flow), self-consistent models of the geomagnetic tail (Schindler, 1975) ($\underset{\sim}{B}$ calculated from the vector potential which is obtained as a solution of an equation derived from stress balance), and much of computer simulations of MHD processes ($\underset{\sim}{B}$ obtained by direct numerical solution of the governing equations, from which $\underset{\sim}{J}$ has been eliminated by the use of (2)). The only important cases in magnetospheric theory where $\underset{\sim}{J}$ is calculated independently of $\nabla \times \underset{\sim}{B}$ are the ring current (Carovillano and Siscoe, 1973), where $\underset{\sim}{J}$ is obtained directly from stress balance (with the use of equation (6) discussed further on), and ionospheric currents which are of course calculated from the electric field $\underset{\sim}{E}$ by means of Ohm's law.

In view of the fact that $\underset{\sim}{B}$ is the primary quantity and $\underset{\sim}{J}$ is derived from it in virtually all observational work to date in the magnetosphere as well as in much theoretical work, it is perhaps somewhat curious that empirical modeling (e.g. Walker, 1979; Olson et al., 1979) has mostly used the reverse approach, $\underset{\sim}{J}$ being treated as the primary quantity to be represented by the model, with $\underset{\sim}{B}$ calculated from it.

The continuity of current, expressed by the equation

$$\nabla \cdot \underset{\sim}{J} = 0 \qquad (3)$$

when the displacement current $\partial \underset{\sim}{E}/\partial t$ is negligible, follows from (2) by virtue of the mathematical theorem $\nabla \cdot \nabla \times \underset{\sim}{B} = 0$ for any $\underset{\sim}{B}$. Thus, when $\underset{\sim}{J}$ has been obtained from $\underset{\sim}{B}$, the closure of the current is guaranteed no matter what model for $\underset{\sim}{B}$ has been adopted - current closure is not a requirement on the model but a property of it. When $\underset{\sim}{J}$ is decomposed into components perpendicular and parallel to $\underset{\sim}{B}$, (3) can be rewritten in the useful form

$$\nabla \cdot \underset{\sim}{J}_\perp = -\nabla \cdot \underset{\sim}{J}_\| = -B(\partial/\partial s)(J_\|/B) \qquad (4)$$

where $\partial/\partial s \equiv \underset{\sim}{b} \cdot \nabla$ is the gradient operator along a magnetic field line ($\underset{\sim}{b} \equiv \underset{\sim}{B}/B$).

Relation to Stress and Flow

Associated with the electric current there is a Lorentz force which must be balanced either by mechanical forces or by the acceleration of the plasma. The current density must thus satisfy the stress balance or momentum equation

$$(1/c)\underset{\sim}{J} \times \underset{\sim}{B} = \rho d\underset{\sim}{V}/dt + \nabla \cdot \underset{\sim}{P} - \rho \underset{\sim}{g} \qquad (5)$$

$$(d/dt) \equiv (\partial/\partial t) + \underset{\sim}{V} \cdot \nabla$$

where the magnetic force density on the LH side is equated to the mass density ρ times the acceleration of the plasma bulk flow plus the divergence of the pressure tensor $\underset{\sim}{P}$ (which includes any viscosity effects as terms proportional to velocity gradients) minus the gravitational force density (negligible in most magnetospheric applications). The electric force does not appear in (5); it is negligibly small by virtue of the electrical charge quasineutrality of the plasma (except in the here irrelevant case of relativistic flow speeds). Equation (5) can be solved explicitly for the components of $\underset{\sim}{J}$ perpendicular to $\underset{\sim}{B}$:

$$\underset{\sim}{J}_\perp = (c\underset{\sim}{B}/B^2) \times \underset{\sim}{f} \qquad (6)$$

where $\underset{\sim}{f}$ stands for the expression on the RH side of (5). The Birkeland current density, the component of $\underset{\sim}{J}$ parallel to $\underset{\sim}{B}$, does not appear in the momentum equation and thus cannot be determined directly from (5); however, its derivative along a field line can be obtained from current continuity (4) and the divergence of (6):

$$B(\partial/\partial s)(J_\|/B) = 2\underset{\sim}{J}_\perp \cdot \nabla B/B + c(\underset{\sim}{B}/B^2) \cdot \nabla \times \underset{\sim}{f} \qquad (7)$$

The net Birkeland current out of a magnetic flux tube can be calculated by integrating (7) along the length of the tube.

The pivotal role of the momentum equation, determining the perpendicular current as well as the field-aligned gradient of the parallel current, is another aspect of magnetospheric physics not represented in models based on circuit analogies, where the electromagnetic forces are simply presumed to be taken up by the mechanical stiffness of the wire structure without any further effects (Lorentz forces in circuits need to be explicitly treated only if moving elements, e.g. electric motors, are involved, or else if the currents are so intense that the stress limits of the wires or their supporting structure might be approached or exceeded; to my knowledge, neither circumstance has yet been invoked in any circuit analogs for the magnetosphere).

Explicit forms of (6) and (7) can be given when the pressure tensor is gyrotropic,

$$\underset{\sim}{P} = P_\perp \underset{\sim}{1} + (P_\| - P_\perp) \underset{\sim}{bb} \qquad (8)$$

It is usually convenient to characterize the magnetic field geometry either by its curvature $\underset{\sim}{b} \cdot \nabla \underset{\sim}{b}$ or by its magnitude gradient, the two being related by the vector identity

$$\underset{\sim}{B} \cdot \nabla \underset{\sim}{B} - B\nabla B = (\nabla \times \underset{\sim}{B}) \times \underset{\sim}{B} = (4\pi/c) \underset{\sim}{J} \times \underset{\sim}{B} \qquad (9)$$

(note also that $\underset{\sim}{B} \cdot \nabla \underset{\sim}{B} = B^2 \underset{\sim}{b} \cdot \nabla \underset{\sim}{b} + B \partial \underset{\sim}{B}/\partial s$). We have then the two equivalent versions of (6):

$$\underset{\sim}{J}_\perp = (c\underset{\sim}{B}/B^2) \times [\nabla P_\perp + (P_\|-P_\perp)\underset{\sim}{b} \cdot \nabla \underset{\sim}{b} + \rho d\underset{\sim}{V}/dt] \qquad (10a)$$

$$\underset{\sim}{J}_\perp \xi = (c\underset{\sim}{B}/B^2) \times [\nabla P_\perp + (P_\|-P_\perp)\nabla B/B + \rho d\underset{\sim}{V}/dt] \qquad (10b)$$

where $\xi \equiv 1 - 4\pi(P_\| - P_\perp)/B^2$ and the gravitational terms have been omitted for simplicity. A considerable amount of vector manipulation yields the two counterparts of (7):

$$B \frac{\partial}{\partial s}\left(\frac{J_\| \xi}{B}\right) = 2\underset{\sim}{J}_\perp \cdot (\underset{\sim}{b} \cdot \nabla \underset{\sim}{b}) + \frac{c\underset{\sim}{B}}{B^2} \cdot \left| \nabla (P_\| - P_\perp) \times \underset{\sim}{b} \cdot \nabla \underset{\sim}{b} \right.$$

$$\left. + \nabla \times \rho \frac{d\underset{\sim}{V}}{dt} \right| \qquad (11a)$$

$$= \underset{\sim}{J}_\perp \cdot \frac{\nabla(B^2 \xi)}{B^2} + \frac{c\underset{\sim}{B}}{B^2} \cdot \left| \nabla (P_\|-P_\perp) \times \frac{\nabla B}{B} \right.$$

$$\left. + \nabla \times \rho \frac{d\underset{\sim}{V}}{dt} \right| \qquad (11b)$$

to which we may add a third version

$$B \frac{\partial}{\partial s}\left(\frac{J_\| \xi}{B}\right) = \frac{\nabla(B^2 \xi)}{B^2 \xi} \cdot \frac{c\underset{\sim}{B}}{B^2} \times \left[\nabla\left(\frac{P_\|+P_\perp}{2}\right) + \rho \frac{d\underset{\sim}{V}}{dt}\right]$$

$$+ \frac{c\underset{\sim}{B}}{B^2} \cdot \nabla \times \rho \frac{d\underset{\sim}{V}}{dt} \qquad (11c)$$

where $J_\perp$ has been eliminated and the whole cast into a form as close as possible to that in the isotropic limit $P_\| = P_\perp$, $\xi = 1$. The acceleration terms in (11) can be shown to be given by

$$\frac{c\underset{\sim}{B}}{B^2} \cdot \nabla \times \rho \frac{d\underset{\sim}{V}}{dt} = -\nabla \rho \cdot \frac{c\underset{\sim}{B}}{B^2} \times \frac{d\underset{\sim}{V}}{dt} + \rho c \frac{d}{dt}\left(\frac{\Omega_\|}{B}\right)$$

$$- \frac{\rho c}{B} \left[\underset{\sim}{\Omega}_\| \cdot (\underset{\sim}{b} \cdot \nabla)\underset{\sim}{V} + \underset{\sim}{b} \cdot (\underset{\sim}{\Omega}_\perp \cdot \nabla)\underset{\sim}{V}\right] \qquad (12)$$

where $\underset{\sim}{\Omega} \equiv \nabla \times \underset{\sim}{V}$ is the vorticity; in deriving (12), the MHD approximation

$$c\underset{\sim}{E} + \underset{\sim}{V} \times \underset{\sim}{B} = 0 \qquad (13)$$

and the mass continuity equation

$$\partial \rho/\partial t + \nabla \cdot \rho \underset{\sim}{V} = 0 \qquad (14)$$

have been used.

When $P_\| = P_\perp$ and $\underset{\sim}{\Omega}_\perp = 0$, equations (11b) and (12) give the expression for $\partial J_\|/\partial s$ derived by Hasegawa and Sato (1979) and Sato and Iijima (1979)(except that Hasegawa and Sato include a finite Larmor radius correction neglected here, and $2\nabla B/B$ is misprinted in Sato and Iijima). The assumption that $\underset{\sim}{\Omega}$ and $\underset{\sim}{B}$ are aligned, adopted by Hasegawa and Sato as "reasonable," is surely not valid in general except in the equatorial plane. As a simple counterexample, consider a corotational flow: the vorticity is everywhere parallel to the rotation axis, regardless of the orientation of $\underset{\sim}{B}$, and its magnitude is constant along a field line, as required by Ferraro's isorotation theorem (whereas if Ω and B were aligned, it is readily shown that Ω/B would have to be constant along a field line).

Another approach is to write the perpendicular current as

$$\underset{\sim}{J}_\perp = -c[\nabla \times (P_\perp \underset{\sim}{B}/B^2)]_\perp + c\underset{\sim}{B}/B^2 \times [P_\perp \nabla B/B + P_\| \underset{\sim}{b} \cdot \nabla \underset{\sim}{b}$$

$$+ \rho d\underset{\sim}{V}/dt] \qquad (15)$$

where the successive terms can be recognized as the contributions from gyromotion, gradient drift, curvature drift, and polarization drift, summed over all energies, pitch angles, and particle species. (The perpendicular sign on the first term is needed, since

$$-c\underset{\sim}{b} \cdot \nabla \times (P_\perp \underset{\sim}{B}/B^2) = -(4\pi P_\perp/B^2) J_\| \qquad (16)$$

and it is instructive to visualize how gyration about the guiding center acquires a parallel component when a Birkeland current is present.) By dint of vector identities, equation (15) can be shown to be rigorously equivalent to (10). Thus, the combined effect of all the particle motions in the first-order guiding center description implies the same $\underset{\sim}{J}_\perp$ as that required by the momentum equation with gyrotropic pressure, a result stated by Parker (1957) and afterwards repeatedly rediscovered. For all the importance of this result in showing the equivalence of two seemingly quite different physical approaches, one should not forget that equation (15), like the guiding center theory from which it is derived, is no more than a calculational device, and its various terms, even if sometimes dignified by the names "magnetization current," "gradient drift current," etc., have no independent physical existence: no one of them, taken by itself, represents a current that could be measured as such either by a magnetometer via equation (2) or a particle detector via (1) - both equations refer only to the total current density $\underset{\sim}{J}$ summed over any and all drift effects.

The so-called polarization or inertial current, in particular, defined by

$$\underset{\sim}{J}_p = c(\underset{\sim}{B}/B^2) \times \rho d\underset{\sim}{V}/dt \qquad (17)$$

has received considerable attention. It can be rewritten, with the use of (14) and Faraday's law, as

$$J_p = \frac{\rho c^2}{B} \frac{d}{dt}\left(\frac{\tilde{E}}{B}\right) + \frac{\rho c}{B} \tilde{V} \times \left|(\tilde{b}\cdot\nabla\tilde{V})\right|_\perp$$

$$- \frac{\rho c}{B} \frac{d}{dt}\left|\frac{c\tilde{E}+\tilde{V}\times\tilde{B}}{B}\right| - \frac{\rho c}{B^2} \times [\nabla\times(c\tilde{E}+\tilde{V}\times\tilde{B})]_\perp \quad (18)$$

with no approximations so far. The terms in the second line of (18) vanish when the MHD approximation (13) is assumed. If one in addition assumes that

$$\tilde{V}\times[(\tilde{b}\cdot\nabla)\tilde{V}]_\perp = 0 \quad (19)$$

then J_p, now given by the first term of (18) alone, bears a certain resemblance to the polarization current of a linear dielectric medium, given by

$$\tilde{J}_p = [(\varepsilon-1)/4\pi]\partial\tilde{E}^*/\partial t \quad (20)$$

and on the basis of this resemblance it is sometimes stated that the plasma can be treated as a dielectric, with dielectric constant $\varepsilon = (4\pi\rho c^2/B^2)+1$. Note, however, that the first term of (18) is not really of the same form as (20); the polarization of a dielectric medium depends on the electric field $\tilde{E}^*$ measured in the frame of reference where the medium is locally at rest, whereas for a plasma $\tilde{E}^* = 0$ whenever the MHD approximation holds. Furthermore, assumption (19) is quite restrictive and unlikely to be valid for other than a few types of simple, mostly two-dimensional, geometries. For these reasons among other (see Vasyliunas, 1983, for a detailed discussion), the so-called dielectric description of a plasma is no more than a calculational device, based on formal analogies of restricted scope and useful only in limited areas (chiefly the theory of plasma waves); attempts to treat the plasma as a true dielectric (e.g. Cole, 1983 - this volume) may lead to errors (see Vasyliunas, 1983).

There exists another connection between the configuration of electric currents and the plasma bulk flow, distinct from the flow contribution to the mechanical stresses balanced by the Lorentz force. The MHD approximation (13), valid for most large-scale structures in the magnetosphere (see, e.g., Vasyliunas, 1980; Siscoe, 1983), implies that $\tilde{B}$ and $\tilde{V}$ are closely connected, so that the magnetic flux through any loop moving with the plasma bulk flow remains constant and points on a magnetic field line moving with the flow continue to form a field line - the well-known flux-preserving and line-preserving properties of hydromagnetic flow. Equation (13) can be combined with Faraday's law to yield

$$\partial\tilde{B}/\partial t = \nabla\times(\tilde{V}\times\tilde{B}) \quad (21)$$

from which $\tilde{B}(\tilde{r}, t)$ can be calculated if the flow configuration $\tilde{V}(\tilde{r}, t)$ is known. When the magnetic field consistent with a given flow has thus been determined, the current density can be obtained simply from $\nabla\times\tilde{B} = (4\pi/c)\tilde{J}$. This MHD connection between $\tilde{J}$ and $\tilde{V}$ has a particularly apparent significance in the case of time-varying systems: a change in $\tilde{J}$ presupposes an appropriate change of $\tilde{B}$, which in turn can only occur if there is a plasma flow pattern that will deform the field lines into their new configuration. The rate at which a current system can change is limited by the speeds of the associated flows which in turn depend on the stresses that are trying to bring about the change; it is not always appreciated that a reduction of the cross-magnetotail current, for example, is thus subject to definite dynamical constraints.

References

Carovillano, R.L., and G.L. Siscoe, Energy and momentum theorems in magnetospheric processes, Rev. Geophys. Space Phys., 11, 289-353, 1973.

Cole, K.D., this volume, 1983.

Frank, L.A., C.Y. Huang, and T.E. Eastman, this volume, 1983.

Hasegawa, A., and T. Sato, Generation of field aligned current during substorm, in Dynamics of the Magnetosphere, edited by S.-I. Akasofu, pp. 529-542, D. Reidel, Dordrecht-Holland, 1979.

Mead, G.D., and D.B. Beard, Shape of the geomagnetic field solar wind boundary, J. Geophys. Res., 69, 1169-1179, 1964.

Olson, W.P., K.A. Pfitzer, and G.J. Mroz, Modeling the magnetospheric magnetic field, in Quantitative Modeling of Magnetospheric Processes, edited by W.P. Olson, pp. 77-85, AGU Geophysical Monograph 21, 1979.

Parker, E.N., Newtonian development of the dynamic properties of ionized gases of low density, Phys. Rev., 107, 924-933, 1957.

Parker, E.N., Interplanetary Dynamical Processes, Chapter X, Wiley-Interscience, New York, 1963.

Sato, T., and T. Iijima, Primary sources of large-scale Birkeland currents, Space Sci. Rev., 24, 347-366, 1979.

Schindler, K., Plasma and fields in the magnetospheric tail, Space Sci. Rev., 17, 589-614, 1975.

Siscoe, G.L., Solar system magnetohydrodynamics, in Solar-Terrestrial Physics, edited by R.L. Carovillano and J.M. Forbes, pp. 11-100, D. Reidel, Dordrecht-Holland, 1983.

Spreiter, J.R., A.Y. Alksne, and A.L. Summers, External aerodynamics of the magnetosphere, in Physics of the Magnetosphere, edited by R.L. Carovillano, J.F. McClay, and H.R. Radoski, pp. 301-375, D. Reidel, Dordrecht-Holland, 1968.

Vasyliunas, V.M., Plasma sheet dynamics: effects on, and feedback from, the polar ionosphere, in Exploration of the Polar Upper Atmosphere, edited by C.S. Deehr and J.A. Holtet, pp. 229-244, D. Reidel, Dordrecht-Holland, 1980.

Vasyliunas, V.M., Dielectric effects in magnetospheric plasma: a critical examination, to be submitted to J. Geophys. Res., 1983.

Walker, R.J., Quantitative modeling of planetary magnetospheric magnetic fields, in Quantitative Modeling of Magnetospheric Processes, edited by W.P. Olson, pp. 9-34, AGU Geophysical Monograph 21, 1979.

ESTIMATION OF ELECTRIC FIELDS AND CURRENTS FROM GROUND-BASED MAGNETOMETER DATA

Y. Kamide[1] and A. D. Richmond

NOAA/ERL Space Environment Laboratory, Boulder, Colorado 80303

Abstract. We review and evaluate recent advances in numerical algorithms for estimating ionospheric electric fields and currents from groundbased magnetometer data. Tests of the adequacy of one such algorithm in reproducing large-scale patterns of electrodynamic parameters in the high-latitude ionosphere have yielded generally positive results, at least for some simple cases. We point out some encouraging advances in producing realistic conductivity models, which are a critical input. When the algorithms are applied to extensive data sets, such as the ones from meridian chain magnetometer networks during the IMS, together with refined conductivity models, unique information on instantaneous electric field and current patterns can be obtained. Examples of electric potentials, ionospheric currents, field-aligned currents, and Joule heating distributions derived from ground magnetic data are presented. Possible directions for future improvements are also pointed out.

Introduction

Worldwide collections of magnetic records obtained on the earth's surface have been used widely to diagnose the state of the ionosphere and magnetosphere and their electrical coupling. In particular, studies of the ground-based magnetic records have proven important in understanding substorm processes in terms of the growth and decay of the three-dimensional current system over the entire polar region. However, because the ground-based magnetometer data include the effects of a variety of source currents flowing in the ionosphere and magnetosphere as well as within the earth, a serious problem arises in attempting to deduce the relative contributions of these different source currents to particular patterns of global magnetic perturbations under study. An extremely simplified but popular method of analysis has been through calculations of "equivalent" currents which are determined by making an unrealistic assumption that all overhead currents flow in a spherical shell concentric to the earth, namely, the ionosphere.

The increase in both quantity and quality of ground magnetic observations in recent years has provided the incentive to develop computational techniques designed to estimate the "true", not equivalent, pattern of ionospheric and field-aligned currents at high latitudes based on magnetic observations made on the earth's surface [Kisabeth, 1979, Hughes et al., 1979; Mishin et al., 1979; Kamide et al., 1981; Levitin et al., 1982]. It is required that the derived current models not only be capable of reproducing the original magnetic records, but also be consistent with other information on electric fields and ionospheric conductances obtained from more direct measurements, such as satellites and radars. The purpose of this paper is to review recent progress in the technique of inferring ionospheric electrodynamics from ground magnetic data, focusing on studies in which the present authors have been involved. For this particular purpose we follow here the algorithm developed by Kamide et al., [1981].

Outline of the Method and its Numerical Tests

The technique used by Kamide et al. [1981] requires that the ionospheric conductivity is available from other sources. The flow chart of the practical steps involved in the algorithm is given in Figure 1. Although the concept and the procedure is described in detail elsewhere, it will be outlined here.

The horizontal, height-integrated ionospheric current $\underline{J}$ can be expressed as the sum of toroidal and potential currents, $\underline{J} = \underline{J}_T + \underline{J}_P$. The potential (irrotational) current $\underline{J}_P$ is connected to the field-aligned current j^P (assumed radial for simplicity) such that $j^P = \mathrm{div}\, \underline{J}_P$. The total poloidal current comprised of j^P and $\underline{J}_P$ produces no ground magnetic effect, and the toroidal (solenoidal) current $\underline{J}_T$ therefore is identical with the equivalent current derivable from ground magnetometers. The total height-integrated ionospheric current $\underline{J}$ is related to the electrostatic field $\underline{E} = -\mathrm{grad}\, \Phi$ by Ohm's Law, $\underline{J} = \Sigma_P \underline{E} + \Sigma_H \underline{E} \times \underline{n}_r$, where Σ_P and Σ_H are the height-integrated Pedersen and Hall conductivities and $\underline{n}_r$ is a unit radial vector. Equating the two expressions for $\underline{J}$ and taking the curl then gives an elliptic second-order partial differential equation for Φ in terms of the known $\underline{J}_T$:

$$\mathrm{curl}\, (-\Sigma_P\, \mathrm{grad}\, \Phi - \Sigma_H\, \mathrm{grad}\, \Phi \times \underline{n}_r) = \mathrm{curl}\, \underline{J}_T$$

---

[1]Permanent affiliation: Kyoto Sangyo University Kyoto 603, Japan.

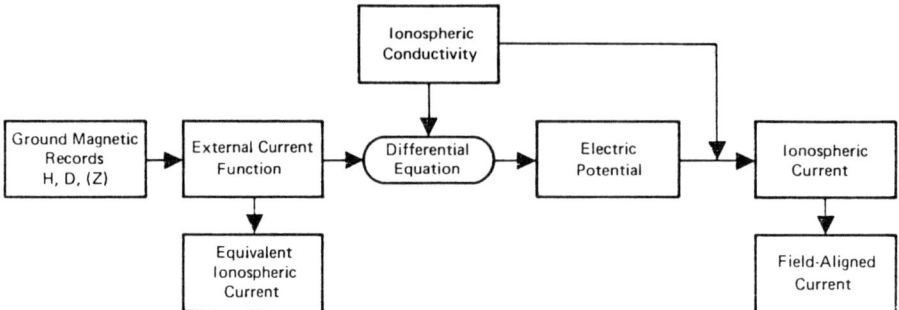

Fig. 1. Schematic diagram showing the steps used in the algorithm of Kamide et al. [1981]. Inputs are the measured ground magnetic variations and a model of the ionospheric conductivity.

This differential equation is numerically solved over a network of points spaced 1° in latitude and 15° in longitude. Having solved the electrostatic potential Φ at all the grid points, the ionospheric current $\underline{J}$ can be obtained from Ohm's Law. One can then derive the field-aligned current div $\underline{J}$ as well as the Joule heating rate $\underline{J} \cdot \underline{E}$.

One important assumption in this algorithm is that all currents above the ionosphere flow radially. However, as discussed quantitatively by Fukushima and Kamide [1973], the effects of field-line curvature influence the low-latitude magnetic perturbations by up to 15%. Furthermore, our neglect of ground-level magnetic effects due to magnetospheric closing currents, such as the ring and tail currents, can be expected to affect the accuracy of the derived parameters, since it is known that on the nightside at low and middle latitudes, where the ionospheric conductivitity is quite low, the partial ring current and the associated field-aligned current make a major contribution to ground magnetic variations [Chen et al., 1982]. Most of the magnetic perturbations that are in fact generated by the ring current system are interpreted in our algorithm as being due to overhead ionospheric currents. Since the conductivity is so low, significant equivalent current densities correspond to unrealistically large electric fields.

In order to increase our confidence that we can rely on the magnetogram-inversion scheme to deduce the global patterns of the ionospheric electric field and current and to study the extent to which the magnetosphere and ionosphere are electrically coupled, it is important to make sure that the numerical accuracy of the algorithm is adequate. The first test of the inversion scheme along this line has been carried out by Akasofu et al. [1981], comparing the deduced three-dimensional current systems from algorithms developed from different principles by Kisabeth [1979] and Kamide et al. [1981]. The two algoithms were applied to the same ground magnetometer data (daily average data from the Alaska meridian chain), and the computed field-aligned currents were in good agreement. The method of Kisabeth [1979] does not yield electric fields directly, so these could not be compared between the two methods.

In the second test, Kamide and Richmond [1982] have conducted a crucial check of the numerical accuracy. Starting with observed ground magnetic records, they computed the field-aligned currents using the algorithm of Kamide et al. [1981]. These field-aligned currents were then used as inputs to a reversed calculation with the model of Kamide and Matsushita [1979a] to attempt to reproduce the original magnetic perturbations. Figure 2 displays the results, where the order of the calculation steps is indicated with the aid of arrows. Panel A shows the equivalent currents obtained from magnetic records for average conditions. Panel B shows the calculated potential distribution, used to deduce the ionospheric current and field-aligned current distributions. Panel C on the right hand side is the potential distribution calculated from these field-aligned currents with the same conductivity model used in earlier steps. It is assuring to see that the two potential patterns are nearly identical. Panel D shows equivalent current vectors which are derived from the electric field of Panel C. It is encouraging that this pattern and Panel A (original input from the ground magnetic perturbations) are very similar.

The third test has been conducted by Wolf and Kamide [1983], who used "theoretical" magnetograms which are estimated from an extensive simulation model for magnetospheric plasma processes [Harel et al., 1981a, b; Spiro et al., 1981]. The simulation model, called the Rice Convection Model (RCM), provides a more theoretical, less observation-based representation of the magnetosphere-ionosphere current system. One can pretend that the theoretical magnetograms at arbitrary grid points calculated by integrating the Biot-Savart law over the current distribution are actually observed, and can input them into the magnetogram-inversion scheme to reproduce the electric field under which the divergences of horizontal ionospheric currents and of magnetospheric gradient/curvature drift currents are balanced by field-aligned currents. In the RCM, a model of the ionospheric conductivity must be

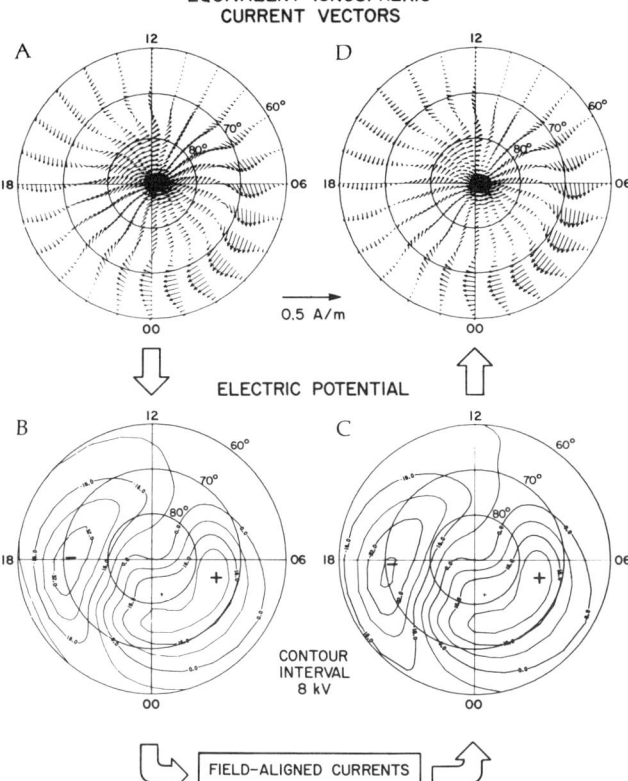

Fig. 2. Results of a test of self-consistency of two algorithms. (a) Equivalent current vectors obtained from the average observed ground magnetic perturbations. (b) Electric potential calculated from the equivalent currents in (a). (c) Electric potential calculated from the algorithm of Kamide and Matsushita [1979a], using as input the field-aligned currents derived as output from the algorithm of Kamide et al. [1981] using the same conductivity model. (d) Equivalent current vectors calculated from the field-aligned currents and ionospheric currents.

flow velocities are too large on the nightside below 65° latitude. Total polar-cap potential drops are found to agree to better than 10%.

Overall, the tests confirm the ability of the inversion altorithm to handle quite accurately the the estimation of auroral ionospheric conductances are given.

Conductivity Models

It should be emphasized that it is not possible to determine uniquely the distribution of ionospheric and magnetospheric currents only from ground magnetic observations without having information on either ionospheric electric fields or conductivities. In other words, the main difficulty in applying the numerical scheme of Kamide et al., [1981] is to determine a suitable distribution of the conductivities for use in the computations. Several different conductivity models developed on the basis of different concepts, with a number of different parameters specifying the conductivity distribution, have been tested. In most of these models, it is assumed that the height-integrated conductivity has two components: one is a "background" conductance of solar ultraviolet origin and the other simulates an enhance-

given, which has been used also in the inversion test. Figure 3 compares horizontal flow velocities $\underline{E} \times \underline{B}/B^2$ computed by the Kamide et al. [1981] algorithm with those in the original RCM simulation for two cases representing times just after a substorm onset and at the maximum epoch of the substorm. There is very good agreement with regard to overall convection patterns at high latitudes, although the following two points must be noted. First, since only the processes in the equatorial plane in the magnetosphere are dealt with in RCM, the velocities in the polar cap, where field lines are not closed in the magnetosphere, cannot be properly compared. Second, as pointed out earlier, because of the assumption of radial currents above the ionosphere in the inversion technique, the inverted

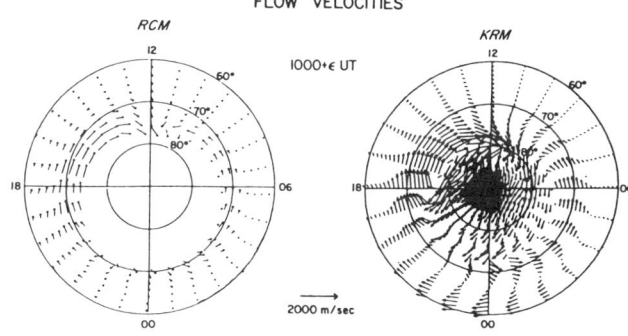

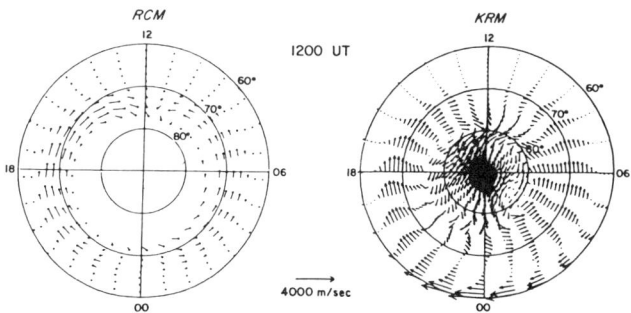

Fig. 3. Comparison of E x B ionospheric flow velocities at two times simulated with the Rice Convection Model (RCM) and deduced from the corresponding simulated ground magnetic variations by application of the Kamide et al. [1981] algorithm (KRM).

ment due to auroral particle bombardment. We may call the former the quiet-time conductance and the latter the auroral enhancement conductance.

As a simple mathematical representation, Kamide et al. [1981] and Akasofu et al. [1981] used a distribution which was devised by Kamide and Matsushita [1979b]. The auroral enhancement is assumed to have a Gaussian distribution in both the latitudinal and longitudinal directions. It is practically possible to put a number of such Gaussian forms within the auroral oval representing local auroral enhancements. The ratio of the Hall to Pedersen conductances must be specified. A major problem of the use of this conductivity model lies in the difficulty in estimating accurately the location of the peak conductivity and its value. Kamide et al. [1981] suggested that it may not be too incorrect to assign the peak conductivity location being coincident with the peak location of the observed magnetic perturbation, since the maximum auroral electrojet is presumably colocated with the highest conductivity. Friis-Christensen et al. [1983] determined the peak conductivity latitude by referring to earlier statistical studies on the relationship between the interplanetary magnetic field and the size of the auroral oval. Kamide et al. [1981] and Kamide and Richmond [1982] examined the sensitivity of the electric fields and currents to the conductivity distribution. It was found that the calculated ionospheric and field-aligned currents are only weakly dependent on the magnitude of the auroral conductivity enhancement, but are affected by the location of the enhancement. The electric field intensity, on the other hand, was found to be sensitive to the assumed conductivity.

Spiro et al. [1982] constructed an empirical model representing the statistical distribution of the height-integrated Hall and Pedersen conductivities, called the Rice University model. Data of precipitating particle energy flux and average electron energy observed by the AE-C and D Satellites were sorted out in terms of different levels of geomagnetic activity. Spiro et al. then used the dependence of the conductances on the characteristic energy of precipitating electrons obtained by Vickrey et al. [1981]. The global distribution of the estimated conductances is similar to that developed separately by Wallis and Budzinski [1981] who used ISIS 2 measurements of precipitating electrons for relatively quiet and active geomagnetic conditions. The Rice University conductivity model has been actively used in the algorithm of Kamide et al. [1981] for several substorm intervals during the International Magnetosphere Study (IMS).

It should be noted that caution must be exercized in applying a statistically obtained, empirical conductivity model to the global modeling of dynamic processes. Kamide et al. [1982a] employed the Rice model with updated improvements in which the Hall and Pedersen conductances are continuous functions of the AE index. However, the auroral precipitation distribution and thus the conductivity distribution may be significantly different even at two instants with the same value of the AE index. Kamide et al. [1982b] and Kamide et al. [1983] subsequently made the following adjustment in the use of the conductivity model: By assuming that the latitude of the maximum equivalent current strength coincides with the latitude of the highest Hall conductivity, a latitudinal shift is made for the entire conductivity distribution whenever a difference between the two latitudes of the maxima is found. This adjustment may partially accomodate conductivity variability, although it does not account for local time shifts in the maxima.

The lack of accurate conductivity models that can reflect conditions on an instantaneous basis has become more and more acute as simultaneous magnetic records from a large number of observatories which represent complex current patterns over the entire polar region with a high time resolution have become available. As a first step toward improving this situation, Ahn et al. [1983] attempted to use the magnitude of ground magnetic disturbances to estimate quantitatively the conductivity distribution. By comparing the height-integrated conductivities deduced from Chatanika variation (ΔH) at College, the following empirical relationships were reached:

$\Sigma_P = 2.4 \, (\Delta H)^{0.24}$ for $\Delta H < 0$

$\Sigma_P = 0.7 \, (\Delta H)^{0.43}$ for $\Delta H > 0$

$\Sigma_H = 4.4 \, (\Delta H)^{0.34}$ for $\Delta H < 0$

$\Sigma_H = 2.1 \, (\Delta H)^{0.36}$ for $\Delta H > 0$

where the conductances are in mhos and ΔH is in nanotesla. These formulas indicate that the ionospheric conductivity is normally larger in the westward electrojet ($\Delta H<0$) than in the eastward electrojet ($\Delta H>0$). Since these empirical formulas would hold only near 65° magnetic latitude, Ahn et al. [1983] devised an appropriate weighting function for both the polar cap and the subauroral regions. Figure 4 compares the global height-integrated Hall conductivity pattern estimated from the ΔH distribution with that from the Rice University model for the maximum epoch of an intense substorm. The AE index level for this particular time was used to display the Rice model. It is noticed that the gross features are similar, but the ΔH-based conductivity includes many local structures which cannot be seen in the average Rice model. While such a method of constructing instantaneous conductivity models is extremely useful, the empirical relationships can probably be improved significantly by including their local-time dependence. It is also expected that data from the new Greenland radar facility could improve the formulas for the polar cap region.

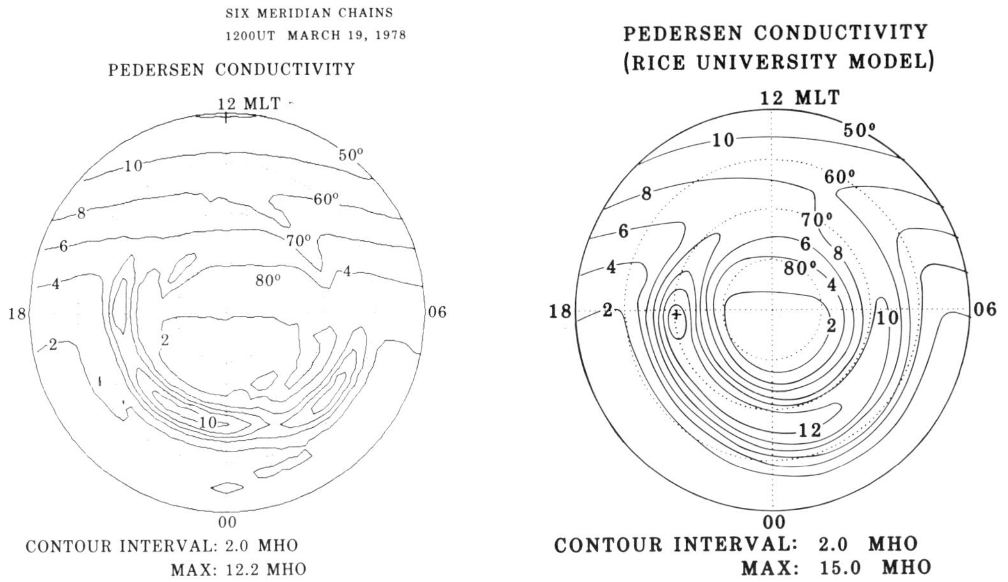

SIX MERIDIAN CHAINS
1200UT MARCH 19, 1978
PEDERSEN CONDUCTIVITY

CONTOUR INTERVAL: 2.0 MHO
MAX: 12.2 MHO

PEDERSEN CONDUCTIVITY
(RICE UNIVERSITY MODEL)

CONTOUR INTERVAL: 2.0 MHO
MAX: 15.0 MHO

Fig. 4. Comparison of two conductivity models, showing the height-integrated Pedersen conductivity at latitudes above 50° magnetic. On the left are values estimated by the procedure of Ahn et al. [1983] using observed ground magnetic variations at the instant shown. On the right are values from the statistical Rice University model.

Most recently, efforts are being made to estimate an "instantaneous" distribution of the ionospheric conductivities over the entire polar region by using available UV auroral imagery from Dynamics Explorer (DE) [Frank et al. 1981].

Most recently, efforts are being made to estimate an "instantaneous" distribution of the ionospheric conductivities over the entire polar region by using available UV auroral imagery from Dynamics Explorer (DE) [Frank et al. 1981]. Kamide et al. [1983] have assumed that the increase in the ionospheric conductivities, thus in the electron density, is proportional to the square root of the auroral luminosity observed by DE. This implies that the auroral luminosity and the ionization production rate are correlated. The proportionality factors depend on the wavelength of the auroral imagery. As Craven et al. [1983] indicated, it may be possible to discuss dynamics of current and aurora patterns in the entire polar region including local structures with the availability of these images.

Electric Potential

Although we have noted the snesitivity of the calculated electric fields to the conductivity model chosen, the general pattern of calculated electrostatic potential, as shown, for example, in Figure 2, reproduces quite satisfactorily known features determined by independent measurements. The two-cell pattern gives a dawn-to-dusk electric field across the polar cap with northward and southward electric fields in the evening and morning auroral oval, respectively. This pattern agrees in general with features deduced from satellite and radar measurements [Heppner, 1977; Evans et al., 1980; Foster et al., 1982; Heelis et al., 1982; Oliver et al., 1983]. The total potential drop across the polar cap is 59 kV in Frame B of Figure 2, which is a reasonable value for conditions of moderate magnetic activity, although it is specific to the particular conductivity model used.

The conductivity distribution is reasonably well determined in the summer polar cap where solar ultraviolet radiation dominates the production of ionization. The polar cap is also the region best characterized by radial geomagnetic field lines. Friis-Christensen et al. [1983] have made use of these advantages to evaluate electric fields and currents around the summer polar cusp for different directions of the interplanetary magnetic field (IMF). The ground magnetic data came from the Greenland magnetometer chain, and were processed to produce average variations for an IMF strength of 5 nT in the solar-magnetospheric y-z plane. The conductivity model included a nightside auroral enhancement, increasing as the algebraic value of the IMF B_z component decreased. Figure 5 shows the calculated electric potentials for four directions of the IMF. The two-cell, electric potential pattern appears consistently in each case, but its shape and magnitude change. The electric field intensity increases as B_z becomes negative. Furthermore, the overall pattern is shifted towards dawn (or dusk) for B_y

ELECTRIC POTENTIAL

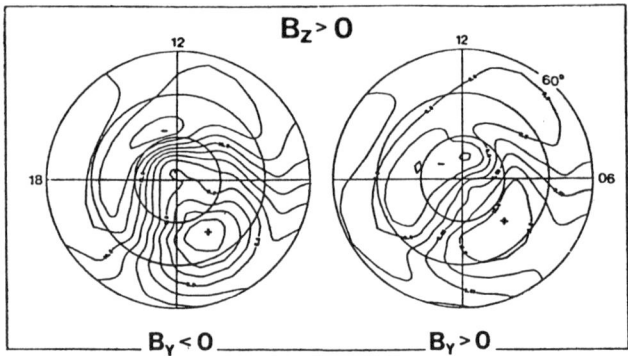

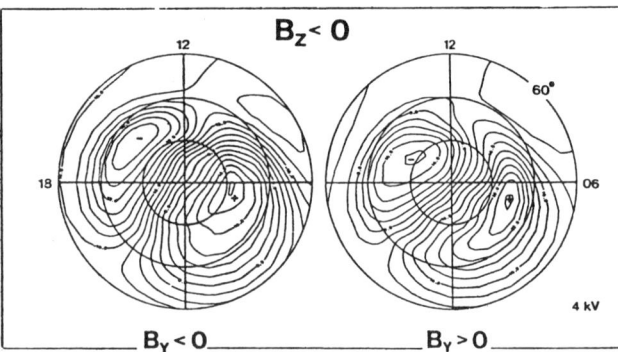

Fig. 5. Electric potential distributions calculated by Friis-Christensen et al. [1983] for different interplanetary magnetic field conditions. The magnitudes of the IMF B_Y and B_Z components are 3.5 nT in each case, with different combinations of signs.

> 0 (or < 0) in agreement with more direct observations by Heppner [1972] and Mozer et al. [1974]. Strong changes in the electric field in the dayside cusp region are evident for changing IMF conditions.

The analysis of substorm events is considerably more difficult than analyses of average conditions. Although the dense coverage of the polar region with magnetometers during the IMS provides the most nearly complete spatial information on instantaneous magnetic variations available to date, this coverage has gaps and is not as comprehensive as the effective area covered by a single magnetometer chain rotating daily under the polar current system. Furthermore, instantaneous patterns of auroral conductivity enhancements can differ substantially from the averaged models so far used in calculations. In spite of these difficulties, Kamide et al. [1982a] have analyzed individual substorm cases using data from the world IMS meridian chains and the statistical conductivity distribution of Spiro et al. [1982]. The resulting potential patterns tended to have a basic twin-vortex pattern, but many local deformations appeared. At present it is difficult to say whether these local features faithfully represent the actual ionospheric electric fields or whether they are artifacts of the calculations caused by inadequate magnetometer coverage and conductivity information.

Ionospheric Currents

By using a number of examples, both of statistical and individual cases, it has been found that the ionospheric current distribution obtained through the inversion scheme is not sensitive to the choice of the ionospheric conductivity model, unless the auroral enhancement is mislocated. This is encouraging in the sense that although for a long history of geomagnetism, our discussions have been limited in scope by the availability of only "equivalent" ionospheric currents derivable directly from ground-based magnetic records, it is now possible to describe the growth and decay of polar magnetic substorms in terms of the global distribution of "true" ionospheric currents. In fact, by comparing in Figure 6 the equivalent and "true" current distributions, several points of interest can be noticed. First, the equivalent currents in the auroral zone flow nearly in the east-west direction, whereas the ionospheric currents have a considerable north-south component as well; they flow southwestward in the morning sector and northeastward in the evening sector. The sense of the north-south component coincides with that of the Pedersen current driven by the large-scale electric field. Second, the magnitude of the equivalent currents is smaller than that of the ionospheric currents by a factor of 0.7-0.8. These facts indicate that the effects of the field-aligned currents in both the ground H and D components are significant.

Recently, it has been postulated that there are possibly two different types of current systems prevailing at different phases of a substorm.

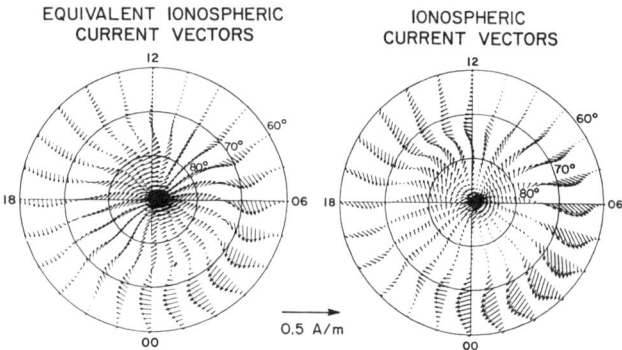

Fig. 6. Comparison of "equivalent" (left) and "true" (right) ionospheric current vectors [from Kamide and Richmond, 1982].

Figure 7 compares two representative cases taken from the CDAW-6 interval [March 22, 1979]. The first case occurred just before the onset of a major substorm, while the second case corresponds to the maximum epoch of the intense substorm. The major difference between these two current patterns lies in the locations of their electrojet maxima. The first example is characterized by the eastward electrojet in the evening sector and the less intense westward electrojet in the morning sector. In the premidnight and midnight sectors, the ionospheric currents are very weak. The maximum eastward and westward currents occur near dusk and dawn, respectively. On the other hand, the second example is dominated by the intense westward electrojet flowing from the prenoon to premidnight sectors.

Field-aligned Currents

The distribution of field-aligned currents can be obtained by taking the divergence of the ionospheric current vectors. Akasofu et al. [1981] have shown by using a data base from the IMS Alaska chain of magnetometers representing average values at every MLT hour, that the numerical scheme could reproduce quite satisfactorily the statistical features of the region 1/region 2 field-aligned currents in auroral latitudes. Friis-Christensen et al. [1983] have argued that the field-aligned current system in the dayside cusp region, whose current direction is controlled by the IMF B_y component, can be reproduced successfully.

It is important to point out, however, that a far greater variability exists during individual substorms. By analyzing data from the six IMS meridian chains of magnetometers, Kamide et al. [1982a] have shown that the field-aligned current regions may not be of the well-known auroral oval shape at particular instants, particularly during the developing stage of substorms, and have many local deformations with high current densities. Such local enhancements were found not only during intense substorms, but also during comparatively periods.

The left frame of Figure 8 is taken from one of of the substorm maximum phases during the three-day interval [March 17-19, 1978] compiled by Kamide et al. [1982a]. It is noticed that the field-aligned current patterns are very complicated, having many local structures. The right frame shows the similar contour plots of the field-aligned current density averaged for the entire three-day interval. This pattern is similar to the statistical pattern of satellite-observed field-aligned currents [Iijima and Potemra, 1976], in that there are two pairs of field-aligned currents, i.e., the region 1 and region 2 currents in the auroral oval. The current distribution during individual substorms demonstrates not only that the field-aligned currents penetrate into local times not seen in the statistical distribution, but also that there are multiple peaks in some regions associated with local enhancements of the substorm-disturbed auroral electrojets.

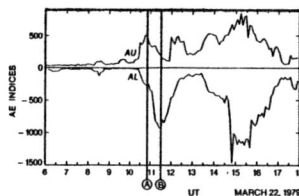

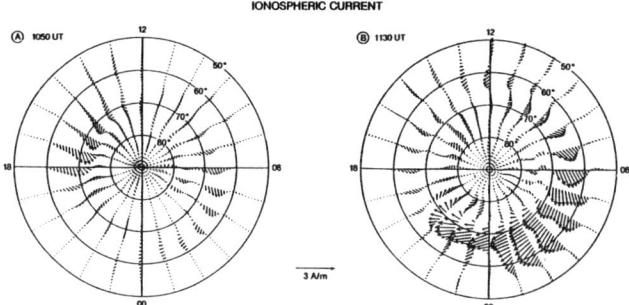

Fig. 7. Ionospheric current vectors at two phases of a substorm. The AU and AL indices are shown at the top.

Joule Heating

Having found the ionospheric electric fields and currents in the numerical scheme, one can calculate the height-integrated Joule heating rate $u_J = \underline{J} \cdot \underline{E} = \Sigma_p E^2$. The total Joule heating in the entire hemispheric ionosphere can then be obtained by integrating u_J both latitudinally and longitudinally.

As a continuation of the joint project of deducing the distribution of electric fields and currents using the six meridian chains of magnetometers for a three-day interval in March 1979, Kamide et al. [1982a, b] have constructed hourly maps of the Joule heat production rate. Figure 9 shows an example of the Joule heating along with the distribution of ionospheric currents for 1300 UT March 18, 1978, which was the maximum epoch of an intense substorm. It appears that there are several separated regions of Joule heating. It is also interesting to see that significant Joule heating results from the eastward electrojet in the evening sector, although the eastward electrojet is less intense than the westward electrojet in the midnight and morning sectors. This agrees with the measurements by Vickrey et al. [1982] who used electric fields and conductivities deduced from Chatanika radar observations. This tendency of having a relatively large heating rate from the eastward electrojet results from the strong electric field in this region. The westward electrojet, by contrast, has a weaker electric field but

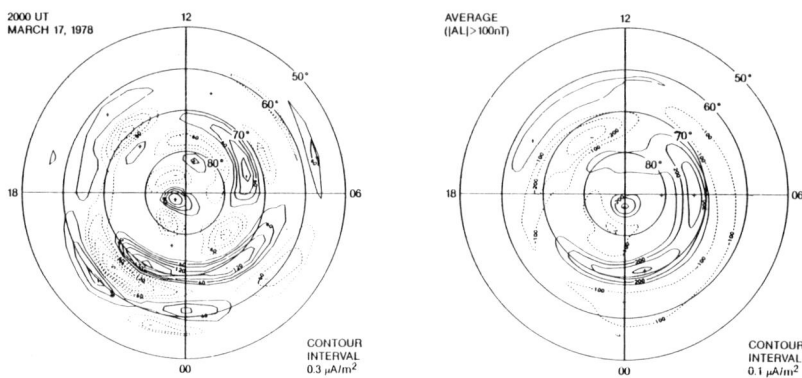

Fig. 8. Instantaneous pattern of estimated field-aligned current density at the maximum phase of a substorm. Solid contours represent downward currents, dashed contours represent upward currents. (Right) Average field-aligned current distribution for March 17-19, 1978. Note the difference in contour intervals.

stronger conductivity enhancement resulting in strong currents but less heating than in the eastward electrojet. The error in determining $\underline{J} \cdot \underline{E}$ is primarily influenced by errors in $\underline{E}$, which are much greater than errors in $\underline{J}$.

Conclusion

The ability to infer ionospheric electric fields and currents from careful analyses of ground magnetometer data is a valuable addition to our set of tools for understanding ionospheric-magnetospheric electrodynamics. Tests of the method on statistically averaged data indicate that the technique gives reasonable estimates of the currents and fields. An exciting prospect is the ability to obtain instantaneous patterns of electrodynamic parameters using a large network of magnetometers, as demonstrated for events during the IMS. Because such continuous large-scale coverage is feasible with relatively inexpensive instruments, the inversion of ground magnetometer data can provide unique information unavailable with satellite or radar instruments. In particular, the improvement of the present magnetometer network at high latitudes will lead to full spatial/temporal coverage of the estimated electric fields and currents and will enable much more complete studies of convection and substorm processes, and other related phenomena, whose study up to now has been limited by the lack of simultaneous monitoring of electromagnetic parameters at multiple locations.

It is clear that further improvements in the analysis technique are possible and desirable. We have noted some of the ongoing studies designed to improve the conductivity models needed in the algorithm. More careful account of the ef-

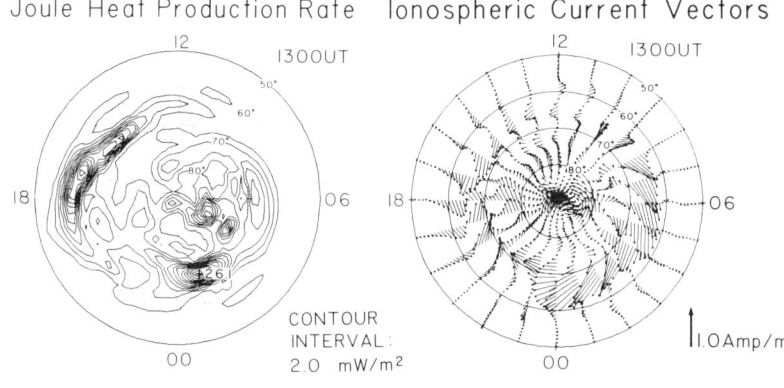

Fig. 9. Distributions of estimated Joule heating rates (left) and corresponding ionospheric current vectors (right) for 1300 UT, March 18, 1978.

fects of nonradial currents above the ionosphere should reduce errors presently arising in regions of low ionospheric conductivity. It would be useful to modify the algorithm so that simultaneous direct measurements of electric fields, conductivities, and field-aligned currents taken by satellite and radar could be directly incorporated into the analysis, effectively placing additional constraints on the calculated patterns. It is desirable for this purpose that radar or satellite data from more than one site be available, because our goal is to obtain the global electric field distribution, not to estimate local electric fields. On a more ambitious scale, a closer coupling between theoretical modeling approaches to magnetosphere-ionosphere interactions and empirical modeling techniques would place even stricter constraints on the patterns and increase their correspondence with reality. We believe that the advancement in our understanding of auroral electrodynamic processes will strongly benefit from these developments.

Acknowledgments. This research was partly supported by NASA Order Number W-15, 347. Y. Kamide was a NRC/NOAA Senior Research Associate during the course of this work.

References

Ahn, B.-H, R. M. Robinson, Y. Kamide, and S.-I. Akasofu, The electric conductivities, electric fields, and auroral particle energy injection rate in the auroral ionosphere and their empirical relations to the horizontal magnetic disturbances, Planet. Space Sci., 31, 641, 1983.

Akasofu, S.-I., Y. Kamide, and J. L. Kisabeth, Comparison of two modeling methods for three dimensional current systems, J. Geophys. Res., 86, 3389, 1981.

Chen, C.-K., R. A. Wolf, M. Harel, and J. L. Karty, Theoretical magnetograms based on quantitative simulation of a magnetospheric substorm, J. Geophys. Res., 87, 6137, 1982.

Craven, J. D., Y. Kamide, L. A. Frank, S.-I. Akasofu, and M. Sugiura, Distributions of aurora and ionospheric currents observed simultaneously on a global scale, this volume, 1983.

Evans, J. V., J. M. Holt, W. L. Oliver, and R. H. Wand, Millstone Hill incoherent scatter observations of auroral convection over $60° \leq G \leq 75°$, 2. Initial results, J. Geophys. Res., 85, 41, 1980.

Foster, J. C., P. M. Banks, and J. R. Doupnik, Electrostatic potential in the auroral ionosphere derived from Chatanika radar observations, J. Geophys. Res., 87, 7513, 1982.

Frank, L. A., J. D. Craven, K. L. Ackerson, M. R. English, R. H. Eather, and R. L. Carovillano, Global auroral imaging instrumentation for the Dynamics Explorer mission, Sp. Sci. Instr., 5, 369, 1981.

Friis-Christensen, E., Y. Kamide, A. D. Richmond, and S. Matsushita, IMF control of high latitude electric fields, ionospheric and Birkeland currents determined from Greenland magnetometer data, in preparation, 1983.

Fukushima, N., and Y. Kamide, Partial ring current models for worldwide geomagnetic disturbances, Rev. Geophys. Space Phys., 11, 795, 1973.

Harel, M., R. A. Wolf, P. H. Reiff, and R. W. Spiro, Quantitative simulation of a magnetospheric substorm, 1. Model logic and overview, J. Geophys. Res., 86, 2217, 1981a.

Harel, M., R. A. Wolf, R. W. Spiro, P. H. Reiff, C.-K. Chen, W. J. Burke, F. J. Rich, and M. Smiddy, Quantitative simulation of a magnetospheric substorm, 2. Comparison with observations, J. Geophys. Res., 86, 2242, 1981b.

Heelis, R. A., J. K. Lowell, and R. W. Spiro, A model of the high-latitude ionospheric convection pattern, J. Geophys. Res., 87, 6339, 1982.

Heppner, J. P., Polar cap electric field distributions related to the interplanetary magnetic field direction, J. Geophys. Res., 77, 4877, 1972.

Heppner, J. P., Empirical models of high-latitude electric fields, J. Geophys. Res., 82, 1115, 1977.

Hughes, T. J., D. W. Oldenburg, and G. Rostoker, Interpretatation of auroral oval equivalent current flow near dusk using inversion techniques, J. Geophys. Res., 84, 450, 1979.

Iijima, T., and T. A. Potemra, The amplitude distribution of field-aligned currents at northern high latitudes observed by TRIAD, J. Geophys. Res., 81, 2165, 1976.

Kamide, Y., and S. Matsushita, Simulation studies of ionospheric electric fields and currents in relation to field-aligned currents, 1. Quiet periods, J. Geophys. Res., 84, 4083, 1979a.

Kamide, Y., and S. Matsushita, Simulation studies of ionospheric electric fields and currents in relation to field-aligned currents, 2. Substorms, J. Geophys. Res., 84, 4099, 1979b.

Kamide, Y., and A. D. Richmond, Ionospheric conductivity dependence of electric fields and currents estimated from ground magnetic observations, J. Geophy. Res., 87, 8331, 1982.

Kamide, Y., A. D. Richmond, and S. Matsushita, Estimation of ionospheric electric fields, ionospheric currents and field-aligned currents from ground magnetic records, J. Geophys. Res., 86, 801, 1981.

Kamide, Y., B.-H. Ahn, S.-I. Akasofu, W. Baumjohann, E. Friis-Christensen, H. W. Kroehl, H. Mauer, A. D. Richmond, G. Rostoker, R. W. Spiro, J. K. Walker, and A. N. Zaitzev, Global distribution of ionospheric and field-aligned currents during substorms determined using magnetic data from six IMS meridian chains: initial results, J. Geophys. Res., 87, 8228, 1982a.

Kamide, Y., H. W. Kroehl, A. D. Richmond, B.-H. Ahn, S.-I. Akasofu, W. Baumjohann, E. Friis-Christensen, S. Matsushita, H. Mauer, G. Rostoker, R. W. Spiro, J. K. Walker, and A. N. Zaitzev, Changes in the global electric fields and currents for March 17-19, 1978 from six IMS meridian chains of magnetometers, Rept. UAG-87, WDC-A, NOAA/EDIS, Boulder, CO 1982b.

Kamide, Y., J. D. Craven, L. A. Frank, and S.-I. Akasofu, Modeling ionospheric current systems using the conductivity distribution inferred from DE auroral images, Geophys. Res. Lett., submitted, 1983.

Kisabeth, J. L., On calculating magnetic and vector potential fields due to large scale magnetospheric current systems and induced currents in an infinitely conducting earth, Quantitative Modeling of Magnetospheric Processes, Geophys. Monogr Ser., vol. 21, edited by W. P. Olson, p. 473, Am. Geophys. Union, Washington, D. C., 1979.

Levitin, A. E., R. G. Afonina, B. A. Belov, and Ya. I. Feldstein, Geomagnetic variation and field-aligned currents at northern high-latitudes, and their relations to the solar wind parameters, Phil. Trans. R. Soc. London., A304, 253, 1982.

Mishin, V. M., A. D. Bazarzhapov, and G. B. Shpynev, Electric fields and currents in the earth's magnetosphere, in Dynamics of the Magnetosphere, edited by S.-I. Akosofu, p. 249, D. Reidel, Hingham, Mass., 1979.

Mozer, F. S., W. D. Gonzalez, F. Bogott, M. C. Kelley, and S. Schutz, High-latitude electric fields and the three-dimensional interactions between the interplanetary and terrestrial magnetic fields, J. Geophys. Res., 79, 56, 1974.

Oliver, W. L., J. M. Holt, R. H. Wand, and J. V. Evans, Millstone Hill observations of auroral convection over $60° \leq \Lambda \leq 75°$, 3. Average patterns vs. K_p, J. Geophys. Res., 88, 5505, 1983.

Spiro, R. W., M. Harel, R. A. Wolf, and P.H. Reiff, Quantitative simulation of a magnetospheric substorm, 3. Plasmaspheric electric fields and evolution of the plasmapause, J. Geophys. Res., 86, 2261, 1981.

Spiro, R. W., P. H. Reiff, and L. J. Maher, Jr., Precipitating electron energy flux and auroral zone conductances - an empirical model, J. Geophys. Res., 87, 8215, 1982.

Vickrey, J. F., R. R. Vondrak, and S. J. Matthews, The diurnal and latitudinal variation of auroral zone ionospheric conductivity, J. Geophys. Res., 86, 65, 1981.

Vickrey, J. F., R. R. Vondrak, and S. J. Matthews, Energy deposition by precipitating particles and Joule dissipation in the auroral ionosphere, J. Geophys. Res., 87, 5184, 1982.

Wallis, D. D., and E. E. Budzinski, Empirical models of height integrated conductivities, J. Geophys. Res., 86, 125, 1981.

Wolf, R. A., and Y. Kamide, Inferring electric fields and currents from ground magnetometer data: a test with theoretically derived inputs, J. Geophys. Res., submitted, 1983.

ELECTRIC FIELDS AND CURRENTS ASSOCIATED WITH ACTIVE AURORA

W. Baumjohann[1]

Max-Planck-Institut für Physik und Astrophysik,
Institut für extraterrestrische Physik,
D-8046 Garching, Fed. Rep. Germany

H.J. Opgenoorth

Uppsala Ionospheric Observatory, S-755 90 Uppsala, Sweden

Abstract. The presence of active aurorae (i.e., auroral break-ups, westward travelling surges and omega bands) in the nighttime auroral oval and the associated spatially confined highly conducting regions severely alter and distort the current flow of the large scale convection-driven electrojet system. Modelling of the current flow within and near to active aurorae based on simultaneous two-dimensional observations of optical aurora, ionospheric electric fields, and magnetic fields on the ground revealed among others three important features:

(1) Strong conductivity gradients at the boundaries of break-up aurorae and westward travelling surges lead to the generation of polarization electric fields and enhanced westward current flow.

(2) The strong westward current within break-up aurora and behind the surge is at the western edge continued upward along the magnetic field lines by a strong and localized line-type field-aligned current.

(3) Omega bands are associated with east-west orientated pairs of up- and downward directed field-aligned currents and drift eastward with $\underline{E}x\underline{B}$ velocity under the influence of the southward directed convection electric field.

1. Introduction

During the International Magnetospheric Study (IMS) our knowledge on the electrodynamics of the auroral ionosphere and near-earth magnetosphere has been significantly increased [see, for example, the reviews by Kamide, 1982, and Baumjohann, 1983]. The first problem adressed was the three-dimensional current flow, electric field, and conductivity structure associated with the large-scale and slowly varying electrojets. By combining averaged global conductivity or electric field patterns with mostly also averaged global magnetic disturbance field distributions on the ground a number of authors [e.g. Hughes and Rostoker, 1979; Mishin et al., 1979; Kamide and Akasofu, 1981; Kamide et al., 1982] have presented a fairly consistent and complete picture of the electrodynamics of the global auroral electrojets. Only some phenomena in the Harang discontinuity region seem so far not to be fully understood [cf. Baumjohann, 1983].

During disturbed times the precipitation of energetic particles into the auroral oval and consequently the ionospheric conductivity tensor is not as uniform as assumed in the aforementioned electrojet models. The enhanced particle precipitation associated with discrete aurora often results in conductivity enhancements within relatively localized regions. A typical distribution of substorm aurora can be seen in the DMSP photograph in Figure 1: break-up aurora and westward travelling surge around midnight and in the late evening sector, respectively, and eastward drifting omega bands in the morning sector. The spatially confined highly conducting regions associated with such substorm aurorae severely alter and distort the two-cell convection electric field pattern and associated electrojet current flow.

In contrast to the slowly varying electrojet system, which exhibits only weak gradients in the azimuthal direction, these distortions have strong gradients along both the meridional and azimuthal axis. The typical scale size of such localized conductivity enhancements can range from several 100 to 1000 km and is furthermore highly variable in time. Thus one needs simultaneous two-dimensional measurements of, for example, ionospheric electric fields and magnetic perturbations on the ground for a unique determination of the associated three-dimensional current system. Such observations have been possible in northern Scandinavia, where the STARE radars [Greenwald et al., 1978] and the Scandinavian Magnetometer Array (SMA) [Küppers et al., 1979] allowed the simultaneous measurement of the two-dimensional distribution of both electron irregularity drift (from which the ionospheric electric field can be deduced; cf. Greenwald,

---
[1] Formerly at Institut für Geophysik der Universität Münster, Corrensstr. 24, D-4400 Munster, Fed. Rep. Germany.

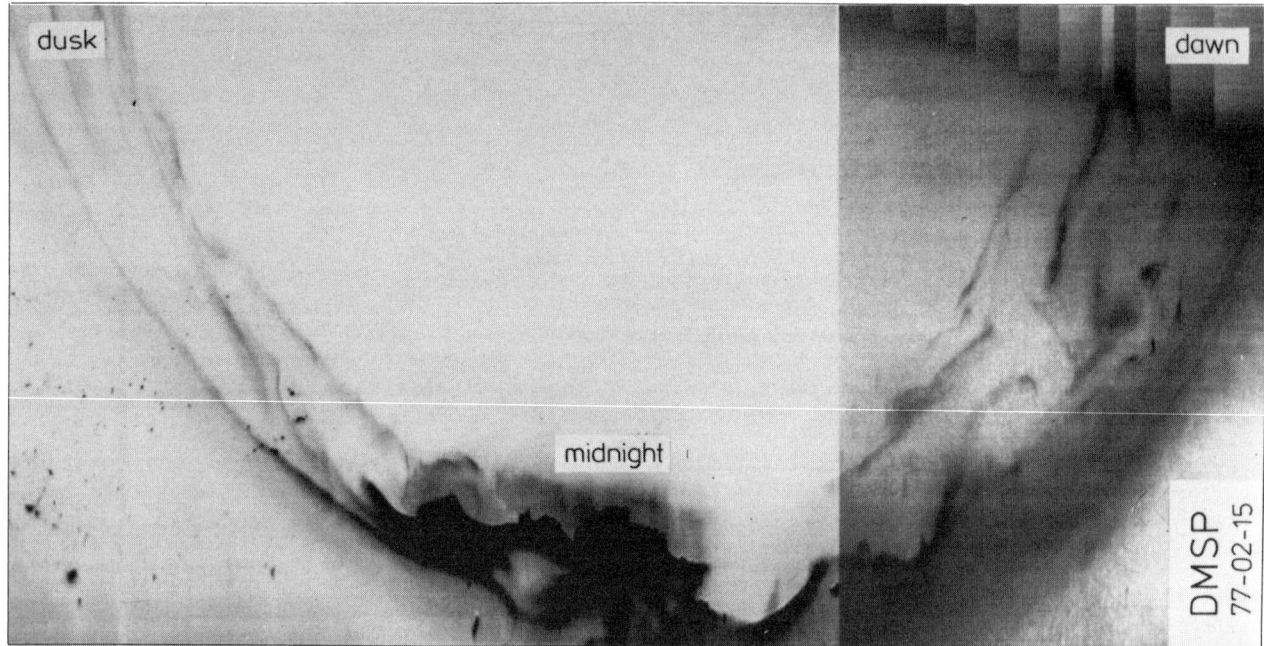

Fig. 1. DMSP photograph of a typical distribution of discrete aurora in the nighttime auroral oval. (Courtesy of AFGL, Hanscom, Mass.)

[1979]) and magnetic perturbations on the ground. Additionally, the Finnish all-sky camera chain documented visual substorm aurora with good temporal resolution.

In the following we will review a number of studies in which the aforementioned data sets have been used to model the three-dimensional current flow in and near active substorm aurorae. The modelling was done by combining the two-dimensional electric field distribution derived from the STARE measurements with an assumed conductivity distribution (basically an enhancement within the auroral forms). The latter was then varied until the magnetic field of the three-dimensional current system associated with a particular electric field/conductivity distribution reproduced the two-dimensional distribution of magnetic perturbations measured by the SMA.

2. Auroral Break-Up and Westward Travelling Surges

During periods of enhanced magnetospheric activity, i.e., during the occurrence of magnetospheric substorms, the ionospheric current flow is affected in two ways [cf. Pellinen et al., 1982; Akasofu, 1981; Rostoker et al., 1983]: During the growth phase or convection bay the Hall current flow in the auroral electrojets increases in direct relation to the energy input from the solar wind into the magnetosphere (so-called driven process). Additionally, sporadic releases of energy previously stored in the magnetotail lead to the formation of substorm current wedges with strongly enhanced westward current flow in the region of active aurora in the midnight sector (so-called unloading process). This substorm electrojet may eventually intrude deeply into the evening sector along with the westward travelling surge and is superimposed on the convection electrojets [cf. Figure 6 of Baumjohann, 1983; see also Rostoker and Hughes, 1979].

In order to explore the electrodynamics of the substorm electrojet and its development during the expansion into the evening sector, Baumjohann et al. [1981], Opgenoorth et al. [1983a] and Inhester et al. [1981] have modelled the three-dimensional current system associated with break-up aurora around 2400 MLT and with westward travelling surges observed around 2230 and 2030 MLT, respectively. Their observations and models will be summarized in this section.

Baumjohann et al. [1981] have analysed two-dimensional magnetic and electric field observations made by the SMA and STARE during three successive shortlived auroral break-ups around magnetic midnight. All three break-ups were moderate and remained relatively localized. In Figure 2 we summarize for one of the three break-ups the ground-based observations of equivalent current vector distributions (i.e., magnetic disturbance vectors rotated clockwise by $90°$), ionospheric electric fields, and auroral structures before the breakup (T = -1 min), at the initial brightening (T = 0 min), and during the maximum expansion of substorm aurora (T = 2 min). The observed features were alike for all three break-ups.

Before the break-up, the electric field distri-

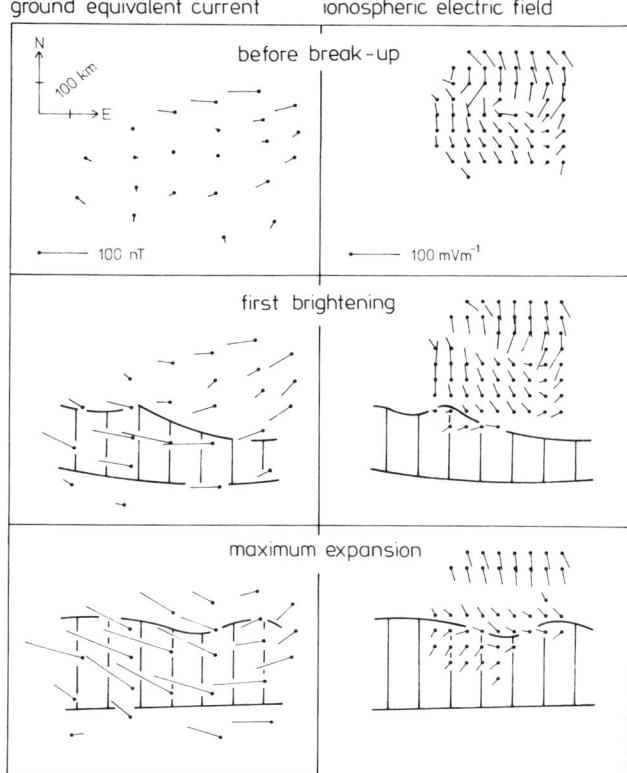

Fig. 2. Spatial distribution of equivalent current vectors on the ground, ionospheric field vectors, and auroral structures before, at the start of, and during the peak development of the auroral break-up [after Baumjohann et al., 1981].

bution displayed a pattern typical for the Harang discontinuity region. That hardly any current flow could be detected indicates a very low ionospheric conductivity. During the initial brightening energetic auroral electrons precipitated into a local region south of the Harang discontinuity. In this region the westward electric field stayed at the same level, while the northward component decreased significantly and even turned southward in some areas, especially during the maximum phase. The magnetometers indicated that a strong westward equivalent current was developing in the same region. These observations agree with those of, for example, Carlson and Kelley [1977] and Marklund et al. [1983].

The described observations match in detail the features to be expected during the generation of an (incomplete) Cowling channel in the highly conducting active regions as envisaged by Boström [1975] and summarized in Figure 3: Due to the conductivity enhancement the westward component of the primary electric field (E^o) drives an enhanced northward Hall current component (J_H^o). The excess Hall current will bring positive charges to the northern border and negative charges develop at the southern boundary. To some extent these are discharged by Birkeland current sheets, but they also give rise to a southward polarization electric field E^p. E^p drives a southward Pedersen current (J_p^p) which balances that part of J_H^o which is not continued via Birkeland currents. The westward currents due to the primary (convection) and secondary (polarization) electric field add to give an intensified westward Cowling current. Note that the westward Cowling current, i.e., the substorm current wedge, dissipates energy in the auroral ionosphere (in contrast to the convection electrojets where energy is dissipated only by the meridional Pedersen currents).

The model constructed by Baumjohann et al. [1981] for the maximum expansion of active aurora is shown in Figure 4. Within the active region Hall and Pedersen conductivities reach peak values of about 30 S and 10 S close to the western edge; outside the active region the conductivity is assumebly zero (see above). The pre-break-up electric field configuration is distorted by the superposition of a southward polarization electric field of up to 50 mV/m in the area covered by break-up aurora. The ionospheric current flows mainly westward but also has a northward component. The height-integrated current densities of 600 - 700 mA/m are comparable with typical current densities within the Hall current westward electrojet [cf. Baumjohann, 1983]. The westward component of the current is closed by very localized and intense upward field-aligned current at the western edge (about 7 µA/m²) and more wide-spread downward field-aligned current of lower density (1-2 µA/m²) in the eastern third of the break-up region (thus indicating that the substorm current wedge is not azimuthally symmetric). The northward current component is closed by field-aligned current sheets (1-2 µA/m²) at the southern and northern boundaries of the active region.

In order to explore, if a similar current structure is found when the substorm electrojet intrudes along with the westward travelling surge (WTS) into the evening sector, Opgenoorth et al. [1983a] have used simultaneous measurements made by the SMA, the STARE-radars and onboard the Barium-GEOS rocket to model the three-dimensional current system in the vicinity of a WTS around 2230 MLT. Their observations and model results are compared in Figures 5 and 6.

The westward equivalent current distribu-

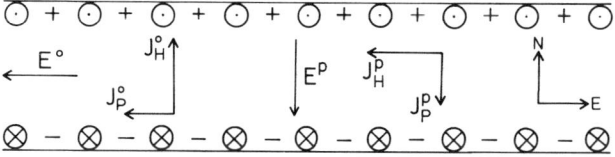

Fig. 3. Generation of a polarization electric field and an complete Cowling channel in a region of enhanced ionization [after Boström, 1975].

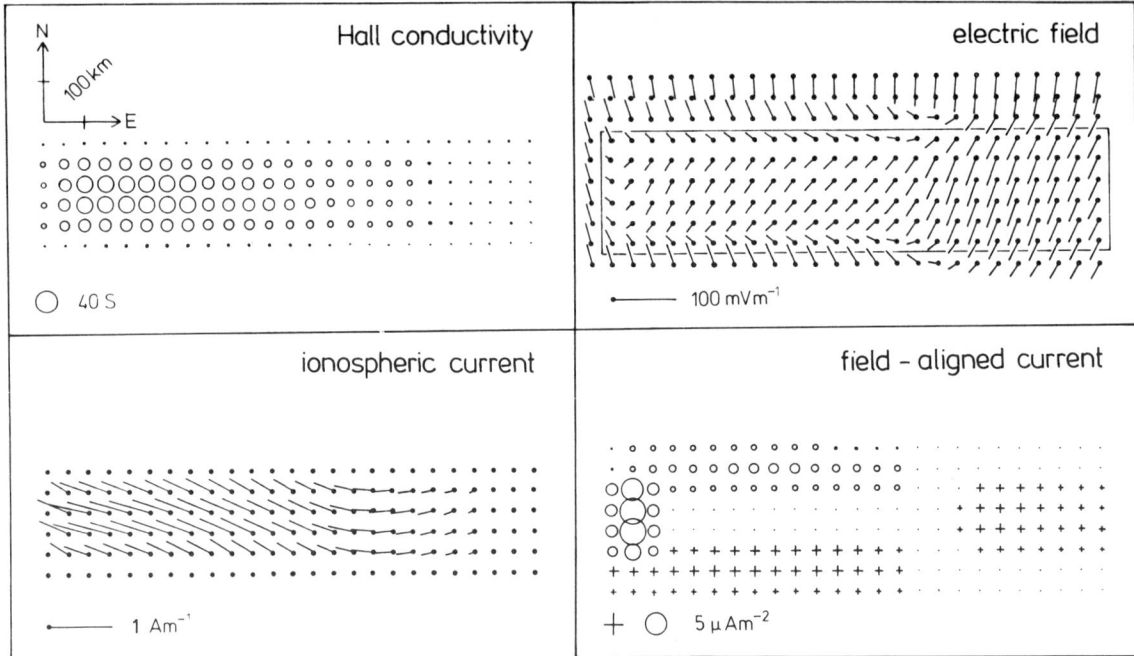

Fig. 4. Spatial distribution of Hall conductivity, horizontal electric field vectors, and ionospheric and field-aligned current in the region of active break-up aurora around local magnetic midnight [after Baumjohann et al., 1981]. The rectangle in the electric field panel frames the region of active aurora. Circles and crosses in the field-aligned current panel denote, respectively, upward and downward direction of the field-aligned current flow.

tions deduced from the SMA observations exhibit a clear longitudinal gradient, i.e., the westward current decreases strongly in the westward direction. The associated southward equivalent current at subauroral latitudes is indicative of net upward field-aligned current near the surge's head [cf. Tighe and Rostoker, 1981]. Electric field observations obtained by the rocket and STARE, during the 2.5 min when the surge passed overhead, are plotted with respect to their location relative to the WTS. A clear radial component directed toward the center of the WTS can be recognized. To the north of the aurora the electric field is southward directed and south of it the rocket data indicates a northward direction. Inside the surge (but close to its southern boundary) the rocket measured a very small electric field (below STARE's threshold level [Cahill et al., 1978]) with a clear westward component.

The modelling effort yielded the result that only a weak southwestward directed electric field east of the surge's head was consistent with all observed features. This field may again be explained by the generation of polarization charges at the northern and southern boundaries of the higher conducting region behind the surge's head (Hall and Pedersen conductivities maximized at 25 S and 20 S, respectively). Accordingly, the westward current east of the surge (of the order of 500 mA/m) might again be a Cowling current. Continuity of this current is conserved by localized and intense upward field-aligned current in the head of the surge (about 6 $\mu A/m^2$). No downward field-aligned current could be detected within the observation area.

Comparing Figure 5 with the observations of Inhester et al. [1981] made at another occasion during the passage of a WTS around 2030 MLT and shown in Figure 7 (here both electric field and equivalent current vectors deduced from STARE and SMA measurements made during an 8 min interval are plotted with respect to their location relative to the surge), it becomes evident that the surge-associated features are essentially the same and that only the ambient convection electrojet environment is different according to its normal MLT dependence: The surge travels westward along the Harang discontinuity just north of a still very prominent eastward electrojet (which buries the typical southward equivalent current). This is substantiated by the modelling results shown in Figure 8 which exhibit essentially the same structure as shown in Figure 5 within the surge (but with smaller amplitudes) and an additional region of strong northward electric field and eastward electrojet current south of the surge.

In conclusion, it can be said that conductivity structure, electric field distribution, and current flow associated with break-up aurora in the midnight sector and westward travelling surges in the late and early evening sector are essentially the same: Strong conductivity gradi-

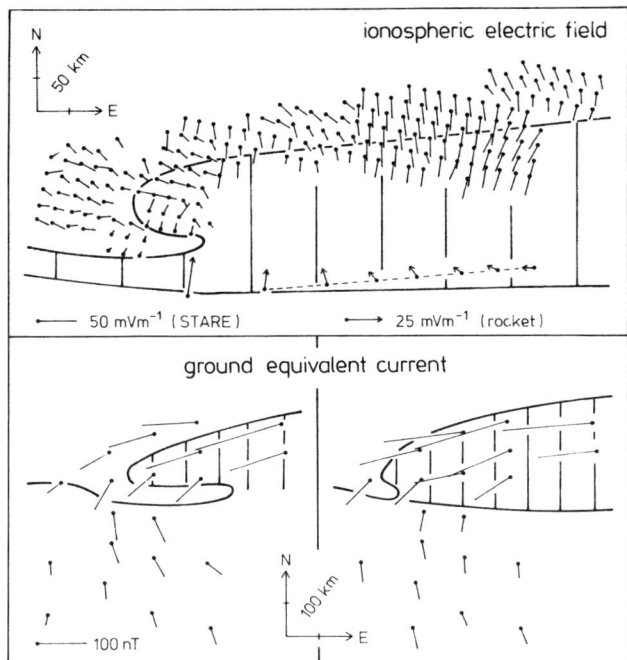

Fig. 5. Spatial distribution of equivalent current vectors on the ground (for two moments 2 min apart), ionospheric electric field vectors, and auroral structure during the passage of a WTS in the premidnight sector [after Opgenoorth et al., 1983a].

ents at the boundaries of the active aurora (mostly found in the vicinity of the Harang discontinuity) lead to the generation of polarization electric fields and enhanced westward current flow. This strong westward current is continued along the magnetic field lines at the western edge by intense and localized line-type upward field-aligned current. The models suggest that the major differences in the observed patterns appear to be dependent 1) on the MLT sector, i.e., if eastward or westward electrojet are prominent, 2) on the general activity, which determines the background conductivity, and 3) on the individual strength of the event, since the southward polarization electric field is weaker for lesser conductivity enhancements.

3. Eastward Drifting Omega Bands

While auroral break-up and westward travelling surge prevail in the premidnight sector during substorms, eastward propagating sequences of omega bands are the dominant feature in the postmidnight sector during disturbed times. The auroral omega bands are associated with localized disturbances in the large-scale electrojet flow [e.g. Baumjohann, 1979]. The magnetic signature on the ground has been denoted as Ps6 pulsations due to the periodic nature of the recorded disturbance when a sequence of omega bands passes over a magnetometer. Several authors (using mainly ground-based magnetic data) came to the conclusion that the Ps6 pulsations must be linked

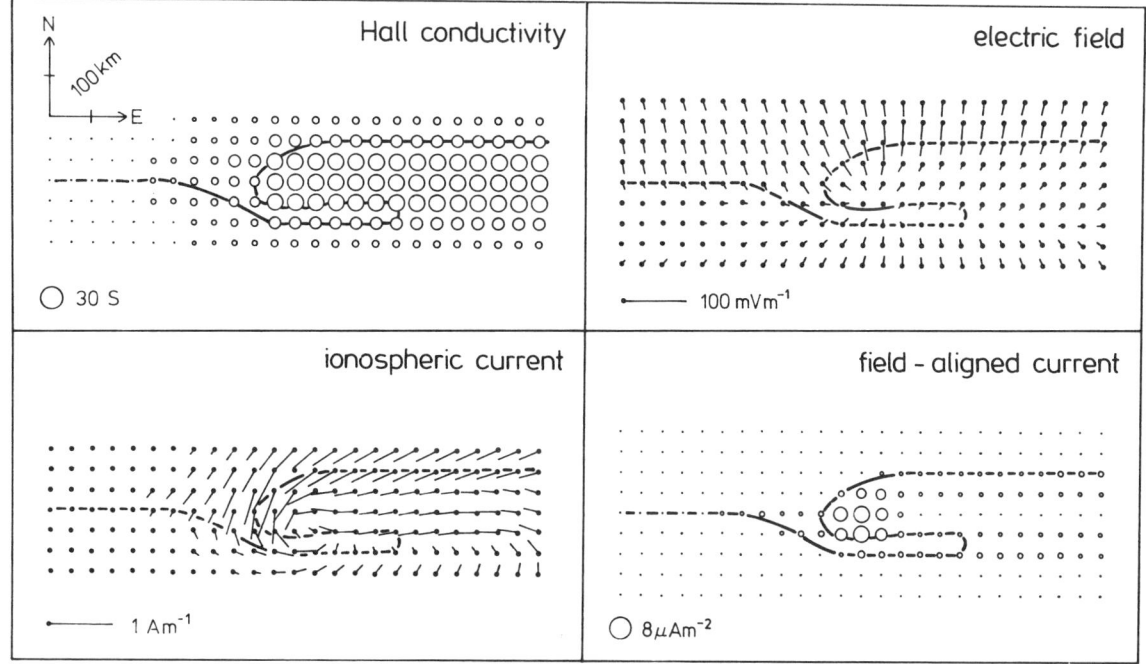

Fig. 6. Same as Figure 4, but for a WTS in the premidnight sector [after Opgenoorth et al., 1983a].

82 ELECTRIC FIELDS AND CURRENTS

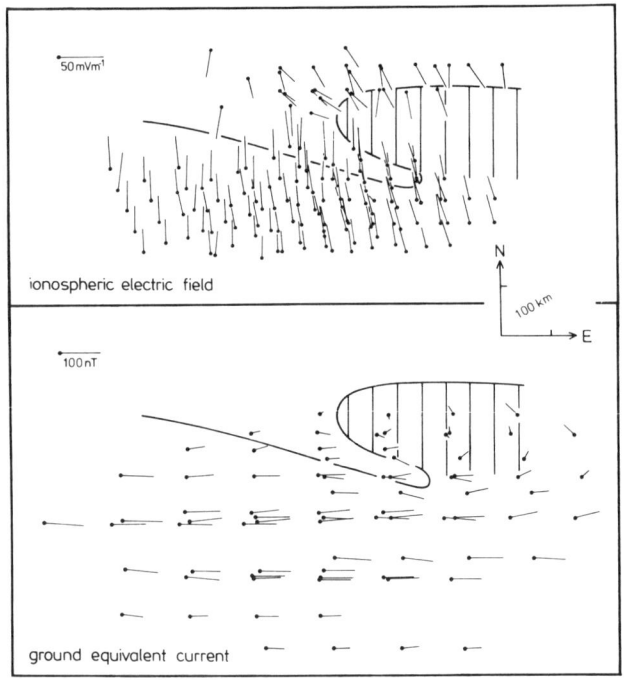

Fig. 7. Same as Figure 5, but during the passage of a WTS in the early evening sector [after Inhester et al., 1981].

to some localized structure in the field-aligned current system, which does not change with time but moves eastward with the omega band [e.g. Kawasaki and Rostoker, 1979, Gustafsson et al., 1981]. But while Kawasaki and Rostoker 1979] interpreted the magnetic Ps6 pulsations as being due to an eastward travelling north-south oriented field-aligned current pair which is closed by southward ionospheric current, Gustafsson et al. [1981] rather advocated an eastward drifting east-west oriented pair of field-aligned current surrounded by clockwise and counterclockwise ionospheric Hall current vortices. Since a decision for one of the two different models could be made only by including simultaneous electric field measurements, André and Baumjohann [1982] and Opgenoorth et al. [1983b] have analysed SMA and STARE measurements for two postmidnight intervals when omega bands were drifting eastward over the Finnish all-sky camera chain. Figure 9 summarizes the observations of André and Baumjohann [1982] made during a 30 min interval around 0330 MLT. Both electric field and equivalent current vectors were superimposed and averaged in 50 x 50 km^2 cells assuming a stationary pattern to drift with the omega band. The equivalent current vectors clearly resemble the "snake-like" electrojet pattern described by Saito [1978]. The pattern is obviously caused by a longitudinally alternating azimuthal electric

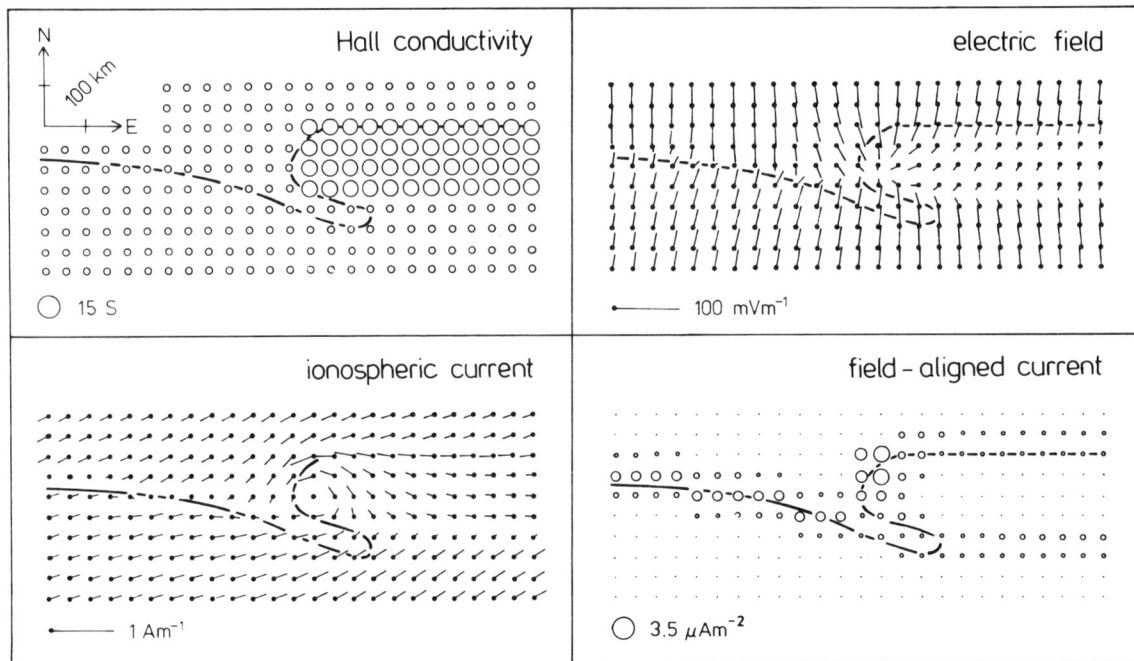

Fig. 8. Same as Figure 4, but for a WTS in the early evening sector [after Inhester et al., 1981].

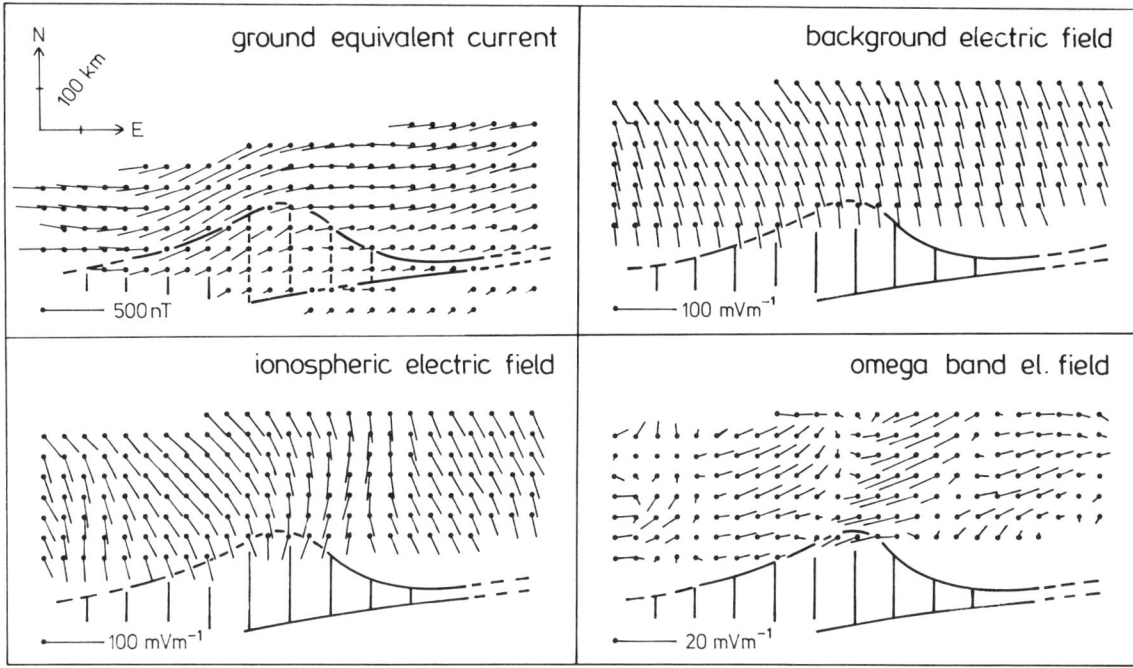

Fig. 9. Same as Figure 5, but during the passage of an east-ward drifting omega band in the morning sector. Additionally, two different components into which the measured electric field may be decomposed are shown: a rather uniform background electric field and the electric field associated with the omega band [after André and Baumjohann, 1982].

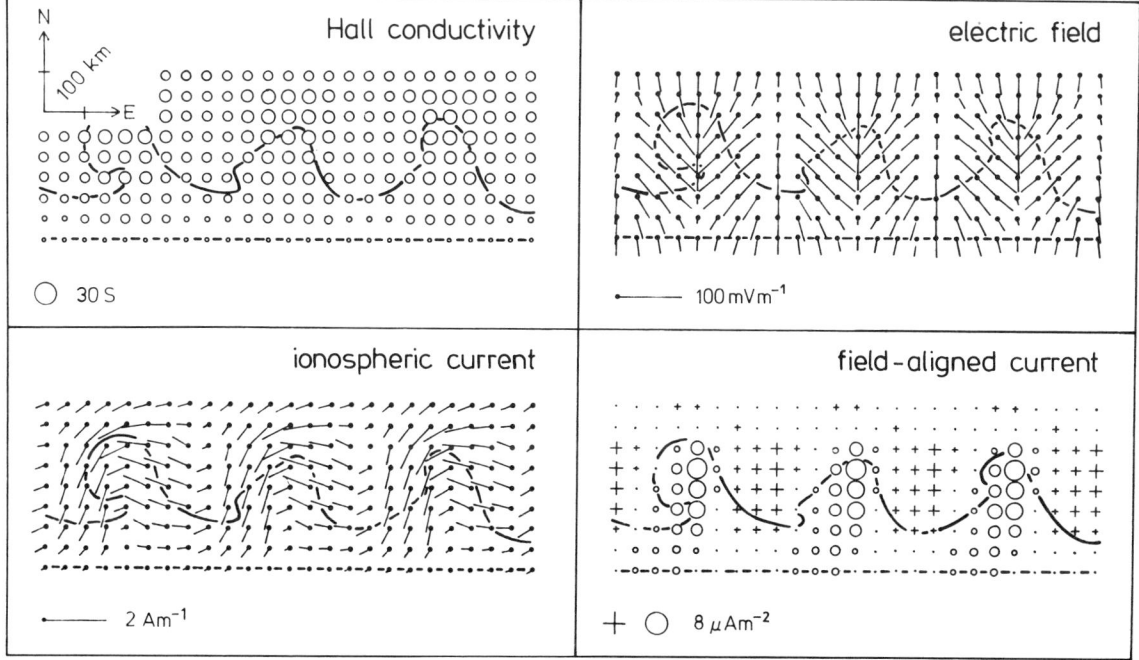

Fig. 10. Same as Figure 4, but for a sequence of eastward drifting omega bands in the morning sector [after Opgenoorth et al., 1983b].

field component of about 10 mV/m added to the southward convection electric field component. This becomes very clear when the ionospheric electric field is separated (by filtering) into a uniform and steady background electric field and a highly structured electric field pattern, which is associated with the omega band and drifts eastward with the same velocity. Moreover, by comparing the drift velocity of the omega band and its associated electric and magnetic field disturbances (in both cited cases typically 700 m/s) with the southward directed electric background field of about 35 mV/m, it is found that the eastward motion of the omega bands must be due to an $\underline{ExB}$ drift of the precipitating particles.

Opgenoorth et al. [1983b] have modelled the three-dimensional current system associated with a series of omega bands. Their observations of electric fields and equivalent current distributions corroborate those of André and Baumjohann [1982]; only the distance between individual omega bands was smaller and thus the periodic nature of these auroral features becomes more apparent in their modelling results (Figure 10). The conductivity panel shows the superposition of regions of enhanced conductivity, within and slightly west of the tongues of the omega bands, on the ambient conductivity structure. The electric field panel displays the superposition of a southward background field and an alternating azimuthal component directed toward the bright tongues and away from the darker zones of the omega band (note the meridional elongation of this structure). The electric field structure leads to a meandering Hall current, which is composed of a westward electrojet Hall current and clockwise and counterclockwise Hall current vortices associated with the alternating azimuthal electric field component, and to a mainly southward directed Pedersen current, which converges toward the bright tongues. This results in an ionospheric current flow which is principally directed southward on the western side and westward on the eastern side of the bright tongues. The symmetry of the electric field pattern also causes the strong upward and downward directed Birkeland current flow of up to 7 $\mu A/m^2$ in the bright tongues and dark zones, respectively, which is mainly fed by Pedersen current. A closer look at the model results also reveals a north-south aligned current wedge as proposed by Kawasaki and Rostoker [1979], but the current flow in this wedge is much weaker than in the main circuit constituted by the east-west aligned pairs of field-aligned current.

In conclusion, the omega bands are associated with east-west orientated pairs of up- and downward directed field-aligned currents which drift eastward together with the omega bands with $\underline{ExB}$ velocity. Thus the present model comes closer to the model of Gustafsson et al. [1981] with its azimuthally alternating Birkeland currents than to the Kawasaki and Rostoker [1979] model with its meridionally aligned current wedge (note that both models only describe the current flow causing the Ps6 disturbance and do not include the background electrojet current).

Acknowledgements. The work of W. Baumjohann was partly supported by grants from the Deutsche Forschungsgemeinschaft. H.J. Opgenoorth appreciates support from the Swedish Natural Science Research Council.

References

Akasofu, S.-I., Energy coupling between the solar wind and the magnetosphere, Space Sci. Rev.,28, 121-190, 1981.

André, D., and W. Baumjohann, Joint two-dimensional observations of ground magnetic and ionospheric electric fields associated with auroral currents, 5, Current system associated with eastward drifting omega bands, J. Geophys., 50, 194-201, 1982.

Baumjohann, W., Spatially inhomogeneous current configurations as seen by the Scandinavian Magnetometer Array, in Proceedings of the International Workshop on Selected Topics of Magnetospheric Physics, edited by the Japanese IMS Committee, pp. 34-40, Tokyo, 1979.

Baumjohann, W., Ionospheric and field - aligned current systems in the auroral zone: A concise review, Adv. Space Res., in press, 1983.

Baumjohann, W., R. J. Pellinen, H. J. Opgenoorth, and E. Nielsen, Joint two-dimensional observations of ground magnetic and ionospheric electric fields associated with auroral zone currents: Current systems associated with local auroral break-ups, Planet. Space Sci., 29, 431-447, 1981.

Boström, R., Mechanisms for driving Birkeland currents, in Physics of the Hot Plasma in the Magnetosphere, edited by B. Hultquist and L. Stenflo, pp.431-447, Plenum Press, New York, 1975.

Cahill, L. J., R. A. Greenwald, and E.Nielsen, Auroral radar and rocket double-probe observations of the electric field across the Harang discontinuity, Geophys. Res. Lett., 5, 687-690, 1978.

Carlson, C. W., and M. C. Kelley, Observation and interpretation of particle and electric field measurements inside and adjacent to an active auroral arc, J. Geophys. Res., 82, 2349-2360, 1977.

Greenwald, R. A., Studies of currents and electric fields in the auroral zone ionosphere using radar auroral backscatter, in Dynamics of the Magnetosphere, edited by S.- I. Akasofu, D. Reidel Publ. Comp., Hingham, Mass., 1979.

Greenwald, R. A., W. Weiss, E. Nielsen, N. R. Thomson, STARE: A new radar auroral backscatter experiment in northern Scandinavia, Radio Sci., 13, 1021-1039, 1978.

Gustafsson, G., W. Baumjohann, and I. Iversen,

Multi-method observations and modelling of the three-dimensional currents associated with a very strong Ps6 event, J. Geophys., 49, 138-145, 1981.

Hughes, T. J., and G. Rostoker, A comprehensive model current system for high-latitude magnetic activity, 1, The steady state system, Geophys. J. R. Astr. Soc., 58, 525-569, 1979.

Inhester, B., W. Baumjohann, R. A. Greenwald, and E. Nielsen, Joint two-dimensional observations of ground magnetic and ionospheric electric fields associated with auroral zone currents, 3, Auroral zone currents during the passage of a westward traveling surge, J. Geophys., 49, 155-162, 1981.

Kamide, Y., The relationship between field-aligned currents and the auroral electrojets, A review, Space Sci. Rev., 31, 127-243, 1982.

Kamide, Y., and S.-I. Akasofu, Global distribution of the Pedersen and Hall Currents and the electric potential pattern during a moderately disturbed period, J. Geophys. Res., 86, 3665-3668, 1981.

Kamide, Y., B.-H. Ahn, S.-I. Akasofu, W. Baumjohann, E. Friis-Christensen, H. W. Kroehl, H. Maurer, A. D. Richmond, G. Rostoker, R. W. Spiro, J. K. Walker, And A. N. Zaitzev, Global distribution of ionospheric and field-aligned currents during substorms as determined from six IMS meridian chains of magnetometers, Initial results, J. Geophys. Res., 87, 8228-8240, 1982.

Kawasaki, K., and G. Rostoker, Perturbation magnetic fields and current systems associated with eastward drifting auroral structures, J. Geophys. Res., 84, 1464-1480, 1979.

Küppers, F., J. Untiedt, W. Baumjohann, K. Lange, and A. G. Jones, A two-dimensional magnetometer array for ground-based observations of auroral zone electric currents during the International Magnetospheric Study (IMS), J. Geophys., 46, 429-459, 1979.

Marklund, G., W. Baumjohann, and I. Sandahl, Rocket and ground-based study of an auroral break-up event, Planet. Space Sci., 31, 207-220, 1983.

Mishin, V. M., A. D. Barzarzhapov, and G. B. Shpynev, Electric fields and currents in the earth magnetosphere, in Dynamics of the Magnetosphere, edited by S.-I. Akasofu, D. Reidel Publ. Comp., Hingham, Mass., 1979.

Opgenoorth, H. J., R. J. Pellinen, W. Baumjohann, E. Nielsen, G. Marklund, and L.Eliasson, Three-dimensional current flow and particle precipitation in a westward travelling surge (observed during the Barium-GEOS rocket experiment), J. Geophys. Res., 88, 3138-3152, 1983a.

Opgenoorth, H. J., J. Oksman, K. U. Kaila, E. Nielsen, and W. Baumjohann, Characteristics of eastward drifting omega bands in the morning sector of the auroral oval, J. Geophys. Res., 88, in press, 1983b.

Pellinen, R. J., W. Baumjohann, W. J. Heikkila, V. A. Sergeev, A. Yahnin, G. Marklund, and A. O. Melnikov, Event study on pre-substorm phases and their relation to the energy coupling between solar wind and magnetosphere, Planet. Space Sci., 30, 371-388, 1982.

Rostoker, G., and T. J. Hughes, A comprehensive model current system for high latitude magnetic activity - II. The substorm component. Geophys. J. R. Astr. Soc., 58, 571-581, 1979.

Rostoker, G., S.-I. Akasofu, W. Baumjohann, Y. Kamide, R. L. McPherron, Magnetospheric substorms - The respective roles of the solar wind and the magnetotail in providing energy for substorm expansion phases, J. Geophys. Res., 88, in press, 1983.

Saito, T., Long-period irregular magnetic pulsations, Pi3, Space Sci. Rev., 21, 427-467, 1978.

Tighe, W. G., and G. Rostoker, Characteristics of westward traveling surges during magnetospheric substorms. J. Geophys., 50, 51-67, 1981.

POLAR CAP CURRENT SYSTEMS

E. Friis-Christensen

Division of Geophysics, Danish Meteorological Institute

Abstract. It is the purpose of this paper to review the present understanding of polar cap current systems.

During the last decade there has been an increasing interest in the very high latitude currents particularly in the polar cap and the polar cusp. These regions are connected to the boundaries of the magnetosphere and are therefore showing immediate response to variations in the solar wind, especially in the interplanetary magnetic field. Ground based measurements as well as polar orbiting satellites have revealed a lot of features of the distribution of currents and electric fields in the polar cap. Individual observations have demonstrated the high degree of variability in the polar cap and the polar cusp. Statistical investigations have revealed some of the general and repeatable features of the polar cap currents. A number of primary current systems, ionospheric and field-aligned, have been defined. The relationship between the different current systems, in particular near the dayside cusp, is still not fully understood because this region is characterized by rapid temporal and spatial variations. The elucidation of these variations demands a lot of simultaneous observations. Different interpretations of the polar cusp currents and their relationship to the auroral electrojets and to the Birkeland currents are discussed.

Introduction

Currents in the polar cap may be divided into two main categories. One which comprises the three-dimensional current system associated with the substorm and which has its main part in the auroral electrojets. The other part, which will be the main topic of this paper, consists of the currents associated with the always present convection in the polar cap.

While the substorm current system is intimately connected to processes in the magnetospheric tail, the convection currents are characterized by a high degree of variability - even at quiet times - and by a clear controlling influence of the solar wind, in particular the interplanetary magnetic field.

The controlling influence is most clearly recognizable near the polar cusp, where the geomagnetic field-lines are connected to the magnetospheric boundaries and, according to some ideas, directly connected to the solar wind.

It is therefore crucial for the understanding of the interaction between the solar wind and the magnetosphere to obtain an adequate description of the distribution of electric fields and currents in this region, which has often been called the "zone of confusion".

Ground-based measurements gave the first hints that the interplanetary magnetic field, IMF, was an important ordering parameter (Svalgaard, 1969; Mansurov, 1969; Friis-Christensen et al., 1972; Berthelier et al., 1974). Electric field observations in this region have been sparse until recently, but from OGO-6 measurements Heppner (1972) found dawn-dusk asymmetries in the electric field distributions which seemed to agree with the observed magnetic perturbations near noon, assuming that the B_y related current was part of the eastward or westward auroral electrojets respectively, according to the sign of B_y. This interpretation was questioned by Friis-Christensen and Wilhjelm (1975) who examined the combined effects of B_y and B_z on equivalent currents in the polar cap. According to their results the B_y-related current called DPY was situated at higher latitudes than the auroral electrojets. Furthermore the DPY current was shown to vary with B_y quite independently of the magnitudes of the auroral electrojets, and might even be larger than the auroral electrojets for quiet magnetospheric conditions which occured for $B_z>0$.

The identification of large-scale field-aligned currents by Zmuda and Armstrong (1974) and the detailed description of the statistical behaviour of these currents by Iijima and Potemra (1976) showed that they were the missing link in coupling the auroral electrojets to the outer magnetosphere. Initially the effect of B_y was not considered when deriving statistical distributions of the field-aligned or Birkeland currents. In a search for the Birkeland current sheets corresponding to the DPY current, Wilhjelm et al. (1978) examined TRIAD data measured over the Greenland IMS magnetometer chain near local magnetic noon and found a one to one correspondence between the direction of the DPY current and the directions of the associated Birkeland current sheets on the equatorward and poleward side of the DPY current indicating that the DPY equivalent current is an ionospheric Hall current.

These observations were confirmed by observations of magnetic field deflections from the ISIS-2 satellite (McDiarmid et al. 1978), indicating that the Birkeland current sheets near noon were influenced by the B_y component of the IMF in a systematic manner.

Whereas the existence and direction of the region 1 and region 2 current systems (Iijima and Potemra, 1976) are well established and independent of the sign of B_y, a number of different interpretations of the cusp Birkeland current have been put forward. Since the Birkeland currents can be regarded as the driving mechanisms of ionospheric electric fields and currents these different interpretations reflect fundamentally different views regarding the source of the currents.

According to Iijima et al. (1978), Saflekos et al. (1979) and Friis-Christensen (1981), the cusp Birkeland current is part of a separate current system in the dayside part of the polar cap.

According to McDiarmid et al. (1979), Rostoker (1980), and Doyle et al. (1981), the morning and afternoon auroral current regions extend across the magnetic local noon meridian and overlap.

The overlap configuration is determined by the IMF B_y component. The cusp Birkeland currents are then not part of a separate system but part of the corresponding region 1 currents. Since single satellite observations are not sufficient to determine the actual pattern, a number of simulation studies (Gizler et al., 1979; Troshichev et al., 1979; Rich and Kamide, 1983) have examined the consequences of different distributions of field-aligned currents near the cusp although no definitive answer has yet been found. In the discussion part of this paper the different arguments favouring one or another pattern will be reviewed.

Solar Wind Control of Polar Cap Currents

Due to their continuous operation magnetic observatories were used in the initial investigations of solar wind influence on polar cap currents. A number of studies were performed which revealed the basic patterns of currents for different conditions in the solar wind. An extensive review of these studies has been given by Nishida (1978).

Direct satellite observations of magnetic perturbations due to Birkeland currents have been more difficult to use on a statistical basis because of the more discontinuous nature of these measurements together with the insufficient spatial coverage at any given time. In spite of these difficulties it has been possible to derive a pattern of the distribution of Birkeland currents which is shown to be ordered in a systematic way by the interplanetary magnetic field components (McDiarmid et al., 1979).

The realization of the important contribution of the field-aligned currents to the real three-dimensional current systems in the polar regions has resulted in the derivation of mathematical and numerical methods to separate the ionospheric and field-aligned part from the horizontal equivalent current system which is the only unique output from ground based magnetic observations.

Using such numerical models which are all based upon certain assumptions regarding the ionospheric conductivities, more realistic current systems for the polar regions have been derived (e.g. Mishin et al., 1980; Kamide et al., 1981; Akasofu and Ahn, 1981; Matsushita and Xu, 1982; Levitin et al., 1982). Since these studies were either performed without taking into consideration the IMF or were based upon a limited spatial coverage of observations in the polar cap and cusp, no definite result regarding the detailed distribution of ionospheric and field-aligned currents was achieved.

For this reason a study was initiated by Friis-Christensen et al. (1983) to use Greenland magnetometer data together with high resolution interplanetary magnetic field data (20 minute resolution) and the numerical model described by Kamide et al. (1981) to obtain statistical relationships between the IMF components and ionospheric- and field-aligned currents especially in the polar cap and polar cusp where the Greenland magnetometer chain offers a good spatial resolution.

Used in this study were digitized magnetograms for approximately 70 days during the summer months of 1972 and 1973. To make efficient use of the limited data a regression technique was used which requires the assumption of linear relationships between the IMF-components and the perturbation vectors for a given observatory, magnetic local time, and season. The validity of this assumption in a statistical sense has been demonstrated by Friis-Christensen and Wilhjelm (1975), Maezawa (1976) and Levitin et al. (1982) using hourly data from permanent observatories in the northern polar cap. It was also demonstrated in these works that the linear approximation is excellent for the B_y component. However, as far as the B_z component is concerned, an improvement is achieved when the data are divided into two classes corresponding to $B_z>0$ and $B_z<0$. The regression analysis resulted in a set of predictors as a function of station and local time which could be used as an empirical model for the magnetic perturbations and their dependences on the IMF B_y and B_z components.

The empirical model was used to predict distributions of magnetic perturbation vectors corresponding to selected and representative conditions of the IMF.

Using these magnetic perturbation vectors, corresponding equivalent current functions

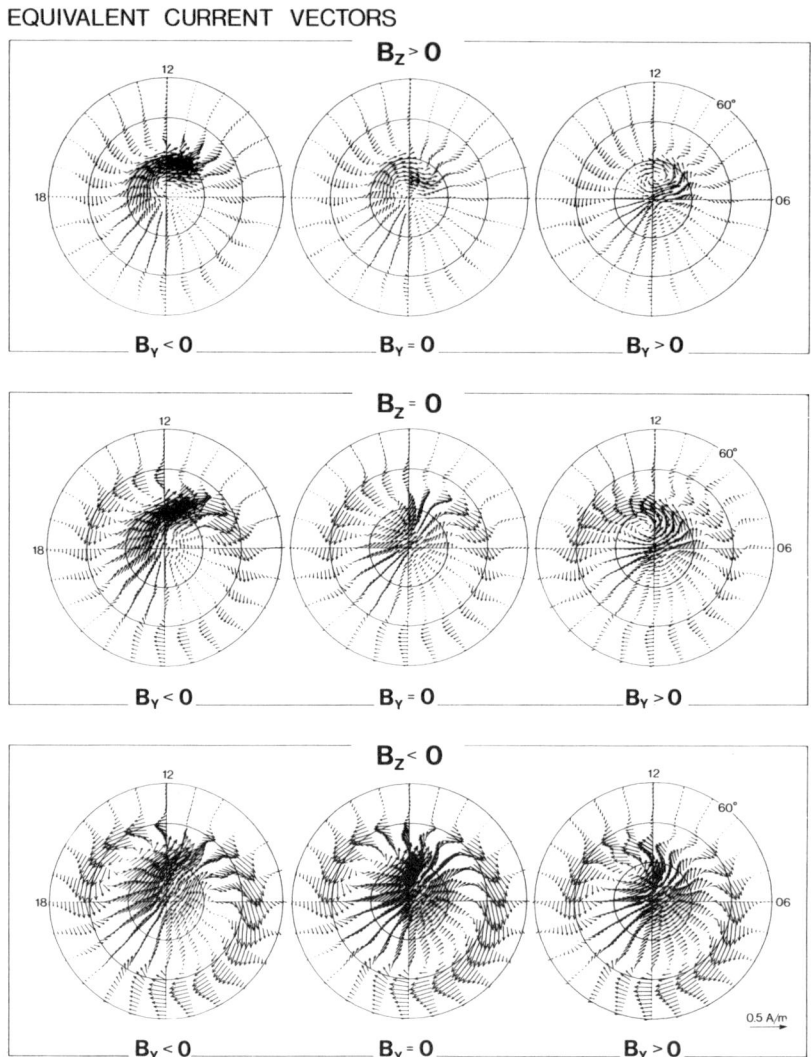

Fig. 1. Equivalent current vectors corresponding to different directions of the interplanetary magnetic field in the Y-Z plane (after Friis-Christensen et al., 1983).

were derived and used together with simple and analytical ionospheric conductivity models to compute the electric potential distribution in the polar region. From the potential distribution and the model conductivity distribution it is possible to derive the patterns of ionospheric Hall and Pedersen currents as well as the field-aligned current distribution. Input to the numerical model is the computed equivalent current functions for various combinations of IMF components.

Figure 1 shows the equivalent current vectors which have been derived from the equivalent current functions. It is interesting to compare the equivalent current vectors with the derived "true" ionospheric currents shown in Figure 2 and the Hall current vectors shown in Figure 3. It is seen that in the dayside (sunlit) polar cap the equivalent currents are predominantly caused by Hall currents in agreement with the fact that in areas where the conductivity is homogeneous the ground based magnetic perturbations caused by the Pedersen currents will cancel the contribution from the field-aligned currents (Fukushima, 1976).

The model results also confirmed that the main part of the equivalent currents in the auroral electrojets is caused by effect from Hall currents. Only the nightside part of the polar cap shows a large difference between the Hall current vectors and the equivalent current vectors indicating a non-cancelled effect from field-aligned currents. It should be noted that the current diagrams show the total currents without any attempt to remove the S_q^p two-cell current pattern assumed to be independent of B_y. The DPY cur-

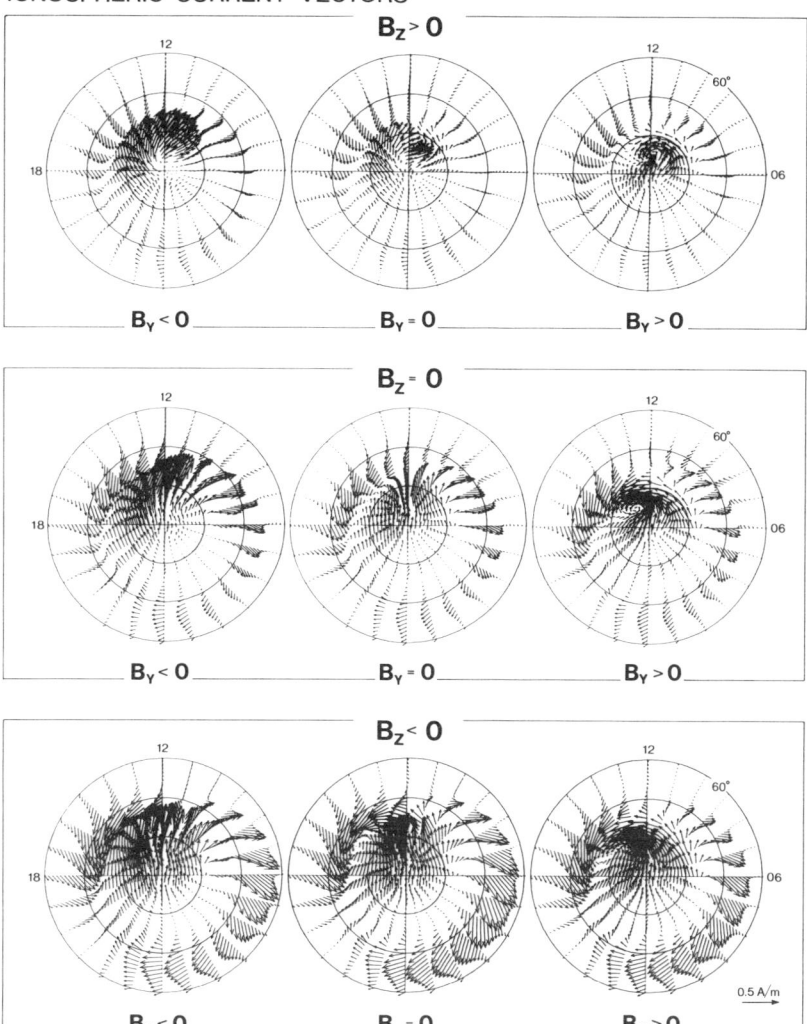

Fig. 2. Total ionospheric current vectors derived from the equivalent currents of Figure 1 and analytical ionospheric conductivity distribution models (after Friis-Christensen et al., 1983).

rent is therefore imbedded in the total currents which means that it is most purely seen as a zonal current for $B_z > 0$ where the two-cell current system has decreased considerably.

The statistical diagrams of Hall current vectors show that the polar cap current has a sunward component in all cases except for $B_z > 0$ and $B_y = 0$. If the S_q^p system is subtracted the residual current system for $B_z > 0$, $B_y = 0$ correspond to a "reversed" two-cell current system earlier reported by Maezawa (1976) and in agreement with convection patterns for northward IMF derived from satellite electric field measurements (Burke et al., 1979). The kind of statistical treatment used in the paper by Friis-Christensen et al. (1983) does not, however, permit an evaluation of the probability that a "reversed" current system really would exist for northward IMF. It might well be that the result is just a combined statistical effect of two DPY-currents of opposite polarity.

Relationship with Birkeland Currents

The numerical method allowed Friis-Christensen et al. (1983) to derive statistical distributions of polar cap field-aligned currents as a function of the IMF. The result is shown in Figure 4 in which the main features are the Birkeland currents located poleward and equatorward of the auroral electrojets and known as the region 1 and region 2 currents (Iijima and Potemra, 1976).

The results of Friis-Christensen are focussed upon the day side of the polar cap where the Birkeland currents near the cusp show the expected asymmetry according to the sign of B_y, re-

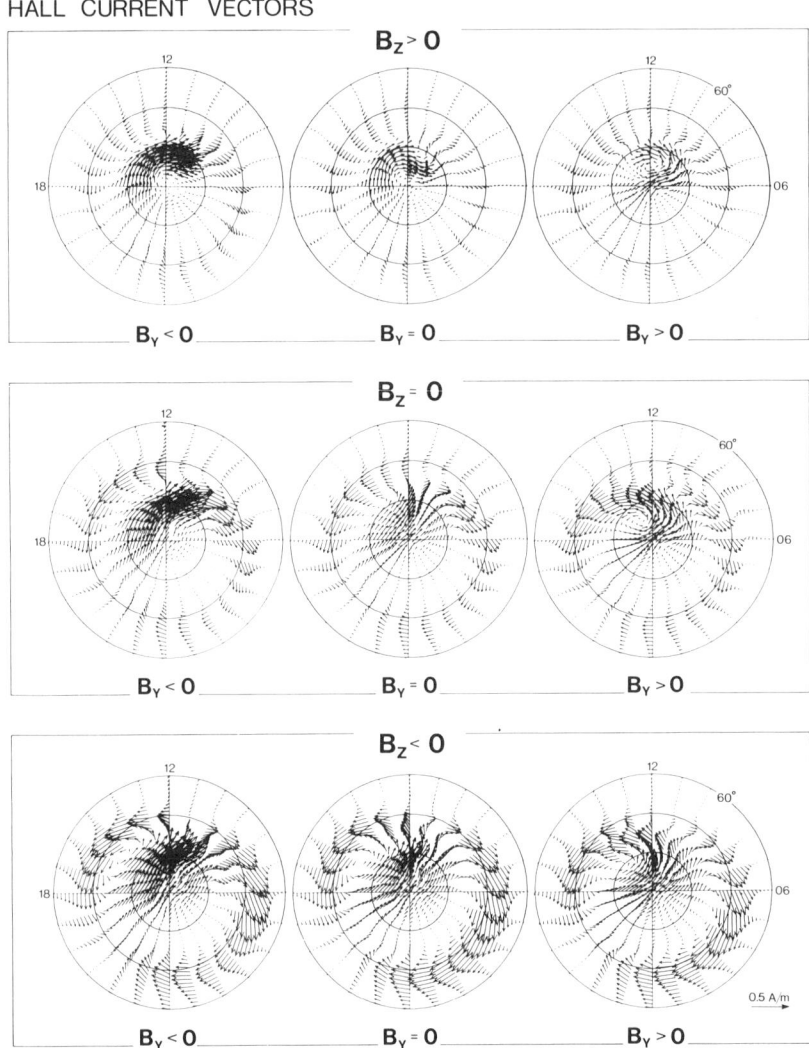

Fig. 3. The Hall-current part of the ionospheric current distribution of Figure 2 (after Friis-Christensen et al., 1983).

gardless of whether $B_z<0$ or $B_z>0$. It is interesting that although the results are statistical and one therefore would expect some smoothing of the distributions, seen for instance in the latitudinal extension of the current regions, which is larger than reported from direct individual measurements, a surprisingly distinct and systematic behaviour of the currents near noon is observed. Compared to other published distributions of field-aligned currents derived by means of ground-based magnetometer observations e.g. Akasofu et al. (1981), Kamide et al. (1981) the distributions show a far better spatial resolution which is due to the good latitudinal coverage of the Greenland magnetometer chain and the fact that pure and idealized states of the IMF have been considered.

The region 2 currents are seen to diminish drastically for $B_z>0$ while the region 1 currents are concentrating in the dayside. The cusp field-aligned currents are nearly vanishing for $B_y=0$ and $B_z=0$ and seem to be statistically detached from the region 1 current of the same sign. For $B_z>0$ and $B_y=0$ the pattern is similar to the original results from the TRIAD satellite (Iijima and Potemra, 1976).

Although Figure 4 basically contains the gross features of the satellite observations summarized by Iijima and Potemra (1976) there are some differences in the details in particular around magnetic noon and near magnetic midnight. The differences around magnetic noon are explained by the importance of the IMF B_y component in averaging and ordering the data. Around magnetic midnight, however,

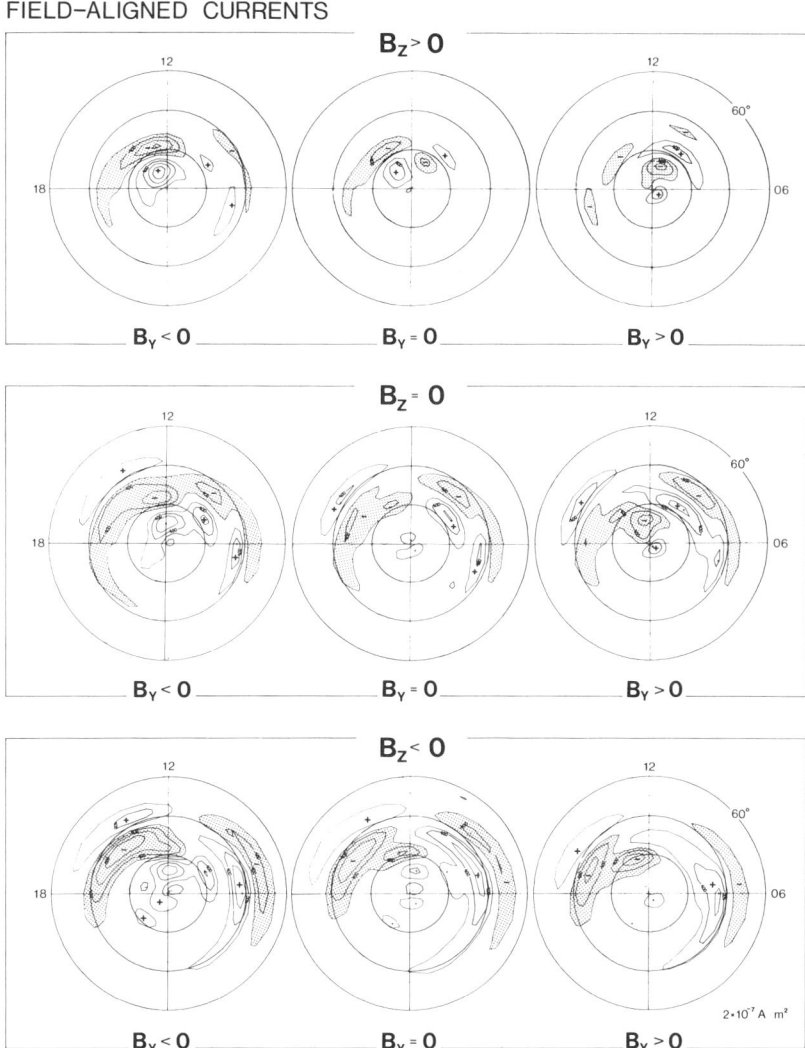

Fig. 4. The field-aligned current distribution corresponding to Figures 1 to 3. Shaded regions mean upward currents. The contour interval is 2×10^{-7} A/m^2 (after Friis-Christensen et al., 1983).

the result by Friis-Christensen et al. (1983) shows no field-aligned currents greater than the contour threshold. The reason for this is that the analytical average conductivity model does not have large gradients contrary to what is expected in the individual cases. The region of upward field-aligned currents, which evidently must exist because of the large gradients in the horizontal currents, must therefore be spread in a larger area implying a density lower than the threshold given by the contour interval. It should be noted that the zero contour lines have been deleted for simplicity reasons.

Polar Cap Currents and Ionospheric Convection

Figure 5 taken from Friis-Christensen et al. (1983) show the electric potential which is computed using the empirical equivalent current systems and the selected conductivity model. Also here the most striking features are located in the dayside polar cap near the cusp. In this region there is still lacking a detailed description of the electric field distributions and the dependences on the interplanetary magnetic field. Overall features of the two-cell convection pattern which is shifted towards the dawn (dusk) for $B_y>0$ ($B_y<0$) (Heppner, 1972; Mozer et al., 1974) are reproduced. For $B_z>0$ a one-cell system controlled by B_y is dominating over the IMF-independent and persistent two-cell system seen in the center of the diagram. For $B_y=0$ and $B_z>0$ near local noon in the dayside part of the polar cap the electric field has a dawnward component in agreement with satellite measurements by Burke et al. (1979), and corresponding to the "reversed" ionospheric current. In general

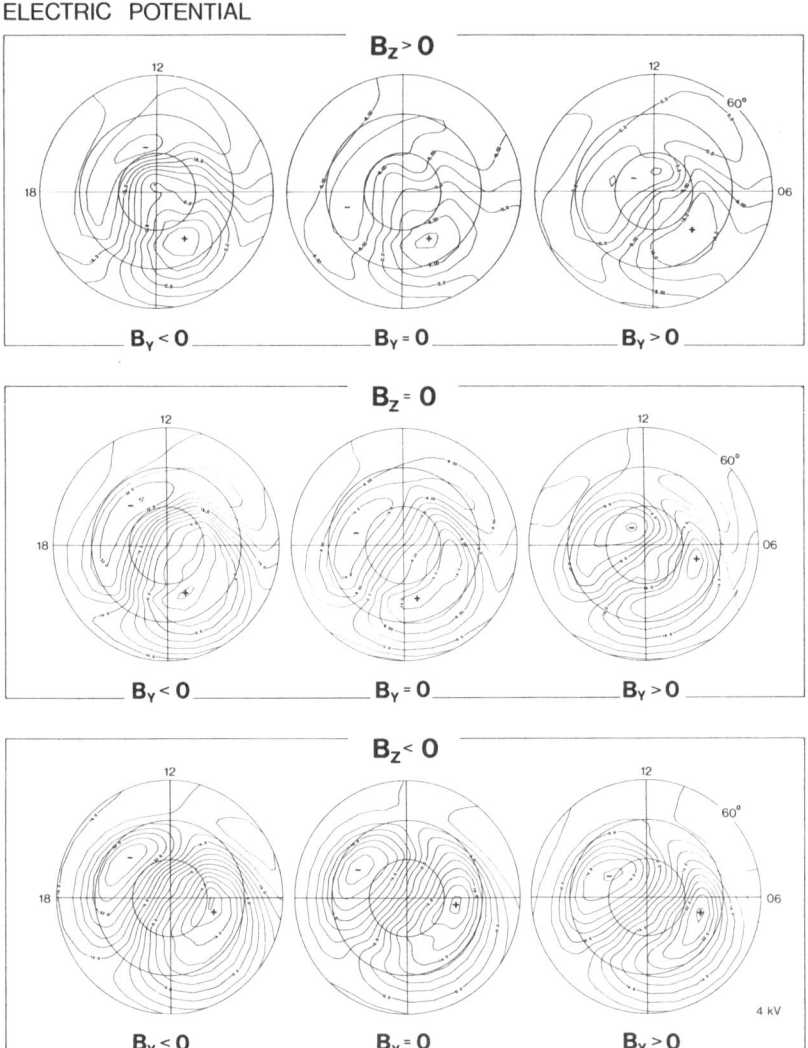

Fig. 5. Electric potential contours corresponding to Figures 1 to 4. The contour interval is 4 KV (after Friis-Christensen et al., 1983).

the dayside part of the polar cap shows potential contours which fairly well are represented by the equivalent current system because of the homogeneous conductivity distribution in this part of the ionosphere where solar EUV radiation is supposed to produce the main part of the ionization. It is worth nothing that the empirical model used by Friis-Christensen et al. (1983) is assumed antisymmetric with respect to B_y when the B_y effect alone is regarded. However, when combining the B_y effect with the background S_q^p current system which is rotated with respect to magnetic local noon, the resulting pattern is neither symmetric nor antisymmetric. This fact is clearly recognizable also in the potential distribution and is probably related to the day-night gradient in the conductivity (Atkinson and Hutchinson, 1978).

Due to B_y some similarity of the dayside electric potential contours with those observed in the Harang discontinuity may be noticed in Figure 5, in the region where the DPY ionospheric current is located. This is just another confirmation of the interpretation of the DPY current as a Hall current with a poleward (equatorward) directed electric field for $B_y>0$ ($B_y<0$) in the northern polar cap.

Discussion and Outstanding Problems

Considerable knowledge has been gained the last five years regarding the distribution of polar cap electric fields and currents. Except for the boundary between the polar cap and the polar cusp the large-scale features of ionospheric and field-aligned currents are well established.

Near local noon electric fields and currents are to a high degree determined by the B_y component of the IMF. Lack of simultaneous measurements with a good spatial resolution in this region has made it difficult in a unique way to relate the currents in the cusp with the large-scale region 1 and region 2 currents.

Wilhjelm et al. (1978) found by a comparison between ground-based measurements of the ionospheric currents and TRIAD measurements of the field-aligned currents that the DPY current is a Hall current sandwiched between a pair of oppositely directed field-aligned current sheets. The poleward current sheet corresponds to the cusp field-aligned currents whereas the equatorward current sheet according to the latitude would correspond to the region 1 currents. Since, however, both current sheets changed direction according to the sign of the B_y component of the IMF, it was not possible to decide from the analysis by Wilhjelm et al. (1978) if the equatorward current sheet around magnetic noon was indeed part of the region 1 current system or if it was a separate current system.

McDiarmid et al. (1979) regarded the cusp field-aligned current sheets as part of the prenoon downward region 1 current when $B_y<0$ and as part of the postnoon upward region 1 current for $B_y>0$. In this way the currents around noon could be described as two oppositely directed region 1 currents which overlapped in a pattern determined by the B_y component of the IMF, (see Figure 6 taken from McDiarmid et al. (1979)).

The morphology of the different polarity regions of the dayside field-aligned currents summarized by McDiarmid et al. (1979) has been taken as an indication that the DPY current is just a continuation across magnetic local noon of the eastward (westward) electrojet for $B_y>0$ ($B_y<0$) (Rostoker, 1980).

This interpretation, however, would require that the DPY current is placed at approximately the same latitude as the auroral electrojets. Statistical studies (Friis-Christensen and Wilhjelm, 1975) and individual case studies by Friis-Christensen (1981) show that the DPY current is always located poleward of an often simultaneously existing auroral electrojet. In Figure 7 taken from Friis-Christensen (1981) latitude profiles for different signs of B_y at local times before and after magnetic noon show that at least for $B_z>0$, where the conditions are not so complicated, it is possible to identify the poleward DPY current as well as the auroral electrojet. This is also confirmed by the statistical studies by Friis-Christensen et al. (1983) of the Hall- and field-aligned current patterns near local noon. Figure 3 and Figure 4 indicate that the DPY current should rather be regarded as the convection current just inside the polar cap. When $B_z>0$ a considerable decrease of the region 1 and region 2 currents along the electrojets is seen. The field-aligned currents near noon, the "cusp-field-aligned current" and

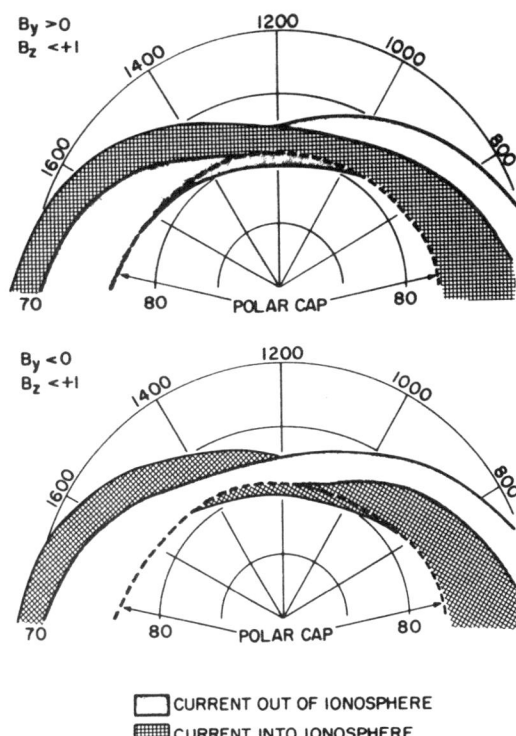

Fig. 6. Field-aligned current directions across the dayside auroral oval inferred from ISIS-2 polar orbiter magnetometer data (after McDiarmid et al., 1979).

the noon part of the region 1 current remain without a considerable decrease.

Bythrow et al. (1982) have compared field-aligned currents measured with the TRIAD satellite with AE-C measurements of plasmas and fields.

They suggest that the region 1 currents near magnetic noon may have a source substantially different from that of region 1 currents observed at other local times. These observations obtained from single events are in agreement with the result by Friis-Christensen et al. (1983) indicating that the "cusp-field-aligned current" is statistically independent of the region 1 currents along the auroral electrojets but is associated with a current sheet of opposite polarity equatorward of the "cusp" sheet near noon.

These two different interpretations have of course quite different implications regarding the origin of the field-aligned currents.

If the "cusp-field-aligned current" is a continuation across noon of the corresponding region 1 current it is implied that these currents are of the same origin. It is however hard to imagine a source in the magnetosphere or in the boundary layer which could be responsible for all region 1 and "cusp-field-

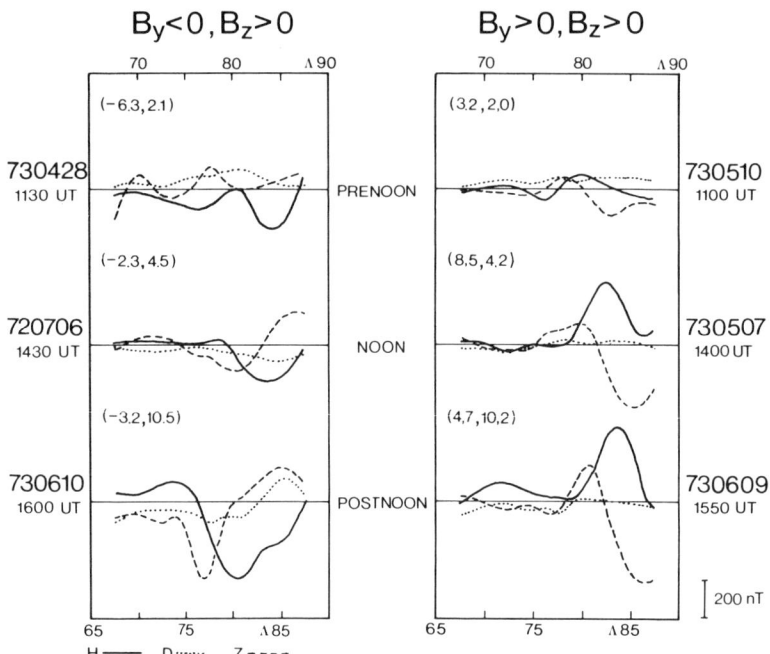

Fig. 7. Latitude profiles of the H, D, and Z perturbations along the Greenland magnetometer chain for different conditions of the interplanetary magnetic field (after Friis-Christensen, 1981).

aligned currents". If on the other hand the region 1 currents have a different source than the "cusp-field-aligned currents" there should be some kind of discontinuity between these currents and an overlap should not be seen. The final solution to this problem has not yet been found. It is hoped that investigations using the extensive IMS database will lead to an improved understanding which will have a great impact on the interpretation of geophysical observations in the polar cap.

References

Akasofu, S.I., and B.H. Ahn, Distribution of the field-aligned currents, ionospheric currents, and electric fields in the polar region on a very quiet day and a moderately disturbed day, J. Geophys. Res., 86, 753, 1981.

Atkinson, G., and G. Hutchison, Effect of the day night ionospheric conductivity gradient on polar cap convection flow, J. Geophys. Res. 83, 725, 1978.

Berthelier, A., J.J. Berthelier, and C. Guérin, The effect of the east-west component of the interplanetary magnetic field on magnetospheric convection as deduced from magnetic perturbations at high latitudes, J. Geophys. Res., 87, 3187, 1974.

Burke, W.J., M.C. Kelley, R.C. Sagalyn, M. Smiddy and S.T. Lai, Polar cap electric field structures with a northward interplanetary magnetic field, Geophys. Res. Lett., 6, 21, 1979.

Bythrow, P.E., T.A. Potemra, and R.A. Hoffman, Observations of field aligned currents, particles and plasma drift in the polar cusps near solstice, J. Geophys. Res., 87, 5131, 1982.

Doyle, M.A., F.J. Rich, W.J. Burke, and M. Smiddy, Field-aligned currents and electric fields observed in the region of the dayside cusp, J. Geophys. Res., 86, 5656, 1981.

Friis-Christensen, E., High latitude ionospheric currents. In Exploration of the Polar Upper Atmosphere, ed. C.S. Deehr and J.A. Holtet, D. Reidel Publishing Company, 1981.

Friis-Christensen, E., K. Lassen, J. Wilhjelm, J.M. Wilcox, W. Gonzalez, and D.S. Colburn, Critical component of the interplanetary magnetic field responsible for large geomagnetic effects in the polar cap, J. Geophys. Res., 77, 3371, 1972.

Friis-Christensen, E., and J. Wilhjelm, Polar cap currents for different directions of the interplanetary magnetic field in the Y-Z plane, J. Geophys. Res., 80, 1248, 1975.

Friis-Christensen, E., Y. Kamide, A.D. Richmond, and S. Matsushita, Interplanetary magnetic field control of high-latitude electric fields and currents determined from Greenland magnetometer data, to be submitted to J. Geophys. Res., 1983.

Fukushima, N., Generalized theorem for no ground magnetic effect of vertical currents connected with Pedersen currents in the uniform-conductivity ionosphere, Rep. Ionos. Space Res. Jpn., 30, 35-40, 1976.

Gizler, V.A., V.S. Semenov, and O.A. Troshichev, Electric fields and currents in the ionosphere generated by field-aligned currents observed by TRIAD, Planet. Space Sci., 27, 223, 1979.

Heppner, J.P., Polar-cap electric field distributions related to the interplanetary magnetic field direction, J. Geophys. Res., 77, 4877 1972.

Iijima, T., and T.A. Potemra, The amplitude distribution of field-aligned currents at northern high latitudes observed by TRIAD, J. Geophys. Res., 81, 2165, 1976.

Iijima, T., R. Fujii, T.A. Potemra, and N.A. Saflekos, Field-aligned currents in the south polar cusp and their relationship to the interplanetary magnetic field, J. Geophys. Res., 83, 5595, 1978.

Kamide, Y., A.D. Richmond, and S. Matsushita, Estimation of ionospheric electric fields, ionospheric currents, and field-aligned currents from ground magnetic records, J. Geophys. Res. 86, 801, 1981.

Levitin, A.E., R.G. Afonina, B.A. Belov, and Ya.I. Feldstein, Geomagnetic variation and field-aligned currents at northern high-latitudes, and their relations to the solar wind parameters, Phil. Trans., R. Soc. Lond., A 304, 253, 1982.

Maezawa, K., Magnetic convection induced by the positive and negative z components of the interplanetary magnetic field: Quantitative analysis using polar cap magnetic records, J. Geophys. Res., 81, 2289, 1976.

Mansurov, S.M., New evidence of a relationship between magnetic fields in space and on earth, Geomagn. Aeron., 9, 622, 1969.

Matsushita, S., and W.-Y. Xu, Equivalent ionospheric current systems representing IMF sector effects on the polar geomagnetic field, Planet. Sci., 30, 641, 1982.

McDiarmid, I.B., J.R. Burrows, and Margaret D. Wilson, Magnetic field perturbations in the dayside cleft and their relationship to the IMF, J. Geophys. Res., 83, 5753, 1978.

McDiarmid, I.B., J.R. Burrows, and M.D. Wilson, Large-scale magnetic field perturbations and particle measurements at 1400 km on the dayside, J. Geophys. Res., 84, 1431, 1979.

Mishin, V.M., A.D. Bazarzhapov, and G.B. Shpynev, Electric fields and currents in the earth magnetosphere, in Dynamics of the magnetosphere, edited by S.I. Akasofu, pp. 249-268, D. Reidel, Hingam, Mass., 1980.

Mozer, F.S., W.D. Gonzalez, F. Bogott, M.C. Kelley, and S. Schutz, High-latitude electric fields and the three-dimensional interaction between the interplanetary and terrestrial magnetic fields, J. Geophys. Res., 79, 56, 1974.

Nishida, A., Geomagnetic diagnosis of the magnetosphere, Springer Verlag, Berlin 1978.

Rich, F.J., and Y. Kamide, Convection electric fields and ionospheric currents derived from model field-aligned currents at high latitudes, J. Geophys. Res., 88, 271, 1983.

Rostoker, G., Magnetospheric and ionospheric currents in the polar cusp and their dependence on the By component of the interplanetary magnetic field, J. Geophys. Res., 85, 4167, 1980.

Svalgaard, L., Sector structure of the interplanetary magnetic field and daily variation of the geomagnetic field at high latitudes, Dan. Met. Inst. Geophys. Pap. R-16, 1969.

Saflekos, N.A., T.A. Potemra, P.M. Kintner, and J.L. Green, Field aligned currents in the South Pole cusp and their relationship to the interplanetary magnetic field, and ULF-ELF waves in the cusp, J. Geophys. Res., 84, 1391, 1979.

Troshichev, O.A., V.A. Gizler, I.A. Ivanova, and A.Y. Merkuryeva, Role of field-aligned currents in generation at high-latitude magnetic disturbances, Planet. Space Sci., 27, 1451, 1979.

Wilhjelm, J., E. Friis-Christensen and T.A. Potemra, The relationship between ionospheric and field-aligned currents in the dayside cusp, J. Geophys. Res., 83, 5586, 1978.

Zmuda, A.J., and J.C. Armstrong, The diurnal variation of the region with vector magnetic field changes associated with field-aligned currents, J. Geophys. Res., 79, 2501, 1974.

RELATIONSHIPS BETWEEN FIELD-ALIGNED CURRENTS, ELECTRIC FIELDS, AND
PARTICLE PRECIPITATION AS OBSERVED BY DYNAMICS EXPLORER-2

M. Sugiura, T. Iyemori, R. A. Hoffman, N. C. Maynard

Goddard Space Flight Center, Greenbelt, Maryland 20771

J. L. Burch, J. D. Winningham

Southwest Research Institute, San Antonio, Texas 78284

Abstract. The relationships between field-aligned currents, electric fields, and particle fluxes are determined using observations from the polar orbiting low-altitude satellite Dynamics Explorer-2. It is shown that the north-south electric field and the east-west magnetic field components are usually highly correlated in the field-aligned current regions. This proportionality observationally proves that the field-aligned current equals the divergence of the height-integrated ionospheric Pedersen current in the meridional plane to a high degree of approximation. As a general rule, in the evening sector the upward field-aligned currents flow in the boundary plasma sheet region and the downward currents flow in the central plasma sheet region. The current densities determined independently from the plasma and magnetic field measurements are compared. Although the current densities deduced from the two methods are in general agreement, the degree and extent of the agreement vary in individual cases.

Introduction

Field-aligned currents, electric fields, and precipitating particle fluxes constitute key physical parameters in the magnetosphere-ionosphere coupling processes. Theoretical models for a self-consistent system of this coupling mechanism have been presented by Vasyliunas (1970, 1972), Wolf (1970, 1974), Harel et al. (1979) and many others. Extensive studies have been made on the relationships between field-aligned currents, electric fields, and particle precipitation using data from satellites, sounding rockets, and ground-based facilities. Bostrom (1974) theoretically recognized the proportionality between the electric and magnetic field components associated with field-aligned currents and pointed out the resemblance between some of the magnetic field signatures of field-aligned currents observed by Triad (Armstrong, 1974) and some of the electric field variations observed from Injun 5 (Gurnett and Frank, 1973), although the comparison was not between simultaneously observed data. Comparisons of simultaneous magnetic and electric field signatures of field-aligned currents have been made using the electric and magnetic field observations on the S3-2 and S3-3 satellites (e.g., Burke et al., 1980; Smiddy et al., 1980; Rich et al., 1981), the ion drift and magnetic field measurements on AE-C (Bythrow et al., 1980), the Chatanika radar observations (e.g., de la Beaujardiere et al., 1977), and a combination of the Chatanika radar observations and the magnetometer measurements from Triad, (e.g., Robinson et al., 1982).

Early investigations of the relations between auroral particle precipitation and field-aligned currents (and also electric fields) are reviewed by Arnoldy (1974) and Anderson and Vondrak (1975). Numerous papers have been written on this subject since then, using particle flux measurements or optical observations of the aurora and the field-aligned current data deduced from magnetic field measurements or from incoherent scatter radar observations (e.g., Klumpar, 1979; Burke et al., 1980; Robinson et al., 1982; Senior et al., 1982).

One of the objectives of the Dynamics Explorer (DE) mission is to investigate quantitatively the relationships between field-aligned currents, electric fields, and particle fluxes with simultaneous observations. Descriptions of the magnetometer, the electric field probe, and the plasma experiment (LAPI) have been given by Farthing et al. (1981), Maynard et al. (1981), and Winningham et al, (1981), respectively.

Correlation Between Electric and Magnetic Fields

Sugiura et al. (1982) have presented initial DE-2 results on the correlation between the magnetic and electric fields in the field-aligned current regions. They showed that the traces of the north-south component of the electric field and of the east-west component of the magnetic field are usually very similar and that the correlation coefficient between these parameters is often as high as 0.99. Figure 1 shows the north-south component of the electric field and the geomagnetic north-south and east-west components of the magnetic field observed from DE-2 during a pass on July 29, 1982, 0606-0610UT, crossing the dayside cusp. The x component of

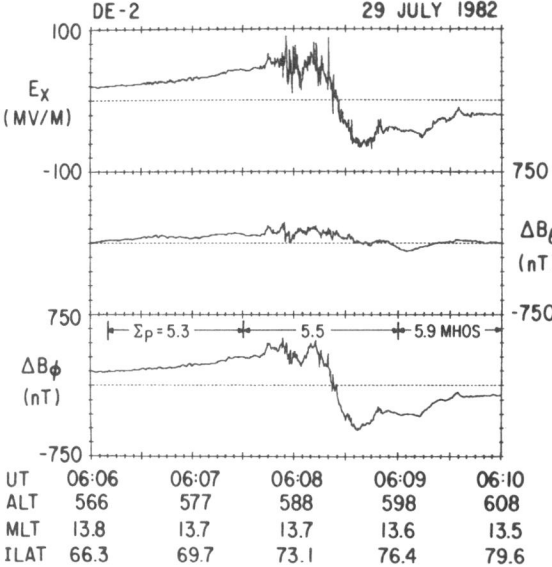

Fig. 1. The component of the electric field along the spacecraft velocity vector (northward on this pass) and the southward and east-ward components of the perturbation magnetic field in geomagnetic diple coordinates, on a dayside cusp pass on July 29, 1982. The height-integrated Pedersen conductivity, Σ_P, is given for the three segments indicated.

the electric field is along the satellite velocity vector and in this case it is northward. The magnetic field data plotted are the differences between the observed field and a reference field based on the latest Magsat model field (see Langel et al., 1980, for a description of the Magsat results). The sampling rate of the measurement is 16 samples per second both for the electric and magnetic field observations. There is a striking similarity between the north-south electric field and east-west magnetic field components except that the electric field data show considerably more rapid fluctuations than the magnetic field data. Using 1/2 second averages, the coefficient of correlation between these two parameters for the 6-minute period 0605 to 0611 (containing the 4-minute interval shown in the figure) is 0.992. To see if the correlation varies significantly between the regions of weak current (equatorward and poleward of the cusp) and the vicinity of the cusp where the current is more intense, the above interval was divided into three segments: 0606:10-0607:30, 0607:30-0609:00, and 0609:00-0610:00, and the correlation coefficient was calculated for each of these segments, using the high time resolution data (at 16 samples per second). The correlation coefficients so calculated are 0.996, 0.98, and 0.98 respectively. This result shows that the high correlation is not limited to the regions where the current densities are large.

We now examine the relation between the electric and magnetic fields analytically. Let the x, y and z axes be toward the north, east and downward, respectively. For simplicity, let the ambient magnetic field be vertical and downward. We assume that $\partial b_z/\partial t = 0$ and that b_x is independent of y (i.e. $j_\parallel = -(1/\mu_o)\partial b_y/\partial x$), where b is the perturbation magnetic field, the y component of which corresponds to ΔB_ϕ in our data presentation. Then by equating $j_\parallel$ to the divergence of the height-integrated horizontal ionospheric current, we obtain

$$(1/\mu_o)\partial b_y/\partial x = \partial(\Sigma_P E_x)/\partial x + \partial(\Sigma_P E_y)/\partial y + E_x \partial \Sigma_H/\partial y - E_y \partial \Sigma_H/\partial x \quad (1)$$

where Σ_P and Σ_H are the height-integrated Pedersen and Hall conductivities. If E_y is independent of x, and if the conductivities and E_y are independent of y, eq.(1) can be integrated to give

$$b_y = \mu_o(\Sigma_P E_x - \Sigma_H E_y) + \text{const} \quad (2)$$

If the observed values of b_y are linearly correlated with those of E_x with a coefficient of correlation near unity over distances of several hundred or even several thousand kilometers along the x axis, we must conclude either (a) that the north-south Hall current is zero and that

$$b_y = \mu_o \Sigma_P E_x + \text{const} \quad (3)$$

or (b) that E_y is always proportional to E_x. However, E_y was assumed to be independent of x in deriving eq.(2). Since E_x varies with x, the proportionality of E_y to E_x contradicts with this assumption. Hence (a), i.e. eq.(3), is the only choice.

If eq.(3) holds, then the slope of the b_y (corresponding to ΔB_ϕ in Figure 1) vs. E_x enables us to determine Σ_P. We have made several simplifying assumptions in deriving the relationship expressed by eq.(3). However, considering the complexity of eq.(1), the probability that the proportionality between b_y and E_x holds over large distances on many passes at different local times would appear near null if the terms other than the first term are not negligible. We thus conclude that when the correlation between the north-south component of the electric field and the east-west component of the magnetic field is high, the field-aligned current equals the divergence of the height-integrated Pedersen current within the meridional plane, and that the Hall current from an east-west electric field, if any, makes no significant contribution to the field-aligned current under this condition.

For the three subintervals mentioned above in reference to the July 29, 1982 example, the value of Σ_p determined by eq.(3) is 5.3, 5.5 and 5.9 mhos, respectively. The variation in Σ_p is relatively small even though the current densities vary between these sections. This feature is consistent with the interpretation that the primary source of ionization is the solar EUV, and is in strong contrast with the condition on the nightside where the ionization is mainly from precipitating particles. Sugiura et al.(1982) have shown an example near 21 hours MLT in which the conductivity is largest where the current is most intense and in which the conductivity decreases away from the main current region.

Relations With Particle Precipitation in the Evening Sector

From an extensive inspection of preliminary energy-time spectrograms obtained from LAPI observations, it has become clear that in no local time sectors do the precipitating positive ions consistently play a primary role in carrying downward field-aligned currents. Therefore in this section we only discuss the relations between electron precipitation and field-aligned currents. The morphological relationships between these parameters are complex and depend strongly on local time. In this paper we limit our discussion to the evening sector. Figure 2 shows from top to bottom, plots of the north-

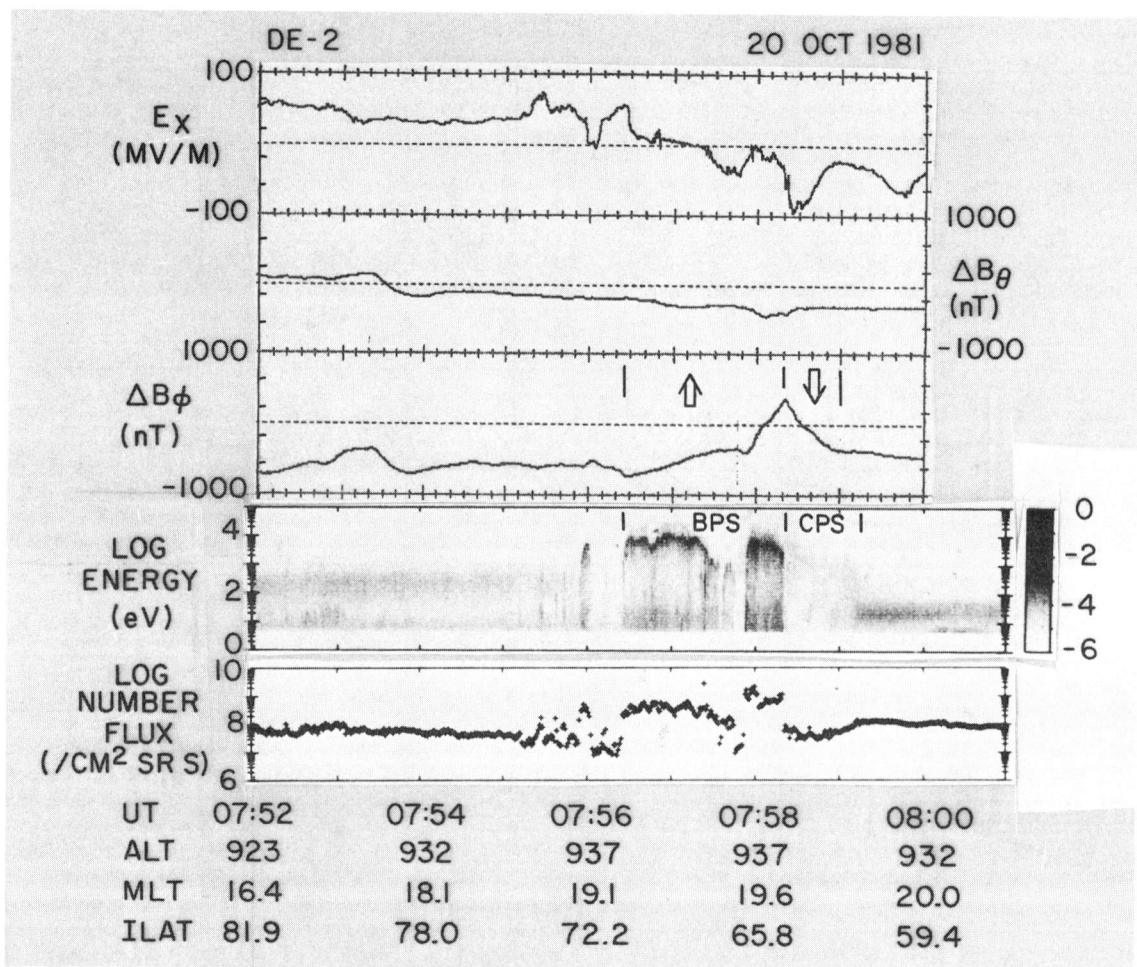

Fig. 2. The component of the electric field along the spacecraft velocity vector (southward on this pass), the southward and eastward components of the perturbation magnetic field in geomagnetic dipole coordinates, and the energy-time spectrogram and the number flux for electrons, on an evening pass on October 20, 1981. Arrows in the third panel schematically indicate field-aligned current directions.

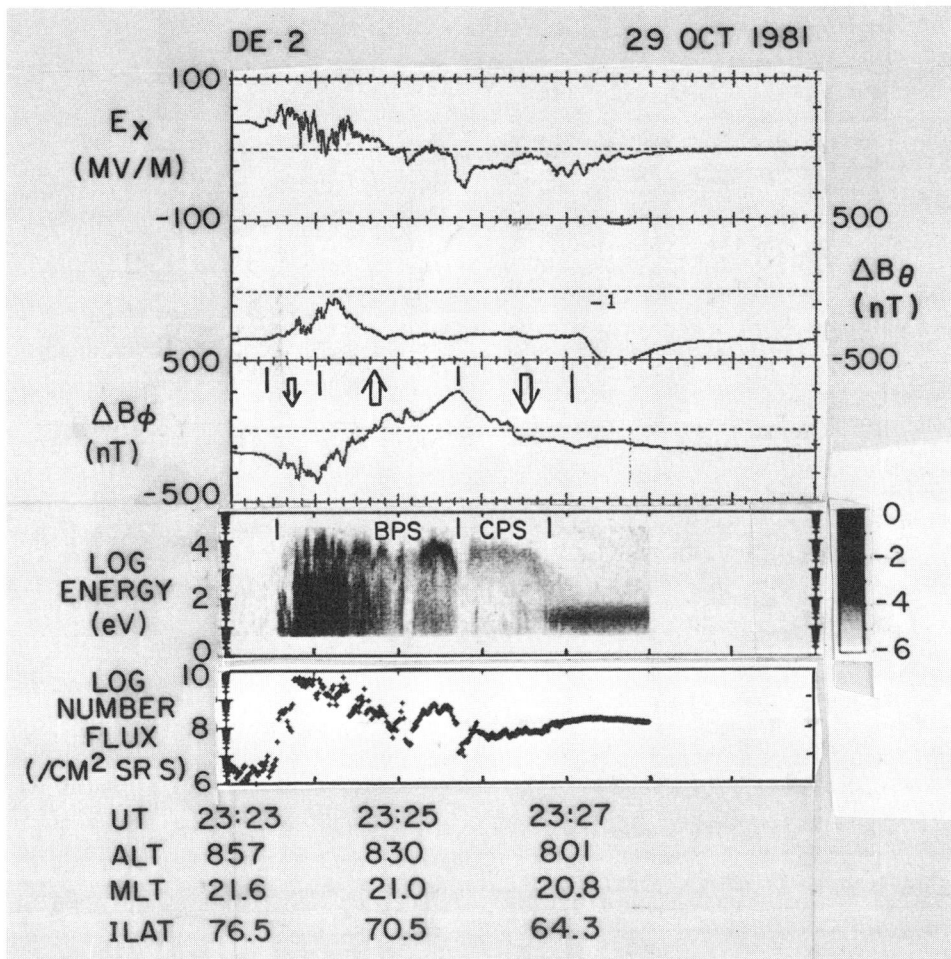

Fig. 3. The components of the electric and magnetic fields and the electron spectrogram and number flux (see Fig. 2 caption) for an evening pass on October 29, 1981.

south component of the electric field, the dipole north-south and east-west components of the magnetic field, and the electron energy-time spectrogram at zero degree pitch angle and the electron number flux obtained from LAPI, for a high latitutde pass near magnetic local time (MLT) 19 hours on October 20, 1981. Electrons measured are in the energy range 5 eV to 32 keV. In the basic mode of operation a 32 point energy spectrum is obtained every second. The electric and magnetic field data are 1/2 second averages. Plots of ΔB_ϕ indicate that the upward field-aligned current region extends approximately from 0756:30 to 0758:20, and that a downward current is observed from 0758:20 to about 0759:00. The current directions are schematically indicated by arrows in the figure. The x component of the electric field in Figure 2 (and also in Figure 3 below) is southward. Therefore the coefficient of correlation between E_x and ΔB_ϕ has an opposite sign to the case shown in Figure 1. The terminology of Winningham et al. (1975) has been found helpful in determining the relationships between the characteristics of electrons and field-aligned currents. According to Winningham et al. (1975), the electron flux precipitating from the central plasma sheet (CPS) is relatively stable with respect to 'substorm time' and spatially uniform, and its variation, if any, is a uniform increase or decrease in intensity. The boundary plasma sheet (BPS) precipitation is characterized by highly variable plasma precipitation poleward of the CPS region. It is in the BPS region that structures such as the 'inverted V' are observed, as seen in the figure. As to the field-aligned currents, it is more convenient for the present purpose to categorize the field-aligned current regions simply by upward and downward current regions rather than by regions 1 and 2 as is usually done following the nomenclature of Iijima and Potemra (1976).

In Figure 2 the upward field-aligned current region coincides with the inverted V's in the boundary plasma sheet (BPS) precipitation and the downward current region is inside the central plasma sheet (CPS). In this example there is a sharp transition between the two main current regions. The spectrogram and the number flux show that the higher electron flux (roughly from 0757:55 to 0758:20) gives a steeper slope in ΔB_ϕ and hence a greater current density. The current density distribution for this pass is given in the next section (Figure 7).

Another example is presented in Figure 3 for an evening pass near 21 hours MLT on October 29, 1981. From ΔB_ϕ, the upward field-aligned current region is approximately from 2324:00 to 2325:40 and is in the BPS electron precipitation region. However, in this example the BPS extends poleward beyond the high latitude edge of the upward current region and into a region of weak predominantly downward current. As in the previous case of October 20, the abrupt (equatorward) termination of the BPS electons clearly coincides with the boundary between the regions of upward and downward field-aligned current. Again the downward current region roughly coincides with the CPS region.

In the October 29 case there are fine structures in both upward and downward current regions. (The current density profile is shown in Figure 5 in the following section.) Corresponding to each structure of intensified electron flux, an indication of upward current is discernible in the magnetic field data. This feature is not limited to the evening sector but is common at all local times.

In the electron spectrograms, narrow gaps of varying width are often found which have appearances of 'holes' in the electron precipitation. In these 'holes' the currents are always downward. As an example, referring to Figure 2, a small downward current exists in the BPS region for about 5 seconds after 0757:50, and the electron precipitation nearly vanishes. Another

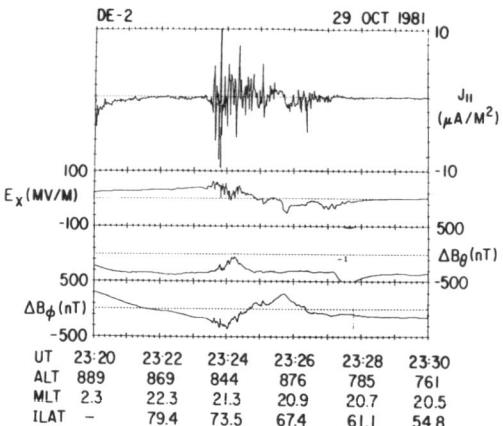

Fig. 5. The current density (1-second averages) and the electric and magnetic field components (1/2-second averages) for the evening pass on October 29, 1981, shown in Fig. 3.

example is seen near 2325:10 in the October 29 event shown in Figure 3.

In these two examples the north-south component of the electric field is generally similar to the east-west component of the magnetic field on large scales but there are appreciable deviations from the proportionality in small scale structures in the evening sector. However, generally speaking, the electric and magnetic field variations are consistent in the manner discussed in the preceding section and in our previous paper (Sugiura et al., 1982).

Current Densities

The density of field-aligned current is calculated from the east-west component of the magnetic field assuming a series of field-aligned infinite current layers that are locally normal to the dipole meridian plane. The differentiation of the magnetic field component is made with respect to the distance normal to the current sheets. The current density profile for the cusp pass on July 29, 1982 (Figure 1) based on 2-second averages of ΔB_ϕ is shown in Figure 4. Even with these smoothed data the current density has large fluctuations. Through a careful examination of the current densities derived from high time-resolution magnetic field data that are not averaged and from averages over various time lengths (1/2 sec, 1 sec, 2 sec, etc.) we have concluded that in terms of current density the field-aligned currents have much more detailed structures than the human eye can resolve in the magnetic field plots. Figure 5 shows the current density plotted with the electric and magnetic field components for the October 29, 1981 pass (Figure 3). The current density given here is based on 1-second averages of ΔB_ϕ, while the

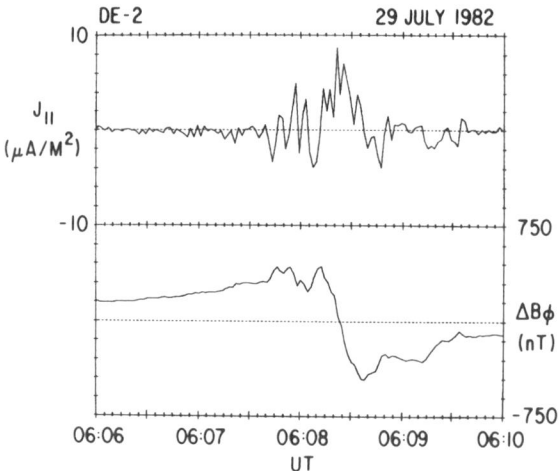

Fig. 4. The current density (upper panel) deduced from 2-second averages of the magnetic field data (lower panel) for the cusp pass on July 29, 1981, shown in Fig. 1.

electric and magnetic field data are 1/2 second averages as in Figure 3. The maximum current density in the time interval shown is about 10 $\mu A/m^2$.

Figure 6 gives the current density (1-second averages) and the electric and magnetic field components (1/2-second averages) for a pass near 20 hours MLT on October 7, 1981. In this case there was a very thin upward current layer coinciding with a narrow intense BPS electron flux region (electron spectrogram not shown in this paper). The region of irregular current roughly coincides with the CPS region. However, a detailed analysis of plasma data is required to specify the nature of the electron population that is related to the irregular field-aligned currents. There is a region of high electric field with a peak intensity of 150 mV/m near 1029:35, which is an example of a rapid subauroral ion drift (SAID) region (see Spiro et al., 1978). Because of low conductivity, however, the field-aligned current is weak with a current density of only a few $\mu A/m^2$.

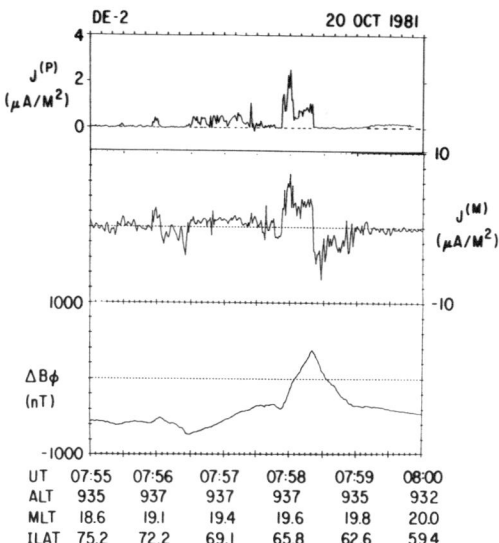

Fig. 7. Comparison of the current density, $J^{(P)}$, deduced from the plasma data and the current density, $J^{(M)}$, deduced from the magnetic field data; these values are computed once in each second. The bottom panel gives the magnetic field component from which the current density was deduced.

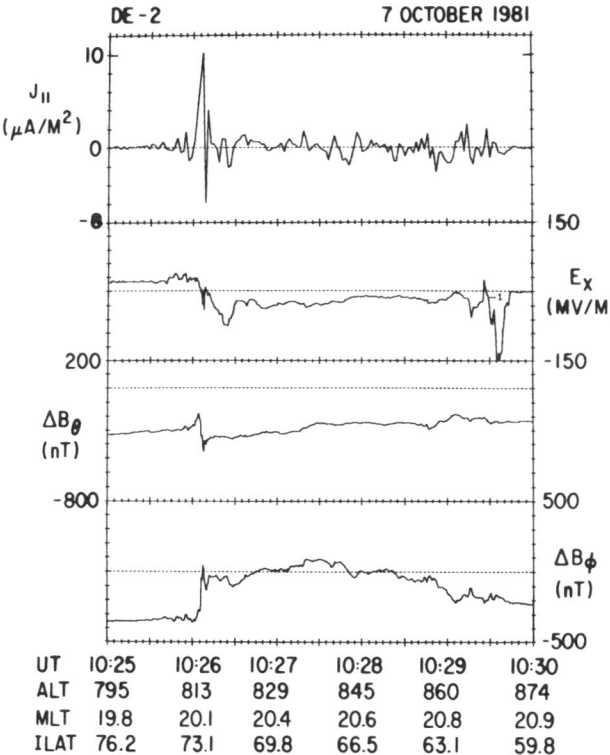

Fig. 6. The current density (1-second averages) and the electric and magnetic field components (1/2-second averages) for an evening pass on October 7, 1981. The electric field, E_x, plot shows a large electric field in a rapid subauroral ion drift convection region near 60° invariant latitude.

A method has been developed to deduce the current density from the plasma measurements on DE-1 and -2 (Burch et al., 1983) on a routine basis. An example of comparison between the current density distributions derived from the LAPI data and from the magnetic field data is shown in Figure 7 for the October 20, 1981 pass (Figure 2). The energy sweep in LAPI takes 1 second; therefore 1 second is the time resolution in the current density calculation using the LAPI data. Correspondingly, the current density from the magnetic field data is based on 1-second averages of the field. The overall features are remarkably similar between the two current profiles in the region of upward current, even for small variations. In the region of negative (downward) current, the LAPI currents go nearly to zero. The boundaries of the upward and downward current regions derived from the two methods agree exactly. In this example the current densities determined from the plasma data are generally smaller than those determined from the magnetic field data by factors of 2 to 4. However, this is not always the case; there are examples in which the current densities derived from the magnetic field are smaller than those from the plasma data. There are also cases where the agreement is nearly perfect in some portions of a pass but is rather poor in other portions. These facts indicate an extreme complexity of the relationships between field-aligned currents and particle precipitation. In deriving the current density from the magnetic field data we assumed infinite current sheets. However, this is an idealization and in reality, structured field-aligned currents are more likely to be filamentary (or of irregular shape) than being longitudinally uniform, as for instance the Ogo 4 observations suggested (Berko et al., 1975). Also the time unit of 1-second used is the time required for the plasma instrument to sweep the 32 energy steps, while the 1-second magnetic field data are true averages. The differences between the values of current density derived from the plasma data and from the magnetic field data may partially stem from the different nature of the data. Thus, future studies would require more detailed investigations of the current distribution taking these factors into consideration.

Conclusions

With the DE-2 observations the relationships between electric fields, field-aligned currents, and precipitating particles were examined. The north-south component of the electric field and the east-west component of the magnetic field are generally highly correlated in the field-aligned current regions. This high correlation gives an observational proof that the field-aligned current equals the divergence of the height-integrated Pedersen current within the meridional plane to a high degree of approximation. The height-integrated Pedersen conductivity can be determined from the constant of proportionality between these parameters. In the evening sector the upward field-aligned currents are usually in the boundary plasma sheet region and the downward currents flow in the central plasma sheet region. The current densities calculated from the plasma data and the magnetic field data agree in general features but the magnitudes of the current density determined by the two methods are not necessarily in agreement. The current density distribution has much more detailed structures than is visible to the eye in the magnetic field plots.

Acknowledegements. We wish to thank K. Babst, J. Byrnes, B. Carroll, S. Kempler, A. Meyers, L. Salter, and J. R. Thieman for their assistance in the preparation of this paper. The work at the Southwest Research Institute was supported by NASA Contracts NAS5-26363 and NAS5-2693. T. Iyemori was supported by a NASA/NRC Resident Research Associateship on leave from the Kyoto University.

References

Anderson, H. R., and R. R. Vondrak, Observations of Birkeland currents at auroral latitudes, Rev. Geophys. Space Phys., 13, 243, 1975.

Armstrong, J. C., Field aligned currents in the magnetosphere, in Magnetospheric Physics, B. M. McCormac, ed., D. Reidel Publishing Company, Dordrecht-Holland, p.155, 1974.

Arnoldy, R. L., Auroral particle precipitation and Birkeland currents, Rev. Geophys. Space Phys., 12, 217, 1974.

Berko, F. W., R. A. Hoffman, R. K. Burton, and R. E. Holzer, Simultaneous particle and field observations of field-aligned currents, J. Geophys. Res., 80, 37, 1975.

Bostrom, R., Ionosphere-magnetosphere coupling, in Magnetospheric Physics, B. M. McCormac, ed., D. Reidel Publishing Company, Dordrecht-Holland, p.45, 1974.

Burch, J. L., P. H. Reiff, and M. Sugiura, Upward electron beams measured by DE-1: A primary source of dayside region-1 Birkeland currents, Geophys. Res. Lett., 10, 753, 1983.

Burke, W. J., D. A. Hardy, F. J. Rich, M. C. Kelley, M. Smiddy, B. Schuman, R. C. Sagalyn, R. P. Vancour, P. J. L. Widman, and S. T. Lai: Electrodynamic structure of the late evening sector of the auroral zone, J. Geophys. Res., 85, 1179, 1980.

Bythrow, P. F., R. A. Heelis, W. B. Hanson, and R. A. Power, Simultaneous observations of field-aligned currents and plasma drift velocities by Atmosphere Explorer C, J. Geophys. Res., 85, 151, 1980.

de la Beaujardiere, O., R. R. Vondrak, and M. Baron, Radar observations of electric fields and currents associated with auroral arcs, J. Geophys. Res., 82, 5051, 1977.

Farthing, W. H., M. Sugiura, B. G. Ledley, and L. J. Cahill, Jr., Magnetic field observations on DE-A and -B, Space Sci. Instr., 5, 551, 1981.

Gurnett, D. A., and L. A. Frank, Observed relationships between electric fields and auroral particle precipitation, J. Geophys. Res., 78, 145, 1973.

Harel, M., R. A. Wolf, P. H. Reiff, and M. Smiddy, Computer modeling of events in the inner magnetosphere, in Quantitative Modeling of Magnetospheric Processes, W. P. Olson, ed., American Geophysical Union, Washington, D. C., p.499, 1979.

Iijima, T., and T. A. Potemra, The amplitude distribution of field-aligned currents at northern high latitudes observed by Triad, J. Geophys. Res., 81, 2165, 1976.

Klumpar, D. M., Relationships between auroral particle distributions and magnetic field perturbations associated with field-aligned currents, J. Geophys. Res., 84, 6524, 1979.

Langel, R. A., R. H. Estes, G. D. Mead, E. B. Fabiano, and E. R. Lancaster, Initial geomagnetic field model from MAGSAT vector data, Geophys. Res. Lett., 7, 793, 1980.

Maynard, N. C., E. A. Bielecki, and H. F. Burdick, Instrumentation for vector electric field measurements from DE-B, Space Sci. Instr., 5, 523, 1981.

Rich, F. J., C. A. Cattell, M. C. Kelley, and W. J. Burke, Simultaneous observations of auroral zone electrodynamics by two satellites: Evidence for height variations in the topside ionosphere, J. Geophys. Res., 86, 8929, 1981.

Robinson, R. M., R. R. Vondrak, and T. A. Potemra, Electrodynamic properties of the evening sector ionosphere within the region 2 field-aligned current sheet, J. Geophys. Res., 87, 731, 1982.

Senior, C., R. M. Robinson, and T. A. Potemra, Relationship between field-aligned currents, diffuse auroral precipitation and the westward electrojet in the early morning sector, J. Geophys. Res., 87, 10469, 1982.

Smiddy, M., W. J. Burke, M. C. Kelley, N. A. Saflekos, M. S. Gussenhoven, D. A. Hardy, and F. J. Rich, Effects of high-latitude conductivity on observed convection electric fields and Birkeland currents, J. Geophys. Res., 85, 6811, 1980.

Spiro, R. W., R. A. Heelis, and W. B. Hanson, Ion convection and the formation of the mid-latitude F region ionization trough, J. Geophys. Res., 83, 4255, 1978.

Sugiura, M., N. C. Maynard, W. H. Farthing, J. P. Heppner, B. G. Ledley, and L. J. Cahill, Jr., Initial results on the correlation between the magnetic and electric fields observed from the DE-2 satellite in the field-aligned current regions, Geophys. Res. Lett., 9, 985, 1982.

Vasyliunas, V. M., Mathematical models of magnetospheric convection and its coupling to the ionosphere, in Particles and Fields in the Magnetosphere, B. M. McCormac, ed., D. Reidel Publishing Company, Dordrecht-Holland, p.60, 1970.

Vasyliunas, V. M., The interrelationship of magnetospheric processes, in Earth's Magnetospheric Processes, B. M. McCormac, ed., D. Reidel Publishing Company, Dordrecht-Holland, p.29, 1972.

Winningham, J. D., F. Yasuhara, S.-I. Akasofu, and W. J. Heikkila, The latitudinal morphology of 10-eV to 10-keV electron fluxes during magnetically quiet and disturbed times in the 2100-0300 MLT sector, J. Geophys. Res., 80., 3148, 1975.

Winningham, J. D., J. L. Burch, N. Eaker, V. A. Blevins, and R. A. Hoffman, The low altitude plasma instrument (LAPI), Space Sci. Instr., 5, 465, 1981.

Wolf, R. A., Effects of ionospheric conductivity on convective flow of plasma in the magnetosphere, J. Geophys. Res., 75, 4677, 1970.

Wolf, R. A., Calculations of magnetospheric electric fields, in Magnetospheric Physics, B. M. McCormac, ed., D. Reidel Publishing Company, Dordrecht-Holland, p.167, 1974.

A STUDY OF HIGH LATITUDE CURRENT SYSTEMS DURING QUIET GEOMAGNETIC CONDITIONS USING MAGSAT DATA

J.R. Burrows, T.J. Hughes and Margaret D. Wilson

Herzberg Institute of Astrophysics, National Research Council of Canada, Ottawa, Canada K1A 0R6

Abstract. Magsat has provided precise vector measurements of the geomagnetic field which have confirmed the large scale Birkeland current pattern observed earlier by higher altitude satellites such as TRIAD and ISIS 2. In addition to detecting Birkeland currents, MAGSAT's lower orbit ($\approx$300-500 km) permits measurement of both ionospheric currents and crustal magnetic anomalies. The large scale Region 1 and 2 Birkeland currents, which are the dominant feature in the dawn and dusk sectors during moderate magnetic activity, tend to be replaced, during quiet magnetic periods, by structured, small scale current sheets throughout the polar region and by correspondingly reduced ionospheric electrojets. Several Magsat orbits across the dark, winter polar cap are quantitatively modelled. Total electrojet current intensities ranging from 1.8×10^4 A to 1.9×10^5 A are inferred along with ratios of zonal to meridional current densities ranging from 0.8 to 1.7. The extension of these modelling procedures to more structured current systems during very quiet magnetic condition is considered and a method of achieving better separation of crustal anomaly fields from external fields is proposed.

Introduction

Measurements of the geomagnetic field by the flux-gate vector magnetometer on Magsat have provided an opportunity to model fields due to the three-dimensional current system flowing in the ionosphere and low altitude magnetosphere as well as the main core field of the earth and fields due to crustal magnetic anomalies of scale size $\approx$1000 km. Various studies have produced a generally good separation of the fields due to these three sources and numerical models of the main field (Langel et al., 1981) and maps of the crustal fields (Langel et al., 1982; Coles et al., 1982) have recently been published. Models have also been developed to represent various sources of the external field such as the equatorial electrojet (Sugiura and Poros, 1969; Maeda et al., 1972), the ring current (Langel and Sweeney, 1971; Langel and Estes, 1983) and the high latitude Birkeland current system (Iijima and Potemra, 1976).

The separation of fields due to crustal anomalies from those due to external currents is least satisfactory at high latitudes where the polar ionospheric and field-aligned currents (FAC) are continually present over a wide area with scale sizes comparable to the crustal anomaly fields and with extreme spatial and temporal variations of the current intensity.

The purpose of this paper is to describe current modelling techniques and apply them to the Magsat data to achieve a better separation of external fields from crustal fields.

Orbits selected during relatively low levels of magnetic disturbance index ($K_p \leq 1+$) are analyzed in order to identify commonality and differences in the spatial distribution and scale size of currents. The data from sample orbits with large scale size and small scale size are fitted with numerical models using the cell method developed by Kisabeth (1979) and Hughes et al. (1982). Fit parameters include ionospheric meridional current densities, the average ratio of the zonal (east-west) to meridional (north-south) current densities in the electrojet and estimates of the azimuthal extent of the current system and the contribution of earth induction currents. The analysis method shows promise in determining the electrojet contribution to the field component in the direction of the main geomagnetic field using the magnetic east-directed (D) component perturbation as initial data.

Modelling Procedures

The cell method is a forward modelling procedure which starts by specifying the southward and eastward components of height-integrated ionospheric current density in 'source cells' located in an infinitesimally thin ionosphere at an altitude of 115 km. Their dimensions are $\frac{1}{2}°$ latitude by $4°$ longitude in a centred dipole coordinate frame. Current continuity for each cell is satisfied by field-aligned current flowing along dipole field lines at the cell boundaries. The resulting field is calculated by summing the contributions from all source cells in the specified current system at observation points located in the centre of 'observation cells' of the same $\frac{1}{2}° \times 4°$ size. Sets of observation cells are located on shells of constant predetermined altitude (e.g., 300 km, 350 km, 500 km). Thus one can calculate vector magnetic perturbations for a specified latitude profile along a meridian at a fixed altitude which approximates the Magsat orbit. Source cells are presently limited to dipole latitudes between $58°$ and $85°$ while field components can be calculated from $50°$ to $88°$. Induced earth currents are modelled by placing a perfect conductor at a depth of 250 km.

In the present work, the ionospheric current system is specified as follows. First, a trial latitude profile for the meridional (N-S) component is derived by scaling the D' component of the residual field at $\frac{1}{2}°$ intervals along the orbit. The FAC causing the D' perturbation is assumed to close by meridional flow in the ionosphere. Current intensity is inferred by using the infinite sheet current approximation. The zonal (E-W) current intensity is then derived by assuming a constant ratio, σ, of zonal to meridional current.

$$\sigma = \frac{j_{EASTWARD} \text{ (amps/m)}}{j_{SOUTHWARD} \text{ (amps/m)}}$$

Note that $|\sigma|$ would be the height-integrated Hall to Pedersen current ratio if the ionospheric electric field were meridional. The current system is assumed to have a uniform latitude profile as a function of longitude and to extend in the east-west direction along circles of magnetic dipole latitude. This uniformity implies that the zonal current is closed by FAC which flow only at the eastern and western ends of the current system.

Next, a trial current system length and a trial meridian for the observation cells (i.e., satellite track) are selected by comparing the data with typical model profiles. The meridian profiles of the three field components (H',D',U') are then computed and compared to the satellite perturbations observed between 50° and 88° eccentric dipole latitude. Adjustments by inspection are made to the latitude profile of the N-S current and to the parameters σ, system length, and track meridian. The profiles are recomputed until a satisfactory fit is obtained.

Data Processing Procedures

Magsat is in a sun-synchronous orbit near the dawn-dusk plane with an 83.4° retrograde inclination of its orbit plane. The data were selected from an epoch in early February 1980 when perigee was at high northern latitudes and the decaying orbital period yielded orbits with almost exactly the same ground track every second day. Pairs of passes, which were selected from magnetically quiet days, typically had longitude differences of less than 0.5°. The north polar ionosphere at E region altitudes was in darkness although the spacecraft remained sunlit at perigee (329 km altitude). Thus one would expect ionospheric currents to be principally confined to regions of ionization created by auroral precipitation along the orbit track.

The magnetometer data are filtered to 1 second resolution (≈8 km spatial resolution) and the MGST 4/81-2 model field is subtracted from it. This model is a 13th degree and order spherical harmonic expansion, epoch 1980.0 with linear time terms but no external field terms (Langel et al., 1981). The vector residuals are resolved into the eccentric dipole (ED) coordinate frame, H',D',U'

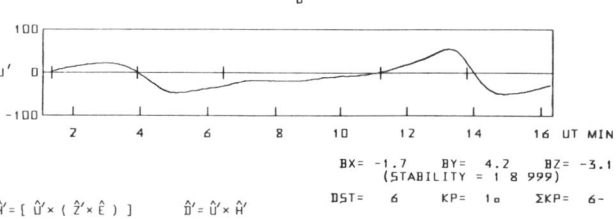

Fig. 1. Magnetic field vector residuals for the Magsat orbit on February 5, 1980 from 1301 to 1316 h UT. The component in the plane perpendicular to the local model field direction is plotted along the orbit track in a polar coordinate representation of eccentric dipole (ED) latitude and ED magnetic local time. The field aligned component, U', is plotted below.

(Wallis et al., 1982) in which U' is defined to be parallel to the model field direction, H' is orthogonal to U' and lies in the ED meridian plane, pointing northward, and D' completes the right-handed triad, being approximately easterly. A particular property of this coordinate frame is that nearby FAC give perturbations only in the H' and D' components, whereas the U' residual will be due principally to ionospheric currents with smaller contributions from more distant FAC due to the diverging dipolar geometry.

Case I - Classic Profiles

The first three figures illustrate the 'classic' latitude profiles which are observed when flying through balanced pairs of Region 1/Region 2 FAC current sheets in the evening and morning sectors in the dark winter hemisphere. Figure 1 shows the

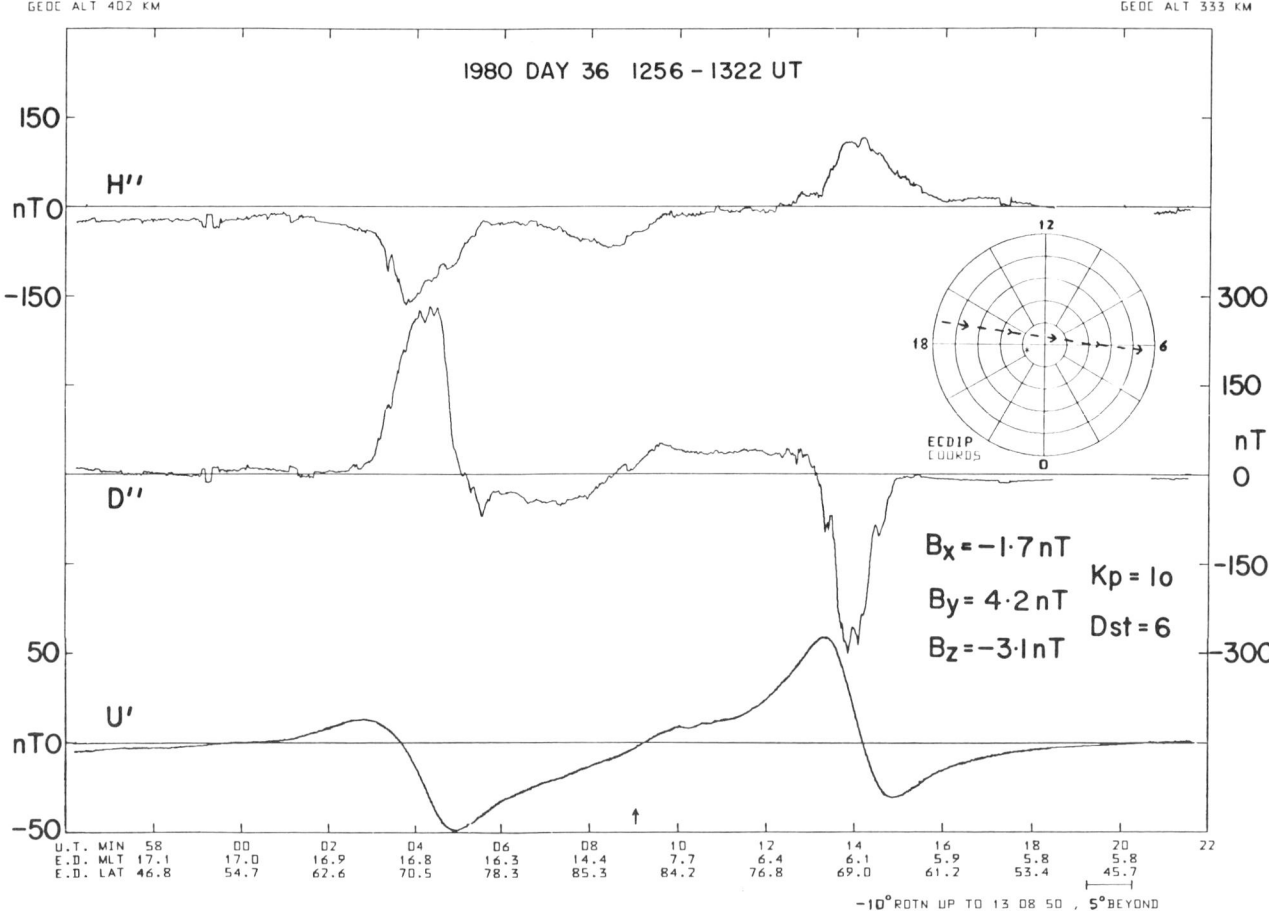

Fig. 2. Same orbit as figure 1. Components of the residual field are plotted versus UT for ED latitudes >40°. Crustal anomaly fields have been removed from the U' component.

perturbation vectors in the H'D' plane along the orbit track plotted in an ED latitude/ED MLT polar plot. The well-ordered antisunward vectors marking the FAC regions are separated by well-ordered sunward vectors in the polar cap which can be thought of as a potential field due to the return flux from the distant solenoidal FAC regions.

Figure 2 shows the three perturbation components for the same pass poleward of 40° ED latitude. In order to remove the significant contribution of crustal anomalies to the U' residual, the plotted U' is the difference between day 36, when K_p = 1o, and the much quieter pass (K_p = 0+) on day 34 along the 'same' ground track (longitude difference <0.2°). Differences are not plotted for H' and D' components, since their amplitudes in the major current regions during the pass on day 34 are ≲10% of those on the day 36 pass. The H" and D" residuals plotted in figure 2 are actually components which have been rotated in the H'D' plane so that H" is 10° east of the ED meridian on the evening side of the pole and 5° west of the ED meridian on the morning side. These rotation angles provide better alignment of D" axis with the FAC sheets as indicated by the separation of the FAC-related perturbation into the rotated D" component and the ionospheric electrojet perturbation into the rotated H" component. However, some small-scale structure attributable to FAC remains in the H' component. The smooth U' component shows the characteristic signatures of eastward and westward electrojets associated with the FAC regions.

On both days 34 and 36, ΣK_p = 6-. On day 36, there were three consecutive crossings of the northern polar cap between 13 h and 17 h UT which had latitude profiles on the morning side which were remarkably similar to figure 2. During this interval, K_p remained at 1o and the hourly-averaged IMF, measured by ISEE-3, remained stable in the away sector with $-3.1 < B_z < -1.6$ nT in GSM coordinates.

Using the procedure described in section 2, modelling is presented in detail for the second pass on day 36 at 1441-1451 UT which is at 330-343 km altitude. Figure 3 shows four profiles (points every 0.5°) fitted to the data (solid lines). The H'D'U' data components are all difference profiles

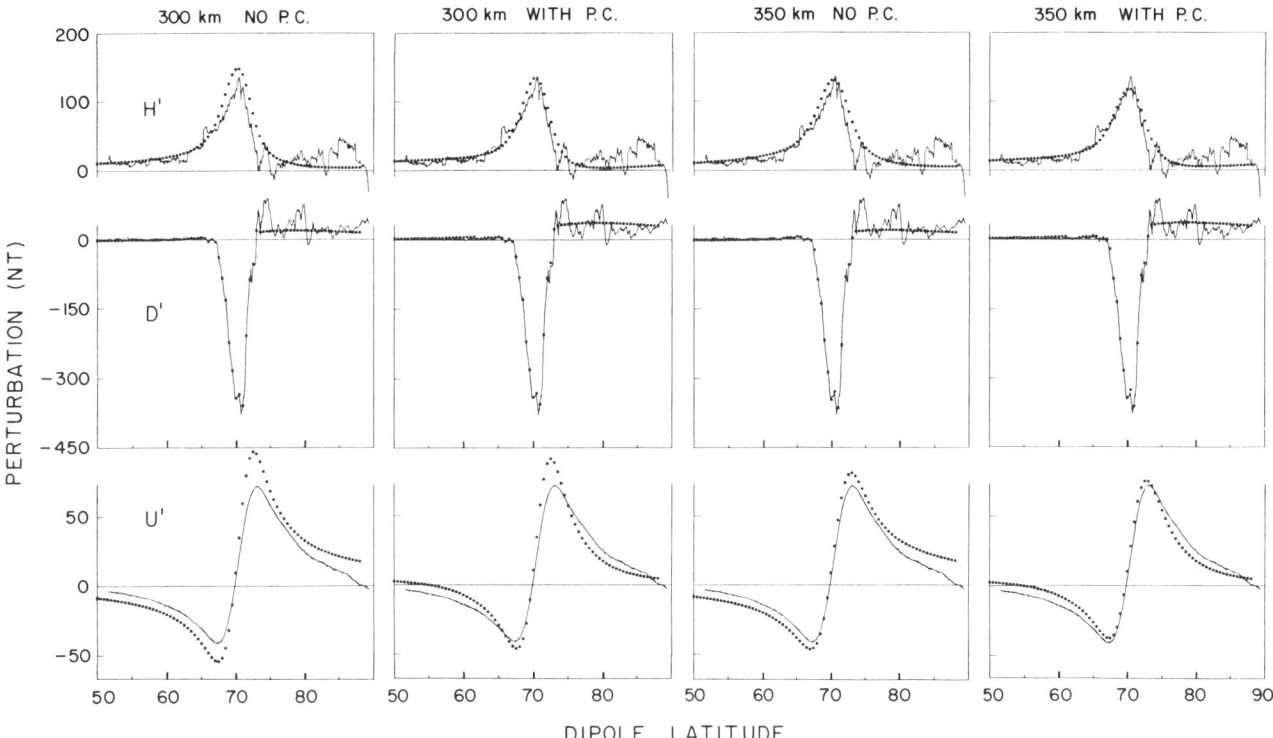

Fig. 3. Magsat residual field components with crustal anomaly fields removed (solid lines) compared to model fields (dots) computed for the same ionospheric current model under four different conditions. See text for details.

using the corresponding quiet pass from day 34 along the same ground track. The four cases all use the same model current system to provide a basis for comparison. Its parameters are: σ -1.70, longitudinal extent 124° (≈8 hours in local time). The westward electrojet extends from 67.5° to 73.5° dipole latitude with a total current of 1.92×10^5 A and a peak southward current intensity of 0.322 A/m at 71° to 71.5°, giving a D' peak of ≈370 nT. The meridian of the satellite crossing is 30° from the eastern (day) end. The four profiles differ in the observation altitude (300 or 350 km) and the presence or absence of a perfect conductor (PC) at 250 km depth to simulate induction.

Reviewing first the general agreement in shape between the computed profiles and the data, it should be noted that somewhat better agreement in peak amplitudes of H' and U' could have been obtained if optimum values of parameters (e.g., σ) had been chosen for each profile instead of keeping them unchanged. All four D' profiles give good fits to the negative perturbation and the U' profiles give excellent agreement in the location of the peaks and the cross-over slope between them. The H' profile shapes also agree well although the narrower width of the induction models seems marginally better.

The differences between computed profiles and data occur mostly in the more distant field. Comparisons are difficult for H' above 80° ED latitude because the data are contaminated by positive steps due to pitch errors of the order of 1 arc min in the attitude determination. The distant field wings of the U' profiles attenuate too slowly if no induction is included and attenuate too quickly for a 250 km deep perfect conductor. A shorter current system (e.g., 60° longitude extent) would accentuate this attenuation. Although the wing attenuation is relatively independent of the satellite meridian, the amplitude ratio of the positive and negative U' peaks is modified by it. This ratio increases monotonically from the eastern end to the western end of the westward electrojet. The choice of satellite meridian also influences the peak amplitude of H', and the distant field in the H' and D' components due to the field from the FACs which provide current continuity at either end of the electrojet. A central meridian crossing would yield minimum H' peak and positive D' wings of equal amplitude on the high and low latitude sides. A crossing displaced toward the eastern end, as selected for these profiles, makes the D' profile in the polar cap more positive and depresses the low latitude part. The field from an equally strong eastward

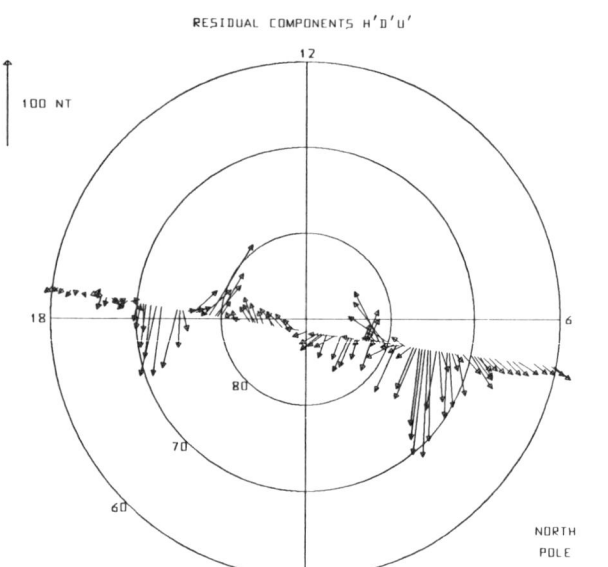
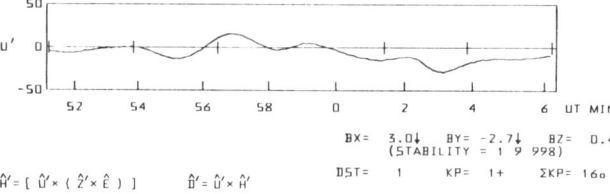

Fig. 4. Magnetic field vector residuals for a Magsat orbit on February 2, 1980 from 1651 to 1706 h UT in the same format as figure 1.

electrojet observed on the afternoon crossing has not been included in the computed profiles. It will tend to push the computed D' more positive in the polar cap and bring the U' component down to zero near the centre of the polar cap. A crossing meridian displaced toward the eastern end of the current system also unbalances the poleward and equatorward baselines of the H' component as figure 3 illustrates. An evaluation of all these factors suggests that a profile calculated for 350 km altitude with a perfect conductor located at ≈600 km depth and σ = -1.60 would give a remarkably close fit to all three components over the full range of 50° to 90° ED latitude.

Case II - Structured Profiles

The next six figures demonstrate the situation in which FAC are much weaker and more structured and the perturbation in the U' component, due to the electrojets, is comparable in magnitude to the crustal anomaly field. Figure 4 is in a format similar to figure 1 but with a reduced vector scale of 100 nT. More disordered vectors are apparent in the polar cap and the poleward borders of the electrojet region suggest subsidiary counter-flowing electrojets directed eastward on the morning side and westward in the afternoon.

Figure 5 shows the H'D'U' components poleward of 40° ED latitude for the same pass on day 33, 1646-1711 UT and its paired pass along the same ground track on day 35. Similarities in the U' components due to crustal anomalies are evident. At ED latitudes less than ≈65°, similarities are also present in the pairs of H' and D' traces, although attitude errors cause step discontinuities, most prominently in H'. The FACs in the latter pass are much more structured. They suggest that the ionospheric currents will also have small scale structure with multiple reversals, resulting in a much weaker magnetic signature in the U' component at Magsat's altitude. This structure extends across the whole polar cap with only a suggestion of stronger large scale currents at 'classical' electrojet latitudes. A typical condition for the weak, highly structured current systems is the northward B_z component of the IMF which occurs during both these passes.

Figure 6 plots the difference for all three components. Note that the vertical scales are reduced compared to figure 2. The crustal anomalies are completely cancelled in the U' component, which is least sensitive to attitude error. The smooth low latitude baselines have been joined through the perturbation region by an upward-convex curve. The convex curvature is attributed to the different ring current strengths and other distant global currents during the two passes (note the difference in Dst index in figure 5). The scale size of the perturbations due to ionospheric currents is comparable to those due to the crustal anomalies.

Detailed modelling of the U' perturbations is carried out for the morning and evening portions of the pass. Since FAC appear to occur throughout the polar cap, the modelling is somewhat handicapped by the lack of source cells at ED latitudes greater than 85°. Even if source cells were available, however, it is probable that the current sheets are not aligned along constant latitude circles, since auroral arcs in the polar cap tend to be sun-aligned.

Figure 7 shows a polar plot of the assumed model current system with a σ = -0.8 and a longitudinal extent of 60° (62°-122°). The satellite track is taken to be on the central meridian at 92°. The latitude profile of the model current extends from 67.5° to 85°. Its computed D' profile (points at 0.5° intervals) is compared to the data's D' profile. The total model current is 2.36×10^4 A in the westward electrojet and 2.30×10^4 A in the eastward current which lies poleward of the westward electrojet. The satellite altitude varied from 351 km near the pole to 329 km at 50° ED latitude.

Figures 8 a) to f) show the U' data profile compared to different calculated U' profiles using the current profile described above. The data

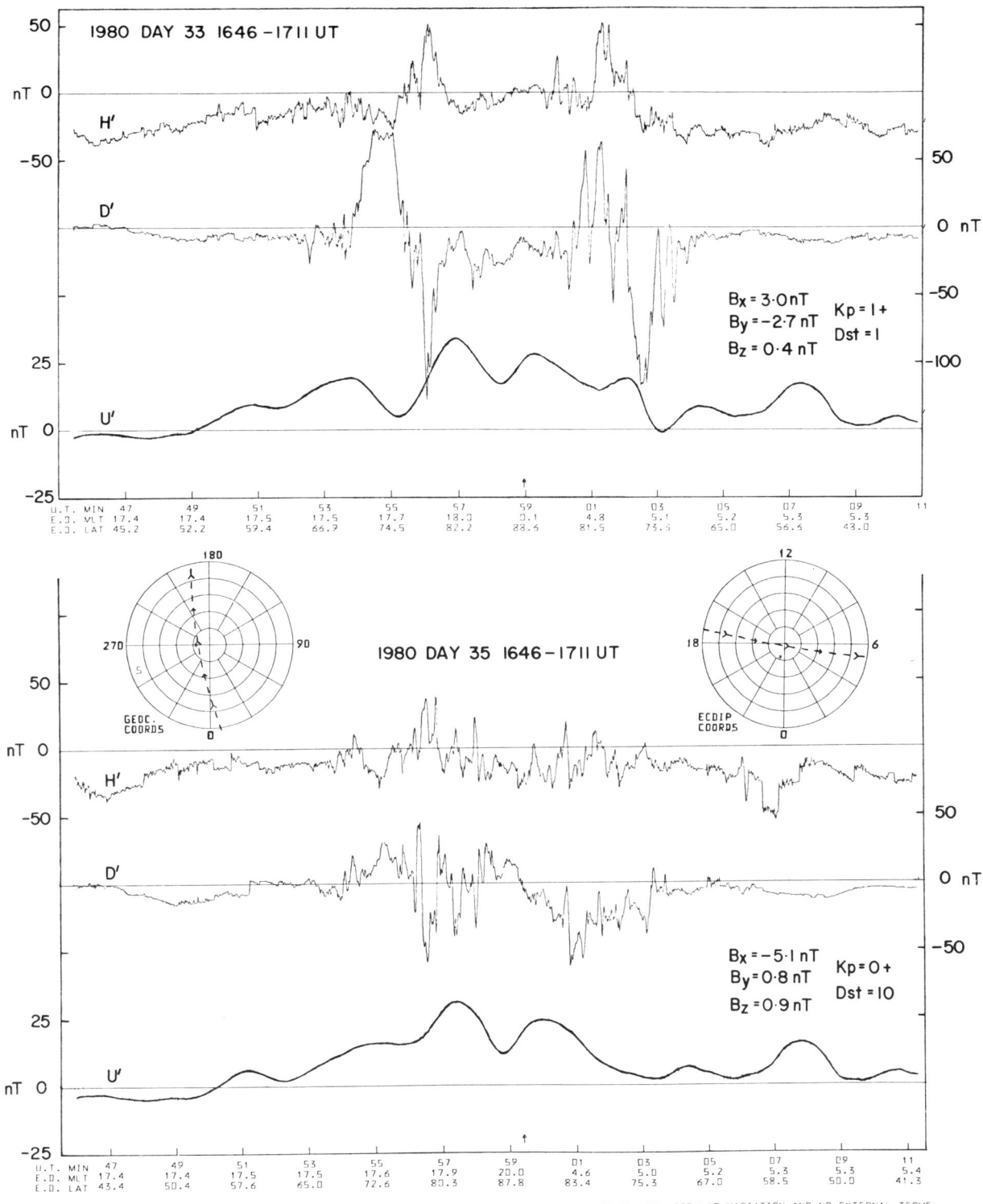

Fig. 5. Components of the residual vector magnetic field for a pair of Magsat orbits along the same ground track on February 2 and 4, 1980 at ∿17 h UT for ED latitudes >40°. Common crustal anomaly features are particularly evident at mid-latitudes in the U' component profiles.

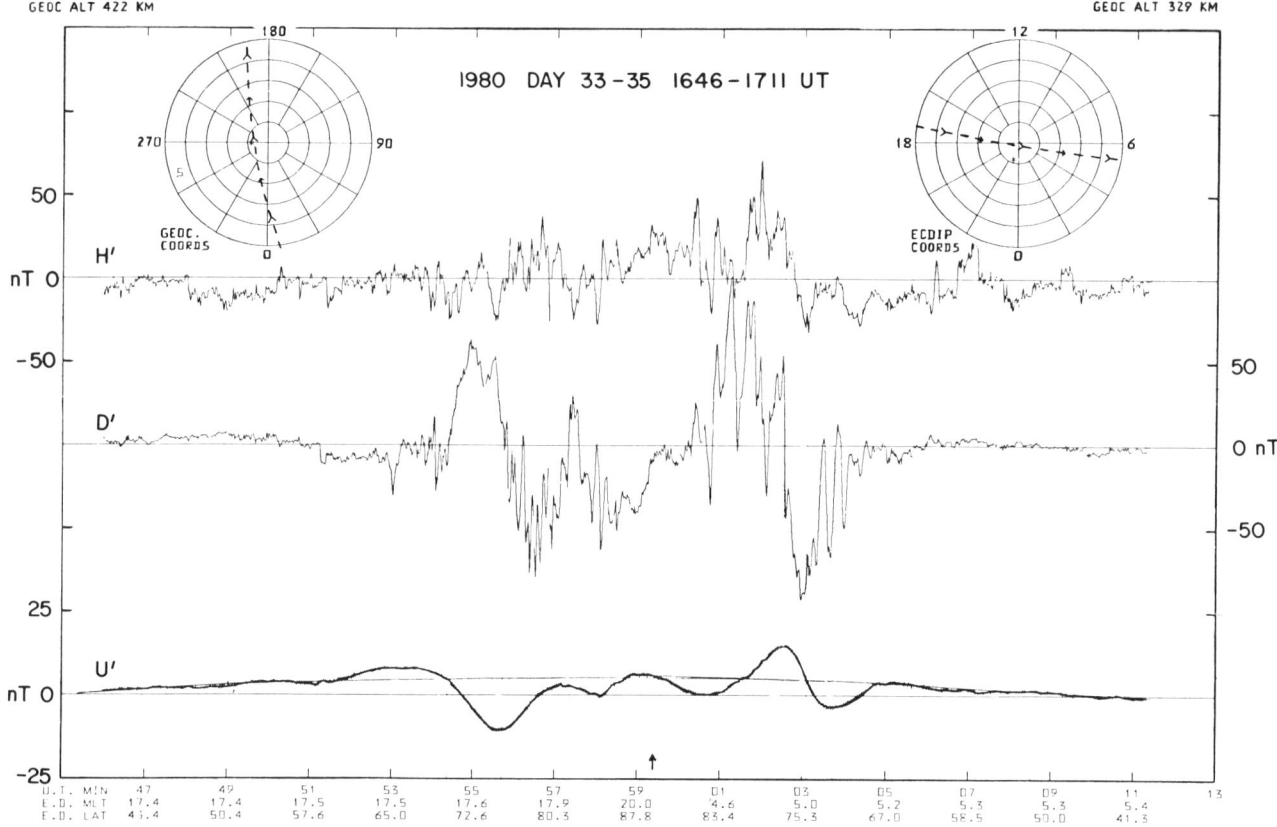

Fig. 6. The differences of magnetic field residuals for the two orbits shown in figure 5 provide an example of external fields with crustal fields removed, when magnetic activity is low.

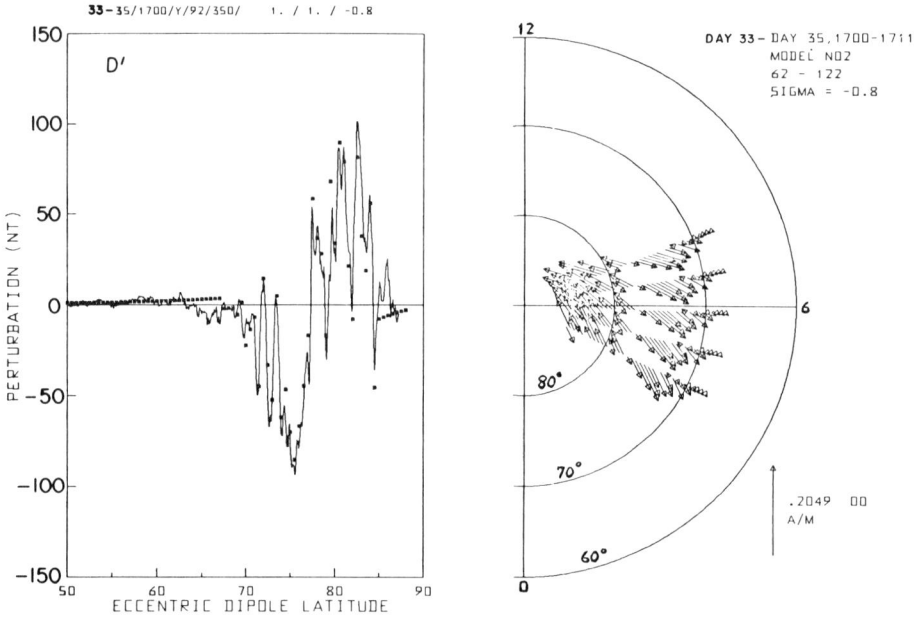

Fig. 7. The D' component data (solid line) from the morning side of figure 6 are fitted by the field computed from a three-dimensional model current (dots); the ionospheric part of this current is represented in the polar plot by sample meridional profiles at longitude intervals of 16°.

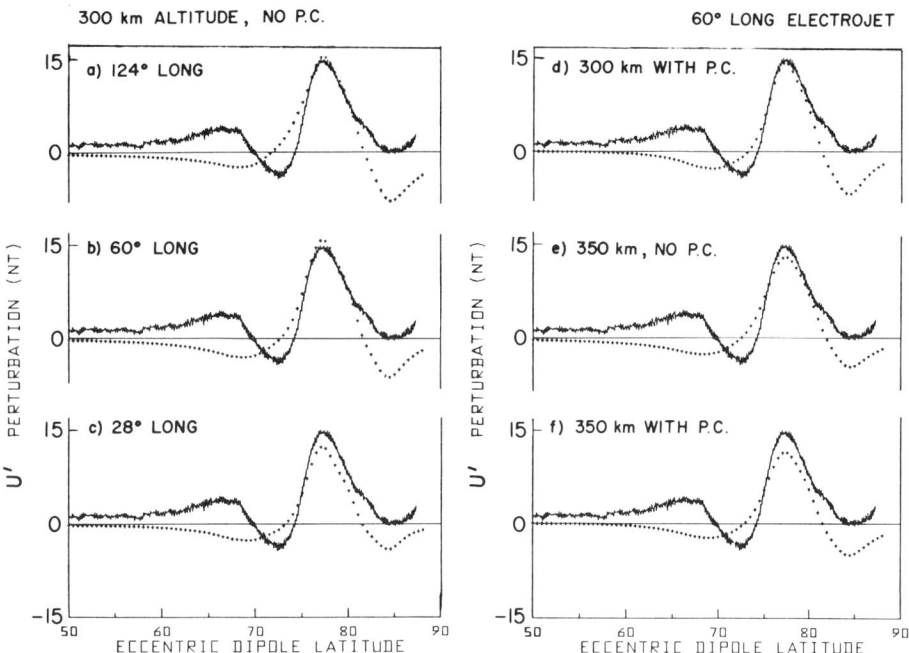

Fig. 8. The U' component data (solid lines) from the morning side of figure 6 are compared to computed fields from several variants of the current system represented in figure 7. See text for details.

profile has not been corrected for the ring current field. Subtraction of the convex baseline sketched in figure 6 would displace the U' profile about 2 nT and 5 nT more negative at 50° and 80°, respectively. Panels 8 a) to 8 c) are calculated for a satellite altitude of 300 km, a current system with no induction and various lengths of 124°, 60° and 28°, respectively. Panel 8 d) is also at 300 km for 60° long system with induction represented by a perfect conductor at 250 km depth. Panels 8 e) and 8 f) show the same 60° system at 350 km without and with induction, respectively.

One can note first some of the common features of the computed profiles. The positive perturbation at 75° to 80°, which straddles the boundary between westward and eastward electrojets, fits well in amplitude and width. The portion from 65° to 70° is never fitted well and would need an additional weaker eastward current at those latitudes to improve the fit. The most variable part of the profiles is poleward of 82° where the fit is better at 350 km than at 300 km and the longest current system (124°) gives the poorest fit. If one allows for the ring current correction, the 8 f) profile gives the best fit.

Figure 9 illustrates the best fit obtained for the evening half of the pass pair. The current system on the left has a ratio of σ = -1.1 and consists of 60° long eastward electrojet with a westward electrojet poleward of it. Its total eastward current is 1.82×10^4 A with 2.24×10^4 A of westward current. The profiles on the right are computed at 350 km along the central meridian of the current system, with induction simulated by a perfect conductor at 250 km depth. The actual satellite altitude varies from 405 km at 50° to

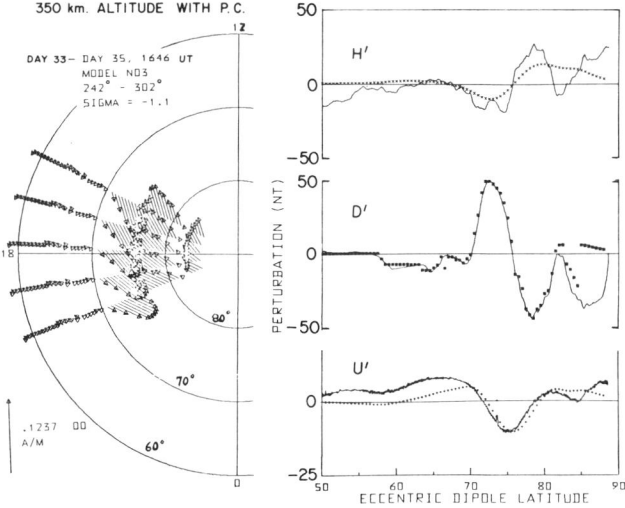

Fig. 9. Filtered data (solid lines) from the evening side of figure 6 are compared to computed field components (dots) due to a three-dimensional model current; the ionospheric part of this current model is represented in the polar plot by sample meridional profiles at longitude intervals of 16°.

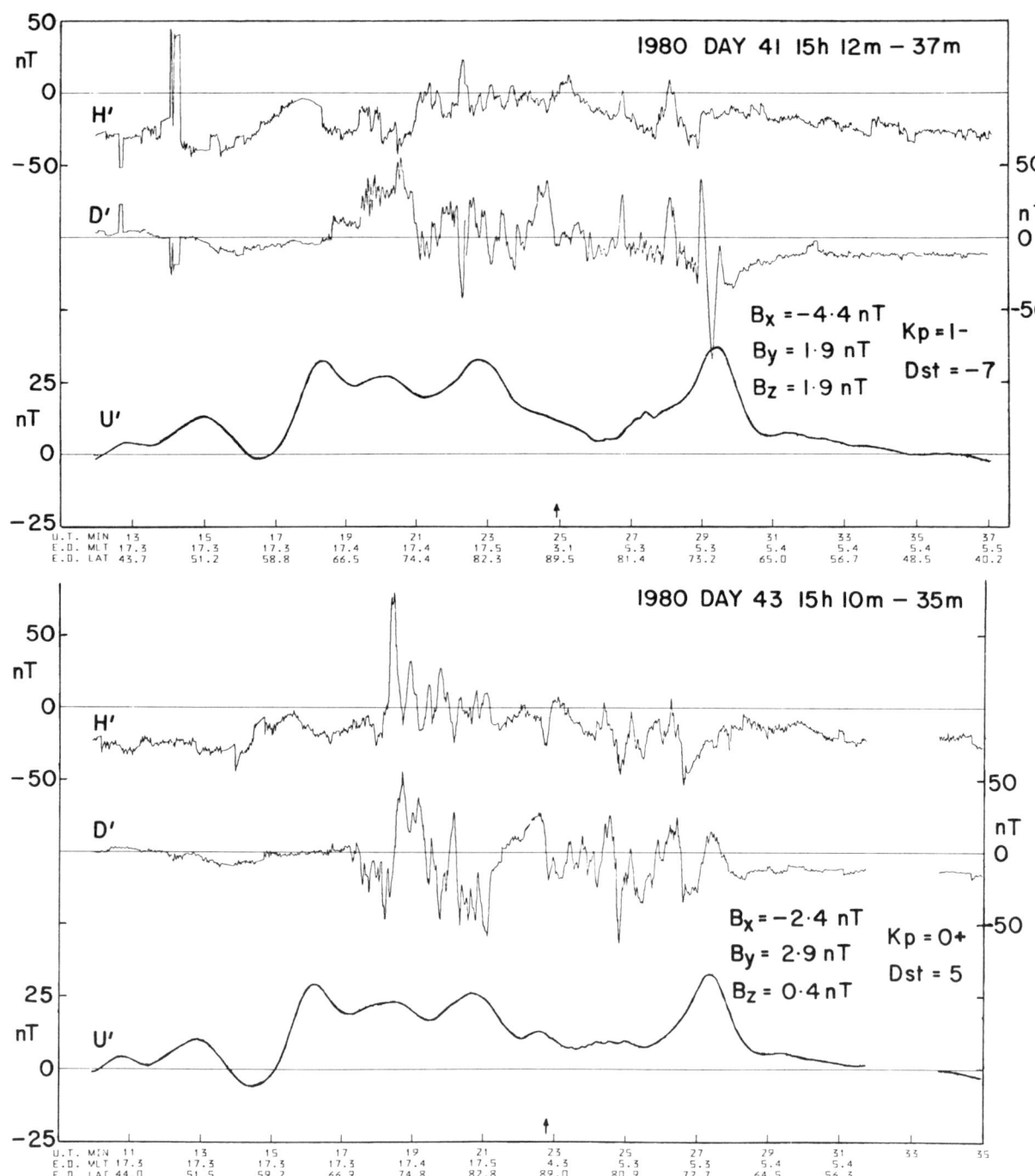

Fig. 10. Components of the residual vector magnetic field for a pair of Magsat orbits along the same ground track on February 10 and 12, 1980 at ∼15 h UT for ED latitudes >40°.

351 km near the pole. The H' and D' data profiles have been passed through a 31 s. boxcar low-pass filter in the time domain. The computed D' profile generally fits well except poleward of 85° due to the absence of source cells. The rather poor fit of the H' profile probably indicates the presence of nearby inhomogeneities in the current sheets. As in figure 8, the U' fit is good at the peak but is less satisfactory equatorward of 70°. It could be improved by a latitude variation in σ that enhanced the eastward electrojet relative to the westward electrojet. Westward current in

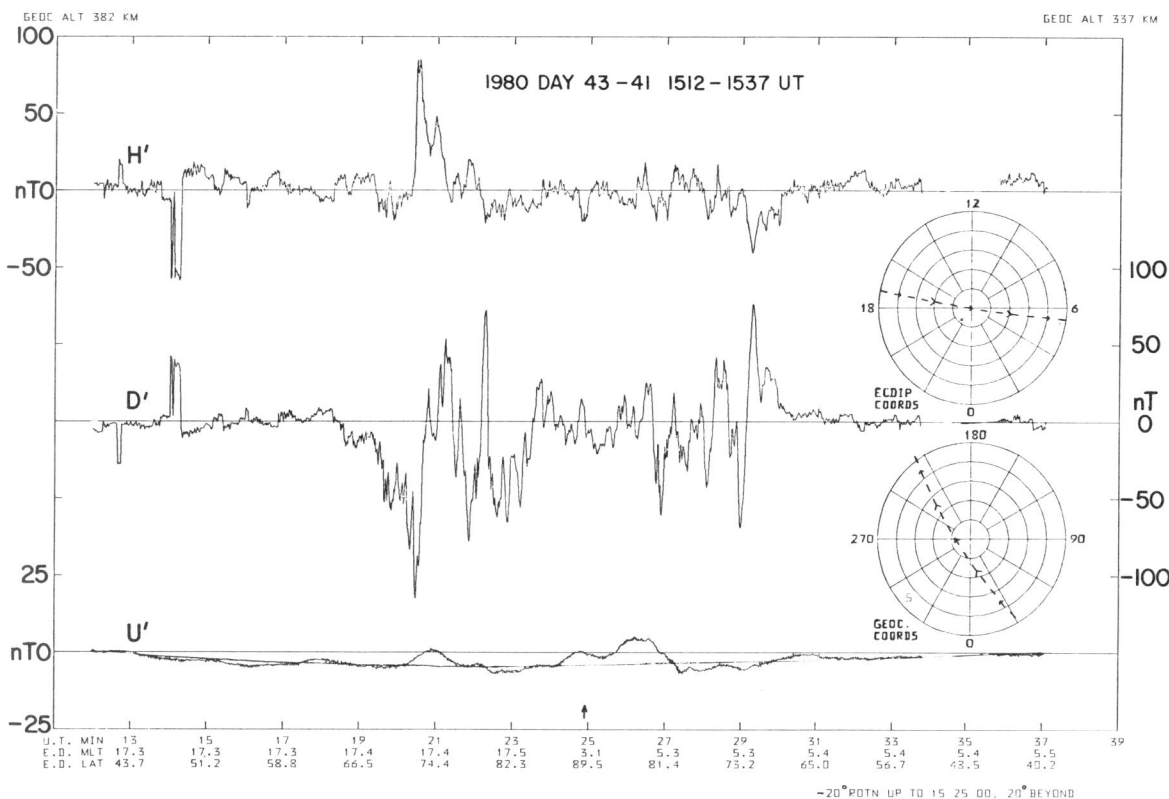

Fig. 11. The differences of magnetic field residuals for the two orbits shown in figure 10 provide an example of external fields with crustal fields removed, when magnetic activity is very low.

cells poleward of 85° would tend to improve the fit there. Applying the ring current correction would tend to displace the U' data negatively which would improve the fit at latitudes below 70° but would require a somewhat larger σ to fit the peak region around 75°.

Case III - Very Weak Current Systems

Figure 10 shows a further pair of passes when the IMF B_z was positive. Highly structured small scale perturbations extend across the polar cap with no well-developed region of large scale currents. In the U' component, the crustal anomalies dominate the fields due to electrojet currents. Figure 11 shows the three components of the difference plot, in which the H'D' plane has been rotated about the U' vector by 20° east of north on the evening side of the pole and 20° west of north on the morning side. These rotation angles resolve the majority of the structure into the D' component, indicating thereby that there is preferential alignment of current sheets along the rotated D' axis. The U' profile shows a concave-upward shape across the polar cap which is consistent with the different ring current strengths indicated by the Dst index. At present, no attempt has been made to use the modelling procedure to fit the D' and U' profiles. The structure in the U' profile is smaller in both scale and amplitude than that in figure 6.

Discussion and Conclusions

The purpose of this study is to apply quantitative modelling techniques to the Magsat data to infer parameters of the high latitude current system in the winter hemisphere near the dawn-dusk plane at times of low magnetic activity.

It is demonstrated that a useful separation of the external field from the crustal field can be achieved for a limited subset of Magsat passes by comparing pairs of passes which have nearly identical ground tracks. This separation is most necessary for the field aligned component which contains the signature of the ionospheric electrojet, since the field amplitudes from the two sources are comparable in this component. Ionospheric fields as small as ≈20 nT have been observed in data taken at 350 km altitude and modelled to yield total electrojet currents ≈2 × 10^4 A.

Confidence in the modelling procedure has been established by studying favourable cases of large scale, relatively unstructured current systems which conform to the 'classical' Region 1/Region 2 model. Good quantitative agreement with measurements has been obtained with a relatively simple current system in which the ionospheric part is confined to the latitude range bounded by the FAC sheets and characterized by a constant eastward to southward current ratio. The value of -1.7 derived for this ratio is somewhat larger than the Hall/Pedersen conductivity of 1.4 inferred by Wallis and Budzinski (1981) for the unilluminated ionosphere in the predawn sector. Their value is derived from an average spectrum of precipitating electrons for $0 \leq K_p \leq 3+$. When the current system is longitudinally uniform, subtle characteristics of the data profiles can be used to estimate the longitudinal extent of the current system and the contribution of induction currents.

In the weaker, more structured profiles analyzed in case II, one might expect more longitudinal inhomogeneities in the current sheets, which would give rise to field contributions to the H' component from both the ionosphere and the nearby FAC. Thus the poor fit to the H' component is not unexpected. Fits to the D' and U' components are sufficiently good to allow estimates of average current ratios of -1.1 and -0.8 to be derived for the evening and morning sides, respectively. These lower values suggest Hall/Pedersen conductivity ratios which are caused by softer precipitation spectra than the 'classical' case I. The eastward (westward) electrojet's latitude profile may not, in reality, duplicate all the structure inferred for the northward (southward) current. However, the use of both filtered and unfiltered D' profiles in case II illustrates that the computed U' profile is smooth and relatively insensitive to any assumed electrojet structure which has a scale size comparable to the 225 km distance between the satellite and the E region. This is expected because small scale structure in the field will be highly damped when observed from this distance.

One drawback to analyzing difference fields from a pair of passes is that the fit parameters are a mixture from the two passes unless, as in case I, the current system from one pass dominates. This is still a reasonable approximation for case II but would be invalid for case III where the two passes have comparable perturbation amplitudes. The profiles computed in figure 8 for different lengths of current system provide some insight into the effect of this parameter. The preference for a shorter current system is tentative since the effect on the profile is small.

If a weak eastward electrojet does exist on the morning side at latitudes near 70°, as suggested in section 5, it would have to be continuous in the ionosphere at longitudes near the satellite track since there is no evidence of FAC closure currents in the D' profile.

At present, modelling of weak structured current systems at high latitudes has been completed for only a few passes like case II. No highly structured pass such as case III has been attempted. The procedure does appear to provide a method of obtaining average ratios of zonal to meridional current during periods of low magnetic disturbance. It is planned to extend the study to provide a better statistical base for this parameter. If typical values of the ratio can be established from studying pairs of passes, it is proposed to use the D' profile from individual passes to estimate the contribution of external currents to the field aligned (U') component. Subtraction of this external field from the U' residual will yield a better estimate of the crustal anomaly field for a larger subset of Magsat passes in polar regions.

References

Coles, R.L., G.V. Haines, G. Jansen van Beek, A. Nandi and J.K. Walker, Magnetic anomaly maps from 40°N to 83°N derived from MAGSAT satellite data, Geophys. Res. Letters, 9, 281, 1982.

Hughes, T.J., D.D. Wallis, J.R. Burrows and M.D. Wilson, Model predictions of magnetic perturbations observed by MAGSAT in dawn-dusk orbit, Geophys. Res. Letters, 9, 357, 1982.

Iijima, T. and T.A. Potemra, The amplitude distribution of field-aligned currents at northern high latitudes observed by TRIAD, J. Geophys. Res., 81, 2165, 1976.

Kisabeth, J.L., On calculating magnetic and vector potential field due to large-scale magnetospheric current systems and induced currents in an infinitely conducting earth, in Quantitative Modelling Magnetospheric Processes, Geophysical Monograph 21, W.P. Olson, Editor, AGU, Washington, 1979.

Langel, R.A. and R.H. Estes, Large-scale, near-earth, magnetic fields from external sources and the corresponding induced internal field, NASA TM85012, April 1983.

Langel, R.A. and R.A. Sweeney, Asymmetric ring current at twilight local time, J. Geophys. Res., 76, 4420, 1971.

Langel, R.A., J. Berbert, T. Jennings and R. Horner, MAGSAT Data Processing: An Interim Report for Investigators, NASA TM82160, November, 1981.

Langel, R.A., J.D. Phillips and R.J. Horner, Initial scalar magnetic anomaly map from MAGSAT, Geophys. Res. Letters, 9, 269, 1982.

Maeda, H., T. Iyemori, T. Araki and T. Kamei, New evidence of a meridional current system in the equatorial ionosphere, Geophys. Res. Letters, 9, 337, 1982.

Sugiura, M. and D.J. Poros, An improved model equatorial electrojet with a meridional current system, J. Geophys. Res., 74, 4025, 1969.

Wallis, D.D. and E.E. Budzinski, Empirical models of height integrated conductivities, J. Geophys. Res., 86, 125, 1981.

Wallis, D.D., J.R. Burrows, T.J. Hughes and M.D. Wilson, Eccentric dipole coordinates for MAGSAT data presentation and analysis of external current effects, Geophys. Res. Letters, 9, 353, 1982.

FIELD-ALIGNED CURRENTS DURING NORTHWARD IMF

Takesi Iijima

Faculty of Science, The University of Tokyo, Tokyo 113, Japan

Abstract. Studies of field-aligned currents have advanced a great deal during the International Magnetospheric Study (IMS) period. However, there still remain many areas that need further examination. One of them is examined here and involves a case study of field-aligned currents and ionospheric currents during a period of strong northward interplanetary magnetic field (IMF). The present results provide the first evidence for reversed two-cell convection-type horizontal current at high latitudes in the ionosphere which is associated with large-scale field-aligned currents. These field-aligned currents are located poleward of region 1 currents and have the opposite flow directions of region 1 currents.

Introduction

It is now accepted that field-aligned currents are a manifestation of magnetosphere-ionosphere coupling. The coupling via field-aligned currents may be classified into two categories: (1) perfect coupling model without significant field-aligned electric field ($E_\parallel \approx 0$) and (2) imperfect coupling with a significant $E_\parallel$. For the perfect coupling an electric field disturbance ($\vec{\nabla} \cdot \vec{E}$) that occurs in the source region of the magnetosphere (or ionosphere) propagates to the terminating region of the ionosphere (or magnetosphere) and is adjusted there by transverse electric current and also by a newly created electric field disturbance that feeds back to the source region. All of these electric field disturbances are transmitted by localized shear Alfvén waves associated with field-aligned currents, similar to power transmission line theory. In this case, field-aligned currents are thought to be principally carried by cold plasma of ionospheric origin [e.g., Mallinckrodt and Carlson, 1978; Sato and Iijima, 1979]. For the imperfect coupling case, the hot plasmas that have been accelerated via field-aligned electric potential difference have become important carriers of field-aligned currents. In this case, transverse electric field disturbances that appear in the magnetosphere and the ionosphere cannot simply be connected via transmission and adjustment of shear Alfvén wave disturbances between these regions. The transverse electric field disturbances, field-aligned currents, energy transport, and ionospheric Pedersen and Hall conductivities directly depend on field-aligned electric potential difference [e.g., Lyons, 1980; Tamao, 1980; Kan and Lee, 1980].

The observations of field-aligned currents and associated phenomena (electric field, plasma convection, particles, ionospheric conductivity, auroras, plasma waves, etc.) have been reviewed by Potemra et al. [1979], Saflekos et al. [1982], Burke [1982], and a number of references therein. As described in these recent reviews, studies of field-aligned currents have advanced significantly during the International Magnetospheric Study (IMS) period. However, there still remain many problems that need further examination. These include the following three areas:

1. <u>Current Closure at Ionospheric Level</u>. Even if sources of field-aligned curents (space charge polarizations) are of magnetospheric origin, the inhomogeneity of ionospheric Pedersen and Hall conductivities produces space charge polarizations in the ionosphere. The polarization electric fields are set up so that increments of enhanced Pedersen current and Hall current are compensated partly by secondary field-aligned currents of ionospheric origin and also by Pedersen currents associated with created polarization electric fields in the ionosphere. Except outside the region of enhanced conductivity regions, the observed field-aligned currents will not simply be continuous to ionospheric Pedersen current but are to be connected with both Pedersen and Hall currents in a complex manner. Thus eastward and westward auroral electrojets are not wholly ascribable to enhanced Hall currents but also possibly include enhanced Pedersen current plus secondary Hall current, which has often been denoted as ionospheric Cowling current. Therefore, current closure between region 1 and region 2 field-aligned currents and ionospheric currents should be determined two-dimensionally, not only in geomagnetic meridian but also in longitudinal extent.

2. <u>Particle Carriers of Field-Aligned Currents and Field-Aligned Electric Potential Drop</u>. Intense field-aligned currents (> $1 \, \mu A/m^2$) flowing away from the ionosphere are believed to be primarily carried by hot plasmas (mostly electrons) that have been accelerated by field-aligned electric fields. In this case the source mechanisms of field-aligned currents must be determined in terms of the generation of transverse current divergences in the magnetosphere and the active response of the iono-

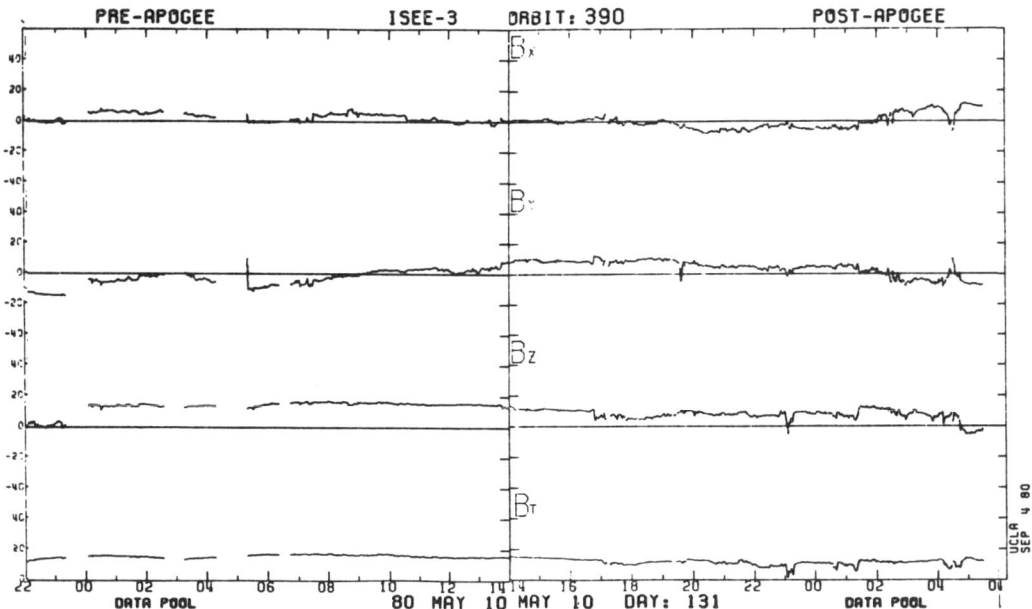

Fig. 1. Interplanetary magnetic field in solar ecliptic coordinate system acquired with the ISEE 3 satellite (provided by C. T. Russell). The prolonged strong northward IMF was observed on May 10, 1980.

spheric conductivity that modifies the magnetospheric convection via field-aligned electric potential drop (see also Greenwald [1982]).

3. <u>Field-Aligned Currents During Strong Northward Interplanetary Magnetic Field (IMF)</u>. Only a few papers have been concerned with this problem [e.g., Saflekos and Potemra, 1980; Zanetti et al., 1982]. This paper is mainly intended to report briefly a case study of field-aligned currents and ionospheric currents for a special period (from 0500 UT, May 10, 1980, to 0500 UT, May 11, 1980) associated with strong northward IMF as shown in Figure 1. (The ISEE 3 magnetic field data has been kindly provided to our laboratory by C. T. Russell.) The geomagnetic data acquired with the MAGSAT satellite at low altitudes (~ 300 km) and with high-latitude ground observatories were used.

Observed Results During Strong Northward IMF

Reversed Field-Aligned Currents Poleward of Region 1 Currents

The vector residual (measured magnetic field from MAGSAT minus MGST 4/81 model field) was separated into $\Delta B_\parallel$ and $\Delta \vec{B}_\perp$ magnetic perturbations parallel and transverse to the reference field). The $\Delta \vec{B}_\perp$ component was further decomposed into sunward (ΔS) and dusk to dawn (ΔD) components. Figure 2 shows examples of the latitude profile of geomagnetic perturbations along dusk to dawn paths in the northern hemisphere. Following the previous analysis of the MAGSAT data [Iijima et al., 1982], these examples indicate the following: as is usual with quiet days, the region 1 field-aligned currents were detected several degrees higher in comparison with disturbed days [Iijima et al., 1982]; in the top panel of Figure 2, an upward current exists at dusk, 73.2°-76.9° magnetic latitude (ML) (ΔS increase, magnitude ~ 290 nT), and a downward current exists at dawn 79.4°-73.2° (ΔS decrease, ~ 330 nT). The most striking feature is the presence of intense field-aligned currents observed poleward of, and with reversed polarities to, region 1 currents. These currents are also associated with a fairly large amplitude of $\Delta B_\parallel$. For example, in the upper panel of Figure 2 there is a downward current at 76.9°-81.3° on the postnoon side (ΔS decrease, ~ 980 nT) and an upward current at 81.3°-79.4° on the prenoon side (ΔS increase, ~ 1330 nT). In the bottom panel the weak region 2 field-aligned currents are also discernible; they are a downward current at dusk 65.3°-74.1° (ΔS decrease, ~ 180 nT) and an upward current at dawn 75.7°-66.9° (ΔS increase, ~ 150 nT).

Figure 3 shows a distribution of transverse geomagnetic disturbance vectors for eight consecutive MAGSAT orbits over the north polar region (beginning 0553:34 and ending 1634:46 UT, May 10, 1980). By using a filtering technique, disturbances with a time scale larger than 30 s (~ 2° in horizontal wavelength) were used. If this composite distribution from different paths

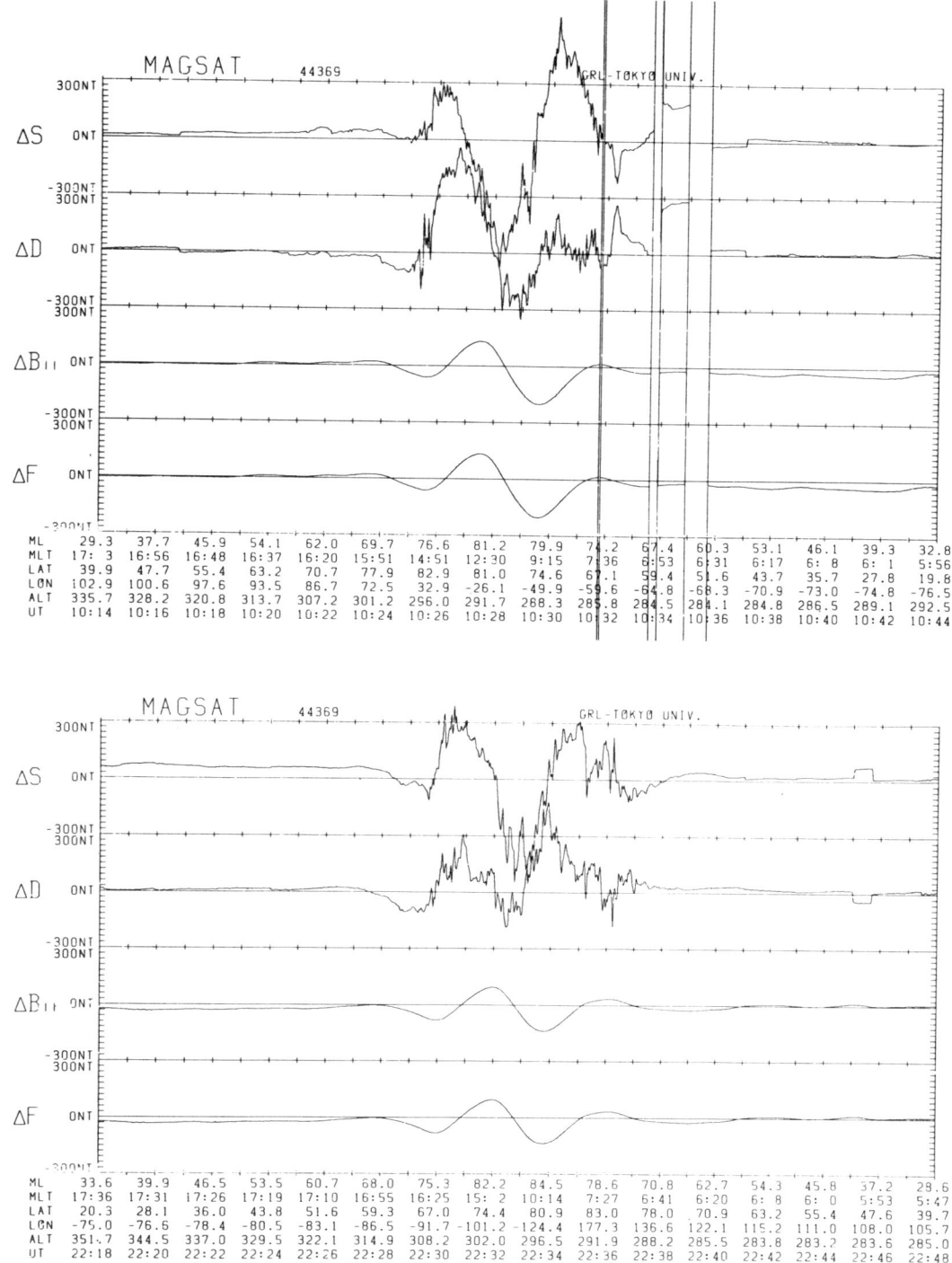

Fig. 2. Geomagnetic vector residuals separated into ΔS (perturbation transverse to the main geomagnetic field and directed toward the sun), ΔD (perturbation transverse to the main field and in the dusk to dawn direction) and $\Delta B_\parallel$ (perturbation parallel to the main field), obtained from the MAGSAT data for two passes over the northern hemisphere on a very quiet day of auroral electrojet activity, May 10, 1980, associated with strong northward IMF.

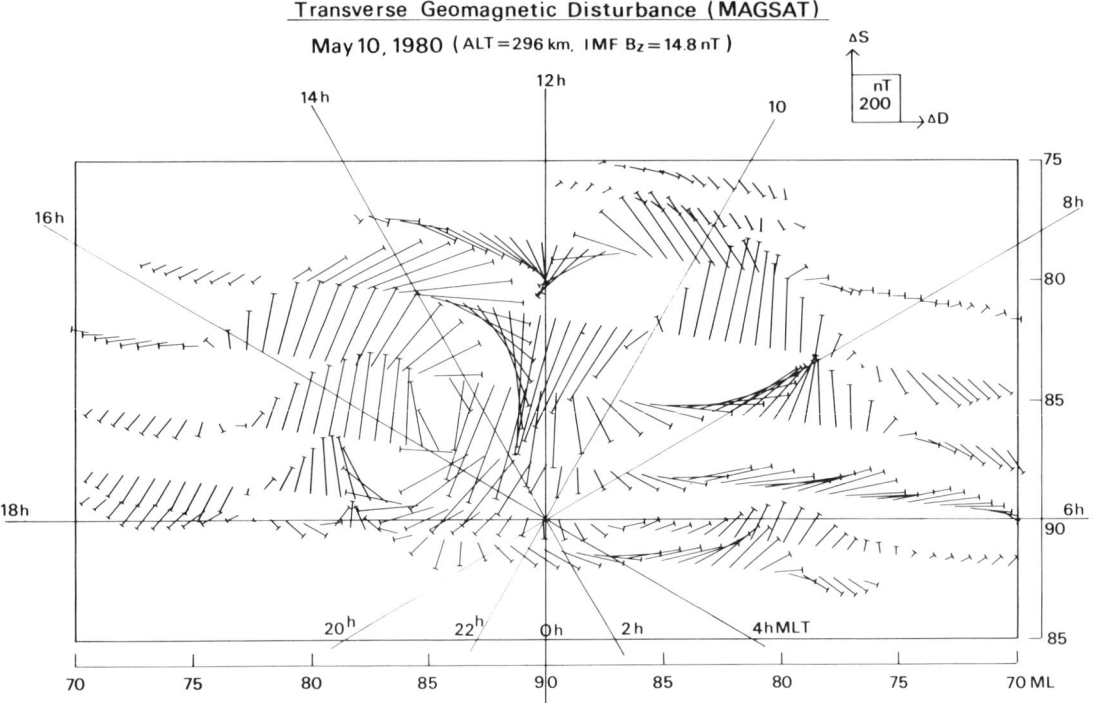

Fig. 3. A composite diagram of transverse geomagnetic disturbance vectors for eight consecutive paths of the MAGSAT (beginning 0553:34 UT and ending 1634:46 UT, May 10, 1980).

was assumed to show a spatial distribution pattern that exists relatively stably with a time scale longer than ~ 100 min (a time scale between consecutive paths), the pattern seems to conform well to a clockwise rotation of $\Delta \vec{B}_\perp$ (indicative of a downward field-aligned current) at high latitudes on the postnoon side and a counterclockwise rotation of $\Delta \vec{B}_\perp$ (indicative of an upward field-aligned current) at high latitudes on the prenoon side. A counterclockwise $\Delta \vec{B}_\perp$ rotation indicative of an upward dusk region 1 current and a clockwise $\Delta \vec{B}_\perp$ rotation indicative of a downward dawn region 1 current are also discernible at lower latitudes on the leftmost and rightmost sides of the figure, respectively. Figure 4 gives a summary of distribution of upward and downward field-aligned currents for eight consecutive paths that were used in Figure 3. In a determination of field-aligned current by $\frac{1}{\mu_0} [-\partial \Delta S/\partial d + \partial \Delta D/\partial s]$, upward and downward currents were approximated here only by the first term. It is confirmed that large-scale upward current and large-scale downward current exist at high latitudes poleward of region 1 currents in the morning and in the afternoon region, respectively, with flow direction showing opposite polarity to the region 1 current in the same MLT sector. For the present case, the intensity of these field-aligned currents at high latitudes was ~ 1.7 times the region 1 current intensity.

Reversed Convection-Type Ionospheric Currents at High Latitudes

As previously discussed by Iijima et al. [1982], $\Delta B_\parallel$ is primarily ascribable to a poloidal magnetic field disturbance associated with ionospheric currents that close two-dimensionally by themselves in the ionosphere, and then a $\Delta B_\parallel$ maximum (or minimum) manifests an ionospheric horizontal current circulating clockwise (or counterclockwise).

As seen in Figure 2 and Figure 4, the $\Delta B_\parallel$ definitely shows a primary maximum corresponding to a downward current at highest latitudes and a primary minimum to an upward current at highest latitudes and also exhibits a secondary minimum and a secondary maximum corresponding to an upward region 1 current and a downward region 1 current, respectively. These indicate, at least, the presence of a two-cell horizontal current confined to the highest latitudes, with an antisunward current over the geomagnetic pole that is opposite to the well-known ordinary current associated with magnetospheric convection.

The results of analysis of ground magnetograms at eight high-latitude stations also support the above conclusion. The overhead current arrows for the horizontal geomagnetic variation in high latitudes in Figure 5 clearly show the existence of a reversed convection current near the geomagnetic pole. This is further confirmed by

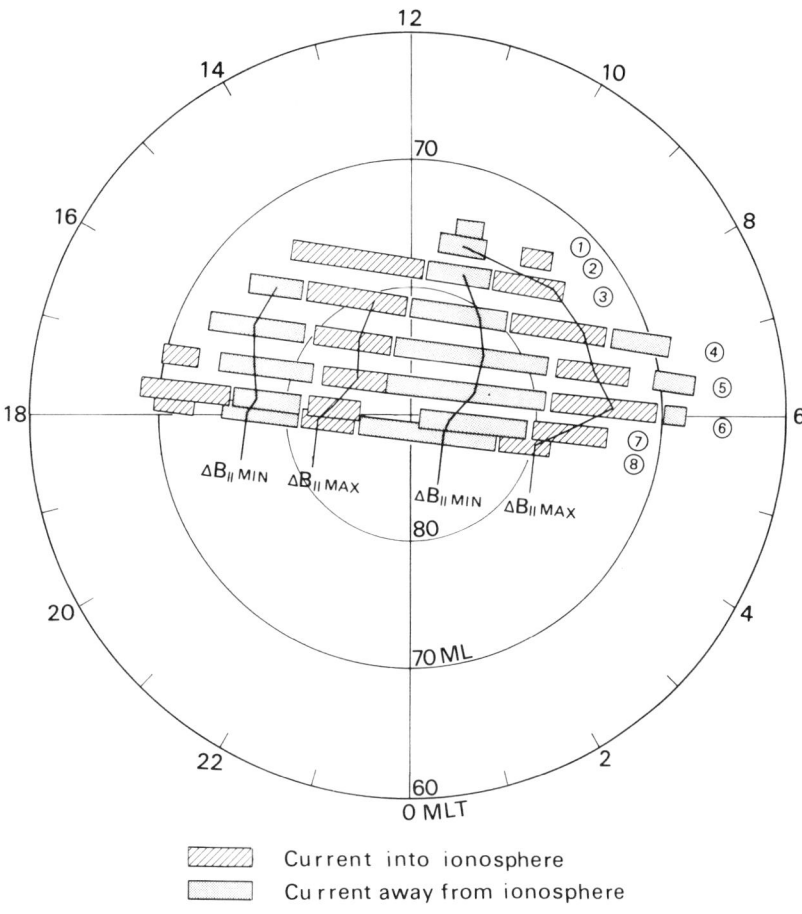

Fig. 4. Spatial distribution of field-aligned currents and maximum-minimum of $\Delta B_\parallel$ over the northern polar region, for same eight consecutive paths as in Figure 3.

Figure 6, which shows the simultaneous geomagnetic variation of ΔZ. Figure 6 shows that ΔZ is positive in the region of a counterclockwise ionospheric current and negative in the region of a clockwise ionospheric current, such as is shown in Figure 5.

The occasional appearance of a reversed convection current near the geomagnetic pole has been advocated by Maezawa [1976] and others from an analysis of ground magnetograms at high latitudes during the period of northward IMF. The present case study provides the first evidence that reversed two-cell convection-type current, which appears at the highest latitudes in the ionosphere, is associated with the large-scale field-aligned currents that appear poleward of region 1 currents and flow in directions opposite to the region 1 currents.

Conclusions

The presence of space charge polarizations and associated passive and active responses between the ionosphere and magnetosphere determine the properties of field-aligned currents. As a final remark, the forces associated with field-aligned currents will be described. When field-aligned current is flowing in a magnetic flux tube, it inevitably causes bending of field lines. Consequently, changes in Maxwell stress occur and magnetic forces appear. As long as field-aligned current persists, there must be certain forces in the magnetospheric plasma that balance the magnetic forces associated with these field-aligned currents. These forces just balance the original forces that are required for producing space charge polarizations in the

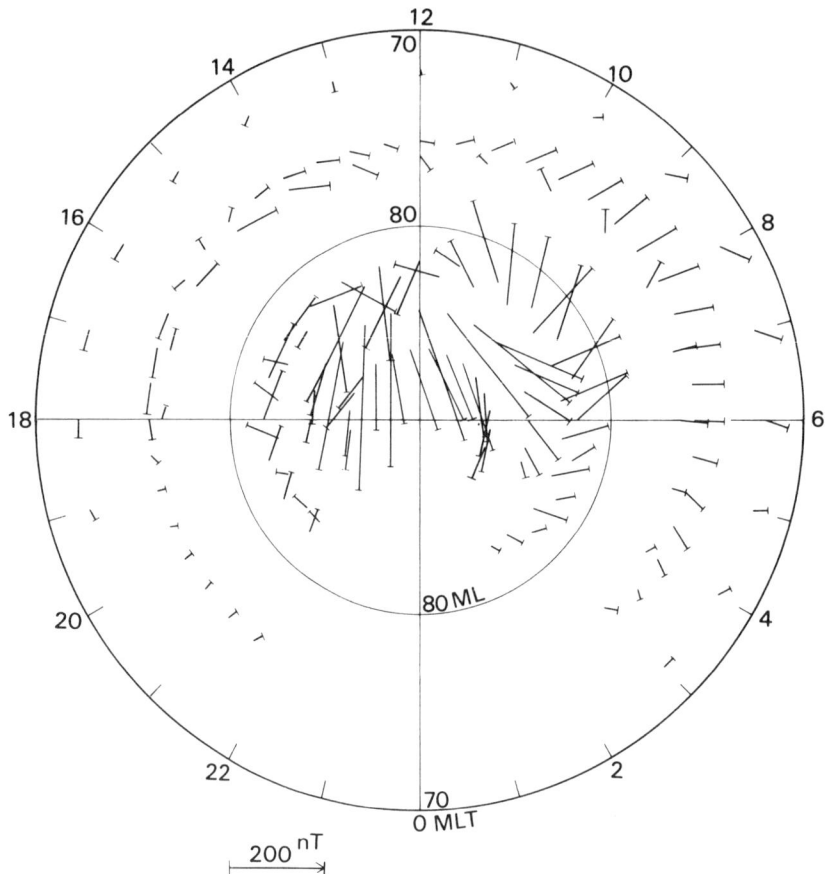

Fig. 5. Spatial distribution of equivalent current vectors derived from horizontal magnetic disturbance vectors at Alert (86.3°), Thule (84.3°), Resolute (82.6°), Mould Bay (81.9°), Cambridge Bay (76.5°), Godhaon (75.8°), and Baker Lake (72.2°) in eccentric dipole latitude. The base line values were defined as average values between 2100 and 0300 MLT at each station.

magnetospheric plasma. Various theoretical and computational studies have estimated or numerically simulated large-scale field-aligned currents, assuming the pressure gradient force and/or the inertia force in magnetospheric plasma (see for example, Greenwald, 1982). Source mechanisms of field-aligned currents must be deduced to determine the physical processes that create them. These mechanisms are not presently known although magnetic reconnection, viscous like interaction, and magnetospheric convection have been advocated.

For the case of field-aligned currents observed during strong northward IMF reported here, it is suggested that sunward forces must be exerting on the magnetospheric plasma and magnetic flux tubes at high latitudes in the ionosphere and antisunward forces must dominate at the dawn and dusk flanks of the magnetosphere. These forces will produce positive space charges at high latitudes on the postnoon side and negative space charges at high latitudes on the prenoon side of the magnetosphere. This space charge distribution is responsible for large-scale field-aligned currents that appear poleward of and with reversed flow directions of region 1 currents. At present, we have no exact knowledge of the basic configuration of the magnetosphere and its dynamic changes during strong northward IMF condition. It has been suggested that causes of field-aligned currents should be systematically determined in observational and theoretical studies by reexamining the problem of formation of the magnetosphere and its changes that depend on the IMF and solar wind conditions. Also, causes of field-aligned currents can be determined by taking acocunt of imperfect magnetosphere-ionosphere coupling that essentially involves the field-aligned electric potential

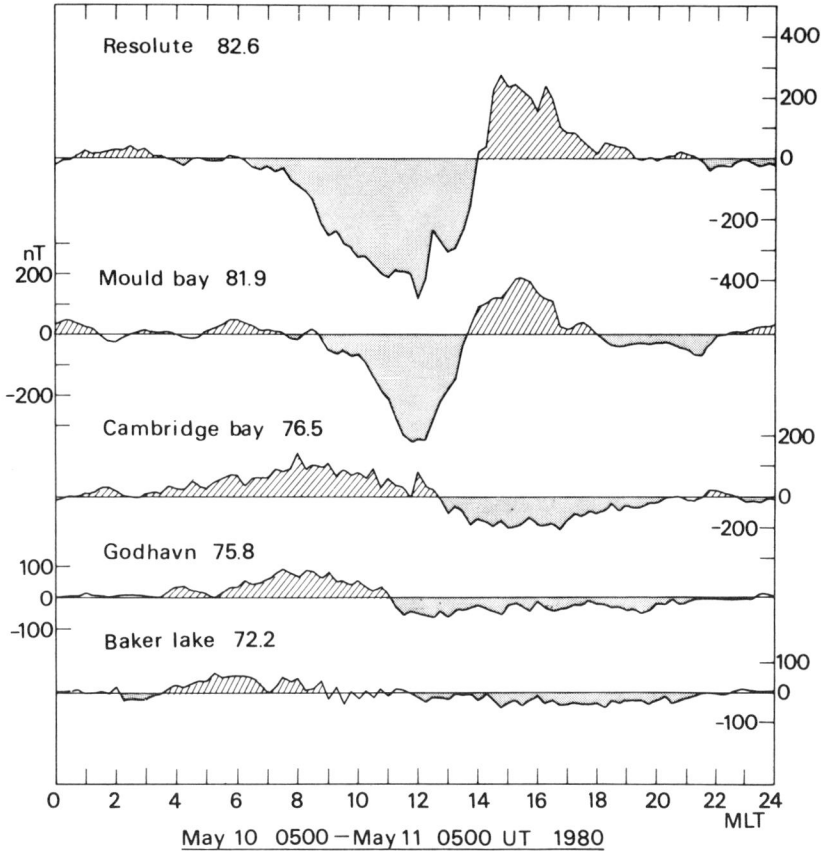

Fig. 6. Geomagnetic perturbations of ΔZ at 5 high-latitude stations.

drop and the active response of the ionosphere to the magnetospheric convection.

References

Burke, W.J., Magnetosphere-ionosphere coupling: Contributions from IMS satellite observations, Rev. Geophys. Space Phys., 20, 685-708, 1982.

Greenwald, R.A., Recent advances in magnetosphere-ionosphere coupling, Rev. Geophys. Space Phys., 20, 577-588, 1982.

Iijima, T., N. Fukushima, and R. Fujii, Transverse and parallel geomagnetic perturbations over the polar regions observed by MAGSAT, Geophys. Res. Lett., 9, 369-372, 1982.

Kan, J.R., and L.C. Lee, Theory of imperfect magnetosphere-ionosphere coupling, Geophys. Res. Lett., 7, 633-636, 1980.

Lyons, L.R., Generation of large-scale regions of auroral currents, electric potentials, and precipitation by the divergence of the convection electric field, J. Geophys. Res., 85, 17-24, 1980.

Mallinckrodt, A.J., and C.W. Carlson, Relations between transverse electric fields and field-aligned currents, J. Geophys. Res., 83, 1426-1432, 1978.

Maezawa, K., Magnetospheric convection induced by the positive and negative Z components of the interplanetary magnetic field: Quantitative analysis using polar cap magnetic records, J. Geophys. Res., 81, 2289-2303, 1976.

Potemra, T.A., T. Iijima, and N.A. Saflekos, Large-scale characteristics of Birkeland currents, in Dynamics of the Magnetosphere, pp. 165-199, edited by S.-I. Akasofu, D. Reidel, New York, 1979.

Saflekos, N.A., and T.A. Potemra, The orientation of Birkeland current sheets in the dayside polar region and its relationship to the IMF, J. Geophys. Res., 85, 1987-1994, 1980.

Saflekos, N.A., R.E. Sheehan, and R.L. Carovillano, Global nature of field-aligned currents and their relation to auroral phenomena, Rev. Geophys. Space Phys., 20, 709-734, 1982.

Sato, T., and T. Iijima, Primary sources of large-scale Birkeland currents, Space Sci. Rev., 24, 347-366, 1979.

Tamao, T., An adiabatic model of stationary field-aligned currents, preprint, Sept. 1980.

Zanetti, L.J., T.A. Potemra, J.P. Doering, J.S. Lee, J.F. Fennell, and R.A. Hoffman, Interplanetary magnetic field control of high-latitude activity on July 29, 1977, J. Geophys. Res., 87, 5963-5975, 1982.

THREE-DIMENSIONAL BIRKELAND-IONOSPHERIC CURRENT SYSTEM, DETERMINED FROM MAGSAT

L. J. Zanetti
The Johns Hopkins University Applied Physics Laboratory, Laurel, Maryland

W. Baumjohann
Max-Planck Institute für Extraterrestrsche Physik, Garching, W. Germany

T. A. Potemra and P. F. Bythrow
The Johns Hopkins University Applied Physics Laboratory, Laurel, Maryland

Abstract. The relationship of Birkeland currents and ionospheric currents has been determined from MAGSAT magnetic field observations. A two-dimensional ionospheric current analysis using a Fourier technique has been used in this study of several MAGSAT orbits on March 21, 1980. The use of the MAGSAT data allows an extensive ionospheric current system to be matched directly to the inferred Birkeland current system without the aid of a conductivity model (this is not the case with Birkeland currents deduced from surface magnetic field measurements). The results include electrojet current distributions that are generally triangular with peak intensity at the center and, in some cases, more complicated distributions associated with equally complicated Birkeland currents. The Birkeland currents generally occupy the same latitudes as do the electrojets, and the Region 1–Region 2 boundary is located near the center of the electrojet.

Introduction

The two major current systems of the near-earth environment, the Birkeland or field-aligned currents and the electrojet currents, are joined in the ionosphere. An increasing amount of effort is being expended in order to understand the electrical connection that takes place in the auroral zone regions. Birkeland current patterns are calculated from equivalent ionospheric current systems that are inferred from ground-based magnetometer data [Kamide et al., 1981; Kamide et al., 1982; Kisabeth, 1979; Hughes and Rostoker, 1979; Hughes et al., 1982; Akasofu and Ahn, 1981; Inhester et al., 1981; Baumjohann et al., 1980; and others]. Birkeland currents have also been matched to electrojets for specific events as well as statistical comparisons [see Potemra, 1979, and references therein]. The purpose of this paper is to communicate new information on this problem; whereas most of the above methods require the input of a conductivity model, the following analysis is free of such constraints.

At our disposal are measurements of disturbances in the magnetic field resulting from both Birkeland currents and ionospheric Hall currents from the same satellite, i.e., MAGSAT. The following will describe briefly the technique and limitations of this preliminary inference of the three-dimensional ionospheric-Birkeland current system. The resulting ionospheric current pattern is at present a composite of two-dimensional analyses on individual MAGSAT orbits as they cover the local times of the auroral oval.

Data and Results

Two features have made the MAGSAT magnetometer data set unique among the huge bank of magnetic field data: altitude and attitude. The altitude of MAGSAT was between 300 and 500 km above the earth, within the ionosphere, such that the magnetometer responded to horizontal ionospheric currents. Since these magnetic disturbances result from distant currents, they appear as slowly varying functions. Oscillations in the attitude of a satellite can introduce similar variations in the response of the magnetometer sensors, making it difficult to assess the actual cause. The attitude of the MAGSAT magnetometer sensor was measured to within 15 seconds of arc (equivalent to 2.5 nT in a 35,000 nT field) by using a combination of star cameras, precision sun sensors, and laser ranging for the boom [Acuna, 1980; Fountain et al., 1980; Heffernan et al., 1980; and Mobley et al., 1980] and the implementation of the resulting information [Langel et al., 1982].

Figure 1 briefly illustrates the technique used to analyze effects in the MAGSAT data caused by ionospheric currents. Detailed discussions of this method can be found in Zanetti et al. [1983] with references to Mersmann et al. [1979]. The view of Figure 1a is from the magnetotail toward the sun. Two electrojet Hall current distributions flowing tailward, at about 110 km above the earth's surface, represent the eastward electrojet at dusk and the westward electrojet at dawn. These current systems produce distant magnetic disturbances in the radial and northward directions, as illustrated in Figure 1a. The method is summarized as follows:

1. Project the MAGSAT disturbance field data onto a Cartesian coordinate system tangent to the earth at the invariant pole at an altitude of 110 km (z = 0).
2. Fourier analyze the B_z disturbance profile, which is presumably related only to the ionospheric current distribution. In most of the examples, the radial component (B_z) is assumed to be free from transverse disturbances, the angle to the magnetic field being less than 20°.
3. Produce the disturbance profile (B_x) at MAGSAT altitudes. This profile represents the one resulting from the ionospheric current distribution.
4. Check to see if the B_x component agrees with the measured B_x MAGSAT data. If it does not, rotate the coordinate system about the z-axis to x′ and y′ and repeat steps 2 and 3 (Figure 1b).
5. The $B_{y'}$ disturbance is due only to the Birkeland-Pedersen poloidal current loop; thus, B_y is independent of height and a measure of the total current in the loop, assuming that the Birkeland currents are the divergence of the Pedersen currents.
6. Calculate the B_x at 110 km (z = 0, the assumed source current location); this is directly convertible to the ionospheric Hall current distribution $J_y(x')$.

The three panels of Figure 2 illustrate the steps in the Birkeland-ionospheric current analysis procedure, this example being one of the orbits contributing to the final two-dimensional horizontal com-

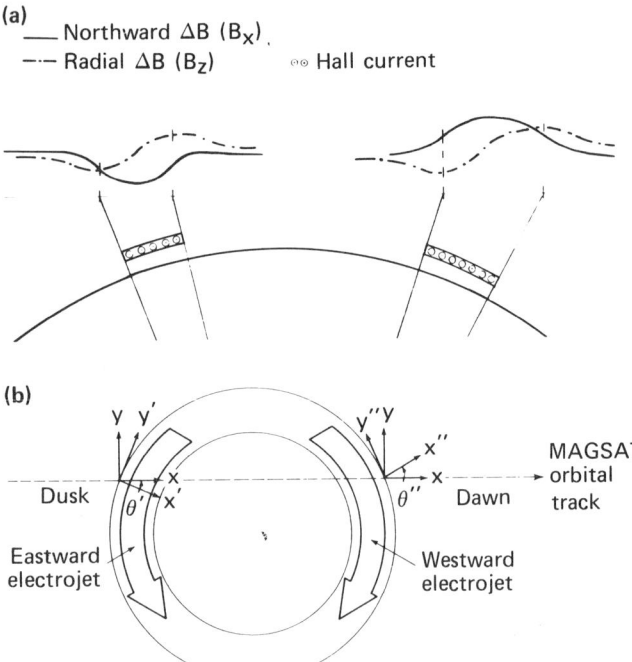

Fig. 1. Hall current analysis schematic. (a) If $+\hat{x}$ is the direction of the MAGSAT orbit over the northern polar region, the dot-dash trace will be the radial magnetic field disturbance (ΔB) and the solid line the north ΔB resulting from ionospheric Hall electrojet currents shown ≈ 110 km above the earth's surface. (b) View looking down on the northern hemisphere of the orbital track across the auroral oval. The analysis coordinate system is adjusted so that it is parallel to the local ionospheric current.

posite of ionospheric Hall currents. Northern and southern data have been combined during this unusually steady but active (Kp ≈ 3 to 4) equinox day. Figure 2a graphically positions the MAGSAT orbit at 19:40 UT on March 21, 1980 (day 81) relative to a magnetic reference frame. The diagram represents the statistical pattern of the Region 1–Region 2 Birkeland currents [Iijima and Potemra, 1978] plotted on an invariant latitude (INV), magnetic local time (MLT) grid. The MAGSAT orbit goes from dusk to dawn, and the ground stations Roberval (R), Chatanika (C), Thule (T), and Kiruna (K) as well as the north geographic pole (P) are indicated. The slanted tick marks correspond to the abscissa information in Figure 2b: universal time (UT), geographic latitude (LAT), geographic longitude (LONG), MLT, INV, and altitude (ALT). The three plotted data sets are the geographic south, east, and radial magnetic field components with the MAGSAT 4/81 [Langel et al., 1981] internal field model subtracted. Zero disturbance is in the middle of each section of Figure 2b with full scales of ±1000 nT. One may notice signatures in the radial disturbance field similar to the schematics of Figure 1a, indicating distant ionospheric Hall currents. The east-west disturbances are primarily due to Birkeland or field-aligned currents and are more structured because of the penetration of local currents. Note the expansion of the large-scale Birkeland current regions relative to the Iijima and Potemra [1978] statistical pattern; the dusk Region 2 downward current is evident from about 19:49 to 19:51 UT and Region 1 from 19:51 to about 19:53 UT. Similarly, the dawn Region 1 downward current starts after 20:02 to 20:03 UT, from which point Region 2 continues to about 20:04:30 UT.

Figure 2c contains the results of the Hall current analysis as described above for the MAGSAT orbit starting at 19:40 UT. The abscissa is now the MAGSAT orbit projected with a conformal mapping onto a Cartesian coordinate system expressed in kilometers from the invariant pole; the negative direction is toward dusk. The solid lines are the y′ or y″ (Figure 1b) disturbances resulting from Birkeland currents and are plotted at the same resolution as the Hall current analysis (250 km). The solid line is very similar to the east disturbance graph of Figure 2b although there is no obvious mapping between the two abscissas of 2b and 2c. The dotted line is a plot of the magnetic disturbance at 110 km calculated from the Hall current analysis, assuming this altitude is just outside the location of the source current distribution. Thus, this quantity is a measure of Hall current density from $J_{Hall} = (2/\mu_o)B_x$. Note the general coincidence of the Birkeland and Hall electrojet current distributions, particularly the coincidence of their edges. The equal amplitudes of ΔB in Figure 2c indicate (assuming that Birkeland currents are closed meridionally by Pedersen currents) a Hall to Pedersen conductivity ratio of 2.0. This ratio is about 1.0 at dawn and 1.5 at dusk in this example. The sign of ΔB in Figure 2c (in fact both c panels of this format) is irrelevant; the dotted trace at dusk indicates the eastward electrojet current distribution whereas the dotted trace at dawn indicates the westward electrojet. From -2000 to $+2000$ km, there appear to be some small Hall currents across the polar cap as well as Pedersen currents connecting the net poleward Region 1 currents.

Figure 3 illustrates results from a pass over the southern hemisphere, sampling a very interesting region in the early morning hours. The format is the same as that in Figure 2 except that Figure 3c is plotted from dusk to dawn (as are Figures 2b and 2c) and Figure 3b is from dawn to dusk. Figure 3a shows the orbital track across the southern auroral zone at about 4:00 MLT and again at about 20:00 MLT. Because of the excessive dip angle (>30°) at these latitudes in the southern hemisphere, eccentric dipole coordinates have been used. In the parallel component one may note what appears to be a complex addition of two "sine waves" between 6:51 and 6:56 UT. The Hall current analysis, depicted in Figure 3c, indicates two peaks in the westward electrojet distribution across the early morning auroral zone. The dotted trace in Figure 3c has two maxima centered at +2100 and +3000 km. Similar observations have been made previously [McDiarmid and Harris, 1976] during a substorm time. In a brief, limited survey of the MAGSAT data set, such a signature in the radial component is not uncommon during times of fairly active Kp. The resolution of the Hall current analysis using the radial component is limited to 800 km wavelengths, which translates into 400 km structures in the electrojet current. This limitation is due to high frequency power in the Fourier analysis that grows exponentially (e^{-kz}, where k is the wave number) as we continue toward the source [see Zanetti et al., 1983, for details]. The corresponding Birkeland current distributions are indicated by solid lines in Figure 3c from +1600 to 3400 km in the dawn sector. The resolution of these data is 250 km as is the Hall current plot, and this belies the finer structure evident in the east and south component graphs of Figure 3b. Once again the boundaries of the electrojet and Birkeland currents coincide. In fact, it appears as if both westward electrojets have individual Birkeland current sheets, a Region 2 starting at the equatorward edge (+3400 to 3000 km), another Region 2 starting at +2400 to 2000 km, and a Region 1 continuing poleward from this point. The remarkable similarities presented above might even suggest finer scale structure in the westward electrojets commensurate with the structure in the Birkeland currents. The dusk ionospheric-magnetospheric current system is similar to that discussed in Figure 2: coincident Birkeland and eastward electrojet Hall currents, the peak Hall current intensity corresponding to the boundary between Region 1 and Region 2, and a Hall to Pedersen conductivity ratio of 1.2.

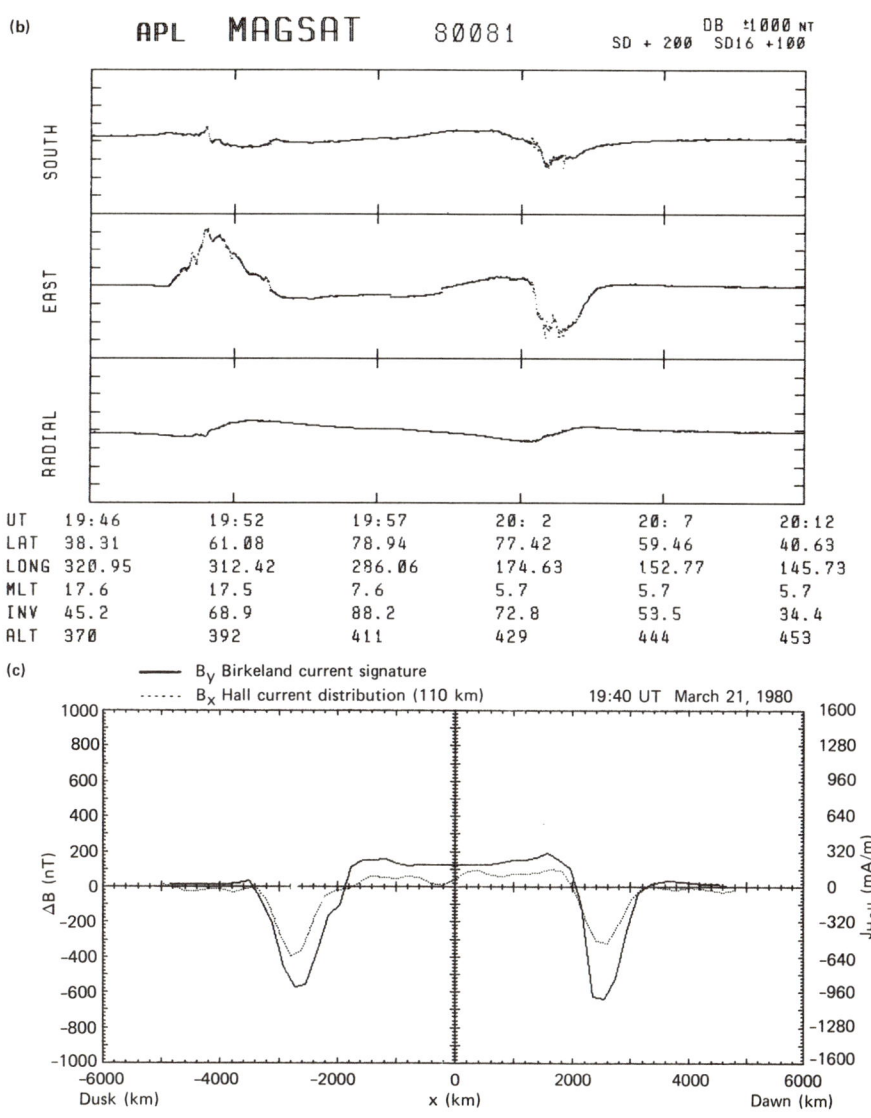

Fig. 2. Three steps in the two-dimensional current analysis procedure, for example, on March 21, 1980 at 19:40 UT. (a) Locate the orbit relative to geomagnetic coordinates and the statistical pattern of Birkeland currents [Iijima and Potemra, 1978]. (b) Magnetic disturbance data (geographic south, east, and radial) plotted versus time, etc. Find electrojet signatures in the radial component (see Figure 1a). (c) Magnetic field disturbances versus Cartesian coordinate x centered on the invariant pole. Reproduced is the Birkeland current signature (ΔB in the y' direction) as a solid line; the dotted line represents ΔB in the x' direction at ≈ 110 km resulting from a Hall current distribution.

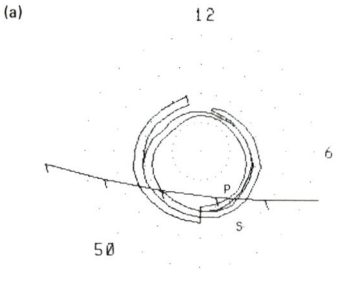

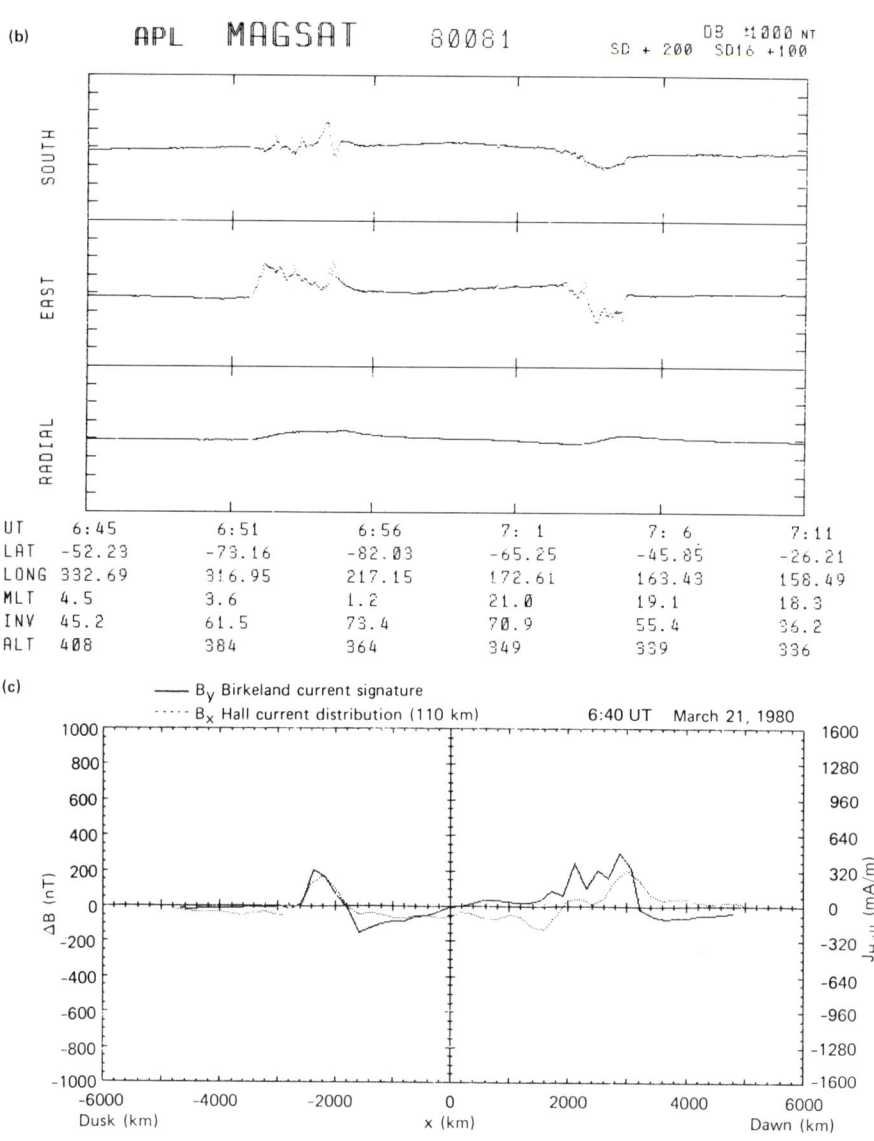

Fig. 3. Same as Figure 2 except that this example is from the southern hemisphere at 6:40 UT. Eccentric dipole coordinates have been used: parallel to B, north (ED North), and east (ED East). Note that the sign of the disturbances in panel (c) is irrelevant; the eastward electrojet is evident at dusk ($-x$) and the westward electrojet is evident at dawn ($+x$).

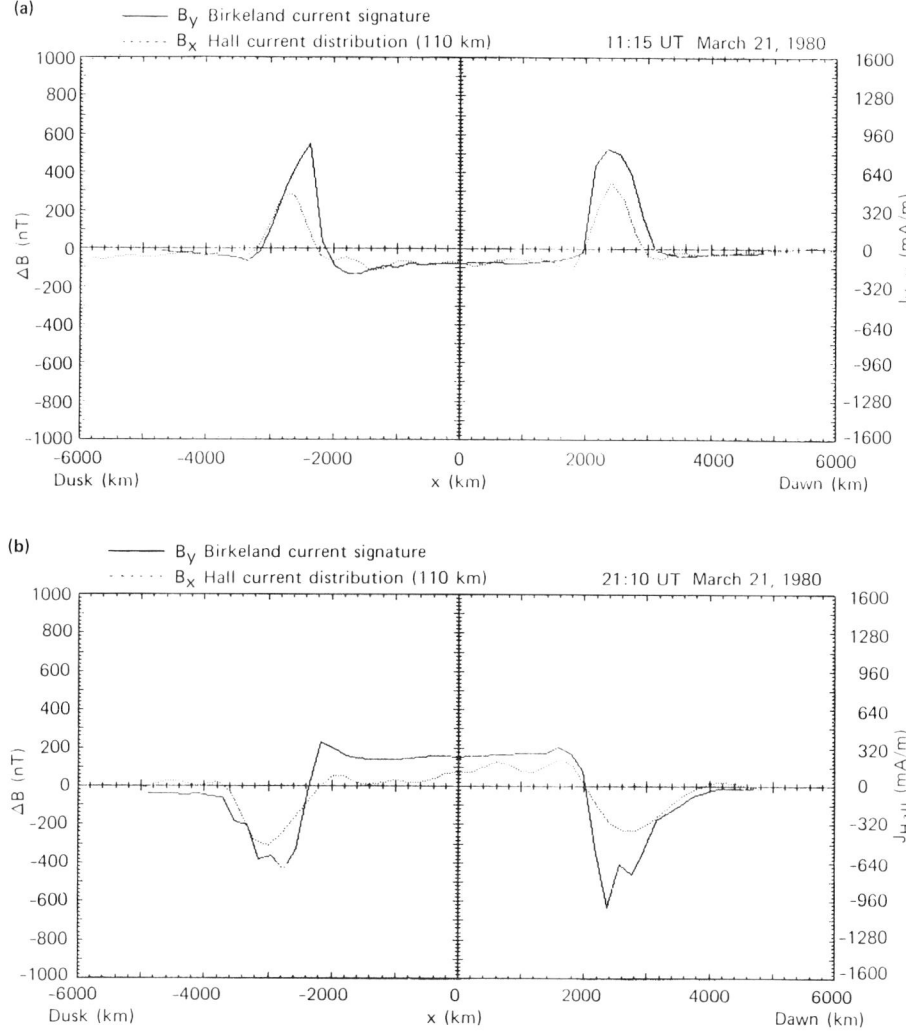

Fig. 4. Comparison of Birkeland current signatures (solid line) and Hall current signatures calculated for 110 km (dotted line) for two cases: (a) 11:15 UT, March 21, 1980, and (b) 21:10 UT, March 21, 1980.

A preliminary attempt at analyzing the horizontal plane of the ionosphere in terms of current distributions is displayed in Plate 1. The background grid of both panels is MLT versus INV, with noon MLT toward the top and 6:00 MLT toward the right. Two circles of INV (50° and 70°) are indicated. On this graph, the orbital tracks are plotted with two sets of information: Plate 1a – Birkeland current distributions, and Plate 1b – ionospheric Hall current distribution. Along each orbital track in Plate 1a, the intensity of color signifies the gradient of the geomagnetic east-west magnetic field disturbance as a first approximation to the field-aligned currents. From black to the brightest blue we indicate the intensity of Birkeland currents into the ionosphere, that is Region 2 at dusk and Region 1 at dawn. Equivalently, the Birkeland currents away from the ionosphere are indicated by the intensity of red-yellow; these would be the Region 1 at dusk and the Region 2 at dawn. One must be cautioned at the outset that this figure is a collection of data taken over a good fraction of a day; thus, there certainly may be time effects in what appears to be a strictly spatial pattern. March 21, 1980 was a particularly steady day and was also near equinox, and we have combined north and south polar data as shown in the composite Plates 1a and 1b. The 11° offset of the magnetic pole allows such coverage in magnetic coordinates from an inertially fixed orbit. Plate 1a shows the general Region 1-Region 2 statistical Birkeland current pattern put forth by Iijima and Potemra [1978] although expanded because of higher magnetic activity.

Plate 1b displays a collection of two-dimensional Hall current analyses equivalent to the dotted trace in Figure 2c or 3c. The amplitude of such a trace (Hall current) is scaled from dark blue to white regardless of sign. Thus, the eastward electrojet is centered at 1800 and just below 70° INV whereas the westward electrojet is near 70° INV at dawn and the current distribution is doubly peaked from about 4:00 MLT toward midnight. Note the general coincidence of Plates 1a and 1b. The lower Hall current intensity as one deviates from the dawn-dusk diameter of this auroral oval toward the sun or toward midnight may not be real. This effect may be a systematic problem from the two-dimensional analysis; that is, the method depends on variations in the radial component of ΔB that disappear as the orbits get progressively more parallel to the ionospheric current

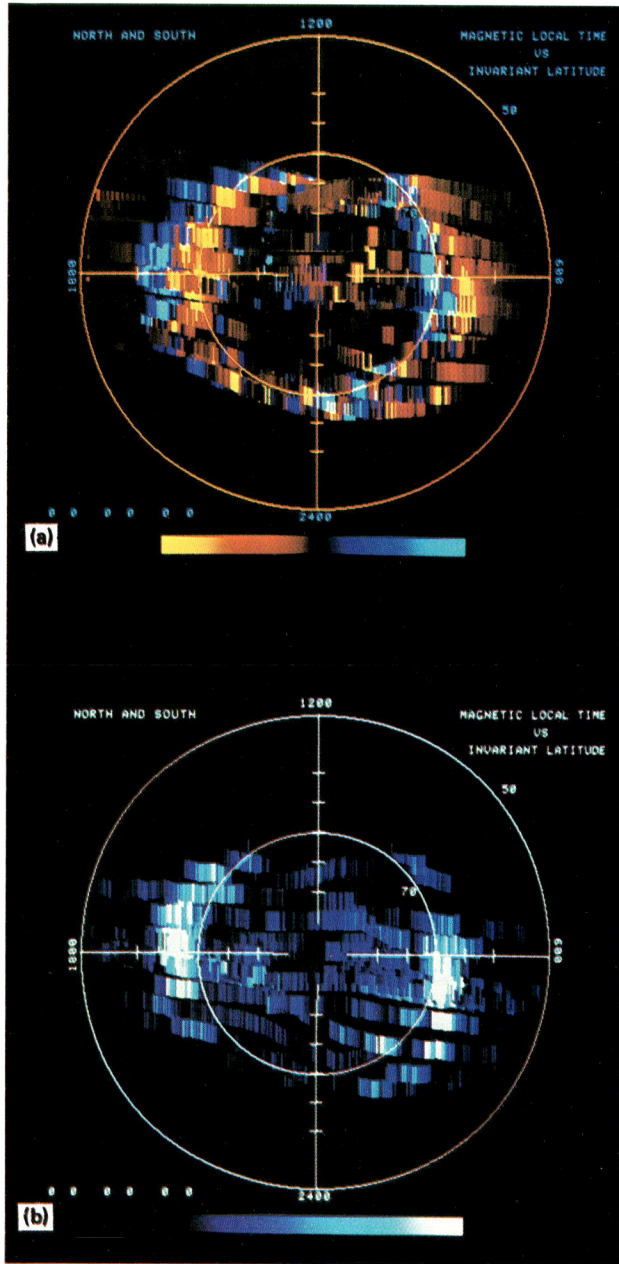

Plate 1. Composite of examples shown in Figures 2 and 3 with the rest of the data for March 21, 1980. The Birkeland currents are shown in (a) plotted versus invariant latitude and magnetic local time; blue is into the ionosphere and red-yellow is away from the ionosphere. Panel (b) is a composite of the electrojet current analyses, also for March 21, 1980. The magnitude of the current intensity is shown by blue to white; the eastward electrojet is at dusk and the westward electrojet is at dawn.

direction. There is, of course, still a DC level in the radial component that will contribute when a proper three-dimensional Fourier analysis is used. The results will unfortunately be degraded because of the lack of data at the noon and midnight auroral oval.

Discussion

We have presented simultaneous observations of both Birkeland currents and the distribution of ionospheric Hall currents as observed in the MAGSAT magnetometer data. A Fourier analysis technique has been applied to the radial component of the disturbance magnetic field in order to determine not only boundaries but the distribution of Hall current intensity in the ionosphere as well. The intensity of Hall current usually has a triangular rather than uniform distribution with the intensity of the peak at the center. Assuming a constant Hall to Pedersen conductivity across the oval, and further assuming that the Birkeland currents are linked meridionally by Pedersen currents in the ionosphere, a triangular Hall current distribution is consistent with a uniform density of adjacent up-down field-aligned currents. The signature of such a Birkeland current system would be a triangular signature in the east-west disturbance magnetic field, evident in the majority of magnetic field-aligned current signatures.

One of the subjects of recent interest has been the relationship of the locations of Birkeland currents and Hall currents in the ionosphere. A recent review [Potemra, 1979] contains a discussion of referenced work on just this subject, and specific events have also been addressed by Senior et al. [1982] and Kamide et al. [1983]. All results have been drawn on statistical or close-in-time event measurements from ground-based magnetometers and/or measurements from satellite magnetometers. We may resolve some questions with very specific data since, at least for a pass through the current systems, both Birkeland and ionospheric currents are inferred simultaneously. For the approximately two dozen cases so far analyzed, there has been an amazing collocation in latitude of the boundaries of those currents. Specifically, the eastward and westward electrojet current intensities at dusk and dawn, respectively, begin increasing in amplitude at the equatorward edge of the Region 2 Birkeland current system and go to zero at the poleward edge of the Region 1 system. For about 75% of the cases, the peak of the electrojet current intensity coincides with the boundary between Region 1 and Region 2 (see Figures 2 and 3). Figure 4 has been included to show examples of the other cases, the format being the same as Figure 2c or 3c. In Figure 4a at 11:15 UT on March 21, 1980 the Region 1–Region 2 boundary (peak of solid line) is poleward of the peak of the dusk eastward Hall electrojet current intensity (dotted trace). This Region 1–Region 2 Birkeland current boundary is poleward of the dawn westward electrojet peak intensity at 21:10 UT (Figure 4b).

A final statement concerns the fact that the entire preceding analysis was done without the input of a conductivity model. In fact, if one assumes that the Birkeland currents are closed along a meridian by Pedersen currents, then the ratio of Hall to Pedersen conductivity is actually an output of this method of current detection. Inferences of field-aligned currents from horizontal Hall currents deduced from surface magnetometers do have advantages, particularly in determining temporal variations [Kamide et al., 1982]. However, in such cases many steps must be taken before the final field-aligned current patterns emerge, the input of a conductivity model being one of the most critical steps. These models are by nature statistical and should be applied cautiously to specific events [Greenwald, 1982; Zanetti et al., 1983]. The MAGSAT data are unique in the vast collection of magnetometer data sets in avoiding this problem. Finally, the statistical patterns of Birkeland currents of Iijima and Potemra [1978] have been substantiated, and the emerging ionospheric Hall current patterns and distributions appear to be self-consistent.

Acknowledgments. We would like to thank MAGSAT teams at JHU/APL and NASA/GSFC for the successful mission from which these data resulted. We appreciate the assistance in data processing contributed by S. Favin, L. L. Suther, C. A. Koontz, and J. S. O'Donnell and the various contributions of S1P personnel. This work was supported by the National Science Foundation, the Office of Naval Research, and the NASA MAGSAT Program.

References

Acuna, M. H., The MAGSAT precision vector magnetometer, *Johns Hopkins APL Tech. Dig., 1,* 210, 1980.

Akasofu, S.-I., and B.-H. Ahn, Distribution of field-aligned currents, ionospheric currents, and electric fields in the polar region on a very quiet day and a moderately disturbed day, *J. Geophys. Res., 86,* 753, 1981.

Baumjohann, W., J. Untiedt, and R. A. Greenwald, Joint two-dimensional observations of ground magnetic and ionospheric electric fields associated with auroral zone currents – 1. Three-dimensional current flows associated with a substorm-intensified eastward electrojet, *J. Geophy. Res., 85,* 1963, 1980.

Fountain, G. H., F. W. Schenkel, T. B. Coughlin, and C. A. Wingate, The MAGSAT attitude determination system, *Johns Hopkins APL Tech. Dig., 1,* 194, 1980.

Greenwald, R. A., Recent advances in magnetosphere-ionosphere coupling, *Rev. Geophys. Space Phys., 20,* 577, 1982.

Heffernan, K. J., G. H. Fountain, B. E. Tossman, and F. F. Mobley, The MAGSAT attitude control system, *Johns Hopkins APL Tech. Dig., 1,* 188, 1980.

Hughes, T. J., and G. Rostoker, A comprehensive model current system for high latitude magnetic activity – I. The steady-state system, *J. R. Astr. Soc., 58,* 525, 1979.

Hughes, T. J., D. D. Wallis, J. R. Burrows, and M. D. Wilson, Model predictions of magnetic perturbations observed by MAGSAT in dawn-dusk orbit, *Geophys. Res. Lett., 9,* 357, 1982.

Iijima, T., and T. A. Potemra, Large-scale characteristics of field-aligned currents associated with substorms, *J. Geophys. Res., 83,* 599, 1978.

Inhester, B., W. Baumjohann, R. A. Greenwald, and E. Nielsen, Joint two-dimensional observations of ground magnetic and ionospheric electric fields associated with auroral zone currents, 3. Auroral zone currents during the passage of a westward travelling surge, *J. Geophys., 49,* 155, 1981.

Kamide, Y., A. D. Richmond, and S. Matushita, Estimation of ionospheric electric fields, ionospheric currents, and field-aligned currents from ground magnetic records, *J. Geophys. Res., 86,* 801, 1981.

Kamide, Y., B.-H. Ahn, S.-I. Akasofu, W. Baumjohann, E. Friis-Christensen, H. W. Kroehl, H. Maurer, A. D. Richmond, G. Rostoker, R. W. Spiro, J. K. Walker, and A. N. Zaitzev, Global distribution of ionospheric and field-aligned currents during substorms as determined from six IMS meridian chains of magnetometers: Initial results, *J. Geophys. Res., 87,* 8228, 1982.

Kamide, Y., R. M. Robinson, S.-I. Akasofu, and T. A. Potemra, Aurora and electrojet configuration in the early morning sector, submitted to *J. Geophys. Res.,* 1983.

Kisabeth, J. L., On calculating magnetic and vector potential fields due to large-scale magnetospheric current systems and induced currents in an infinitely conducting earth, in *Quantitative Modelling of Magnetospheric Processes,* Geophysical Monograph 21, W. P. Olson, Editor, AGU, Washington, 1979.

Langel, R. A., J. Bervert, T. Jennings, and R. Horner, *MAGSAT data processing: An interim report for investigators,* NASA TM 82160, November 1981.

Langel, R. A., G. Ousley, and J. Berbert, The MAGSAT mission, *Geophys. Res. Lett., 9,* 243, 1982.

McDiarmid, D. R., and F. R. Harris, The structure of a connected sequence of magnetospheric substorms, *Planet. Space Sci., 24,* 269, 1976.

Mersmann, U., W. Baumjohann, F. Kuppers, and K. Lange, Analysis of an eastward electrojet by means of upward continuation of ground-based magnetometer data, *J. Geophys., 45,* 281, 1979.

Mobley, F. F., L. D. Eckard, G. H. Fountain, and G. W. Ousley, MAGSAT – A new satellite to survey the earth's magnetic field, *IEEE Trans. Magn., MAG-16,* 758, 1980.

Potemera, T. A., Current systems in the earth's magnetosphere, *Rev. Geophys. Space Phys. 17,* 640, 1979.

Senior, C., R. M. Robinson, and T. A. Potemra, Relationship between field-aligned currents, diffuse auroral precipitation and the westward electrojet in the early morning sector, *J. Geophys. Res., 87,* 10469, 1982.

Zanetti, L. J., W. Baumjohann, and T. A. Potemra, Ionospheric and Birkeland current distributions inferred from the MAGSAT magnetometer data, *J. Geophys. Res., 88,* 4875, 1983.

VARIATION OF THE AURORAL BIRKELAND CURRENT PATTERN ASSOCIATED WITH THE NORTH-SOUTH COMPONENT OF THE IMF

P. F. Bythrow, T. A. Potemra, and L. J. Zanetti

The Johns Hopkins University Applied Physics Laboratory, Laurel, Maryland

Abstract. In this study we have examined the data set acquired from the vector magnetometer on board MAGSAT during the period October 1979 to June 1980. Auroral zone crossings were sorted by hourly averaged values of the Interplanetary Magnetic Field. The result is a catalog of sorted data files from which we have generated color coded magnetic local time invariant latitude plots that image the auroral Birkeland current pattern, over each prescribed parameter range. Using these images, we have statistically examined the dynamics of the current pattern.

Our results indicate that during the periods of $B_z < 0$, the region 1 and 2 current system expands toward lower latitudes accompanied by an expansion of the auroral zone. When $B_z > 0$, we find that the region 1 and 2 currents continue to flow with greatly reduced amplitude in the presence of extensive small-scale structure.

Introduction

An understanding of the role played by the Interplanetary Magnetic Field (IMF) in the dynamics of auroral phenomena has been the goal of extensive satellite and ground-based investigations conducted throughout the past decade [see Cowley, 1982, and references therein]. It is the objective of this paper to identify how the poleward and equatorward boundaries of the Birkeland current patterns are displaced in response to the variation of the IMF. It has been suggested that the location of these boundaries and how they react to the IMF are intimately involved with proposed current-generating mechanisms [Bythrow et al., 1981; Stern, 1983] and with substorm events via the coupling of one current system to the other [Siscoe, 1982a and 1982b; Siscoe and Crooker, 1983]. The geometry of the polar cap as defined by the region of open field lines has been examined for its dependence on the IMF [Akasofu et al., 1981], as has the location of the poleward and equatorward boundaries of auroral electron precipitation [Makita et al., 1983; Hardy et al., 1981]. Such studies have indicated that as B_z becomes more negative these boundaries move equatorward, resulting in an expansion of the polar cap and the auroral zone, while as B_z becomes more positive the polar cap contracts. With respect to the field-aligned currents, it has been shown by Klumpar [1979] and Bythrow et al. [1981] that the boundaries of magnetic field perturbations resulting from field-aligned currents do not necessarily coincide with those of precipitating electrons and therefore are not necessarily subject to the same variations resulting from changes in the IMF.

The location of the statistical Birkeland current pattern has been mapped by Iijima and Potemra [1976]. The pattern reflects a clear dependence on geomagnetic conditions and substorm occurrence [Iijima and Potemra, 1978]. Recent investigations using data from Triad [Iijima and Potemra, 1982] and MAGSAT [Bythrow and Potemra, 1983] have shown that a strong correlation exists among interplanetary parameters, the Birkeland current density, and the total current carried by the region 1 and 2 current system. The latter of these studies also suggests that the poleward and equatorward limits of the Birkeland currents display behavior similar to that of the electron precipitation boundaries expanding and contracting with B_z [Hardy et al., 1981; Makita et al., 1983]. In this paper, we confirm this suspicion by means of an extensive search of the MAGSAT data set in order to generate images of the Birkeland current pattern for various conditions of B_z and by examining 102 dawn-dusk crossings of the northern auroral zone that lie very near the magnetic dawn-dusk meridian.

Data Management

MAGSAT was launched into a low altitude, sun-synchronous, dawn-dusk aligned orbit in October 1979. It provided continuous, high-resolution vector magnetic field data with a resolution of ±0.5 nT and a sampling rate of 16 samples/axis second [Acuna, 1980] until its planned reentry in June 1980. Because of the diurnal rotation of the magnetic poles about the geographic pole, the dayside (nightside) auroral zone in the northern (southern) hemisphere was swept beneath the orbiting spacecraft. The relative motion of the auroral zone with respect to MAGSAT's inertially fixed orbit plane provided a way to make magnetic field measurements that included the entire dayside and nightside Birkeland current pattern twice daily. The data set from 6 months' operation therefore includes some 6000 dawn-dusk crossings of the auroral zone at nearly all latitudes and local times. (Measurements in the Cusp and Harang regions are relatively sparse, particularly during magnetically disturbed periods, because of this dawn-dusk orbital configuration.)

In order to perform rapid statistical and correlative studies of the large-scale Birkeland currents with interplanetary conditions, it was necessary to be able to access randomly any individual MAGSAT polar crossing. This goal was achieved by limiting data to those taken above 50° invariant latitude, reducing spatial resolution (by selecting every 32nd point) from about 0.5 to about 16 km, rotating the magnetic field vector from geographic to geomagnetic coordinates, and storing this reduced record on magnetic disk. In this manner, the entire data set was reduced and stored on a single 500 megabyte disk, thus providing a way to sort rapidly according to any desired parameter.

To examine the macroscopic properties of the region 1 and 2 current system on a global scale, we have introduced an imaging technique that represents gradients in the east-west component of ΔB as variations in color and intensity. (ΔB is the difference between the observed magnetic field and the field computed from the MAGSAT model field.) Colorations from blue through white represent current flow into the ionosphere, while outward flow is represented by red through yellow. The result of combining data from both the northern and southern hemispheres is a color image in invariant latitude and magnetic local time (MLT) coordinates of the complete (dayside and nightside) large-scale Birkeland current pattern. When generated from data acquired during a single day or from data sorted by the z component of the IMF, this pattern bears a marked resemblance to the statistical patterns derived from Triad magnetic field measurements by Iijima and Potemra [1976].

For this study, files were generated by sorting MAGSAT data according to hourly averaged values of the IMF B_z and B_y compo-

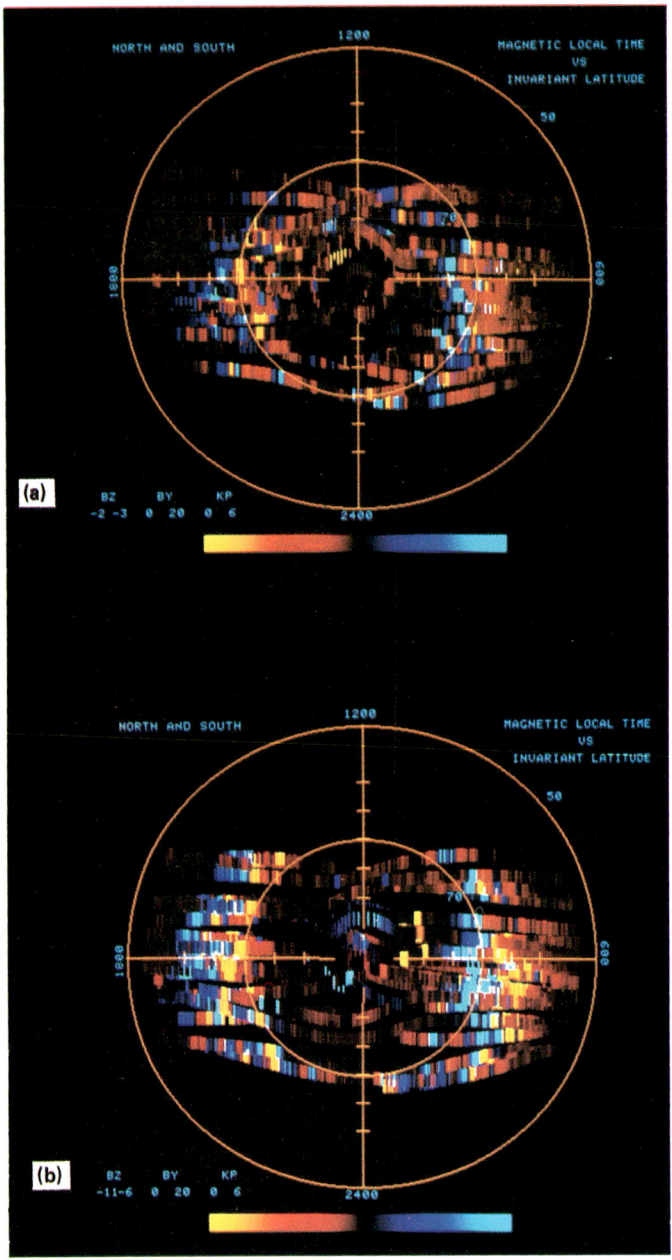

Plate 1. Color images of the large-scale Birkeland current pattern generated from gradients of magnetic field measurements taken during multiple crossings of the northern and southern hemispheres by MAGSAT. (a) $-2 \leq B_z \leq -3$. (b) $-6 \leq B_z \leq -11$.

nents. The IMF data were from NASA/GSFC IMP-8 OMNI Tapes [King, 1982]. The values of B_z and B_y used in the sorting procedure were hourly values that preceded by 1 hr the median time of each auroral zone crossing.

Observations: $B_z < 0$

We have conducted both qualitative and quantitative examinations of the variation of the Birkeland current pattern in response to changes in the z component of the IMF during conditions of positive and negative B_z. The color images in Plates 1a and 1b map out the system of region 1 and 2 currents during periods of negative B_z. For these images, the intensity saturates at about ± 3 $\mu A/m^2$ and cutoff (black) is at about 0.2 $\mu A/m^2$. These images are a composite of about 50 crossings of the northern and southern hemispheres in the range $-2 \geq B_z \geq -3$ and $-6 \geq B_z \geq -11$.

The resultant image represents a statistical distribution in magnetic local time and invarient latitude of the large-scale auroral zone field-aligned currents for the indicated conditions of B_z and B_y. In the range of B_z used to generate Plate 1a, the auroral zone is apparently contracted enough for MAGSAT to provide a nearly complete coverage of both the region 1 and 2 current systems over the entire oval. The pre-noon sector region 1 currents flowing into the ionosphere are represented by the blue arc, about 3° wide, that extends from about 1130 MLT around to about 0100 MLT. The poleward edge of this arc lies at about 77° on the dayside and about 72° toward midnight. Equatorward, the region 2 current flowing out of the ionosphere extends over approximately the same magnetic local time as region 1 and spans about 4° down to about 70° near noon and about 63° at about 0200 MLT. On the evening side between noon and midnight, the region 1 current flows out of the ionosphere and extends from about 1230 to 2300 MLT. The red arc, about 3° wide, depicts this region of outwardly directed current flow. Region 2 currents in this sector flow into the ionosphere and are represented by the faint blue arc from about 1700 to about 2200 MLT. In general, the region 1 and 2 current densities observed here are about 1 $\mu A/m^2$. The image indicates that the region 1 current systems on both sides of the noon-midnight meridian are nearly equal in density, location, and latitudinal extent. The region 2 currents both begin to fade toward noon, but the inwardly directed current in the post noon sector fades nearer to 2400 than does that in the pre-noon sector.

Plate 1b was generated by selecting about 50 MAGSAT crossings for the range $-6 \leq B_z \leq -11$ and $B_y > 0$. The region 1 and 2 current systems are readily apparent in this figure, but because of the overall expansion of the pattern to lower latitudes MAGSAT does not provide coverage of the noon and midnight sectors. The region 1 currents in both the pre-noon and post-noon sectors show a peak in current density near 0600 and 1800 MLT of about 3 $\mu A/m^2$; they are about 3° to 5° wide and begin at about 74° near 1330 and 1030 MLT, while in the sectors centered around midnight they begin lower in latitude, at about 70°, apparently because of the offset of the auroral zone toward midnight. Region 2 currents extend down to about 60° and have a maximum current density of about 3 $\mu A/m^2$ in the pre-noon sector and about 2 $\mu A/m^2$ in the post-noon sector.

A comparison of these two images provides an insight into the control exerted by B_z over the large-scale Birkeland currents. As B_z becomes more negative, three changes become obvious:
1. The current density over the entire auroral zone increases.
2. The poleward edge of the region 1 current, the equatorward edge of the region 2 current, and the interface between these currents move equatorward.
3. The equatorward motion of all boundaries is such that, on the average, the overall pattern undergoes an expansion.

In an attempt to better quantify these results, we selected 102

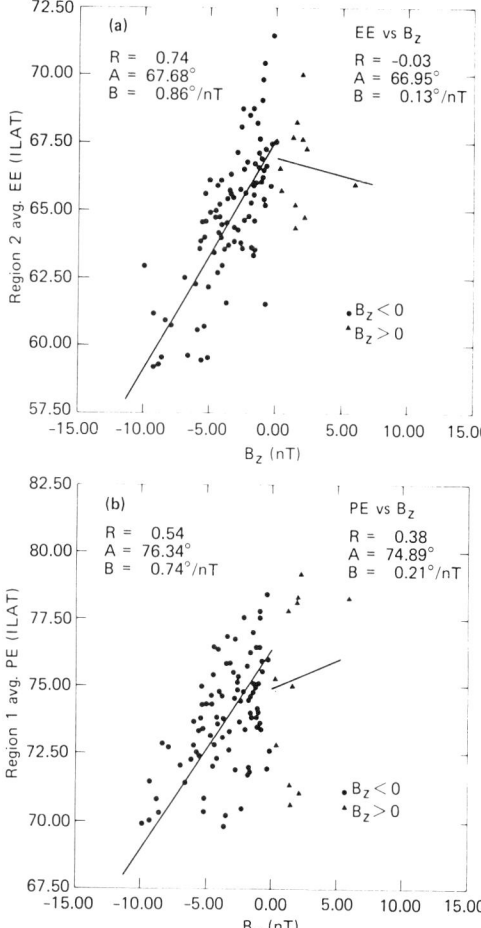

Fig. 1. Scatter plots of the (a) equatorward (EE) and (b) poleward (PE) boundaries of the region 1 and 2 Birkeland currents versus B_z. A linear least-squares fit was performed separately on the data for $B_z > 0$ and $B_z < 0$. Correlation coefficients, intercepts, and slopes are indicated by R, A, and B, respectively.

crossings of the northern hemisphere auroral zone. We required that the crossings lie within about 8° of the dawn-dusk meridian, slightly offset toward midnight, and that they exhibit a somewhat triangular magnetic field disturbance on both sides of the auroral zone. The invariant latitudes of the edge of these disturbances were tabulated and compared with hourly averaged values of B_z. In Figures 1a and 1b we show the pre-noon and post-noon averaged location in invariant latitude (ILAT) of the (a) equatorward (EE) and (b) poleward (PE) edges of the region 1 and 2 currents plotted versus B_z. We also performed a linear orthogonal least-squares fit over the data. This fit assumes error in both the ordinate and the abscissa and therefore minimizes the orthogonal distance between the points and the fit. The results of these fits are in agreement with the indications of the images in Plate 1; that is, when B_z is negative both the equatorward edge of region 2 and the poleward edge of region 1 move equatorward. The slope of region 2 (0.86°/nT) is greater than that of region 1 (0.74°/nT), resulting in an expansion of the overall pattern with $B_z < 0$. When $B_z > 0$, the linear fits are quite poor and indicate very little correlation of the field-aligned current boundaries.

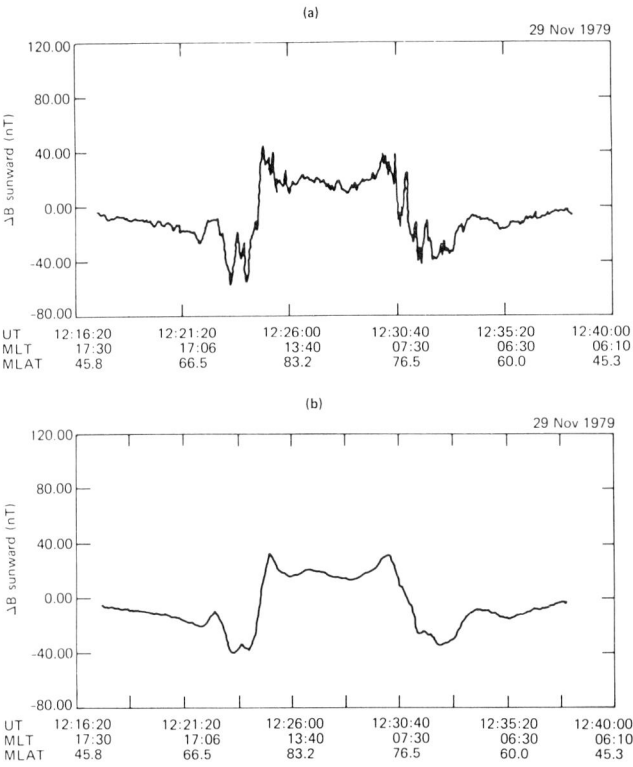

Fig. 2. (a) The sunward/antisunward ΔB component of a single crossing of the auroral zone for $B_z > 0$. The hourly average of B_z was +1 and had been greater than 0 for the preceding 14 hours. Note the extensive small-scale spatial structure. (b) Same data as in (a) but after 150 km spatial filtering.

Observations: $B_z > 0$

During periods of $B_z > 0$, our investigation of field-aligned currents has been carried out in the same fashion as for $B_z < 0$. In Plate 2a we have plotted about 50 MAGSAT passes of the northern and southern hemispheres for conditions of B_z weakly positive, $1 \leq B_z \leq 2$. The range of current density is from about ± 0.2 to ± 3 $\mu A/m^2$. We observed very little evidence of significant magnetic field disturbance below about 70° invariant latitude (ILAT). Above that latitude, the disturbance in the east-west component of B_z increases in a dramatic but chaotic fashion up to about 82°, with little indication of a coherent pattern of large-scale current systems.

Initial examination of this image gives the impression that under the conditions of $B_z > 0$, the region 1 and 2 currents have disappeared and have been replaced by random pairs of small-scale currents, but a closer study of individual orbits and of a spatially smoothed image reveals that this, in fact, is not the case. Plate 2b is a spatially smoothed (about 150 km) reconstruction of Plate 2a in which the intense small-scale (about 15 km) structure has been filtered by means of a 9 point running average and the amplitude scale has been decreased by a factor of 4, so that maximum intensity is now produced by current densities of about 0.8 $\mu A/m^2$. The resultant image indicates the presence of the region 1 and 2 currents with greatly reduced intensity displaced toward much higher latitudes. The smoothing technique used was simply a running average, hence the poleward and equatorward boundaries were smeared over about 0.5°.

In Figures 2a and 2b we have demonstrated the results of smoothing on an individual dawn-dusk crossing of the northern auroral zone with B_z about +1.0. For this pass on 29 November 1979, the hourly average of B_z had been >0 for the previous 14 hours. With B_z at this weakly positive value, there is extensive structure in ΔB (Figure 2a) having a spatial scale of about 30 to 60 km, possibly indicating multiple, intense, and oppositely directed field-aligned currents throughout the high-latitude portion of the pass. When the data were smoothed with a 9 point (about 150 km) running average in Figure 2b, the structure was removed and the underlying large-scale (200 km) pattern of region 1 and 2 currents was clearly reestablished. In this sunward/antisunward component, there is also an indication of two currents poleward and adjacent to the region 1 currents but directed opposite to that of region 1.

Conclusions

Over the past few years, the evidence for the existence of a solar wind/IMF dynamo that drives currents and convection during periods of $B_z < 0$ has been mounting [see Stern, 1983, and references therein]. A precise description of the means by which electric fields generated by this dynamo couple to the ionosphere is still lacking, but nonetheless these fields apparently do couple and persist in driving Birkeland currents and ionospheric convection. The concept of a background potential that is supplied by a continuously operating viscous mechanism has also been suggested [see again Cowley, 1982]. The results of Reiff et al. [1981] and Iijima and Potemra [1982] show that, as B_z turns positive, terminating dayside merging, both the polar cap potential and the field-aligned currents maintain a residual level, indicating the presence of another source. The observations discussed in this paper provide additional

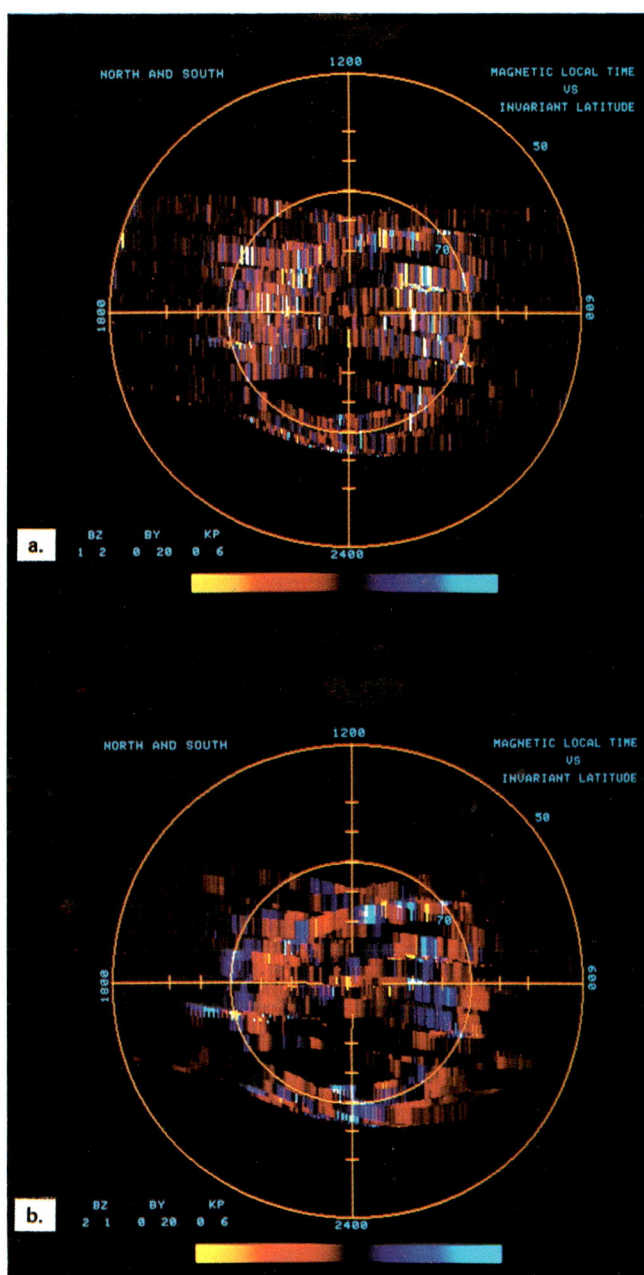

Plate 2. (a) Color image of the Birkeland current pattern generated from multiple crossings of the northern and southern hemispheres with $B_z > 0$. (b) Color image produced from the same data as in (a) but after 150 km spatial filtering.

evidence that the large-scale Birkeland current system is driven simultaneously by at least both of these processes.

In the first section of this paper, we presented observations of the Birkeland current pattern with $B_z < 0$. The expansion of the current system, its motion toward lower latitude, and the overall intensification of the current density with increasingly negative B_z strongly suggest the following. A dayside merging process that transmits the solar wind/IMF electric field (modulated by B_z) to the ionosphere is responsible for driving much of the large-scale Birkeland currents. The linear fits of the poleward and equatorward edges of the region 1 and 2 currents are not completely destroyed as B_z becomes positive but are limited to latitudes of about 70° and 80°, respectively. This suggests that these currents persist when dayside merging does not occur. In the second set of observations, we examined cases for $1 \leq B_z \leq 2$ and found that the region 1 and 2 current pattern retained its identity, albeit at a much reduced amplitude and masked by the superposition of small-scale structure. This small-scale structure has amplitudes comparable to those produced by the spatially larger-scale currents.

The principal conclusions of this study are:
1. The region 1 and 2 current system appears to be a permanent feature of the auroral zone.
2. When $B_z < 0$, the dominant source of large-scale Birkeland currents is the solar wind/IMF dynamo.
3. A mechanism operating independently of dayside merging drives region 1 and 2 Birkeland currents at a background level, even during periods of positive B_z.

In order to better understand these interactions, a more comprehensive study of Birkeland currents is required, particularly during periods of $B_z > 0$ and with consideration of the effects induced by the x and y components of the IMF.

Acknowledgments. We would like to thank the MAGSAT teams at JHU/APL and NASA/GSFC for the successful mission from which these data result. We appreciate the assistance in data processing contributed by S. Favin, L. L. Suther, and C. A. Koontz and the various work of S1P personnel. This work was supported by the National Science Foundation, the Office of Naval Research, and the NASA MAGSAT Program.

REFERENCES

Acuna, M. H., The MAGSAT precision vector magnetometer, *Johns Hopkins APL Tech. Dig., 1,* Jul-Sep 1980.

Akasofu, S.-I., D. N. Covey, and C.-I. Meng, Dependence of the geometry of the region of open field lines on the interplanetary magnetic field, *Planet. Space Sci., 29,* 803, 1981.

Bythrow, P. F., R. A. Heelis, W. B. Hanson, and R. A. Power, Observational evidence for a boundary layer source of dayside region 1 field-aligned currents, *J. Geophys. Res., 87,* 5577, 1981.

Bythrow, P. F., and T. A. Potemra, The relationship of total Birkeland currents to the merging electric field, *Geophys. Res. Lett., 10,* 573, 1983.

Cowley, S. W. H., The causes of convection in the Earth's magnetosphere: A review of developments during the IMS, *Rev. Geophys. Space Phys. 20,* 531, 1982.

Hardy, D. A., W. J. Burke, M. S. Gussenhoven, N. Heinemann, and E. Holeman, DMSP/F2 electron observations of equatorward auroral boundaries and their relationship to the solar wind velocity and the north-south component of the interplanetary magnetic field, *J. Geophys. Res., 86,* 9961, 1981.

Iijima, T., and T. A. Potemra, The amplitude distribution of field-aligned currents at northern high latitudes, *J. Geophys. Res., 81,* 2165, 1976.

Iijima, T., and T. A. Potemra, Large scale characteristics of field-aligned currents associated with substorms, *J. Geophys. Res., 83,* 599, 1978.

Iijima, T., and T. A. Potemra, The relationship between interplanetary quantities and Birkeland current densities, *Geophys. Res. Lett., 9,* 442, 1982

King, J., *NASA/GSFC IMP-8 Omni Tapes,* 1982.

Klumpar, D. M., Relationships between auroral particle distributions and magnetic field perturbations associated with field-aligned currents, *J. Geophys. Res., 84,* 6524, 1979.

Makita, K., C.-I. Meng, and S.-I. Akasofu, The shift of the auroral electron precipitation boundaries in the dawn-dusk sector in association with geomagnetic activity and interplanetary magnetic field, *J. Geophys. Res.* (in press, 1983).

Reiff, P. H., R. W. Spiro, and T. W. Hill, Dependence of polar cap potential drop on interplanetary parameters, *J. Geophys. Res., 86,* 7639, 1981.

Siscoe, G. L., Polar cap size and potential, a predicted relationship, *Geophys. Res. Lett., 9,* 672, 1982a.

Siscoe, G. L., Energy coupling between Regions 1 and 2 Birkeland current systems, *J. Geophys. Res., 87,* 5124, 1982b.

Siscoe, G. L., and N. V. Crooker, Coupling of Birkeland current rings, Chapman Conf. on Magnetospheric Currents, April 1983.

Stern, D. P., The origins of Birkeland currents, *Rev. Geophys. Space Phys., 21,* 125, 1983.

DISTRIBUTION OF AURORA AND IONOSPHERIC CURRENTS
OBSERVED SIMULTANEOUSLY ON A GLOBAL SCALE

J. D. Craven[1], Y. Kamide[2], L. A. Frank[1], S.-I. Akasofu[3], and M. Sugiura[4]

Abstract. The instantaneous spatial distribution of auroral emissions is observed with auroral imaging photometers on board the spacecraft Dynamics Explorer 1 (DE 1) as ground-based meridian chains of magnetometers simultaneously detect the magnetic signatures of ionospheric and field-aligned currents flowing at northern polar latitudes. Ionospheric conductivities at nighttime auroral latitudes are estimated from the measured auroral luminosities and used with the measured polar magnetic variations to compute model distributions of ionospheric and field-aligned currents. Temporal resolution for the coordinated observations and model calculations is 12 minutes. Model ionospheric and field-aligned current distributions are overlayed on global auroral images to illustrate spatial relations on a global scale at the maximum epoch of an auroral substorm. Eccentric-dipole-latitude and magnetic-local-time coordinates are used. A model field-aligned current distribution is compared quantitatively with the distribution of field-aligned currents inferred from simultaneous observations by the DE-2 magnetometer.

Introduction

A small number of meteorological spacecraft are able to monitor significant portions of the terrestrial atmosphere on a continuous basis through the use of remote-sensing instrumentation at visible and infrared wavelengths. When combined with local and remote observations gained by a distribution of ground- and aircraft-based instruments, an extensive data base is available for investigations of atmospheric dynamics on a global scale. A similar capability does not exist in magnetospheric studies as significant portions of the magnetosphere can not be sampled by remote-sensing instruments and because the fiscal and technical requirements would be immense to install and operate a widely distributed spacecraft network within and about the extensive magnetospheric cavity. A program less comprehensive than that in meteorology is in place, however. The magnetosphere is sampled routinely by a small number of earth-orbiting spacecraft (e.g., IMP 8; ISEE 1,2; Dynamics Explorer 1) through measurements of local plasma and plasma-wave environments, and of local electric and magnetic fields. Distant plasma-wave source regions are identified by direction finding techniques and images of the aurora are gained at visible and ultraviolet wavelengths. Remote sensing of the aurora on a global scale provides direct observations of the inner boundary of the magnetospheric system at the ionosphere. Below this interface at ground level widely distributed observations are (or can be) made to monitor physical parameters of that boundary (e.g., electron densities, drift speeds, ionospheric currents and auroral emissions).

The geometric configuration of the near-earth magnetosphere is determined primarily by the earth's magnetic field while the configuration of the more distant magnetosphere is determined in large part by current systems established in response to the incident solar-wind plasma and the induced electric field. A principal objective of magnetospheric physics is the identification of these current systems and their responses to variations in the interplanetary environment. Surprisingly, gross current patterns within the distant magnetosphere identified to date are largely inferred from the magnetic field configuration, and not from direct measurements of the current carriers. Recent new measurements within the distant magnetosphere [Frank et al., 1983] provide direct observations of these currents.

At much lower altitudes (typically $\lesssim 4\ R_E$) magnetically field-aligned currents are routinely identified at high latitudes by direct detection of the charge carriers and indirectly by means of magnetometers [e.g., Sugiura et al., 1983]. The principal charge carriers for outwardly directed currents appear to be the energetic electrons responsible for auroral optical radiation and enhanced ionization within the ionosphere. Typical energies vary from hundreds of electronvolts to tens of kiloelectronvolts. Direct detection of charge carriers responsible for the inwardly directed currents continues to challenge investiga-

---
[1] Department of Physics and Astronomy, The University of Iowa, Iowa City, Iowa 52242.
[2] Kyoto Sangyo University, Kamigamo, Kita-ku Kyoto 603, Japan.
[3] Geophysical Institute, University of Alaska, Fairbanks, Alaska 99701.
[4] Laboratory for Extraterrestrial Physics, Goddard Space Flight Center, Greenbelt, Maryland 20771.

tors. The spatial distributions of ionospheric electric fields and the enhanced ionospheric conductivities direct ionospheric currents along and across the auroral regions. At ground level the horizontal component of this current (the auroral electrojet) is readily detected by its magnetic signature. Ground level detection of magnetic fields arising from magnetically field-aligned closure currents to and from the distant magnetosphere is more difficult. Recently developed advanced computer codes make possible the approximate determination of the large-scale distributions of these field-aligned currents [Akasofu et al., 1981].

Analysis of ground-based and low-altitude spacecraft observations provided the early average global views of the high-latitude regions. The distributions of magnetic perturbations from the auroral electrojets [Silsbee and Vestine, 1942; Harang, 1946], of the optical aurora [Feldstein, 1963] and of field-aligned currents above the auroral zones [Zmuda and Armstrong, 1974] demonstrate the presence of intense current systems with a complex continuity in local time and a confinement to a small range of latitudes generally referred to as the auroral latitudes. Imaging instrumentation on board the ISIS-2 and DMSP spacecraft have confirmed the complex distributions of auroral emissions in local time at visible and near-infrared wavelengths by providing views of significant portions of the auroral zones [e.g., Anger et al., 1973; Snyder et al., 1974] and of the full auroral region [Murphree and Anger, 1980]. Orbital periods of ~ 100 minutes determine the basic temporal resolution for observations with these optical instruments. This time interval is long in comparison to time scales for most temporal variations in auroral phenomena.

A dynamical description of auroral phenomena on a global scale was developed in the early 1960's from the extensive library of all-sky auroral photographs obtained during the International Geophysical Year [Akasofu, 1964]. While the initial models were based upon collections of observations, each covering a small portion of the auroral region, more refined later models have been based upon individual DMSP and ISIS-2 images of significant fractions of the auroral regions. Photo-mosaics of all-sky and DMSP photographs are used frequently to illustrate the several stages of auroral activity in the substorm model [see Chapter 2 of Akasofu, 1977].

Dynamics of the auroral electrojet current systems have been investigated with chains of magnetometers distributed in latitude along magnetic meridians [e.g., Kisabeth, 1979]. The six meridian chains of more than 70 magnetometers assembled for the International Magnetospheric Study (IMS) represent the most comprehensive investigation from the ground to date of auroral magnetic phenomena. Kamide et al. [1982] have developed computer algorithms which use these extensive observations to determine ionospheric electric fields, currents and Joule-heating rates, and the magnitudes of field-aligned currents towards and away from the ionosphere. Temporal resolutions of several minutes are obtained by Kamide and coworkers.

This greatly increased temporal resolution exceeds that of nearly all other available global magnetospheric observations, but effective utilization has been hampered in part by the lack of simultaneous determinations of ionospheric conductivities with comparable temporal resolution. What have been used are conductivity distributions based upon average particle precipitation patterns as a function of magnetic activity. The application of these average conductivity models in the algorithim of Kamide et al. [1981] is discussed by Kamide et al. [1982]. Conductivities arising from ultraviolet insolation are calculated for standard model upper atmospheres.

With the launch of the spacecraft Dynamics Explorer 1 in August 1981 [Hoffman et al., 1981] global auroral imaging has become available at visible and ultraviolet wavelengths with temporal resolutions of 12 minutes [Frank et al., 1981]. This ability to determine the distributions of auroral emissions with good spatial and temporal resolution at three wavelengths, simultaneously, provides a unique opportunity to obtain improved ionospheric conductivities on a global scale. Additionally, auroral imaging provides indirect observations of field-aligned currents directed out of the ionosphere at auroral and polar-cap latitudes due to precipitating auroral electrons.

The availability of accurate models for the instantaneous ionospheric and field-aligned current systems at the ionospheric boundary of the magnetosphere can provide an additional, significant tool for studies of the dynamics of magnetospheric current systems. Several correlative investigations outlined in Figure 1 are being developed, with the triad of investigations using ground and near-earth observations the subject of this paper.

Observations

For this investigation magnetic records are presently available from 34 ground magnetometer stations. These stations are identified in Table 1 and their eccentric-dipole coordinates [Cole, 1963] tabulated. The spatial distribution of the stations in an eccentric-dipole-latitude, magnetic-local-time format is illustrated in Figure 2 to identify those portions of the polar region for which ground observations are available. The local-time distribution is given for 0610 UT. Good spatial coverage is available from the magnetic pole to ~ 65° latitude over a 12-hour interval of local time, and then to ~ 55° latitude over a 6-hour interval, each centered near local midnight. In each case the largest concentration of stations is found at the earlier local times. The Scandinavian chain of stations

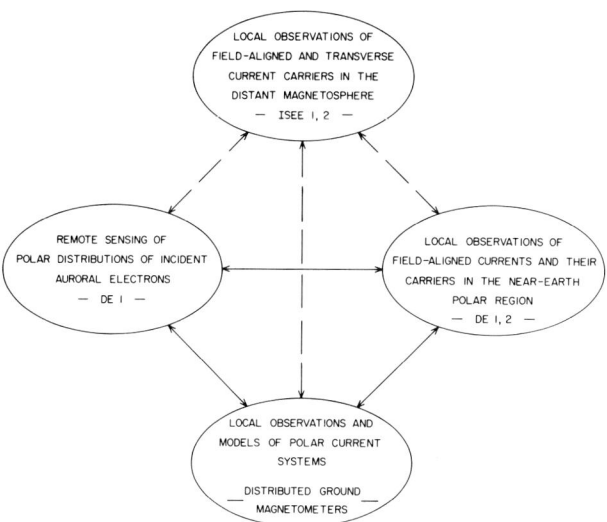

Fig. 1. Four ground and spacecraft investigations for studies of magnetospheric current systems. The three correlative investigations for observations at ground level and at near-earth distances are the subject of this work.

is located at approximately 0800 hours local time.

Magnetic records from analog-recording stations are digitized at the University of Alaska's Geophysical Institute and combined with digitally recorded data to provide a data base of magnetic perturbation vectors referenced to quiet auroral conditions with one minute temporal resolution. These data are used by Kamide et al. [1981] to compute model ionospheric and field-aligned current systems. The perturbation vectors are also used to compute the auroral electrojet index AE(34) with which magnetic substorm activity is characterized on a global scale.

Optical images of the northern polar region are gained simultaneously at visible and ultraviolet wavelengths with auroral imaging instrumentation carried on the polar-orbiting spacecraft Dynamics Explorer 1 [Frank et al., 1981]. A combination of high apogee altitude (3.65 R_E) and an initial apogee latitude near 90° provides up to 5 hours of continuous auroral imaging for each 6.83-hour orbit. By means of a method described by Kamide et al. [1983] these auroral images can be used to infer height-integrated ionospheric conductivities over the entire polar region with 12-minute time resolution.

Ionospheric Currents

Ionospheric currents are computed for an intense auroral substorm with an onset time of ~ 0700 UT on 8 November 1981. The AE index is ~ 220 nT at the onset and increases to ~ 1000 nT by the maximum epoch of the substorm at ~ 0850 UT. A steady decrease in the AE index follows, reaching ~ 150 nT by ~ 1020 UT.

Global auroral activity during this magnetic substorm displays a corresponding increase to maximum intensities by ~ 0850 UT, followed by decreases in the following hour. This temporal behavior is displayed in Plate 1 by a sequence of 12 consecutive auroral images in false color spanning the time interval 0719 to 0945 UT. The first image at upper left in the plate presents the spacecraft view of earth from 69° north latitude in the early-morning sector. With time the spacecraft proceeds from apogee inbound across the north pole (fifth frame at 0807 UT) and into the mid-evening sector. The universal time (UT) assigned to an image identifies the time at which the 12-minute image period begins. For this se-

Table 1. Eccentric-Dipole Coordinates of the Ground Magnetometers

No.	Station Name	Longitude	Co-Latitude
1	Resolute Bay	311.52	6.70
2	Yellowknife	294.13	20.38
3	Fort Churchill	323.00	22.47
4	Great Whale River	344.79	25.74
5	Cambridge Bay	301.74	12.96
6	Baker Lake	318.76	17.07
7	Meanook	299.66	28.07
8	St. John's	13.97	34.64
9	Ottawa	347.05	35.41
10	Mould Bay	262.58	8.27
11	Victoria	289.96	35.32
12	Alert	101.04	3.53
13	Glenlea	320.80	31.75
14	Igloolik	340.49	11.68
15	Tromso	98.44	24.12
16	Barrow	236.07	18.35
17	Ny Alesund	107.29	15.18
18	Bjornoya	104.11	19.80
19	Thule	22.55	4.54
20	Odessa	96.98	47.61
21	Leirvogur	56.90	22.97
22	Narssarssuaq	27.93	22.58
23	Cape Chelyuskin	159.85	21.62
24	Cape Wellen	230.37	25.17
25	Leningrad	101.57	34.96
26	Godhavn	25.46	13.78
27	Newport	297.26	34.75
28	Sachs Harbor	268.04	12.74
29	Norman Wells	276.32	19.21
30	Fort Simpson	286.19	21.97
31	Inuvik	264.86	17.34
32	Fort Yukon	253.82	20.83
33	Fort Smith	298.88	22.58
34	College	252.90	23.08

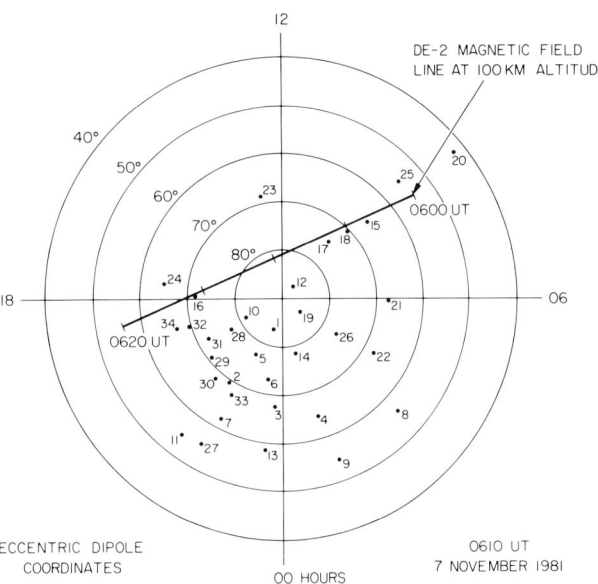

Fig. 2. The spatial distribution of ground magnetometer stations at 0610 UT in an eccentric-dipole-latitude, magnetic-local-time format. The trajectory of DE 2 in the interval 0600 - 0620 UT on 7 November 1981 is mapped along the geomagnetic field to an altitude of 100 km.

quence each 30° X 30° nadir-centered image is a record of the emission intensities of ultraviolet light observed at wavelengths 120 to 155 nm. The geocorona is visible in the scattered light of solar Lyman-α radiation at upper left in the images, sunward of the dayside limb. Antisunward of the terminator the entire auroral region is detected, principally in the light of neutral atomic oxygen at 130.4 and 135.6 nm.

Acquisition of the first frame in the plate is initiated at 0719 UT as magnetic activity is just beginning to increase. The AE index is ~ 240 nT. By the fifth frame at 0807 UT the AE index exceeds 600 nT, and is ~ 1000 nT by the maximum epoch of the substorm at ~ 0850 UT (eighth frame). Throughout this interval auroral emissions increase steadily in brightness and latitudinal width. There is a marked poleward displacement at local midnight during the maximum epoch of the substorm. Images at 0856, 0908 and 0920 UT (ninth through eleventh frames) reveal that during the declining phase of magnetic activity a discrete arc along the nightside edge of the polar cap is replaced by a diffuse band of emissions. A system of discrete arcs reappers at 0932 UT (twelfth frame). Diffuse aurora are visible at more equatorward latitudes throughout the period.

Height-integrated conductivities are inferred from these auroral images in the manner described by Kamide et al. [1983]. A smoothed Pedersen conductivity distribution for this substorm at maximum epoch (0844 - 0856 UT) is shown in Plate 2 superposed over the original auroral image which is displayed in a false-color format. Light blue denotes emission intensities of ~ 2 kR, with greater intensities coded by green, yellow, red and white ($\geq$ 32 kR). The conductivity distribution and auroral image are presented in eccentric-dipole-latitude, magnetic-local-time coordinates. A latitude of 50° is identified by the outer circle and concentric circles of smaller radii denote latitudes of 60°, 70° and 80°. The maximum conductivity of ~ 32 mhos is located at ~ 63° latitude and ~ 0100 hours local time. Hall conductivities are assumed to be twice the Pedersen conductivities. Magnetic perturbation vectors for the 12-minute interval are obtained for each of the 34 ground stations by averaging the 12 samples obtained the 0844 to 0856 UT time interval. A model ionospheric current system is derived with the computer algorithm of Kamide et al. [1981] using the inferred conductivity distribution and the 34 magnetic perturbation vectors. This current system is displayed in Plate 3 using the format of Plate 2. Recalling the spatial distribution of ground stations illustrated in Figure 1 for 0610 UT, the 12-hour range of local times to 65° latitude and the 6-hour range to 55° are centered near 0230 hours local time. Within this 12-hour local time sector the latitude of maximum model westward electrojet current decreases monotonically from 69° eccentric dipole latitude at ~ 0830 hours local time to 66° at ~ 2030 hours local time. The latitude of maximum current lies near the poleward edge of maximum auroral intensities associated with the diffuse aurora (excluding the local enhancement in intensities at ~ 0030 hours and 69°). Width of the westward electrojet decreases from $\geq$ 10° in the morning sector to ~ 5° in the late-evening sector. Maximum ionospheric currents range from 1.1 A/m at 0600 hours to 0.75 A/m at 2200 hours local time. A second westward electrojet near local midnight and ~ 76° is coincident with the most poleward arc of the expanding auroral bulge. A weak eastward electrojet is computed near local midnight and equatorward of ~ 60°. Westward currents in the morning sector at latitudes below 60° are coincident with aurora of intensities less than the 2-kR threshold used in displaying the image.

Field-Aligned Currents

A model field-aligned current distribution is obtained by computing the divergence of the ionospheric current distribution. The field-aligned current distribution for maximum epoch in the substorm is presented in Plate 4. The format is identical to that used for the two previous plates. The most striking features of this cur-

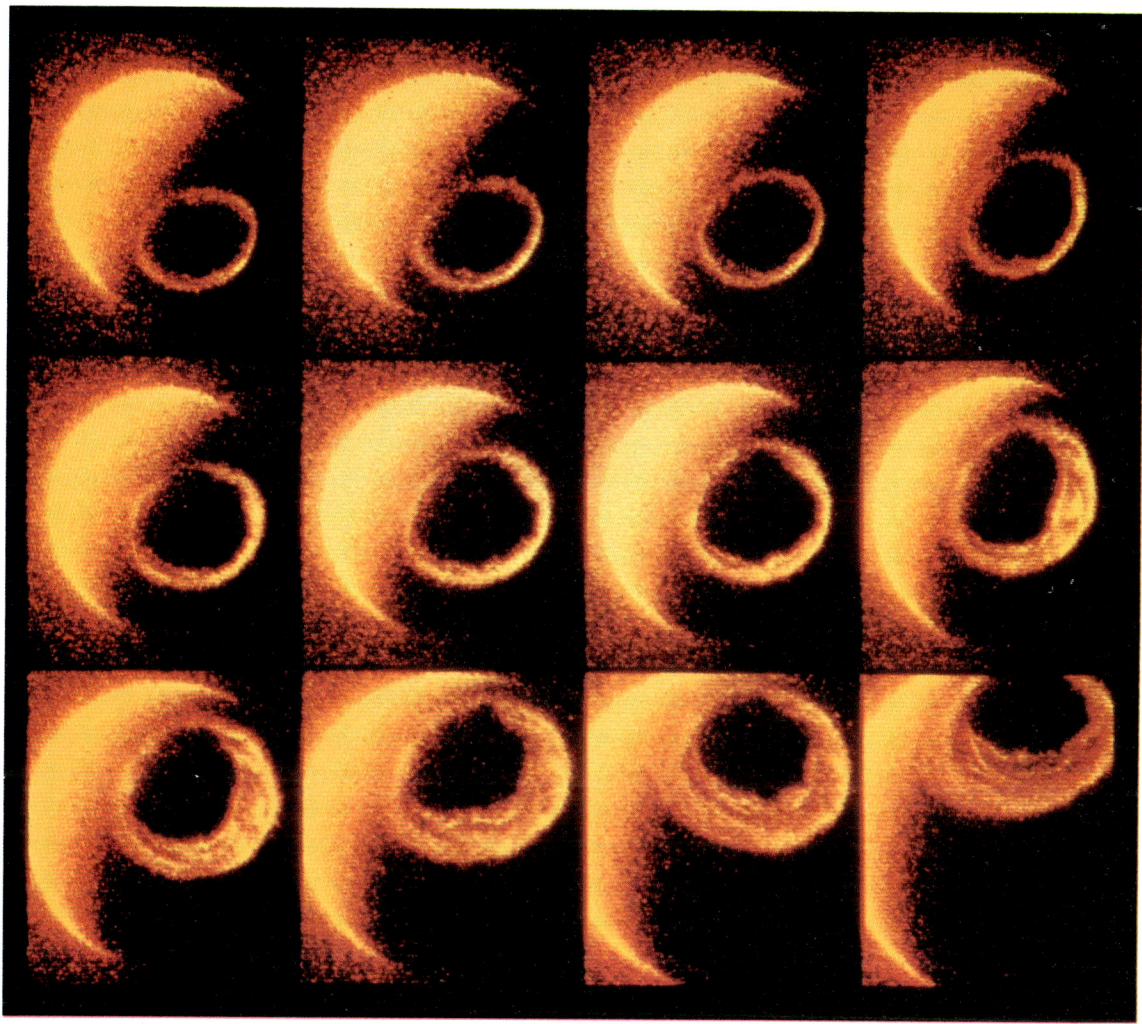

Plate 1. Imaging sequence of 12 consecutive frames displaying global auroral activity at ultraviolet wavelengths in the interval 0719 - 0945 UT on 8 November 1981. Onset of a substorm occurs at ~ 0719 UT (first frame), with maximum epoch at ~ 0850 UT (eighth frame). The geocorona is visible in the scatterd light of solar Lyman-α radiation sunward of the dayside limb, at upper left in each frame. Antisunward of the terminator the entire auroral region is detected in the light of OI at 130.4 and 135.6 nm.

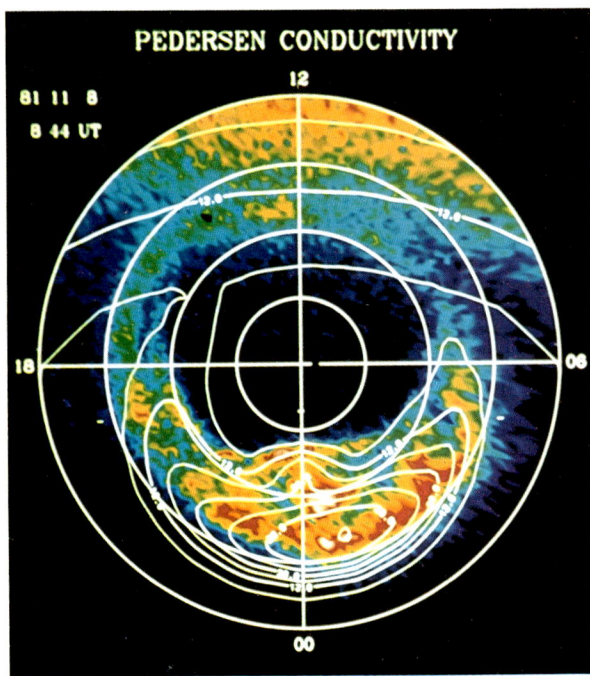

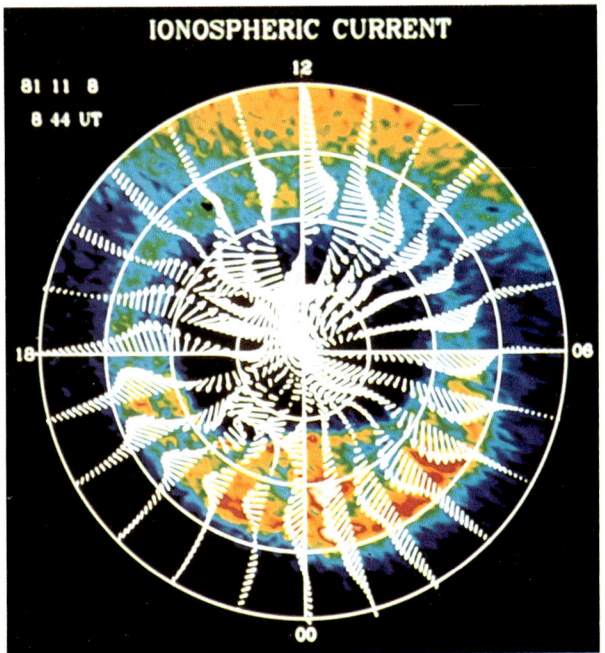

Plate 3.

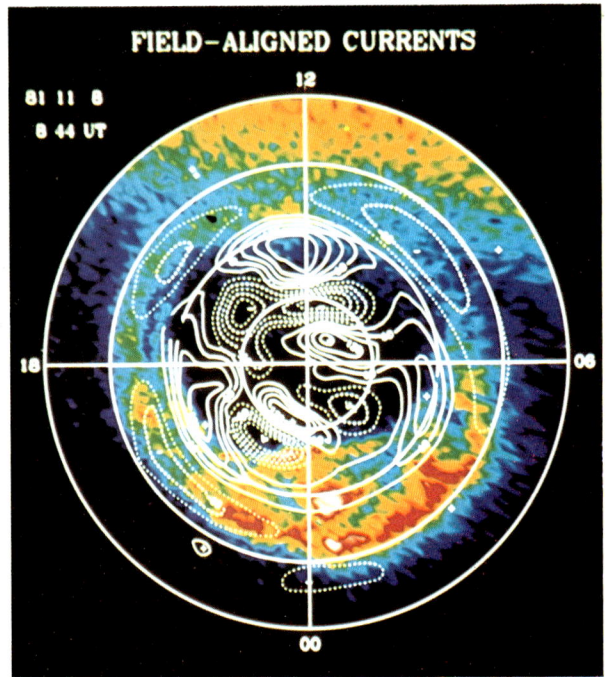

Plate 4.

Plate 2. Smoothed Pedersen conductivity distribution for maximum epoch in the substorm overlayed on the image used to deduce the conductivity distribution. The maximum conductivity of ~ 32 mhos is located at ~ 63° latitude and ~ 0100 hours local time. The increment in conductivity between contours is 4 mhos. For the false-color image, light blue denotes emission intensities of ~ 2 kR, with greater intensities coded by green, yellow, red and white ($\gtrsim$ 32 kR). An eccentric-dipole-latitude, magnetic-local-time coordinate system is used. Concentric circles denote latitudes of 80°, 70°, 60° and 50°.

Plate 3. Modeled ionospheric current distribution at maximum epoch in the substorm overlayed on the image used to deduce the conductivity distribution (Plate 2). Peak ionospheric currents are ~ 1 A/m.

Plate 4. A continuation of Plate 3 for the modeled field-aligned current distribution. Solid lines denote contours for currents directed towards the ionosphere. Dashed lines denote contours for outward currents.

rent distribution are the numerous cells of inwardly (solid lines) and outwardly (dashed lines) directed currents concentrated within the polar cap. Maximum current densities are ~ 1.5 µA/m². No aurora of intensities greater than ~ 1 kR are detected within the polar cap which can be associated with these model cells of field-aligned currents.

At the latitudes of maximum intensities in the diffuse aurora the predominant feature of the model current system is a ~ 5° - 8° wide distribution of outwardly directed currents with typical densities ~ 0.6 µA/m² in the pre-midnight and post-dawn local-time sectors. These distributions are coincident in latitude with the diffuse aurora but are not coincident in local time with the locations of maximum auroral intensities. The minimum current density contour in this plate is 0.3 µA/m². The current distribution for the previous 12-minute interval, with a minimum current density contour of 0.1 µA/m², demonstrates more clearly that outward currents in excess of 0.2 µA/m² are present at latitudes 60° - 70° over a large range of local times. Within the 0000 to 0300 hour local time sector inward currents of the order of 0.2 µA/m² are computed for these latitudes. This current distribution is a low-latitude extension to earlier local times of a more intense model current distribution centered at 71° and 0500 hours. In the morning sector the distributions of outward currents are displaced equatorward of the westward electrojet by ~ 3 degrees, and are nearly coincident in the late-evening sector.

Comparison with DE-2 Observations

Further analysis of the model field-aligned current distributions is gained through comparisons with simultaneous observations from low-altitude spacecraft magnetometers. An example of available simultaneous observations is found in the interval from 0603 to 0619 UT on 7 November 1981 when the DE-2 spacecraft passed over the IMS Scandinavian magnetometer chain, proceeded across the polar cap and then crossed over Alaska. A mapping of this portion of the DE-2 orbit along the magnetic field to an altitude of 100 km is displayed in Figure 2. Local times for this spatial distribution of ground magnetometers and the orbit mapping are determined for 0610 UT, when the spacecraft is traversing the polar-cap region midway between auroral zone crossings at 0800 and 1800 hours local time. An auroral substorm occurring in the period 0440 - 0600 UT has been discussed previously by Kamide et al. [1983]. The DE-2 polar crossing occurs at the end of this substorm as a system of intense arcs is brightening in the evening sector and advancing poleward. Auroral intensities in the morning sector near 0800 hours local time remain diffuse and weak by comparison (see Plate 1 of Kamide et al., [1983]).

Summaries of the observations with the DE-2

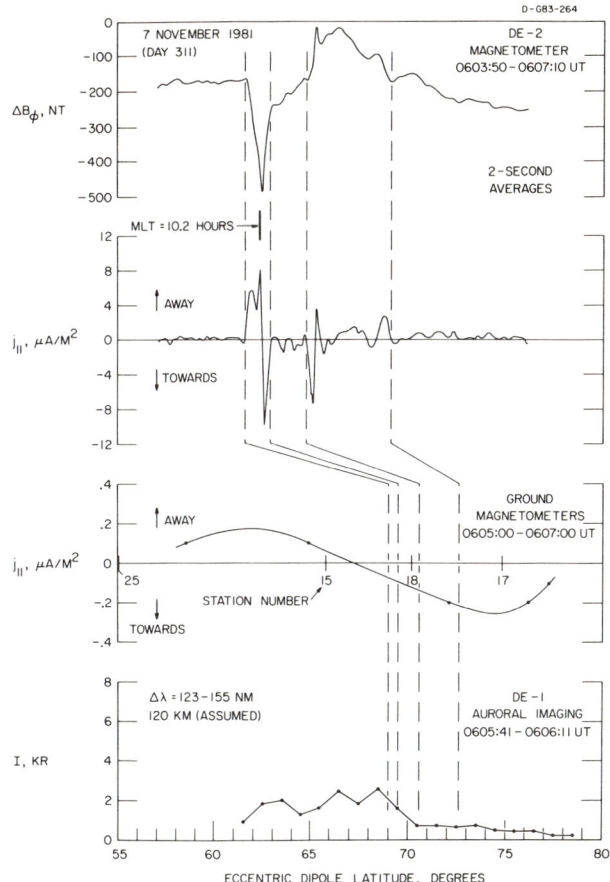

Fig. 3. Simultaneous observations of aurora and magnetic perturbations within the interval 0603:50 - 0607:10 UT on 7 November 1981 at ~ 0800 hours local time. Magnetic perturbations observed with the DE-2 spacecraft and the inferred field-aligned current densities are displayed in the upper and second panels, respectively. Model field-aligned currents deduced from magnetic perturbations observed at ground level are displayed in the third panel. The intensities of auroral OI emissions at 130.4 and 135.6 nm are presented in the lower panel.

magnetometer, the 34 ground magnetometers and the DE-1 auroral imaging instrumentation are presented in Figures 3 and 4. Observations in the morning (evening) sector are presented in Figures 3 (4). Observed deviations of the horizontal component of the geomagnetic field perpendicular to the DE-2 orbit plane are shown in the upper panel of each figure. Smaller deviations parallel to the orbit plane are ignored in this analysis. Assuming a distribution of infinite field-aligned current sheets tangential to the geomagnetic dipole L shells, current distributions are computed from the observed magnetic perturba-

144 AURORA AND IONOSPHERIC CURRENTS

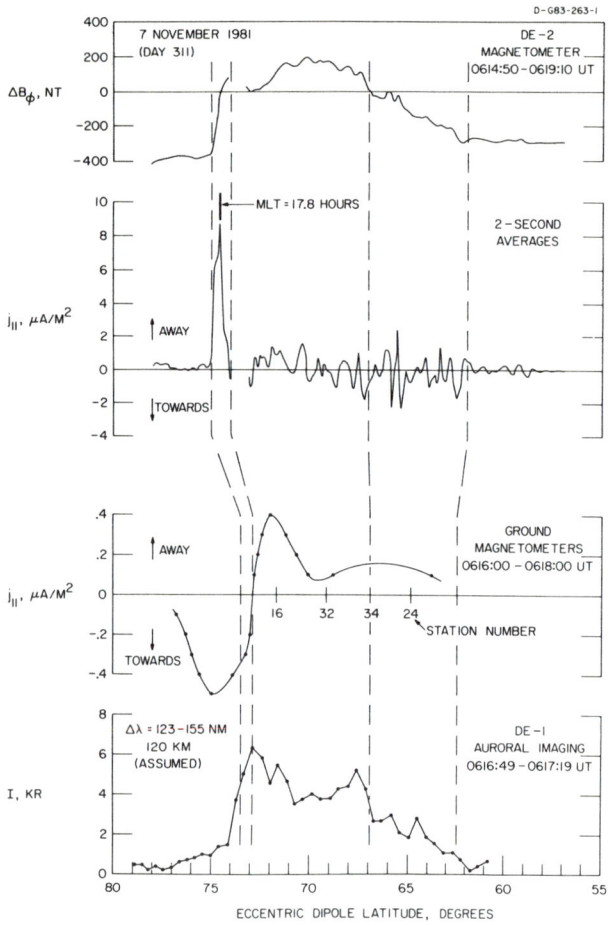

Fig. 4. A continuation of Figure 3 for the interval 0614:50 - 0619:10 UT at ~ 1800 hours local time.

tions. The magnitudes and directions of these computed current distributions are shown in the second panel of each figure. Current directions are specified with respect to the ionosphere. The magnetic perturbations and inferred current densities are plotted as functions of time. Eccentric dipole latitudes at significant locations are mapped from the abscissas by dashed lines.

Simultaneously with the DE-2 crossings of the auroral field lines in the morning and then the evening sectors, the DE-1 auroral imagers measured auroral emission intensities at the ionospheric intersections of the DE-2 magnetic field lines. A slant angle of ~ 40° and a large radial distance to the weaker morning-side aurora results in separations of ~ 1° in latitude between consecutive pixels in an image scan line. In the evening sector the separations are ~ 0.4°. The auroral luminosities are plotted in the lower panel of each figure. The errors in determining the geographic and then geomagnetic coordinates of a pixel are less than the angular width of a pixel.

Model field-aligned current distributions have been determined from ground magnetometer observations in the time intervals 0605 - 0607 UT and 0616 - 0618 UT for concurrent observations with the two DE spacecraft in the morning and evening sectors, respectively. From the resulting current density coutours of the form illustrated in Plate 4, current densities are determined along the DE-2 orbit track. These current-density profiles are presented in the third panels of the two figures. Ground magnetometer stations near the DE-2 ground track are identified by number from Table 1.

In the morning sector (Figure 3), field-aligned currents with densities in excess of ~ 1 $\mu A/m^2$ are detected with DE-2 at latitudes 69° to 72.6°. The lowest latitude current sheet is located at 69°, with current densities of 6-8 $\mu A/m^2$ directed out of the ionosphere. Poleward at 69.5° and 71° inward currents with peak current densities of 10 and 8 $\mu A/m^2$ are detected. In the evening sector (Figure 4) an upward current is located at the poleward edge of the auroral zone coincident with greatly increased optical emissions of ~ 6 kR. An untimely data gap prohibits accurate computation of currents densities equatorward of this intense (~ 9 $\mu A/m^2$) outward current, but the magnetic perturbations on either side of the data gap establish the presence of an inward current equatorward of the outward current. Multiple current sheets with smaller densities are observed at more equatorward latitudes. This spatial distribution in the morning (evening) sector of inward (outward) currents poleward of outward (inward) currents is in agreement with previous spacecraft observations [e.g., Iijima and Potemra, 1976].

This direct comparison with spacecraft observations illustrates clearly a difficulty inherent in models of global field-aligned current distributions which are based upon a widely spaced distribution of ground magnetometers: the field-aligned current sheets have latitudinal widths and separation distances which are small compared to the resolutions provided by the distributions of ground magnetometers and the computational schemes. These modeled current distributions may provide smoothed, average values for the large scale features, but cannot resolve the observed multiple current sheets with adjacent, oppositely directed currents.

In the evening sector (Figure 4) four ground stations are located at latitudes for which 1-2 $\mu A/m^2$ field-aligned current densities are detected with DE 2. The most poleward station (Barrow, #16) is located ~ 1.5° equatorward of the intense upward current sheet at ~ 73° for which the current sheet intensity is ~ 0.3 A/m, and ~ 1° from the inferred downward current sheet for which the estimated current sheet intensity is ~ 0.1 A/m. These current sheets provide a net upward current intensity of ~ 0.2 A/m which is the principal source of field-aligned currents near the Barrow ground station. The current sheet intensity computed from the model current densities at latitudes 70° - 73° is ~ 0.1 A/m, upward, in reasonable agreement with the DE-2 observation, but the sheet is located ~ 1.5° equatorward of the current sheets observed with DE 2. The absence of a ground station poleward of 73° precludes improved spatial agreement. The large inwardly directed currents poleward of 73° are nonphysical artifacts of the model solution.

In the morning sector (Figure 3) only one ground station (Bjornoya, #18) is located within the latitudinal range of significant field-aligned currents. Intensities of the four principal current sheets nearest the ground station are approximately 0.3 A/m outward and 0.2 A/m inward at ~ 69.2° and 0.15 A/m inward and 0.05 A/m outward at ~ 70.6°. The net current intensities at the two latitudes are approximately 0.1 A/m outward and 0.1 A/m inward, respectively. The magnitudes of the current sheet intensities computed by the model for outward currents (57° - 67°) and inward currents (67° - 77°) are each ~ 0.1 A/m. While these values for average current sheet intensities compare favorably with the observed values, a comparison of the spatial distributions indicates little agreement.

Conclusions

Determination of the simultaneous distributions of the aurora and ionospheric currents on a global scale as a function of time with good temporal resolution will provide a new tool for studies of magnetospheric current systems. The auroral images provide a means of obtaining improved conductivity models for polar latitudes which can be used with advanced computer codes to obtain the distributions of ionospheric and field-aligned currents. The images also provide an indirect monitor of currents carried by precipitating auroral electrons.

Spatial resolutions with the network of ground magnetometers and the computational schemes are too coarse to resolve discrete, multiple current sheets present at auroral latitudes. These discrete current structures are observed only by in situ spacecraft observations. However, such observations reveal only local conditions at one instant in time, while the objective is to develop global-scale dynamic models which describe the average current systems at the magnetosphere's ionospheric boundary [e.g., Harel et al., 1981]. These models are a necessary part of the larger goal to obtain comprehensive dynamic models for the magnetospheric current systems. The in situ spacecraft observations are important to the development of these models by determining absolute magnitudes and latitudinal distributions of current sheet intensities whose averages should be described correctly by the models. The comparisons with DE-2 observations presented here demonstrate that the average sheet intensities can be obtained with the models. The latitudinal widths are comparable to the statistically determined distributions [e.g., Iijima and Potemra, 1976]. Further improvements to the mathematical models may require the a posteriori inclusion of multiple current sheets. Correlative investigations with the two DE spacecraft of the spatial distributions of current sheets and aurora can provide necessary new information.

Large-scale current structures appear to be resolved with the present distribution of ground stations. Typical latitudinal widths for the auroral electrojets are comparable or greater than the differences in latitude between many adjacent ground stations. The electrojet's great extent in longitude and slow variation in latitude with local time further improve the determination of average magnitude, width and latitude. For the example provided in Plate 2 at maximum epoch in the substorm, the auroral electrojet follows the poleward edge of the diffuse aurora for the entire 12-hour local time interval covered by ground stations. Resolution is adequte to identify an electrojet at the poleward edge of the expanding auroral bulge approximately 10° poleward of the diffuse aurora.

Large-scale current structures of the polar cap should also be resolved, in particular during periods of enhanced auroral activity when discrete polar-cap aurora are not present. Present difficulties include polar-cap conductivity models, the smaller current densities of the polar-cap current systems and nonphysical model solutions in areas not adequately sampled by ground stations.

Acknowledgements. This research was supported at the University of Iowa in part by the National Aeronautics and Space Administration under contract NAS5-25689 and grant NGL 16-001-002 and by the Office of Naval Research under grant N00014-76-C-0016. Contributions to this research by the University of Alaska were supported by the National Science Foundation under grant ATM81.

References

Akasofu, S. -I., The development of the auroral substorm, Planet. Space Sci., 12, 273, 1964.

Akasofu, S. -I., Physics of Magnetospheric Substorms, D. Reidel, Dordrecht, Netherlands, 1977.

Akasofu, S. -I., Y. Kamide and J. Kisabeth, Comparison of two modeling methods for three-dimensional current systems, J. Geophys. Res., 86, 3389, 1981.

Anger, C. D., A. T. Y. Lui and S. -I. Akasofu, Observations of the auroral oval and a westward traveling surge from the Isis 2 satellite and the Alaskan meridian all-sky cameras, J. Geophys. Res., 78, 3020, 1973.

Cole, K. D., Eccentric dipole coordinates, Australian J. Phys., 16, 423, 1963.

Feldstein, Y. I., Some problems concerning the

morphology of auroras and magnetic disturbances at high latitudes, Geomagn. Aeron., 3, 183, 1963.

Frank, L. A., J. D. Craven, K. L. Ackerson, M. R. English, R. H. Eather and R. L. Carovillano, Global auroral imaging instrumentation for the Dynamics Explorer mission, Sp. Sci. Instr., 5, 369, 1981.

Frank, L. A., C. Y Huang and T. E. Eastman, Currents in the earth's magnetotail, in Magnetospheric Currents, ed. by T. A. Potemra, AGU Geophysical Monograph 29, 1983.

Harang, L., The mean field of disturbance of polar geomagnetic storms, Terr. Magn. Atmos. Elec., 51, 353, 1946.

Harel, M., R. A. Wolf, P. H. Reiff, R. W. Spiro, W. J. Burke, F. J. Rich and M. Smiddy, Quantitative simulation of a magnetospheric substorm, 1. model logic and overview, J. Geophys. Res., 86, 2217, 1981.

Hoffman, R. A., G. D. Hogan and R. C. Maehl, Dynamics Explorer spacecraft and ground operations system, Sp. Sci. Instr., 5, 349, 1981.

Iijima, T., and T. A. Potemra, The amplitude distribution of field-aligned currents at northern high latitudes observed by TRIAD, J. Geophys. Res., 81, 2165, 1976.

Kamide, Y., A. D. Richmond and S. Matsushita, Estimation of ionospheric electric fields, ionospheric currents and field-aligned currents from ground magnetic records, J. Geophys. Res., 86, 801, 1981.

Kamide, Y., B. -H., Ahn, S. -I. Akasofu, W. Baumjohann, E. Friis-Christensen, H. W. Kroehl, H. Maurer, A. D. Richmond, G. Rostoker, R. W. Spiro, J. K. Walker and A. N. Zaitzev, Global distribution of ionospheric and field-aligned currents during substorms as determined from six IMS meridian chains of magnetometers: initial results, J. Geophys. Res., 87, 8228, 1982.

Kamide, Y, J. D. Craven, L. A. Frank and S. -I. Akasofu, Modeling substorm current systems using conductivity distributions inferred from DE auroral images, Geophys. Res. Lett.,(to be submitted), 1983.

Kisabeth, J. L., On calculating magnetic and vector potential fields due to large-scale magnetospheric current systems and induced currents in a infinitely conducting earth, in Quantitative Modeling of Magnetospheric Processes, ed. by W. P. Olson, AGU Geophysical Monograph 21, pp. 473, 1979.

Murphee, J. S. and C. D. Anger, An observation of the instantaneous optical auroral distribution, Can. J. Phys., 58, 214, 1980.

Silsbee, H. C., and E. H. Vestine, Geomagnetic bays, their frequency and current systems, Terr. Magn. Atmosph. Elect., 47, 195, 1942.

Snyder, A. L., S. -I. Akasofu and T. N. Davis, Auroral substorms observed from above the North Pole region by a satellite, J. Geophys. Res., 79, 1393, 1974.

Sugiura, M., T. Iyemori, R. A. Hoffman, N. C. Maynard, J. L. Burch and J. D. Winningham, Relationships between field-aligned currents, electric fields, and particle precipitation as observed by Dynamics Explorer-2, in Magnetospheric Currents, ed. by T. A. Potemra, AGU Geophysical Monograph 29, 1983.

Zmuda, A. J. and J. C. Armstrong, The diurnal flow pattern of field-aligned currents, J. Geophys. Res., 79, 4611, 1974.

CURRENTS IN THE EARTH'S MAGNETOTAIL

L. A. Frank, C. Y. Huang and T. E. Eastman

Department of Physics and Astronomy, The University of Iowa, Iowa City, Iowa 52242

Abstract. Currents in the earth's magnetotail are detected with the plasma instrumentation on board the ISEE-1 spacecraft. Field-aligned currents directed into and out of the ionosphere are found in the boundary layer of the plasma sheet. Typical current densities are in the range of 5×10^{-9} to 5×10^{-8} A/m^2. These currents are associated with the Region 1 current system that is observed previously at low altitudes. An intense current sheet is shown to exist at a discontinuity in convection electric fields ($\vec{\nabla} \cdot \vec{E} < 0$) during a period of great magnetic activity. Electron acceleration, similar to that for electron inverted-V precipitation regions at low altitudes, occurs in this current sheet. Examination of the plasma velocity distributions at a neutral sheet crossing reveals that the neutral sheet current is carried by electrons and protons. The relative directions for the bulk flows of the proton and electron plasmas indicate that the first adiabatic invariant is not conserved for the protons, and quite possibly for the electrons also. These latter findings are in substantial agreement with a neutral sheet model for acceleration of charged particles in the presence of a weak electric field.

Introduction

Direct observation of the currents in the earth's magnetotail is a newly developing research activity. Decisive determinations of the small current densities in the magnetotail are possible with comprehensive measurements of the plasma velocity distributions. With the plasma instrumentation on board recently launched spacecraft it is now possible to investigate the velocity distributions and spatial configuration of the current-carrying plasmas in the magnetotail. Even so, the perturbations in the velocity distributions that correspond to the signatures of currents are generally weak and considerable analyses are required to evaluate the currents. Our present knowledge of current systems in the magnetotail is derived largely from spacecraft-borne magnetometers. For example, the existence of field-aligned current sheets has been established for over a decade (Aubry et al., 1972; Fairfield, 1973; Sugiura, 1975). With such measurements the spatial configuration and current sheet intensities are gained. Typical current sheet intensities, or more precisely the equivalent intensities of an infinite current sheet, are in the range of tens of mA/m in the magnetotail. The direct detection of these currents provides further unique information concerning the physical processes participating in the formation and dynamics of the magnetotail, e.g., the identification of the current carriers, the magnitudes of the current densities and their source regions. To date few reports of findings of currents via measurements of plasma velocity distributions are available in the published literature. The first findings of field-aligned, or Birkeland currents in the magnetotail are given by Frank et al. (1981). These field-aligned currents are positioned at the boundary between the plasma sheet and the magnetotail lobe. Typical current densities are $\sim 10^{-8}$ A/m^2 and can be directed either into or out of the ionosphere. Further examples of field-aligned current densities are given by Huang et al. (1983a, b). No published literature is available for cross-field currents detected with plasma measurements in the magnetotail.

Our efforts in the direct evaluation of magnetotail currents are summarized in the following sections. The three areas of major effort are (1) field-aligned currents at the boundary of the plasma sheet, (2) a field-aligned current associated with a large discontinuity in the convection electric fields and (3) the cross-field, or perpendicular currents in the vicinity of the neutral sheet. We begin our discussion with several comments on the feasibility and limitations of such measurements of weak current densities.

Direct Detection of Currents

It is useful to review briefly several factors in the direct evaluation of current densities from plasma velocity distributions in the magnetotail. In order to find the current density it is necessary to measure the entire velocity distribution $f(\vec{v})$ of each major species and compute subsequently the bulk velocities $\vec{V}_o$,

$$\vec{V}_o = \frac{1}{n} \int \vec{v} f(\vec{v}) d^3v \qquad (1)$$

where n is the number density.
Suppose for the purposes of illustration that the velocity distribution is Maxwellian with tempera-

ture T and drifting with velocity $\vec{V}_o$, then

$$\vec{V}_o = K \int \vec{v} \exp\left(-\frac{m(\vec{v}-\vec{V}_o)^2}{2kT}\right) d^3v \quad , \qquad (2)$$

where $K = (m/2\pi kT)^{3/2}$. Thus from an observational viewpoint the determination of $\vec{V}_o$ is extremely dependent upon the ratio of bulk flow speed and the characteristic thermal speed $v_t = (2kT/m)^{1/2}$. For ions in the magnetotail these values are quite favorable for the detection of bulk flows, v_t is in the range of 100 to 1000 km/sec and V_o can be reliably determined with a threshold of tens of km/sec with present plasma instrumentation. This range of V_o is sufficient for detecting most ion bulk flows in the magnetotail. Numerous investigations of magnetotail dynamics have been based upon such determinations of ion bulk motions (cf. Hones, 1976, Frank, 1976). On the other hand, the electron bulk velocity must be evaluated also in order to find the current density. The detection of $\vec{V}_o$ for the electron velocity distributions is generally much more difficult since the thermal speeds are greater. The electron characteristic thermal speeds are $\sim 5 \times 10^3$ to 5×10^4 km/sec, and bulk speeds of < 100 km/sec are generally undetectable in the plasma sheet with present-day plasma instrumentation.

An example of a typical electron velocity distribution in the plasma sheet is shown in Figure 1. The velocity distribution is isotropic within the accuracy of the measurement. The positive $v_\parallel$ axis is directed parallel to the local magnetic field $\vec{B}$. Electron number density and temperature are 0.12 cm^{-3} and 7.9 $\times$ 10^6 °K, respectively. If the electron distribution is drifting along $\vec{B}$, for example, the bulk speed displaces the isodensity contours along the $v_\parallel$-axis. A speed of -107 km/sec, the instantaneous proton bulk speed, is shown for reference in Figure 1. It is seen that electron drift speeds of the order of typical ion bulk speeds are not detectable with presently operating plasma instrumentation. By numerical integration of the electron velocity distribution, an upper limit on the drift speed is found to be 120 km/sec and the corresponding current density is $< 2.1 \times 10^{-9}$ amp/m^2. Note that this upper limit on the electron drift speed is similar to the observed ion drift speed. Thus the direct detection of currents in the magnetotail is limited generally to regions where the electron drift speeds are > 100-200 km/sec. This situation is ameliorated somewhat for field-aligned currents since the current-carrying electrons favor small pitch angles with respect to $\vec{B}$.

In view of the above restriction it is worthwhile to consider the expectations for the magnitudes of these magnetotail current densities from considerations of the low-altitude measurements of field-aligned currents above the auroral zones. These low-altitude current densities are in the range of $\sim 10^{-5}$ to 10^{-4} A/m^2 (Cloutier and Anderson, 1975; Casserly and Cloutier, 1975; Potemra, 1979). The corresponding current densities as mapped into the magnetotail, B $\approx$ 30 nT, are 10^{-8} to 10^{-7} A/m^2. Thus these auroral currents should be detectable in the magnetotail unless the current system closes at lower altitudes.

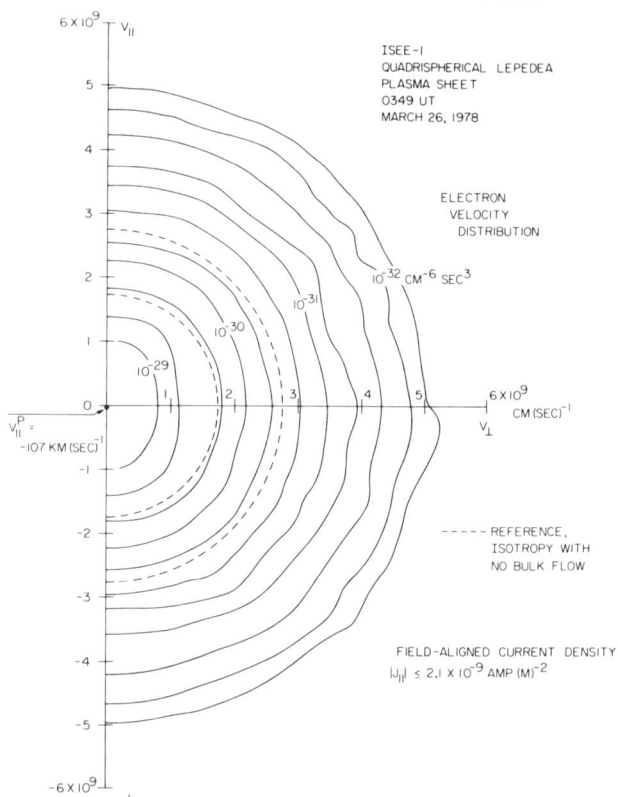

Fig. 1. Isodensity contours for an isotropic electron velocity distribution observed in the plasma sheet. Densities are plotted as functions of $v_\parallel$ and $v_\perp$, the velocity components parallel and perpendicular to the magnetic field vector. Also shown is the parallel component of the proton bulk flow $V_\parallel^P$. The upper limit for the field-aligned current density is $|j_\parallel| < 2.1 \times 10^{-9}$ A/m^2 (after Frank et al., 1981).

Field-aligned Currents at the Plasma Sheet Boundary

Surveys of the electron velocity distributions with the ISEE-1 plasma instrument reveal that the largest field-aligned current densities in the magnetotail occur most frequently at the boundary of the plasma sheet. The existence of persistent and distinct boundary layers of plasma at the interfaces of the magnetotail lobes and the plasma sheet is shown by DeCoster and Frank (1979).

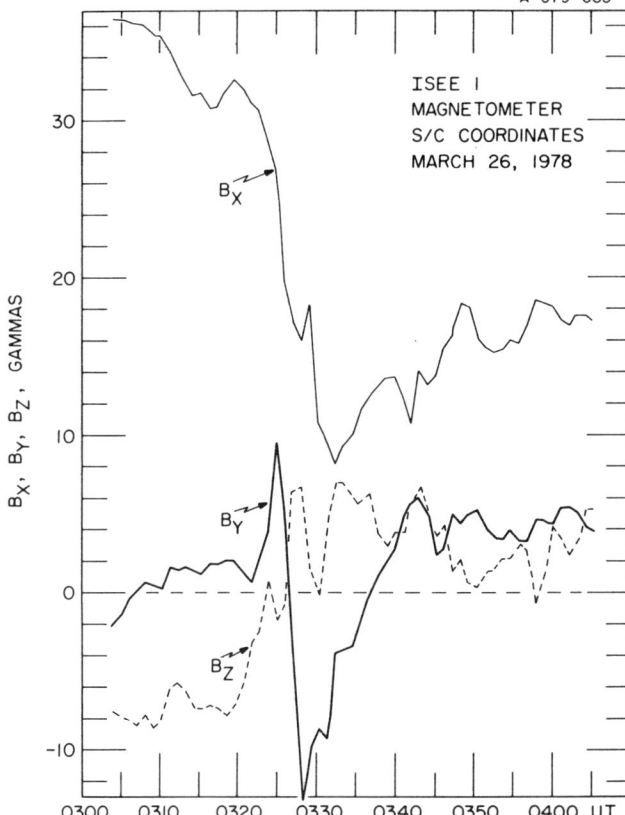

Fig. 2. Magnetic field components B_x, B_y, and B_z in spacecraft-referenced coordinates for the period spanning the crossing of the boundary of the plasma sheet. The spacecraft coordinate system is nearly aligned along the solar-ecliptic coordinate axes. The component B_y displays the signature of a multiple-sheet current system at ~ 0323 to 0340 UT (after Frank et al., 1981).

An extensive study of magnetotail plasmas by Eastman et al. (1983) finds that these boundary layers of the plasma sheet are among the primary plasma transport regions within the earth's magnetotail. These boundary layers are present during all phases of magnetic activity and are characterized by large field-aligned flows of plasma ions. Positive ions from both magnetosheath and ionosphere are found in these plasma sheet boundary layers. Large field-aligned current densities which are carried by electrons are detected also in the boundary layers.

We summarize here several principal findings of observations of field-aligned currents in the boundary layer from the analysis of the examples given by Frank et al. (1981). On March 26, 1978 at ~ 0330 UT the northern boundary layer of the plasma sheet is intercepted by the ISEE spacecraft as the plasma sheet expands during substorm recovery. The spacecraft coordinates (X_{SM}, Y_{SM}, Z_{SM}) are (-19.8, -6.1, 5.2 R_E), i.e., in the postmidnight sector of the magnetotail. The magnetic signature as provided by the UCLA magnetometer is shown in Figure 2. The magnetic field is directed generally sunward as expected for this position in the magnetotail and the fluctuations in the B_y component indicate the presence of field-aligned currents during the period 0320-0340 UT. Currents flowing into and away from the ionosphere are indicated by these magnetometer measurements.

These field-aligned currents are detected directly by the ISEE-1 plasma instrumentation. As the spacecraft traverses the boundary layer due to the expansion of the plasma sheet, three current sheets are detected with current densities directed into, out of and again into the ionosphere in order of decreasing distance to the plasma sheet. An example of the electron velocity distributions for the current sheet intercepted first is shown in Figure 3. The velocity

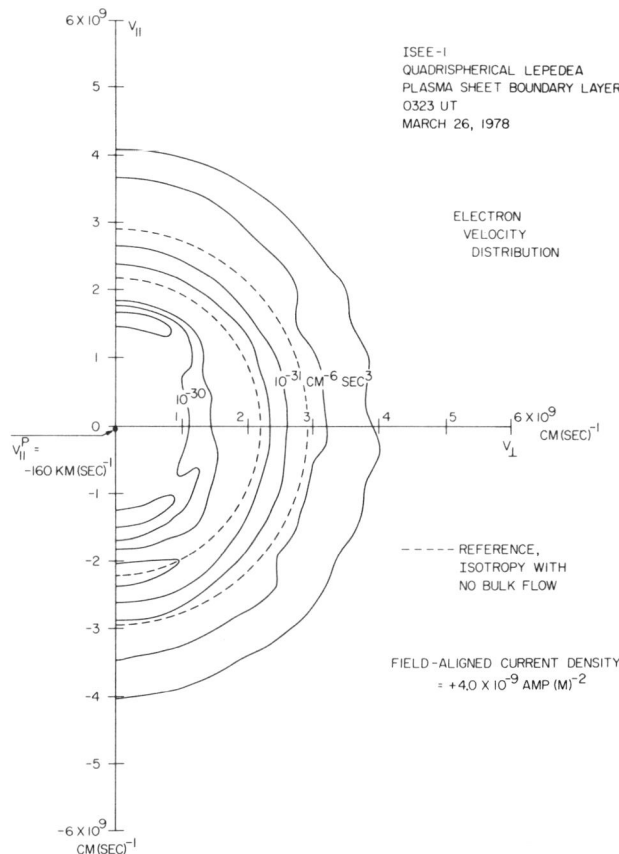

Fig. 3. Isodensity contours for electron velocity distributions within the first field-aligned current sheet encountered at 0323 UT by ISEE 1 as the plasma sheet expands over the spacecraft position. The current density is directed into the ionosphere (after Frank et al., 1981).

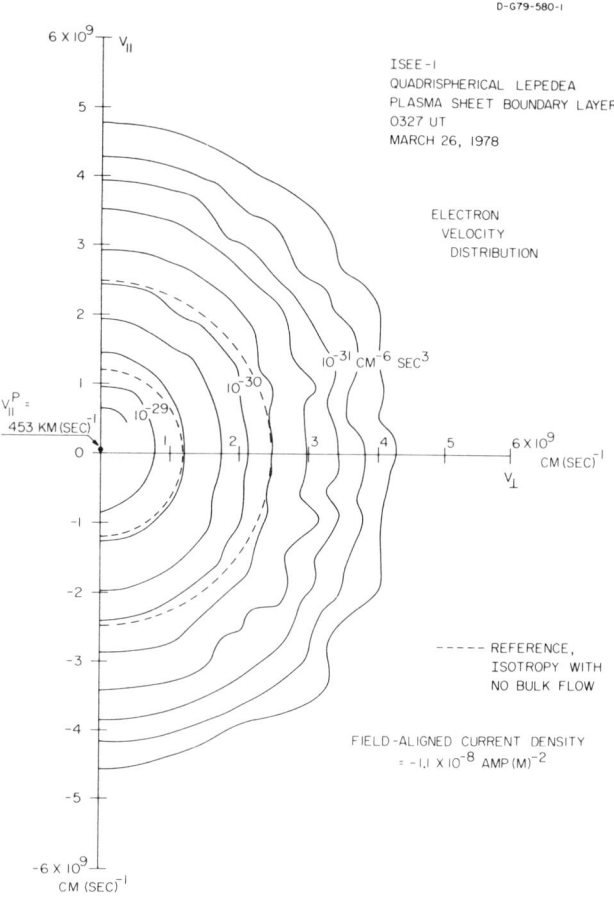

Fig. 4. Electron velocity distribution within the second, or center field-aligned current sheet at 0327 UT. The field-aligned current is directed antiparallel to $\vec{B}$ (net electron flux directed toward the ionosphere). The current is carried dominantly by electrons with energies < 1 keV (after Frank et al., 1981).

distribution $f(\vec{v})$ is plotted as functions of the parallel and perpendicular velocities, $v_\parallel$, $v_\perp$, relative to the local magnetic field vector $\vec{B}$. Note that this velocity distribution is not a Maxwellian at speeds less than $\sim 2 \times 10^9$ cm/sec and that the current-carrying electrons are found within $\sim 25°$ of the $v_\parallel$ axis. The field-aligned current density is numerically integrated, $d_{j\parallel} = e\, v_\parallel\, f(\vec{v}) d^3v$, from the isodensity contours of Figure 3. This current density, $+4.0 \times 10^{-9}$ A/m², is directed along $\vec{B}(+)$ and hence into the ionosphere. Since the current carriers favor considerably the pitch angles nearly parallel or antiparallel to $\vec{B}$ the detection of the current is considerably easier relative to that for a convecting Maxwellian distribution.

The electron temperature of the current-carrying electrons for the first current sheet is in the range of $\sim 10^6$ °K. These electrons are drifting away from the ionosphere into the magnetotail. Since electron temperatures at altitudes ~ 1000 km are only several thousand °K, heating and acceleration of these electrons occur at some position along the magnetic field lines between the spacecraft altitude at 20.4 R_E and the ionosphere.

An electron velocity distribution for the second current sheet encountered in the plasma sheet boundary layer is shown in Figure 4. The corresponding field-aligned current density is -1.1×10^{-8} A/m² and is directed out of the ionosphere. For such electron velocity distributions that flow toward the ionosphere the isodensity contours are anisotropic but without the structured features near the $v_\parallel$ axis. The third current sheet exhibits similar velocity distributions as those for the first sheet, i.e., current densities directed into the ionosphere and structured isodensity contours near the $v_\parallel$ axis. Thus three separate current sheets are detected during this particular traversal of the plasma sheet boundary layer.

A comparison of the current sheet intensities, A/m, as determined with the magnetometer and the direct measurements of the current densities and sheet thicknesses with the plasma analyzer is given in Figure 5. In order to compute the current sheet intensities from the plasma measurements the thickness of each sheet must be evaluated. The expansion speed of the plasma sheet in the Z_{SM} direction is evaluated using measurements of the proton convective velocity and independently verified with two-spacecraft observations of boundary motion, the latter with the Berkeley particle instruments. The corresponding expansion speed of the plasma sheet, and hence of the motion of the boundary layer, is 14 km/sec. The thickness of each current sheet is then 0.36, 0.55 and 0.74 R_E in chronological order of their crossing. As shown in the table of Figure 5, the agreement for current sheet strengths between magnetometer and plasma observations is good.

These current sheets are to be associated with the high-latitude Region 1 current system as identified at low altitudes by Iijima and Potemra (1978). The average Region 1 current in the local morning sector is directed into the ionosphere. Two of the above current sheets encountered at large distances in this local time sector with ISEE are directed toward the ionosphere. The current density in the center sheet is directed out of the ionosphere. This situation poses no inherent disagreement since Region 1 current patterns represent the average result of a large number of low-altitude measurements whereas the ISEE observations are of the detailed current sheet structure for a single example. However the ISEE results indicate an expectation for complex interleaving of positive and negative current sheets at auroral altitudes. If global magnetic field models (cf. Walker, 1979) are em-

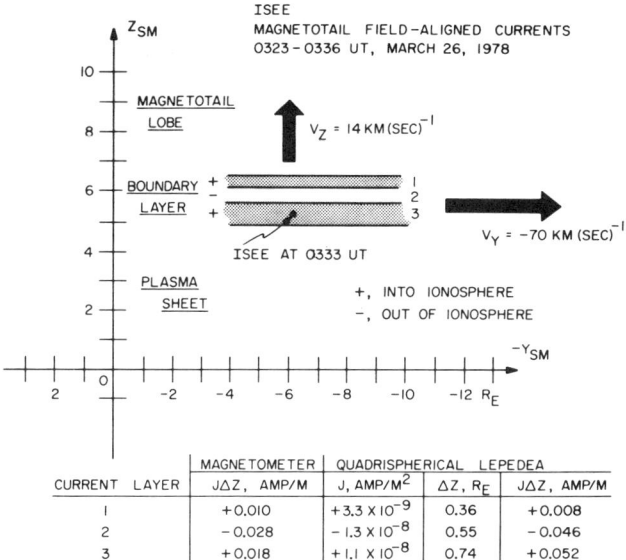

Fig. 5. Summary of the locations and thicknesses of the three field-aligned current sheets in the boundary layer of the plasma sheet. The spacecraft positional coordinate $X_{SM} = -19.8\ R_E$. Sheet intensities, A/m, as observed simultaneously with the magnetometer are tabulated at the bottom of the figure, along with current densities, sheet thicknesses and sheet intensities as computed from the plasma measurements (after Frank et al., 1981).

ployed to estimate the widths of the current sheets as projected to auroral altitudes, these widths are ~ 25, 39 and 52 km, in order of decreasing latitude.

Thus it is possible to directly detect the field-aligned currents in the magnetotail that are associated with the low-altitude Region 1 current system. These findings establish one more element in the circuit for the global ionospheric-magnetospheric current system. Direct observations of the lower latitude Region 2 currents are not available. These currents are expected to be positioned in the plasma sheet and its earthward extension, where detection of field-aligned currents may be considerably more difficult.

Field-aligned Currents at an Electric Field Discontinuity

One of the most striking examples of charged particle acceleration in the earth's magnetosphere is the well-known 'inverted-V' precipitation event that occurs at low altitudes, about one to several thousands of kilometers, over the evening sector of the auroral zone (Frank and Ackerson, 1972). As a low-altitude spacecraft traverses this electron beam incident on the earth's atmosphere, a dramatic increase in the average electron energies is observed. This electron precipitation region is associated usually with a reversal of convection electric fields and is believed to be the signature of magnetically field-aligned electric fields at higher altitudes (Frank and Gurnett, 1971). In order to establish the high-altitude boundary conditions for this phenomenon, e.g., plasma parameters and electric fields, it is an ongoing effort to search for the high-altitude signatures of this acceleration mechanism with the ISEE spacecraft. Since these precipitation regions associated with discrete arcs are most likely mapped into the boundary layer of the plasma sheet, and the boundary layer thickness is only < 1 R_E, the observational opportunities are limited. The identification requirements for a high-altitude encounter are a substantial reversal in convection electric fields, current densities directed out of the ionosphere and coincident heating or acceleration of the electron plasma. Only one such example is available at this time.

At about 1204 UT on May 1, 1978 the ISEE-1 spacecraft passed through a large discontinuity in convection electric fields coincidentally in the presence of field-aligned currents and electron heating (Huang et al., 1983b). This period is characterized by unusually great magnetic activity, the AE index increases to ~ 2500 nT. The solar-magnetospheric coordinates for the spacecraft position are (-13.2, 2.1, 3.9 R_E), i.e., at an altitude of 12.9 R_E in the premidnight sector of the magnetotail. Components of the magnetic field and of the convection electric fields are shown in Figure 6. The values for convection electric fields are computed from determinations of the proton drift velocities perpendicular to $\vec{B}$ from the velocity distributions. A large gradient in the z component of $\vec{E}_\perp$ is observed during the period 1202-1208 UT. The change in this component is 5.1 mV/m. Examination of the proton velocity distributions (not shown here) locates the spacecraft within the plasma sheet boundary layer during the discontinuity in $\vec{E}$. Proton bulk speeds along the magnetic field direction range from ~ 20 to 150 km/sec and the proton average kinetic energy is ~ 4 keV. The magnetosphere exhibits a large distortion. This distortion is noted in the change in sign of B_x at 1200-1205 UT in Figure 6, when the magnetotail is deflected sufficiently northward that the spacecraft is in the southern magnetotail.

A large field-aligned current density is detected at about 1204 UT and is coincident with the rapid change in $E_{\perp z}$. The electron velocity distribution in this current sheet is shown in Figure 7. The isodensity contours are qualitatively similar to those for electrons drifting toward the ionosphere in the boundary layer (cf. Figure 4). At 1204 UT the field-aligned current density is $+5.4 \times 10^{-8}$ A/m^2, i.e., the electron

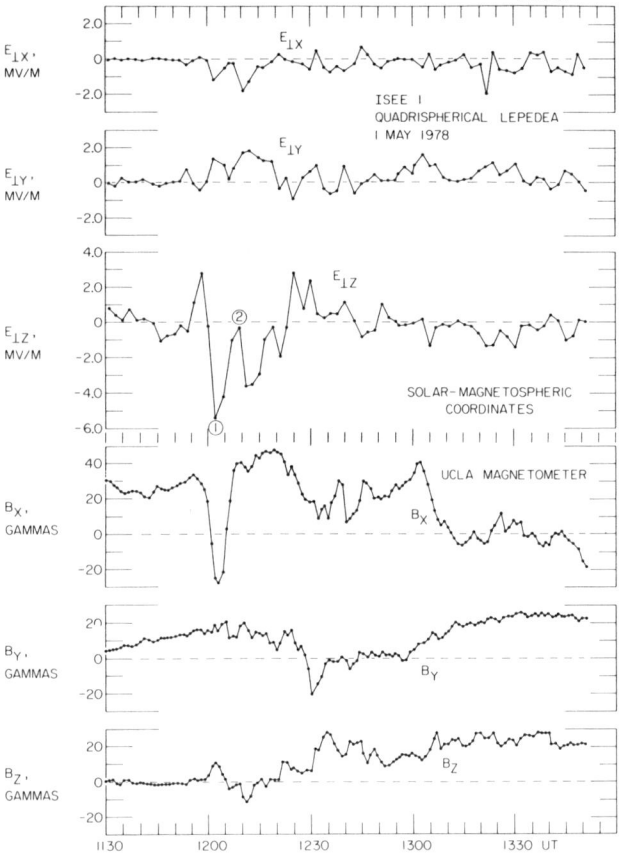

Fig. 6. Convection electric fields and magnetic fields as observed with the ISEE-1 spacecraft on May 1, 1978. The electric fields are calculated from determination of the proton $\vec{E} \times \vec{B}$ drifts with the plasma instrument. An unusual crossing of a high-altitude inverted-V structure is detected during the period centered at 1204 UT, coincident with a large gradient in $E_{\perp z}$. The spacecraft altitude is 12.9 R_E (after Huang et al., 1983b).

velocity component is small, < 50 km/sec, V_z is estimated from time delay of magnetic features at the two ISEE spacecraft. It is found that V_z = +25 (±5) km/sec. The thickness of the current sheet is based upon V_z and the magnetic field profile and is estimated at ~ 1000 km. The positive value of V_z yields $\vec{\nabla} \cdot \vec{E} < 0$. The major features of the magnetospheric convection and the field-aligned current sheet that are associated with this high-altitude electric field discontinuity are summarized in Figure 8.

These observations of an inverted-V acceleration region extending to altitudes of ~ 12 R_E can be compared quantitatively with the boundary-value model given by Lyons (1980). This model is derived from previous work by Atkinson (1970), Knight (1973), Lemaire and Scherer (1974), Chiu and Shultz (1978) and Chiu and Cornwall (1979). Specific mechanisms for electron acceleration are reviewed by Swift (1981). A discontinuity in the

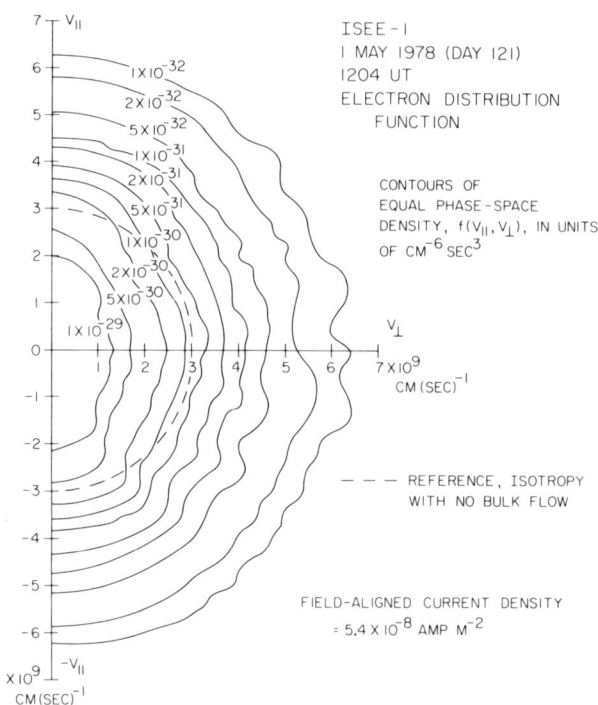

Fig. 7. Electron velocity distribution in the intense current sheet detected at 1204 UT on May 1, 1978. The field-aligned current density is +5.4 × 10^{-8} A/m². Ancillary observations are used to determine that this current is a net electron flux directed toward the southern auroral ionosphere. The signature of electron heating, similar to that observed for inverted-V precipitation zones at low altitudes, occurs coincidentally with the appearance of the current sheet at the spacecraft position (after Huang et al., 1983b).

flux is directed antiparallel to $\vec{B}$. Since the spacecraft is located in the southern magnetotail at this time, this net electron flux is directed toward the southern auroral ionosphere. Electron heating to thermal energies ~ 1 keV that is similar in character, if not magnitude, to that of the 'inverted-V' phenomena at lower altitudes is present in the current sheet.

In order to evaluate the thicknesses of the current sheet and convection zones from the time history of the events at the spacecraft it is necessary to determine the z component of the velocity of these features past the spacecraft. The sign of V_z determines also the sign of $\vec{\nabla} \cdot \vec{E}$ at the discontinuity in convection electric fields. Since the plasma measurements indicate that this

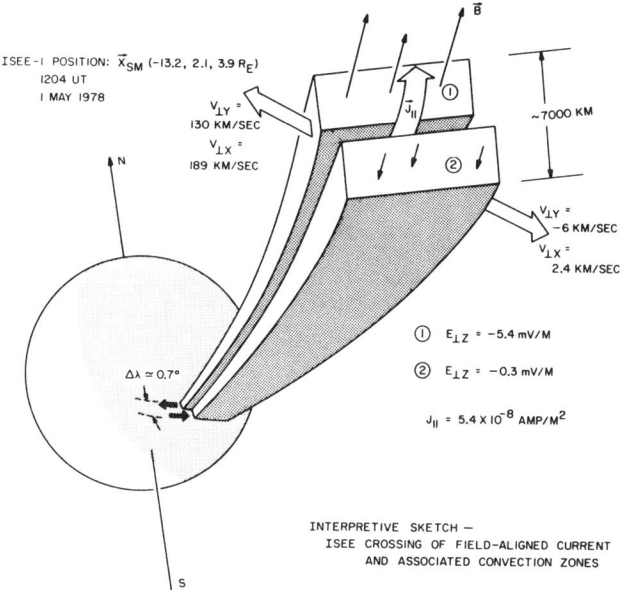

Fig. 8. Interpretive diagram showing the gross relationship of the plasma convection and field-aligned current sheet associated with the crossing of the high-altitude signature of an electron inverted-V region. The magnetic field directions, plasma bulk flow perpendicular to $\vec{B}$, and the electric field component $E_{\perp Z}$ are given for the boundaries of the large gradient in $E_{\perp Z}$ (after Huang et al., 1983).

convection electric fields at the upper boundary in the magnetosphere is assumed such that $\nabla \cdot \vec{E} < 0$ at the discontinuity. At this upper boundary the electron thermal energy K_{th} and number density n are specified. Electrons are the dominant current carriers. At the lower boundary in the ionosphere a uniform, height-integrated Pedersen conductivity Σ_p is assumed. For our calculations we use $\Sigma_p = 40$ mhos (Evans et al., 1977). The current density is assumed to be divergence-free, $\nabla \cdot \vec{j} = 0$, and the geometry is reduced to two orthogonal dimensions, the distance x along the ionosphere and the distance along the magnetic field lines. Thus Hall currents are not considered in the model. The ionospheric potential $\phi_i(x)$ can be found as the solution to the equation

$$j_\parallel = \frac{\partial}{\partial x}\left(\Sigma_p \frac{\partial \phi_i}{\partial x}\right) \quad (3)$$

where $j_\parallel$ is the field-aligned current density. Since the potential ϕ is specified at the upper boundary, the field-aligned potential is $\phi_i(x) - \phi$. The electrons are assumed to move freely in the time-independent $\vec{E}$ and $\vec{B}$ fields. The current density $j_\parallel$ is given by Knight (1973)

$$j_\parallel = e\, n \left(\frac{K_{th}}{2\pi m}\right)^{1/2} \cdot$$
$$\left\{\frac{B_i}{B_v}\left[1 - \left(1 - \frac{B_v}{B_i}\right) \exp -\left(\frac{eV_\parallel}{K_{th}\,(B_i/B_v - 1)}\right)\right]\right\} \quad (4)$$

where e is the electron charge, m is the electron mass, and B_i and B_v are the magnetic field strengths in the ionosphere and at the acceleration region, respectively. The field-aligned potential difference is $V_\parallel$. The upper, or magnetospheric boundary is assumed to be at the top of the acceleration region. The values for B_v and K_{th} at the ISEE position during the observation of the field-aligned current are 25 nT and 1.6 keV respectively. Thus the assumption that $B_i/B_v \gg eV_\parallel/K_{th}$ is valid and equation (4) reduces to

$$j_\parallel = en\left(\frac{K_{th}}{2\pi m}\right)^{1/2}\left(1 + \frac{eV_\parallel}{K_{th}}\right) \quad . \quad (5)$$

Also it is reasonable to assume that $eV_\parallel/K_{th} \gg 1$ for intense inverted-V events. Then

$$j_\parallel = K(\phi_i - \phi) \quad (6)$$

where $K = e^2 n/(2\pi m K_{th})^{1/2}$. The solutions of equation (3) are of the general form $C \exp(-x_i/x_w)$ where the ionospheric half-width is x_w (Lyons, 1980),

$$x_w = \left(\Sigma_p/K\right)^{1/2} \quad . \quad (7)$$

Thus this boundary-value model predicts the maximum field-aligned potential, the field-aligned current density and a half-width for the potential distribution at the ionosphere.

For comparison the ISEE plasma measurements provide the magnetospheric boundary values, the current density and the width of the current sheet at the outer boundary. The electron density and thermal energy at 12.9 R_E are 0.5 cm^{-3} and 1.6 keV, respectively. The measured value for $\Delta E_\perp$ across the discontinuity is -5.1 mV/m. The computed maximum current density and halfwidth x_w are 1.3 × 10^{-5} A/m^2 and 360 km, respectively, at the ionosphere. The observed current density as mapped to the ionosphere is 9.0 × 10^{-5} A/m^2 and is in moderate agreement with the above calculated current density. The comparison of the calculated width, $2x_w$, of the ionospheric current system with the observed width of the current sheet is more difficult to consider. The predicted width is ~ 720 km. The observed current sheet thickness as projected into the ionosphere is only ~ 20-30 km. However the plasma measurement is sufficiently sensitive to detect only the maximum current densities. The actual width of the region of parallel currents is probably much

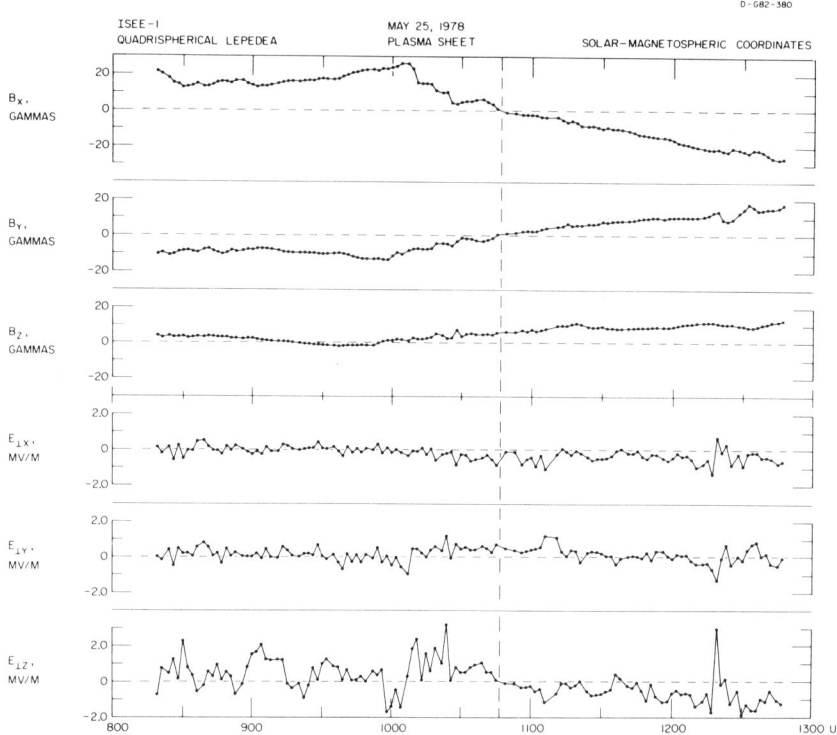

Fig.9. The components of the magnetic field $\vec{B}$ and the convection electric field $\vec{E}_\perp$ for a time interval centered on the crossing of the neutral sheet at ~ 1046 UT on May 25, 1978. The components of $\vec{E}_\perp$ are computed from the proton velocity distributions (after Frank et al., 1983).

larger. On the other hand the model assumes uniform ionospheric conductivity and a point discontinuity in the electric fields. The actual conductivities and distribution of convection electric fields are expected to impact the calculated widths (Lyons, 1981). The calculated maximum field-aligned potential is 43 kV and is in agreement with previous observations for intense inverted-V regions at low altitudes. For the present high-altitude observation it is quite likely that the potential difference is distributed along the magnetic field to much lower altitudes.

Cross-field Currents near the Neutral Sheet

Perhaps the most difficult observational task for direct detection of currents in the magnetotail is that for determination of cross-field, or perpendicular current densities in the hot plasmas near the neutral sheet. For our exploratory analyses we choose crossings of the neutral sheet for which a measurable convection electric field is present, a well-behaved reversal of the magnetic field component B_x occurs, and magnetic quiescent conditions prevail. Such an example is shown in Figure 9 for May 25, 1978 (Frank et al., 1983). The components of the magnetic and convection electric fields are given. The ISEE spacecraft position at the time of reversal in the magnetic component B_x at 1046 UT is (-10.7, 6.5, 2.6 R_E) in solar-magnetospheric coordinates. Again $\vec{E}_\perp$ is computed from observations of the proton bulk flow velocity. The AE index is < 50 nT for the period 0900-1200 UT. At the neutral sheet crossing, B_z = 6 nT and $B_x \simeq B_y \simeq$ 0 nT. The magnitudes of the components of $\vec{E}_\perp$ are in the range of 0.5 mV/m during the period 1030-1130 UT. The reversal of the sign of $E_{\perp z}$ is due to a similar reversal of B_x; the direction of the ion convection velocities does not reverse.

The lack of any major signature of the neutral sheet in the plasma observations, $\vec{E}_\perp$ in the case of Figure 9, is typical of the insensitivity of macroscopic plasma parameters in the identification of the neutral sheet position. These plasma moments include convection velocities, electron and ion temperatures, and number densities. However a detailed examination of the electron and proton velocity distributions reveals a remarkable situation. The proton velocity distribution perpendicular to the magnetic field, $f_\perp(\phi)$, at the neutral sheet is shown in Figure 10. The angle ϕ is the ion velocity direction in solar-ecliptic coordinates, i.e., at ϕ = 180° the velocity vector is directed antisunward. The velocity distribution is shown in Figure 10 for v = 1.70 X 10^8 cm/sec. Proton temper-

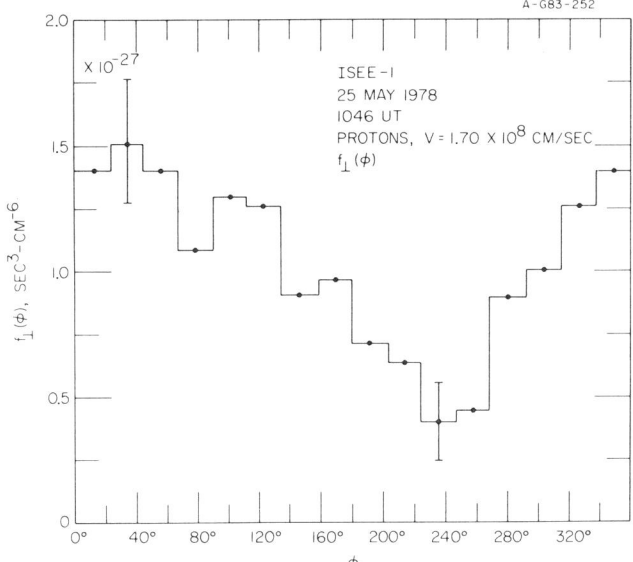

Fig. 10. The proton velocity distribution perpendicular to $\vec{B}$, $f_\perp(\phi)$, during the crossing of the neutral sheet at 1046 UT. The angle ϕ is the azimuthal angle about $\vec{B}$ and the proton speed is 1.70×10^8 cm/sec (after Frank et al., 1983).

ature is 9.9×10^7 °K and the number density is 0.19 protons/cm^3. A moments analysis for the entire velocity distribution yields a proton bulk flow velocity (122, 150, -35 km/sec) in solar-magnetospheric coordinates. This bulk flow corresponds to the maximum densities at $\phi \simeq 30°-50°$ in Figure 10. On the other hand, the electron velocity distribution at $v = 6.15 \times 10^9$ cm/sec shown in Figure 11 reveals that the electron bulk motion is in the direction $\phi \simeq 210°-270°$. This result is similar for the ion and electron velocity distributions that are taken just prior to and after 1046 UT. Thus the perpendicular current density $\vec{j}_\perp$ is measurable at the neutral sheet. The electron and proton bulk velocities show clearly that the plasmas are not convecting solely under the influence of $\vec{E} \times \vec{B}$ drifts.

The moments analysis of the entire electron velocity distribution for 1046 UT overestimates the electron bulk velocity due to the presence of secondary and photoelectrons at lower speeds in the measured velocity distribution. In order to obtain a more precise estimate of the electron bulk speed we assume a convecting Maxwellian distribution

$$f(\vec{v}) = C \exp(-m\vec{v}'^2/2kT) \quad (8)$$

where $\vec{v}' = \vec{v} - \vec{V}_o$ and $\vec{V}_o$ is the bulk velocity. If we define the quantity

$$\varepsilon(v) \equiv [f(v)]_{max}/[f(v)]_{min} = \exp(2mvV_o/kT) \quad (9)$$

then the drift speed is

$$V_o = 7.5 \times 10^{10} \frac{T}{v} \ln(\varepsilon) \quad (10)$$

where T is in °K and v, V_o are in units of cm/sec. Thus the electrons at higher energies in the velocity distribution can be used to estimate the drift speed, a technique which circumvents the effects of secondary and photoelectrons and utilizes the most sensitive portion of the observed velocity distribution. The relation (10) is limited to electron kinetic energies < 3-4 kT. At higher energies the electron energy spectrum is usually a power law.

The current density at the neutral sheet can be computed from the above analyses. The electron temperature for the measurements displayed in Figure 11 is 3.1×10^7 °K and the anisotropy ratio $\varepsilon \simeq 1.20$. The calculated electron bulk speed from equation (10) is 690 km/sec. The magnitudes of the current densities $j_\perp = enV_\perp$ due to the proton and electron plasmas are 5.7×10^{-9} A/m^2 and 2.0×10^{-8} A/m^2, respectively. Vector addition of the current densities for the proton and electron plasmas gives a total current density $j_\perp = 2.4 \times 10^{-8}$ A/m^2 directed along $\phi \simeq 60°$ (generally dawn to dusk).

The directions of the drift velocities of protons and electrons near the neutral sheet differ by $\Delta\phi \simeq 130°-190°$. Thus the fluid motions of the two plasmas cannot be attributed entirely to $\vec{E} \times \vec{B}$ drifts. The equation for fluid motion is

$$ne(\vec{E} + \vec{V} \times \vec{B}/c) = \nabla p \quad (11)$$

if we assume a time-independent Boltzmann equation for a collisionless plasma and an isotropic pressure p. The role of pressure gradients can

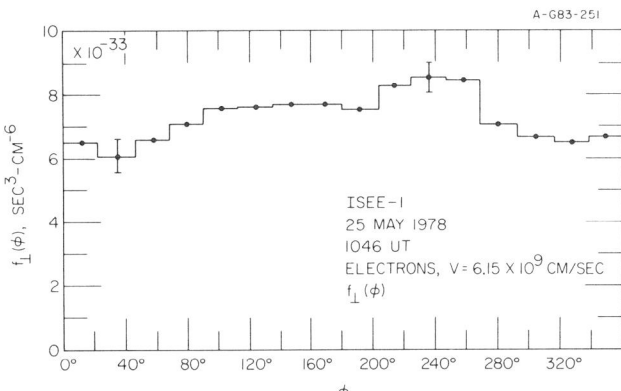

Fig. 11. The electron velocity distribution $f_\perp(\phi)$ measured at 1046 UT and to be compared with that for the proton distribution shown in Figure 10. The electron speed for this cut of the velocity distribution is 6.15×10^9 cm/sec (after Frank et al., 1983).

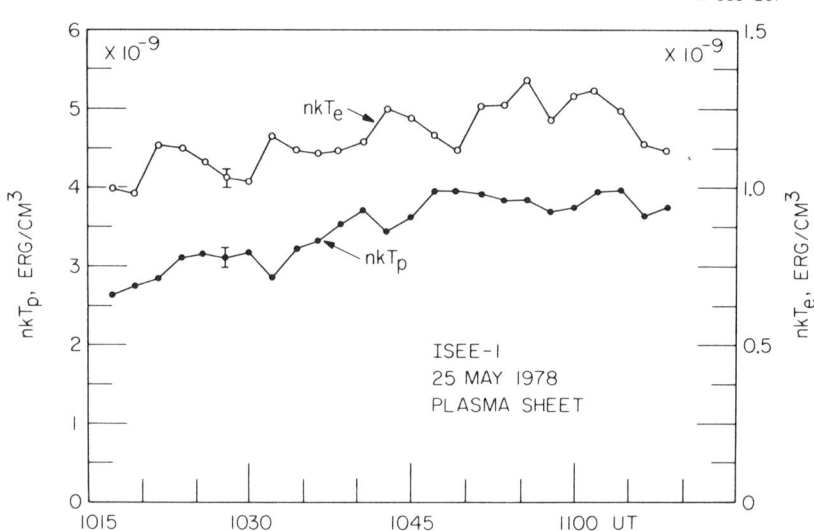

Fig. 12. Electron (e) and proton (p) plasma pressures observed in the vicinity of the neutral sheet (after Frank et al., 1983).

be examined with the electron and ion pressure profiles that are shown in Figure 12. The electron pressures exhibit the largest fluctuations, $\Delta p \simeq 2 \times 10^{-10}$ erg/cm^3, on time scales ~ 500 sec. We can estimate the scale lengths ΔL if pressure gradients are important for the observed electron fluid motion

$$\Delta L \simeq \frac{c\ \Delta p}{neVB} \ . \quad (12)$$

This scale length $\Delta L \simeq 150$ km if we attribute the entire electron fluid velocity to pressure gradients. This magnitude is inconsistent with the observed scale lengths, $\Delta L \simeq (50$ km/sec$)(500$ sec$) = 2.5 \times 10^4$ km, computed from the averaged bulk motion of the proton plasma relative to the spacecraft position. The distance the spacecraft traverses in 500 seconds is > 1000 km. The observed fluid motions are not due to pressure gradients.

It is evident from the fluid motion equation (11) that curvature and gradient drifts do not contribute to the observed plasma motions or current densities (Spitzer, 1952). Thus it appears that the proton plasma is weakly magnetized and the convective derivative, $\vec{V} \cdot \nabla \vec{V}$, in the Boltzmann equation should be included. The typical gyroradii for the protons and electrons at the neutral sheet for the example given above are 3000 km and 60 km, respectively. At least for the protons, the conditions for conserving the first adiabatic invariant of motion are violated by the magnetic scale lengths near the neutral sheet. From the measurements of the magnetic fields shown in Figure 9 gross estimates of scale lengths are < 1000 km. Speiser (1965, 1967) calculates single-particle trajectories in models of the magnetotail neutral sheet and finds that particles are accelerated in a weak dawn-to-dusk electric field. The normal component, B_z, in the neutral sheet is assumed typically to be 1 nT and the electric field strength ~ 0.25 mV/m. For our present example, the observed B_z is larger, ~ 6 nT, and the ambient convection electric fields are ~ 0.5 mV/m. The observed change in the proton energy corresponding to the bulk flow speed is $\sim 2-3$ keV. Lyons and Speiser (1982) report further calculations for the neutral sheet model and find that the particle acceleration is nearly independent of the $|\vec{E}|/B_z$ ratio. For our example, $|\vec{E}| \simeq 0.5$ mV/m, $B_z \simeq 6$ nT, the modeled results give an energy gain of 1-2 keV for a 20-keV proton. This result is to be compared with the observed bulk flow energy above. The agreement between observed and calculated values is good.

If the electron plasma is fully magnetized the electron $\vec{E} \times \vec{B}$ drift is expected to be closely perpendicular to the proton bulk velocity near the neutral sheet. The observed direction is in the range of 160° ($\pm$ 30°). Thus the electron plasma is probably not fully magnetized at the neutral sheet. This relative flow direction and the large fluid speed for electrons is consistent qualitatively with those given by the Lyons-Speiser model if the electron motion violates the first adiabatic invariant. In any case, the plasma phenomena of the neutral sheet are not readily seen in the macroscopic plasma parameters such as bulk flows and temperatures, but are found with examination of the velocity distributions.

Acknowledgments. We wish to thank N. D'Angelo for several useful discussions concerning the interpretation of observations. In our analyses measurements with other ISEE instruments are used: UCLA magnetometer (C. T. Russell), JHU/APL medium energy detector (D. J. Williams), UC/Berkeley fast plasma instrument (G K. Parks) and the LANL/MPI plasma instrument (E. W. Hones). This research was supported in part by the National Aeronautics and Space Administration under contract NAS5-26257 and grant NGL-16-001-002 and by the Office of Naval Research under grant N00014-76-C-0016.

References

Atkinson, G., Auroral Arcs: Result of the interaction of a dynamic magnetosphere with the ionosphere, J. Geophys. Res., 75, 4796, 1970.

Aubry, M. P., M. G. Kivelson, R. L. McPherron, and C. T. Russell, Outer magnetosphere near midnight at quiet and disturbed times, J. Geophys. Res., 77, 5487, 1972.

Casserly, R. T., and P. A. Cloutier, Rocket-based magnetic observations of auroral Birkeland currents in association with a structured auroral arc, J. Geophys. Res., 80, 2165, 1975.

Chiu, Y. T., and J. M. Cornwall, Electrostatic model of a quiet auroral arc, J. Geophys. Res., 85, 543, 1980.

Chiu, Y. T., and M. Schulz, Self-consistent particle and parallel electrostatic field distributions in the magnetospheric-ionospheric auroral regions, J. Geophys. Res., 83, 629, 1978.

Cloutier, P. A., and H. R. Anderson, Observations of Birkeland currents, Space Sci. Rev., 17, 563, 1975.

DeCoster, R. J., and L. A. Frank, Observations pertaining to the dynamics of the plasma sheet. J. Geophys. Res., 84, 5099, 1979.

Eastman, T. E., L. A. Frank, and C. Y. Huang, The boundary layers as the primary transport regions of the earth's magnetotail, (submitted for publication), J. Geophys. Res., 1983.

Evans, D. S., N. C. Maynard, J. Troim, T. Jacobsen, and A. Egelund, Auroral vector electric field and particle comparisons, 2, Electrodynamics of an arc, J. Geophys. Res., 82, 2235, 1977.

Fairfield, D. H., Magnetic field signatures of substorms on high-latitude field lines in the nighttime magnetosphere, J. Geophys. Res., 78, 1553, 1973.

Frank, L. A., and K. L. Ackerson, Local-time survey of plasma at low altitudes over the auroral zones, J. Geophys. Res., 77, 4116, 1972.

Frank, L. A., and K. L. Ackerson, and R. P. Lepping, On hot tenuous plasmas, fireballs, and boundary layers in the earth's magnetotail, J. Geophys. Res., 81, 5859, 1976.

Frank, L. A., and D. A. Gurnett, Distributions of plasmas and electric fields over the auroral zones and polar caps, J. Geophys. Res., 76, 6829, 1971.

Frank, L. A., C. Y. Huang, and T. E. Eastman, Currents in the neutral sheet of the earth's magnetotail, (to be submitted for publication), J. Geophys. Res., 1983.

Frank, L. A., R. L. McPherron, R. J. DeCoster, B. G. Burek, K. L. Ackerson, and C. T. Russell, Field-aligned currents in the earth's magnetotail, J. Geophys. Res., 86, 687, 1981.

Hones, E. W., Jr., S. J. Bame, and J. R. Asbridge, Proton flow measurements in the magnetotail plasma sheet made with Imp 6, J. Geophys. Res., 81, 227, 1976.

Huang, C. Y., T. E. Eastman, and L. A. Frank, Periodic substorm activity in the geomagnetic tail, (submitted for publication), J. Geophys. Res., 1983a.

Huang, C. Y., L. A. Frank, and T. E. Eastman, High-altitude observations of an intense inverted-V event, (submitted for publication), J. Geophys. Res., 1983b.

Iijima, T., and T. A. Potemra, Large-scale characteristics of field-aligned currents associated with substorms, J. Geophys. Res., 83, 599, 1978.

Knight, S., Parallel electric fields, Planet. Space Sci., 21, 741, 1973.

Lemaire, J., and M. Scherer, Ionosphere-plasmasheet field-aligned currents and parallel electric fields, Planet. Space Sci., 22, 1485, 1974.

Lyons, L. R., Generation of large-scale regions of auroral currents, electric potentials, and precipitation by the divergence of the convection electric field, J. Geophys. Res., 85, 17, 1980.

Lyons, L. R., The field-aligned current versus electric potential relation and auroral electrodynamics, Physics of Auroral Arc Formation, AGU monograph, 1981.

Lyons, L. R., and T. W. Speiser, Evidence for current sheet acceleration in the geomagnetic tail, J. Geophys. Res., 87, 2276, 1982.

Potemra, T. A., Current systems in the earth's magnetosphere, Rev. Geophys. Space Physics, 17, 640, 1979.

Speiser, T. W., Particle trajectories in model current sheets, 1, Analytical solutions, J. Geophys. Res., 70, 4219, 1965.

Speiser, T. W., Particle trajectories in model current sheets, 2, Applications to auroras using a geomagnetic tail model, J. Geophys. Res., 72, 3919, 1967.

Spitzer, L., Jr., Equations of motion for an ideal plasma, Ap. J., 116, 299, 1952.

Sugiura, M., Identification of the polar cap boundary and the auroral belt in the high-altitude magnetosphere: A model for field-aligned currents, J. Geophys. Res., 80, 2057, 1975.

Swift, D. W., Mechanisms for auroral precipitation: A review, Rev. Geophys. Space Phys., 19, 185, 1981.

Walker, R. J., Quantitative modeling of planetary magnetospheric magnetic fields, Quantitative Modeling of Magnetospheric Processes, ed. by W. P. Olson, published by American Geophysical Union, p. 9, 1979.

CHARACTERISTICS OF THE CROSS-TAIL CURRENT IN THE EARTH'S MAGNETOTAIL

A. T. Y. Lui

Applied Physics Laboratory, Johns Hopkins University, Laurel, Maryland 20707

Abstract. The basic features and substorm changes of the neutral sheet within the downstream distance of ~ 30 R_e in the magnetotail are briefly reviewed. Different signatures of the neutral sheet are presented. A classification scheme for the varieties in neutral sheet signatures is proposed. Based on this classification scheme, the most common features observed at the neutral sheet are identified to be those anticipated for a classical neutral sheet crossing. Detailed study of the varieties in neutral sheet signatures reveals (1) wave profiles on the neutral sheet surface, (2) magnetic islands embedded in the neutral sheet, (3) disordered magnetic field regions, (4) magnetic structures with large dawn-dusk field components, and (5) partial penetration ($\sim 50\%$) of the dawn-dusk component of the interplanetary magnetic field into the neutral sheet. The large-scale substorm changes of the cross-tail current are examined in one case study. The general observed features are consistent with the idea of diversion of the near-earth cross-tail current into the ionosphere at substorm onset.

Introduction

It is well established that the auroral electrojets are parts of a three-dimensional current system which links the ionosphere with the magnetosphere via magnetic-field-aligned currents (Zmuda and Armstrong, 1974; Bostrom, 1974; Potemra, 1979). The ionospheric currents are dissipative portions of the circuit. The energy and dynamo required to drive these currents reside above the ionosphere, such as in the magnetospheric boundary layer (Eastman et al., 1976) and the plasma sheet (Rostoker and Bostrom, 1976). The cross-tail current is a magnetospheric component of the current system and its basic features and dynamic characteristics constitute to an integral understanding of the entire current circuit in the magnetosphere-ionosphere medium.

In this paper, the neutral sheet which carries a significant portion of the cross-tail current is examined in detail. The observed varieties in the signature of the neutral sheet are first reviewed in Section 2 and are classified in Section 3. The dawn-dusk component of the magnetic field at the neutral sheet is investigated in Section 4. Several inferred features of the neutral sheet are then discussed in Section 5, followed by inferred substorm changes of the cross-tail current in Section 6. The observations presented below are primarily from IMP 5 and 6. Both spacecraft were in highly eccentric orbits with IMP 5 apogee at ~ 29 R_e and IMP 6 apogee at ~ 32 R_e.

Observed Signatures of the Neutral Sheet

The neutral sheet in the earth's magnetotail was discovered almost two decades ago by magnetic field measurements from IMP 1 (Ness, 1965). The neutral sheet was recognized as the region in which the magnetic field orientation reverses from pointing sunward to tailward or vice versa. Figure 1 shows examples of neutral sheet crossings by IMP 6 in the dusk portion of the magnetotail at a downstream distance of ~ 23 R_e on Nov. 16, 1971. The measurements are averages over ~ 15.36s intervals. The magnetic field parameters in Figure 1 are the field magnitude (B), elevation angle (θ_{sm}) and azimuthal angle (ϕ_{sm}) in solar magnetospheric system, and field variances parallel ($\sigma_\parallel$) and perpendicular ($\sigma_\perp$) to the averaged magnetic field direction. The field variances are defined by

$$\sigma_\parallel^2 = [\sigma_{xx}^2 B_x^2 + \sigma_{yy}^2 B_y^2 + \sigma_{zz}^2 B_z^2 + 2(\sigma_{xy}^2 B_x B_y + \sigma_{yz}^2 B_y B_z + \sigma_{zx}^2 B_z B_x)]/B^2$$

$$\sigma_\perp^2 = \sigma_{xx}^2 + \sigma_{yy}^2 + \sigma_{zz}^2 - \sigma_\parallel^2$$

$$\sigma_{xy}^2 = \frac{N \sum_{i=1}^{N} B_{yi} B_{xi} - \sum_{i=1}^{N} B_{xi} \sum_{i=1}^{N} B_{yi}}{N(N-1)}$$

where (B_{xi}, B_{yi}, B_{zi}) and (B_x, B_y, B_z) are the components of individual (40 ms) field measurement and of the averaged field, respectively. The variances σ_{yz}, σ_{zx}, σ_{xx}, σ_{yy}, σ_{zz} are defined in a similar way as σ_{xy}.

In the beginning of the interval at 05:40 UT, the field is pointing tailward, suggesting that IMP 6 is in the southern half of the tail. Near 05:50 UT, the field orientation changes from tailward to sunward, suggesting that IMP 6 crosses the neutral sheet to the northern half of the tail. Two other crossings of the neutral sheet are detected near 05:55 UT and 05:59 UT. Coincident with these three field reversals are decreases in the field magnitude to ~ 4 nT and increases in the field elevation angle to above 80°. The field fluctuations are very small during these crossings, as indicated by both $\sigma_\|$ and $\sigma_\perp$ being less than 1 nT. These crossings shown in Figure 1 may be considered as classical examples of neutral sheet crossings.

The decrease in field magnitude and increase

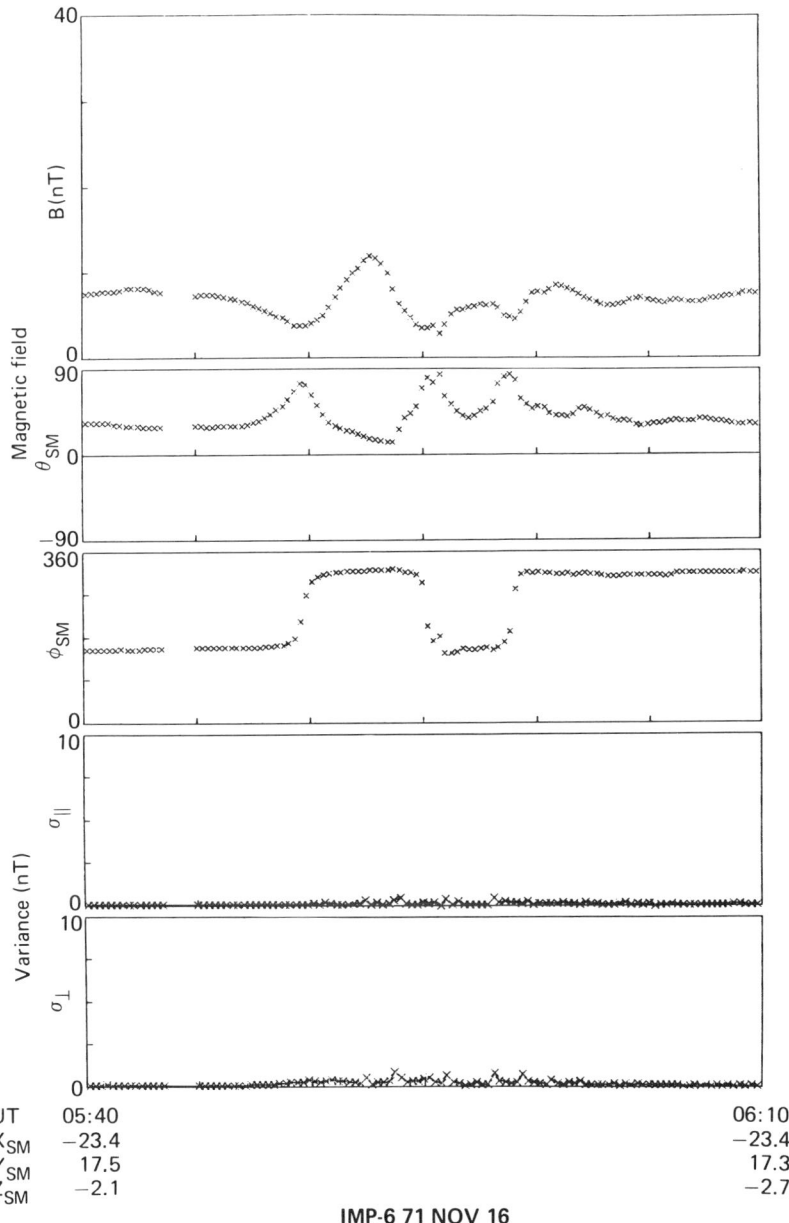

Fig. 1. Three crossings of the neutral sheet on Nov 16, 1971 are shown by the IMP-6 magnetic field measurements averaged over 15.36 sec. The field fluctuations parallel ($\sigma_\|$) and perpendicular ($\sigma_\perp$) to the averaged magnetic field direction are defined in the text. The spacecraft location is given in solar magnetospheric coordinates below the bottom panel. These crossings are "classical" examples of neutral sheet crossings which show decreases in the field magnitude and increases in the field elevation angle.

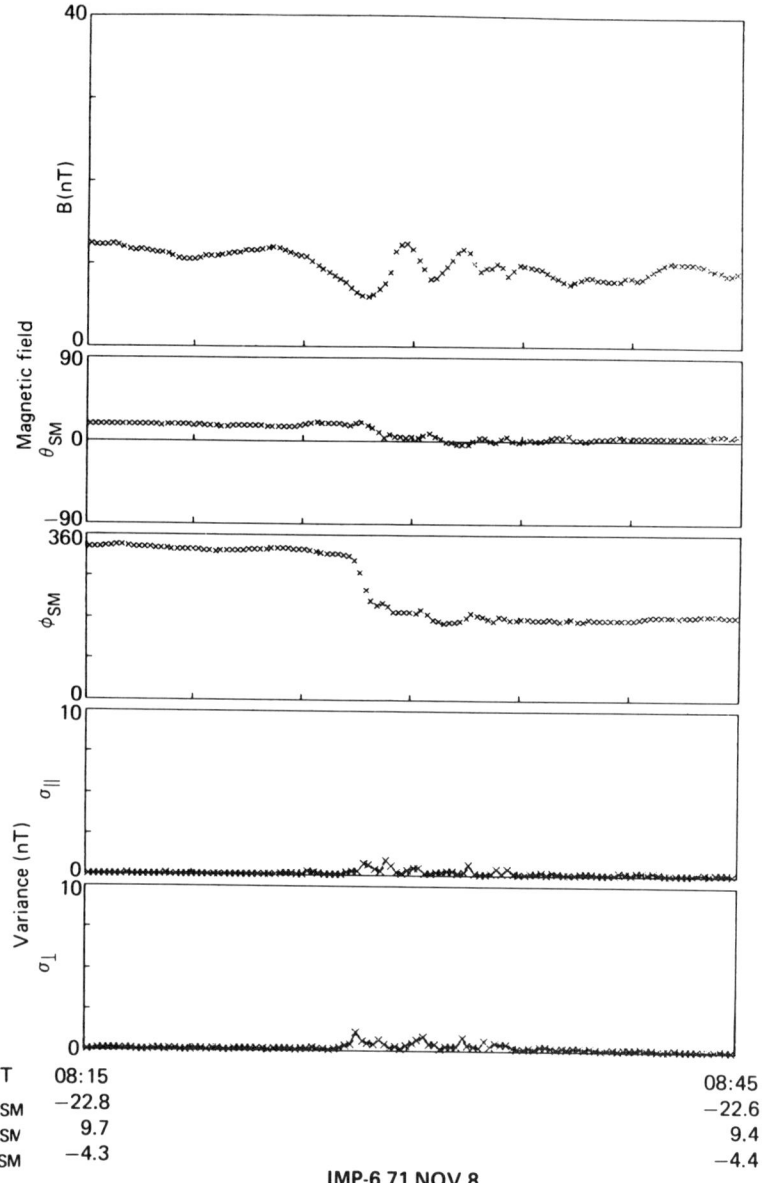

Fig. 2. An example of neutral sheet crossing which deviates from the classical signatures of the neutral sheet in that the elevation angle (θ_{sm}) shows a step-like transition at the time of crossing.

in field elevation angle associated with the field reversal in azimuth are anticipated signatures of the neutral sheet. However, all these features are not necessarily duplicated in many other neutral sheet crossings. Figure 2 shows a neutral sheet crossing in which the variation of the field elevation angle θ_{sm} deviates from the previous examples in Figure 1. The neutral sheet is encountered near 08:27 UT in Figure 2 when the field direction reverses from sunward to tailward. In this case, no large increase of θ_{sm} is detected at the crossing. When IMP 6 is in the northern half of the tail before 08:27 UT (indicated by the magnetic field pointing sunward), the angle θ_{sm} is small at $\sim 20°$. After the neutral sheet is crossed, the angle θ_{sm} decreases to $\sim 0°$.

Figure 3 shows another deviation of the features anticipated for the neutral sheet. In this example, the transition of the field azimuthal direction from sunward to tailward occurs very gradually over an interval of ~ 8

min from 20:49 to 20:57 UT. This is relatively long in comparison with the short transition of ~ 1 min or less for the previous examples of neutral sheet crossings. In addition, the field magnitude decreases to a minimum of ~ 3 nT at the start of the transition, but increases in the middle of the transition to ~ 6 nT, a value which is quite comparable with the field magnitude outside the neutral sheet crossing. In other words, the field minimum is not found near the mid-transition of the crossing.

Whereas the neutral sheet crossings in previous examples are associated with very low field fluctuations, there are occasions when neutral sheet crossings are very complicated and are difficult to be isolated. Figure 4 shows an interval containing this complex field signature. During this period, IMP 6 is at a downstream distance of ~ 19 R_e not far from the tail axis. It is seen that the field magnitude varies from ~ 1 nT to 15 nT. The angle θ_{sm} fluctuates within -70° to 80° while θ_{sm} often

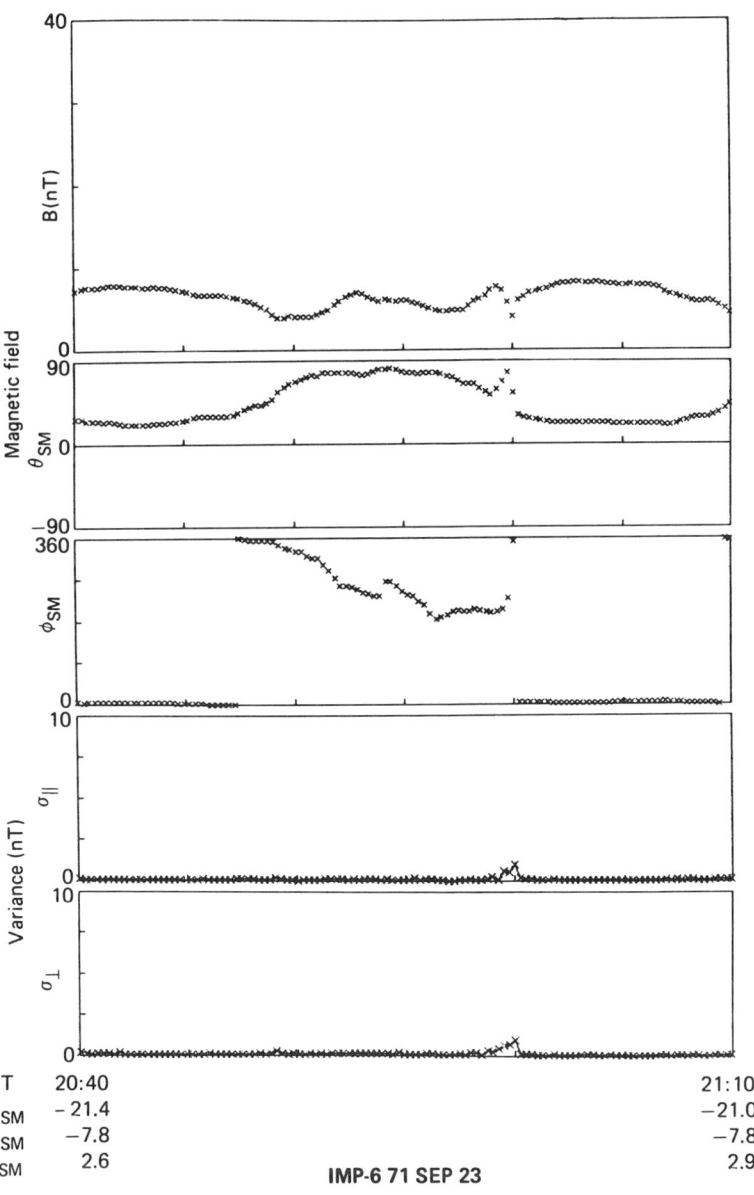

Fig. 3. A neutral sheet crossing from 20:49 to 20:57 UT in which the minimum field magnitude appears near the beginning and not near the middle of the transition.

varies irregularly at field reversals. The accompanying field variances $\sigma_\parallel$ and $\sigma_\perp$ are often large, exceeding 5 nT during some crossings.

Classification of Neutral Sheet Signatures

In section 2, four different signatures of neutral sheet crossings are illustrated. Each of the examples show differences which cannot be accounted for by a relative motion between the spacecraft and the neutral sheet. The result that all neutral sheet crossings are not similar raises the question of how often each type of crossing is observed. To address this question, a simple one-parameter classification scheme is adopted here as a first approach. Three years (1971 to 1973 inclusive) of magnetic field measurements from IMP 6 are used. Only neutral

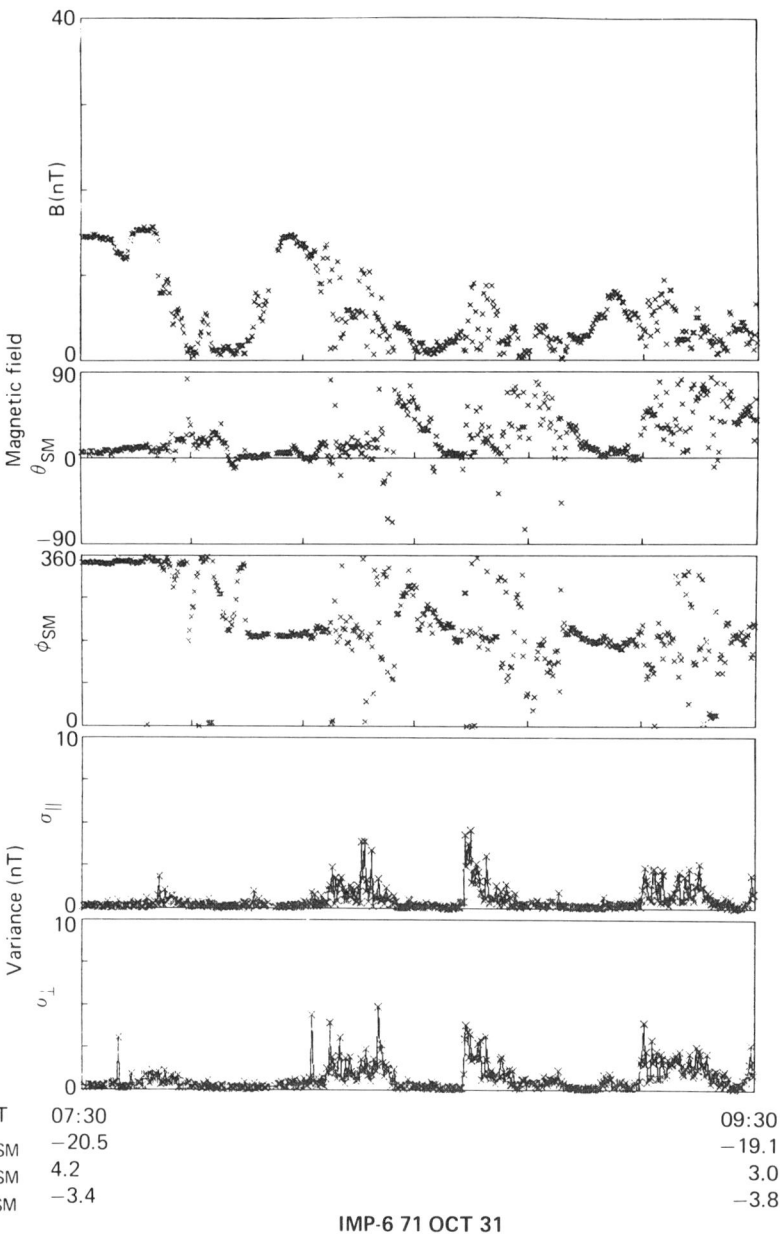

Fig. 4. An interval with many complicated neutral sheet crossings which are difficult to be isolated. All the field parameters fluctuate considerably and the field variances $\sigma_\parallel$ and $\sigma_\perp$ are high during some crossings.

sheet crossings identified at locations within $|Y_{sm}| \leq 20 R_e$ and $X_{sm} < -10 R_e$ are included. Each group of multiple crossings are sorted by judging the behavior of the field magnitude, field elevation angle, field azimuthal angle, field variances separately at the field reversals indicated by the change in ϕ_{sm}. Neutral sheet encounters with the same classification category and are separated by less than 0.5 hr from neighboring neutral sheet encounters are treated to be within the same group of multiple crossings. Table 1 lists the result of the classification, based on a total of 363 sets of multiple neutral sheet crossings. Three types of field magnitude variations across the neutral sheet are used. The field magnitude shows (1) a significant decrease as illustrated by the neutral sheet crossings in Figure 1, or (2) a highly structured variation as the first neutral sheet crossing in Figure 3, or (3) a relatively uniform field magnitude across the neutral sheet. It is found that about 58% of neutral sheet crossings are in the first category, 33% in the second and 9% in the third. The most common field magnitude variation across the neutral sheet is a significant decrease, as anticipated. In terms of the field elevation angle, it may increase (as the neutral sheet crossings in Figure 1), remain unchanged, decrease (as in some neutral sheet crossings in Figure 4), fluctuate or change in a stepwise fashion at the field reversal (as the example in Figure 2). From Table 1, it is seen that the most common θ_{sm} variation (56%) is an increase at the field reversal, again as anticipated. Under the field azimuthal angle classification, the ϕ_{sm} transition may be rather monotonic such as the neutral sheet crossings shown in Figures 1, 2, and 3, or be rather unorderly as those in Figure 4. The monotonic ϕ_{sm} transition may be completed within 1 minute, tens of minutes or on the order of hours. The result indicates that the majority of neutral sheet crossings (57%) show monotonic ϕ_{sm} transition. Finally, in terms of field variances $\delta B = \sqrt{\sigma_\perp^2 + \sigma_\parallel^2}$, the neutral sheet may be associated with low ($\delta B < 5$ nT) or high ($\delta B \geq 5$ nT) fluctuations. The statistics indicate that about 77% of neutral sheet crossings are those with low field fluctuations. It is indeed reassuring to know that all the most common characteristics of the neutral sheet in this 1-parameter scheme are those in classical examples of neutral sheet crossings such as the three multiple crossings in Figure 1.

The Dawn-Dusk Magnetic Field Component at the Neutral Sheet

For a group of multiple neutral sheet crossings, it is often found that the B_y component is nonzero and maintains the same sign throughout all crossings. The case shown in Figure 1 is an example, in which the B_y component is negative for all three crossings. Fairfield (1980) has examined the dawn-dusk field component in the earth's magnetotail based on IMP 6 measurements. When the effects of tail flaring and aberration of the tail axis are subtracted, it is found that the residual B_y component is correlated statistically with the interplanetary B_y component in the following way:

$$B_y \text{ (tail)} = 0.13 \, B_y \text{ (interplanetary)} - 0.30$$

The dependence of B_y in the tail on B_y in the solar wind suggests that the dawn-dusk component of the interplanetary magnetic field penetrates partially (~ 13%) into the magnetotail.

At equal distance from the neutral sheet, the B_y components due to tail flaring and aberration of the tail axis are equal and opposite for the northern and southern halves of the tail. Therefore, these two effects do not contribute any B_y component at the neutral sheet. This allows one to investigate the relationship between B_y in the solar wind and B_y in the tail at the neutral sheet without the need to remove the two effects discussed above. Figure 5 shows a scatter plot between the averaged B_y component in multiple neutral sheet crossings ($\bar{B}_{yt}$) versus the hourly averaged B_y of the interplanetary magnetic field (B_{yip}) taken from King (1977). Data from two years (1971 - 1972) of IMP 6 tail crossings are used. There is indication from this scatter plot that these two parameters are

TABLE 1: Classification of Neutral Sheet Signatures

FIELD MAGNITUDE		
Decrease	57.9%	
Structured	33.3%	
Uniform	8.8%	
FIELD ELEVATION ANGLE		
Increase	55.9%	
Unchanged	6.3%	
Decrease	3.6%	
Fluctuate	34.2%	
FIELD AZIMUTHAL ANGLE		
Monotonic Transition 1 min	25.3%	
10's min	39.1%	57.2%
hour	2.8%	
Transition Unorderly	42.8%	
FIELD FLUCTUATIONS		
Low Variance ($\delta B < 5$ nT)	77.1%	
High Variance ($\delta B \geq 5$ nT)	22.9%	

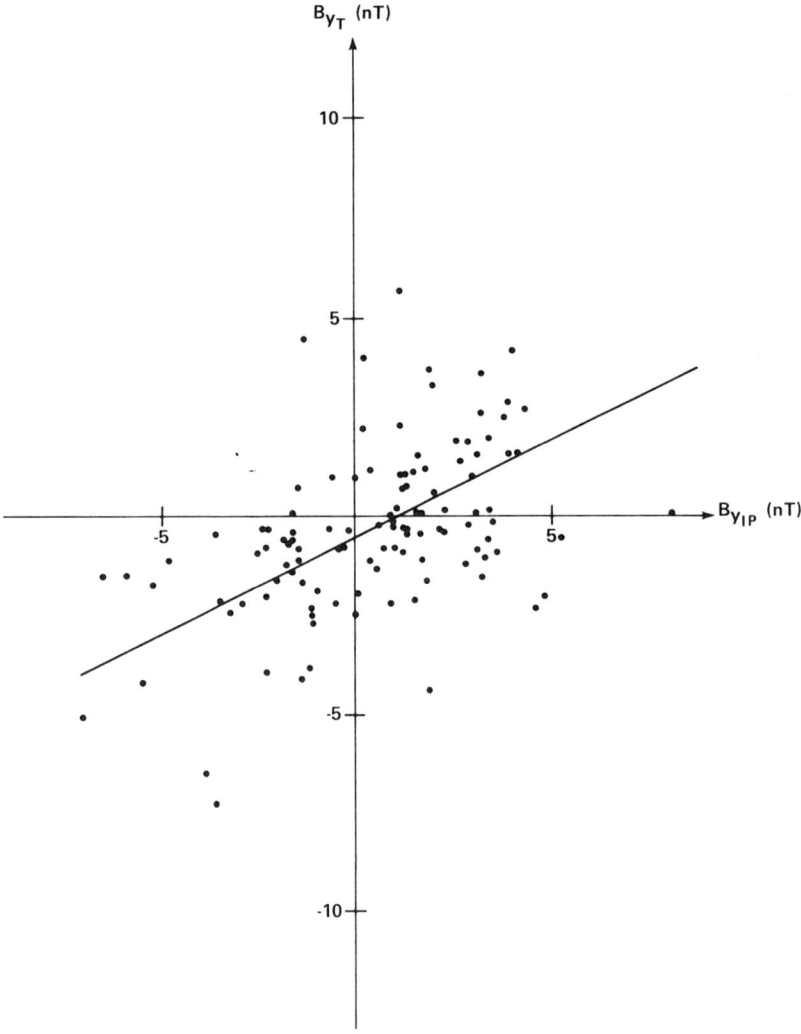

Fig. 5. A plot of the averaged B_y component in multiple neutral sheet crossings (B_{yT}) versus the hourly B_y component of the interplanetary magnetic field (B_{yIP}) at the corresponding time. A linear correlation is indicated, suggesting that about 50% of the B_y component of the interplanetary magnetic field penetrates into the neutral sheet region.

related. A linear regression fit gives

$$B_y \text{ (tail)} = (0.5 \pm 0.1) B_y \text{ (interplanetary)} - (0.5 \pm 0.2)$$

with a correlation coefficient of 0.6. The scatter in data points may be attributed to the contribution from local structures of the neutral sheet. The slope obtained here is slightly larger than that given by Fairfield (1980) while the y-intercepts are about the same within the uncertainty range. The difference in the slope does not necessarily indicate that the present result contradicts with that of Fairfield. The slope derived in Fairfield's analysis is based on magnetic field measurements over the entire magnetotail sampled by IMP 6 and thus include tail regions outside the neutral sheet, whereas the present result is pertinent to the neutral sheet region only. Therefore, it is possible that the result in Figure 5 may indicate that the dawn-dusk component of the interplanetary magnetic field is shielded less in the neutral sheet region of the tail.

Several Features of the Neutral Sheet

In this section, we address briefly the question of how different signatures presented in Section 2 may reveal structures in the neutral sheet.

Wave Profile of the Neutral Sheet Surface. Referring back to Figure 2, the unusual feature in the observed crossing is a significant decrease in the field elevation angle resembling a step function change. Figure 6 shows how this type of neutral sheet crossing can be modeled by a wave profile of the neutral sheet. This type of wave profile has been suggested by Speiser (1973). The diagram at the right top corner illustrates schematically the geometry of the situation. The one-dimensional equilibrium current sheet solution (Harris, 1962) is modified to allow a wave profile of the neutral sheet along the tail axis. The magnetic field configuration is constructed by the vector potential

$$\vec{A} = (B_{yo}Z, -B_oL \ln[\cosh(\frac{Z-A \sin kx}{L})]+B_nX, 0)$$

where

$B_o = 10$ nT, $B_{yo} = -5$ nT, $B_n = 2$ nT, $L = 2\ R_e$, $A = 0.5\ R_e$ and $k = 0.6\ R_e^{-1}$

are used. The k value corresponds to a wavelength of about 10 R_e and A is the amplitude of the wave. It can be seen from Fig. 6 that the modeled magnetic field variation across the neutral sheet portrays very well the observed features in Figure 2, i.e. step-like change in θ_{sm} and a decrease in B.

The wave profile is not confined only along the tail axis. Features indicating a wave profile of the neutral sheet along the dawn-dusk direction have also been presented (Lui et al., 1978). Therefore, the neutral sheet may deviate from a plane surface. Since the features suggestive of a warped neutral sheet surface are not the most commonly seen (Table 1), the deviation of the neutral sheet from a plane surface is therefore present occasionally or only in localized regions.

Magnetic Island Structure. Another feature which is detected occasionally is a decrease in the elevation angle at neutral sheet crossings. This feature has been studied by many workers (Mihalov et al., 1968; Schindler and Ness, 1972; Speiser, 1973; Bowling, 1975; Lui and Meng, 1979) and is consistent with a magnetic structure named variously as magnetic bubble, magnetic loop or magnetic island. This structure which gives rise to a southward magnetic field component in the neutral sheet

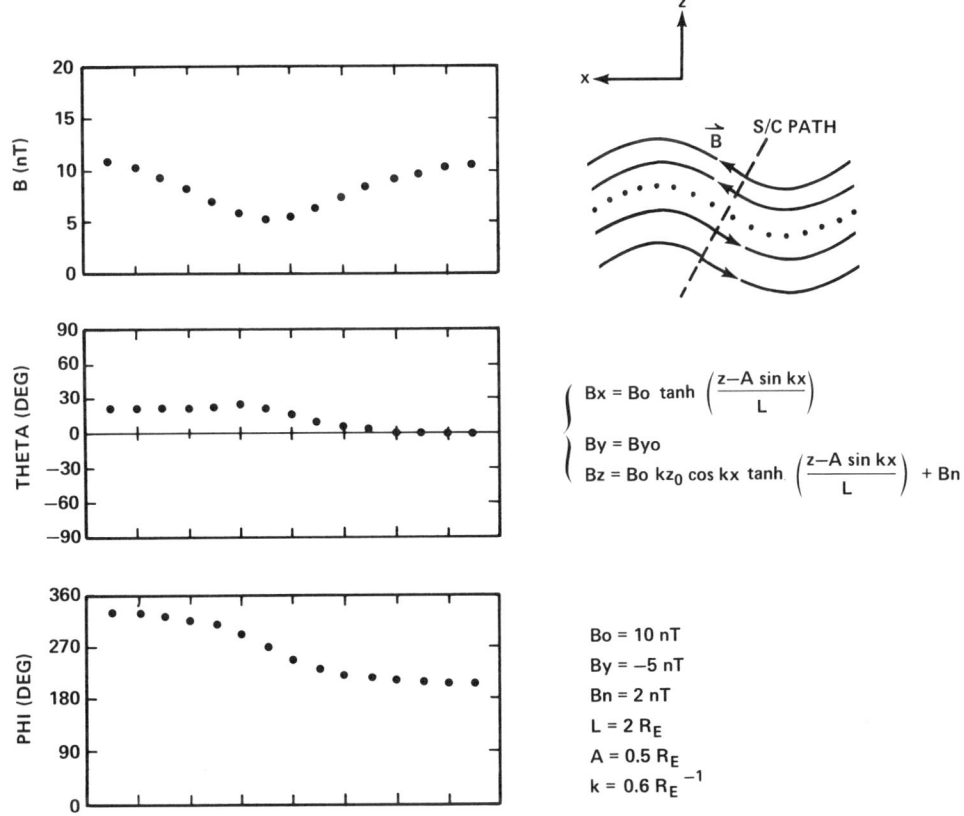

Fig. 6. The observed signatures of the neutral sheet shown in Figure 2 are modeled very well by a wavy neutral sheet surface along the tail axis.

region is often regarded as an indication of magnetic reconnection associated with magnetospheric substorms. However, from a statistical survey of this feature at neutral sheet crossings, it is found that magnetic bubbles occur during both quiet and substorm times (Lui and Meng, 1979). The quiet magnetic conditions in these cases are verified with global auroral images from DMSP satellites. Therefore, the magnetic bubbles can be considered as intrinsic features of the neutral sheet. However, it is conceivable that some of the magnetic bubbles may be activated during substorms to play a role in energizing particles.

Summary of Magnetic Structures of the Neutral Sheet. The structures of the neutral sheet are recapitulated in Figure 7 which encompasses the essential ingredients to account for all the observed features of the neutral sheet from this survey. At the top of the figure is a diagram of the field configuration for a simple neutral sheet which shows the typical signatures, i.e. a decrease in the field magnitude, an increase in the elevation angle θ, a smooth transition in the azimuthal angle ϕ, and low field fluctuations ($\delta B < 5$ nT). From frequent multiple crossings of the neutral sheet, it is deduced that flapping and tilting of the neutral sheet occur very commonly (Speiser and Ness, 1967; Mihalov et al., 1968). This feature is not considered as a structure of the neutral sheet but rather as an effect related to the choice of a particular coordinate system. This simple planar model of the neutral sheet may be deviated by the formation of wave profiles on the neutral sheet surface along the tail axis or the dawn-dusk direction. These wave profiles are illustrated in the third and fourth diagrams from the top. For a wave profile along the tail axis, the ϕ angle may exhibit a step-like variation across the neutral sheet. For a wave profile in the dawn-dusk direction, a characteristic feature is a 360° rotation in the ϕ-angle for adjacent crossings of the neutral sheet. Another structure shown by the fifth diagram is the magnetic island embedded in the neutral sheet. This may be indicated by the θ-angle showing a decrease or relatively constant value as the neutral sheet is crossed. The sixth diagram in Figure 7 illustrates a complex field geometry indicated only rarely by magnetic field observations. In this disordered field region, the field fluctuations are high and the transition of the ϕ-angle tends to be irregular. Time variations of the magnetic field are probably significant. Large B_y variations across the neutral sheet are detected at times, particularly during intervals with a substantial southward magnetic field component (Akasofu et al., 1978; Hones et al., 1982). These variations are rare and tend to be found during substorms which suggest that the magnetic structures are probably some substorm-associated 3-dimensional local features of the neutral sheet. Finally, the feature that the minimum field magnitude does not occur near the middle of a smooth transition in the ϕ-angle (Figure 3) seems to indicate that time variations of the magnetic field are also dominate at these times.

Large-Scale Substorm Changes in the Cross-Tail Current

Large-scale changes in the cross-tail current during substorms have been suggested and inferred from observed magnetic field perturbations (Atkinson, 1967; Bostrom, 1974; McPherron et al., 1973; Lui, 1978). At the geosynchronous distance and in the near-earth tail, observation indicates that a portion of the nightside magnetosphere relaxes to a more dipolar-like configuration during substorm expansion. This phenomenon has been interpreted as the interruption or diversion of the near-earth portion of the cross-tail current into the ionosphere at substorm onset.

A case study of substorm changes of magnetic field in the near-earth tail is given in Figure 8. The top diagram shows the IMP 6 trajectory on the XZ-plane, with the magnetic field model from Mead and Fairfield (1975) appropriate to the condition at the time of observation (MF 730; $K_p = 0$, 0^+, tilt = 30°). The magnetic field variation in the XZ plane from 08:25 UT to 08:55 UT are plotted below the diagram along the magnified IMP 6 trajectory. The time progresses from right to left since IMP 6 is inbound at this time. At 08:36 UT, within a minute of substorm onset from ground all-sky-camera observation (Lui et al., 1976), the magnetic field vector starts to swing northward, reaching its maximum swing at 08:41 UT. The simultaneous magnetic field variations projected on the equatorial plane is given at the bottom of Figure 8. Changes in the B_y component starts also at 08:36 UT, reaching the largest perturbation at ~ 08:44 UT.

The magnetic field variations in this event has been modeled to provide a crude estimate of the substorm changes in the cross-tail current (Lui, 1978). The observed magnetic field is first subtracted by a model field which simulates magnetic field contributions from the earth's dipole field and the magnetopause currents. The residual magnetic field is taken to be due to the cross-tail current. For a current sheet with current density j_0 at the inner edge and decreases linearly with downstream distance over a length of L, i.e.

$$j = j_0 \left(1 - \frac{X}{L}\right)$$

the magnetic field perturbation at point (X,Z)

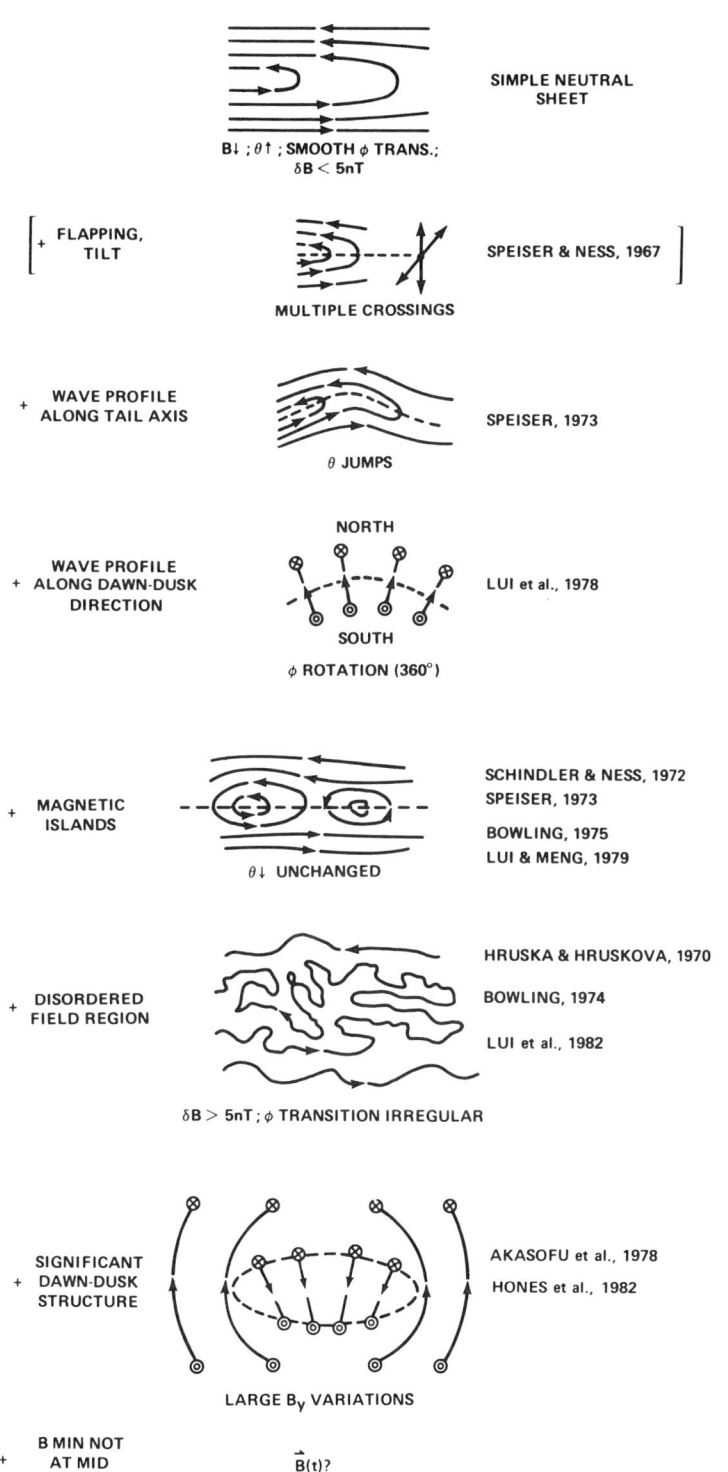

Fig. 7. The structures revealed by the different signatures of neutral sheet crossings are illustrated schematically. The features included are (1) flapping and tilting of the neutral sheet, (2) wave profiles on the neutral sheet surface, (3) magnetic islands embedded in the neutral sheet region, (4) three-dimensional disordered magnetic field regions, (5) three-dimensional structures with large dawn-dusk magnetic field components, and (6) significant time variations.

is given by

$$\Delta B_x \simeq -\frac{\mu_0 j_0}{2\pi}\left(\frac{\pi}{2} + \tan^{-1}\frac{X}{Z}\right)$$

$$\Delta B_z \simeq -\frac{\mu_0 j_0}{2\pi}\left[\ln\left(\frac{L}{\sqrt{x^2+z^2}}\right) - 1\right]$$

which may be rewritten as

$$j_0 \simeq \frac{-\Delta B_z}{\frac{\mu_0}{2\pi}\left[\ln\left(\frac{L}{\sqrt{x^2+z^2}}\right) - 1\right]}$$

$$X \simeq Z \cot\left(\frac{2\pi \Delta B_x}{\mu_0 j_0}\right)$$

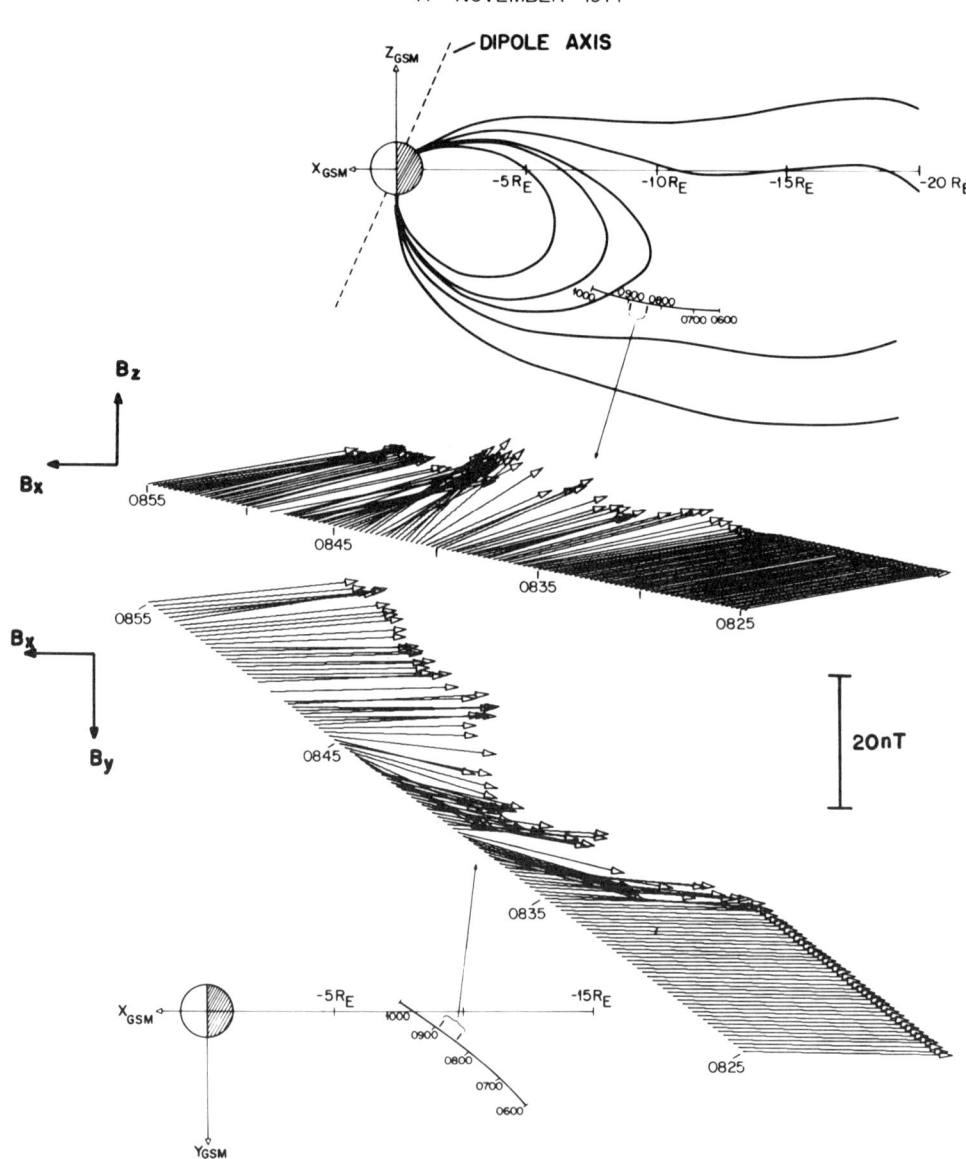

Fig. 8. Magnetic field variations observed in the near-earth tail region during a substorm on Nov 17, 1971. The relaxation of the magnetic field from a tail-like to a more dipolar-like configuration is modeled to indicate that the observed field changes are consistent with a disruption or diversion of the near-earth cross-tail current into the ionosphere at substorm onset.

Letting $L = 100 \, R_e$ and Z to be the estimated distance of the spacecraft to the neutral sheet, the above two equations are solved numerically for the current density j_o and the location of the inner edge of the cross-tail current. The result indicates that the observed magnetic field perturbation may be interpreted as a tailward retreat of the cross-tail current sheet location by 0.4 to 3.5 R_e and an overall reduction of the current density by 5 to 25%. These changes are coincident with B_y changes which are indicative of field-aligned currents. Earthward plasma flows of ~ 500 km/s coincident with the observed magnetic field changes have also been observed and reported (Lui et al., 1976).

Summary

An attempt is made to give a brief review of the basic and substorm features of the neutral sheet which carries a significant portion of the cross-tail current. The main points covered here may be summarized as follows:

(1) Different signatures of the neutral sheet are found.
(2) A simple one-parameter classification scheme is introduced to distinguish the varieties in the signature of the neutral sheet.
(3) The most common features associated with a neutral sheet crossing are identified to be those anticipated for a classical neutral sheet crossing, namely, a decrease in field magnitude, an increase in field elevation angle, a smooth rotation of the field azimuthal angle, and low field fluctuations.
(4) The dawn-dusk component of the magnetic field at the neutral sheet is statistically related to the dawn-dusk component of the interplanetary magnetic field, suggesting a partial penetration (~ 50%) of the B_y component of the interplanetary magnetic field into the neutral sheet of the magnetotail.
(5) The neutral sheet may depart from a plane surface by the development of wave profiles.
(6) Magnetic islands are intrinsic features of the neutral sheet and are present during substorms as well as quiet periods.
(7) Substorm changes of the cross-tail current sheet examined in one case study are consistent with the idea that the near-earth cross-tail current is being interrupted and diverted into the ionosphere at the substorm onset.

Acknowledgments. This research is supported by the Atmospheric Sciences Section of the National Science Foundation, Grant ATM83-05537 to the Johns Hopkins University.

References

Akasofu, S.-I., A. T. Y. Lui, C.-I. Meng, and M. Haurwitz, Need for a three-dimensional analysis of magnetic fields in the magnetotail during substorms, Geophys. Res. Lett., 5, 283, 1978.

Atkinson, G. J., An approximate flow equation for geomagnetic flux tubes and its application to polar substorms, J. Geophys. Res., 72, 5373, 1967.

Bostrom, R., Ionosphere-magnetosphere coupling, in Magnetospheric Physics, edited by B. M. McCormac, D. Reidel Pub. Co., Hingham, MA, p. 45, 1974.

Bowling, S. B., Transient occurrence of magnetic loops in the magnetotail, J. Geophys. Res., 80, 4741, 1975.

Eastman, T. E., E. W. Hones, Jr., S. J. Bame, and J. R. Asbridge, The magnetospheric boundary layer: site of plasma, momentum and energy transfer from the magnetosheath into the magnetosphere, Geophys. Res. Lett., 3, 685, 1976.

Fairfield, D. H., On the average configuration of the geomagnetic tail, J. Geophys. Res., 84, 1950, 1979.

Harris, E. G., On a plasma sheath separting regions of oppositely directed magnetic field, Niovo Cimento, 23, 116, 1962.

Hones, E. W., Jr., J. Birn, S. J. Bame, G. Paschmann, C. T. Russell, On the three-dimensional magnetic structure of the plasmoid created in the magnetotail at substorm onset, Geophys. Res. Lett., 9, 203, 1982.

King, J. H., Interplanetary medium data book - Appendix, NSSDC/WDC-A-R&S 77-04a, September, 1977.

Lui, A. T. Y., Estimates of current changes in the geomagnetotail associated with a substorm, Geophys. Res. Lett., 5, 853, 1978.

Lui, A. T. Y., S.-I. Akasofu, E. W. Hones, Jr., S. J. Bame, and C. E. McIlwain, Observation of the plasma sheet during a contracted oval substorm in a prolonged quiet period, J. Geophys. Res., 81, 1415, 1976.

Lui, A. T. Y., C.-I. Meng, S.-I. Akasofu, Wavy nature of the magnetotail neutral sheet, Geophys. Res. Lett., 5, 1978.

Lui, A. T. Y., and C.-I. Meng, Relevance of southward magnetic fields in the nuetral sheet to anisotropic distribution of energetic electrons and substorm activity, J. Geophys. Res., 84, 5817, 1979.

McPherron, R. L., M. P. Aubry, C. T. Russell, and P. J. Coleman, Jr., Satellite studies of magnetospheric substorms on August 15, 1968, 4, Ogo 5 magnetic field observations, J. Geophys. Res., 78, 3068, 1973.

Mead, G. D., and D. H. Fairfield, A quantitative

magnetosphere model derived from spacecraft magnetometer data, J. Geophys. Res., 80, 523, 1975.

Mihalov, J. D., D. S. Colburn, R. G. Currie, and C. P. Sonnett, Configuration and reconnection of the geomagnetic tail, J. Geophys. Res., 73, 943, 1968.

Ness, N. F., The earth's magnetic tail, J. Geophys. Res., 70, 2989, 1965.

Potemra, T. A., Current systems in the Earth's magnetosphere, Res. Geophys. Space Phys., 17, 640, 1979.

Rostoker, G., and R. Bostrom, A mechanism for driving the gross Birkeland current configuration in the auroral oval, J. Geophys. Res., 81, 235, 1976.

Schindler, K., and N. F. Ness, Internal structure of the geomagnetic neutral sheet, J. Geophys. Res., 77, 91, 1972.

Speiser, T. W., Magnetospheric current sheets, Radio Sci., 8, 973, 1973.

Speiser, T. W., and N. F. Ness, The neutral sheet in the geomagnetic tail: its motion, equivalent currents, and field line connection through it, J. Geophys. Res., 72, 131, 1967.

Zmuda, A. J., and J. C. Armstrong, The diurnal flow pattern of field-aligned currents, J. Geophys. Res., 79, 4611, 1974.

FIELD-ALIGNED CURRENTS NEAR THE MAGNETOSPHERE BOUNDARY

Edward W. Hones, Jr.

University of California, Los Alamos National Laboratory
Los Alamos, NM 87545

Abstract. This paper describes present thinking about the structure of magnetospheric boundary layers and their roles in the generation of the field-aligned currents that are observed in the polar regions. A principal effect of the momentum loss by magnetosheath plasma to the magnetosphere boundary regions just within the magnetopause, whether it be by a diffusive process or by magnetic reconnection, is the tailward pulling of the surface flux tubes relative to those deeper below the surface. The dayside region 1 currents at low altitudes flow along field lines in the resulting regions of magnetic shear. The direction of the shear and its magnitude, actually measured in the boundary region, confirm that the polarities and intensities of the dayside region 1 currents can be accounted for by this process. The low latitude boundary layer, formerly thought to be threaded entirely by closed field lines, now appears to contain at least some open field lines, newly reconnected, that are in the process of being swept into the high latitude tail to form the plasma mantle. The open flux tubes of the flux transfer events, thought to be the product of "patchy reconnection" have a spiral magnetic structure whose helicity is such as to suggest currents having the polarities of the region 1 currents.

Introduction

The primary interaction of magnetosheath plasma with the magnetosphere is reflected in a boundary region hundreds to thousands of kilometers thick that lies just within the magnetopause (see Figure 1a). This boundary region is characterized by plasma, having density, temperature, and flow properties similar to those of the magnetosheath, but generally thought to be threaded by earth-tied magnetic field lines. While it is essentially continuous over the surface of the magnetosphere the boundary region is thought to consist of two major parts that can be recognized to some extent by differences in their observed properties but that are differentiated more clearly in the concepts of their means of formation. The first of these, the low latitude boundary layer (LLBL), covers the sunward surface of the magnetosphere generally equatorward of the polar cusps and extends, at low to moderate latitudes, to some unknown distance (perhaps 50-100 R_E) along the dawn and dusk flanks of the magnetotail. It is thought to contain closed magnetic field lines onto which magnetosheath plasma has penetrated by a diffusive process. It is often identified with the outermost region of the Axford-Hines (1961) model where magnetospheric plasma was set into motion by momentum transferred from the solar wind through a "viscous-like" interaction. The second part, the plasma mantle, covers the higher latitude surfaces of the magnetosphere from the polar cusps back to very large distances along the tail. It is thought to contain field lines that have recently become open by reconnection with the magnetosheath field and that are populated by magnetosheath plasma. The mantle is usually identified with the regime of open field lines, resulting from reconnection, pictured in the Dungey (1961) model of the magnetosphere. A third separately identified part of the boundary region comprises the two entry layers (Paschmann et al., 1978), small areas lying near and equatorward of the two polar cusps. They separate the LLBL from the mantle near the noon meridian and perhaps are avenues for entry of some magnetosheath plasma to both.

Figure 1b is a projection, along magnetic field lines to the ionosphere, of the various boundary layers. Figures 1a and 1b show, also a "plasma boundary layer" (also referred to as the plasma sheet boundary layer) that is essentially the boundary of the plasma sheet. Although this region, too, is an important source of field-aligned currents, it is primarily a nightside phenomenon, not associated with the magnetosphere boundary, and will not be addressed directly in this paper. It must be said, regarding Figure 1b, that the mapping of field lines from the boundary regions to the inosphere is subject to considerable uncertainty. That is because the boundary regions, containing greatly stretched and distorted field lines as they do, are the very regions to which existing magnetospheric models are least applicable. There is, in fact, no accepted theoretical model of these regions. Thus the apportioning of the ionospheric projection, in Figure 1b, to the various boundary regions should be regarded as only semi-quantitative, indicating that around local noon there is some mixture or arrangement of entry layer, LLBL, and mantle field lines, while at earlier and later local times there is increasing tendency to encounter LLBL and mantle field lines and finally plasma boundary layer field lines.

It is primarily along the magnetic flux tubes of the LLBL and the mantle that momentum of the solar wind is imparted to the magnetosphere, stressing the magnetotail and providing the energy for internal magnetospheric phenomena. The process is entirely analogous to the functioning of a magnetohydrodynamic (MHD) dynamo although the geometry is different in the two regions. For the LLBL the plasma on closed boundary layer field lines plays the role of the flowing conducting fluid of the dynamo; for the mantle, the magnetosheath plasma outside the magnetopause (i.e., on the extensions of the mantle field lines) largely plays this role. For both, the ionosphere plays the role of the dynamo's resistive load.

Electic fields, generated in the boundary region and conveyed along field lines to the ionosphere, drive Pedersen currents there. These force convective motion of the ionospheric plasma and where there is shear in this imposed flow (i.e., where the Pederson currents are divergent) field-aligned currents flow so as to maintain current continuity. And since the flow of ionospheric plasma is a mapping along field lines of the boundary region flows, the field-aligned currents are associated with flow shears in the boundary regions. These shears are mainly in the sense that the flow speed increases as the magnetopause is approached from within and result in enhanced tailward draping of those field lines nearest the magnetopause. Such enhanced draping of the outer field lines implies the existence of field-aligned currents having the polarity of the region 1 currents of Iijima and Potemra (1976). (See Vasyliunas (1979) and McDiarmid et al. (1978) for other discussions of the relationship of field aligned currents to stresses in the outer geomagnetic field.)

Boundary Layer Field Aligned Currents and Reverse Draping of Field Lines

We now turn to satellite observations of the boundary regions where large flow shear occurs and where, from the above introduction, the

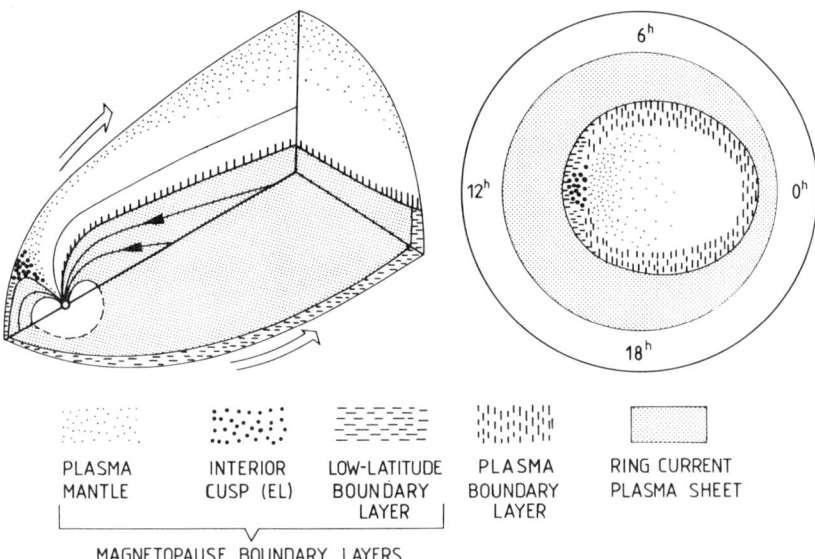

Fig. 1. (a) (left) Schematic diagram of various observed magnetospheric boundary layers. The great extent of the low-latitude boundary layer into the night side is speculative. Open arrows indicate the flow of magnetosheath plasma.

(b) (right) Schematic diagram of the regions occupied by the various boundary layers, mapped down to the ionosphere along magnetic field lines. Hours indicate local time. The local time of the transition from the low-latitude boundary layer to the plasma boundary layer is rather uncertain. (Adapted from Vasyliunas, 1979).

region 1 field-aligned currents originate. Information regarding direction and/or intensities of currents at these places is derived from the behavior of the local magnetic field. We shall find that field-aligned currents of intensities adequate to be identified with the region 1 currents are encountered there. But in these observations there remains some uncertainty as to whether the currents were driven by boundary layer or magnetosheath flow, i.e., whether they were on closed or open field lines.

Hones et al. (1982) conducted a survey of magnetic field variations measured with ISEE 1 during crossings of the dawn boundary of the magnetosphere, mostly at moderate northern solar magnetospheric latitudes where encounters with the LLBL were expected. The object was to identify the anticipated enhanced tailward draping of boundary layer field lines and to derive a measure of the associated field-aligned currents there. They identified the boundary layer by the density, temperature, and flow properties of the plasma. To their surprise they found no cases of obviously enchanced tailward draping of the field as they crossed from the plasma sheet into the boundary layer. But they found several cases of "reverse draping" of the field. That is, the boundary layer field rotated to more tailward orientation rather than to a more sunward orientation as had been anticipated (for the northern hemisphere). Using data from both ISEE 1 and ISEE 2, Hones et al. made a detailed study of one such dawn boundary crossing in which repeated encounters with LLBL, plasma sheet, and magnetosheath were experienced by the outward-bound ISEE satellites for a five-hour interval.

Figure 2 shows the components of the magnetic field at both satellites for one 8-minute interval of that crossing on January 5, 1978. The components are given in a boundary-normal system of coordinates, L, M, N, where N is outward normal to the magnetopause; L is tangent to the magnetopause, and approximately northward; M is tangent and approximately westward (tailward in this instance). The satellites were on the north dawn shoulder of the magnetosphere about 9 R_E above the midplane and were separated by 315 km along the N-axis. From 0929 to 0930 UT the satellites were in the plasma sheet where the field pointed northward (+L) and earthward (–M) as is expected in that sector of the outer magnetosphere. From ~09:30:30 to 09:32:30 and again from ~0936 to 0937 they were in the LLBL where the field was again northward but now tailward (+M), the signature of "reverse draping". From ~09:32:30 to ~09:33:30 and from ~09:34:00 to ~09:34:20 the satellites were in the magnetosheath where B_L was negative (southward).

The observed spatial rate of shear of the magnetic field through the boundary (assumed quasi-two-dimensional) provided a measure of the density of current flowing in the L-M plane. Its components parallel and perpendicular to $\bar{B}$ are plotted at the bottom of Figure 2. The current was mostly anti-parallel to $\bar{B}$. At the crossings from the plasma sheet to the LLBL at ~09:30:30 UT and ~09:35:30 UT j reached values of -3×10^{-2} µA/m². At crossings into and out of the magnetosheath j reached -7×10^{-2} µA/m². If the currents at the plasma sheet –LLBL interface flow undiverted along field lines to the (southern) polar cap where the field strength is ~2000 times greater, their intensity at the ionosphere would be ~60 µA/m², a value consistent with some region 1 current intensitites measured bythe DE-2 satellite (M. Sugiura, private communication, 1982).

The remarkable reverse draping of the LLBL field found on this and other occasions has several possible explanations, three of which are illustrated in Figure 3. (Top panel) Plasma enters the closed LLBL flux tubes primarily at the entry layers near the north and south polar cusps. Its expansion tailward around the magnetosphere stretches the field lines preferentially at high latitudes, leaving the equatorial segments to follow more slowly. Possibly the equatorial segments are still embedded in the plasma sheet whose plasma is moving more slowly or, perphaps, even moving sunward. (Middle panel) The field lines are pulled tailward but with their single apex pulled up to high northern latitudes. Such an asymmetry with respect to the equatorial plane might be a seasonal effect, the much larger Pedersen conductivity in the sunlit southern

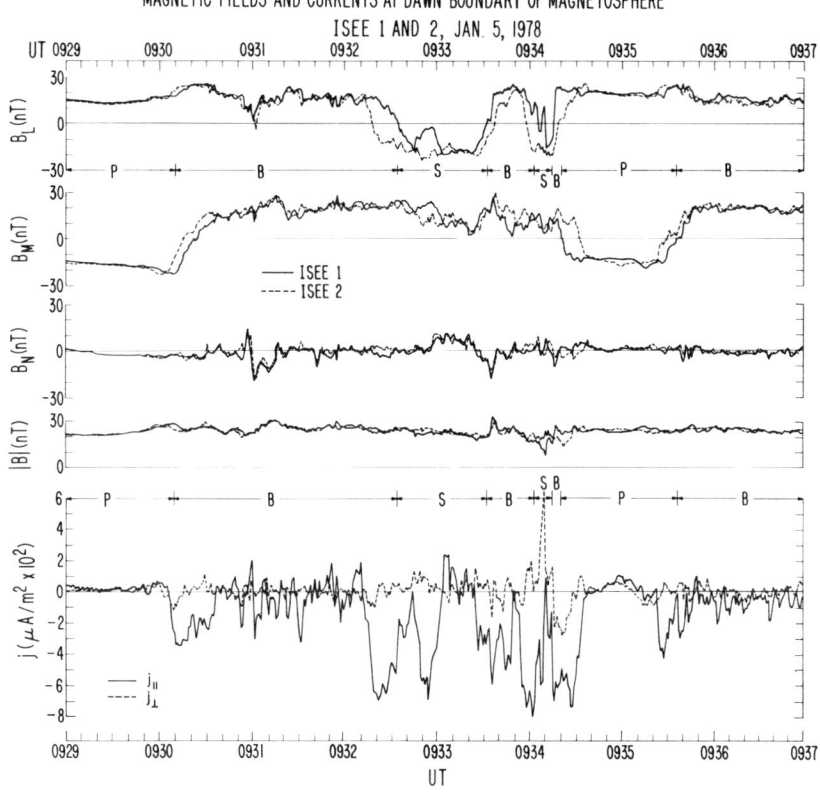

Fig. 2. (top three panels) L, M, and N components of the magnetic field measured by ISEE 1 (solid curves) and ISEE 2 (dashed curves) from 0929 UT to 0937 UT, January 5, 1978. (Fourth panel) Total field measurements. (Fifth panel) Electric current density flowing parallel (solid curve) and perpendicular (dashed curve) to the magnetic field (and in the LM plane). Bars under the first and fourth panels indicate boundary layer. Dots indicate magnetosheath and blanks indicate plasma sheet. (From Hones et al., 1982).

polar cap diverting a larger portion of $j_\parallel$ into the southern hemisphere than into the northern one. (Bottom panel) Magnetic reconnection occurs between magnetospheric field lines on the north-dawn surface of the magnetosphere and magnetosheath field lines. (Data from IMP 8 satellite showed the IMF to have a negative B_y component during the crossing discussed above, appropriate for reconnection in the location suggested). The field lines at ISEE 1 and 2 might then be segments from the southern polar cap that are being pulled tailward by their reconnected magnetosheath extensions, and that are on their way to becoming plasma mantle field lines. If the situation as depicted in the bottom panel is the true explanation for the reverse draping, the characterization of the LLBL as flowing plasma giving up momentum to closed field lines on that occasion is incorrect; at least part of the LLBL was on open field lines and the driving agent was the magnetosheath flow instead. (Cowley et al. (1983) investigated the tangential stress balance condition across the magnetopause for the January 5, 1978 event, using the theoretical procedure of Sonnerup et al. (1981) and found it to be quantitatively consistent with the explanation in the bottom panel of Figure 3 but not inconsistent with the explanation in the top panel.)

Quasi-Steady Reconnection and Flux Transfer Events

In the previous section we found that the reverse draping of field lines sometimes observed in the LLBL might indicate that the LLBL is threaded with open, rather than closed, field lines on those occasions. In this section we shall examine some of the evidence for magnetic reconnection at the magnetopause that would result in an open LLBL and consider some of its implications with regard to field-aligned currents.

High resolution plasma, magnetic field, and energetic particle data that have become available through the ISEE 1 and 2 satellite program provide the first direct evidence that magnetic reconnection occurs at the magnetopause. This evidence grew from two paths of investigation that were followed concurrently and independently. One revealed that reconnection can occur in a quasi-steady manner, enduring on time scales of tens of minutes, and the other revealed that it can occur as a transient process, localized in both spatial and temporal extent. Future research may show that the two processes are not sharply distinct but simply represent different regions of a continuous spectrum of space and time scales (Cowley, 1982).

The situation that may prevail at the magnetopause during quasi-steady reconnection is depicted schematically in Figure 4. The reader is referred to the paper by Sonnerup et al. (1981) for a detailed discussion of those aspects of reconnection, illustrated in the figure, that have been important in identifying instances of quasi-steady reconnection in the ISEE satellite data. Briefly, it was found, in a substantial number of magnetopause crossings, that (a) the plasma velocity (V_2) in the magnetopause and boundary layer was significantly larger than that (V_1) in the magnetosheath; (b) the difference in tangential plasma

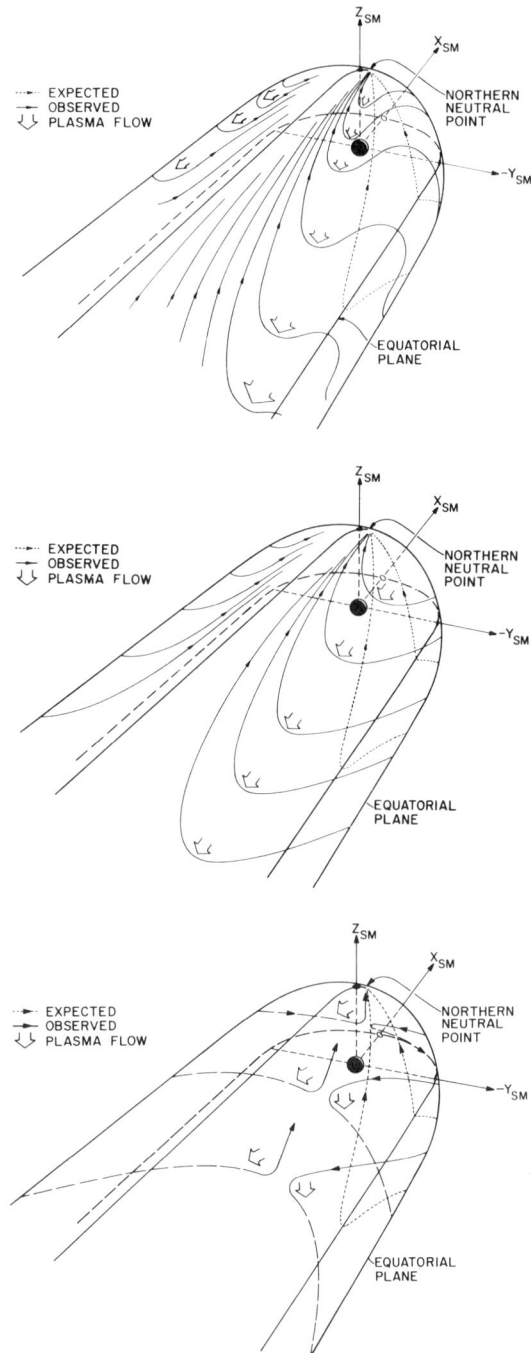

velocity ($V_{2t} - V_{1t}$) had the direction and magnitude appropriate for the Maxwell stress derived from the change of the tangential magnetic field across the magnetopause; (c) the normal component of the magnetic field (B_n) had the signs appropriate for the observed boundary layer flow ($B_n < 0$, i.e., inward, for northward flow, $B_n > 0$ for southward flow); (d) the normal component of the flow (V_n) at the magnetopause was inward; (e) energetic magnetospheric particles had directional anisotropies indicating outward escape along open flux tubes; (f) the outer separatrix, S_1, could be identified in the flux of escaping magnetospheric particles.

A flux transfer event (FTE) is thought to result when a bundle of magnetic field lines (perhaps several thousand kilometers in cross-section) of the magnetosheath reconnects to a comparable flux tube in the surface of the magnetosphere. The magnetic tension of the reconnected flux tubes causes them to contract, as illustrated in Figure 5, while their magnetosheath sector continues to be carrried tailward by the magnetosheath plasma flow. The feature in the ISEE magnetic data that first attracted attention to FTEs was a quasi-sinusoidal variation of the magnetosheath field component normal to the magnetopause in periods of ~1 minute that was quite often seen when the satellite was short distance outside the magnetopause. Russell and Elphic (1979)

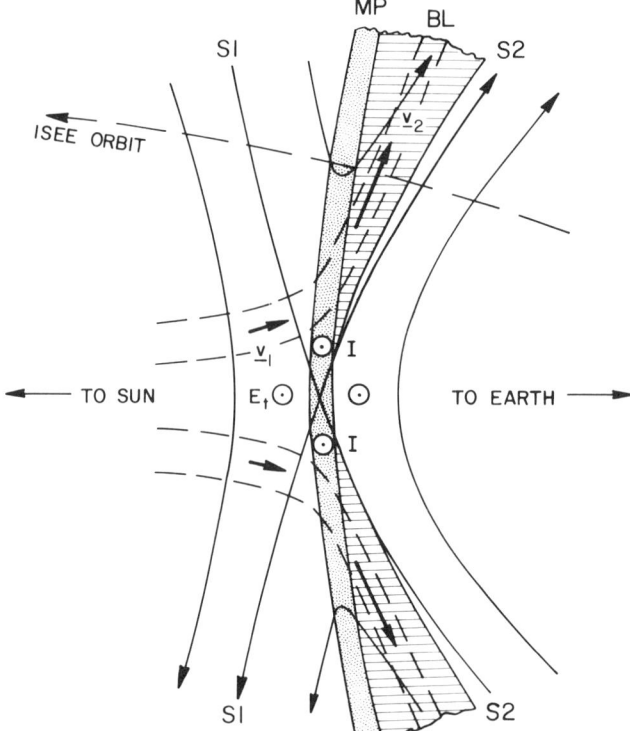

Fig. 3. Three tentative interpretations of the observed reverse draping of boundary layer magnetic field lines. (Top) Plasma entry around polar cusps causes high northern and southern segments of field lines to be pulled tailward faster than equatorial segments. Dotted field lines show line shape for "normal draping" anticipated for equatorial or uniform entry of plasma to boundary layer. (Middle) Unequal conductivity in northern and southern ionospheres may cause northward displacement of single apices. (Bottom) Magnetosheath field lines (dashed) reconnect with magnetospheric field lines on north dawn shoulder of magnetosphere. (From Hones, 1983).

Fig. 4. Meridional view of the reconnection configuration for antiparallel external and internal magnetic fields. The magnetic field lines are shown as solid lines. The magnetopause (MP) is shown as a current layer of finite thickness, with an adjoining boundary layer (BL) of comparable thickness. Those magnetosheath and magnetospheric field lines connected to the separator (or X line) form the outer (S_1) and inner (S_2) separatrices. Dashed lines are stream lines and the heavy arrows indicate plasma flow speed outside and inside the magnetopause. The reconnection electric field, E_t is aligned with the magnetopause current, I. (From Sonnerup et al., 1981).

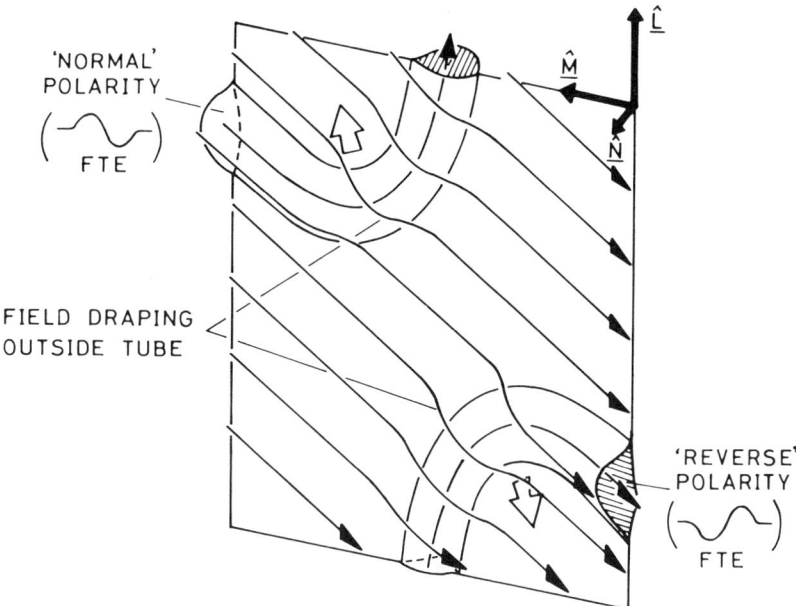

Fig. 5. Schematic view (from the magnetosheath) of two open FTE flux tubes shortly after a localized reconnection event at the dayside magnetopause. The LMN corrdinate directions are included for reference in the top right hand part of the sketch. Magnetic field lines are designated by black arrowheads. The white arrows show the direction of contraction of the newly-reconnected flux tubes away from the site of reconnection (at the center of the figure). (From Saunders, 1983).

interpreted this as the draping of the overlying magnetosheath field as the reconnected flux tube moved by underneath. Because of the orientation of the ISEE orbit early observation of FTEs were mostly at northern latitudes where the variation of the normal component was observed to be outward (+) then inward (−) and these were called normal polarity FTEs. Figure 5 illustrates how this variation occurs and shows that if reconnection occurs near the equator, the field variation at southern latitudes should be inward-then-outward. Examples of this latter variation have been found (Rijnbeek et al., 1982) and are called reverse polarity FTEs.

Within the open flux tubes of the FTEs deflections of the plasma flow and the field direction are expected and are found in the data. In particular, the field in the magnetosheath portion of the tube will turn toward the direction of the earth's field, while the field in the magnetospheric portion of the tube will turn toward the direction of the magnetosheath field. The plasma flow inside the northerly contracting open tube will be deflected northward and westward relative to the ambient magnetosheath flow while the plasma flow within the southerly contracting open tube should be deflected southward and eastward.

It is to be expected that the moving bulge associated with the contracting reconnected flux tube might be sensed in the underlying magnetospheric magnetic field just inside the magnetopause as well as in the overlying magnetosheath field. Indeed this is found to be the case and so there are magnetosheath FTEs and magnetospheric FTEs. As a moment's reflection will show, the polarity of the B_N change in magnetospheric FTEs will be the same as for magnetosheath FTEs (outward-inward in the northern hemisphere, inward-outward in the southern hemisphere). (The reader is referred to the paper by Paschmann et al. (1982) for a detailed account of plasma and magnetic properties of several magnetosheath and magnetospheric FTEs).

Sample Observations of Quasi-Steady Reconnection, Magnetosheath FTEs, and a Magnetospheric FTE

Figure 6 shows plasma and magnetic field data for an ISEE 1 passage from the magnetosheath into the low-latitude dusk sector of the magnetosphere. ISEE 1 was in the LLBL between 1934 UT and 1953 UT, as is evidenced by the irregular flow velocity (V_p) and density (N_p) and by the northward orientation of the magnetic field (B_L positive). This is one of the intervals studied by Sonnerup et al. (1981). They deduced that quasi-steady reconnection was taking place near the spacecraft and, specifically, that the magnetopause/boundary layer region observed between 1934 and 1953 UT was located on open field lines southward of the reconnection region. The substantially increased plasma flow speed and the southward (V_z negative) deflection of its direction during this interval were quantitatively consistent with the requirements of tangential stress balance across an open magnetopause with B_N positive (that is, on open tubes connected to the southern hemisphere of the earth). Note that during this interval the magnetic field pointed generally northward and tailward (B_M negative), whereas in the plasma sheet (after ∼2011 UT) the field pointed northward and earthward, appropriate for the satellite's location at northern latitudes. This is the signature of "reverse draping" of field lines reported by Hones et al. (1982) and discussed in an earlier section of this paper. The tailward draping of the LLBL field lines (relative to those in the neighboring plasma sheet) implies a shear at their interface (which interface was not actually encountered in the 1934-1953 UT interval) requiring a field-aligned current parallel to $\bar{B}$ (i.e., away from the south polar region of earth). This is compatible with the direction of the low-altitude region 1 field-aligned currents at the post-noon sector of the polar cap. Note that the example of reverse draping of the dawn LLBL field reported by Hones et al. (1982) required FAC flowing anti-parallel

176 FIELD-ALIGNED CURRENTS

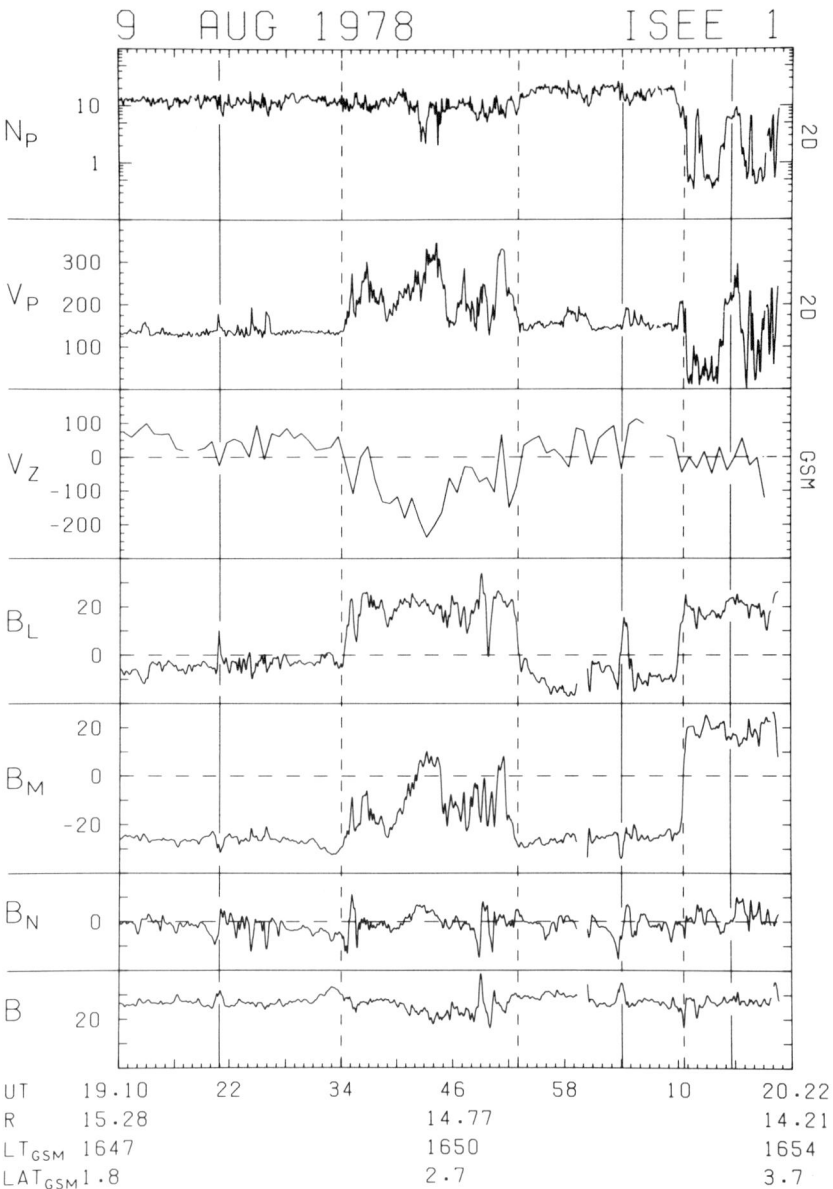

Fig. 6. ISEE 1 plasma and magnetic field measurement of the magnetopause region for the inbound crossing on August 9, 1978. The top three graphs show the plasma number density (cm^{-3}), plasma flow speed in the ecliptic plane (km/sec) and plasma flow speed in the GSM z-direction (km/sec). The bottom four graphs show the L, M, and N components of B and its total intensity (nanoteslas). Reverse polarity FTEs are denoted by vertical guidelines at ~19:21 and ~20:04 UT. The dashed line at ~20:11 UT identifies the magnetopause. (From Saunders, 1983).

to $\bar{B}$ which is also compatible with region 1 FAC if reconnection (Figure 1c) is the correct explanation of that event also.

Reverse polarity magnetosheath FTEs are seen at 1921 UT and 2004 UT in Figure 6, evidenced by the inward-then-outward perturbation of B_N. Accompanying this signature in each case was a northward rotation (increasing B_L) of the magnetosheath field toward the direction of the magnetosphere field. The LLBL is encountered several times again after the first entry into the plasma sheet at 2011 UT. In the encounter at ~2015 there is a reverse polarity (inward-then-outward) perturbation of B_N together with a rotation of the field (increasing B_L, decreasing B_M) toward a less earthward orientation. This is possibly a magnetospheric FTE signature.

Field-Aligned Currents in FTE Flux Tubes

Several investigators (Paschmann et al., 1982; Cowley, 1982; Rijnbeek et al., 1982; Saunders, 1983) have noted that the inward-then-outward (or vice versa) perturbation of B_N that identifies an FTE often appears continuous even within the open flux tube, possibly indicating

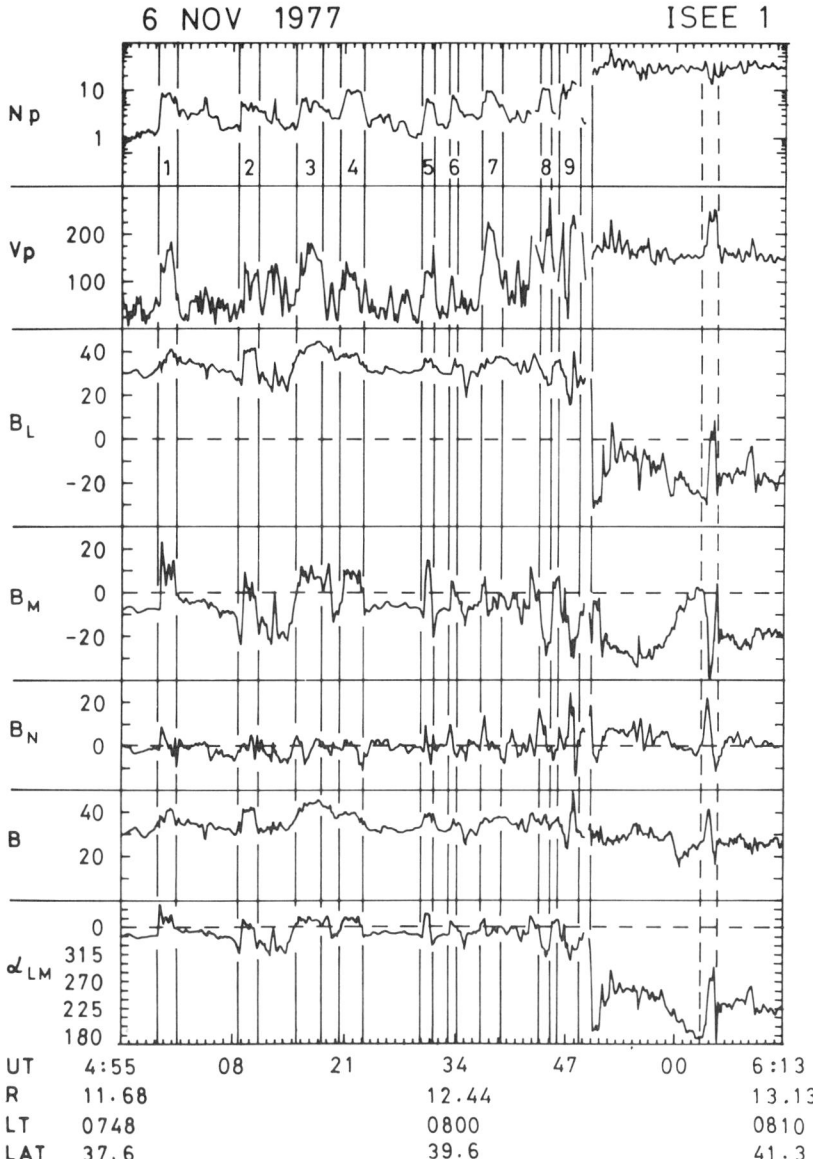

Fig. 7. ISEE-1 plasma and magnetometer recordings for the November 6, 1977 outbound crossing of the low latitude boundary layer and magnetopause. Plotted parameters are the same as in Figure 6 except α_{LM}, which is the field angle in the LM plane (see text). The main intervals (or "pulses") of flowing boundary layer plasma are marked by vertical lines and numbered near the top of the diagram. (From Saunders, 1983).

helicity of these flux tubes caused by field-aligned current flowing within them. Rijnbeek et al. (1982) deduced that there was field-aligned current flowing parallel to the field (i.e., from the magnetosphere to the magnetosheath) in the two magnetosheath FTEs in Figure 6, consistent with the low-altitude region 1 currents. Cowley (1982) has likewise noted that dawn-side FTEs are twisted in such a direction as to imply field-aligned currents in the magnetosheath-to-magnetosphere sense, again consistent with region 1 currents.

Saunders (1983) has analyzed the ISEE-1 crossing of the north, morning sector of the megnetopause that Sckopke et al. (1981) earlier described in detail. The crossing involved a ~50-minute passage through the LLBL that revealed a number of square-wave "pulses" of the plasma density and of the (tailward) flow speed (Figure 7). Saunders draws attention to the variations of B_N and of the angle, α_{LM} (= $\tan^{-1} B_M/B_L$), that accompany the nine identified square-wave plasma variations. The plasma pulses are accompanied by well-defined rotations of the magnetic field. In pulses 1-6, these are caused principally by positive (tailward) excursions of B_M that cause α_{LM} to change from its plasma sheet value (~345°) to ~15°. This change of direction constitutes reverse draping of the field such as we have discussed earlier in connection with two other LLBL encounters. Pulses 8 and 9, which also display higher plasma flow velocity, are accompanied by negative (sunward) excursions of B_M. Variations of B_N resembling FTE signatures accompany the individual pulses. Saunders regards the pulses as passages of FTEs; the variations of B_M then reflect samplings of the field within the open flux tubes as they pass over the satellite. The

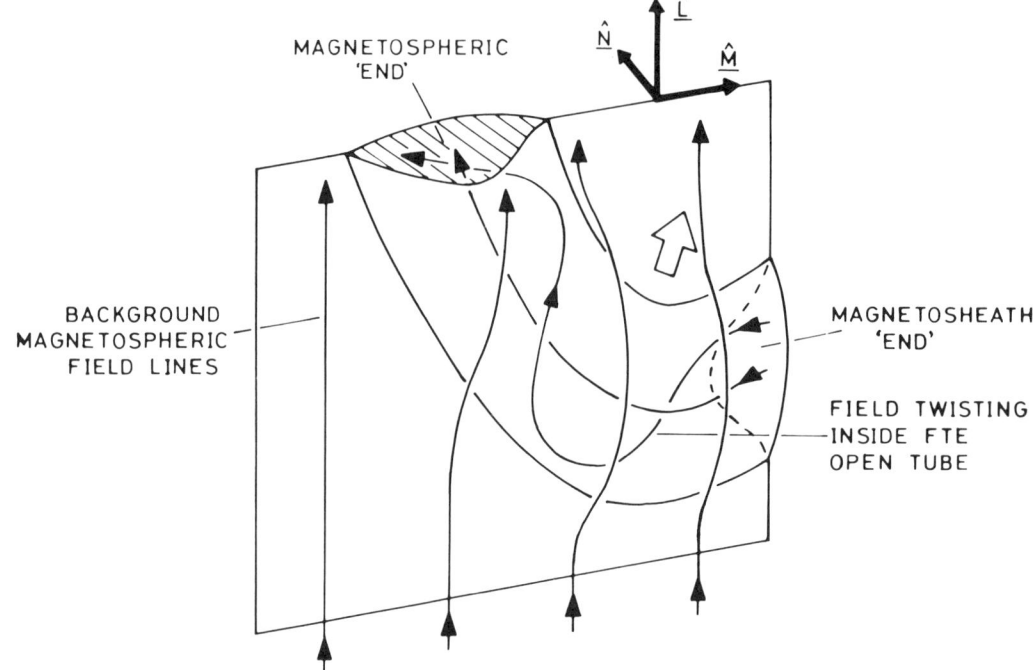

Fig. 8. Sketch of a tailward and northward propagating FTE observed from inside the magnetosphere. The field within the open tube is shown twisted as implied by the observed B_N perturbations. (From Saunders, 1983).

quasi-sinusoidal perturbation of B_N in each pulse, together with the change of the B_M signature from tailward excursions in the early pulses (1-6), when the satellite was near the inner edge of the LLBL, to sunward excursions in the later pulses (8,9) when the satellite was near the outer excursions edge, are all taken as signatures of field-aligned currents within the FTE open tubes. The situation is shown schematically in Figure 8. The northward contraction of the reconnected flux tube (connected to earth's northern polar cap) and tailward dragging of its magnetosheath end tilt the magnetosphere end toward the sun causing a negative B_M variation at its axis. But the twist caused by a field aligned current toward the earth (consistent with region 1 currents) can cause a net tailward tilt at the inner surface of the tube, along with an enhanced sunward tilt at its outer surface.

Saunders and Russell (1983) report that they have examined the spatial properties of FTEs with simultaneous magnetic data from ISEE 1 and 2 when they were separated by many thousands of kilometers. They find that the field perturbations inside FTEs are often oppositely directed at the two satellites, implying a spiral geometry of the FTE internal magnetic structure. The direction of field twisting is systematic both for FTEs seen in the magnetosheath and in the boundary layer, with the field in the FTE flux tube section furthest from (nearest) the magnetopause always tilting away from (towards) the ambient field direction within the neighboring plasma sheet.

Conclusions

The region 1 field aligned currents on the dayside of the polar cap are associated with magnetic shears in the plasma flowing just within the magnetopause at the sunward surface of the magnetosphere and along the flanks of the magnetotail. The magnetic shear results when tailward flowing magnetosheath plasma becomes entrained on earth-tied magnetic field lines and loses its momentum, imparting tangential stresses to the magnetic field. The entrainment of the magnetosheath plasma can take place by diffusion of plasma through the magnetopause or by reconnection, at the magnetopause, of magnetosheath flux tubes to those of the magnetosphere. In the latter case the entrainment of magnetosheath plasma occurs both by flow of plasma through the magnetopause to the magnetospheric sector of the open flux tubes and by constraint of motion of the magnetosheath sector of the tubes that occurs by virtue of their earth-connection. Whatever the means, the net effect of the momentum transfer is to pull the boundary region flux tubes tailward with respect to flux lying deeper beneath the surface. The resulting magnetic shear is in the correct sense to explain the polarity of the region 1 field aligned currents at low altitudes (toward the earth in the pre-noon sector; away from the earth in the post-noon sector). Also, the intensity of field-aligned currents in the LLBL and at the LLBL-plasma sheet interface deduced from the magnetic shear is of the correct order to account for observed intensities of the region 1 currents.

The boundary region is customarily considered to comprise two parts—the low latitude boundary layer, commonly thought to consist of closed flux tubes into which magnetosheath plasma penetrates by diffusion, and the plasma mantle, thought to consist of open field lines into whose magnetospheric sectors (i.e., earthward of the magnetopause) magnetosheath plasma flows directly after reconnection. But now there are suggestions in the ISEE data that the LLBL may, when the IMF is southward (or at least not strongly northward), consist partly, or even largely, of open flux tubes. Among these suggestions are (a) the fairly common occurrence of reverse draping of LLBL field lines, (b) convincing demonstrations of quasi-steady reconnection occurring in the area considered LLBL, and (c) observations, within the LLBL, of FTE signatures. It appears that the LLBL may be a transitional stage of magnetospheric topology constaining (at least some) newly reconnected flux tubes that are being pulled into the plasma mantle. The most severe magnetic shears occur in this stage and are the cause of the dayside region 1 currents.

The magnetic field within FTEs is found to have a twisted structure, indicative of field-aligned currents. Remarkably, the deduced field-aligned currents in the several cases analyzed have been in the appropriate direction (magnetosheath to magnetosphere, or vice versa) to account for region 1 currents at the location where the FTE is thought to connect to the earth. It is not obvious why this should be true and further investigation is required.

Acknowledgments. I am grateful to S. W. H. Cowley for helpful discussions and for sending me several reports of work at Imperial College before their publication. This work was done under the auspices of the U.S. Department of Energy.

References

Cowley, S.W.H., The causes of convection in the earth's magnetosphere: a review of developments during the IMS, Rev. Geophys. Space Phys. 20, 531, 1982.

Cowley, S.W.H., D. J. Southwood, and M. A. Saunders, Interpretation of magnetic field perturbations in the earth's magnetopause boundary layers, Planet. Space Sci., in press, 1983.

Hones, E. W., Jr., Magnetic structure of the boundary layer, Space Sci. Rev., 34, 201, 1983.

Hones, E. W. Jr., B.U.O. Sonnerup, S. J. Bame, G. Paschmann, and C. T. Russell, "Reverse draping" of magnetic field lines in the boundary layer, Geophys. Res. Lett. 9, 523, 1982.

Iijima, T. and T. A. Potemra, The amplitude distribution of field-aligned currents at northern high latitudes observed by Triad, J. Geophys. Res. 81, 2165, 1976.

McDiamrid, I. B., J. R. Burrows, and M. D. Wilson, Comparison of magnetic field perturbations at high latitudes with charged particle and IMF measurements, J. Geophys. Res. 83, 681, 1978.

Paschmann, G., G. Haerendel, I. Papamastorakis, N. Sckopke, S. J. Bame, J. T. Gosling, and C. T. Russell, Plasma and magnetic field characteristics of magnetic flux transfer events, J. Geophys. Res. 87, 2159, 1982.

Rijnbeek, R. P., S.W.H. Cowley, D. J. Southwood, and C. T. Russell, Observations of reverse polarity flux transfer events at the earth's dayside magnetopause, Nature, 300, 23, 1982.

Russell, C. T. and R. C. Elphic, ISEE Observations of flux transfer events at the dayside magnetopause, Geophys. Res. Lett. 6, 33, 1979.

Saunders, M. A., Recent ISEE observations of the magnetopause and low latitude boundary layer: a review, Journal of Geophysics, in press, 1983.

Saunders, M. A. and C. T. Russell, Flux transfer events: scale size and magnetic structure, Geophys. Res. Lett., submitted, 1983.

Sonnerup, B.U.O., G. Paschmann, I. Papamastorakis, N. Sckopke, G. Haerendel, S. J. Bame, J. R. Asbridge, J. T. Gosling, and C. T. Russell, Evidence for magnetic field reconnection at the earth's magnetopause, J. Geophys. Res. 86, 10,049, 1981.

Vasyliunas, V. M., Interaction between the magnetospheric boundary layers and the ionosphere, in Magnetospheric Boundary Layers, ESA SP-148, Bruce Battrick, ed., p. 387, published by ESTEC, Noordwijk, The Netherlands, 1979.

MODELS OF AURORAL-ZONE CONDUCTANCES

Patricia H. Reiff

Department of Space Physics and Astronomy, Rice University, Houston, TX 77251

Abstract. The magnetosphere-ionosphere system is strongly coupled, with magnetospheric Birkeland currents feeding ionospheric Pedersen and Hall currents. Central to any computer simulation of this system is a detailed, valid conductivity model. An accurate conductivity model is also vital in order to infer Birkeland currents and electric field patterns from inversions of magnetometer chain data.

Several recent attempts at constructing conductivity models are presented and their strengths and weaknesses discussed. Incoherent scatter radar measurements can determine height profiles of electron content, from which Pedersen and Hall conductances may be calculated. These yield excellent spatial and good temporal resolution; however, they are limited in field of view. A global pattern requires either 24 hours of data or a chain of stations. Synoptic empirical models (quantized by indices such as Kp or AE) typically are limited by their large bin size (1° invariant latitude × 1 hour MLT), and cannot reproduce arcs. Estimating conductivity globally from Dynamics Explorer auroral images is promising, and can yield reasonable time scales (~10 minutes); however, this procedure is still only now being tested.

Introduction

The magnetosphere-ionosphere-atmosphere system is strongly coupled, with the ionosphere, because of its drag with the neutral atmosphere, acting as a load on magnetospheric current (or voltage) generators. This load may be characterized by the height-integrated Pedersen and Hall conductivities [Wolf, 1983]. Modeling of the complex physical interaction has progressed to the point where the lack of an accurate conductivity model is a serious impediment to further progress (and physical insight). This paper examines the existing conductivity models, and how their strengths and weakness can profoundly affect theoretical models of the magnetosphere-ionosphere-atmosphere interaction.

Sources of Ionospheric Conductivity

Solar Illumination

On the dayside, the bulk of the ionospheric conductivity arises from solar EUV ionization of the neutral atmosphere. Hanson [1965] calculated height profiles of ionospheric electron density from solar UV fluxes and chemical recombination and combined these ionospheric densities with a neutral atmosphere model to calculate the height profile of the conductivity, using two values of the solar spectrum (solar maximum and solar minimum), but he did not show height-integrated values.

Two studies have calculated solar-induced conductances based on experimentally-determined height profiles. Mehta [1978], using Chatanika data, calculated the Pedersen conductance $\Sigma_p = 12.579 - 0.122\ (\chi)$ (mho), where χ is the solar zenith angle in degrees (valid for $45° \leq \chi \leq 95°$), and the Hall conductance $\Sigma_H = 24.421 - 0.239\ \chi$. Vickrey et al. [1981], also using Chatanika electron concentration data, found $\Sigma_p = 5\cos^{1/2}(\chi)$ (mho) and $\Sigma_H \approx 2\Sigma_p$ (Fig. 1). The ratio $\Sigma_H/\Sigma_p \approx 2$ appears to be nearly constant for sunlight-induced conductivity [Brekke et al., 1974]. The factor-of-two difference between the two studies may be caused by differences in the collision models used [Vickrey et al., 1981] or by different EUV fluxes on the two dates used.

During the conference talk for which this is the written version, I stressed the usefulness of developing a solar conductance model parameterized by the magnitude of the solar ionizing (EUV) radiation. The solar F10.7 cm flux, although not a measure of the EUV flux, is nevertheless readily available and is correlated with the EUV flux [Kockarts, 1981; Donnelly, 1982; Donnelly et al., 1983]. In the intervening weeks, Robinson [private communication, 1983] took my suggestion and performed just such a study, finding that in fact the solar-induced conductivity is well ordered by F10.7, varying by nearly a factor of two as the solar F10.7 flux increases from 80 to 200. These results should be quite helpful to modelers.

Background Radiation

Cosmic radiation and galactic EUV both contribute weakly to the ionospheric conductance: roughly 0.1 and 0.2 mho for Σ_p and Σ_H, respectively [Wallis and Budzinski, 1981]. As pointed out by those authors, conductances Σ_1 and Σ_2 arising from two different processes are not simply additive; rather, the ionization rates must be added at each altitude and then appropriately integrated over altitude. If the sources are strictly separated in altitude, one may simply

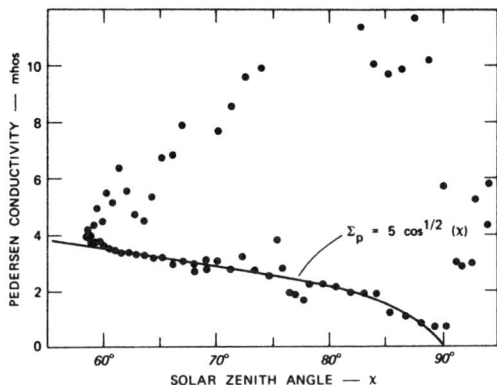

Fig. 1. Scatter plot of height-integrated Pedersen conductivity versus solar zenith angle on April 6, 1977 [from Vickrey et al., 1981].

add the conductances. If the source altitude distributions are identical, however, one can add the ionization rates at each altitude, yielding $\Sigma_T = (\Sigma_1^2 + \Sigma_2^2)^{1/2}$. For combining solar-induced and particle-induced conductances, for example, the latter estimate (root-sum-square) is far more accurate (to within 10%) than a simple sum [Wallis and Budzinski, 1981]).

Precipitating Electron Flux

The dominant nightside conductivity source is ionization from precipitating particle fluxes. Since the ionization rate is proportional to the incident energy flux Φ_E, with about 35 eV required for each electron-ion pair [Rees, 1963], the precipitating ion contribution to the conductivity is generally negligible. This ionization occurs at progressively lower altitudes for progressively more energetic electrons, with the altitude of maximum ionization occurring at 72 km for 300 keV electrons; 95 km for 40 keV; ~107 km for 10 keV, ~170 km for 1 keV, and > 216 km for .4 keV electrons (for isotropic incident distributions; field-aligned fluxes penetrate to lower altitudes) [Rees, 1963].

One can therefore take the energy distribution of precipitating electrons, calculate the resulting altitude distribution of ionospheric electron density, add a neutral atmosphere model [e.g., Banks and Kockarts, 1973; Hedin, 1977a,b] and collisional conductivity formulas, and calculate the height profile of the conductivity. This profile can then be integrated with altitude to obtain the height-integrated ionospheric conductivity (which has units of conductance), so long as the flux is steady over several recombination times (tens of seconds) [Jones and Rees, 1973]. Thus one may develop an empirical conductivity model by first developing a global electron flux model and then calculating the conductances from the fluxes (Fig. 2) [Wallis and Budzinski, 1981;

see also Harel et al., 1981; Spiro et al., 1982; Hardy et al., 1983].

In cases where the ion precipitation is substantial, the ionization produced by the ions must be included in calculation of the conductances. The ions also require about 35 eV per electron-ion pair, and thus the ionization depends on the precipitating ion energy flux; however, since precipitating ions also charge exchange, their ionization efficiency is less than that for precipitating electrons of the same energy [Rees, 1982]. The ionization created by proton precipitation occurs at a higher altitude than for electrons of the same energy. For energies less than about 10 keV, the difference is only about 10-20 km; as the ion energy increases to $\gtrsim 100$ keV the altitude of the deposition levels off at ~110 km, whereas the electrons continue to lower altitudes [Rees, 1982].

To include ion fluxes in the conductivity calculations, one could then calculate an 'equivalent' electron energy flux (equal to the ion energy flux times the ratio of efficiencies), and an 'equivalent' electron mean energy (equal to 80% of the ion energy for energies < 10 keV, and ≈ 8 keV for ion energies > 10 keV), add these equivalent electron fluxes to the measured electron fluxes, and calculate the conductivities.

The SRI group has developed a computer model,

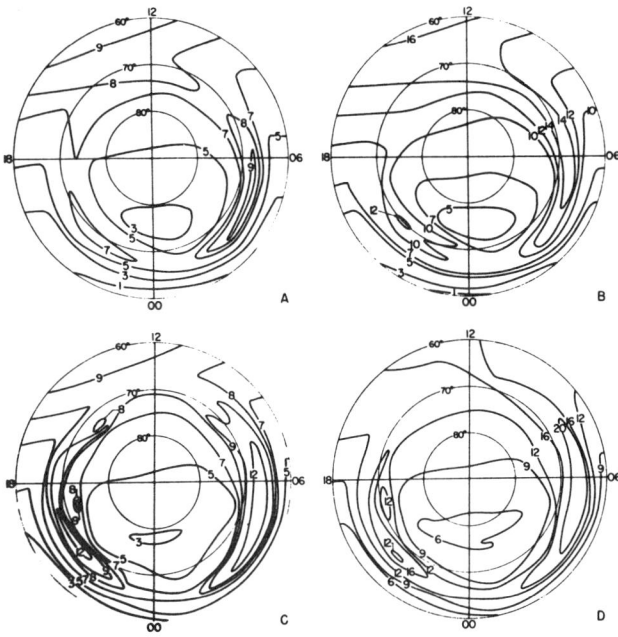

Fig. 2. Contour plots of Σ_P and Σ_H (mhos) for the two K_p cases produced by particle, background, and solar sources for 1700 UT on April 1. (a) Σ_P for $0 \leq K_p \leq 3_0$, (b) Σ_H for $0 \leq K_p \leq 3_0$, (c) Σ_P for $3_0 < K_p$, (d) Σ_H for $3_0 < K_p$ [from Wallis and Budzinski, 1981].

TANGLE, to calculate conductances from detailed electron flux spectra [Vondrak and Baron, 1976, 1977; Vickrey et al., 1981]. To simplify our calculations of conductivities, we have fitted simple formulae to their results: for an isotropic Maxwellian distribution with average energy E_0, $\Sigma_P/(\Phi_E)^{1/2} \approx 20 E_0/(4 + E_0^2)$ (mho), where Φ_E is in ergs/cm^2-s and E_0 in keV; and $\Sigma_H/\Sigma_P \approx E_0^{5/8}$ [Spiro et al., 1982] (Fig. 3). Similar calculations by Wallis and Budzinski [1981] and Rees et al. [1983] have yielded similar results.

Non-Maxwellian spectra are often encountered in the auroral zone, however, and our more recent work divides up the electron distribution into energy bands, taking the energy flux Φ_i (erg/cm^2-s) in a band centered at energy E_i, (keV) and a fit to the monoenergetic conductance plots of Wallis and Budzinski [1981] or TANGLE [Robinson,

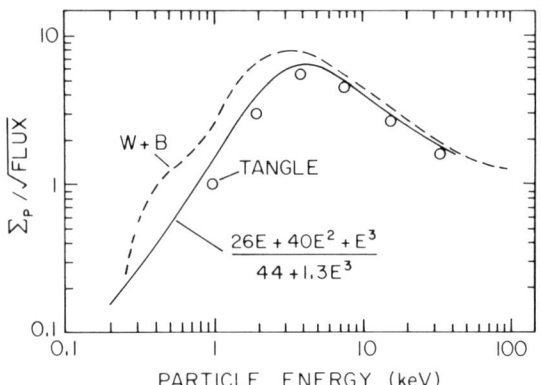

Fig. 4a.

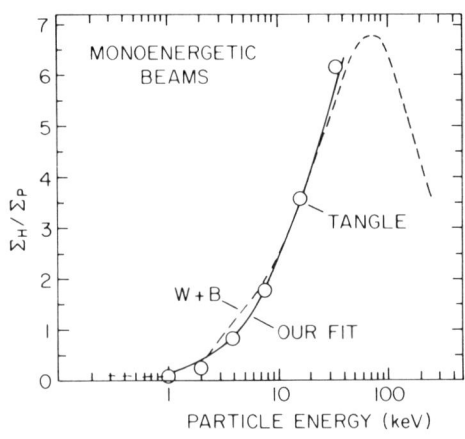

Fig. 4b.

Fig. 4. Similar to Figure 3 but for monoenergetic spectra. TANGLE results are courtesy of Bob Robinson [personal communication, 1983]. The fit to 4b is $\Sigma_H/\Sigma_P = -.04 + .24 E_i - 0.0022 E_i^2$ for $E_i > 0.6$ keV.

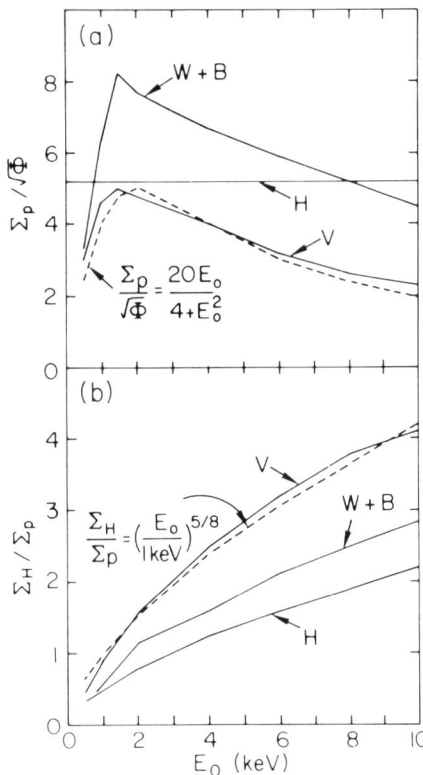

Fig. 3. (a) Pedersen conductance (mho) per unit square root energy flux (erg cm^{-2} s^{-1}) as a function of average electron energy in keV for Maxwellian spectra. (b) Ratio of Hall to Pedersen conductances, Σ_H/Σ_P, as function of average electron energy. The curves labeled 'W + B,' and 'V,' and 'H' refer to the results of Wallis and Budzinski [1981], Vickrey et al. [1981], and Harel et al. [1981a], respectively. The dashed curves represent empirical fits to the results of Vickrey et al. [1981] [from Spiro et al., 1982].

private communication, 1983] (Fig. 4), yielding

$$\Sigma_P \approx \left(\sum_i \Sigma_{Pi}^2\right)^{1/2} \quad (1)$$

where

$$\Sigma_{Pi} = (\Phi_i)^{1/2}(26 E_i + 40E_i^2 + E_i^3)/(44 + 1.3E_i^3)$$

and

$$\Sigma_H \approx \left(\sum_i \Sigma_{Hi}^2\right)^{1/2} \quad (2)$$

where

$$\Sigma_{Hi}/\Sigma_{Pi} = \begin{cases} 0.10, & E_i < 0.6 \\ -0.04 + .2416E_i - 0.0022E_i^2, & E_i \geq 0.6 \end{cases}$$

Although these formulas are more complicated than before, they are still considerably simpler than TANGLE and can reproduce the conductance to within about 20%.

Incoherent Radar Techniques

Conductances from Observations

Incoherent radar in a meridian scanning mode can produce two-dimensional (height and north-south distance) contours of ionospheric electron density. These can be combined with neutral atmosphere models, as above, to produce conductivity models [Brekke et al, 1974; Baron and Vondrak, 1976, 1977; Vickrey et al., 1981; Robinson and Potemra, 1983, this volume]. The conductances so calculated are not too sensitive to the choice of neutral atmosphere model [Brekke et al., 1974].

Electron Fluxes from Ionization Profiles

An interesting inverse technique has also been developed at SRI: UNTANGLE, whereby the height profiles of ionospheric density are deconvolved to yield the incoming electron distribution [Baron and Vondrak, 1976, 1977; Robinson and Potemra, this volume]. This technique, which was suggested by the nomograms of Chestnut et al. [1971], has been shown to be both internally consistent with TANGLE and with in situ electron flux measurements, when available.

Inference from Electric and Magnetic Fields

A fascinating new technique for remotely estimating height-integrated conductivities has been developed by Smiddy et al. [1980]. In regions of space where the Pedersen current flows meridionally (with $\hat{x}$ = northward), and the Birkeland currents feeding it consist of effectively infinite east-west sheets ($\hat{y}$ = eastward), then the current conservation law $\nabla_h \cdot \underline{J}_P = j_\parallel$ and Maxwell's equation $\nabla \times \underline{B} = \mu_0 \underline{j}$ can be combined (recalling Ohm's law $\underline{J}_P = \overline{\Sigma}_P \underline{E}_x$ for this geometry) to give

$$\frac{\partial}{\partial x}(\Delta B_y - \mu_0 \Sigma_P E_x) = 0 \qquad (3)$$

In other words, in regions of nearly constant Σ_P, the electric field E_x and transverse magnetic perturbations from the main field ΔB_y should be proportional, with a constant of proportionality = $\mu_0 \Sigma_P$. For ΔB in nT, E in mV/m, and Σ_P in mhos, $\Delta B_y - 1.256 \Sigma_P E_x$ = const [Smiddy et al., 1980]. This technique ignores the effects of neutral winds and the fact that gradients in Hall conductivities can also contribute to the Birkeland currents.

Naturally, the fields will only be proportional in regions of nearly constant Σ_P (most likely on the dayside), whereas on the nightside, the latitudinally varying fluxes require shorter and shorter time segments to analyze. Nevertheless, Sugiura et al. [1982] found regions of space

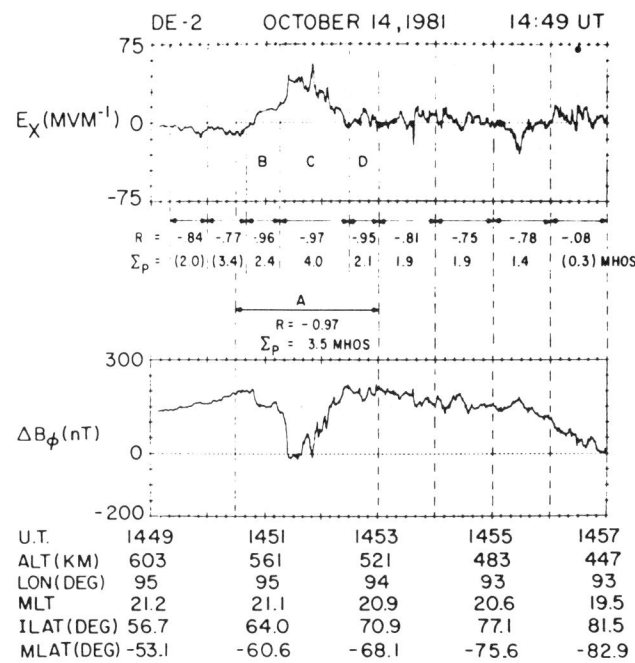

Fig. 5. Highly correlated electric and magnetic field data from which the Pedersen conductance may be calculated [from Sugiura et al., 1982].

where the correlation between ΔB and E was very large, with a correlation coefficient approaching unity (Fig. 5). We examined the in-situ electron fluxes for this case, to determine what the conductivity inferred from the particles alone would be. Unfortunately, in this case, the electron fluxes were relatively weak and cool (100-500 eV). In this case (high altitude absorption), the recombination time constants are large (tens

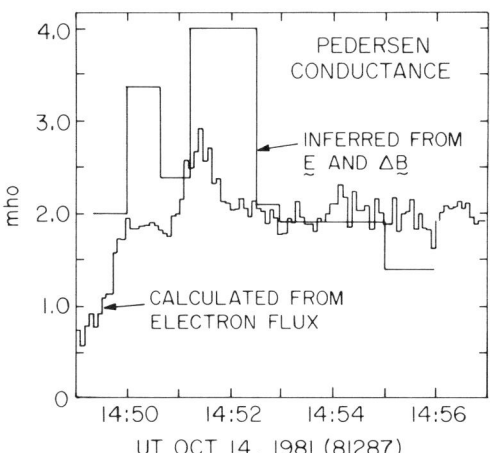

Fig. 6. Pedersen conductance inferred from E and ΔB (Fig. 5) compared to conductance inferred directly from particle fluxes.

of minutes) and convective history becomes important, with ionization transport from the dayside being comparable to in-situ ionization. Thus it is not too surprising that the two conductivities disagree somewhat (Fig. 6). We hope to test a more intense, more energetic case in the near future.

Another clever empirical technique to estimate directly the conductivity from magnetic perturbations was given by Ahn et al. [1983]. In this study, the conductivity calculated from Chatanika radar data is compared to simultaneous local ΔH perturbations. Although the scatter in the data is very large (Fig. 7), and the power-law fit requires Σ = 0 when ΔH = 0, nevertheless the method is interesting because it allows a global small-scale conductivity model to be constructed instantaneously. A comparison with the Gaussian form of our empirical model shows reasonable agreement (Fig. 8).

Global Optical Techniques

When we first required global time-dependent conductivity estimates for our convection model [Harel et al., 1981], we first hoped that DMSP photographs could supply our need. With several spacecraft photographing the aurora each 45 minutes, a model could conceivably be developed on ~15 minute time scales. Although the DMSP images can show accurately auroral positions on a large (two-dimensional) scale [Meng, 1976], nevertheless they are of limited usefulness in developing a global conductivity model, because (1) calibration is difficult [Eather, 1979]; (2) the entire oval is seldom imaged; and (3) the imager responds to a single broad visible band, whereas to get both the energy flux and characteristic energy of the precipitating electrons,

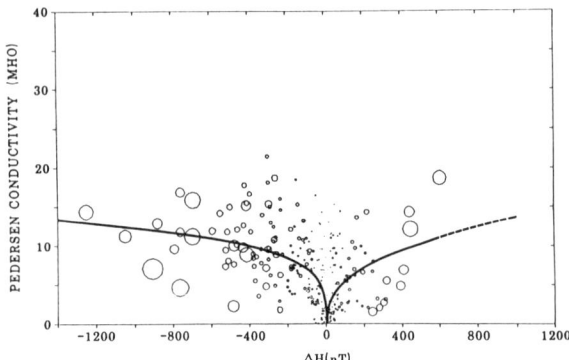

Fig. 7. Height-integrated Pedersen conductivity measured at Chatanika plotted against the College Alaska horizontal magnetic disturbance. The radius of the circles indicate the weight of each event. The fitted values are E_p = 2.4 $(\Delta H)^{.24}$ for ΔH < 0 and E_p = 0.7 $(\Delta H)^{.43}$ for ΔH > 0 [from Ahn et al., 1983].

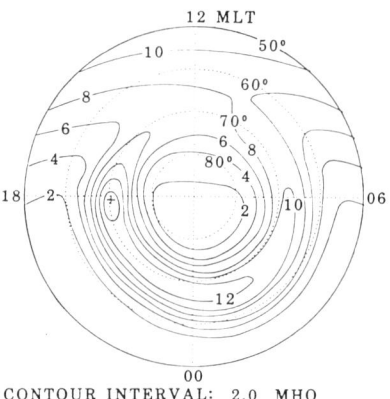

Fig. 8a.

Fig. 8b.

Fig. 8. Conductance on 1200 UT March 19, 1978 from the Ahn technique (a) compared to the Gaussian conductance model of Simons et al. [1983] (b) [from Ahn et al., 1983].

one needs at least two wavelength ranges, preferably line emissions for which the excitation mechanisms are readily quantifiable (e.g., 6300 and 4278 Å) [Rees and Luckey, 1974; Rees et al., 1983]. Thus, although the Auroral Scanning Photometer on ISIS-2 did not have the spatial resolution of DMSP, its ability to image in spectral lines [Murphree and Anger, 1978] made it more useful for this kind of study (the above referenced study, however, used only a single line, 4278 Å, which was reasonably accurate in determining precipitating energy fluxes).

Global images in ultraviolet lines, as are now

measured by Dynamics Explorer [Frank et al., 1982] can also be used to determine conductances. However, the detailed dependence of UV lines on the precipitating electron distribution has not yet been theoretically calculated, but work is in progress [Rees, private communication, 1983; Kamide et al, in preparation, 1983; Eather et al., in preparation, 1983]. Again, at least two lines must be used, and the most satisfactory pair (UV-UV or UV-visible or visible-visible) is yet to be determined.

Another recent technique involves imaging the aurora using X-ray Bremsstrahlung radiation [Mauk et al., 1981; Imhof et al., 1982]. These have the advantage of being usable even in daylight; however, at present their disadvantages are two fold: first, the spatial resolution on the images are rather poor, because of the low X-ray fluxes; secondly (and most critically), the Bremsstrahlung, being continuum radiation, cannot uniquely determine the electron spectrum, although it can be used as an indicator of the hardest electron fluxes. Nevertheless, it is likely that the X-ray images, when used in conjunction with two or more UV and/or visible lines, can be significant help in remotely sensing the ionospheric conductivity.

The real advantage of distant optical techniques is the ten-minute time scales for which one can determine a global conductivity model. With DE and the proposed OPEN mission, this technique may be feasible in the near future.

Strengths and Weaknesses of the Models

Incoherent radar systems are excellent at obtaining high spatial resolution (5 km), high time resolution (5 minutes) cuts through the ionospheric conductivity. However, their field of view is limited in north-south extent (~4° in latitude), and, since they are fixed in geographic longitude, they require 24 hours to build up an ionospheric conductance model. Joint radar studies, such as MITHRAS [de la Beaujardière et al., 1983] comprising Chatanika, EISCAT, and Millstone Hill, can build models more quickly, but still much slower than typical substorm times. Radar finds its best use when operated in conjunction with rocket or spacecraft overflights, when detailed patterns of ionospheric and Birkeland currents can be inferred, as well as both ionospheric flows and neutral winds [e.g., Robinson, 1982; Senior et al., 1982].

Conductivity models based on spacecraft electron flux measurements [Wallis and Budzinski, 1981; Spiro et al., 1982; Hardy et al., 1983] require at least six months of data (because of the slow drift in local time of the orbital plane). Thus these models must be parameterized by an activity index of some kind. Wallis and Budzinski [1981] used two ranges of Kp; Spiro et al. [1982] using four ranges of AE and five of Kp, showed that the AE index organized their data better than Kp (Fig. 9). It is likely that AL parameterizes the duskside observations, and AU the dawnside observations, better than AE alone, but that has not been tested. Also, an indicator of substorm phase would be helpful (since AE = 200 could be the peak expansion phase of a small substorm, or the recovery phase of a large substorm). Hardy et al. [1983] found that Kp was better for organizing their data. Possibly their data included more storm times, when ring current inflation causes expansion of the auroral zone [Siscoe, 1979]. Thus, an AE plus Dst index should be useful. All these improvements, of course, make the models more complicated and more cumbersome to use.

The major difficulty with such global empirical

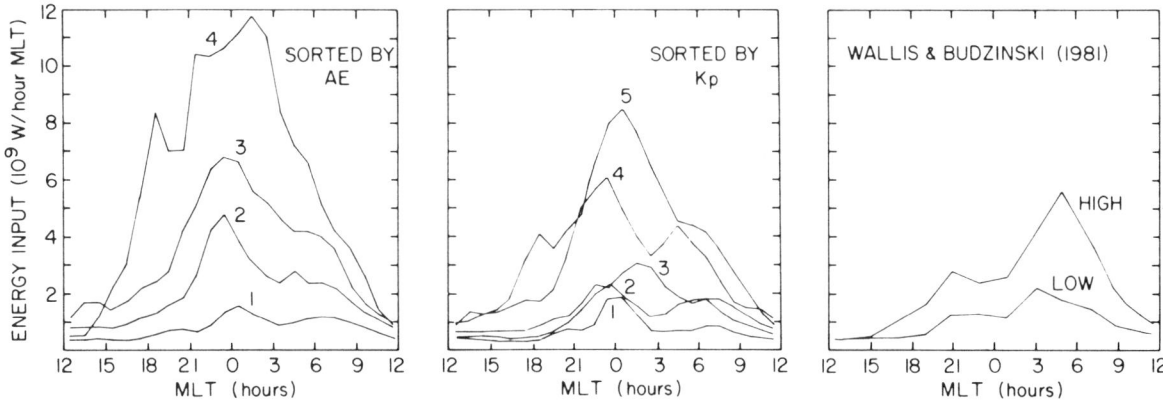

Fig. 9. Total precipitating electron energy input (both hemispheres) per hour magnetic local time. (a) Data sorted according to AE index, (b) data sorted according to Kp index, and (c) results calculated from data given in Table A1 and Table A2 of Wallis and Budzinski [1981] [from Spiro et al., 1982].

models is that they smear out the conductivity distribution in latitude because of the constantly changing location of the aurora. This can have serious consequences to the modeler who uses these models as inputs to his computer programs. Small-scale structures, obviously, cannot be reproduced; for these, incoherent scatter or in-situ electron flux measurements must be used. In addition, individual events may have much larger or smaller electron fluxes than the mean values [Murphree and Anger, 1978]; thus one must use the models with due caution.

Test of the Models

The recent CDAW-6 event presented an excellent opportunity to check the accuracy of an empirical model. In this event (March 22, 1979), data from the DMSP, P78-1, and TIROS spacecraft were used, both on a pass-by-pass basis and on a global basis, to test the accuracy of the model. Figure 10 shows four examples of pass-by-pass comparisons between the data, the binned model (histogram: BCM) and the gaussian fit to the empirical model (GCM: smooth curve). The binned model is determined by a straightforward interpolation (or extrapolation) of the model of Spiro et al. [1982]; the gaussian fit is given by

$$\Sigma = A_1 \exp\left\{-(\Lambda - A_2)^2/2A_3^2\right\} \quad (4)$$

with Λ being invariant latitude, $A_1 = a_{10} + a_{11}AE$, $A_2 = a_{20} + a_{21}AE$, and A_3 only a function of MLT [Simons et al., 1983]. The values a_{10}, a_{11}, a_{20}, a_{21}, and A_3 for both Σ_P and Σ_H are tabulated in Table 1.

We see that although the conductance is approximately correct when integrated over latitude, the latitude structure is generally broader in the model than in the data. This is also apparent in Figure 11, which shows the data from 1400-2300 UT (AE = 1085 and 190), both simply binned by ILAT and MLT (top), and smoothed in ILAT and MLT (middle), in a similar fashion to that performed in the bin scheme of Spiro et al. [1982] (bottom). Although the data show considerably more structure than the model, nevertheless, the model appears adequate on the large scale. The latitude shift between data and model is most easily explained by ring current inflation of the magnetosphere [Siscoe, 1979]. For other and more detailed comparisons, see Simons et al. [1983].

Effects of Conductivity Models on Other Modeling Efforts

Kamide and Richmond [1982] have shown, for their magnetogram-inversion (KRM) technique, that doubling the ionospheric conductivity will halve the inferred electric fields. What effect does

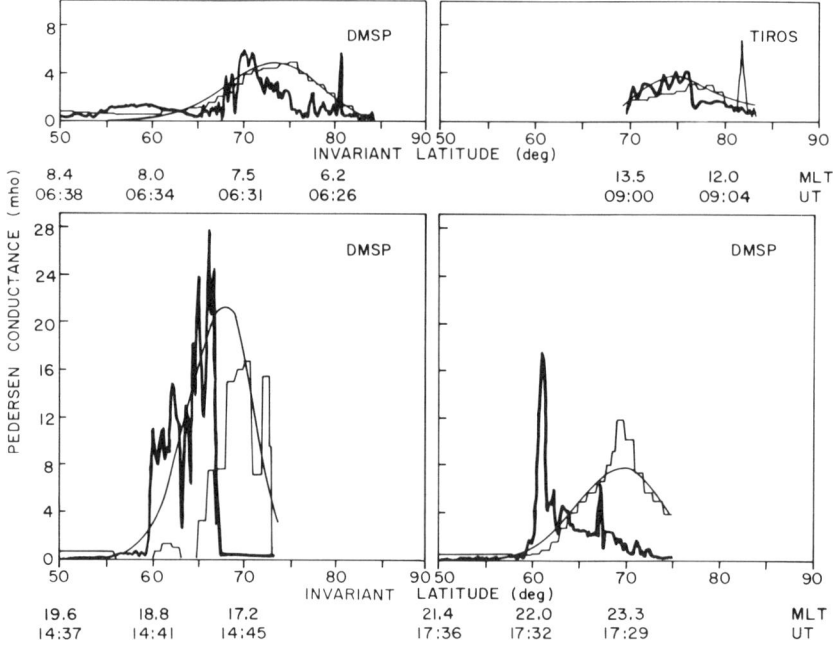

Fig. 10. Pass-by-pass comparisons of conductances calculated from particle fluxes on March 22, 1979 (heavy line), the gaussian conductivity model (GCM, smooth line) and from the binned conductivity model (BCM, histogram). Despite a factor-of-seven range in conductance peaks, the models fit reasonably well, although a latitude shift is evident at later UT's. The data were not used in the construction of the model.

TABLE 1. Constants of Gaussian Fits

MLT	Pedersen Conductance					Hall Conductance				
	a_{10}	a_{11}	a_{20}	a_{21}	A_3	a_{10}	a_{11}	a_{20}	a_{21}	A_3
0- 1	5.90	0.0090	70.30	-0.0047	5.40	10.60	0.0190	69.90	-0.0046	4.90
1- 2	4.90	0.0100	69.80	-0.0040	5.40	8.50	0.0230	69.30	-0.0042	4.90
2- 3	4.00	0.0110	69.40	-0.0033	5.50	6.70	0.0260	68.70	-0.0032	4.90
3- 4	4.10	0.0089	70.30	-0.0047	5.80	6.90	0.0220	69.30	-0.0045	4.90
4- 5	4.60	0.0081	70.90	-0.0055	5.40	8.00	0.0200	69.80	-0.0048	4.60
5- 6	4.70	0.0076	71.70	-0.0054	5.40	8.70	0.0180	70.50	-0.0048	4.60
6- 7	4.90	0.0064	72.80	-0.0057	5.40	9.50	0.0130	71.40	-0.0049	4.90
7- 8	4.50	0.0060	73.80	-0.0063	5.60	8.90	0.0120	72.10	-0.0050	5.20
8- 9	3.90	0.0055	74.50	-0.0062	5.70	7.80	0.0110	72.70	-0.0050	5.40
9-10	3.50	0.0044	74.90	-0.0060	5.90	7.10	0.0083	73.20	-0.0053	5.60
10-11	3.20	0.0023	74.50	-0.0047	6.40	6.40	0.0048	72.70	-0.0038	5.80
11-12	2.90	0.0013	74.30	-0.0031	6.60	6.20	0.0009	72.30	-0.0023	6.20
12-13	3.20	0.0012	74.40	-0.0010	6.00	6.50	0.0002	72.60	-0.0008	5.90
13-14	4.00	0.0016	76.30	-0.0043	4.50	6.60	0.0021	72.80	-0.0010	5.80
14-15	4.40	0.0027	76.70	-0.0060	3.90	5.20	0.0053	72.50	-0.0004	6.10
15-16	4.30	0.0045	76.50	-0.0060	3.50	4.30	0.0089	73.50	-0.0006	5.30
16-17	3.40	0.0074	75.40	-0.0054	4.00	4.20	0.0120	73.00	-0.0011	5.10
17-18	2.00	0.0150	74.00	-0.0045	4.50	1.80	0.0290	72.10	-0.0010	4.90
18-19	1.60	0.0190	72.80	-0.0037	4.60	0.67	0.0420	71.40	-0.0009	4.60
19-20	2.80	0.0130	71.40	-0.0021	4.60	4.30	0.0300	70.20	0.0004	4.30
20-21	3.60	0.0110	70.80	-0.0011	4.80	6.70	0.0250	70.10	0.0005	4.20
21-22	4.00	0.0120	69.90	0.0004	5.10	6.40	0.0310	69.50	0.0021	4.40
22-23	5.80	0.0093	69.40	-0.0002	5.00	10.20	0.0230	69.30	0.0007	4.60
23-24	6.90	0.0073	69.80	-0.0026	5.00	12.50	0.0170	69.70	-0.0024	4.60

redistributing the conductivity (i.e., by smoothing out arcs) have on modeling efforts?

The KRM technique [Kamide et al., 1982, 1983] uses a two-dimensional conductance model to invert H and D perturbations on a two-dimensional plane to infer a two-dimensional ionospheric electric potential pattern. This, coupled again with the conductance model, yields true (not merely "equivalent") ionospheric current vectors whose divergence yields the Birkeland current pattern.

Similarly, the Rice Convection Model (RCM) solves a two-dimensional differential equation in the ionosphere, conserving current, given a two-dimensional conductance model [Harel et al., 1981a; Wolf et al., 1982]. Here, however, the Birkeland currents are calculated from magnetospheric current divergences, and the ionospheric currents and electric fields are calculated self-consistently, using a conductivity model.

In both types of models, the inferred electric field is roughly inversely proportional to the conductivity. In the KRM model, the magnetic perturbations are used to calculate ionospheric currents; if the conductance is too large, the electric fields that will be inferred will be too small and vice-versa [Kamide and Richmond, 1982]. In the RCM model, sudden increases in the conductance will be accompanied by decreases in the electric field (to preserve Pedersen currents), at least until a new shielding pattern is established.

Let us examine an (oversimplified) RCM-type calculation to see what our errors in conductance may cause. For the example of Figure 10 (1440 UT), the inferred conductance from the DMSP data shows a sharp conductivity gradient at $\Lambda \approx 60°$ and at 67°, whereas the Gaussian conductivity model (GCM) used by RCM is quite smooth. Averaging both the DMSP-derived conductances and the GCM model to the latitude bins of the RCM simulation yields Figure 12a.

In both KRM and RCM modeling techniques, the latitude of the conductivity is shifted: in the KRM technique, the shift is to make the latitude of the strongest electrojet coincide with the latitude of the largest conductivity. In the RCM technique, the conductivity is shifted so that the equatorward edge (10% of the peak) of the conductivity coincides with the inner edge of the electron plasma sheet. It is gratifying that both techniques shifted the conductivity equatorward, as the actual data show is necessary [Simons et al., 1983].

Figure 12b shows the latitude profile of Birkeland current (mostly downward) and Pedersen current (northward) for the 1440 UT dusk-side crossing of Figure 10 (RCM model). The Birkeland cur-

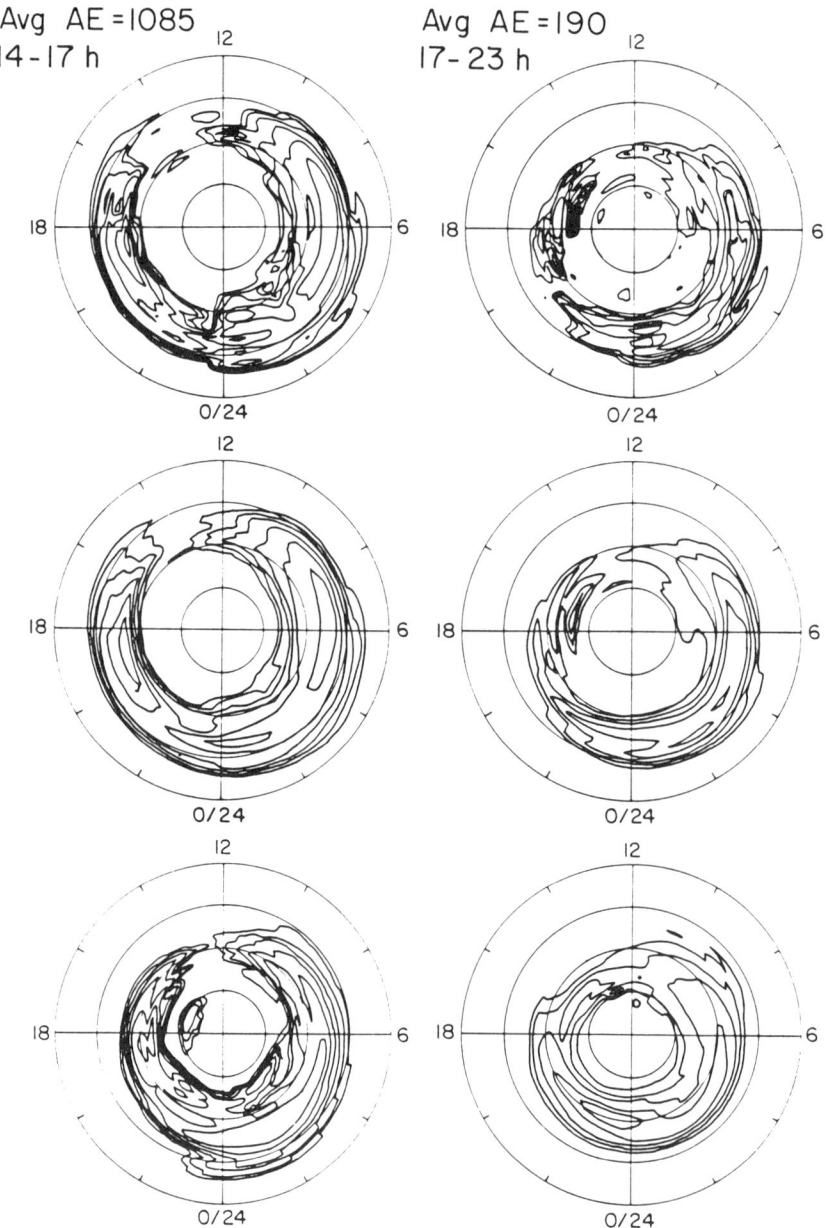

Fig. 11. Contour plots of precipitting electron fluxes based on (top) Data, (middle) Smoothed data, and (bottom) binned model (BCM) for the UT range 1400-1700 UT (left) and 1700-2300 UT (right) March 22, 1979. The bottom contour is 0.25 ergs/cm^2-s and successive contours factors of two apart [from Simons et al., 1983]. General features agree reasonably well.

rent is calculated from the horizontal divergence of magnetospheric drift currents, and the Pedersen current (and convection electric fields) are solved self-consistently on a two-dimensional ionospheric grid, given the Birkeland currents, the ionospheric height-integrated conductivity, and the polar boundary electrostatic potential distribution as inputs [Harel et al., 1981a]. For the sake of illustration, let us assume that the electric field is exactly northward (not a bad assumption at dusk), and examine how that calculated electric field varies with the conductivity model (i.e., similar assumptions as used by Smiddy et al. (above)).

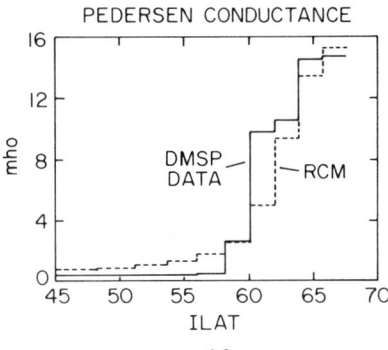

Fig. 12a.

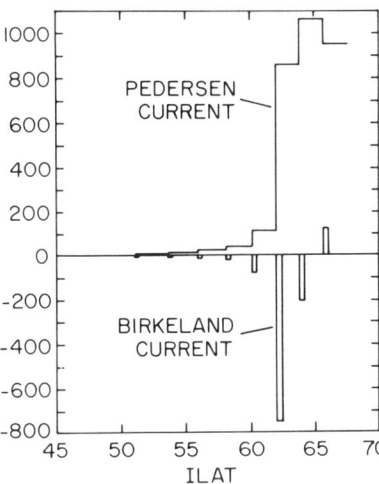

Fig. 12b.

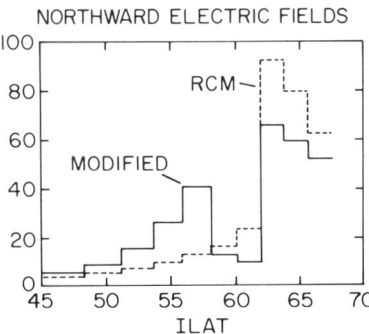

Fig. 12c.

Fig. 12. (a) Conductance from DMSP data (solid) and gaussian model (dashed) binned to the latitude ranges of the Rice Convection Model (RCM) for 1440 UT March 22, 1979 event. (b) Pedersen and Birkeland currents at same time as (a). (c) Comparison of electric fields calculated using RCM conductance (dashed) and DMSP data (solid).

Figure 12c shows, for the Pedersen current of Figure 12b, how the electric fields vary with the smoothness of the conductivity model. The dashed curve shows the actual RCM result. The solid curve is calculated from the Pedersen current of Figure 12b and the DMSP data conductance of Figure 10a (solid line), and $E_N = J_{PN}/\Sigma_P$. Since the RCM technique requires that the total potential drop $\int E_N dx$ remain constant (as predicted from IMF measurements), the resulting electric fields are here multiplied by the ratio of the actual potential drop to the inferred potential drop (roughly .75 in this case).

Thus, a relatively minor inaccuracy of the conductivity model (mostly due to smoothing) can cause a fairly severe change in the detailed electric field structure, although the total potential difference was not too far off (25%). Thus, odd electric field patterns such as occasionally arise in modeling efforts [e.g., Kamide et al., 1982], may well be attributed to the conductivity model. It should be noted, however, that the RCM technique would then use the inferred electric fields to move the magnetospheric particles around, generating new Birkeland currents, etc., and a strong low-latitude electric field such as that in Figure 12c may or may not persist in the self-consistent RCM magnetosphere model. Computer experiments performed by adding induction electric fields in the equatorial plane at midnight, for example, generally make no significant change in the electric field pattern, once the magnetospheric particle population responds to it self-consistently [Harel et al., 1981b].

Conclusions

A global conductance model is vitally important for several kinds of models of the magnetosphere-ionosphere-atmosphere system. Global empirical models have recently been developed that can be quite useful for large-scale modeling. However, small-scale structures are smoothed out in the empirical models, and so are not reproduced well.

New global models of ionospheric conductance on ~10 minute time scales may be feasible in the future by examining Dynamics Explorer photographs of the auroral zone in two wavelength ranges, one (4278 Å) to infer the precipitating electron energy flux, and the ratio with the other (6300 Å) to infer the average energy of the precipitating particles. Similarly, X-ray or UV images of the aurora can also be used to make conductivity maps from the inferred energetic particle precipitation.

Acknowledgments. The author thanks R. A. Frahm, D. A. Hardy, R. A. Hoffman, Y. Kamide, H. Kroehl, R. M. Robinson, R. W. Spiro, and R. A. Wolf for data and insight. This research was supported by the Atmospheric Sciences section of the National Science Foundation under grant ATM-8306772 and by NASA under grant NGR-44-006-137.

References

Ahn, B.-H., R. M. Robinson, Y. Kamide, and S.-I. Akasofu, Electric conductivities, electric fields, and auroral particle energy injection rate in the auroral ionosphere and their empirical relations to the horizontal magnetic disturbance, Planet. Space Sci., (in press), 1983.

Banks, P. M., and A. Kockarts, Aeronomy, Academic Press, New York, 1973.

Brekke, A., J. R. Doupnik, and P. M. Banks, Incoherent scatter measurements of E region conductivities and currents in the auroral zone, J. Geophys. Res., 79, 3773-3790, 1974.

Chestnut, W. G., J. C. Hodges, and R. L. Leadabrand, Correlation of radar echoes from aurora with satellite measured particle precipitation, Rep. DNA-2825F, Stanford Res. Inst., Menlo Park, CA, Dec. 1971.

de la Beaujardière, O., V. B. Wickwar, M. J. Baron, J. Holt, R. M. Wand, P. Baver, M. Blanc, C. Senior, D. Alcaydé, G. Caudal, J. Foster, E. Nielsen, and R. Heelis, Mithras: A brief description, Radio Sci., (in press), 1983.

Donnelly, R. F., Comparison of non-flare solar soft X-ray flux with 10.7-cm radio flux, J. Geophys. Res., 87, 633-6334, 1982.

Donnelly, R. F., D. F. Heath, and J. L. Lean, Active-region evolution and solar rotation variations in solar UV irradiance, total solar irradiance, and soft X-rays, J. Geophys. Res., 87, 10318-10324, 1982.

Eather, R. H., DMSP calibration, J. Geophys. Res., 84, 4134-4144, 1979.

Hanson, W. B., Structure of the ionosphere, in Satellite Environment Handbook, F. S. Johnson, ed., pp. 23-49, Stanford University Press, Stanford, CA, 1965.

Hardy, D. A., M. S. Gussenhoven, and E. Holeman, High latitude maps of precipitating auroral electrons, (abstract), EOS, Trans. AGU, 64, 300, 1983.

Harel, M., R. A. Wolf, P. H. Reiff, R. W. Spiro, W. J. Burke, F. J. Rich, and M. Smiddy, Quantitative simulation of a magnetospheric substorm, 1, Model logic and overview, J. Geophys. Res., 86, 2217-2241, 1981a.

Harel, M., R. A. Wolf, R. W. Spiro, P. H. Reiff, C.-K. Chen, W. J. Burke, F. J. Rich, and M. Smiddy, Quantitative simulation of a magnetospheric substorm, 2, Comparison with observations, J. Geophys. Res., 86, 2242-2260, 1981b.

Hedin, A. E., J. E. Salah, J. V. Evans, C. A. Reder, G. P. Newton, D. C. Kayser, D. Alcayde, P. Bauer, L. L. Cogger, and J. P. McClure, A global thermospheric model based on mass spectrometer and incoherent scatter data: MSIS I. N_2 density and temperature, J. Geophys. Res., 82, 2139-2147, 1977a.

Hedin, A. E., C. A. Reber, G. P. Newton, N. W. Spencer, H. C. Brinton, and H. G. Mayr, A global thermospheric model based on mass spectrometer and incoherent scatter data: MSIS 2. composition, J. Geophys. Res., 82, 2148-2156, 1977b.

Imhof, W. L., J. Stadsnes, J. R. Kilner, D.W. Datlowe, G. H. Nakano, J. B. Reagan, and P. Stauning, Mappings of energetic electron precipitation following substorms using the satellite bremsstrahlung technique, J. Geophys. Res., 87, 671-780, 1982.

Jones, R. A., and M. H. Rees, Time-dependent studies of the aurora, 1, Ion density and composition, Planet. Space Sci., 21, 537-557, 1973.

Kamide, Y., and A. D. Richmond, Ionospheric conductivity dependence of electric field and currents estimated from ground magnetic observations, J. Geophys. Res., 87, 8331-8337, 1982.

Kamide, Y., B.-H. Ahn, S.-I. Akasofu, W. Baumjohann, E. Friis-Christensen, H. W. Kroehl, H. Maurer, A. D. Richmond, G. Rostoker, R. W. Spiro, J. K. Walker, and A. N. Zaitzev, Global distribution of ionospheric and field-aligned currents during substorms as determined from six IMS meridian chains of magnetometers: Initial results, J. Geophys. Res., 87, 8228-8240, 1982a.

Kamide, Y., H. W. Kroehl, B. A. Hausman, R. L. McPherron, S.-I. Akasofu, A. D. Richmond, P. H. Reiff, and S. Matsushita, Numerical modeling of ionospheric parameters from global IMS magnetometer data for the CDAW-6 intervals, NCAR Technical Note (in press), 1983.

Kockarts, G., Effects of solar variations on the upper atmosphere, Solar Phys., 74(2), 295-320, 1981.

Mauk, B. H., Chin, J., and Parks, G., Auroral X-ray images, J. Geophys. Res., 86, 6827-6835, 1981.

Mehta, N. C., Ionospheric electrodynamics and its coupling to the magnetosphere, Ph.D. thesis, University of California, San Diego, CA., 1978.

Meng, C.-I., Simultaneous observations of low-energy electron precipitation and optical auroral arcs in the evening sector by the DMSP 32 satellite, J. Geophys. Res., 81, 2771-2785, 1976.

Murphree, J. S., and C. D. Anger, Instantaneous auroral particle energy deposition by optical means, Geophys. Res. Lett., 5, 551-554, 1978.

Rees, M. H., Auroral ionization and excitation by incident energetic electrons, Planet. Space Sci., 11, 1209-1218, 1963.

Rees, M. H., On the interaction of auroral protons with the Earth's atmosphere, Planet. Space Sci., 30, 463-472, 1982.

Rees, M. H., and D. Luckey, Auroral electron energy derived from ration of spectroscopic emissions: 1. Model computations, J. Geophys. Res., 79, 5181-5186, 1974.

Rees, M. H., B. A. Emery, R. G. Roble, and K. Stamnes, Neutral and ion gas heating by auroral electron precipitation, J. Geophys. Res., (in press), 1983.

Robinson, R. M., and T. A. Potemra, Auroral zone conductivities within the field-aligned current sheets, in Magnetospheric Currents, this volume, edited by T. A. Potemra and J. N. Barfield, AGU, (in press), 1983.

Robinson, R. M., R. R. Vondrak, and T. A. Potemra, Electrodynamic properties of the evening sector ionosphere within the region 2 field-aligned current sheet, J. Geophys. Res., 87, 731-741, 1982.

Senior, C., R. M. Richmond, and T. A. Potemra, Relationship between field-aligned currents, diffuse auroral precipitation, and the westward electrojet in the early morning sector, J. Geophys. Res., 87, 10469-10477, 1982.

Simons, S. L., P. H. Reiff, R. W. Spiro, D. A. Hardy, and H. W. Kroehl, A comparison of precipitating electron energy flux on March 22, 1979 with an empirical model, J. Geophys. Res., (submitted), 1983.

Siscoe, G. L., A Dst contribution to the equatorial shift of the aurora, Planet. Space Sci., 27, 997-1000, 1979.

Smiddy, M., W. J. Burke, M. C. Kelly, N. A. Saflekos, M. S. Gussenhoven, D. A. Gussenhoven, D. A. Hardy, and F. J. Rich, Effects of high-latitude conductivity on observed convection electric fields and Birkeland currents, J. Geophys. Res., 65, 6811-6818, 1980.

Spiro, R. W., P. H. Reiff, and L. J. Maher, Jr., Precipitating electron energy flux and auroral zone conductances: An empirical model, J. Geophys. Res., 87, 8215-8227, 1982.

Sugiura, M., N. C. Maynard, W. H. Farthing, J. P. Heppner, and B. G. Ledley, Initial results on the correlation between the magnetic and electric fields observed from the DE-2 satellite in the field-aligned current regions, Geophys. Res. Lett., 9, 985-988, 1982.

Vickrey, J. F., R. R. Vondrak, and S. J. Matthews, The diurnal and latitudinal variation of auroral zone ionospheric conductivity, J. Geophys. Res., 86, 65-75, 1981.

Vondrak, R. R., and M. J. Baron, Radar measurements of the latitudinal variation of auroral ionization, Radio Sci., 11, 939-946, 1976.

Vondrak, R. R., and M. J. Baron, A method of obtaining the energy distribution of auroral electrons from incoherent scatter radar measurements, in Radar Probing of the Auroral Plasma, Procedings of the EISCAT summer school, Tromso, Norway, 1975, edited by A. Brekke, pp. 315-330, Universitetforlaget, Tromso-Oslo-Bergen, 1977.

Wallis, D. D., and E. E. Budzinski, Empirical models of height-integrated conductivities, J. Geophys. Res., 86, 125-137, 1981.

Wolf, R. A., Effects of ionospheric conductivity on convective flow of plasma in the magnetosphere, J. Geophys. Res., 75, 4677, 1970.

Wolf, R. A., The quasi-static (slow-flow) region of the magnetosphere, in Solar-Terrestrial Theory, edited by R. M. Carovillano and J. M. Forbes, (in press), AGU, Washington, D.C., 1983.

Wolf, R. A., M. Harel, R. W. Spiro, G.-H. Voigt, P. H. Reiff, and C.-K. Chen, Computer simulation of inner magnetospheric dynamics for the magnetic storm of July 29, 1977, J. Geophys. Res., 87, 5949-5962, 1982.

COORDINATED GROUND AND SATELLITE OBSERVATIONS OF CONDUCTIVITIES,
ELECTRIC FIELDS, AND FIELD-ALIGNED CURRENTS

R. M. Robinson

Radio Physics Laboratory, SRI International, Menlo Park, California 94025

Abstract. Chatanika radar measurements of conductivities and electric fields have been made during coincident passes of several polar orbiting satellites. Observations made in coordination with the Triad satellite have been especially useful because they have yielded information about the electrodynamic coupling between the ionosphere and the magnetosphere. The Triad data are used to determine the field-aligned currents along field lines connected to the ionosphere. The radar measurements of electric fields and conductivities on the same field lines provide information about the closure of the field-aligned currents. Among the auroral features that have been studied in this manner are the morning- and evening-sector diffuse aurora and auroral arcs. In this paper, these observations are reviewed, and the manner in which the ionospheric conductivity influences the coupling between the ionosphere and the magnetosphere is discussed.

Introduction

Much progress has been made recently in understanding the relationships between auroral electrodynamic parameters by examining the data simultaneously obtained by ground and space-borne instrumentation. In particular, over the past several years an increased effort has been made to coordinate space-borne measurements with measurements simultaneously made by the Chatanika radar.

Radar observations have been made in coordination with several satellites including AE-C, Triad, S3-2, DMSP, Isis, Dynamics Explorer, and the NOAA satellites. It would be impossible to review the results of all these coordinated experiments, and much of the work is still in progress. In this review, we restrict our discussion to the results of coordinated experiments with the Triad satellite. These experiments were designed to study magnetosphere-ionosphere coupling by field-aligned currents. The Triad satellite has been orbiting the earth since 1972. It carries a vector magnetometer that measures the magnetic perturbation produced by field-aligned currents. In September 1979, a Triad receiving station was installed at the radar site in Chatanika, Alaska, which facilitated the scheduling of coordinated Triad and Chatanika measurements.

The Chatanika radar has many possible operating modes. One, in particular, proved to be especially useful for coordinated experiments with satellites. In this mode, the radar scans in elevation at a fixed azimuth, usually in the magnetic meridian. Figure 1 shows the geometry of the scan and the trajectory of Triad when the satellite passes directly over the radar. In the elevation scan, the radar can measure ion drift velocity and electron density as a function of both latitude and altitude. The initial measurements obtained when the radar was operating in the elevation scan mode were presented by Vondrak and Baron [1976]. The method by which electric fields are determined by this technique was discussed by de la Beaujardiere et al. [1977] and Robinson et al. [1981]. The electron density measurements are used to compute the height-integrated conductivity, which can be used with the electric-field measurements to calculate the horizontal ionospheric current intensity. Electric fields and conductivities can be measured within about three degrees of latitude from the radar. Each elevation scan takes 10 to 15 min to complete. When operated in coordination with a satellite pass, the scans were made for at least two hours before and two hours after the time of the pass. Examination of radar data obtained before and after the pass reveals to what extent the ionosphere varied during the experiment.

This review is divided into two sections. In the first section we discuss the spatial relationships between field-aligned currents and ionospheric parameters such as conductivity and electric field. In the second section, we describe some of the ways by which the ionospheric conductivity affects the flow of field-aligned currents.

Spatial Relationships Between Field-Aligned
Currents and Ionospheric Electric Field
and Conductivity

Evening Sector

Robinson et al. [1982] combined electric fields and conductivities measured by the Chatanika radar with field-aligned currents measured by Triad within the Region 2 current sheet in the evening sector. As defined by Iijima and Potemra [1976], the Region 2 evening-sector current sheet is downward. The results of the Triad/Chatanika study are summarized in Figure 2, which shows electron density contours in the meridian plane constructed from radar-electron-scan measurements. The arrows show the parallel and perpendicular currents inferred from the Triad and Chatanika data. The

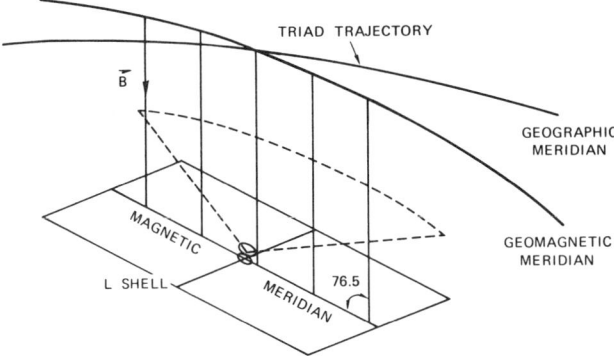

Fig. 1. Operating mode used during Chatanika radar/Triad satellite coordinated experiments.

ionization feature shown in this figure is often observed by the radar in the evening sector. Its occurrence frequency and latitudinal structure suggest that it is associated with electron precipitation from the central plasma sheet. Whalen [1982] studied this ionization, referred to as the "continuous" or "diffuse" aurora, and found that it is distributed in a nearly circular pattern offset from the geomagnetic pole. The latitudinal profile of energy flux from precipitating electrons is usually Gaussian in shape. The center of the structure in Figure 2 probably represents the center of this Gaussian distribution. By combining the radar electron-density measurements with the Triad field-aligned-current measurements, Robinson et al. [1982] showed that the equatorward edge of the Region 2 current sheet coincides with the edge of approximately 1-keV electron precipitation from the central plasma sheet. The poleward edge of the current sheet was situated in the vicinity of the localized maximum in the precipitation pattern. Because the field-aligned current reversal was so closely correlated with the center of the diffuse precipitation, Robinson et al. [1982] referred to this feature as the interface arc. Although not visually as bright as the discrete arcs, interface arcs are clearly visible as diffuse bands, especially when viewed obliquely.

Robinson et al. [1982] also found that the downward current in the Region 2 current sheet was consistent with the radar measurements of the ionospheric current. These measurements showed that the gradient of the northward ionospheric current was directed poleward, as it must be if downward currents are entering the ionosphere. The change in northward current was produced by a combination of electric field and conductivity gradients. The northward electric field drives a moderate eastward electrojet in the ionosphere.

Although electric fields and conductivities in the Region 2 current sheet in the evening sector are consistent with the field-aligned currents present, this is not the case for Region 1. The primary difficulty in studying this region is that it invariably contains very dynamic discrete arcs. Figure 3 shows radar and satellite data obtained during an evening-sector pass on 28 February 1981. The top panel is a contour plot constructed from radar data obtained simultaneously with the Triad pass. The broad E-region ionization enhancement centered between 65° and 66° is the signature of the central plasma sheet described above. The maximum density in this enhancement is about 2×10^5 el/cm^3. North of this interface arc is a discrete arc. The discrete arc is distinguished from the interface arc by its higher density (about 6×10^5 el/cm^3) and sharp latitudinal gradients.

The second panel in Figure 3 shows the radar electric-field measurements. The east-west electric field is small and fairly constant with latitude. The north-south electric field increases with latitude to a peak of 30 mV/m at a latitude of 66.5°, then falls sharply to zero within the arc. The maximizing of the northward electric field between the interface arc and the bright arc is a fairly common feature of the electric-field pattern. This region appears dark relative to the arcs that surround it and has been referred to as the dark band. The dark band warrants special attention because it is always present and because the large electric fields within it also drive the most intense eastward currents. In addition, the Joule heating rate peaks in the dark band.

The Triad measurements of the east-west magnetic perturbations during this pass are shown by the solid line in the bottom panel of Figure 3. Upward slope in these data implies the presence of downward field-aligned current, while negative slope implies upward field-aligned current. The field-aligned current reverses just poleward of the interface arc. The upward (Region 1) current sheet is composed of the dark band in the equatorward portion and the bright arc in the poleward portion. The solid circles in the bottom panel of Figure 3 represent the radar measurements of the northward current density scaled so that the measurements coincide with the magnetic perturbations if the northward ionospheric currents completely

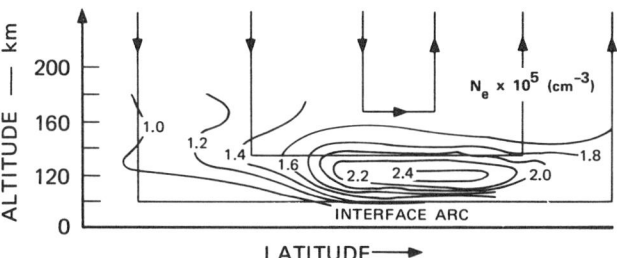

Fig. 2. Typical E-region ionization pattern within the Region 2 evening-sector field-aligned-current sheet. The electric field is predominantly northward, while the horizontal currents are northeastward.

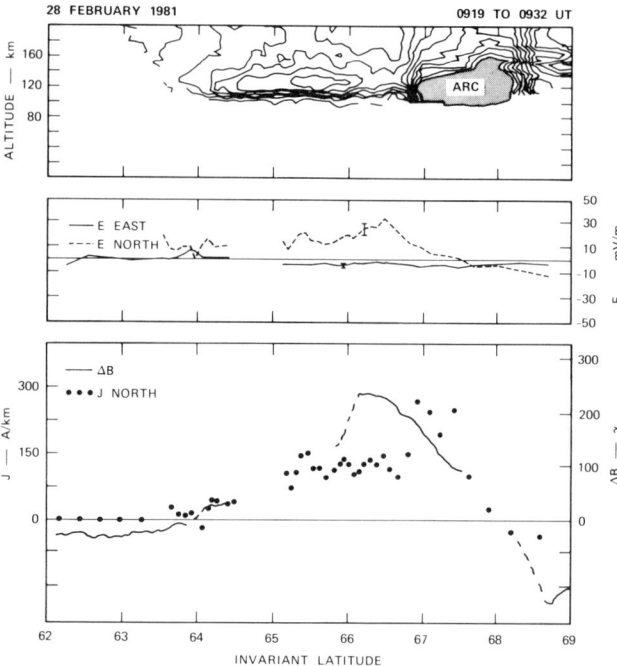

Fig. 3. Ionization, electric fields, and field-aligned currents measured during a coordinated Triad/Chatanika experiment on 28 February 1981.

close the upward currents. It is apparent that this assumption is not valid for this data set. In general, there is very poor agreement between the field-aligned currents in Region 1 and the northward ionospheric currents. As mentioned above, this may be due to the dynamic nature of auroral arcs.

The results presented above are in reasonable agreement with past studies of the spatial relationships among field-aligned currents, electric fields and visible aurora in the evening sector. In particular, Kamide and Rostoker [1977] showed that the bright aurora coincided well with the region of upward field-aligned currents. Armstrong et al. [1975] also related the diffuse aurora to the region of downward current flow. Both of these studies involved coordinated measurements by ground and space-borne techniques. The electrodynamics of the downward current sheet in the evening sector has also been studied using Chatanika radar data obtained during two S3-2 satellite overpasses [Vondrak and Rich, 1982]. Other coordinated experiments have attempted to examine the field-aligned current structure of auroral arcs. For example, de la Beaujardiere et al. [1981] combined Chatanika radar data with measurements made by the AE-C satellite. Robinson et al. [1981] used radar data in conjunction with sounding rocket measurements. The results of these experiments have been reviewed recently by Vondrak [1981].

Morning Sector

Ionospheric electrodynamic properties have been studied within the morning-sector field-aligned current sheets by Senior et al. [1982]. Figure 4 shows data from four morning-sector Triad/Chatanika experiments. The electron density contour plots shown for the four cases are very similar. All contain a latitudinally extended E layer with one or more central maxima. The appearance of the ionosphere in these contour plots is similar to that of the interface arc discussed above although somewhat patchy. This similarity suggests that this ionization is associated with precipitation from the central plasma sheet. The maximum E-layer electron density is somewhat greater in the morning sector than in the evening sector. Also, the height of the E-layer maximum is less in the morning suggesting the presence of a harder precipitation source. These differences are consistent with the differences in central plasma-sheet precipitation measured by polar-orbiting satellites. The shaded regions in Figure 4 show the locations of the Region 2 (upward) and Region 1 (downward) current sheets as determined by the simultaneous Triad data. The upward current sheet encompasses the entire latitudinal range within which central plasma-sheet precipitation is present. This differs from the pattern observed in the evening sector in which the field-aligned current reversal occurs near the center of the central plasma-sheet precipitation. Senior et al. [1982] also showed that the westward electrojet filled the entire upward-current region and extended well into the region of downward-current flow. However, in the upward-current region the southward electric field that drives the westward electrojet is small, the large conductivities producing the intense electrojet. In the downward-current sheet, the conductivities are small, but large southward electric fields are present to sustain the electrojet. These results are consistent with those of Kamide and Vickrey [1982] who combined Chatanika radar data with ground-based magnetometer data to show that the westward electrojet is divided into two regions, one dominated by electric field, the other dominated by conductivity.

Effects of Conductivity on Field-Aligned Currents

We begin our discussion of conductivity effects by looking in more detail at the spatial relationship between the Region 2 field-aligned current sheet in the evening sector and the central plasma sheet precipitation. As shown in Figure 2, Robinson et al. [1982] found a close correspondence between the equatorward edge of the current sheet and the ionization produced by kiloelectron volt electrons from the central plasma sheet. This result contradicts the findings of Klumpar [1979] who used Isis magnetometer and electron spectrometer data to show that the equatorward

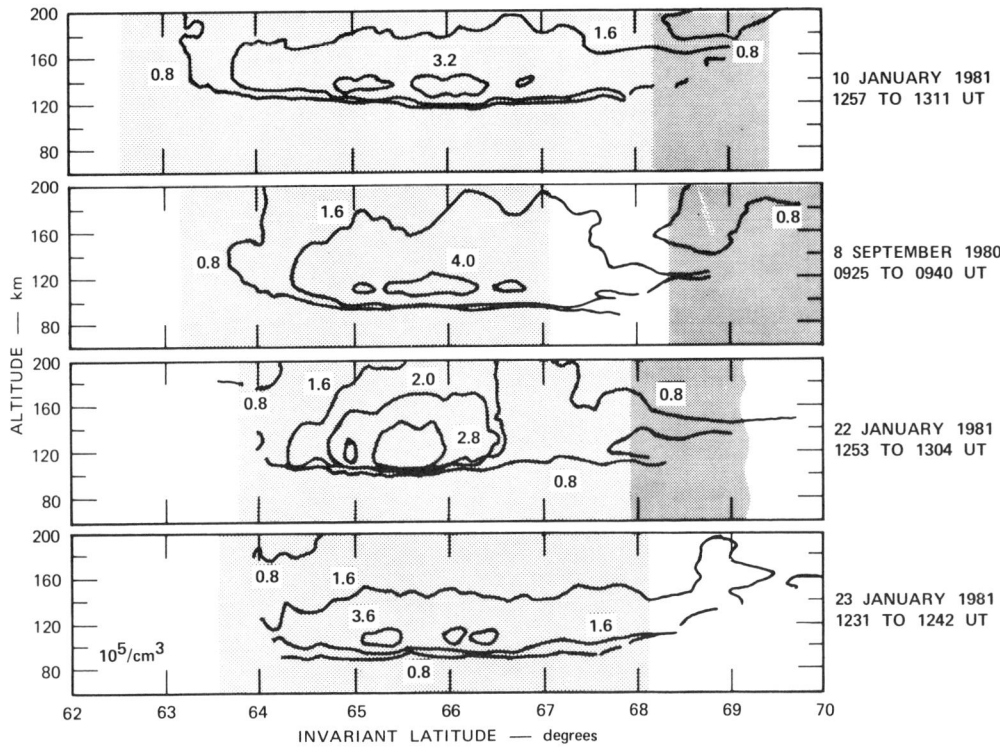

Fig. 4. Spatial relationships between ionization and field-aligned currents measured during four coordinated Triad/Chatanika experiments from Senior et al. [1982]. Lightly shaded regions indicate upward field-aligned current; darker shading indicates downward field-aligned current.

edge of the Region 2 current sheet extended, on the average, two degrees equatorward of the point at which the kiloelectron volt electron precipitation ended. This discrepancy can perhaps be reconciled by noting that Klumpar [1979] used mostly summer passes to reach his conclusion. The Triad/Chatanika data studied by Robinson et al. [1982], on the other hand, were obtained during the winter. This immediately suggests that the presence of ionization produced by solar ultraviolet radiation can alter the field-aligned current distribution in the high-latitude ionosphere. In particular, the conductivity produced equatorward of plasma-sheet precipitation can provide a conducting pathway for closure of field-aligned currents. Thus, the equatorward edge of the Region 2 current sheet tends to be tied to the conductivity distribution in the ionosphere rather than the precipitating particle characteristics. The applicability of this property to other field-aligned current regions remains to be shown.

In addition to affecting the distribution of currents, the ionospheric conductivity can also affect the local intensity of field-aligned currents. This is perhaps best illustrated in the Region 1 current sheet in the evening sector within which auroral arcs occur. In Figure 5, we offer two possible effects on field-aligned currents of the enhanced conductivity associated with arcs. In Case 1, a conductivity enhancement results in an increase in the northward ionospheric current. For the horizontal current to have such a latitudinal profile, field-aligned currents must flow into and out of the edges of the arc. The resulting magnetic signature detected by a satellite traversing the structure is shown in the bottom trace of Figure 5(a). In Case 2, we do not presume to know anything about the current flowing through the enhanced conductivity region. However, we assume that the enhancement is being produced by a stream of electrons into the ionosphere. Thus, there is an upward current coincident with the enhancement. This upward current must be fed by downward currents that we have assumed in Figure 5(b) flow poleward of the arc. This field-aligned current pattern implies a horizontal current structure in the ionosphere shown by the second trace in Figure 5(b) and creates a magnetic perturbation measured by a satellite shown in the fourth trace.

In examining Triad and Chatanika data during passes that included an auroral arc, we conclude that Case 2 best describes the observed patterns. In no case have we seen evidence for field-aligned currents flowing at the edges of auroral arcs in the sense shown by Case 1. An example of a perturbation in field-aligned current produced by an auroral arc is shown in Figure 6. The contour

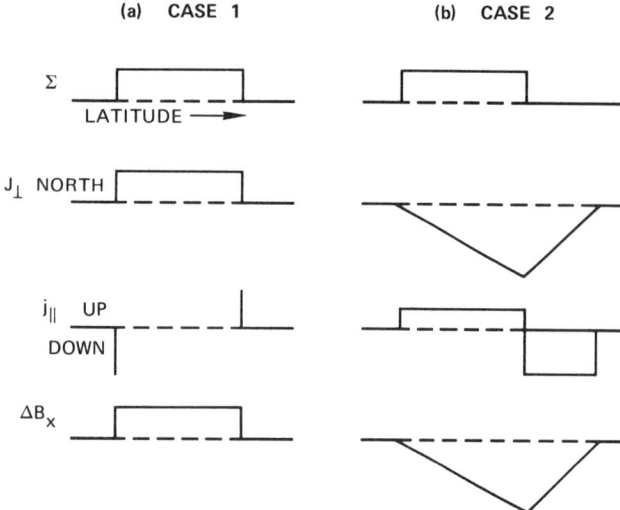

Fig. 5. Two possible field-aligned-current configurations associated with a localized ionospheric conductivity enhancement. The four traces show the latitudinal profiles of conductivity, northward current, parallel currents, and east-west magnetic perturbation.

plot in the lower part of the figure shows the interface arc and a discrete arc to the north. The magnetic deflections measured by Triad are shown in the top panel. Note the perturbation imbedded in the Region 1 current sheet. Between 67° and 68°, the field-aligned current density is higher than that in the surrounding regions as indicated by the larger negative slope. The return current flows at the poleward edge. This current at the poleward edge appears to contradict our earlier statement about currents associated with the edges of auroral arcs. However, the direction of this current is opposite to that predicted by Case 1. In most of the cases we have studied the return current flows to the north of the arc. However, Burke et al. [1980] have shown an example where the return current is situated equatorward of the arc, actually within the Region 2 current sheet. We conclude from these observations that conductivity gradients do not perturb the large-scale field-aligned-current pattern, but enhanced electron precipitation can increase the field-aligned-current density on any given flux tube.

The absence of field-aligned currents at conductivity gradients suggests that the electric field is able to adjust itself to compensate for these gradients, maintaining a nearly constant horizon-

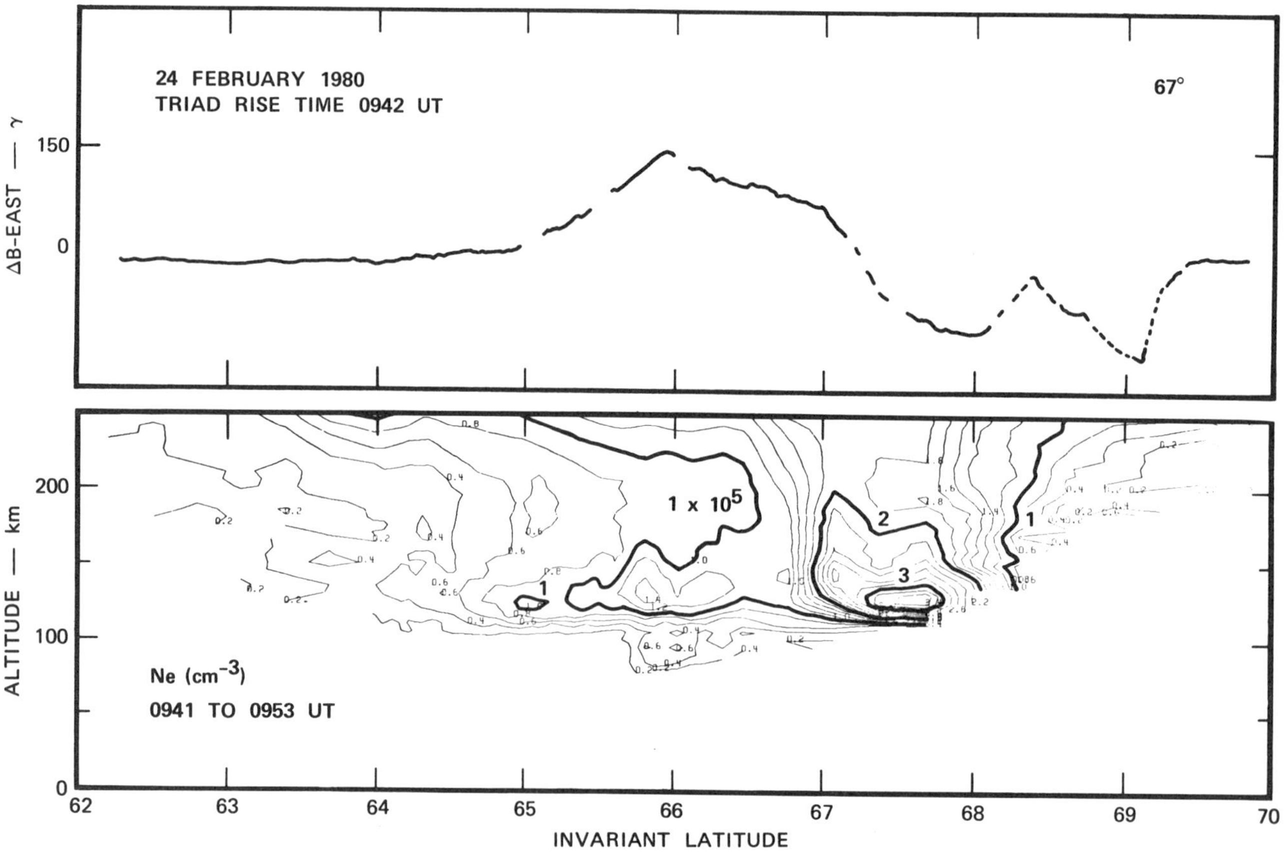

Fig. 6. Triad magnetometer data and Chatanika radar ionization data obtained during an evening-sector pass over a discrete arc.

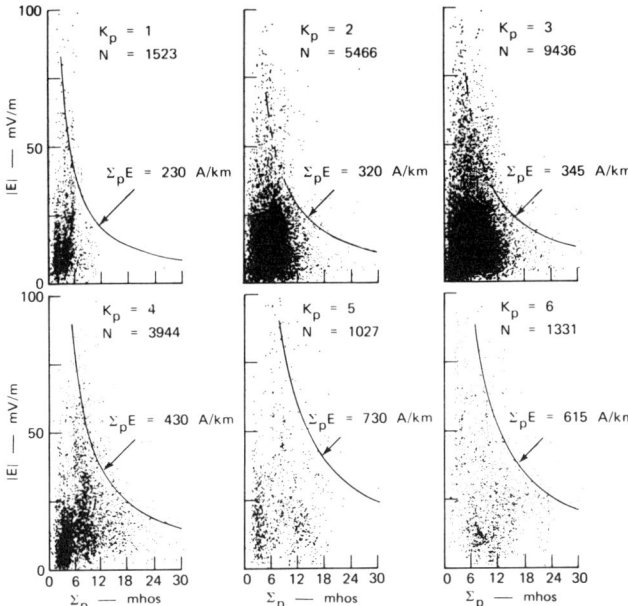

Fig. 7. Electric-field magnitude as a function of height-integrated Pedersen conductivity sorted by Kp. The number of data points in each plot is given by N. The hyperbolas were drawn to contain 95 percent of the data.

tal current in the ionosphere. We have examined the relationship between electric fields and conductivity using data from more than 200 hours of elevation scan experiments [Robinson, 1983]. In Figure 7, we plot the magnitude of the electric field as a function of Pedersen conductivity, sorting the measurements according to Kp. It is apparent in these plots that the electric field tends to be restricted to smaller values for large values of Σ_p. One way to interpret these results is that the product $\Sigma_p E$ is limited. We have drawn curves in Figure 7 each given by an equation of the form $\Sigma_p E = J_p$ where the value of J_p is chosen so that it contains 95 percent of the data in each Kp range. It is clear that as Kp increases the limiting current J_p also increases. The value of J_p as a function of Kp is shown by the solid circles in Figure 8. The triangles represent the extent to which J_p changes if it contains 85 percent of the data.

The importance of the limiting current J_p is that it is equal to the total field-aligned current that can enter the ionosphere along any meridian for a given value of Kp. This is shown in Figure 9 in which we have sketched two field-aligned current sheets in cross-section with their closure in the ionosphere. Measurements of the ionospheric Pedersen current will result in values up to but not exceeding J_p. The maximum Pedersen current measured is equal to the current entering and leaving the ionosphere along field lines, J_o.

Figure 8 can then be interpreted to mean that the total field-aligned current generated in the magnetosphere increases with Kp. Reiff et al. [1981] have used data from the AE-C satellite to correlate Kp with the polar-cap potential drop, Φ_o. We have included the corresponding value of Φ_o in Figure 8. Thus, Figure 8 represents a current-voltage diagram for the magnetospheric generator. The results agree well with those of Iijima and Potemra [1976] who showed that for Kp between 2- and 4+ the intensity of field-aligned currents was between 250 and 300 A/km. Because of the way the data in Figure 7 were selected our results apply primarily to the winter hemisphere currents on the nightside between dusk and dawn.

Figure 8 implies that for a given polar cap potential drop the field-aligned current must assume a certain value. This is probably an oversimplification. Fujii et al. [1981] showed that the field-aligned-current intensity is almost twice as large in the summer hemisphere than in the winter hemisphere. Thus, the currents in Figure 8 must represent the maximum currents that can be generated, the actual current at any given time depending on the total ionospheric conductivity. This may be the most important effect of conductivity on field-aligned currents. Large-scale conductivity enhancements such as those present in the summer polar cap can cause substantial redistribution of the ionospheric potential and may account for the formation of auroral

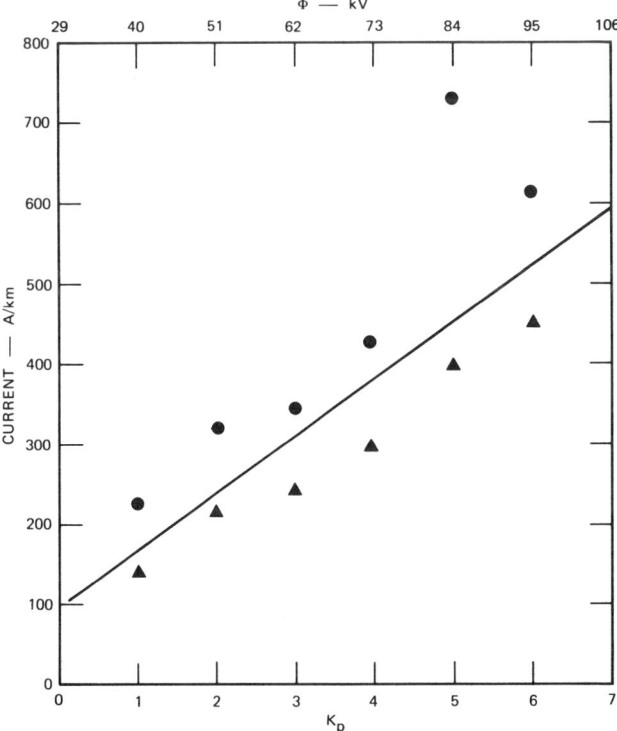

Fig. 8. The limiting Pedersen current in the ionosphere as a function of Kp.

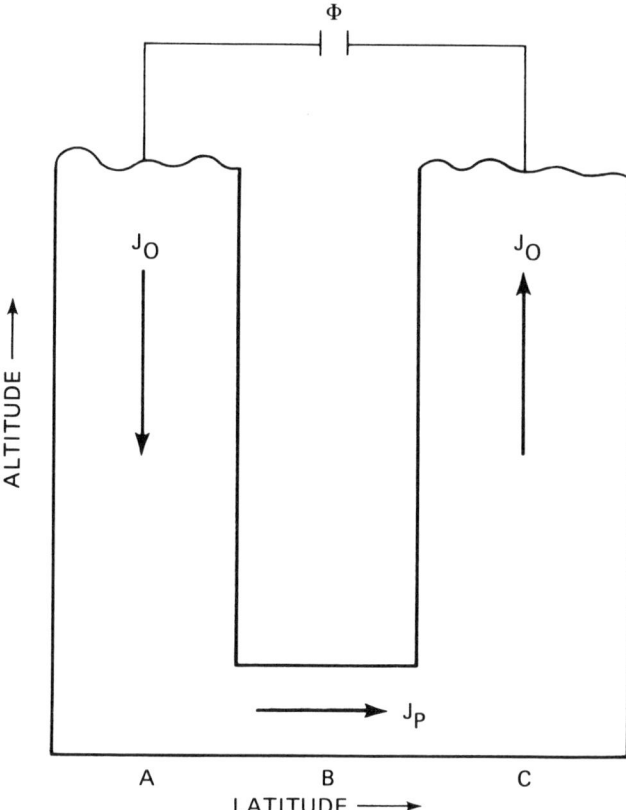

Fig. 9. Simplified model of field-aligned currents and their closure in the ionosphere. The current sheets, shown in cross section, are assumed to be extended along the auroral oval.

arcs in the evening sector. Assume, for example, that the polar cap potential drop is fixed, but that the ionospheric conductivity is low so that the field-aligned-current intensity is somewhat less than the value shown in Figure 8. Any small conductivity enhancement will allow more current to enter the ionosphere. In the evening sector, Region 1 current sheet, this is accomplished by acceleration of electrons into the ionosphere. These accelerated electrons produce a further conductivity enhancement that allows more current to be drawn from the magnetosphere. This feedback mechanism continues until the currents have reached the values shown in Figure 8, at which time the mechanism ceases to operate.

Summary

The spatial relationships between field-aligned currents and ionospheric electrodynamic parameters are summarized in Figure 10 from Senior et al. [1982]. The shaded semicircular area represents the ionospheric signature of central plasma sheet precipitation. In the evening sector the field-aligned current reversal at the Region 1/Region 2 interface lies near the center of this precipitation zone. The center of the central plasma sheet precipitation is represented in the ionosphere by a localized enhancement in E-region ionization called the interface arc. In the morning sector, the Region 2 current sheet encompasses the entire central plasma-sheet precipitation. Auroral arcs in the evening sector occur in the poleward portion of the Region 1 current sheet separated from the interface arc by a dark band. The northward electric field, Joule heating, and eastward electrojet maximize within the dark band. Within the arc the northward electric field is small, but there is usually a small westward component of the electric field that drives a westward electrojet. In the morning sector, the westward electrojet spans the upward-current region and extends into the downward-current region. The poleward portion of the westward electrojet is sustained by large southward electric fields while the equatorward portion is sustained by the high conductivities.

The ionospheric conductivity can influence the intensity and location of field-aligned currents in several ways. Robinson et al. [1982] found that latitudinal gradients in the conductivity are responsible for a significant fraction of the Region 2 field-aligned current in the evening sector. However, steep gradients in conductivity such as those found on the edges of auroral arcs are not associated with corresponding narrow field-aligned-current sheets. Narrow intensifications of upward field-aligned currents can be produced in association with large fluxes of precipitating electrons.

In addition to precipitating particles, significant conductivity can be produced by solar extreme ultraviolet radiation. Iijima and Potemra [1976] showed that the higher conductivity in the summer hemisphere allows more field-aligned current to flow through the ionosphere. The amount of field-aligned current that closes meridionally through the ionosphere in the winter has been

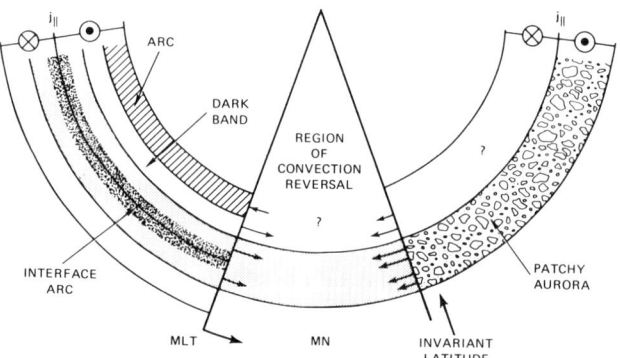

Fig. 10. Schematic representation of the relationship between field-aligned currents, visible aurora, and auroral electrojets in the evening and morning sectors. From Senior et al. [1982].

determined from Chatanika radar measurements of electric field and conductivity. The field-aligned-current intensity increases approximately linearly with Kp. Because Kp also increases with the polar-cap potential drop, these results can be used to derive a current-voltage relationship for the magnetospheric generator.

Conductivity produced by extreme ultraviolet radiation can also affect the location of the field-aligned currents. The results of Klumpar [1979] and Robinson et al. [1982] suggest that the Region 2 current sheet in the evening sector in the summer may extend well equatorward of auroral electron precipitation owing to EUV enhanced conductivity.

Acknowledgments. I wish to acknowledge the contributions made to the Triad/Chatanika experiments by R. Vondrak of the Lockheed Palo Alto Research Laboratory, T. A. Potemra of the Applied Physics Laboratory (APL), the Johns Hopkins University (JHU), and C. Senior of CNET/CRPE. J. Neary and J. Debrul of APL set up the Triad ground station at Chatanika. SRI International's staff at Chatanika maintained the Triad receiver and operated the radar during the coordinated experiments. The entire Triad project has been made possible by the Space Department of APL/JHU. The Triad work was supported at APL by the National Science Foundation and by the Office of Naval Research. The Chatanika radar operations and research at SRI were supported by NSF grant ATM78-23658.

References

Armstrong, J., S.-I. Akasofu, and G. Rostoker, A comparison of satellite observations of Birkeland currents with ground observations of visible aurora and ionospheric currents, J. Geophys. Res., 80(4), 575-585, 1975.

Burke, W., D. Hardy, F. Rich, M. Kelley, M. Smiddy, B. Shuman, R. Sagalyn, R. Vancour, P. Widman, and S. Lai, Electrodynamic structure of the late evening sector of the auroral zone, J. Geophys. Res., 85(A3), 1179-1193, 1980.

de la Beaujardiere, O., R. Vondrak, and M. Baron, Radar observations of electric fields and currents associated with auroral arcs, J. Geophys. Res., 82, 5051-5062, 1977.

de la Beaujardiere, O., R. Vondrak, R. Heelis, W. Hanson, and R. Hoffman, Auroral arc electrodynamic parameters measured by AE-C and the Chatanika radar, J. Geophys. Res., 86, 4671, 1981.

Iijima, T., and T. Potemra, The amplitude distribution of field-aligned currents at northern high latitudes observed by Triad, J. Geophys. Res., 81, 2165, 1976.

Kamide, Y., and G. Rostoker, The spatial relationship of field-aligned currents and auroral electrojets to the distribution of nightside auroras, J. Geophys. Res., 82(35), 5589-5608, 1977.

Kamide, Y., and J. Vickrey, Relative contribution of ionospheric conductivity and electric field to the auroral electrojets, J. Geophys. Res., 88, in press, 1983.

Klumpar, D., Relationships between auroral particle distributions associated with field-aligned currents, J. Geophys. Res., 84, 6524, 1979.

Reiff, P., H. Spiro, and T. Hill, Dependence of polar cap potential drop on interplanetary parameters, J. Geophys. Res., 86, 7639, 1981.

Robinson, R., E. Bering, R. Vondrak, H. Anderson, and P. Cloutier, Simultaneous rocket and radar measurements of currents in an auroral arc, submitted to J. Geophys. Res., 1981.

Robinson, R., R. Vondrak, and T. Potemra, Electrodynamic properties of the evening sector ionosphere within the region 2 field-aligned current sheet, J. Geophys. Res., 87, 731, 1982.

Robinson, R., K_p dependence of field-aligned current intensity, submitted to J. Geophys. Res., 1983.

Senior, C., R. Robinson, and T. Potemra, Relationship between field-aligned currents, diffuse auroral precipitation and the westward electrojet in the early morning sector, J. Geophys. Res., 87(A12), 10,469-10,477, 1982.

Vondrak, R., and M. Baron, Radar measurements of the latitudinal variation of auroral ionization, Radio Sci., 11(11), 939-946, 1976.

Vondrak, R., Chatanika radar measurements of the electrical properties of auroral arcs, In "Physics of Auroral Arc Formation," Geophysical Monograph Series, 25, 1981.

Vondrak, R., and F. Rich, Simultaneous Chatanika radar and S3-2 satellite measurements of ionospheric electrodynamics in the diffuse aurora, J. Geophys. Res., 87(A8), 6173-6185, 1982.

Whalen, J. A., General characteristics of the auroral ionosphere, In "Physics of Space Plasmas," SPI Conference Proceedings and Reprint Series, 4, 85-114, (Scientific Publishers, Inc., Cambridge, MA, 1982).

MAGNETOSPHERIC DYNAMO PROCESSES

David P. Stern

Planetary Magnetospheres Branch, Goddard Space Flight Center, Greenbelt, MD 20771

Abstract. Three processes are examined whereby an effective electromotive force and energy input arise in circuits of magnetospheric currents, even in the absence of time-varying maagnetic fields. The first involves currents on "open" field lines, linking the ionosphere with the solar wind, and it underscores the role of polarization currents. The second may exist on the current filament observed in the vicinity of Jupiter's satellite Io. The third may operate along the high-latitude boundary of the earth's magnetic tail, from where it pumps energy into the plasma sheet.

Electric currents in planetary magnetospheres belong to one of two classes. Some of them—the ring current, for instance—involve no energy loss and therefore, in principle, do not require a mechanism which constantly supplies energy (loss mechanisms remain, but they are not tied to the current flow). The $\underline{j}\times\underline{B}$ force then has to be taken into account in deriving equilibrium condition, but a dynamo is not needed to maintain to the current.

However, when the flow of a current is accompanied by an energy loss, a dynamo process, which constantly adds energy, is needed. This occurs when the circuit passes through a section of the ionosphere, or when energy is given to the participating plasma, by heating, by imparting to it a bulk velocity, or by accelerating some of its particles.

Discussed below are 3 basic types of dynamo which energize currents in solar-system magnetospheres. Commercial generators generally work by induction, i.e. they involve a time-dependent magnetic field, which sets up a net emf around their circuits. Time-dependent magnetic fields are important in transient magnetospheric phenomena such as magnetic storms and substorms, and may play a role in setting up the ring current. However, such dynamos will not be discussed, in part because even their qualitative features are still uncertain. Instead, the 3 examples described involve magnetic configurations which do not change in time, i.e. with

$$\partial \underline{B}/\partial t = 0 \qquad (1)$$

It follows then from Maxwell's equations that in such processes the e.m.f. around any closed circuit is zero:

$$\varepsilon = \oint \underline{E}\cdot\underline{ds} = 0 \qquad (2)$$

With the above restriction, dynamo processes generally depend on the relative motion between one part of the circuit and the rest. This may be illustrated by assuming an ohmic current $\underline{j}=\sigma\underline{E}$; a situation of this type will be encountered later (see eqn. 7), though it should be noted that most magnetospheric currents are not ohmic. The ohmic relation must then be taken in the local frame, in which the electric field equals $\underline{E}^* = \underline{E} + \underline{v}\times\underline{B}$. Thus

$$\underline{j} = \sigma\underline{E}^* = \sigma(\underline{E} + \underline{v}\times\underline{B}) \qquad (3)$$

One can view $\underline{j}$ here as being driven by an "effective e.m.f."

$$\varepsilon_{eff} = \oint (\underline{v}\times\underline{B})\cdot\underline{ds} \qquad (4)$$

If all parts of the circuit are moving with the same constant $\underline{v}$, ε_{eff} will be zero, because $\underline{v}$ then can be transformed away. The same happens if the medium moves like a single rigid body. But if there exists relative motion between different parts of the circuit, ε will not vanish, and one obtains the same value for it in any frame of reference is used. This is what drives the dynamos discussed here.

The notion of such a process goes back to 1832, when Faraday stretched a wire along Waterloo Bridge [Williams, 1965] and connected its ends to plates immersed in the river, so that the flowing water completed the circuit [Figure 1]. With a galvanometer he then tried to detect the induced current: the experiment did not work, but the idea was sound. Viewed from the reference frame of the bridge, $\underline{v}$ only exists in the water, hence the effective e.m.f. is the integral of $\underline{v}\times\underline{B}$ across the river, which is certainly not zero.

If the earth's magnetosphere is open—and there exists a large body of evidence suggesting that it is—then a completely analogous situation exists above the polar caps of the earth [Figure 2]. The wire along the bridge is now replaced by the polar ionosphere, the wires dipping into the river by open field lines at opposite sides of the polar cap, and the river Thames by the flow of the solar wind, in the magnetosheath region. Through the $\underline{j}\times\underline{B}$ force, the solar wind loses momentum and energy, and both these are given to the ionosphere at the other end of the circuit. The directions of current flow along these open field lines, by the way, agree with those of region 1 Birkeland currents, and the observation that such currents peak in the noon quadrant sug-

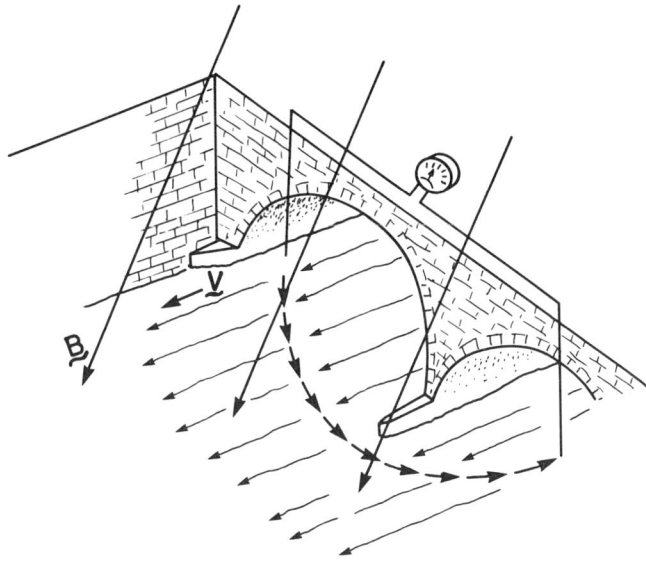

Fig. 1. Faraday's Fluid Dynamo Experiment (1832).

gests that at least some of them indeed come from this type of dynamo process.

This leaves two points to ponder. One, how is the circuit closed in the solar wind? Collisions there are rare--of the order of one per day per ion [Parker, 1962]--so that collisional conductivity is negligible. What one finds instead is a <u>polarization current</u>.

First consider the situation purely formally. Let those two field lines in the figure have a potential difference of 50,000 volts, and let them be treated as two plates of a charged capacitor. Furthermore, everything is viewed from the frame moving with the field lines, which is also the frame of the plasma particles (this may not be an inertial frame, but acceleration effects will be neglected). Because the ionosphere connects those two field lines, their electric charge will quickly begin to drain away and the voltage will drop, and this, in the moving frame, gives a definite $d\underline{E}/dt$. However, there exists a guiding center drift proportional to $d\underline{E}/dt$, the so-called <u>polarization drift</u>, which will drive a "polarization current" trying to restore $\underline{E}$ to its previous value. The polarization current cannot restore $\underline{E}$ completely --if it did, there would remain no $d\underline{E}/dt$ and there would be nothing to drive all those processes--but it comes remarkably close, restoring all but about $1/10^7$ of the drained charge. It is thus the polarization current, essentially, that closes the circuit.

Next, what does all this mean <u>in terms of physical processes</u>? When one withdraws charge from a vacuum capacitor and causes the voltage to drop, the energy obtained is taken from the energy density $\varepsilon_0 E^2/2$ of the vacuum electric field, or if there is a dielectric present, from the stored energy $\varepsilon E^2/2$. In the present case, however, there exists an additional mode of energy storage, because in the space between the plates there also exists a plasma of some density ρ, drifting with velocity

$$\underline{u} = \underline{E} \times \underline{B}/B^2 \tag{5}$$

To remove $\underline{E}$, one must remove not only the vacuum energy density $\varepsilon_0 E^2/2$, but also the kinetic energy density $\rho u^2/2$, which is generally far greater. When the polarization current withdraws from the dynamo a current far greater than the vacuum calculation would allow, this is the account from which the withdrawal is made.

Back to formalism: the vacuum energy density is $\varepsilon_0 E^2/2$, while the kinetic energy density $\rho E^2/2B^2$ is also proportional to E^2. The ratio between the two is $\rho/\varepsilon_0 B^2 = c^2/V_A^2$, with V_A the Alfvén velocity: in the magnetosheath V_A may reach 100-200 km/sec, so this ratio is rather large. One can join the two terms together in the form of $\varepsilon E^2/2$, where ε now is the "plasma dielectric constant"

$$\varepsilon = \varepsilon_0 [1 + (c/V_A)^2] \tag{6}$$

with the "1" representing the vacuum contribution and the rest due to the plasma.

The second point involves <u>energy</u>. The discussion above makes clear that such dynamos are driven by the kinetic energy of the bulk motion of

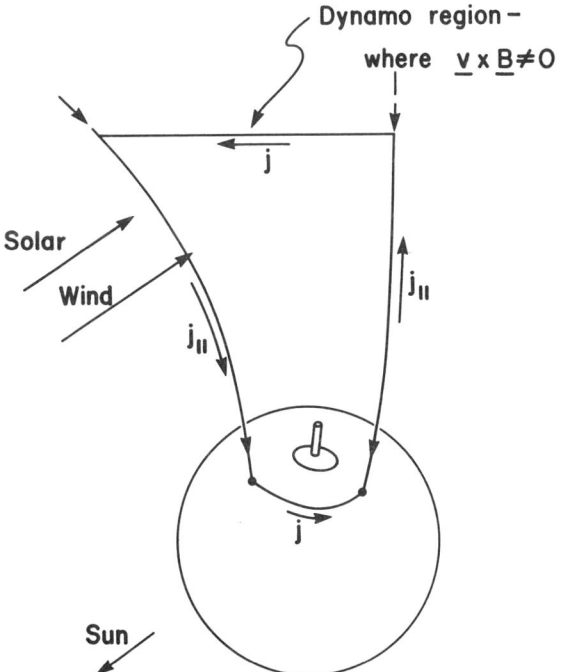

Fig. 2. Dynamo action on open terrestrial field lines.

the plasma: the individual kinetic energy of the particles does not count, only the part of the energy due to the $\underline{E}\times\underline{B}$ drift does. One may ask, is there enough energy available in this form to drive Birkeland currents?

With the solar wind, there is no problem. The total magnetospheric drain has been estimated as 1-2% of the energy of the SW crossing an area equal to the magnetospheric cross section--and open field lines can, in principle, reach out and tap energy even from further away. It thus seems unlikely that this source ever becomes overloaded.

However, proposals have also been made by which similar dynamo processes operate inside the magnetosphere and supply the observed Birkeland currents. Then it is not at all certain that the required energy is available, since both density and convective velocity inside the magnetosphere are typically 1/10 or less of what they are in the solar wind. The one significant exception would be the boundary layer, where the density is about 1/5 and the velocity about 1/2 its value in the adjoining magnetosheath, with great fluctuations. The layer, however, is fairly thin, so that its total flow energy still falls short. Lack of energy is a problem for all such internal dynamos [Stern, 1983]: for the various proposed boundary layer dynamos [e.g. Eastman et al., 1976], for dynamos tapping sunward convection [Akasofu et al., 1981] and for Rostoker's dynamo depending on cross-tail convective flows [Rostoker and Boström, 1976]. One qualification should be made here: the boundary layer might well be the source of some of the region 1 Birkeland currents, but only by putting its own weak dynamo in series with the much more powerful solar wind, which provides the required energy.

The Io-Jupiter Interaction

We proceed to the second example. Pioneer 10 found that the innermost large moon of Jupiter, Io, has an ionosphere and therefore conducts electricity [Kliore et al., 1974]. In retrospect there exists now some question about whether that conductivity might perhaps reside inside Io itself, where an internal heat source exists, and about the effect of the Io plasma torus on the data from which the "ionosphere" was deduced. In any case, there is evidence that a conductivity does exist.

Now the plasma at Io's orbit is observed to co-rotate with Jupiter, while Io's orbital speed is considerably smaller than the corotation velocity. This leaves a relative velocity $\underline{v}_o$=57 km/s between the two, and since Jupiter has a conducting ionosphere, there exists scope for dynamo action [Hill et al., 1983].

Suppose first that Io and Jupiter were quite close to each other, so that the intervening plasma contributed just a minor effect [Figure 3]. Then dynamo action could proceed as follows. In a circuit like ($\alpha\beta\gamma\delta$) there will exist an

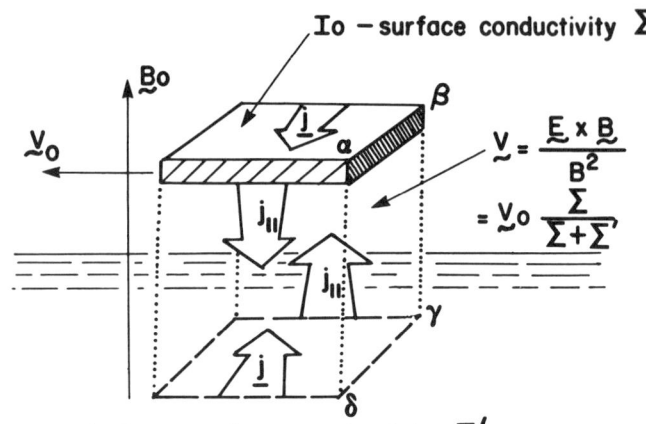

Fig. 3. The Io dyanmo, assuming that only a negligible amount of plasma separates Io from Jupiter.

effective e.m.f., because in the frame of Jupiter, only along ($\alpha\beta$) does $\underline{v}\times\underline{B}$ differ from zero. Assuming the conductivities are ionospheric, one can now calculate the electric field between Io and Jupiter. For simplicity assume that both Io and Jupiter are flat and parallel, with surface conductivities, integrated over the thickness of the ionosphere, of Σ and Σ', respectively. Further assume a rectangular shape for Io, and neglect edge effects. Then from current continuity

$$\underline{E} = (\underline{v}_o \times \underline{B})\ \Sigma/(\Sigma+\Sigma') \qquad (7)$$

and the convection velocity of the plasma is

$$\underline{v} = (\underline{E}\times\underline{B})/B^2 = \underline{v}_o \Sigma/(\Sigma+\Sigma') \qquad (8)$$

The meaning is quite clear. If Io's conductivity Σ is very large, $\underline{v}=\underline{v}_o$, meaning field lines are frozen to Io and the plasma in "Io's shadow" stays glued to it (the rest of the plasma has to split up and make way, presumably). If on the other hand the Jovian conductivity Σ' is very large, $\underline{v}=0$, meaning that the plasma stays frozen to Jupiter and field lines part smoothly to let Io pass between them, without touching it and incidentally, without any dynamo action.

For in-between values of Σ and Σ' a compromise is reached, in which the plasma is dragged in the same direction as $\underline{v}_o$, but only at a fraction of the speed. With the electric field comes a pair of field-aligned current filaments or sheets between Io and Jupiter. The magnetic signature of such currents, suggesting an intensity of a few million amps, has indeed been detected by Voyager 1 [Ness et al., 1979], which is why it was earlier stated that some conductivity of Io is known to exist.

Matters are complicated, however, by the

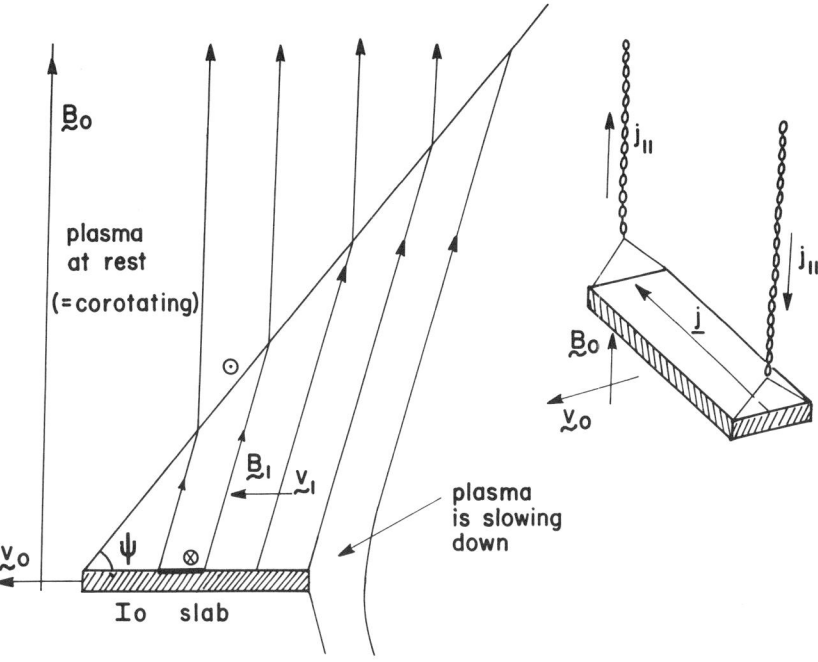

Fig. 4. The plasma wedge around Io (swing illustrates 3-D analogy).

ambient plasma. So far the role of that plasma has been viewed as passive--it flows in accordance with the induced electric field, but does not influence $\underline{E}$. Actually, the plasma surrounding Io is relatively dense, or at least was when Voyager 1 observed it [Belcher, 1983], containing sulfur ions and other ones, from Io's volcanos. In the previously discussed model, as Io overtakes this plasma (viewing things now from the frame of the plasma, which co-rotates with Jupiter), an electric field which is some fraction of $\underline{v}_0 \times \underline{B}$ spreads out and tries to make the plasma acquire at least a fraction of Io's velocity. If the plasma is dense enough, however, this requires an appreciable transfer of momentum from Io to the plasma.

To transfer momentum to the plasma, a current must flow through it and exert a $\underline{j} \times \underline{B}$ force. One expects $\underline{j}$ to be a polarization current, because the plasma must experience an increase in its $\underline{E} \times \underline{B}$ velocity, which suggests a non-zero $d\underline{E}/dt$. And one expects the return flow of $\underline{j}$ to take place through Io.

I have not seen a rigorous solution of the complete problem, taking into account Io's finite conductivity, nor am I able to offer one. Some studies do exist [e.g. Neubauer, 1980; Hill et al., 1983], giving the conditions at some distance from Io in terms of Alfvén waves. They suggest that field lines from Io are deformed into a cone with an angle θ_A, where tg θ_A is the ratio of Io's velocity to the Alfvén velocity. Let me try here to reconstruct the physics of this process, using simplified 2-dimensional MHD in the (x-z) plane, and assuming such a cone indeed exists--except that here everything is reduced to 2 dimensions and the cone becomes a wedge [Figure 4].

Io here is replaced by a very elongated slab in the z=0 plane, extending some great distance along the y axis, perpendicular to the figure--so great that we neglect end effects. A wedge surrounds it, diverging by some angle ψ, inside which the plasma is dragged along by some velocity $\underline{v}_1$, which is less than Io's velocity $\underline{v}_0$. The gradual widening of the wedge represents the gradual widening of the plasma region dragged along by Io, and the surface of the wedge is where plasma is caught up and is given the velocity $\underline{v}_1$, so that the polarization current will flow along that surface, perpendicular to the figure.

That current closes through Io, and the two darker segments in Figure 4 show part of the boundary flow and its return circuit, in profile. One may visualize the slab as resembling a child's swing, sticking out from the plane of the figure. Currents run along its length, and twin filaments connect them to the boundary, flowing at each end along the chains by which the swing is suspended.

Note that Jupiter only becomes involved after the entire plasma separating it from Io is "brought up to speed" (but see note below). For Io (assuming $\psi = \theta_A$ and tg $\theta_A \sim 7$) it could be that this never happens. The dynamo then only involves the planetary body of Io and the moving (co-rotating) plasma, linked to it by field lines, just as the plasma and the earth were linked in the earlier example of open field lines. The full calculation is complicated—there is no assurance $\underline{v}_1$ is a constant and that the field $\underline{B}_1$ inside the wedge is homogeneous—but this will not be followed.

Because $\underline{v}_1$ is smaller than $\underline{v}_o$, the plasma flows through the wedge with a velocity $(\underline{v}_1-\underline{v}_o)$ relative to Io, and after a short while it arrives at the trailing edge and becomes decoupled from Io. Now the situation is completely changed: all that remains is a flux tube where the middle part (initially at least) advances with a velocity $\underline{v}_1$, but where the ends are at rest. As new plasma crosses the boundary of the wedge, it gains momentum, but only at the expense of the momentum of the moving plasma near z=0. The latter is therefore decelerated, its $d\underline{v}_1/dt$ is negative, which means $d\underline{E}/dt$ is non-zero and that a polarization current flows, and this current replaces the one which earlier flowed through Io itself. The final outcome, as may be guessed, is that the tube of moving plasma gradually gets longer and slower, as more and still more plasma shares the initial momentum. Ultimately the momentum is shared by the entire flux tube, all the way down to the Jovian ionosphere, and then a new dynamo starts operating, slowing down the relative motion between the plasma and Jupiter until corotation is restored.

(Added Note: A discussion with Dr. Fritz Neubauer made clear that the above interaction differs markedly from the one proposed by him. Here the behavior of plasma immediately adjacent to Io is studied, while Neubauer's calculation is 3-dimensional (finite width in y) and refers primarily to the distant interaction, which has properties of an Alfvén wave. The conclusions of the preceding paragraph could thus require revision).

The Tail Dynamo

Finally, I would like to speculate on the interaction between the solar wind and the geomagnetic tail, which involves a type of "surface dynamo." If the magnetosphere is open, field lines connected to Earth fall into one of two classes—those that return to earth and are closed, and those that wander off into space and are open.

But from the point of view of magnetospheric currents, one can further divide closed field lines into 2 types. The division reflects the behavior of current flow lines—lines analogous to magnetic field lines, but for the current density $\underline{j}$, which like $\underline{B}$ has zero divergence. If such lines are drawn in the equatorial surface (defined, say, as the locus of minimum B on closed field lines), two regions are found. Near earth the flow lines are closed around the earth, and this is the ring current region, while further away they just cross from one side to the other, and that is the tail.

Now the ring current requires no dynamo, since it does not have to overcome losses. It is just the result of guiding center drifts, moving electrons one way and positive ions in the opposite direction. Neither is a dynamo required for the cross tail current, if the magnetosphere is closed. That current then also reflects merely the reaction of plasma to an inhomogeneous magnetic field, though now some of the motions may be non-adiabatic, and some of them may close along the flanks.

If the magnetosphere is open, however, the situation is different, for then it is linked to interplanetary space by a bundle of open field lines with (say) 50,000 volts across them. The flanks of the tail are on opposite sides of this bundle, hence they span a voltage difference of 50,000 volts, and this difference will be bridged by the cross-tail current. The energy involved in this is rather large. The cross-tail current is typically 100,000 amperes for each 1 R_E of length, and if this is multiplied by 50,000 volts and by a length of 40 R_E (very conservative) one gets $2\;10^{11}$ watt, which is 5-10 times the average auroral input. This immediately raises 2 questions: (1) where does this energy go? (2) where does it come from? Each question will be handled separately below.

One would expect the energy to go to particles drifting across the tail, since both ions and electrons drift in such a direction that they gain energy from a dawn-to-dusk field (in principle, energy could also be gained in the segment of the circuit along the flanks, if the large $E_\perp$ associated with this does not disrupt things). It may then be shown that particles confined near z=0 (those with steep pitch angles) drift slowly and gain energy slowly, but particles which undergo a few irregular gyrations near the middle of the current sheet and then escape to mirror in the dipole-like region near the earth gain an appreciable amount.

Among the studies of this process, there is one by Lyons and Speiser [1982], who considered a narrow current sheet, about 1000-2000 km thick, inside which the particles moved non-adiabatically [Figure 5]. Such thin sheets are occasionally observed [Fairfield et al., 1981], though at other times the current may spread over 3-5 R_E. It then turned out that particles may increase their energy by 50% during a typical encounter with the z=0 plane. As they emerge from near z=0 (and this may occur on either side of the plane), they encounter a stronger magnetic field and may or may not be reflected. If they are reflected they give up most of their gained energy. However, it may be shown (as already noted) that those which make it to the dipolar region near earth, not only get to keep their gains, but actually increase them slightly.

In all cases, the energy gain $\int \underline{E} \cdot \underline{j}\; dV$ ends up in the plasma, there is really no other place for it to go. Two things may now happen to it: (1) The plasma may heat up. Indeed, ions in the plasma sheet are about 2-5 times more energetic

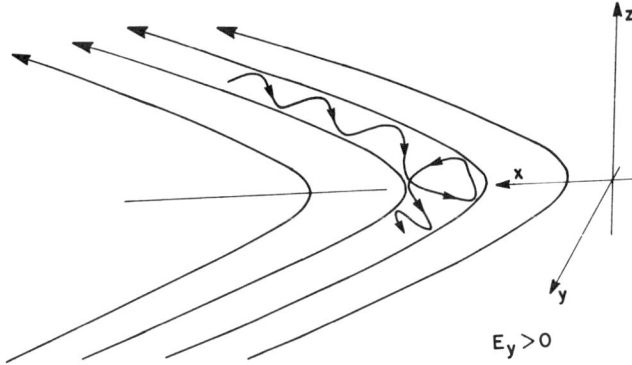

Fig. 5. Orbit of charged particle gaining energy in its encounter with the tail's current sheet.

than those of the solar wind, from which they probably originated (if the source is the ionosphere, the energization is even more profound!). (2) The energy may drive region 1 currents. The argument that region 1 currents come from the tail is rather simple, and it is best stated by looking at a map of the polar cap [Figure 6]. Region 2 currents, according to both observation and theory, come from the ring current region. Therefore, if all region 1 currents came directly along open field lines, there would exist a large gap between the two systems on the night side, corresponding to field lines threading the tail current. Observations only rarely see a gap and in general the two current systems are found to be contiguous, making it likely that region 1 currents come from the tail and are a shunt on the crosstail flow. They may draw several 10^{10} watt,

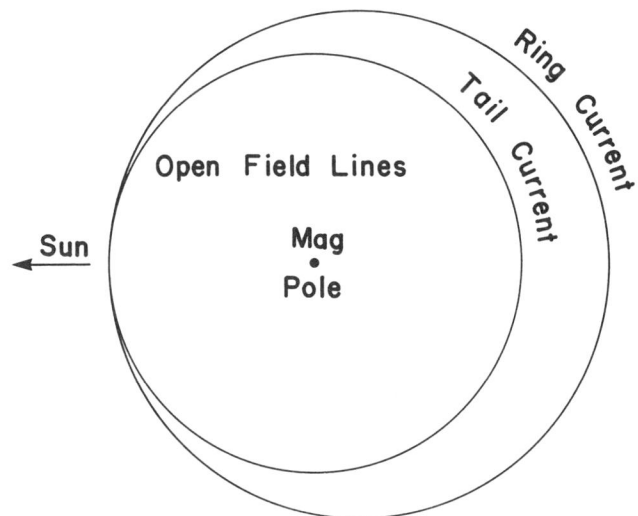

Fig. 6. The footprints of different field line regions on the polar cap.

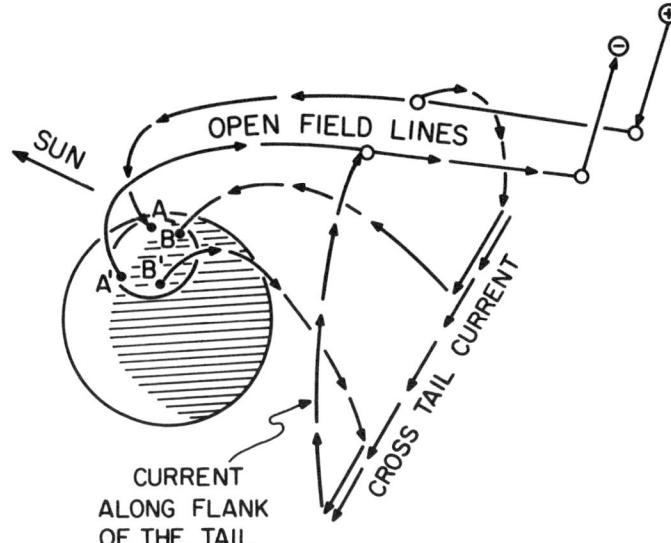

Fig. 7. Schematic view of the way in which region 1 currents may form a shunt of the cross-tail circuit.

an appreciable fraction of the amount listed earlier. Figure 7 shows schematically how this shunt might function.

The really interesting question, however, is where this energy comes from. The cross-tail current closes along the flanks in the "theta configuration" proposed long ago by Axford, Petschek and Siscoe [1965], and it is easy to see that the energy input must take place where the current crosses the bundle of open field lines which emerges from the magnetosphere. In fact, Siscoe and Cummins [1969] calculated what this input amounts to, but their formula does not give any mechanism, it is just a re-affirmation of electrodynamics, equivalent to saying that the energy input equals $\int \underline{E}\cdot\underline{j}\, dV$ in regions where that integral is negative. Let me therefore speculate on how this might happen--no one knows for sure, of course.

At the boundary of the tail lobes the magnetic pressures do not balance--typically, on the inside B = 20 nT and on the outside B = 5 nT, so some extra pressure must exist from the outside. As noted by Sonett et al [1971], a key fact is that the radius of the magnetopause slowly expands with distance down the tail [Figure 8], from about 15 R_E in the dawn-dusk plane to about twice as much 200 R_E down the tail, where ISEE 3 has recently made observations. The solar wind thus impinges on a <u>sloping</u> tail boundary, which deflects some of its flow, and this produces the extra pressure. Notice that, since area increases like r^2, this means that 3/4 of the frontage of the magnetosphere involves the tail.

Now, inside the two tail lobes $\underline{B}$ has roughly opposing directions, which means that in one lobe

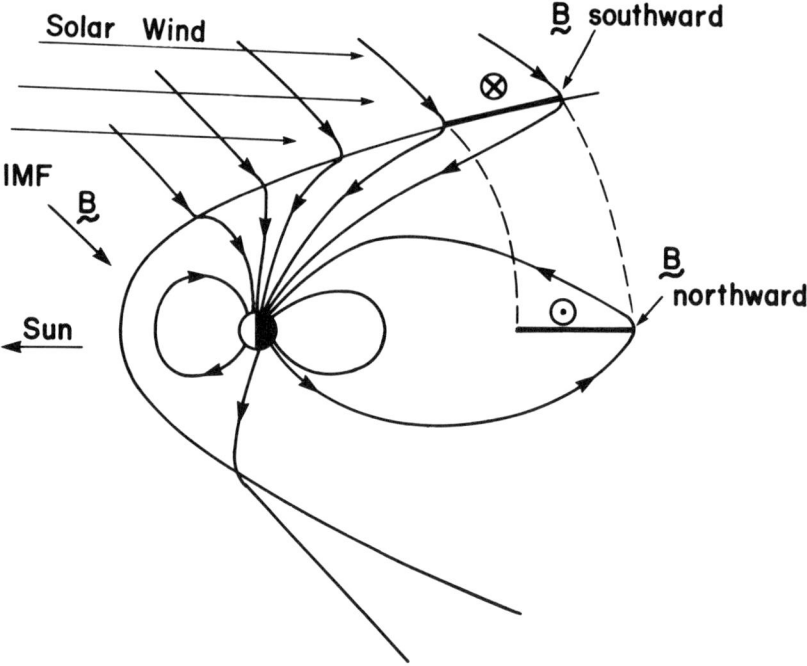

Fig. 8. Schematic view of the surface dynamo of the cross-tail current.

B is relatively anti-parallel to the IMF, and in the other one relatively parallel. By Maxwell's equations, the cross-tail current splits unevenly and its majority flows along the boundary where fields are antiparallel, which in Figure 8 happens to adjoin the northern lobe. As in the illustration of the Io wedge, the two segments marked darker represent two cross-sections of the same current circuit.

Now, I do not know how that extra pressure is applied at the boundary, but let me make a simple assumption, that the solar wind just impinges at an angle: no matter what one assumes, the trajectories of some solar wind particles must intersect the boundary, for if they all flow parallel to it, they exert no force. The motion of an individual ion destined to encounter the magnetopause is then very similar to that of a plasma sheet ion encountering a thin current sheet near z=0, in the model of Lyons and Speiser. As Figure 8 reveals, however, there exists one major difference: B is now reversed. This means that the magnetic drifts are also reversed, that the ion moves against the potential drop, and that it loses energy, which drives the dynamo.

The process may possess some crude analogy with the interaction between a stream of water and the vanes of a turbine. The ion flow hits the boundary, is deflected in a way which causes it to lose energy, carries a portion of the current in the "uphill" section of the circuit, and then generally returns to the solar wind. A few ions will penetrate into the lobes, but not many, and one might speculate on whether they do or do not contribute to the boundary layer. The fact that most ions are reflected is intuitively clear if the motion is adiabatic, because B on the inside is about 4 times stronger, so that many particles will mirror. However, the nonadiabatic reflection ratio at a rotational discontinuity is also about the same as the one deduced from adiabatic theory—this was shown by P.D. Hudson [1974]. Thus most ions return to the solar wind. They react there with the rest of the plasma to regain their bulk speed, but this is no longer of concern here.

There is more than enough energy in this source to supply $2 \cdot 10^{11}$ watts, which is about 0.5% of the solar wind energy impinging upon the frontage of the magnetosphere. As noted, about 3/4 of that frontage occurs in the tail: even if only one lobe is involved, even if open field lines make their exit only over 1/10 of the width of that lobe, and even if impinging ions only lose a fraction of their energy, there still remains enough.

All this is just the bare outline of an idea. It suggests that the high-latitude boundary of the tail lobes is a rather important region, from which the magnetosphere quite possibly derives most of its energy. So far this region was studied mainly by HEOS 1 and 2, whose magnetic field sampling rate was extremely low, once per 32 or 48 seconds. Even though Hawkeye and Prognoz have provided some additional data, perhaps it is time for this very interesting region to be visited again, and to be studied with greater resolution of detail.

Acknowledgment. I am grateful to Dr. Bengt Sonnerup for bringing to my attention the work of P.D. Hudson.

References

Akasofu, S.-I., et al., Power transmission from the solar wind-magnetosphere dynamo to the magnetosphere and to the ionosphere: Analysis of the IMS Alaska meridian chain data, Planet. Space Sci., 29, 721-730, 1981.

Axford, W.I., H.E. Petschek and G.L. Siscoe, Tail of the magnetosphere, J. Geophys. Res., 70, 1231-36, 1965.

Belcher, J.W., The low-energy plasma in the Jovian magnetosphere, pp. 68-105 in Physics of the Jovian Magnetosphere, A.J. Dessler, ed., Cambridge Univ. Press, 1983.

Eastman, T.E. E.W. Hones, S.J. Bame and J.R. Asbridge, The magnetospheric boundary layer: Site of plasma, momentum and energy transfer from the magnetosheath into the magnetosphere, Geophys. Res. Let., 3, 685-688, 1976.

Fairfield, D.H., E.W. Hones, Jr. and C.-I. Meng, Multiple crossings of a very thin plasma sheet in the earth's magnetotail, J. Geophys. Res., 86, 11,189-11,200, 1981.

Hill, T.W., A.J. Dessler and C.K. Goertz, Magnetospheric Models (esp. section 10.5: The Io-Jupiter interaction) pp.353-394 in Physics of the Jovian Magnetosphere, A.J. Dessler, ed., Cambridge Univ. Press, 1983.

Hudson, P.D., The reflection of charged particles by rotational discontinuities, Planet. Space Sci., 22, 1571-77, 1974.

Kliore, A., D.L. Cain, G. Fjeldbo, B.L. Seidel and S.I. Rasool, Preliminary results on the atmospheres of Io and Jupiter from the Pioneer 10 S-band occultation experiment, Science, 183, 323-4, 1974.

Lyons, L.R. and T.W. Speiser, Evidence for current sheet acceleration in the geomagnetic tail, J. Geophys. Res., 87, 2276-86, 1982.

Neubauer, F.M., Nonlinear standing Alfvén wave current system at Io: Theory, J. Geophys. Res., 85, 1171-8, 1980.

Ness, N.F., M.H. Acuña, R.P. Lepping, L.F. Burlaga, K.W. Behannon and F.M. Neubauer, Magnetic field studies at Jupiter by Voyager 1: Preliminary results, Science, 204, 982-87, 1979.

Parker, E.N. Kinetic properties of interplanetary matter, Planet. Space Sci., 9, 461-75, 1962.

Rostoker, G. and R. Boström, A mechanism for driving the gross Birkeland current configuration in the auroral oval, J. Geophys. Res., 81, 235-44, 1976.

Siscoe, G.L. and W.D. Cummins, On the cause of geomagnetic bays, Planet. Space Sci., 17, 1795-1802, 1969.

Sonett, C.P., J.D. Mihalov and J.P. Klozenberg, The flux content and form of the geomagnetic tail, Cosmic Electrodynamics, 2, 22-33, 1971.

Stern, D.P., The origins of Birkeland currents, Rev. Geophys. Space Phys. 21, 125-38, 1983.

Williams, L.P., Michael Faraday, Basic Books, New York 1965.

MAGNETOSPHERIC TOPOLOGY OF FIELDS AND CURRENTS

Walter J. Heikkila

Center for Space Sciences, The University of Texas at Dallas
P.O. Box 688, Richardson, TX 75080

Abstract. This paper proceeds from the view that the existence of the boundary layer inside the magnetopause is crucial to the plasma physics of the magnetosphere. The boundary layer flow is so massive that it can generate its own electric field for continued anti-sunward flow. At very great distances (some 100 to 200 R_e downstream from the earth) the dawn and dusk boundary layers become joined together, and the magnetotail is essentially just boundary layer plasma, on closed magnetic field lines. Evidence concerning closed field lines in the polar cap is the migration of breakup auroras near midnight during the recovery phase. This poleward migration is here interpreted as the ionospheric footprint of the tailward expulsion of a plasmoid from the plasma sheet through the distant magnetotail; the auroral forms move poleward to 75°-80° latitude or even higher, indicating that there is a wide band of closed field lines above the nightside auroral oval. The region of open field lines is smaller than previously thought. The electric field must have some dramatic changes, some in consort with changes of the magnetic field. These changes involve charge separation as well as induction through time-dependence of the magnetic field. Field aligned currents arise due to both of these sources. Our suggestion of how region 1 and 2 currents arise is generally in agreement with others. However, there is still another current, which we name as the region 0 current system at the boundary between open and closed magnetic field lines. The region 0 current may be weak; it is in parallel with the magnetopause current over the lobes of the magnetotail; together, they are the closure current of part of the region 1 current system (region 2 provides the remainder). The main conclusion of these conjectures is that auroral phenomena along the oval are mostly the result of the boundary layer, a form of viscous interaction.

Introduction

Although much has been learned about the magnetosphere in the past three decades, both observationally and theoretically, we are still a long way from a complete understanding. This is evidenced by the fact that we still have doubts as to the nature of the interaction at the magnetopause, and of the reason for a substorm. Thus, for example, Kennel (1980) had to conclude that "...many questions (about them) are not settled even conceptually, much less quantitatively".

In this paper I attempt to provide a total qualitative concept of the interaction of the solar wind with the magnetosphere, based on fundamental processes in physics, and on observations whenever these are available. As such it is an ad hoc concept, but I shall focus on the essential physics, which is basically rather simple. This same elementary physics is equally applicable to other cosmic plasmas, e.g., for solar flares.

As the paper proceeds there is a gradual transition from the familiar to the strange. I do carry the discussion to the strange in order to emphasize that there are many questions that have never been asked before.

Questions and Answers

The concept is based on two key features of the magnetosphere. Their reality is somewhat debatable, but if true each of these poses some interesting and provocative questions which must be answered if we are to understand the magnetosphere as a whole. The first is that some polar cap magnetic field lines may reach out into the interplanetary medium; the other, that there is a boundary layer of magnetosheath plasma inside the magnetopause, at least partly on closed magnetic field lines.

The openness of polar cap field lines is suggested by the easy access of energetic solar flare particles, both electrons and positive ions, to both polar caps (Paulikas, 1974). Also, it is well known that the geomagnetic field lines are deformed into a long magnetotail by the action of the solar wind. These observational facts are represented in Figure 1 for an interplanetary magnetic field (IMF), that is assumed to be strictly southward for simplicity. The regions closest to the earth are well documented, but the more distant parts of the magnetotail are open to conjecture as only a very few spacecraft have sampled these regions. In particular, we have no precise knowledge about the location of the nightside X-line. Some of the

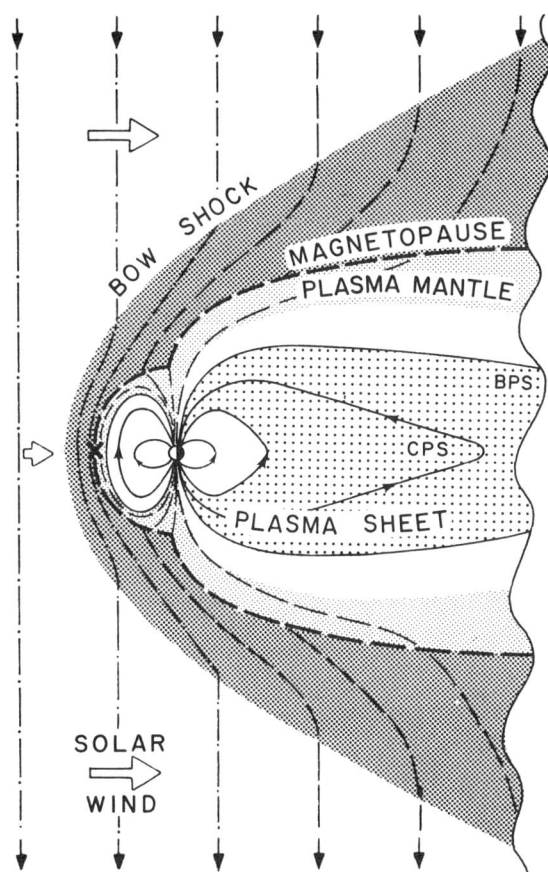

Fig. 1. A southward interplanetary magnetic field favors an open magnetosphere, with an X-line (separator line) at the boundary between closed and open magnetic field lines.

questions which are posed are: Is there a (well defined) nightside magnetopause and X-line; if so, how far away are they; and are they in motion at all times? From the geometry of the magnetic field we can in principle obtain the current distribution (if the precision is adequate), but the question of closure is uncertain, especially of field aligned and distant magnetotail currents, and bow shock currents. What is the flow pattern of plasma, especially in the far-tail? Above all, what is the topology of the electric field that goes along with this open magnetosphere?

The magnetospheric boundary layer at noon is shown enlarged in Figure 2. Apparently, magnetosheath plasma is able to get into the boundary layer very easily, ostensibly on closed field lines, sometimes with no change in energy, or plasma momentum (Eastman and Hones, 1979). How can that happen? The observations show that the boundary layer flow is antisunward; how far does this antisunward flow continue? Does the plasma return toward the earth in the plasma sheet, or does it escape from the magnetotail? If it does escape, what does that imply for the electric and magnetic field topologies? A related question concerns the fate of the plasma in the plasma mantle; if some mantle particles become trapped in the plasma sheet, does this mean that all low energy charged particles going upward from the polar cap become trapped also, for it is difficult to see how their paths could cross.

This paper proposes an ad hoc model that provides a conceptual answer to these questions, answers craved for by the magnetospheric community as evidenced in the quotation by Kennel above. Only when we have such a concept can we with confidence try to provide more quantitative answers.

Magnetopause Boundary Layer

We first take up the question of the magnetospheric boundary layer. It would be satisfying

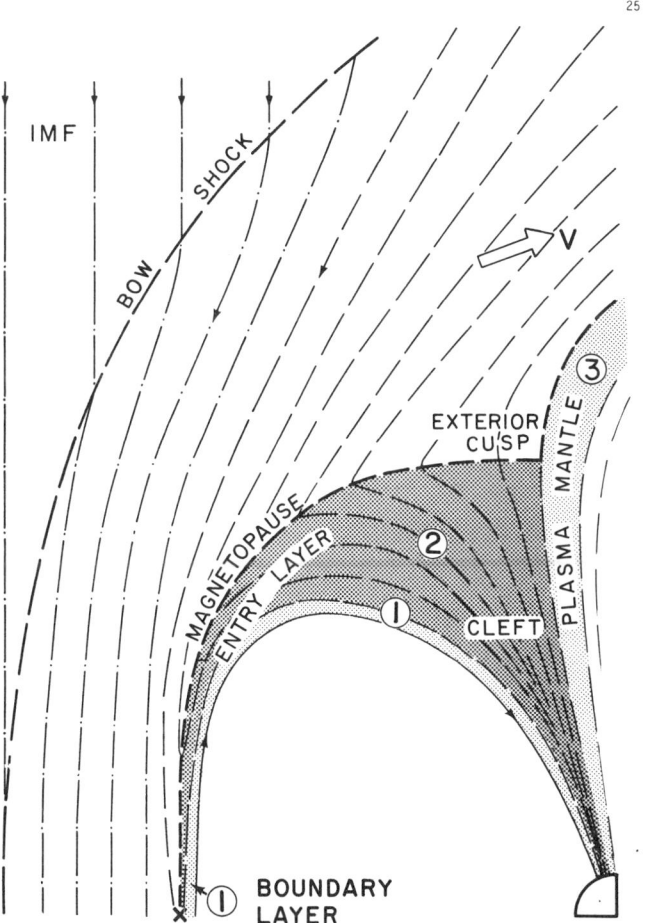

Fig. 2. Observations are consistent with a boundary layer partly on closed magnetic field lines, indicated by 1. Some of the field lines in the cleft are open (2), as well as in the plasma mantle (3).

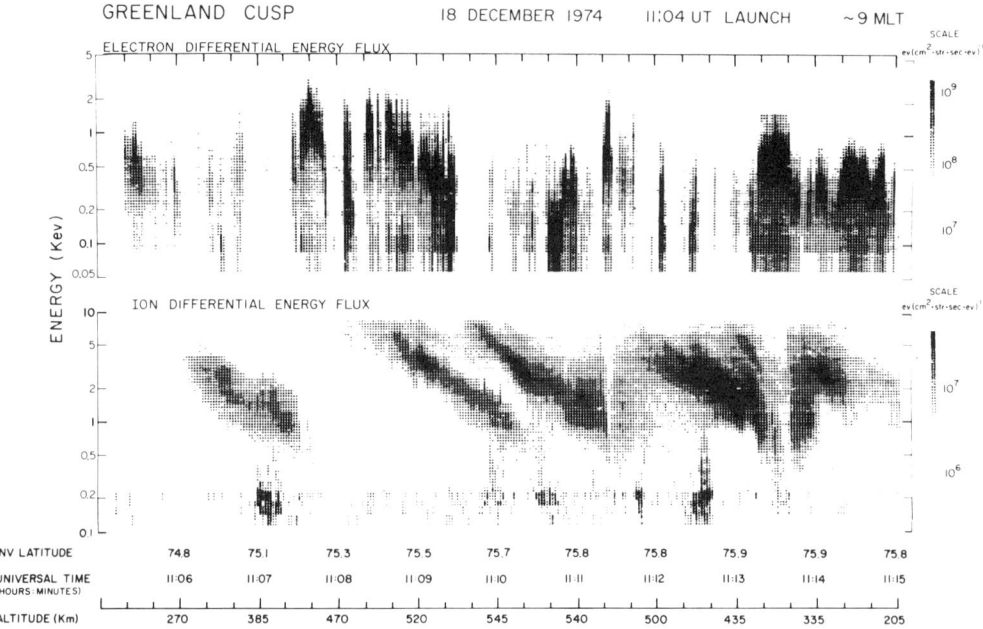

Fig. 3. Rocket evidence for impulsive plasma transport, showing time dispersion of ions in the cleft.

for present theories, especially theories of reconnection, if there were evidence that the boundary layer is entirely on open field lines (see e.g. Heikkila, 1978, and Fritz and Fahnenstiel, 1982). However, any evidence that we now have is better explained by saying that at least part of it is on closed field lines (McDiarmid et al., 1976) as suggested by Figure 2. In this figure the last closed field line is within the cleft, not at its equatorward boundary. There is absolutely no evidence of open field lines equatorward of the cleft; this lack of evidence rules strongly against classical reconnection, according to two-dimensional steady state theories (Heikkila, 1978, page 124). An entirely different argument is the so-called (but not by me) energy crisis, which is equally compelling (Heikkila, 1975); if the energy predicted by reconnection theories is not found we should be prepared to modify the models upon which the theories are based, rather than ignoring observations.

Some other ideas for the processes operating at the magnetopause have been proposed, such as diffusion or direct penetration of magnetosheath particles, or the action of MHD waves on the plasma. At this moment, none of these has reached the state where they are generally regarded as the true answer (see Hill, 1979 for an appraisal).

However, there is one model which deserves to be considered in greater detail, that of impulsive penetration (Lemaire, 1977). Direct observational evidence supporting this idea was obtained in a rocket flight by Torbert and Carlson (1976) in the morning-side cleft in Greenland. Their data are shown in Figure 3 in the form of electron and ion spectrograms. They concluded that these date are consistent with sudden injection at a distance of about 12 earth radii of ions of all energies, at the same time, with the high speed ions preceding low speed ions along the flux tube. These results are consistent with some type of impulsive entry; in fact, they almost demand that.

Heikkila (1982a) has proposed another variation of impulsive entry, drawing explicit attention to an induction electric field; this theory is totally different from that of Lemaire. Figure 4 shows a localized plasma cloud (or flux tube) assumed to possess more momentum than the adjacent magnetosheath plasma directed into the magnetopause; this excess momentum is assumed to deform the magnetopause current as shown here, as is frequently observed. (In actual practice there would also be a large tangential component, which is not shown in the interests of simplicity.) A normal component of $\underline{B}$ (i.e., an open magnetosphere) is assumed; this is similar to what is done in reconnection models, where it is also assumed at the outset. Figure 5 shows an idealization of how the magnetopause current might change, with a clockwise current perturbation as appropriate for entry. This current perturbation will create an induction electric field $\underline{E}^i$ in a counter-clockwise sense, by Lenz's law, opposing the current perturbation everywhere, of the order of 1 mV/m (Heikkila, 1982a); it will be directed in opposite directions on the two sides of the moving magneto-

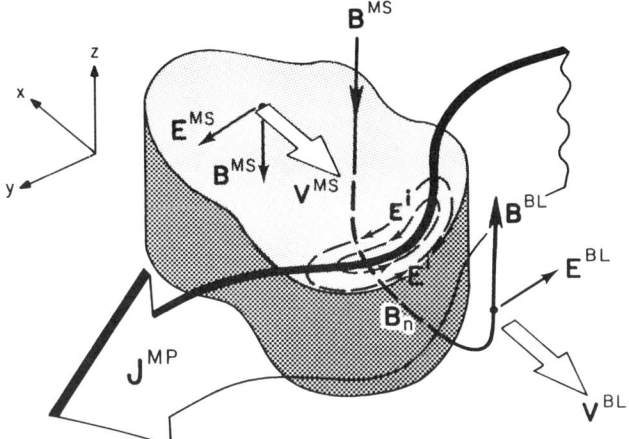

Fig. 4. Model for impulsive plasma transport across an open magnetopause. A plasma cloud (a flux tube) with excess momentum is assumed to distort the magnetopause current locally, creating an inductive electric field, everywhere opposing the current perturbations (by Lenz's law).

pause. This E^i is just what is needed for the plasma to follow the moving magnetopause. Now comes the key to the entire situation; if there is a normal component of the magnetic field through the magnetopause, as appropriate for an open magnetosphere, then the plasma will try to decrease the field aligned component of the induction electric field along B_n by delivering charge, via a small polarization current through the magnetopause; the resulting electrostatic field has the opposite sense to the induction electric field, so the net electric field along the magnetic field lines will be reduced, perhaps even to zero. These same charges will enhance the tangential component of the induction electric field, in opposite directions on the two sides. This enhancement of the induction electric field by the plasma will drive the plasma cloud on to the moving magnetopause via an $\underline{E} \times \underline{B}$ drift. The plasma particles can continue along B_n through the magnetopause current layer, even if $|E|$ is very small. Finally, it can continue flowing earthward (again mostly by $\underline{E} \times \underline{B}$ drift) or tailward in the boundary layer, as long as it still has excess momentum.

In effect, the external plasma (and if you wish, the frozen-in-magnetic field) is convected toward the moving magnetopause locally. Within the current layer it is improper to regard the aggregate of particles as a plasma because the length and time scales necessary for MHD to be satisfied are not obeyed (see Heikkila, 1982a, p. 161). Instead, we should take the view that the individual particles go through the current sheet along B_n, and these then form an entirely new plasma on the other side, with a different orientation of magnetic and electric fields. The particles gain some energy on one side, but they then lose most or even all this gain on the other side. This explains the observed easy access into the boundary layer, in agreement with observations, such as those reported by Eastman and Hones (1979).

It is interesting to note that if there is no normal component of $\underline{B}$ through the magnetopause, as appropriate to an open magnetosphere, this

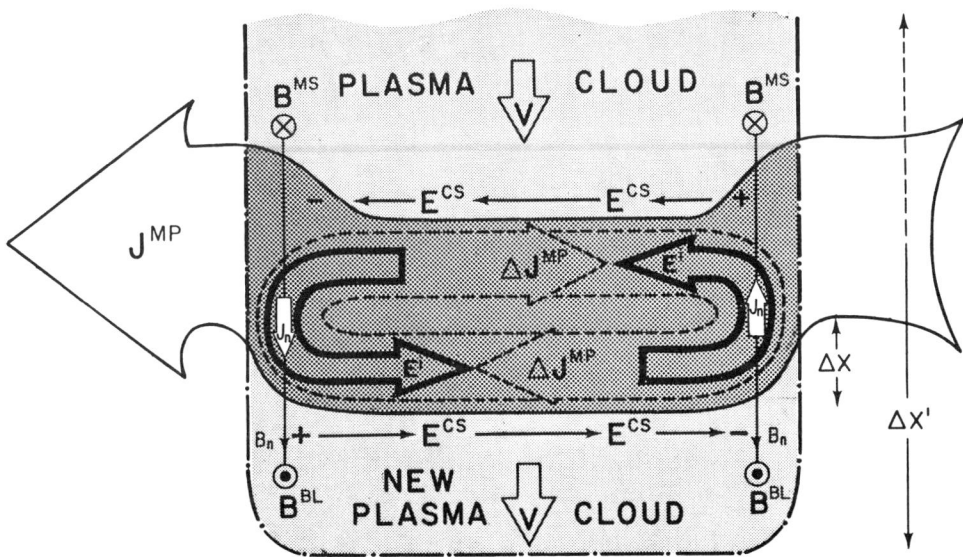

Fig. 5. Impulsive penetration is produced by an electrostatic field due to a charge distribution created by an induction electric field. Charged particles from the old plasma cloud go through the current sheet along B_n and form a new plasma cloud on closed magnetic field lines.

mechanism will not work. This property will lead to a dependence of geomagnetic activity on the interplanetary magnetic field.

In a further article Heikkila (1982b) has shown evidence from Mozer's electric field instrument on ISEE-1 that support the above model. The data, shown in Figure 6, indicate a constant reading for two-thirds of a satellite spin in the middle of the current layer; they indicate that the electric field vector must be rotating, even though there may be temporal changes in magnitude as well. Assuming no temporal changes in magnitude, the electric field vector would be as indicated in the bottom portion of the figure, in complete agreement with the model shown in Figure 5.

Away from local noon, the magnetosheath plasma will have a large antisunward component of drift, shown in Figure 7 (after Heikkila, 1979), whereas at the subsolar point this component may be small. The tangential drift is not likely to be changed much by crossing the magnetopause, so in fact, the boundary layer flow should also be anti-sunward, as is observed. The net result is that the boundary layer flow is created by the barrage of plasmoids, in agreement with the conclusions of Sckopke et. al (1981), shown in Fig. 8. This flow behaves as a voltage generator (Heikkila, 1979) as proposed by Eastman et al in 1976. The physics is quite simple; the plasma flow drives a polarization current to produce the charge separation, and thereby the electrostatic field in the boundary layer. The resulting convection pattern is shown in Figure 9, both in the equatorial plane and in the ionosphere; the convection in the high-latitude ionosphere shows a "throat" such as observed by Heelis et al. (1976).

The Distant Magnetotail

Now we are ready to tackle another aspect of the problem, that of the X-line and the total amount of open magnetic flux. With an open magnetosphere there should be an X-line through the subsolar point, as in Figure 1 and 2 (although an O-line with two X-lines is a possibility, defining a visor according to the model of Podgorny et al., 1978). Now an X-type neutral line must be an extended feature; it cannot end in nothing. It can be a continuous ring, joining back onto itself, or it can join to an O-type neutral line. This becomes obvious when we recall that a neutral line separates regions where the polarity of the magnetic field reverses (for a southward IMF); a physically realistic field must change continuously, and so it must go through zero magnitude before it can change sign. For our model with a strictly southward IMF we can resort to these symmetry arguments, and we can deduce that there should be an X-type neutral line completely around the magnetosphere in the equatorial plane. The X-line (and the separatrix) circumscribes the total amount of open magnetic flux, a point to be dealt with further below. This feature is illustrated by the row of x-s in Figure 7 for the front half of the magnetosphere.

Observations have shown that the boundary layer flow is quite massive, at least 10^{27} ions per second (Eastman, 1979). This flow is inside the magnetopause and X-line, and a question immediately presents itself as to the ultimate fate of this plasma. It cannot all flow into the plasma sheet, since that can be supplied with a flow which is an order of magnitude less (Hill, 1979). On the basis that most of the magnetotail is in fact boundary layer and mantle plasma, as proposed by Heikkila (1983), a cut in the equatorial plane would appear as in Figure 10, and in the noon-midnight plane as in Figure 11. The plasma flow is tailward both in the magnetosheath and in the distant tail as observed by Hardy et al. (1975) at lunar orbit. The plasma flow is still tailward through the moving magnetopause and X-line (similar to what happens on the dayside magnetopause). The X-line is moving also, but with a velocity which is less than the plasma velocity.

Open Field Lines in the Polar Cap

The geomagnetic field lines in the polar caps are shown in Fig. 1 and 11 as going through the polar magnetopause as a rotational discontinuity, into the high latitude magnetosheath, through the bow shock and into the interplanetary medium; this is what is meant by open field lines. Maxwell's equations require that the same amount of open flux appear at both poles.

The total amount of open flux is determined by the size of the polar cap from which this flux emanates. Charged particles show a trapped pitch angle distribution (i.e., with two loss cones indicative of absorption on either end of a closed field line) up to the approximately circular auroral oval. Thus the oval represents an upper limit to the size of the region of open magnetic field lines. However, quite often a trapped pitch angle distribution of a weak particle flux is seen poleward of the auroral oval; we might contemplate a slightly smaller circle as the area of open field lines, with a radius of 1500 km, with an area of about $7 \times 10^{12} m^2$, and deduce that the total open flux is at most $\Phi^M = 4 \times 10^8 Wb$.

An independent line of evidence concerning closed magnetic field lines is provided by recovery phase auroral forms in the polar cap. These sometimes show poleward motion to about 75°-80° in latitude, some 5° to 15° poleward from the breakup arc as shown in Figure 12. It is consistent to interpret these forms as the footprint of a plasmoid that is formed at the onset of the expansion phase (Birn and Hones, 1981; Forbes and Priest, 1983), as it travels tailward to the nightside magnetopause (Heikkila and Treilhou, manuscript under preparation). Based on this interpretation, the surface enclosing the

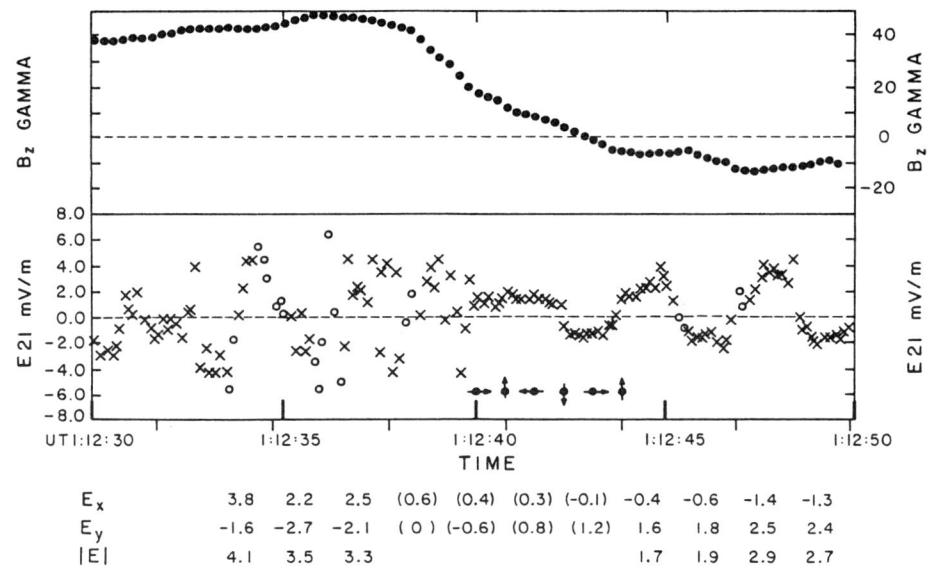

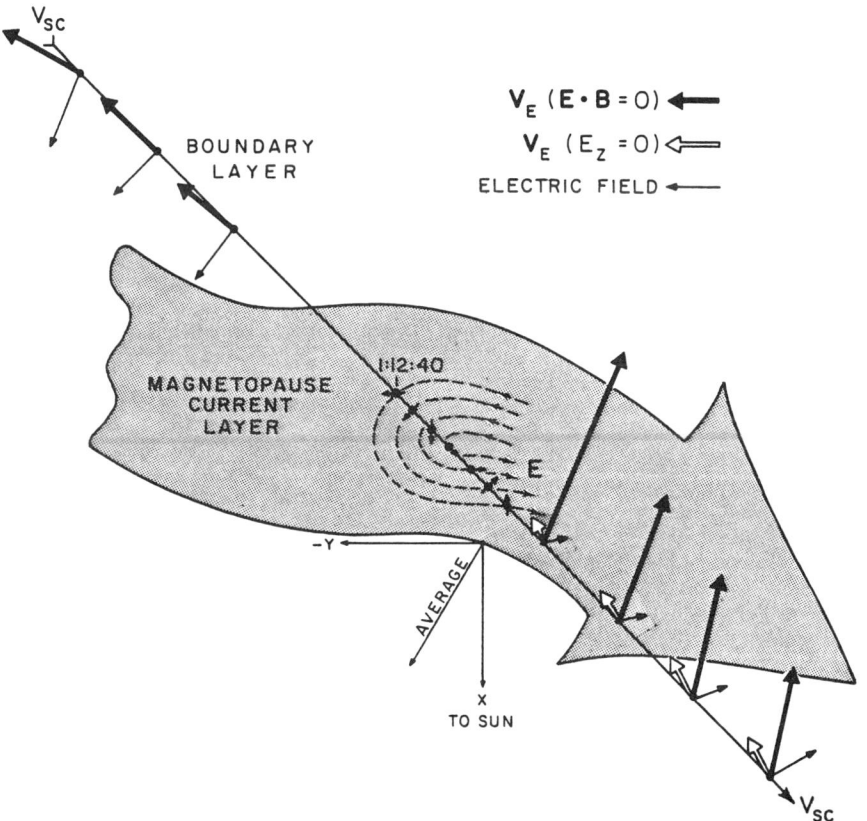

Fig. 6. The constant potential difference for 270° of satellite rotation shows that the electric field vector is rotating in the middle of the current layer, suggesting an induction electric field.

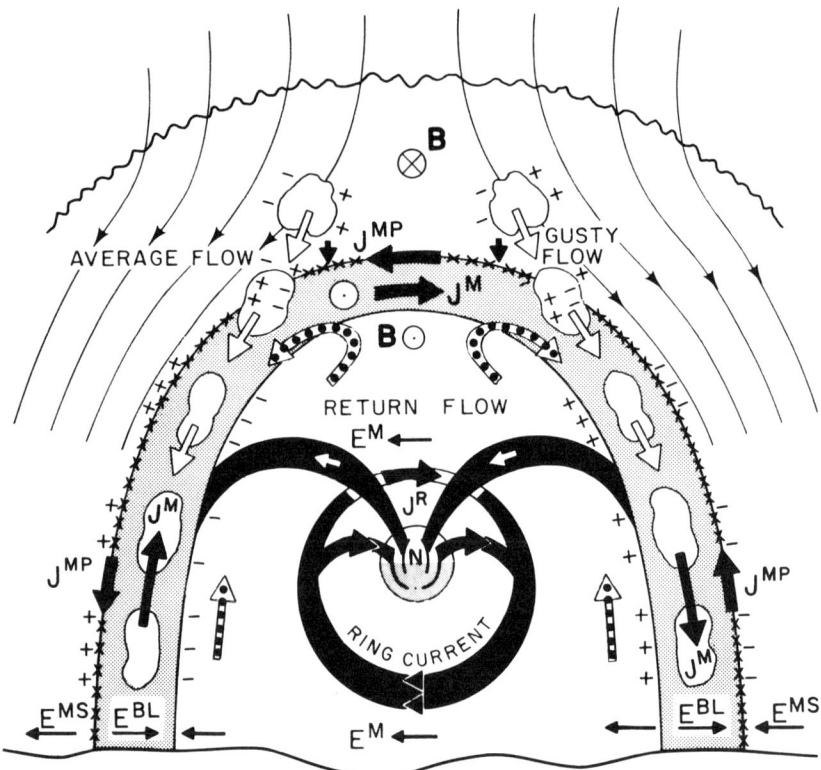

Fig. 7. Away from noon the tangential component of velocity is antisunward, suggesting this pattern of electric fields and currents. The barrage of plasma clouds charges up the sides of the boundary layer as shown, consistent with tailward flow and field aligned currents. The magnetospheric return flow is also caught up by the boundary layer.

plasmoid remains connected to the polar caps by closed magnetic field lines for one or two hours during the recovery phase as it travels tailward with speeds of several hundred km/s, again suggesting an extended feature of closed field lines as shown in Figures 10 and 11. The conclusion is that the region of open magnetic field lines is smaller than previously thought, perhaps a circular region extending from the dayside cleft (at 75° to 80°) to 75° to 80° on the night side. This latter view yields a very crude estimate for the total amount of open flux as being

$$\Phi^M = 2 \text{ to } 3 \times 10^8 \text{Wb}$$

Still another piece of observational evidence concerning polar field lines is shown by the existence of polar rain, essentially at all times. An example of polar rain is shown in Figure 13 for a transpolar crossing of ISIS 1 (Winningham and Heikkila, 1974). It is a flux of electrons with the same energy spectrum as the electron flux in the cleft, but reduced by one or two orders of magnitude in intensity. This great reduction rules out adiabatic processes, such as convection and compression; the origin of polar rain could be pitch angle scattering of electrons at the rotational discontinuity at the lobe magnetopause, shown in Figure 11. Electrons in the magnetosheath flow have a large random component of velocity (i.e., a high temperature) and

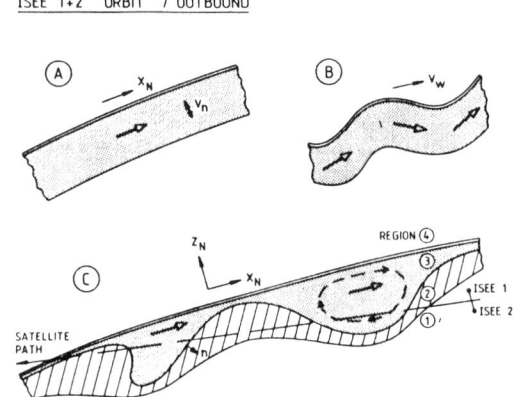

Fig. 8. Satellite observations suggest this plasma configuration in the boundary layer, in agreement with the rocket data shown in Fig. 3, and with impulsive penetration through the magnetopause.

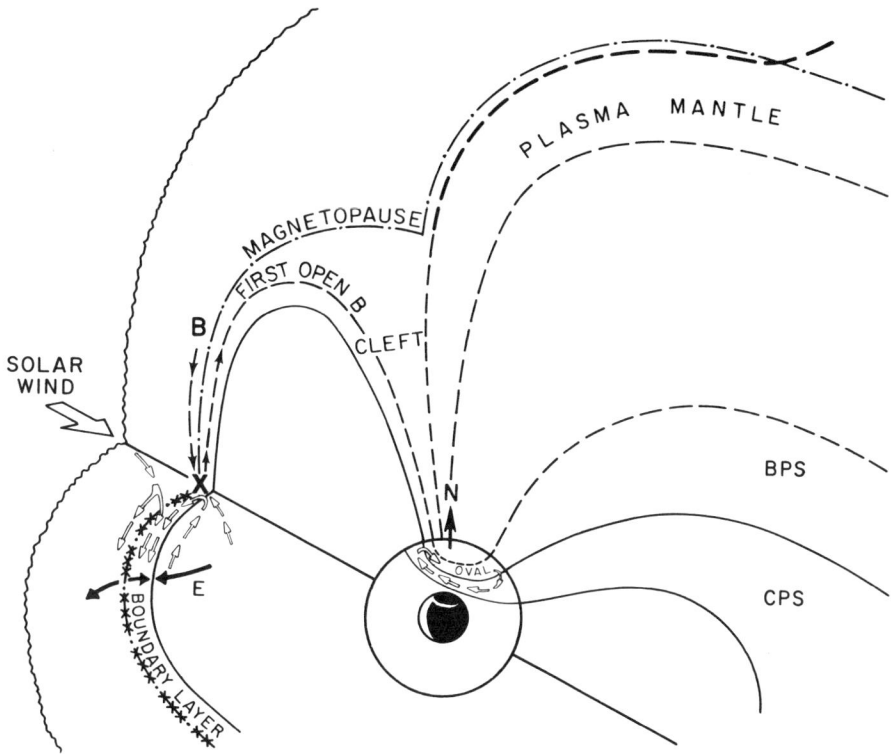

Fig. 9. Three dimensional view showing a throat through which the plasma convects, both in the equatorial plane and in the ionosphere.

with their small mass they can be scattered easily to travel down the open polar cap field lines to show up as polar rain. Protons or other positive ions in the magnetosheath flow have mostly a directed antisunward velocity, are not so easily deflected, and continue their tailward flow. Here we have one more bit of evidence concerning open vs. closed geomagnetic field lines, that of a dropout of polar rain just poleward of auroral fluxes. This is shown in Figure 14 from Meng and Kroehl (1977). This is readily explained by the model of entry of polar rain particles discussed above: this source cuts off at the nightside magnetopause and X-line.

Under the action of the solar wind the cross-section of this flux tube is not likely to be circular on going through the high latitude magnetopause in the lobes and bow shock; instead, it will more than likely be a narrow fin of open field lines, as has been suggested by Stern (1973). We can estimate its width from a measurement of the polar cap electric field. That is observed to be rather small during moderate geomagnetic activity, of the order of 5 to 10 mV/m, providing a total voltage drop of 10 to 20 KV across the (now smaller) polar cap. In addition, there are much larger field strengths within the auroral oval, which in that narrow region (which we assume is on closed field lines) produces a somewhat larger voltage drop; a total drop of 30 to 60 KV across the entire region is consistent with observations. A 20 KV drop across open field lines will be produced by the solar wind of 400 km/s blowing across the IMF of 5 nT over a width

$$w = 10^7 m \text{ or } 1.5 R_e$$

This feature is shown in Figure 15 and also in Figure 16 where the relative size of the earth and boundary layer are shown enlarged (typically, the distance from center of the earth to the subsolar magnetopause is $10 R_e$). The narrow fin of open field lines is bounded by the two sheets of the separatrix S_1 and S_2 extending to the X-line. Figure 15 shows that the open field lines are bent back, intersecting the X-line at right angles.

Having estimated the width of the open flux tube emanating from the polar caps we are now in a position to estimate its length. The above value of the open flux of $3 \times 10^8 M$ yields for the length

$$L = 4 \text{ to } 6 \times 10^9 m, \text{ or } 600 \text{ to } 900 R_e.$$

This agrees roughly with the finding by Evans (1972) using the delay time of cosmic ray protons produced by solar flares as compared with the

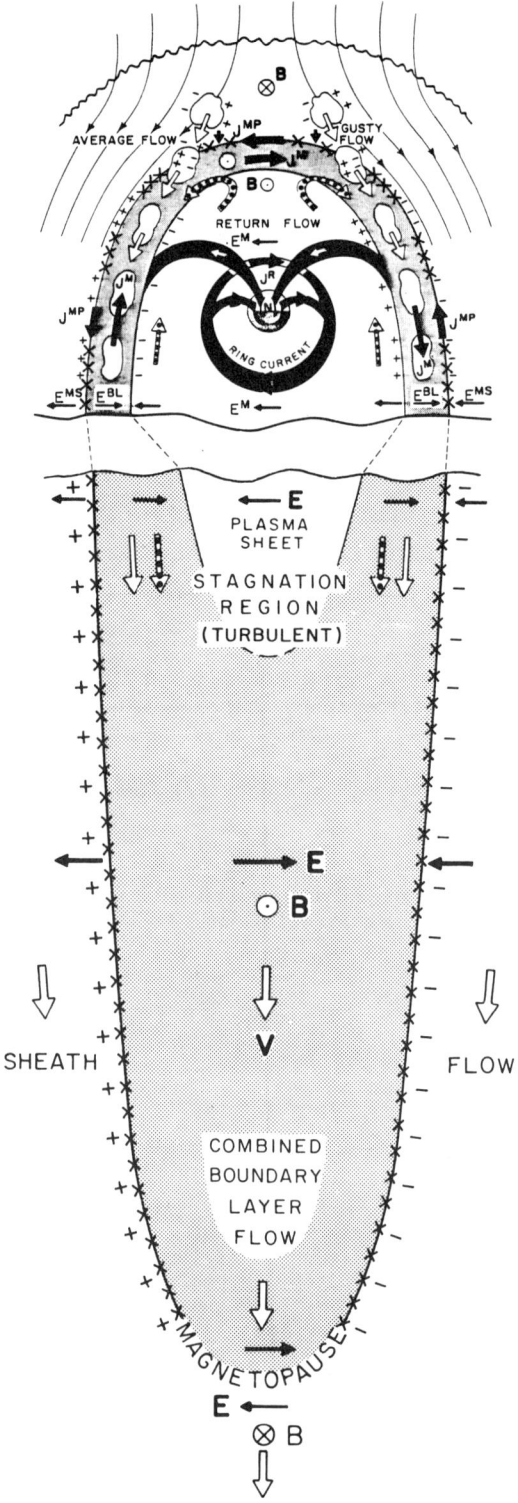

Fig. 10. The boundary layer flow is so massive that it can create its own electric field for tailward flow, as shown in this equatorial cut of the magnetotail.

optical flare; he obtained values ranging from $320R_e$ to $1600R_e$.

Birkeland Currents

Since the boundary layer is partly on closed field lines this layer must extend earthward of the first open field line, or the separatrix. The boundary layer thickness is shown greatly enlarged in Fig. 16 so as to be able to indicate the dynamo or generator produced by the boundary layer plasma flow. As we said before, this dynamo works via a polarization current to provide charges on the sides of the boundary layer. It is a voltage generator, since its main function is to produce the required electric field, so that the rest of the plasma can flow with the proper momentum. The massive boundary layer plasma will generate whatever current is required to maintain this value of the electric field (Heikkila, 1979). The required arrangement of charges is shown in Figures 7 and 16. The former also shows the electric field in the boundary layer, which points from dusk to dawn, the reverse of the magnetotail field in the plasma sheet.

The charges on the sides of the boundary layer are some of the plasma particles themselves. To a first approximation, their motion is governed mostly by the Lorentz force without collisions, and as such they spiral up and down the magnetic field lines; their average motion will be by gradient and curvature drifts in the inhomogeneous magnetic field, as well as $\underline{E} \times \underline{B}/B^2$. Since like charges repel, the charges of the same sign will be accelerated away from the respective charge clouds, while opposite charges will be attracted. The inside surfaces of the dawn and dusk boundary layers (see Figure 16) will try to produce a field-aligned current, since some particles will gain enough energy to counteract the mirror force on closed field lines; in fact, these are the Region 1 currents (Figure 16) described by Iijima and Potemra (1976), indicated in Fig. 7 and 10. The physics has been discussed by Lyons (1980) and others.

It seems likely that most (or at least many) particles can remain trapped on closed field lines on the inside surface, at least for some time, losing and gaining energy as they repeatedly traverse the region of net charge density in the bouncing motion between northern and southern hemispheres. Not so for the outer surfaces of the boundary layers; any particle which is not in the loss cone will mirror on polar cap field lines, but will find after reflection that it is on open field lines, and is consequently lost from the system. Positive ions would be preferentially lost from the evening side, and electrons from the morning side. The closure current in the distant magnetotail would be provided by curvature drift at the rotational discontinuity as the particles leave the nightside magnetopause, as can be deduced from Figure

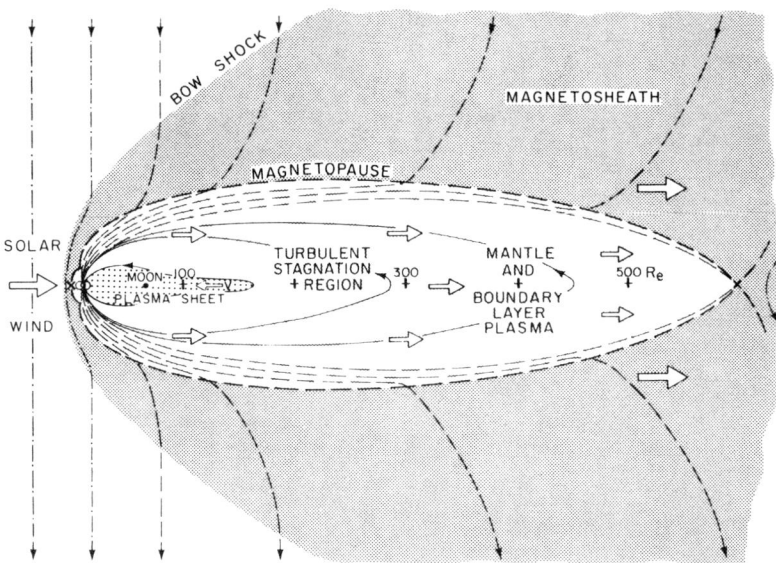

Fig. 11. A cut in the noon-midnight meridian with mantle and boundary layer plasma throughout most of the magnetotail. The plasma sheet is a small cavity of energized plasma close to the earth.

11. The function of the voltage generator is to keep replenishing all the lost charges.

In fact, we can see that such charged particle flow is consistent with the required sign of the current flow over the high latitude lobes of the magnetotail, deduced from the magnetic field topology, i.e., the magnetopause current over the lobes. This magnetopause current, we believe, is part of the closure current of the region 1 system, as it does go to the correct terminal of the boundary layer dynamo; when we monitor region 1 currents we are really monitoring part of the magnetopause current system. In this model, region 2 currents are not the only closure for region 1 since we need to consider both voltage generators (see Fig. 16). The physics is complicated by the fact that these field aligned currents share the same ionosphere.

It seems likely that some particles from the outer surface of the boundary layer may be lost to the atmosphere on the first bounce; therefore, I propose a new current system, called the region 0 current system (Figure 16). Alternatively we can think of 0 as standing for open, since this current system marks the location of the first open field line. Figure 17 indicates a prediction of the region 0 current system by the dashed lines, going tailward from the cusp currents (from Iijima and Potemra, 1976); these currents may be small, for the reasons we have outlined above, but they are significant in that they could provide the best observation of the boundary between closed and open field lines.

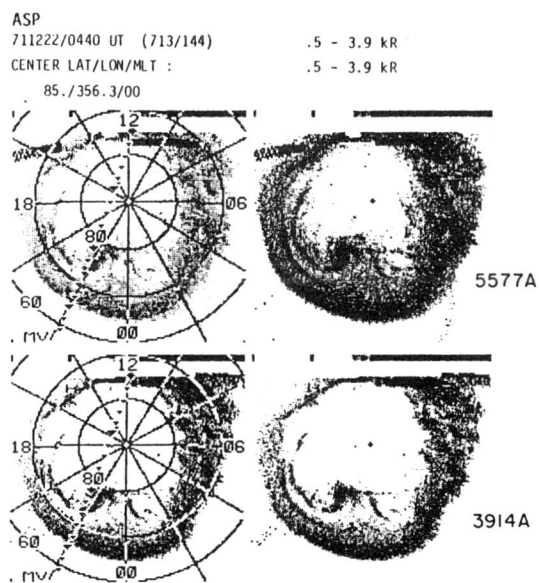

Fig. 12. A satellite view by ISIS-2 of a major auroral breakup, with poleward motion to $80°$ in latitude; this is viewed here as being the footprint of a plasmoid that is being ejected from the magnetotail, with the enclosing surface connected to the earth by closed magnetic field lines.

Field Topology

In this model the magnetospheric electric field due to the solar wind comprises two parts. One is by mapping the interplanetary electric field along open geomagnetic field lines to the polar cap, the other is due to the boundary layer.

The fact that the electron energy spectrum in

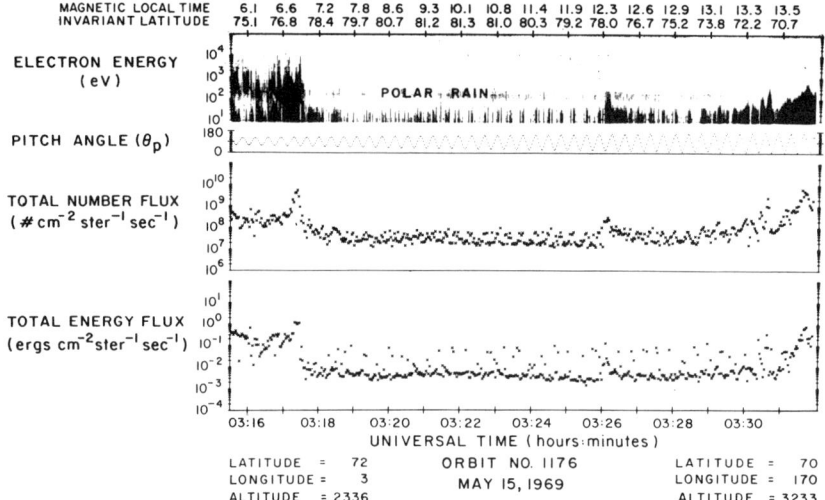

Fig. 13. Recording of polar rain, the weak flux of electrons near 100 eV, present essentially at all times over the polar cap.

the polar rain is the same as the cleft spectrum indicates that there is no (or at least very small) component of the electric field parallel to the magnetic field lines in general. Thus, there is every reason to expect that the interplanetary electric field will be mapped down to the polar cap.

However, this is not the only source of the magnetospheric electric field due to the solar wind. The boundary layer plasma momentum creates a voltage generator to produce the boundary layer electric field, as described above; this can be mapped to the ionosphere (perhaps with some distortion due to $E_{\parallel}$).

A thin polarized boundary layer is a dipole sheet. It has some interesting properties, quite different from those of a charge layer, as shown by the variation of electrostatic potential and field in Fig. 3 of Heikkila (1979). Since a small pillbox enclosing a portion of a perfect infinite dipole layer has no net charge it is clear that there is no discontinuity in electric field from one side of the dipole sheet to the other (by Gauss' law); within the layer there is

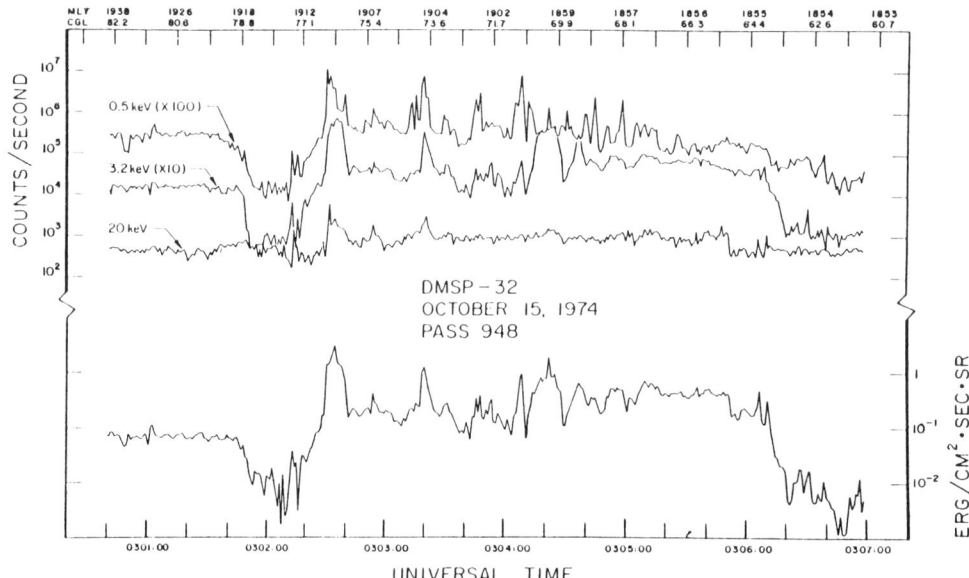

Fig. 14. A dropout of polar rain and higher energy particles just above the auroral precipitation.

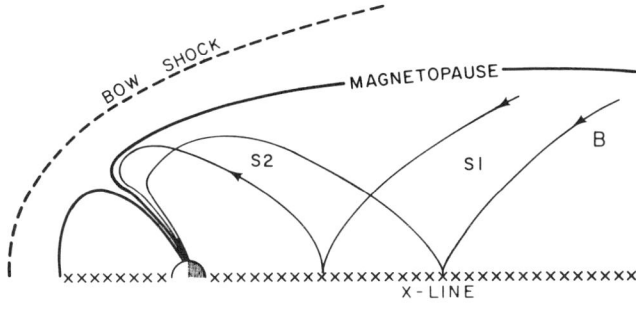

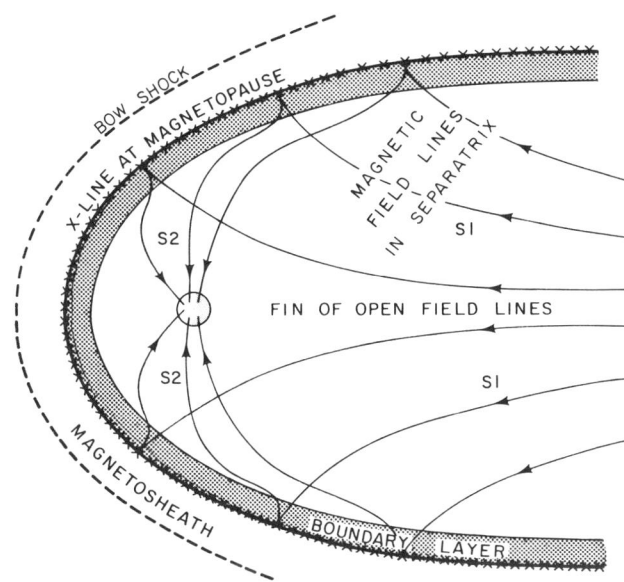

Fig. 15. Magnetic field lines in the two sheets of the separatrix S_1 and S_2.

a reversed field, which introduces a change in potential. In particular, the dipole sheet will transmit the external magnetosheath field into the magnetosphere. This is an important conclusion, since it provides an alternate prescription for the correlation of geomagnetic activity with a southward IMF, an alternate to that suggested by Dungey (1961) involving reconnection. This is a form of viscous interaction, since losses of particles, momentum, and energy of the boundary layer plasma are involved.

At first glance, this idea seems to work out quite well in the dawn-dusk cross-section shown in Figure 16. However, there are some serious difficulties for the overall topology, caused mainly by the observed lack of energization near the subsolar magnetopause. This is a very persistent feature of the observations (Eastman and Hones, 1979); in only a few cases is there some doubt. Even in these cases, the authors (Paschmann et al, 1979; Sonnerup et al, 1981) have concentrated on plasma jets (momentum balance) rather than plasma energization, and similar jets would be produced by the impulsive injection

model of Heikkila (1982a). Consequently, this lack of particle energization implies that the tangential component of the electric field is small; in fact it can reverse within the magnetopause current layer, as indicated by Figure 6.

What this means is that now we have difficulties on the dawn and dusk flanks of the magnetopause, contrary to our first guess. With an entirely closed magnetosphere there would be no problem; it is like having any perfect conductor in a flowing plasma. The difficulty is due to the open polar cap flux; this forces the dawn and dusk magnetopause to a relatively high potential, 10KV and -10KV respectively in the typical magnitudes discussed above.

There are two possible ways out of this dilemma. The first is that there may be a tangential component of the electric field on the dawn and dusk flanks, and therefore some type of classical "reconnection" there, rather than on the dayside magnetopause (Heikkila, 1978, p.128). The long extent of the magnetopause on the flanks implies that the electric field along the X-line can be miniscule. We need to analyze plasma and electric field observations in these regions to see whether indeed this is the case. It should be noted, however, that the plasma velocity is different from that assumed in reconnection theories, being more parallel to the X-line rather than perpendicular, and so it is not obvious whether present theories are applicable.

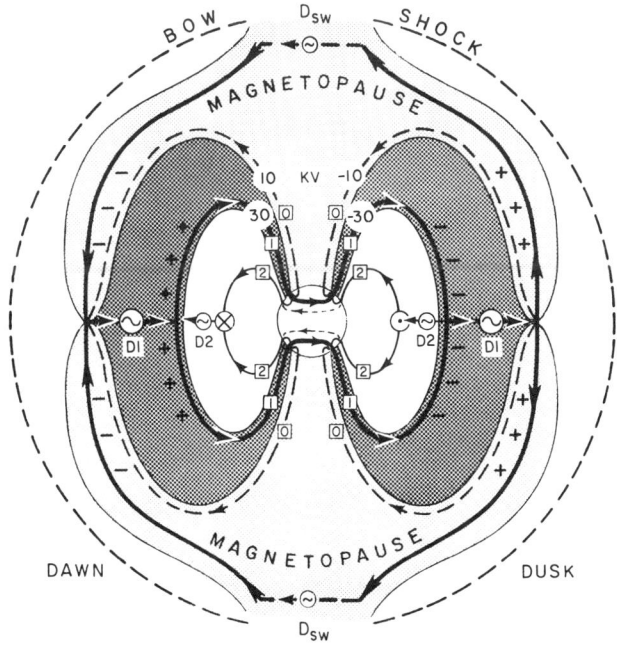

Fig. 16 Voltage dynamo D_1 in the closed field line portion of the boundary layer drives region 0 and 1 field aligned currents. The solar wind dynamo D_{sw} is also indicated.

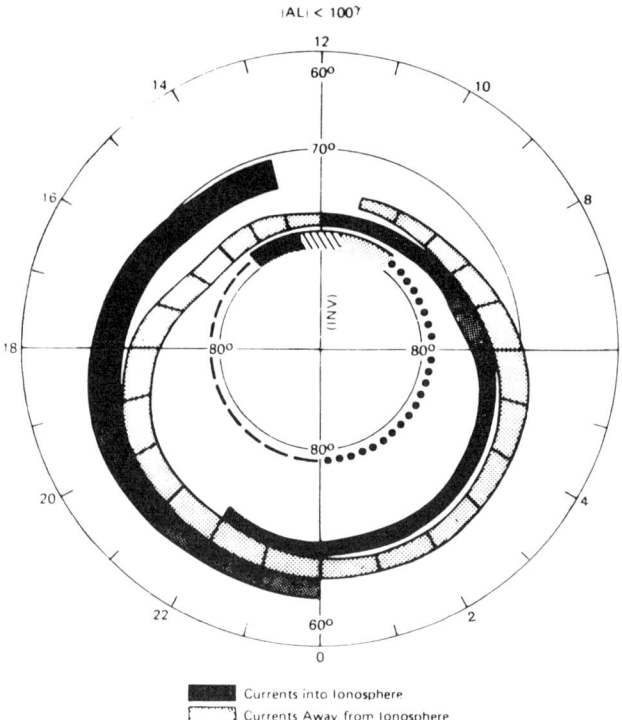

Fig. 17. Observed pattern of field aligned currents, with the suggested continuation of region 0 currents from the so-called cusp currents, defining the region of open field lines. Region 1 currents (between the low latitude Region 2 currents and Region 0) occur within the boundary layer.

The other possibility is even more exotic. A likely possibility is that a truly steady state may never be reached; such a view has been expressed by Erickson and Wolf (1981), and Schindler and Birn (1982). It then becomes very diffi-difficult to predict the electric field topology, as pointed out by Schindler and Birn. This idea is presented here simply to note that with an induction electric field the different topologies of induced and electrostatic fields come into play (Heikkila et al., 1979).

Magnetic Bottle Confining the Plasma Sheet

The plasma in the plasma sheet usually has a high beta, and pressure arguments would lead us to conclude that the plasma would have to escape from the plasma sheet unless it were not contained by some mechanism. Fig. 11 indicates that this mechanism is a magnetic bottle. This containment may continue for some time, but perhaps not forever. The plasma in the plasma sheet is continually energized (see Heikkila et al., 1979); eventually a plasmoid may escape away from the earth from the region of closed magnetic field lines. Such escape involves a magnetospheric substorm, which is the subject of a recent review by Pellinen and Heikkila (1983).

Summary and Conclusions

This paper is based on several features of the magnetosphere which still are open to debate, that of open magnetic field lines in the polar cap, and of a boundary layer just inside the magnetopause which is partly on closed field lines. However, these have not been ruled out conclusively, and in fact I believe they are real; we should try to incorporate them into a consistent model of the magnetosphere. This I have attempted to do, and find that several observations fit into this model quite well, with the single exception of the electric field. Thus, the length of the magnetotail, the existence of the polar rain, the bursts of particles at the edge of the plasma sheet, and the poleward motion of auroral forms during the recovery phase of a magnetospheric substorm all may have natural explanations with this concept.

The one exception is the topology of the electric field. The apparent problems associated with reversal of the electric field within the magnetotail are of three kinds.

At the distant X-line the magnetic field reverses from northward to southward for a southward interplanetary magnetic field (IMF). For continued anti-sunward flow of boundary layer plasma it is necessary that the electric field also reverses; this sudden reversal may mean that the field has a curl, associated with a changing magnetic field by Faraday's law, implying that time-dependent theory must be used. The magnetopause and X-line could either have a wave-like motion, or new X-lines might be formed upstream. On the other hand, it is possible that charge distributions in three dimensions could produce the correct electric field topology.

The second (apparent) difficulty is the reversal within the magnetotail itself. This reversal suggests a (turbulent) stagnation region, in between the anti-solar flow of more dense boundary layer plasma, and the earthward flow of the more tenuous plasma sheet. At first glance, this conjecture appears to be supported by Pioneer 7 and 8 observations (Heikkila, 1983); the data from ISEE-3 in the tail may resolve this point.

This pattern of electric fields in the distant tail should show up at low altitudes in terms of particle precipitation. Since curvature and gradient drift of particles beyond the stagnation region would cause the particles to lose energy, their mirror points will be raised; the result is no precipitation of trapped particles at latitudes above the auroral oval. This feature has been observed; for example, Meng and Kroehl (1977) have observed no precipitation of any particles for a band poleward of auroral zone (see Figure 14 near 0302 UT). Thus, at first glance, there may be no difficulty associated with a turbulent stagnation region in the distant magnetotail.

The third question concerns ionospheric electric fields. There should be an eastward electric field on the open field lines referred to in the preceding paragraph, although parallel electric fields might modify this conclusion. In the case of time dependence, it would be a mistake to try to infer magnetotail electric fields from ground based observations, as pointed out by Heikkila et al. (1979).

It scarcely need be pointed out that this model is in stark contrast with the reconnection models, and is closer to the idea of viscous interaction proposed by Axford and Hines (1961).

Acknowledgements. Much of this work originated as lectures and discussions at the Royal Institute of Technology during the period from 1974 to 1982. I owe Carl-Gunne Fälthammar and his group many thanks for their comments and hospitality. This paper was given in a preliminary form at the IAGA meeting in Edinburg, August, 1981. It was funded under NSF grant ATM-80-25194 and NASA grant NGL44-004-130.

REFERENCES

Axford, W.I., and C.O. Hines, A Unifying Theory of High-Latitude Geophysical Phenomena and Geomagnetic Storms, Can. J. Phys. 39, 1433, 1961.

Birn, J., and E.W. Hones, Three-Dimensional Computer Modeling of Dynamic Reconnection in the Geomagnetic Tail, J. Geophys. Res., 86, 6802, 1981.

Dungey, J.W., Interplanetary Field and the Auroral Zones, Phys. Rev. Lett., 6, 47, 1961.

Eastman, T.E., The Plasma Boundary Layer and Magnetopause Layer of the Earth's Magnetosphere, written in partial fulfillment of requirements for degree of Doctor of Philosophy, Press of Los Alamos Scientific Laboratory, 1979.

Eastman, T.E., and E.W. Hones, Jr., Characteristics of the Magnetospheric Boundary Layer and Magnetopause Layer as Observed by Imp 6, J. Geophys. Res. 84, 2019-2028, 1979.

Eastman, T.E., E.W. Hones, Jr., S.J. Bame, and J.R. Asbridge, The Magnetospheric Boundary Layer: Site of Plasma, Momentum, and Energy Transfer from the Magnetosheath into the Magnetosphere, Geophys. Res. Lett., 3, 685-688, 1976.

Erickson, G.M. and R.A. Wolf, Is Steady Convection Possible in the Earth's Magnetotail?, Geophys. Res. Lett. 7, 897-900, 1980.

Evans, L.C., Magnetospheric Access of Solar Particles and the Configuration of the Distant Geomagnetic Field, Volume I, Thesis written in partial fulfillment of requirements for PhD, California Institute of Technology, 1972.

Forbes, T.G., and E.R. Priest, On Reconnection and Plasmoids in the Geomagnetic Tail, J. Geophys. Res. 88, 863-870, 1983.

Fritz, T.A., and S.C. Fahnenstiel, High Temporal Resolution Energetic Particle Soundings at the Magnetopause on November 8, 1977, Using ISEE 2, J. Geophys. Res., 87, 2125-2131, 1982.

Hardy, D.A., H.K. Hills, and J.W. Freeman, A New Plasma Regime in the Distant Geomagnetic Tail, Geophys. Res. Lett. 2, 169-172, 1975.

Heelis, R.A., W.B. Hanson, and J.L. Burch, Ion Convection Velocity Reversals in the Dayside Cleft, J. Geophys. Res., 81, 3803-3809, 1976.

Heelis, R.A., J.D. Winningham, W.B. Hanson, and J.L. Burch, The Relationships between High-Latitude Convection Reversals and the Energetic Particle Morphology Observed by Atmosphere Explorer, J. Geophys. Res. 85, 3315-3324, 1980.

Heikkila, W.J., Outline of a Magnetospheric Theory, J. Geophys. Res. 79, 2496-2500, 1974.

Heikkila, W.J., Is There an Electrostatic Field Tangential to the Dayside Magnetopause and Neutral Line? Geophys. Res. Lett., 2, 154, 1975.

Heikkila, W.J., Criticism of Reconnection Models of the Magnetosphere," Planet. Space Sci., 26, 121, 1978.

Heikkila, W.J., Impulsive Penetration and Viscous Interaction, published in the Proceedings of the AGU Chapman Conference on Magnetospheric Boundary Layers, p. 375, Alpbach, Austria, June 1979, by the European Space Agency.

Heikkila, W.J., Impulsive Plasma Transport through the Magnetopause, Geophys. Res. Lett., 9, 159-162, 1982a.

Heikkila, W.J., Inductive Electric Field at the Magnetopause, to be published in Geophys. Res. Lett., 1982b.

Heikkila, W.J., Exit of Boundary Layer Plasma from the Distant Magnetotail, Geophys. Res. Lett., 10, 218-220, 1983.

Heikkila, W.J., R.J. Pellinen, C.-G. Falthammar, and L.P. Block, Potential and Inductive Electric Fields in the Magnetosphere during Auroras, Planet. Space Science, 27, 1383, 1979.

Hill, T.W., Rates of Mass, Momentum, and Energy Transfer at the Magnetopause, Proceedings of Magnetospheric Boundary Layers Conference, Alpbach, 11-15 June, 1979, 325-332.

Iijima, T., and T.A. Potemra, Field-aligned currents in the Dayside Cusp Observed by Triad, J. Geophys. Res., 81, 5971-5979, 1976.

Kennell, C., Magnetospheric and Ionospheric Physics, Solar-System Space Physics in the 1980's: A Research Strategy, 1980, p. 55.

Lemaire, J., Impulsive Penetration of Filamentary Plasma Elements into the Magnetospheres of the Earth and Jupiter, Planet. Space Sci., 25, 887-890, 1977.

Lyons, L.R., Generation of Large-Scale Regions of Auroral Currents, Electric Potentials, and Precipitation by the Divergency of the

Convection Electric Field, J. Geophys. Res., 85, 17-24, 1980.

McDiarmid, I.B., J.R. Burrows, and E.E. Budzinski, Particle properties in the day side cleft, J. Geophys. Res. 81, 221.

Meng, C.-I., and H.W. Kroehl, Intense Uniform Precipitation of Low-Energy Electrons over the Polar Cap, J. Geophys. Res. 82, 2305-2313, 1977.

Paschmann, G., B.U.O. Sonnerup, I. Papamastorakis, N. Sckopke, G. Haerendal, S.J. Bame, J.R. Asbridge, J.T. Gosling, C.T. Russell, and R.C. Elphic. Plasma Acceleration at the Earth's Magnetopause: Evidence for Reconnection, Nature, 282, 243-246, 1979.

Paulikas, G.A., Tracing of High Latitude Magnetic Field Lines by Solar Particles, Rev. Geophys. and Space Physics 12, 117-128, 1974.

Pellinen, R.J., and W.J. Heikkila, Review: Inductive Electric Fields, Space Sci. Rev., 1983.

Podgorny, E, M. Dubinin, and Yu. N. Potanin, The Magnetic Field on the Magnetospheric Boundary from Laboratory Simulation Data, Geophys. Res. Lett. 5, 207-210, 1978.

Schindler, K. and J. Birn, Self-Consistent Theory of Time-Dependent Convection in the Earth's Magnetotail, J. Geophys. Res., 87, 2263-2275, 1982.

Sckopke, N., and G. Paschmann, The Plasma Mantle: A survey of Magnetotail Boundary Layer Observations. Journal Atmospheric and Terrestrial Physics 40, 261-278, 1978.

Torbert, R.B. and C.W. Carlson, Impulsive Ion Injection into the Polar Cusp, Magnetospheric Particles and Fields, ed. B.M. McCormac, 47-53, 1976.

Williams, D.J., Energetic Ion Beams at the Edge of the Plasma Sheet: ISEE 1 Observations Plus a Simple Explanatory Model, J. Geophys. Res., 86, 5507-5518, 1981.

Winningham, J.D. and W.J. Heikkila, Polar Cap Auroral Electron Fluxes Observed with ISIS 1, J. Geophys. Res., 79, 949-957, 1974.

A NEW THEORY OF SOURCES OF BIRKELAND CURRENTS

K. D. Cole[1]

NASA/Goddard Space Flight Center, Greenbelt, Maryland 20771

Abstract. A new approach to collisionless plasma (Cole 1983) shows the existence of current orthogonal to $\underline{B}$ along the low latitude boundary layer of the magnetosphere driven by electric field which is orthogonal to both $\underline{B}$ and the layer. In this case the relationship

$$p_\perp + \frac{B^2}{8\pi} - \frac{\varepsilon E_\perp^2}{8\pi} = \text{constant},$$

holds on a line orthogonal to B and the layer, where ε is the dielectric constant of the plasma for electric fields orthogonal to $\underline{B}$. Across the geomagnetic tail there flows a current in the direction of the dawn-dusk electric field, and in this case a relationship

$$p_\perp + \frac{B^2}{8\pi} + \frac{\varepsilon E_\perp^2}{8\pi} = \text{constant},$$

holds along a line orthogonal to $\underline{E}$ and $\underline{B}$. Divergence of both these currents is shown to be a source of Birkeland currents. Also some of the boundary layer current is continuous with current across the tail. Electric currents of physically similar origin flow in interplanetary space, and when the magnetosphere interrupts them, additional Birkeland currents are driven.

Introduction

In a recent paper a general theory of electromagnetic fields in plasmas in a quasi-steady state was proposed (Cole 1983). This theory is now applied to the problem of the sources of electric current responsible for geomagnetic disturbance. In the first part of the paper, some idealized configurations of electromagnetic fields in plasmas are considered. In the second part, these are applied to structures in the geomagnetic field and interplanetary space as part of a new theory of geomagnetic disturbance.

Though many components of earlier theories of geomagnetic disturbance (Axford and Hines, 1961; Dungey, 1961; Piddington, 1960, 1962; Cole, 1960, 1961) remain valid today, none of them is entirely satisfactory and all fail to be quantitative on the precise mechanism of transfer of energy from the solar wind into the magnetosphere and the atmosphere. Later, this author (Cole 1974) illustrated mechanisms whereby solar wind plasma could enter the geomagnetic field. These included "grad B capture" and plasma flow across the magnetopause caused by tangential electric fields and inertial drifts.

The tangential electric fields could be caused either as the result of grad B capture on the day side of the magnetosphere (Cole 1974) or as the result of dynamo action in the high latitude ionosphere (Cole 1976).

In this paper an attempt is made to advance the theory of solar wind interaction with the magnetosphere invoking new understanding of currents generated by plasma flows (Cole, 1983) in interplanetary space, and in various regions of the magnetosphere and showing how the currents couple into the upper atmosphere. The units used are c.g.s., except where otherwise stated.

Idealized Field Configurations. In this paper the theory of the Maxwell stresses in a collisionless, 2-D plasma (Cole, 1983) is applied to discuss the sources of electric current responsible for geomagnetic disturbance. By treating a collisionless plasma as a medium with specifiable dielectric constant and magnetic permeability it has been shown that, in a steady state,

$$(\text{div } \underline{D}) \underline{E} + (\nabla \times \underline{H}) \times \underline{B} - \frac{1}{2} E_\perp^2 \nabla \varepsilon - \frac{1}{2} H^2 \nabla \mu = 0, \quad (1)$$

where $\underline{E}$ = electric field, assumed perpendicular ($\perp$) to $\underline{B}$

$\underline{B}$ = magnetic induction

$\underline{H}$ = magnetic field,

$\mu = \frac{B}{H}$ = magnetic permeability,

---

[1] On leave from La Trobe University, Bundoora, Victoria, Australia, 3083.

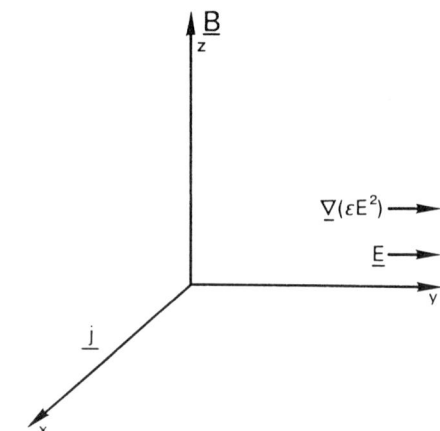

Fig. 1. Geometry for type I current.

$$\underline{D} = \varepsilon \underline{E}_\perp,$$

and
$$\varepsilon = 1 + \sum_s \frac{4\pi n_s m_s c^2}{B^2}, \quad (2)$$

where n_s = number of ions cm^{-3} of species s, and m_s = molecular mass of ions of species s.

It has been shown elsewhere (Cole 1983) that in plasmas such as exist in the low latitude boundary layer and interplanetary space, the force denoted by the term $(\text{div } \underline{D}) \underline{E} - \frac{1}{2} E_\perp^2 \nabla \varepsilon$ in equation 1 drives significant current. This current is referred to as dielectric current and arises principally because of the large value of ε in magnetized plasmas. In this paper divergence of currents of this kind in the geomagnetic environment are invoked as sources of Birkeland currents observed in the magnetosphere. First, some simple models of crossed electric and magnetic fields in plasmas are presented, which define the new currents. Then these models are applied to the plasma environment of the earth in space to estimate the strength of Birkeland currents.

From equation (1)

$$(\text{div } \underline{D}) E_x - \frac{1}{2} E_\perp^2 \frac{\partial \varepsilon}{\partial x} + \frac{1}{2} B^2 \frac{\partial}{\partial x}\left(\frac{1}{\mu}\right)$$
$$- B_z \left(\frac{\partial (B_z/\mu)}{\partial x} - \frac{\partial (B_x/\mu)}{\partial z}\right)$$
$$+ B_y \left(\frac{\partial (B_x/\mu)}{\partial y} - \frac{\partial (B_y/\mu)}{\partial x}\right) = 0, \quad (3)$$

$$(\text{div } \underline{D}) E_y - \frac{1}{2} E_\perp^2 \frac{\partial \varepsilon}{\partial y} + \frac{1}{2} B^2 \frac{\partial}{\partial y}\left(\frac{1}{\mu}\right)$$
$$- B_x \left(\frac{\partial (B_x/\mu)}{\partial y} - \frac{\partial (B_y/\mu)}{\partial x}\right)$$
$$+ B_z \left(\frac{\partial (B_y/\mu)}{\partial z} - \frac{\partial (B_z/\mu)}{\partial y}\right) = 0, \quad (4)$$

Consider now different simple cases of equation 1 which can be integrated and applied to interplanetary space or the geomagnetic field.

Current of Type I:

$$\text{Suppose } \underline{B} = (0, 0, B_z) \quad (5)$$

$$\text{and } \underline{E} = (0, E_y(y), 0) \quad (6)$$

It is shown in Cole (1983) that

$$\mu = \frac{1}{1 + 8\pi p_\perp/B^2} \quad (7)$$

This case (see Fig. 1), discussed earlier (Cole, 1983) shows that a current flows in the x direction, of magnitude given by

$$j_x = \frac{1}{8\pi} \frac{c}{B_z} \frac{\partial}{\partial y} [\varepsilon E_y^2 - 8\pi p_\perp]. \quad (8)$$

In this case $\varepsilon E^2/8\pi$ acts as a tension along $\underline{E}$. Equation 3 integrates to

$$8\pi p_\perp + B_z^2 - \varepsilon E_y^2 = \text{constant}. \quad (9)$$

This corresponds to the circumstances of plasma flows and currents in the x direction. We investigate later the possibility that the flow in the low latitude boundary layer of the earth (Eastman, 1979 Eastman and Hones, 1979) is like this. Also we expect that flows describable by this model exist in interplanetary space (see applications, later in the paper).

Current of Type II:

In this case (see Fig. 2),

$$\text{suppose } \underline{B} = (0, 0, B_z), \quad (10)$$

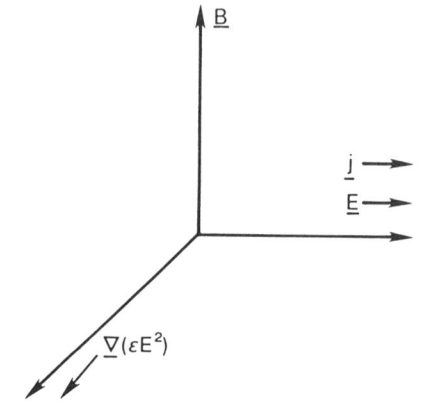

Fig. 2. Geometry for type II current.

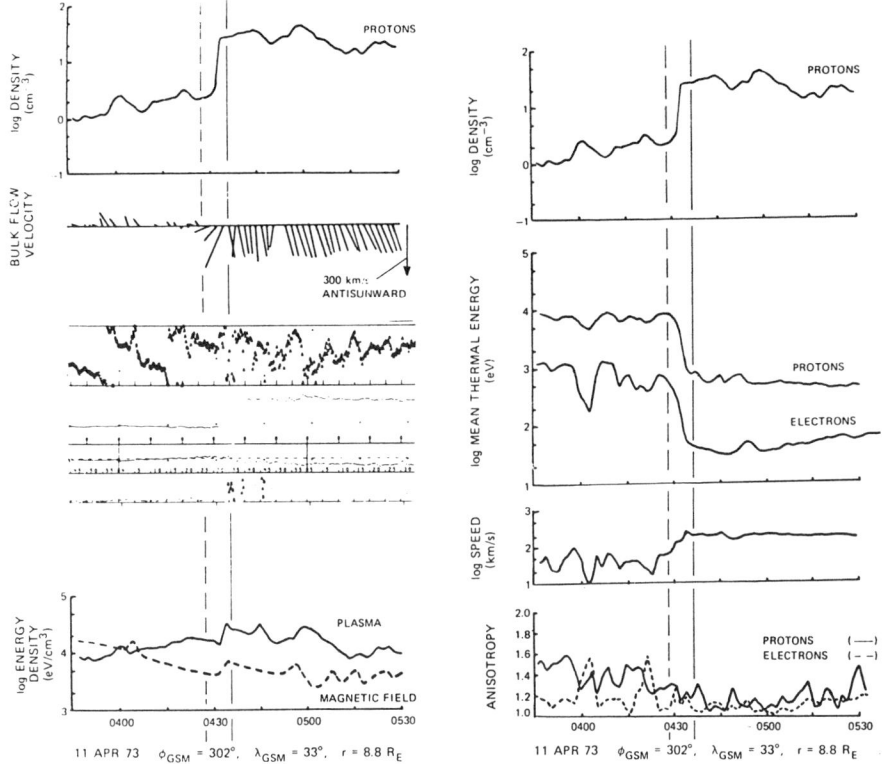

Fig. 3. Example of parameters measured in low latitude boundary layer, taken from Eastman (1979). The boundary layer is identified between the vertical full and dashed lines.

$$\underline{E} = (0, E_y(x), 0). \quad (11)$$

with $\frac{\partial}{\partial y} = 0$, and equation 3 yields

$$\frac{\partial}{\partial x}\left(\frac{B_z^2}{\mu}\right) + E_y^2 \frac{\partial \varepsilon}{\partial x} = 0 \quad (12)$$

Integrating equation 12, yields

$$8\pi p_\perp + B_z^2 + \varepsilon E_y^2 = \text{constant} \quad (13)$$

This case is appropriate to discussions of flow of current across the geomagnetic tail under the influence of a "cross-tail" electric field.

In this case the current in the y-direction driven by electric field is given by

$$j_y = \frac{1}{8\pi} \frac{c}{B} E_y^2 \frac{\partial \varepsilon}{\partial x}. \quad (14)$$

As distinct from case 1, $\varepsilon E^2/8\pi$ acts here as a pressure orthogonal to $\underline{E}$. In the case of cold plasma $p_\perp = 0$, and, equation 12 yields

$$B_z^2 + E_y^2 c^2/V_A^2 = \text{constant}, \quad (15)$$

where $V_A^2 = C^2/\varepsilon$ = square of Alfvén velocity (16)

Equation (15) resembles the results obtained by Alfvén) (1968) and Cowley (1973) in discussing properties of neutral surfaces. While the general solution of the problem of an infinite neutral sheet is contained in equation (13), the Alfvén (1968) type solution is contained as a special case (corresponding to $p_\perp = 0$), while the Harris (1961) type solution is another case, corresponding to $E_y = 0$. In the former case, a balance of forces is maintained by $\nabla(\varepsilon E_y^2)/8\pi$ and $c^{-1} \underline{j} \times \underline{B}$. In the case $E_y = 0$, a balance of $\underline{j} \times \underline{B}$ and $-\nabla p_\perp$ could be maintained.

Case II is, of course, relevant to discussion of magnetic "merging." Equation 13 can be interpreted as implying the existence of static structures of opposed magnetic fields in which plasma is driven towards the merging region by an applied electric field E_y. The plasma is turned away from the neutral line before reaching it generating current in the direction of E_y. The partitioning of energy between thermal energy of the plasma, as manifested by p, magnetic energy and electric field energy density, is dependent on the plasma density, or equivalently, the Alfvén speed (see equation 13). Clearly there is a limitless variety of possible partitions. Experiments looking for merging regions in the geomagnetic tail or at the magnetopause should take into account all three terms of equation

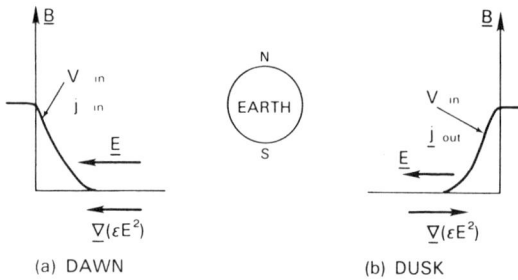

Fig. 4. Schematic representation of velocity, electric field energy density and dielectric current (type I) in low latitude boundary layer.

13. It appears that a steady state neutral sheet can exist in which no merging takes place. Additional factors to those considered here would appear to be necessary in order to generate merging.

Applications

First we consider the regions near the magnetopause. Traditionally, currents at the magnetopause have been considered to be produced solely by a gradient of pressure pointing outwards from the magnetosphere. A Lorentz force $\underline{j} \times \underline{B}$ was considered to oppose the force due to this gradient of pressure (see Willis 1971 for a review). These currents are the Chapman-Ferraro currents. It is considered here that the traditional approach must now be modified. It will be seen to be of limited application in portions of the magnetosphere only, viz., over a small range of longitudes either side of the "nose" of the magnetosphere. In this region a negative pressure gradient acts from the solar wind into the magnetosphere. In vast regions of the magnetosphere, and the low-latitude boundary layer associated with it, there is a flow of solar wind plasma inside and roughly parallel to the magnetosphere in which the speed decreases inwards (see e.g., Fig. 3 taken from Eastman (1979)), and in which the term $\varepsilon E_\perp^2$ is a considerable fraction of p. In making this claim, we assume that $\varepsilon E_\perp^2$ is estimated by $\frac{1}{2} n_i m_i V_{sw}^2$ (c.f., Cole, 1983).

Boundary Layer

It is claimed by the present author that in the boundary layer the current system is different from what might be expected on Chapman-Ferraro theory and the consequences are significantly different. One long recognized difficulty of the Chapman-Ferraro approach is that it predicts no energy or plasma flow into the magnetosphere at all. Nor does it account for the neutral sheet in the geomagnetic tail. The Chapman-Ferraro theory, however, appears applicable in a limited manner to some parts of the magnetopause as will be demonstrated. Let us now apply equations 8 and 9 of Case I to the low latitude boundary layer.

The Boundary Layer Current

Consider an idealized low latitude boundary layer in which the flow of solar wind plasma is laminar and parallel to the boundary. Let the velocity, pressure and density be functions of distance from the magnetosheath. The magnetopause is considered to be a tangential discontinuity of the magnetic field.

Figures 4a and 4b illustrate the situation in cross section in the dawn-side and dusk-side boundary layer. This configuration conforms to Case I. We realize from equation 8 that the dielectric current is sunward in the dusk side and anti-sunward in the dawn side. The presence of these currents has not hitherto been taken into account in the problem of solar wind interaction with the magnetosphere. Figure 5 illustrates the dielectric current in equatorial cross section. The current due to $-\nabla p$ is not illustrated. The dielectric current flows in the opposite direction to the classical Chapman-Ferraro currents which are caused by the gradient of plasma pressure. The dielectric current is a dielectric current of type I. The complete current orthogonal to $\underline{B}$ in such flows leads to equation 9 and is specified by equation 8.

This theory calls into question the widespread view that the geomagnetic field is separated from the interplanetary medium solely by a current layer caused by the gradient pressure of solar wind normal to the current layer. However, the theory appears to be consistent with observations of the outer regions of the geomagnetic field where it interacts with the solar wind. Taking the ion plasma parameters for a typical dawn-dusk meridian in a sequence of steps through plasma

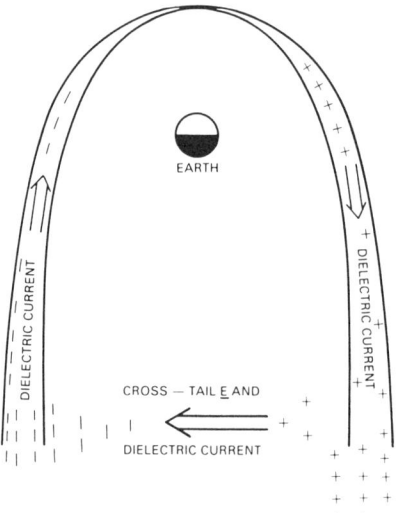

Fig. 5. Scheme of dielectric currents in low latitude boundary layer, and tail. Currents due to ∇p not illustrated.

Table I

Plasma Parameter	Magnetosheath	PBL 1	PBL 2	PBL 3	PBL 4	Magnetosphere
$n(cm^{-3})$	18	14	10	6	2	0.5
$V(km\ s^{-1})$	300	280	200	120	50	<30
$\overline{E}$ (kev)	0.5	1	2.5	4	6.5	7
$\frac{1}{2} nmV^2$ (kev cm^{-3})	9.7	5.3	2.0	0.4	0.02	<0.003
$p\ (=\frac{2}{3} n \overline{E})$ (kev cm^{-3})	6	9.3	16	16	8	2.4

boundary layer (Eastman 1978) Table 1 is constructed.

In this table $\overline{E}$ is the mean energy per particle. Therefore there is a substantial dielectric current to deal with. There is a decrease of $\frac{1}{2} \rho V_{sw}^2$ (where V_{sw} is the plasma velocity and ρ the density) from the solar wind into the boundary layer.

It is evident from Table I (a) that the component of current in the layer due to the gradient of (εE^2) is of comparable magnitude to those due to gradients of pressure; (b) the component due to the gradient of pressure adds to that due to $\nabla(\varepsilon E_\perp^2)$ in the outer section of the plasma boundary layer; (c) in the inner section of the layer the current due to gradient of pressure is in the direction of the classical Chapman-Ferraro current.

The strength of the dielectric current in the boundary layer can be estimated from equation 15 assuming $E = c^{-1} V_{sw} B$.

Then, per cm along B, integrated through the layer it is estimated by

$$j^E\ (cm^{-1}) = \frac{c}{2} \rho V_{sw}^2 B^{-1}. \quad (17)$$

The total dielectric current (J_T) is given by integrating above and below the equatorial plane wherever the boundary layer exists. This is not known, but within a factor of 2 it is estimated to be 10 R_E above and below the equatorial plane. So

$$J_T^E \approx 10\ R_E\ c\rho V_{sw}^2 B^{-1} \quad \text{c.g.s.} \quad (18)$$

Table II shows values of J_T^E for a variety of values of V_{sw}, B and n_s where $\rho = n_s m_s$ and m_s is assumed = mass of H atom.

These currents are geophysically significant. They exert an inwards force on the magnetosphere and would contribute to a decrease of the geomagnetic field earth-ward of them, and an increase of magnetic induction outside the layer in the solar wind. Together with the tail current (see later in the paper) these currents cause a perpetual "ring" current which should manifest itself at the earth's surface as a depression of the geomagnetic field. All currents in the earth environment need to be taken into account in the search for an effect, including field-aligned currents, otherwise known as Birkeland currents. Equation 9 is derived on the assumption that the fields and medium parameters do not vary in the x direction. In the low latitude boundary layer we find that they do vary. This would mean that current generated locally according to equation 18 would have divergence. It is now suggested that the current in the low latitude boundary layer finds continuity partly with Birkeland currents into the polar ionosphere in the day a.m. sector and out in the day p.m. sector. Later the continuity with electric current across the geomagnetic tail will be discussed. Consider the quantity

$$\text{div}\ J_T^E \approx 10\ R_E c V_{sw}^2\ \text{div}(\frac{\rho}{B}), \quad (19)$$

we note that the boundary layer inside the geomagnetic field allows solar wind plasma to expand along the geomagnetic field towards the ionosphere. It follows that (ρB^{-1}) decreases with distance from the "nose" of the magnetosphere. Approximately we can say that

$$\frac{\rho}{B} = \frac{M_T}{V_T B} = \frac{M_T}{\ell_T A_T B}, \quad (20)$$

where M_T, V_T, ℓ_T, A_T, B = mass content, volume,

TABLE II

n_s (cm^{-3})	V_{sw} (kms^{-1})	B (γ)	J_T^E (amps)
10	400	40	4.5×10^6
10	400	10	1.8×10^7
5	800	40	9×10^6

length, cross sectional area and magnetic field (at equatorial plane) in a tube of magnetic field. Given M_T and $A_T B$ (tube flux) constant during drift along the boundary layer, div $\underline{J_T}$ is negative. It follows that Birkeland current must flow along the geomagnetic field to the pre-noon sector of the polar ionosphere to which field lines from the magnetopause are connected (see Fig. 6), there to be continued by ionospheric current. By symmetry Birkeland current would flow away from the polar ionosphere in the post-noon sector. It is suggested here that the part of this dielectric current which remains after flow along the flanks of the magnetosphere becomes continuous with current across the tail (see later). Birkeland current must flow along the geomagnetic field to the pre-noon sector of the polar ionosphere to which field lines from the magnetopause are connected (see Fig. 6), there to be continued by ionospheric current. By symmetry Birkeland current would flow away from the polar ionosphere in the post-noon sector. It is suggested here that the part of this dielectric current which remains after flow along the flanks of the magnetosphere becomes continuous with current across the tail (see later in the paper).

The strength of the current per cm of longitude into the auroral ionosphere can be estimated by

$$j_\parallel \,(\text{cm}^{-1}) = f \, \text{div} \, j_T^E , \qquad (21)$$

where f is the separation at the boundary layer of the lines of force which are 1 cm apart in longitude near the cusp at the ionosphere. Approximately $f = 15$. Estimating the length scale of variation of ρ/B around the boundary using equation 21, a value of 10 R_E is adopted. Then, at the ionosphere,

$$j_\parallel \,(\text{cm}^{-1}) \approx \frac{15 c V_{sw}^2 \rho}{B} , \qquad (22)$$

where the parameter ρ and B are of course boundary layer values. With $B = 40\gamma$, $V_{sw} = 4 \times 10^7$ cm sec^{-1}, $\rho = 10 \times 1.6 \times 10^{-24}$ gm cm^{-3}, $j_\parallel = 1.0 \times 10^{-2}$ amp cm^{-1}. Such currents are in the range actually observed over the polar ionosphere (Sugiura and Potemra, 1976).

Let us return for a moment to the Chapman-Ferraro currents. In the region around the noon-midnight plane of the magnetopause the gradient of pressure of the solar wind may be considered as the dominant non-magnetic force and reflection of solar wind particles would cause a surface current which is towards dusk, equatorwards of the geomagnetic cusps and towards dawn, polewards of the cusps (see Fig. 7). Outside of this region the term $\varepsilon E_y^2/8\pi$ appears to play a progressively more significant role increasing up to a substantial fraction of p. This is suggested by the work of Eastman (1979), who shows that in the numerous low latitude boundary layer crossings reported generally $\frac{1}{2} \rho V_{sw}^2$ is between one half and one third of p. In the present context we estimate

$$\frac{1}{2} \rho V_{sw}^2 \approx \varepsilon E_y^2/8\pi . \qquad (23)$$

It is clear that the magnitude of Chapman-Ferraro currents near the noon plane will dominate the dielectric current elsewhere. This is so because the reflection of solar wind particles in that region is equivalent to a gradient of pressure which, integrated across the region gives an equivalent pressure (p) of approximately $2 \rho V_{sw}^2$. The region in which reflection is the dominant interaction of the solar wind with the magnetosphere is here called the "median" strip (illustration in Fig. 7).

Grad p current exists completely around the magnetosphere in the layer with density given by

$$\underline{j}_{\nabla p} = \frac{\underline{B} \times \underline{\nabla} p}{B^2}$$

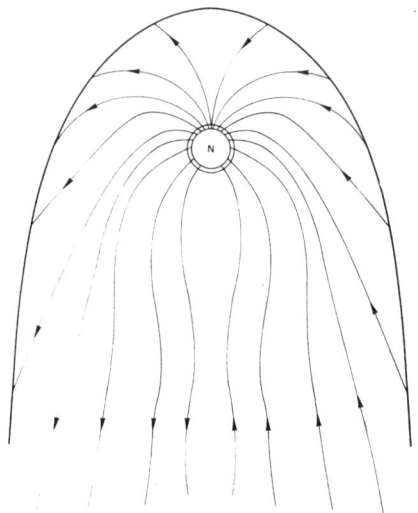

Fig. 6. Schematic representation of Birkeland currents produced by divergence of currents in low latitude boundary layer and geotail.

Integrated throughout the boundary layer it is estimated by

$$j_{T\nabla p} \approx \frac{p_{sw}}{B_{sw}} \quad (24)$$

where p_{sw} and B_{sw} are values in the magnetosheath rather than the boundary layer. In the magnetosheath, along the flanks of the magnetosphere, the flow is expected to be approximately laminar with no variation of p_{sw}/B_{sw} and therefore no divergence of $j_T(\nabla p)$. It is suggested that near the "nose", spatial variation of p_{sw}/B_{sw} may be such that the ratio increases away from the nose. This would form field-aligned (Birkeland) currents towards the ionosphere in the a.m. sector and opposite in the p.m. sector. Also the Birkeland currents would tend to cause a depression of the magnetopause surface in the "median" strip between the cusps. However, polewards of the cusps, the Birkeland current from the Chapman-Ferraro current would reinforce that from the dielectric current, producing a dawn-to-dusk electric field at ionospheric levels and at intermediate heights in the magnetosphere.

Historically, the increase of geomagnetic field at the earth's surface at the commencement of some magnetic storms has been interpreted in terms of an increase of the Chapman-Ferraro currents at the magnetopause particularly on the day side. This explanation would appear to be qualitatively valid in the present theory, however, the effect is complicated by the diminishing effect of dielectric currents and ∇p currents in the opposite direction to the conventional ∇p current of Chapman and Ferraro.

In the region of the cusp of the magnetosphere and the median strip polewards of the cusp it is conceivable that a mixture of Chapman-Ferraro current and dielectric current could flow if there is penetration of solar wind plasma there with a component of velocity orthogonal to $\underline{B}$. In this "zone of confusion" one could expect considerable variability in these currents and associated Birkeland currents. This is suggested as the origin of the irregular Birkeland currents observed at ionospheric altitudes at the site of the dayside cusp. (Ledley and Farthing, 1974).

The "Distant" Cross-Tail Current

Far downstream from the earth at distances greater than about 15 R_E the geotail is known to contain an almost neutral sheet and attempts to explain it have been in two classes. The first, exemplified by the work of Harris (1962) employs a set of "hot" particles only and essentially produces the result, on a line orthogonal to the sheet,

$$8\pi p + B_z^2 = \text{constant}. \quad (25)$$

The second approach exemplified by Alfvén (1968),

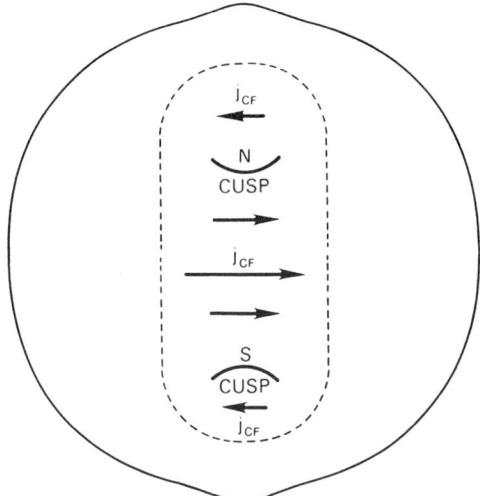

Fig. 7. Illustration of the "median" strip enclosed by dashed curve on day side magnetopause.

works with cold plasma, and essentially produces the result

$$B_z^2 + \varepsilon E_y^2 = \text{constant} \quad (26)$$

In a sense both approaches are valid but the most general approach combines the two, and is embodied in equation 13, viz.,

$$8\pi p + B_z^2 + \varepsilon E_y^2 = \text{constant}, \quad (27)$$

where E_y is now the cross-tail electric field, B_z the geotail magnetic field, and p the pressure of the entrapped plasma perpendicular to B_z.

It is proposed here that the cross-tail electric field is created under "ordinary" conditions by the dielectric current in the flanks of the low latitude boundary layer of the magnetosphere. Any mismatch of the current down the flanks and current across the tail would be taken up by Birkeland current at their junctions discharging through the polar ionosphere.

The current density across the tail from the dawn side to the dusk side is given by

$$j_y = -\frac{c}{B}\frac{\partial p}{\partial x} - \frac{1}{8\pi}\frac{c}{B}E_y^2\frac{\partial \varepsilon}{\partial x}. \quad (28)$$

The component of j_y, per unit of tail length, may be estimated as,

$$j_y (\text{cm}^{-1}) \approx \frac{5B}{2\pi} \text{ amps cm}^{-1}. \quad (29)$$

So that, with $B = 10\ \gamma$, $j_y (\text{cm}^{-1}) = 8 \times 10^{-5}$ amp cm^{-1}. For a tail of scale length 40 R_E, the total cross tail current from this cause would be 2×10^6 amps.

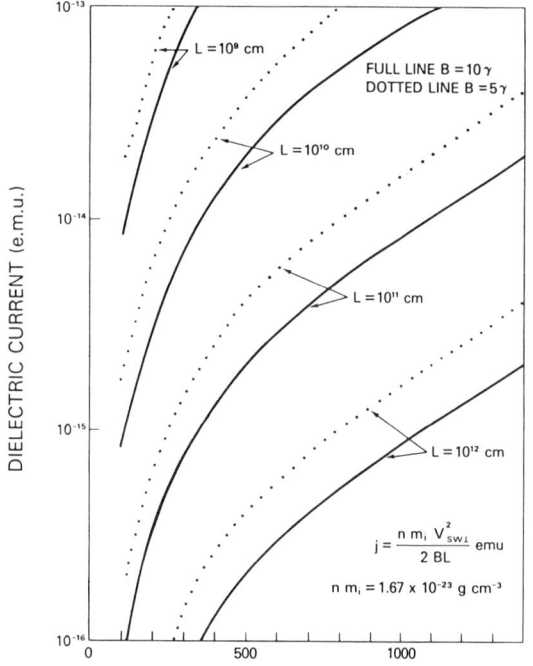

Fig. 8. Curves of measure of dielectric current in interplanetary structures assuming $n\, m_i = 1.67 \times 10^{-23}$ g cm^{-3}. L = scale length across $\underline{B}$ of structure.

By a similar argument made in the case of the divergence of boundary layer currents to obtain Birkeland current, a case is now made for such currents derived from the divergence of cross-tail current using equation 29. This equation gives the locally produced current in the tail. It is assumed that changes of local parameters across the tail in the dawn-dusk (y) direction lead to divergence of j_y which leads to Birkeland current. Firstly, let us note the symmetries of the terms in j_y. Both $\partial p/\partial x$ and $\partial \varepsilon/\partial x$ are expected to point away from the neutral sheet, and since B reverses sign there, divergence of j_y will not change sign with change of position above or below the sheet. However, the divergence of j_y would be expected to change sign depending on whether the position is on the dawn side or the dusk side of the center line of the sheet, if there is symmetry of the local parameters about this line. It is expected on average that at any given distance X from the neutral sheet, $\frac{\partial p}{\partial x}$, $\frac{\partial \varepsilon}{\partial x}$ and B would each increase in magnitude from the central meridian plane of the tail to the magnetopause. The divergence of j_y would then be such as to produce Birkeland current towards the polar ionospheres from the "dawn" half of the tail, and from the ionospheres in the "dusk" half. However, great departures from this simple picture may be expected because of the dependence of div j_y on $\partial p/\partial x$, $\partial \varepsilon/\partial x$ and B. This perhaps accounts for the complexity of Birkeland currents near the "Harang" discontinuity.

The divergence of tail current may be estimated using equation 29, by

$$\text{div } j_y \approx f_T \frac{5BL}{2\pi W} \approx j_\parallel \text{ (amps cm}^{-1}\text{)}, \quad (30)$$

where W, L = half width, length of tail $\approx 15\, R_E$, and f_T = separation in the tail of two lines of force 1 cm apart in longitude near the midnight auroral belt. Therefore, for $B \approx 10\gamma$ and $f_T \approx 20$, the Birkeland current per cm to each polar ionosphere would be $\approx 1.2 \times 10^{-3}$ amps cm^{-1}. Birkeland currents of this magnitude are observed in the night time auroral belt (Potemra et al., 1979).

This distribution of Birkeland electric current would also be compatible with Sugiura's (1975) inference from observations of magnetic fields regarding the connection of Birkeland currents near the polar ionosphere with those from the tail. These currents would cause the magnetic field lines to be splayed outwards away from the earth between them and the neutral sheet and to be oppositely splayed between the currents and the tail mantles of the magnetosphere. There would be a tendency for the field-aligned currents in the tail to produce a component of $\underline{B}$, orthogonal to the neutral sheet, between the neutral sheet and the mantles, pushing the magnetic field out at the middle of the tail mantles.

The electric field across the tail from dawn-side to dusk-side, will cause the drift of plasma towards the earth and the consequent energization of trapped plasma; production of partial ring current and Birkeland currents at places of divergence of the partial ring current (Harel et al., 1981). It is not the purpose of this paper to examine these internal processes of the magnetosphere but to discuss principally the external origin of the Birkeland currents, and the dawn-dusk electric field.

Additional Currents from the Interplanetary Medium

It is clear that equation 1 predicts significant electric currents in the interplanetary medium and that such currents may find conduction paths through the geomagnetic plasmas if the magnetosphere encounters them. The strength of these currents will now be estimated and their interaction with the geomagnetic field discussed. The currents of interest are those associated with shears in the flow of interplanetary plasma and in neutral sheets. These are examples of types I and II treated earlier in this paper. Both of these cases may apply to neutral sheets. In type I the electric field is perpendicular to both B and the neutral sheet. In type II the electric field is perpendicular to B but parallel to the neutral sheet.

In interplanetary space it is generally expected that $p \ll \varepsilon E_y^2$ so the current can be estimated in magnitude in both cases by

$$j \approx \frac{1}{8\pi} \frac{c}{B} \frac{\varepsilon E^2}{L}, \quad (31)$$

where L is the scale of variation of a critical parameter ε or E^2 of the system. When $L > c^2 E m_i/Be_y$, the plasma drift velocity is cE/B (Cole 1983), so we can say, in that case,

$$j \approx \frac{1/2 n m_i V_{sw}^2 c}{B L}, \quad (32)$$

where V_{sw} is the component of the solar wind orthogonal to $\underline{B}$. Fig. 8 shows a range of values of j for a variety of values of V_{sw}, B and L. Assuming $n = 10$ cm^{-3}. There may be occasions, too, when currents due to pressure are significant.

It is assumed that an interplanetary current, if intersected by the magnetosphere, will find continuity across the magnetopause with field aligned current to the polar ionosphere and out again, or may cross the tail of the earth via the neutral sheet. This would be done by a build up of space charge at the boundaries to adjust the electric fields to give current continuity (see Fig. 9). This seems a reasonable proposition.

Suppose the effective area of the magnetosphere accessible to the interplanetary current is $(15 R_E)^2$, if $L > 15 R_E$ and $15 L R_E$ if $L < 15 R_E$. Then we see that for the structures characterized in Fig. 10, may produce currents through the magnetosphere of order $10^6 - 10^7$ amp. Now observations of magnetic bays (Cole 1962a) show that their durations range from about 20 mins. to about 200 ecliptic in the direction $\pm \underline{E} \times \underline{B}$ if $\nabla(\varepsilon E_\perp^2)$ is in the $\pm z$ direction (respectively).

In type II the direction of the current is in the direction of $\underline{E}$ for a gradient of $\varepsilon E_\perp^2$ in the plane of the ecliptic and $\underline{j}$ will be in the $\pm z$ direction depending on whether $\underline{B}$ is away from or towards the sun. In this case we should expect current to flow through the magnetosphere from N to S or vice versa depending on whether $\underline{B}$ is towards or away from the sun, producing significant hemispherical asymmetries in magnetic disturbance as observed at the earth's surface.

Generally for type I currents (i.e., corresponding to Fig. I geometry, if $\nabla(\varepsilon E_\perp^2)$ is in the direction of $\pm \underline{E}$, $\underline{j}$ is in the direction of $\pm \underline{E} \times \underline{B}$ respectively, and if $L \gg mE c^2/Be$, we can say that the direction of $\underline{j}$ is $\pm$ that given by equation 36. Whereas, for type II currents, if $\nabla(\varepsilon E_\perp^2)$ is in the direction of $\pm \underline{E} \times \underline{B}$, $\underline{j}$ is in the direction of $\pm \underline{E}$ respectively, i.e., as specified by $\pm$ that given by equation 35.

The amount of current converted to Birkeland current from the tail would depend upon the angle (α) between the current direction in interplanetary space and the direction defined by the vector cross product of the geomagnetic axis direction and the normal to the geomagnetic neutral sheet. However, Birkeland current from the day sectors of the magnetopause would be

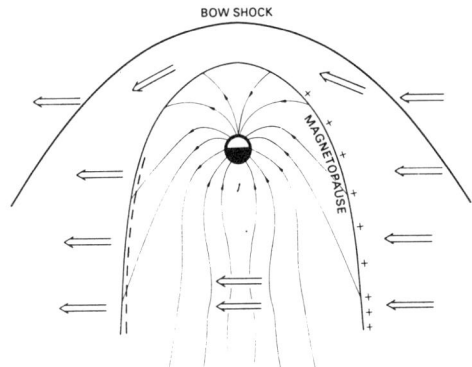

Fig. 9. Illustration of interplanetary currents impacting on the magnetosphere. They are an additional source of Birkeland currents inside the magnetosphere.

mins. though some are outside these values. By assuming these bays are caused by changes to the magnetosphere/ionosphere system brought about by interplanetary currents flowing through the system, it is inferred that structures of scale length greater than about 5×10^{10} cm have sufficient current to account for geomagnetic bays.

Let us now consider the morphology of large scale interplanetary structures corresponding to types I and II. Take a system of cartesian coordinates in which x is radially outwards from the sun, y is azimuthal and z perpendicular to the plane of the ecliptic. Let the solar wind velocity be given by

$$\underline{V}_{sw} = (V_x, V_y, V_z). \quad (33)$$

Let
$$\underline{B} = (B_x, B_y, B_z). \quad (34)$$

For $L \gg m \frac{Ec^2}{Be}$,

$$\underline{E} = (V_y B_z - V_z B_y, V_z B_x - V_x B_z, V_x B_y - V_y B_x),$$

$$\approx 0, -V_x B_z, V_x B_y, \quad (35)$$

and, $\underline{E} \times \underline{B} = V_x(B_x^2 + B_y^2), V_x B_y B_x, V_x B_z B_x. \quad (36)$

The magnitude of the current j is then approximated by

$$j \approx \frac{\varepsilon V_{sw}^2 (B_y^2 + B_z^2)}{8\pi L Bc},$$

$$= \frac{c M^2 (B_y^2 + B_z^2)}{8\pi L B}, \quad (37)$$

where M is the ratio of the solar wind speed to the Alfvén speed.

In type I the direction of the current is given by $\pm \underline{E} \times \underline{B}$ depending on whether $\nabla(\varepsilon E^2)$ is in the same or opposite direction to $\underline{E}_\perp$ (Cole 1976). In this case, $\underline{E}$ is in the + z direction if $\underline{B}$ is away from the sun and $\underline{j}$ will be in the plane of the ecliptic in the direction $\pm \underline{E} \times \underline{B}$ if $\nabla(\varepsilon E_\perp^2)$ is in the $\pm$ z direction (respectively).

In type II the direction of the current is in the direction of $\underline{E}$ for a gradient of $\varepsilon E_\perp^2$ in the plane of the ecliptic and $\underline{j}$ will be in the $\pm$ z direction depending on whether $\underline{B}$ is away from or towards the sun. In this case we should expect current to flow through the magnetosphere from N to S or vice versa depending on whether $\underline{B}$ is towards or away from the sun, producing significant hemispherical asymmetries in magnetic disturbance as observed at the earth's surface.

Generally for type I currents (i.e., corresponding to Fig. 1 geometry, if $\nabla(\varepsilon E_\perp^2)$ is in the direction of $\pm \underline{E}$, $\underline{j}$ is in the direction of $\pm \underline{E} \times \underline{B}$ respectively, and if $L \gg mE c^2/Be$, we can say that the direction of $\underline{j}$ is $\pm$ that given by equation 36. Whereas, for type II currents, if $\nabla(\varepsilon E_\perp^2)$ is in the direction of $\pm \underline{E} \times \underline{B}$, $\underline{j}$ is in the direction of $\pm \underline{E}$ respectively, i.e., as specified by $\pm$ that given by equation 35.

The amount of current converted to Birkeland current from the tail would depend upon the angle (α) between the current direction in interplanetary space and the direction defined by the vector cross product of the geomagnetic axis direction and the normal to the geomagnetic neutral sheet. However, Birkeland current from the day sectors of the magnetopause would be controlled more by the angle (δ) between the current direction and the geomagnetic axis. The precise description of the amount of Birkeland current generated by divergence of interplanetary currents depends on the configuration of the draping of interplanetary magnetic induction lines over the magnetopause. The detailed configurations of the plasma in this coupling process will be the subject of future work.

Recently Iijima and Potemra (1982) correlated a variety (20 in number) of interplanetary quantites with morning and afternoon Birkeland current densities and found that some are better correlated than others. One of the best correlations is with the quantity

$$\left[nV_{sw} (B_y^2 + B_z^2) \sin \theta/2\right]^{1/2}$$

where θ is the angle between the geomagnetic axis and the z-direction. This interplanetary quantity is clearly closely related "mathematically" to the current $\underline{j}$ defined by equation 37, although its origin in physical discussion is different.

The theory of the present paper would explain a permanent Birkeland current system associated with divergence of dielectric current in the boundary layer and the geotail plus a variable component associated with interplanetary parameters via equation 37 and the angles α and δ. To test this in the manner of Iijima and Potemra (1982) it would be preferable to use total Birkeland current per cm of auroral oval rather than current density.

The Bow Shock

That part of the bow shock which is called a quasi-perpendicular shock resembles the conditions for type II in the foregoing; while in the regions in which it is quasi-parallel, the conditions are those like type I. Since the thermal pressure of solar wind plasma is $\ll \frac{1}{2} \rho V_{sw}^2 (\approx \varepsilon E^2)$ it is clear that the electric currents in the bow shock are dielectric currents of the kind discussed in this paper, viz., type II currents in quasi-perpendicular shocks and type I currents in quasi-parallel shocks. The size of the currents in the bow shock are estimated to be of the order of $10^6 - 10^7$ amps and represent another potential source of Birkeland currents from them to and through the magnetosphere to the ionosphere. This could happen if there were field lines connected to the shock, at places of divergence of the dielectric current draping over the magnetosphere.

Discussion

Having applied the unabridged Maxwell equations to the discussion of the electrodynamics of plasmas (Cole 1983) new "dielectric" currents have been predicted in the magnetosphere and interplanetary space to be of geophysical significance. The low latitude boundary layer is the site of one such current in the direction $\underline{E} \times \underline{B}$ (type I) caused by an electric field ($\underline{E}$) orthogonal to $\underline{B}$ accompanied by a gradient of εE^2 in the direction of $\underline{E}$. Divergence of this current produces Birkeland current to the polar ionosphere and the remainder is continuous with current across the geomagnetic tail. In the tail, the cross-tail electric field is accompanied by a gradient of εE^2 perpendicular to $\underline{E}$ and $\underline{B}$ producing a current in the direction of $\underline{E}$. In addition there is cross-tail current caused by $-\nabla p$. Divergence of the cross-tail field appears to be responsible for the Birkeland currents to the auroral region of the ionosphere which were inferred to exist by Sugiura (1975).

In addition to these currents, dielectric currents of origin in interplanetary plasma structures may be conducted across the magnetopause through the magnetosphere by Birkeland currents, and also across the tail by alteration of the cross-tail electric field. The paper does not address the problem of how interplanetary currents negotiate the bow shock to enter the magnetosphere.

The theory so far developed using this new approach to collisionless plasmas (Cole, 1983) has been applied only to situations in which $\underline{B}$ is unidirectional. To develop the theory further one needs to express the "μ" of the plasma in terms appropriate to general magnetic induction $\underline{B}$, i.e., to allow for effects of "parallel" pressure of plasma in addition to "perpendicular" pressure to which this paper is presently restricted.

Acknowledgement. The author is indebted to the Commission on Human Resources of the National Academy of Sciences, U.S.A. for the award of a Research Associateship and Goddard Space Flight Center for hospitality.

References

Alfvén, H., Some properties of magnetospheric neutral surfaces, J. Geophys. Res., 73, 4379, 1968.

Axford, W.I. and C.O. Hines, A unifying theory of high-latitude geophysical phenomena and geomagnetic storms, Can. J. Phys., 39, 1433, 1961.

Cole, K.D., A dynamo theory of the aurora and magnetic disturbance, Aust. J. Phys., 13, 484, 1960.

Cole, K.D., On solar wind generation of polar geomagnetic disturbances, Geophys. J. R.A.S., 6, 103, 1961.

Cole, K.D., Magnetic bays at Macquarie Island, Aust. J. Phys., 15, 277, 1962.

Cole, K.D., Outline of a theory of solar wind interaction with the magnetosphere, Planet. Space Sci., 22, 1075, 1974.

Cole, K.D., Physical argument and hypothesis for sun-weather relationships, Nature, 260, 229, 1976.

Cole, K.D., Dielectric and permeability effects in collisionless plasmas (companion paper, this volume), 1983.

Cowley, S.H.W., A self-consistent model of a simple magnetic neutral sheet system surrounded by cold, collisionless plasma, Cosmic Electrodynamics, 3, 448, 1973.

Dungey, J.W., Interplanetary magnetic field and the auroral zones, Phys. Rev. Letters, 6, 47, 1961.

Eastman, T.E., Ph.D. Thesis, Univ. of California Document LA-7842-T, UC-34b, 1979.

Eastman, T.E., E.W. Hones, Jr., Characterstics of the magnetospheric boundary layer and magnetopause layer as observed by IMP 6, J. Geophys. Res., 84, 2019, 1979.

Harel, M., R.A. Wolf, P.H. Reiff, R.W. Spiro, W.J. Burke, F.J. Rich, and M. Smiddy, Quantitative simulation of a magnetospheric substorm 1. Model logic and overview, J. Geophys. Res., 86, 2217, 1981.

Harris, E.G., On a plasma sheath separating regions of oppositely directed magnetic field, Il. Nuovo Cimento, XXIII, 115, 1962.

Iijima, T. and T.A. Potemra, The relationship between interplanetary quantities and Birkeland current densities, Geophys. Res. Letters, 9, 442, 1982.

Ledley, B.G. and W.H. Farthing, Field-aligned current observations in the polar cusp ionosphere, J. Geophys. Res., 79, 3124, 1974.

Piddington, J.H., A theory of polar geomagnetic storms, Geophys. J. R.A.S., 65, 93, 1960.

Piddington, J.H., A hydromagnetic theory of geomagnetic storms and auroras, Planet. Space Sci., 9, 947, 1962.

Potemra, T.A., T. Iijima and N.A. Saflekos, Large-scale characteristics of Birkeland currents, in "Dynamics of the Magnetosphere," ed. by S.-I. Akasofu, D. Riedel, Dordrecht, Holland, p. 3, 1979.

Sugiura, M. and T.A. Potemra, Net field-aligned currents observed by TRIAD, J. Geophys. Res., 81, 2155, 1976.

Sugiura, M., Identification of the polar cap boundary and the auroral belt in the high-altitude magnetosphere: A model for the field-aligned currents, J. Geophys. Res., 80, 2057, 1975.

Thiele, B. and H.M. Praetorius, Field-aligned currents between 400 and 3000 km in auroral and polar latitudes, Planet. Space Sci., 21, 179, 1973.

Willis, D.M., Structure of the magnetopause, Rev. Geophys., 9, 953, 1971.

DIELECTRIC AND PERMEABILITY EFFECTS IN COLLISIONLESS PLASMAS

K. D. Cole[1]

NASA/Goddard Space Flight Center, Greenbelt, Maryland 20771

Abstract. In a steady state, it is found that spatially varying energy density of the electric field (E) orthogonal to B produces electric current, given under certain conditions, by

$$\underline{j} = [\underline{\nabla}(\varepsilon E^2)] \times \underline{B}/8\pi B^2,$$

where ε is the dielectric constant of the plasma for fields orthogonal to $\underline{B}$. The current is significant in geophysical and astrophysical plasmas.

Introduction

In this paper, macroscopic processes slow (compared to the ion gyro-period) in a collisionless plasma are considered from the standpoint of Maxwell's unabridged equations including all four vectors $\underline{D}$, $\underline{E}$, $\underline{B}$ and $\underline{H}$. This is in contrast to the widespread approach which employs $\underline{E}$ and $\underline{B}$, but not $\underline{D}$ and $\underline{H}$.

The paper starts with Maxwell's unabridged equations and states, as is well known, how these are often abbreviated for use in plasma physics. After uncovering new effects in collisionless plasma using general arguments from Maxwell's equations, the approach of this paper is placed in context in plasma physics, and is shown how it extends our understanding of plasmas.

Maxwell's equations as commonly written are, in c.g.s. units,

$$\underline{\nabla} \times \underline{E} = -\frac{1}{c}\frac{\partial \underline{B}}{\partial t}, \quad (1)$$

$$\underline{\nabla} \times \underline{H} = \frac{1}{c}\frac{\partial \underline{D}}{\partial t} + \frac{4\pi \underline{J}}{c} \quad (2)$$

where $\underline{E}$ = electric field, i.e., electric force per unit charge,
$\underline{H}$ = magnetic field,
$\underline{B}$ = magnetic induction = $\underline{H} + 4\pi\underline{M}$, (3)
$\underline{M}$ = magnetic dipole moment per unit volume (magnetization vector),
$\underline{D}$ = the electric displacement = $\underline{E} + 4\pi\underline{P}$, (4)

---

[1] On leave from La Trobe University, Bundoora, Victoria, Australia, 3083.

$\underline{P}$ = electric dipole moment per unit volume (polarization vector),
$\underline{J}$ = total conduction current excluding displacement current (see Maxwell, 1954).

Following Maxwell,

$$\underline{\nabla} \cdot \underline{D} = 4\pi\rho. \quad (5)$$

From equation 1, $\frac{\partial}{\partial t}(\underline{\nabla} \cdot B) = 0$, from which it is usually assumed that (Stratton, 1941)

$$\underline{\nabla} \cdot B = 0 \quad (6)$$

From equation 2 it follows that $4\pi \,\text{div}\,\underline{J} + \frac{\partial}{\partial t}\,\text{div}\,\underline{D} = 0$, and

$$\underline{\nabla} \cdot \underline{J} + \frac{\partial \rho}{\partial t} = 0 \quad (7)$$

Either one accepts 5 to infer 7 or uses the conservation of charge (equation 7) to infer 5 (Stratton, 1941).

Maxwell's equations may be presented in other forms (Stratton, 1941) e.g.,

$$\underline{\nabla} \times \underline{E} = -\frac{1}{c}\frac{\partial \underline{B}}{\partial t}, \quad (8)$$

$$\underline{\nabla} \times \underline{B} = \frac{1}{c}\frac{\partial \underline{E}}{\partial t} + \left(\frac{4\pi}{c}\frac{\partial \underline{P}}{\partial t} + 4\pi\,\underline{\nabla} \times \underline{M} + \frac{4\pi}{c}\underline{J}\right), \quad (9)$$

$$\underline{\nabla} \cdot \underline{B} = 0 \quad (10)$$

$$\underline{\nabla} \cdot \underline{E} = 4\pi\rho - 4\pi\,\underline{\nabla} \cdot \underline{P} \quad (11)$$

In many plasma physics books and papers (e.g., Spitzer, 1962) instead of equation 9, there is written

$$\underline{\nabla} \times \underline{B} = \frac{1}{c}\frac{\partial \underline{E}}{\partial t} + \frac{4\pi}{c}\underline{j}, \quad (12)$$

where, therefore,

$$\underline{j} = \frac{\partial \underline{P}}{\partial t} + c\,\underline{\nabla} \times \underline{M} + \underline{J} \quad (13)$$

and instead of equation 11 is written

$$\underline{\nabla} \cdot \underline{E} = 4\pi \rho_E, \qquad (14a)$$

where, therefore,

$$\rho_E = \rho - \underline{\nabla} \cdot \underline{P} \qquad (14b)$$

for a plasma consisting of electrons (−) and one kind of ion (+)

$$\underline{j} = q_s(N^+\underline{V}^+ - N^-\underline{V}^-), \; j = \Sigma \; q_s \; N_s \underline{V}_s \qquad (15a)$$

$$\rho_E = (N^+ - N^-)e, \; = \Sigma \; q_s \; N_s$$

$$\underline{V}_s = \frac{1}{N_s} \int \underline{v} \; f_s(\underline{r}, \underline{v}, t) \; d^3v, \qquad (15b)$$

and

$$\underline{N}_s = \int f_s(\underline{r}, \underline{v}, t) \; d^3v. \qquad (16)$$

It is argued in texts on plasma physics that the velocity distribution f takes into account all the motions of all the particles and hence equation 14a and integral 14b gives the quantity $\underline{j}$, then using integral 15 to find N^+ and N^- provides the quantity ρ_E. Sometimes J is taken out of $\underline{j}$ (Clemmow and Dougherty, 1969). As expected

$$\text{div } \underline{j} = \frac{\partial}{\partial t}(\underline{\nabla} \cdot \underline{P}) + \text{div } \underline{J} \qquad (17)$$

$$= -\frac{\partial \rho_p}{\partial t} - \frac{\partial \rho}{\partial t}$$

where ρ_p = "equivalent" charge density due to polarization in agreement with equation 11. To solve problems of phenomena purely internal to plasma, the equations 12, 13, 14(a,b) and 15 are commonly used, putting $\underline{J} = 0$ and $\rho = 0$. Thus by these means one can learn the low frequency dielectric constant (cf. Spitzer 1962).

This paper treats a plasma in quasi-static conditions as a fluid with specifiable dielectric constant and magnetic permeability, using the unabridged Maxwell equations 1 to 7. The plasma is allowed to be driven from the outside by "applied" currents ($\underline{J}$) and charged distributions (ρ). This is in contrast to the present approach to plasma physics referred to above, which, in effect considers the charges and currents to be in a vacuum (Clemmow and Dougherty, 1969), and the equivalent information on ε and μ is contained in velocity distribution functions and appropriate equations of motion.

Consistency between the two approaches is demonstrated (see Section 3); however, this paper brings out simply new effects driven in a plasma by externally applied charge distributions which, in principle, could be quite arbitrary. Suspecting that many phenomena in the ionosphere, magnetosphere and interplanetary space are driven by outside sources we approach these phenomena theoretically in this way in the following discussion.

Electromagnetic Stress on a Dielectric Medium

Discussion in this section is restricted to quasi-static conditions, i.e., to variations slow compared to the ion gyroperiod. For slow variations of electric field orthogonal to $\underline{B}$, it is known that a plasma behaves like a dielectric medium. So that (Stix, 1962) for $\underline{B} = (0, 0, B_z)$

$$\underline{D} = (\varepsilon E_x, \varepsilon E_y, 0) \qquad (18)$$

where

$$\varepsilon = 1 + \frac{4\pi \rho_m c^2}{B^2} \qquad (19)$$

and

$$\rho_m = \sum_s n_s m_s. \qquad (20)$$

n_s, m_s = number density, molecular mass respectively of species s of charged particle in the plasma.

Before discussing the macroscopic forces on a plasma let us refer to the classical discussion of the forces on a volume (V) of fluid with isotropic dielectric constant (Stratton, 1941). He shows that, given $\underline{B} = \mu \underline{H}$,

$$\frac{d}{dt}(\underline{G}_{\text{mech}} + \underline{G}_e) =$$

$$\int_V [\rho \underline{E} + c^{-1} \underline{J} \times \underline{B} - \frac{1}{8\pi} E^2 \nabla \varepsilon - \frac{1}{8\varepsilon} - \frac{1}{8\pi} H^2 \nabla \mu$$

$$+ \frac{1}{4\pi c} \frac{\partial}{\partial t}(\underline{D} \times \underline{B})] \, dV, \qquad (21)$$

where $\underline{G}_{\text{mech}}$ is the mechanical momentum of the volume of fluid and $\underline{G}_e$ is the electromagnetic momentum in the volume (V). The components of the electric force $\rho \underline{E} - \frac{1}{8\pi} E^2 \nabla \varepsilon$ are given, in the case when $\underline{\nabla} \times \underline{E} = 0$ by

$$4\pi \rho E_j - \frac{1}{2} E^2 \frac{\partial \varepsilon}{\partial x_j}$$

$$= \sum_{k=1}^{3} \frac{\partial}{\partial x_k}(E_j D_k) - \frac{1}{2} \frac{\partial}{\partial x_j}(\underline{E} \cdot \underline{D}). \qquad (22)$$

In the condition of constant ($G_e + G_{\text{mech}}$) inside the volume V, equation 21 yields

$$(4\pi)^{-1} (\text{div } \underline{D}) \underline{E} + c^{-1} \underline{J} \times \underline{B}$$

$$- \frac{1}{8\pi} E^2 \underline{\nabla} \varepsilon - \frac{1}{8\pi} H^2 \nabla \mu = 0 \qquad (23)$$

In addition one can say that the component of $\underline{J}$ perpendicular to $\underline{B}$ is

$$\underline{J} = \frac{c \, \underline{F} \times \underline{B}}{B^2}, \quad (24)$$

where

$$4\pi \, \underline{F} = (\text{div } \underline{D}) \, \underline{E} - \frac{1}{2} E^2 \underline{\nabla}\varepsilon - \frac{1}{2} H^2 \underline{\nabla}\mu \quad (25)$$

Taking the stress tensor of equation 22 as general (Jones, 1964) and applying it to a collisionless plasma in which

$$\underline{B} = [0, 0, B_z(y)], \quad (26)$$

$$\underline{E} = [0, E_y(y), 0], \quad (27)$$

$$\varepsilon = \varepsilon(y), \quad (28)$$

it is readily shown, as expected, that the electric force is given by

$$4\pi \, F_y^{el} = (\text{div } \underline{D}) E_y - \frac{1}{2} E_y^2 \frac{\partial \varepsilon}{\partial y}, \quad (29)$$

Stratton (1941) points out that it is difficult to identify the mechanical and electric parts of the RHS of equation 21. However, in the case of plasma this appears straightforward as follows. Firstly let us expand the right hand side as:

$$\int_V [\rho \underline{E} + \frac{1}{c} \underline{J} \times \underline{B} - \frac{1}{8\pi} E^2 \underline{\nabla}\varepsilon$$

$$- \frac{1}{8\pi} H^2 \underline{\nabla}\mu + \frac{1}{c} \frac{\partial P}{\partial t} \times \underline{B} + \frac{1}{c} \underline{P} \times \frac{\partial \underline{B}}{\partial t}$$

$$+ \frac{1}{4\pi c} \frac{\partial}{\partial t} (\underline{E} \times \underline{B})] \, dV. \quad (30)$$

Then $\rho\underline{E} - \frac{1}{8\pi} E^2 \underline{\nabla}\varepsilon$ is the electric force; $\underline{J} \times \underline{B}$ the Lorentz force due to the current $\underline{J}$; $-\frac{1}{8\pi} H^2 \underline{\nabla}\mu$ is $-B^2 \underline{\nabla}(p/B^2)$, part of the force due to $\underline{\nabla}p$, see equation 38; $\frac{\partial P}{\partial t} \times \underline{B}$ is the Lorentz force due to the polarization current, which is responsible for accelerating the plasma; $\frac{1}{c} \underline{P} \times \frac{\partial \underline{B}}{\partial t}$ being $-\underline{P} \times \underline{\nabla} \times \underline{E}$ is an additional force exerted by the field $\underline{E}$ on the medium - this force has the dimensions, interestingly enough, of $(-\underline{\nabla} \cdot \underline{P}) \underline{E}$ but is different; notice that a force $(-\underline{\nabla} \cdot \underline{P}) \underline{E}$ does not appear in this approach, finally $\frac{1}{4\pi c} \frac{\partial}{\partial t} (\underline{E} \times \underline{B})$ is tentatively ascribed to the rate of increase of the momentum of the electromagnetic field (c.f. Chandrasekhar, 1960).

It may be noted that equation 21 was derived (Stratton 1941) under the condition that μ is a function of position but not of field intensity.

This is so in all applications in this paper. The value of μ in a 2-D collisionless plasma (see equation 68a) is independent of B for the reason that p/B^2 is constant as B is changed in such a plasma (Chandrasekhar, 1960).

Let us now investigate the steady state flow of plasma as prescribed by equations 26, 27 and 28.

The equations 9 and 24 yield

$$E_y \left[\frac{\partial}{\partial y} (\varepsilon E_y)\right] - \frac{1}{2} E_y^2 \frac{\partial \varepsilon}{\partial y}$$

$$- \mu H_z \frac{\partial H_z}{\partial y} - \frac{1}{2} H_z^2 \underline{\nabla}\mu = 0, \quad (31)$$

or

$$\frac{\partial}{\partial y} (\varepsilon E_y^2) - \frac{\partial}{\partial y} (\mu H_z^2) = 0. \quad (32)$$

So

$$\varepsilon E_y^2 - \frac{B_z^2}{\mu} = \text{constant}. \quad (33)$$

Equations 68 and 69 show that in the present special case

$$\mu = \frac{B_z}{H_z} = \frac{1}{1 + 8\pi p/B^2} \quad (34)$$

So equation 32 yields

$$\frac{\varepsilon E_y}{8\pi} - \frac{B_z^2}{8\pi} - p = \text{constant} \quad (35)$$

The Current Orthogonal to $\underline{B}$

Expressed generally by equations 25 and 26 the portion of $\underline{J}$ driven by electric field, hereafter called the dielectric current, is given by

$$\underline{J}_{el} = \frac{c[(\text{div } \underline{D}) \underline{E} - \frac{1}{2} E^2 \underline{\nabla}\varepsilon] \times \underline{B}}{4\pi B^2} \quad (36)$$

In the special case given by equations 26, 27 and 28, this current is in the x-direction and of magnitude

$$J_{el} = \frac{c \frac{\partial}{\partial y} (\varepsilon E_y^2)}{8\pi B}. \quad (37)$$

The total current density orthogonal to $\underline{B}$ is, for the same geometry,

$$J_x = \frac{c \frac{\partial}{\partial y} [\varepsilon E_y^2 - 8\pi p]}{8\pi B}. \quad (38)$$

Consistency With Conventional Approach

Earlier, one example of these currents was uncovered (Cole 1976) but their generality and applicability was not fully realized then. Consider the motion of a charged particle in an electric field and magnetic induction which are orthogonal. The equation of motion is

$$m_s \frac{dV_s}{dt} = q_s (\underline{E} + c^{-1} \underline{V} \times \underline{B}), \qquad (39)$$

where m_s, q_s = mass, charge of particles of species s, $\underline{V}_s$ = velocity of molecule of species s. Considering a cold plasma, and a spatially varying electric field which is switched on at time t = 0, it was shown (Cole 1976) that if (in Cartesian coordinates) $\underline{B} = (0, 0, B_z)$ and $\underline{E} = (0, E_o + (\nabla E) y, 0)$ then the particles drift in the x-direction with the speed

$$V_{DS} = \frac{\omega_s^2}{\Omega_s^2} \frac{Ec}{B}, \qquad (40)$$

where

$$\Omega_s^2 = \omega_s^2 - q_s \nabla E/m_s, \qquad (41)$$

and

$$\omega_s = B q_s/m_s c. \qquad (42)$$

This produces an electric current in the plasma in the x-direction given by

$$j_x = \sum_s n_s q_s \frac{\omega_s^2}{\Omega_s^2} \frac{Ec}{B} \qquad (43)$$

In the case where $q_s \nabla E/m_s \ll \omega_s^2$, noting that

$$\sum_s n_s q_s = \underline{\nabla} \cdot \underline{E}/4\pi = (4\pi)^{-1} \partial E/\partial y \qquad (44)$$

(in the present geometry, then (Cole 1983)

$$j_x = c \varepsilon (\nabla E^2)/8\pi B, \qquad (45)$$

where

$$\varepsilon = 1 + 4\pi \rho_m c^2/B, \qquad (46)$$

where

$$\rho_m = \sum_s n_s m_s. \qquad (47)$$

The current is ε times stronger than given by the $\underline{E} \times \underline{B}$ drift alone. It is common to assume $\underline{\nabla} \cdot \underline{E} = 0$ and hence this latter current to be zero. In the magnetosphere and interplanetary space ε is of order 10^5 or 10^6. The current (44) is not present in linearized moments of the Boltzmann equation such as in Spitzer (1962). The collisionless Boltzmann equation for a cold plasma is the same as equation (39). However, putting

$$\frac{dV_s}{dt} = \frac{\partial V_s}{\partial t} + \underline{V}_s \cdot \underline{\nabla} \underline{V}_s, \qquad (48)$$

then linearizing by dropping the term $\underline{V}_s \cdot \underline{\nabla} \underline{V}_s$, and then assuming a steady state ($\frac{\partial}{\partial t} = 0$), consequently $\underline{j} = 0$. This procedure loses all information about the current (43). Whereas, if we solve equation 48 prescribing E in advance, the current manifests itself (Cole 1976). Of course the current (43) is negligible in a host of plasma structures in the laboratory and the ionosphere, but we find it significant in plasmas in which $\nabla(\frac{1}{2}\rho_m V^2)$ is comparable with $\nabla(B^2/8\pi)$, as will be shown later in the paper.

In the circumstances in which $(\nabla \varepsilon)/\varepsilon \ll (\nabla E^2)/E^2$ equation 45 may be written

$$j_x = C\nabla(\varepsilon E^2)/8\pi B \qquad (49)$$

suggesting that j_x is the current driven by the force $\nabla (\varepsilon E^2/8\pi)$ which has been applied to the plasma. It will now be shown how the mechanical force necessary to balance the Lorentz force of the current (45) manifests itself through equation (39). It was shown by Cole (1976) that equation (39) implies, for the electric field $\underline{E} = (0, E_o + (\nabla E) y, 0)$, the velocity distribution function of each species undergoes periodic oscillation at its frequence Ω_s while each special as a whole drifts with the velocity V_{DS} (equation 40). Please note the typographical error in Cole (1976) in the two equations after equation 8 the Jacobian should be inverted so that, if f' is the velocity distribution before the electric field is switched on

$$f(u,v,w) = f'(u',v',w') \frac{\partial(u,v,w)}{\partial(u',v',w')}, \qquad (50)$$

where $\frac{\partial(u,v,w)}{\partial(u',v',w')} = (1 - \frac{\omega^2}{\Omega^2}) \cos \Omega t + \frac{\omega^2}{\Omega^2}.$ (51)

During oscillation a cold positive particle is displaced in the direction of $\underline{E}$ a distance by an amount (Cole (1976)

$$y = \frac{E_o}{B} \frac{\omega}{\Omega^2} (1 - \cos \Omega t),$$

where E_o is the electric field at its origin. The average of the electric force of the positive charge over one gyration is given by

$$\langle m_i \frac{dv_i}{Dt} \rangle = \langle q_i E_{(y)} \rangle$$

$$= \langle q_i [E_o + (\nabla E) \frac{E_o}{B} c \frac{\omega}{\Omega^2} (1-\cos \Omega t)] \rangle$$

(52)

TABLE I

Region	V_E(cms^{-1})	B(gauss)	M_i(amu)	N_i(cm^{-3})	ΔB(gamma)
Low Lat. Ionosphere*	10^5**	0.3	20	5×10^6	4
Auroral Ionosphere*	4×10^5	0.5	30	5×10^6	50
Mid-Magnetosphere	3×10^6	0.05	1.5	10	0.3
Low lat. Boundary	1.5×10^7	4×10^{-4}	1.5	10	17
Layer	4×10^7	4×10^{-4}	1.5	10	$\sim B$
Interplantary Space	4×10^7	B	1.5	10	$\sim B$
Inner Solar Corona	10^7	B	1.5	10	$\sim B$
Base of Corona		1	1.5	10	$\sim B$

*Above altitude at which collision frequency $\nu = \omega_s$.
**Including corotation velocity with earth.

$$= q_i E_o + q_i \frac{E_o}{B} C \frac{\omega}{\Omega^2} (\nabla E)$$

$$= q_i E_o + \frac{m_i c^2}{B^2} \frac{\nabla E^2}{2}, \text{ when } \omega \approx \Omega. \quad (53)$$

Similarly for electrons

$$\langle m_e \frac{dv_e}{dt} \rangle = q_e E_o + \frac{m_e c^2}{B^2} \frac{\nabla E^2}{2} \quad (54)$$

It follows that the net electric force on the plasma per unit volume averaged over the ion gyroperiod is

$$\rho_m \frac{dv_m}{dt} = \frac{\text{div } E}{4\pi} E + \frac{4\pi \rho_m c^2}{B^2} \frac{\nabla E^2}{8\pi} \quad (55)$$

$$= \varepsilon \frac{\nabla E^2}{8\pi} \quad (56)$$

and this exactly balances the Lorentz force due to the current (45). It may be noted that $V_{\nabla E}/V_E$ is approximately the ratio of the gyroradius associated with a velocity V_E to the scale size of the spatial variation of εE_y.

In order to estimate the magnetic change due to the dielectric current, consider equation 37 in the form

$$B \Delta B = E_y \Delta(\varepsilon E_y). \quad (58)$$

So

$$\Delta B = \frac{E_y}{B} \Delta(\varepsilon E_y).$$

The quantity $C E_y/B$ has the dimensions of velocity V_E. So we can write

$$\Delta B \approx \frac{V_E^2}{V_A^2} B, \quad (59)$$

where V_A is the Alfvén speed = $C/\varepsilon^{1/2}$..

Alternatively

$$\Delta B \approx V_E^2 a \pi n_s m_s B^{-1}. \quad (60)$$

where $a = 2$ if $\Delta B/B \ll 1$, and $a = 4$ if $\Delta B \approx B$.

Table I shows estimates of upper limits to ΔB in various regions of interest.
The effects of dielectric currents are greatest in the boundary layer and in interplanetary space and the geophysical consequences of these currents will be dealt with in a separate paper.

It is clear from Table I that the currents must be taken into account in discussions of the physics of the low latitude boundary layer, interplanetary medium and the solar corona where the effects of dielectric current can be large.

The current cm^{-1} integrated in the y direction through the model is, from equation 37,

$$j(\text{cm}^{-1}) = \Delta(\varepsilon E_y^2)/8\pi B \text{ emu}, \quad (61)$$

$$= n_i m_i V_E^2/2B. \qquad (62)$$

In the case of gradient of electric fields above, or in the vicinity of auroras, we assume $n_i = 5 \times 10^6$ cm^{-3}, $m_i = 30$ amu (i.e., NO$^+$), $B = 0.5$ g, $V_E = 4 \times 10^5$ cm/sec. These are extreme values. From equation 45 then $j \approx 1.6 \times 10^4$ amp cm^{-1}. Integrating over the height of the ionosphere where the ions are freely gyrating, i.e., say 180-400 km altitude the total current would be 3.5×10^3 amp. In the first approximation the altitude variation of the current density is like that of the ion mass density. A significant effect should hold from 150 km altitude to the altitude of transition of heavy ions (NO$^+$, O$^+$) to light ones (He$^+$, H$^+$). Sometimes above auroras this transition can be very high in which case the total current could be greater.

In the low latitude boundary layer (Cole, 1974; Eastman and Hones, 1979) in which, e.g., $B = 5 \times 10^{-4}$, $V_{E_3} = 2 \times 10^7$ cm$_s$-1, $n_i = 10$ cm^{-3} of H$^+$ plus 3 cm^{-3} of He$^+$ the current integrated through the layer is $\approx 1.83 \times 10^{-5}$ emu = 1.83×10^{-4} amp cm^{-1}. If the boundary layer extends five earth radii above and below the equatorial plane, then the total dielectric current flowing in the boundary layer would be $\sim 1.2 \times 10^6$ amps. For uncomplicated anti-solar flow and velocity gradient pointing outwards, the current is towards the sun on the dusk side and away from the sun on the dawn side. On both sides the Lorentz force of the current is inwards to the magnetosphere. The currents can be of magnitude comparable to the conventional Chapman-Ferraro currents. Fuller discussions of these problems and applications in the solar wind and other astrophysical situations will be taken up in later papers.

The Momentum Per Unit Mass Orthogonal to $\underline{B}$ (when p = 0)

It is clear from the examination of particle motions (Cole, 1976) that the current (equation 37) is due to ions in proportion to their contribution to the mass density of the plasma. When $\frac{\partial}{\partial y}(\varepsilon E_y^2) = 0$, the drift of the plasma orthogonal to $\underline{B}$, caused by an electric field, is given in a collisionless plasma, by

$$\underline{V} = \frac{c\, \underline{E} \times \underline{B}}{B^2}. \qquad (63)$$

However, in the circumstances given by equations 26, 27 and 28, neglecting the momentum of electrons, the momentum per unit mass is given by

$$V_x = \frac{c}{B}\left(E_y + \frac{1}{en}\frac{\partial(\varepsilon E_y^2)}{8\pi \partial y}\right). \qquad (64)$$

This result means that only if

$$\frac{\partial}{\partial y}(\varepsilon E_y^2) \ll 8\pi\, en\, E_y$$

may it be stated that, approximately

$$c\,\underline{E} + \underline{V} \times \underline{B} = 0 \qquad (65)$$

In this approximation equation 38 becomes, when $\varepsilon \gg 1$,

$$p + \frac{B_z^2}{8\pi} - \frac{1}{2}\rho_m V_x^2 = \text{constant}, \qquad (66)$$

which is a Bernoulli-like relationship, but pertaining to a line orthogonal to the flow.

The Force $-\underline{P} \times \underline{\nabla} \times \underline{E}$

This force which was uncovered in equation 30 makes sense as the electrical counterpart of the force $-\underline{B} \times \underline{\nabla} \times \underline{H}$. Both may be described as forces originating from dipoles. The fact that there is no magnetic counterpart of $(4\pi)^{-1}(\text{div } \underline{D})\underline{E}$ is related to the lack of magnetic monopoles.

The Magnetic Permeability of a 2-D Collisionless Plasma

Consider an infinite two dimensional collisionless plasma in which $\underline{B} = (0, 0, B_z(y))$. In a steady state in the absence of electric field, equation 30 yields,

$$(\nabla \times H) \times B - \frac{1}{2}H^2 \nabla\mu = 0, \qquad (67)$$

which, in the assumed geometry, becomes

$$-\mu H \frac{\partial H}{\partial y} - \frac{1}{2}H^2 \nabla\mu = 0,$$

or

$$\frac{\partial}{\partial y}(\mu H^2) = 0.$$

Therefore,

$$\mu H^2 = K = B^2/\mu. \qquad (68)$$

where K is a constant.

Also $H = K/B$.

In the case of the collisionless plasma let us assume that

$$K = B^2 + 8\pi p.$$

Of course K is a well known constant of 2-D collisionless plasma. Then,

$$\mu = \frac{1}{1 + 8\pi p/B^2}, \qquad (69)$$

because $\mu = 1$, when $p = 0$.

This defines the vectors $\underline{H}$ and $\underline{M}$ by

$$\underline{H} = \left(\frac{1}{1 + 8\pi p/B^2}\right) \underline{B} = \frac{K}{B^2} \underline{B} \qquad (70)$$

and

$$\underline{M} = \left(-\frac{2p}{B^2}\right) \underline{B} = \frac{1}{4\pi}\left(1 - \frac{K}{B^2}\right) \underline{B} \qquad (71)$$

The magnetization vector $\underline{M}$ is twice the "conventional" magnetization vector and presumably incorporates equal components of the diamagnetism due to gyration and due to grad $\underline{B}$ drifting. The total current in the plasma is the sum of

$$(4\pi)^{-1} c \underline{\nabla} \times \underline{H} \text{ and } c(\underline{\nabla} \times \underline{M}),$$

i.e., $(4\pi)^{-1} c \underline{\nabla} \times \underline{B}$. In the present geometry this total plasma current is then

$$\frac{c}{4\pi}\frac{\partial B}{\partial y} = -\frac{c}{B}\frac{\partial p}{\partial y},$$

which is the same as that calculated from particle motions in the field $\underline{B}$ (Chandrasekhar, 1969).

Discussion

The dielectric current orthogonal to a magnetic field in a plasma has been described. An earlier discussion of the movement of charge particles in crossed magnetic and spatially varying electric fields revealed a special case of it (Cole, 1976). The current comes about principally on account of the large value of the dielectric constant of a plasma. Structures exist in plasmas in the earth's environment in which the current plays a significant role, e.g., in the auroral ionosphere and the low latitude boundary laye. In a later paper the significance of the current in interplanetary space will be demonstrated. This new current should be sought experimentally and taken into account theoretically in models of plasma structures.

In the ($\underline{D}$, $\underline{E}$, $\underline{B}$, $\underline{H}$) approach to plasmas, all the internal currents produced by properties of the medium were lumped together in determining μ. Thus, in the model defined by equations 26, 27 and 28, a value of μ was used which combined into the vector $\underline{\nabla} \times \underline{M}$ both the conventional "magnetization" current of plasma physics and the gradient $\underline{B}$ current in a plasma (Chandrasekhar, 1960) i.e., all the currents related to ion and electron pressure in the specified geometry. The value of ε is taken from standard plasma physics. Parameterizing a plasma in this way may facilitate, in some circumstances, the macroscopic description of phenomena in it driven from the outside.

Acknowledgments: The author is indebted to the National Research Council of the National Academy of Science, U.S.A. for the award of a Resident Research Associateship and NASA Goddard Space Flight Center for hospitality.

References

Chandrasekhar, S., "Plasma Physics", Phoenix Books, The University of Chicago Press, 1960.
Cole, K.D., Effects of crossed magnetic and (spatially dependent) electric fields on charged particle motion, Planet. Space Sci., 22, 515-518, 1976.
Cole, K.D., Outline of a theory of solar wind interaction with the magnetosphere, Planet. Space Sci., 22, 1075-1088, 1974.
Cole, K. D., A new theory of sources of Birkeland currents, (companion paper, this volume), 1983.
Clemmow, P.C. and J.P. Dougherty, "Electrodynamics of Particles and Plasmas", Addison-Wesley Publishing Company, London, 1969.
Eastman, T.E. and E.W. Hones, Jr., Characteristics of the magnetospheric boundary layer and magnetopause layer as observed by Imp 6, J. Geophys. Res., 84, 2019, 1979.
Maxwell, J.C., "Treatise on Electricity and Magnetism", Vol. II, 1891, Dover Publications, Inc., New York, 1954.
Møller, C., "The Theory of Relativity", Clarendon Press, Oxford, 2nd Ed., 1972.
Spitzer, L., Jr., "Physics of Fully Ionozed Gases", 2nd Ed. Interscience Tracts, John Wiley, New York, 1962.
Stix, T.H., "Theory of Waves in Plasmas", McGraw Hill, New York, 1962.
Stratton, J.A., "Electromagnetic Theory", McGraw Hill, New York, 1941.

FIELD-ALIGNED CURRENT SHEETS AS TANGENTIAL AND ROTATIONAL DISCONTINUITIES

G. Atkinson

Herzberg Institute of Astrophysics, National Research Council of Canada, 100 Sussex Drive,
Ottawa, Ontario K1A 0R6, Canada

Abstract. Field-aligned current sheets in the magnetosphere should be either tangential or rotational discontinuities. Four types are considered here based on whether they are tangential or rotational and whether they require field-aligned potential drops:

1. A simple tangential discontinuity would be expected if the current is downward or weak upward and there is no convection of flux tubes across the discontinuity.
2. A tangential discontinuity is modified by the U-shaped potential structure if the current is strong upward. Polarization currents, associated with the U-shaped potentials, are an important part of the current system and may determine the thickness.
3. A simple rotational discontinuity occurs for downward and weak upward currents if there is a flow across the discontinuity. Standing Alfven waves flip the field line to the new orientation required by passage through the current sheet. There is an incident and an ionospherically reflected standing wave with an unpaired electrostatic shock in between for thin current sheets. Thick sheets have a more extensive weaker electric field. A current sheet that is smoothly varying in the outer magnetosphere may have fine structure at ionospheric heights due to the time-delayed feedback of ionospheric reflections.
4. If the rotational discontinuity requires a strong upward current, the reflected wave is modified by the U-shaped potential structure.

Analytic solutions to 2 and 4 give thicknesses ~ 10 km.

Introduction

Many current sheets have been observed in the near-earth magnetosphere in which the current is flowing parallel to or nearly parallel to the magnetic field lines. Since the variation across them is much more rapid than in either of the other directions, it is reasonable to expect that they be composed of one or more of the standard discontinuities that can exist in a magnetized plasma. (Even thick one-dimensional structures conform to the same jump conditions and are technically therefore describable as discontinuities.) Since the magnetic field changes in direction, but very little in magnitude across the sheets, they could be either tangential discontinuities or rotational discontinuities (oblique Alfven waves) standing in the magnetospheric convective flow.

Unlike shocks, neither of these has an intrinsic thickness (at least, not in the MHD approximation). If there is no normal component of flow and magnetic field across them, they would be tangential discontinuities, and if there is a normal component of flow and magnetic field, they would be rotational discontinuities (standing Alfven waves).

Whether they are tangential or rotational discontinuities depends on the processes producing them in the outer magnetosphere. A current sheet is essentially a boundary between two regions of flux subjected to different stresses in the outer magnetosphere. That is, there is a spatial discontinuity in the $\underline{J} \times \underline{B}$ forces, and the divergence of the cross-field currents becomes field-aligned current. A number of such boundaries are believed to exist in the magnetosphere. These include regions between open and closed flux tubes, and regions with different convective flows. This was discussed at length in Atkinson (1982). It is also likely that there is a component of convective flow through many of these, so that many of them should be standing Alfven waves. In either case, it can be expected that in the absence of other effects (and much of this paper will be concerned with these) their thickness is determined by the distribution of stresses in the outer magnetosphere. We shall not be concerned with the nature of the stresses in this paper since this is adequately discussed in papers on the large scale convection. The ideas here fit in with most such models.

If the currents are sufficiently strong and upwards, then electron acceleration regions may be required to maintain current flow between the ionosphere and outer magnetosphere. These are believed to occur at an altitude of about one earth radius. The associated U- or V-shaped potential systems would be expected to modify the discontinuity. Thus four types of current sheets will be considered here: tangential discontinuities with and without field-aligned potential drops and rotational discontinuities (standing Alfven waves) with and without potential drops.

In an accompanying paper (Atkinson, 1983), the author proposes a model in which the major current sheets (region I, region II, Harang) have a convective flow through them and should be Alfven waves therefore. It is the author's opinion that most of the magnetospheric currents fall in this class. However since there is a possibility that some of the finer scale structures are tangential discontinuities, they will be discussed briefly. A further reason for brevity is that tangential discontinuities were treated in some detail in Atkinson, 1982.

Tangential Discontinuities Without Electron Acceleration

This is the simplest form of current sheet and is also the classical model for current sheets. Variations in the east-west direction along the sheet are not important since there are no rapid flows involved. There is no convective flow component normal to the sheet (in a reference frame attached to the sheet) and magnetic plus plasma pressure must be the same on both sides. The thickness and current distribution (except for gyroradius and finite conductivity effects) are arbitrary, being determined by the properties of the outer magnetosphere and ionosphere.

Tangential Discontinuities With Electron Acceleration

This type of current sheet is consistent with models using kinetic resistance (Chiu and Schulz, 1978; Chiu and Cornwall, 1980; Fridman and Lemaire, 1980; Lyons, 1981), double layers (Carlquist and Bostrom, 1970) and oblique shocks (Swift, 1981) to maintain parallel electric fields. In a north-south cross-section, U or V shaped potentials of the type shown in figure 1 are apparent. In Atkinson (1982), it was pointed out that, because of the very rapid flow in the east-west direction (north-south electric fields) in the arms of the U-shaped potentials, inflow and outflow (tangential electric fields) had to

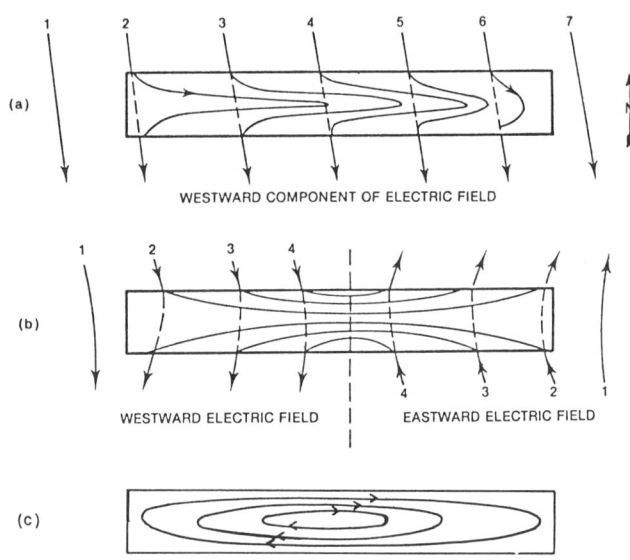

Fig. 2. If the structure of figure 1 is of finite east-west length (limited to the rectangle) east-west electric fields are added as indicated producing the electric potentials shown. View is along field lines. Solid and dashed lines are equipotentials above and below the electron acceleration region respectively. (a) is a rotational discontinuity and (b) and (c) are tangential.

be included in the model so that the divergence of the flow (curl E) was equal to zero. That is the model could be quasi-two-dimensional, but an inflow and hence a variation in the total potential drop must exist along the arc in the east-west direction. Streamlines (or equipotentials) are illustrated in figures 2b and 2c for two simple models. The model in 2a is a rotational discontinuity because there is a net flow through it. Views are looking along magnetic field lines.

In the model in figure 2b, a flux tube convects into the system and the section above the field-aligned potential drop starts to convect eastward or westward at a rapid velocity. As seen in the frame of the convecting flux tube current continuity becomes a problem and parallel electric fields appear first at electron-acceleration heights. That is rapid motion starts at this height and propagates upward as an oblique Alfven wave. In a reference frame fixed with respect to the structure, the boundary between slowly convecting and rapidly convecting regions is seen as a standing (but upward propagating) oblique Alfven wave. Thus, even the tangential discontinuity consists of standing Alfven waves.

In the model in figure 2c, the outer parts of flux tubes are undergoing a steady-state elliptical motion (a very flattened ellipse) presumably

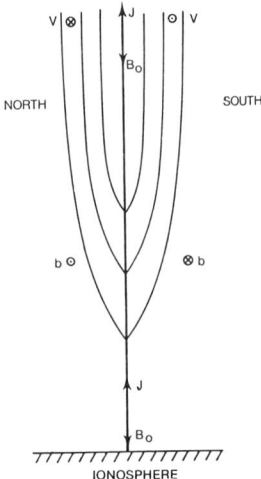

Fig. 1. Typical representation of U- or V-shaped potentials as seen in a north-south cross section through an east-west structure. The lighter lines are electric equipotentials.

driven from the outer magnetosphere and the standing Alfven waves are downward propagating. Since there is not a flow of flux tubes right across the structure, it is still a tangential discontinuity.

In Atkinson (1982), it was assumed that the current in the electron-acceleration region obeyed an Ohmic type law ($J_z = -\Sigma U$) where J_z is the vertical current density, Σ the height-integrated conductivity of the region and U the potential above it. Current continuity at the top of the acceleration region was assumed to be provided by the currents associated with the Alfven waves, and the equation

$$V_A \mu_0 \Sigma U = \pm \partial^2 U/\partial x^2 \quad (1)$$

resulted where x is the variation in the north-south direction and V_A the Alfven velocity. The plus sign applies to upward propagating Alfven waves and the minus sign to downward. The solutions to (1) for the potential U, just above the electron acceleration region are an exponential decay away from the centre of the structure for the model in figure 2b, and a cosine centred on the centre of the structure and extending a quarter of a period each side for the model in 2c. In each case the scale distance is $(\mu_0 V_A \Sigma)^{\frac{1}{2}} \sim 10$ km and the maximum U is $-\frac{1}{2} J_0 (\mu_0 V_A/\Sigma)^{\frac{1}{2}} \sim 1$ to 10 kV where J_0 is the current in the sheets. These solutions are the thin current sheet limit where the sheet thickness would be zero in the absence of the field-aligned potential. The sheets will of course be thicker for other distributions of stresses in the outer magnetosphere.

Rotational Discontinuities Without Electron Acceleration

Throughout this section rotational discontinuities are referred to as standing Alfven waves for aesthetic reasons because we wish to include thick current sheets in the discussion, especially the region I and II and Harang discontinuity currents. Earlier work on MHD Alfven waves and ionospheric reflections as current sheets includes Maltsev et al (1977) and Mallinckrodt and Carlson (1978). We shall be particularly concerned with the situation where a flux tube convects from a region of flux subjected to outer magnetosphere stresses and convective flow in one direction into a region of flux subjected to stresses and flow in a different direction. The information on stress change and flow is propagated downward as a standing Alfven wave in steady-state flow. This is, of course, a current sheet. At the ionosphere, the downcoming wave should be almost perfectly reflected as illustrated in figure 3 for a yz current sheet. The right-hand side shows the incident and reflected current sheet and the left-hand side the shape of a field line at different locations in the sheet. The effect of the wave and reflection is to "flip" the angle of inclination of the flux

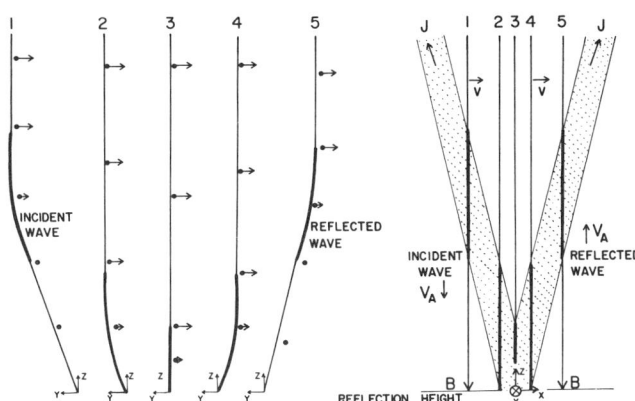

Fig. 3. North-south cross section of current sheets of incident and reflected standing Alfven waves. The shape of field lines in the yz plane is shown for increasing values of x (or t as seen in the convection reference frame). Arrows show plasma velocity. Field lines are flipped from a positive to a negative slope as seen in yz planes.

tube from By < 0 to By > 0 as required by Maxwell's equations as it convects through a current sheet.

This field line "flipping" has associated electric fields. Figure 4a shows the equipotentials for a zero thickness current sheet as seen in a cross-section. The vertical electric fields outside the current sheets are not field-aligned (field lines are equipotentials) but are due to the tilt of field lines through the cross-sectional plane. A compression of a factor of 1000 in the horizontal direction would bring the figure close to real-world geometry. It would of course equally enhance the ratio of horizontal to vertical electric fields. What is clear in figures 3 and 4 is that the total y-directed displacement of the flux tube and hence the electric field associated with "flipping" increases with height. For b ~ 250 nT it is ~ 1 kV at 10,000 km and several times this in the outer magnetosphere. Thus a height variation in the cross-oval potential is expected, even for equipotential field lines. This may be related to the observations of such a variation by Mozer and Torbert (1980). It should be noted that since Δb_y depends only on the total current, the total potential associated with flipping does not depend on the number of reflections between ionosphere and outer magnetosphere or the details of the current sheets.

For thin current sheets, the electric field associated with field-line flipping would resemble an unpaired electrostatic shock. Figure 4a is a solution for the symmetric case where $|b_y|$ is the same on each side of thin current sheets. (These simple cases can be solved by inspection for perfect reflection.) In general, symmetry would not be expected, and even more important, several

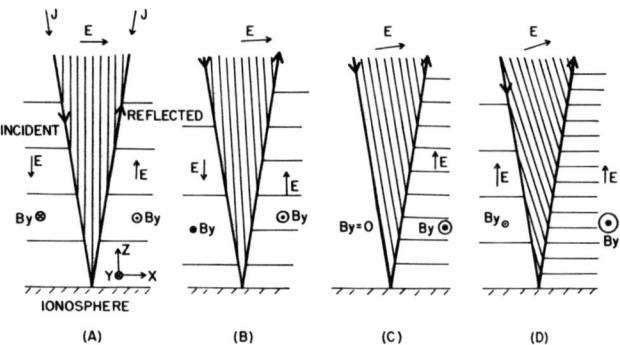

Fig. 4. North-south cross section of electric equipotentials, associated with field-line flipping for thin waves. All the currents are downwards. Field lines are equipotentials. The cases shown are for ΔB_y (across the sheets) constant, but B_y decreasing and going negative. (A) is the symmetric case and the degree of assymetry increases in (B), (C) and (D). Compression by 1000 in the horizontal direction would approximate the real world geometry.

reflections between outer magnetosphere and ionosphere would be expected before equilibrium was established. Figures 4b, c and d show the effect of assymetry. A B_y component in the negative direction has been added, increasing in size in figures 4a to 4b. The equipotentials associated with field-line flipping change as illustrated in the cross-sections in the figure.

The next point to be discussed for standing Alfvén waves (rotational discontinuities) is the effect of multiple reflections. As stated earlier we are interested in the case where a flux tube convects from a region of the magnetosphere subject to stresses in one direction into a region subject to stresses in a different direction. Let us assume east-west stresses for simplicity of discussion, and an east-west aligned current sheet. Consider a surface normal to magnetic field lines in the outer magnetosphere with z outward along the magnetic field, y eastward and x southward and inward (that is the system is the transfer along field lines of the coordinate system used earlier). The Maxwell east-west shear stress across the surface is $b_y B_z/\mu_0$ where b_y is the east-west component of field and B_z the normal component (assumed constant). Thus as a flux tube convects in the x direction between regions subject to different stresses, b_y varies and the field aligned current, $j_z = \partial b_y/\partial x$.

Since we are discussing Alfvén waves we assume j_z is associated with a standing oblique Alfvén wave. Further we shall refer to downward propagating waves as incident waves (on the ionosphere) and upward propagating as reflected. It seems reasonable to assume that the stresses and hence b_y and j_z vary smoothly in the outer magnetosphere. However this is not necessarily true at the ionosphere. j_z in the outer magnetosphere is composed of incident plus reflected wave contributions and the relative phasing of the two is completely different at the two locations due to the transmission delay time for a wave to travel from the ionosphere to the outer magnetosphere. Thus in the outer magnetosphere: $j_z = j_{zi} + j_{zr}$ where i and r stand for incident and reflected. For a time delay of τ for propagation from outer magnetosphere to ionosphere and return, which corresponds to a distance "delay" of $d = \tau V_x$ (v_x is the convective velocity), the above equation becomes:

$$j_z(x) = j_{zi}(x) + j_{zi}(x - d) \qquad (2)$$

This assumes perfect reflection by the ionosphere. It is well known that this difference equation has oscillatory solutions with spatial wavelengths $2d/n$ where n is any odd number (e.g. Cunningham, 1958). Figure 5 is a plot of sample solutions. The top graph shows four assumed forms of j_z in the outer magnetosphere. They are smoothly varying sinusoids fitting to constant values at each

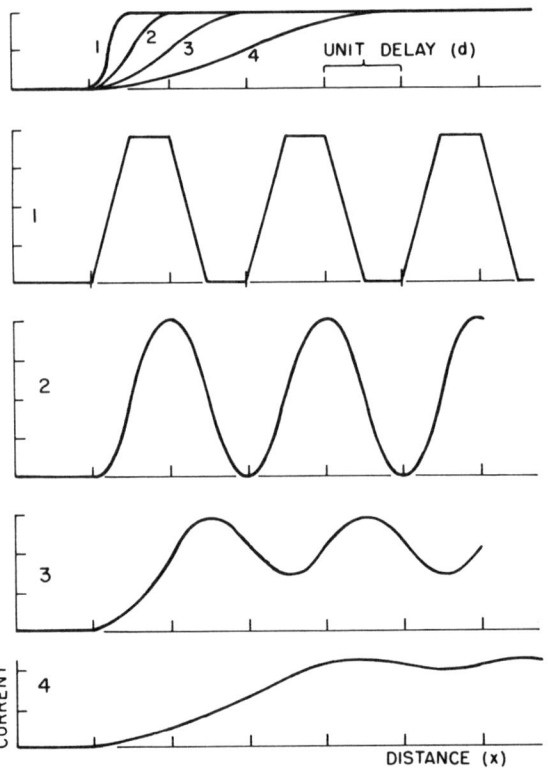

Fig. 5. Plot of field-aligned current against distance perpendicular to the sheet. Smooth variations in the outer magnetosphere (top diagram) produce oscillations in the currents of the incident wave (bottom four diagrams) and hence in current sheets at ionospheric heights.

end. They represent the case where a current increases smoothly from zero to a constant value. The unit of distance is d. The corresponding calculated values of j_{zi} are shown in the four graphs below. In each case the current increases and oscillates about its final values, primarily in the n = 1 harmonic, although for the case where j_z varies on a length scale shorter than d, the presence of higher order harmonics is apparent. The current at ionospheric heights is $2 j_{zi}(x-d/2)$. The factor of two arises because it consists of superimposed incident and reflected waves, and the d/2 from the delay in propagation to the ionosphere. In the real world damping terms in the equation would be expected to cause the oscillation to decrease in amplitude with x.

A decrease in the current from a constant value to zero gives the mirror images in the x axis of the graphs with the oscillation centred on zero. The distance scale (d) is of the order of 10 km when mapped to ionospheric heights. The point of relevance in this section is that smoothly varying stresses in the outer magnetosphere tend to produce sinusoidally-varying current sheets at ionospheric heights with a temporal period equal to twice the time for a wave to propagate from outer magnetosphere to ionosphere and back, or distance scales equal to the above period times the normal component of drift velocity.

Rotational Discontinuities With Electron Acceleration

If the current in the Alfven waves is upwards and sufficiently intense, then electron acceleration regions might be expected. The waves are depicted in the north-south cross-section of figure 6 for a thin incident wave. There is a wave reflected off the acceleration layer, with electric fields consistent with the U-shaped potentials required for acceleration. The transmitted wave contains a similar effect, and the ionospheric reflection tends to cancel the inverted U-shaped potential below the acceleration region and enhance the U-shaped potential above. Note that the angles of incidence and reflection should be very small (equal to the convective velocity divided by the Alfven velocity) and hence the incident and reflected waves overlap much more than indicated in the figure. There are higher order reflections off the acceleration layer of the ionospherically reflected wave. As before it is assumed that the Alfven wave currents feed an Ohm's law acceleration layer. Figure 7 shows a sample solution for a north-south section, ignoring higher order reflections, for a zero thickness incident wave. The acceleration layer is at 5000 km. The insert at 5000 km is a cartoon to show how equipotentials join. Above and below are plots of analytical solutions of the electric potentials. Evidence can be seen for both the field-line "flipping" electric fields and the U-shaped potentials. Equipotentials are field

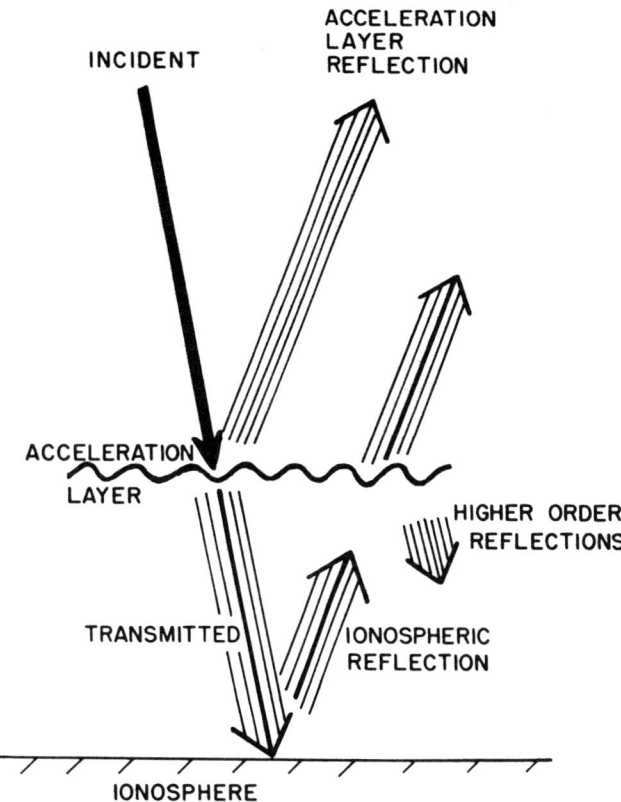

Fig. 6. North-south section showing the waves present if there is an acceleration layer.

aligned everywhere except at 5000 km. Vertical fields elsewhere are due to the field lines tilting so that they pass through the section. It is the symmetric case with the field line tilt equal and opposite on the two sides of the system. A horizontal compression of a factor of 1000 would bring the diagram close to the real world situation.

The derivation of the equations used is very similar to the derivation of equation 1 of this paper except that we have to keep track of three potentials: U_i, U_r, U_t corresponding to incident, reflected and transmitted waves. The potential above the layer is $U_i + U_r$ and the potential below is U_t. The final equations have the form

$$- \mu_o V_A \Sigma (U_i + U_r - U_t) = \frac{\partial^2}{\partial x^2} (U_i - U_r)$$

$$= \frac{\partial^2}{\partial x^2} (U_t) \quad (3)$$

For the zero thickness (step function in electric field) incident wave case which is shown in figure 6, the wave reflected off the acceleration layer consists of oppositely directed electric fields

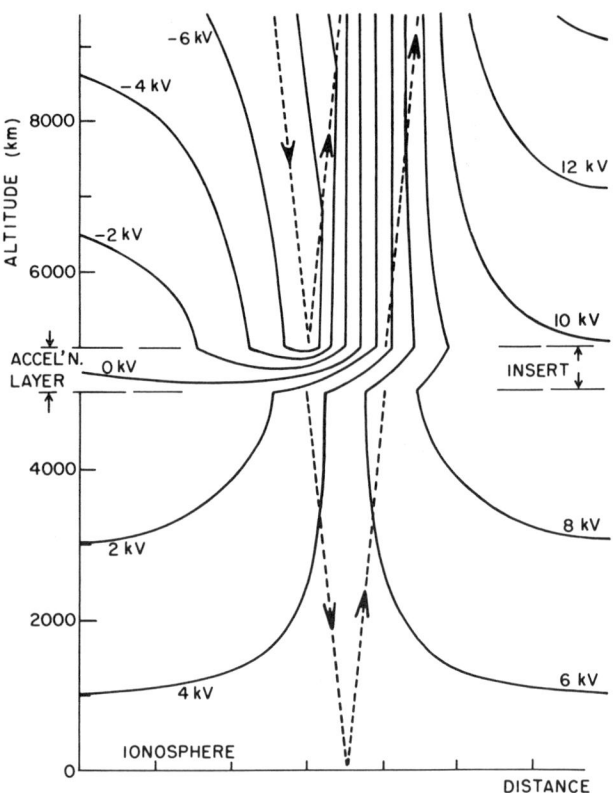

Fig. 7. North-south section of sample solution for case where Ohmic-type acceleration layer exists at 5000 km altitude and currents elsewhere are standing Alfven waves. Equipotentials are shown in kV. Dashed lines are the waves. The incident wave has zero thickness. Equipotentials in the insert at 5000 km are schematic.

decaying exponentially on each side and the transmitted wave and its ionospheric reflection consist of step function plus exponentially decaying fields. The exponential decay distance is $(2\mu_0 V_A \Sigma)^{\frac{1}{2}}$ for all waves, similar to the tangential discontinuity. Similar scale sizes were obtained by Lysak and Carlson (1981) and Lysak and Dum (1983).

References

Atkinson, G., Inverted Vs and/or discrete arcs: a three-dimensional phenomenon at boundaries between magnetic flux tubes, J. Geophys. Res., 87, 1528-1534, 1982.

Atkinson, G., The role of currents in plasma redistribution, paper in this issue, 1983.

Carlqvist, P., and R. Bostrom, Space charge regions above the aurora, J. Geophys. Res., 75, 7140-7146, 1970.

Chiu, Y.T., and M. Schulz, Self-consistent particle and parallel electrostatic field distributions in the magnetospheric-ionospheric auroral region, J. Geophys. Res., 83, 629, 1978.

Chiu, Y.T., and J.M. Cornwall, Electrostatic model of a quiet auroral arc, J. Geophys. Res., 85, 543, 1980.

Cunningham, W.J., Introduction to nonlinear analysis, McGraw-Hill, New York, 1958.

Fridman, M., and J. Lemaire, Relationship between auroral electron fluxes and field-aligned electric potential difference, J. Geophys. Res., 85, 664, 1980.

Goertz, C.K., and R.W. Boswell, Magnetosphere-ionosphere coupling, J. Geophys. Res., 84, 7239, 1979.

Hasegawa, A., Particle acceleration by MHD surface wave and formation of aurora, J. Geophys. Res., 81, 5083.

Lysak, R.L., and C.W. Carlson, Effect of microscopic turbulence on magnetosphere-ionosphere coupling, Geophys. Res. Lett., 8, 169, 1981.

Lysak, R.L., and C.T. Dum, Dynamics of magnetosphere-ionosphere coupling including turbulent transport, J. Geophys. Res., 88, 365-380.

Mallinckrodt, A.J., and C.W. Carlson, Relations between transverse electric fields and field aligned currents, J. Geophys. Res., 83, 1426, 1978.

Maltsev, Y.P., W.G. Lyatsky and A.M. Lyatskaya, Currents over the auroral arc, Planet. Space Sci., 25, 53, 1977.

Mozer, F.S., and R.B. Torbert, An average parallel electric field deduced from the latitude and altitude variations of the perpendicular electric field below 8000 kilometers, Geophys. Res. Lett., 7, 219-221, 1980.

Temerin, M., M.H. Boehm, and F.S. Mozer, Paired Electrostatic Shocks, Geophys. Res. Lett., 8, 799, 1981.

ELECTRODYNAMICS OF CONVECTION IN THE INNER MAGNETOSPHERE

R. W. Spiro and R. A. Wolf

Department of Space Physics and Astronomy, Rice University, Houston, TX 77251

Abstract. During the past ten years, substantial progress has been made in the development of quantitative models of convection in the magnetosphere and of the electrodynamic processes that couple that magnetosphere and ionosphere. These models compute electric fields, electric currents, and magnetospheric plasma distributions within the inner magnetosphere, where the flow speeds associated with convection are slow compared to the propagation speeds of waves in the magnetospheric medium, and where the magnetic field configuration can be estimated with reasonable accuracy from an independent magnetic field model.

Using a computational scheme first proposed by Vasyliunas, the convection models under consideration separate the three-dimensional problem of convection in the inner magnetosphere/ionosphere into a pair of two-dimensional problems coupled by Birkeland currents flowing between the two regions. We review the logic, development, and major results of the inner magnetosphere convection model with emphasis on ionospheric and magnetospheric currents. A major theoretical result of the models has been the clarification of the relationship between the region 1/region 2 picture of field-aligned currents and the older partial ring current/tail current interruption picture of substorm dynamics.

Introduction

Earth's magnetosphere and ionosphere are coupled by magnetic field-aligned (Birkeland) currents that flow between the two regions. In order to explain the observation of field-aligned currents near the inner edge of the plasma sheet, Schield et al. [1969] resurrected an idea originally proposed by Alfvén [1939]. According to this mechanism, as the plasma sheet convects sunward on the nightside of the Earth, ions gradient and curvature drift toward dusk, while plasma sheet electrons drift toward dawn. Charge separation develops near the inner edge of the plasma sheet, leading to a dusk to dawn electric field that opposes the sunward convection of the plasma sheet. This field is partially neutralized by field-aligned currents down into the ionosphere on the dusk side and up out of the ionosphere on the dawn side. The currents that connect the inner edge region of the plasma sheet with the highly conducting ionosphere are the currents that we now associate with the Region 2 system of Iijima and Potemra [1978].

A closed, self-consistent computational scheme to calculate the electric fields, currents, and charged particle distributions of the coupled magnetosphere/ionosphere system was enunciated by Vasyliunas [1970]. The framework and underlying assumptions of this computational approach serve as the basis for many analytic [e.g., Vasyliunas, 1972; Siscoe, 1982; Senior and Blanc, 1983] and computer [Jaggi and Wolf, 1973; Harel et al., 1981a,b; Spiro et al., 1981; Wolf et al., 1982] treatments of convection in the inner magnetosphere. In this review, we concentrate on the computer simulation of inner magnetosphere plasma dynamics, specifically on the Rice Convection Model, a computer model developed over the past several years at Rice University.

Before describing the computational scheme, we first define the limits of the 'inner magnetosphere,' at least in the context of the Rice Convection Model. For computer modeling purposes, the inner magnetosphere is restricted to regions where two theoretical assumptions are valid. We first assume that the plasma flow speed is small relative to the sound speed and that time scales are long compared to MHD wave travel times. Under these assumptions, the inertial term in the MHD momentum equation can be neglected relative to the pressure gradient and $\underline{J} \times \underline{B}$ terms, i.e.,

$$\rho \left(\frac{\partial}{\partial t} + \underline{v} \cdot \underline{\nabla}\right) \underline{v} = -\nabla p + \underline{J} \times \underline{B} \approx 0 \quad (1)$$

This requirement specifically excludes the magnetopause boundary layer and those regions of the plasma sheet and tail where x type neutral lines form.

The second major assumption that limits the region of applicability of the convection calculation involves the use of standard magnetic field models. Presently, the convection model does not self-consistently compute $\underline{B}$, but, instead, assumes that $\underline{B}$ can be estimated from an independent empirical or theoretical magnetic field model [e.g., Olson and Pfitzer, 1974; or Voigt, 1981]. This assumption reduces the three-dimensional convection calculation to the solution of two coupled two-dimensional problems, one in the ionosphere and one in the magnetosphere, with the two regions connected by the magnetic field structure. By assuming the magnetic field mapping from the ionosphere to the magnetospheric equatorial plane, we explicitly exclude the polar cap regions from the model. The two assumptions of subsonic plasma flow and known magnetic field geometry limit the region of applicability of the Rice Convection Model to the sub-polar cap ionosphere and its extension into the closed field line part of the magnetosphere.

Computational Scheme

Figure 1 illustrates the logical structure of the computational scheme used in the Rice Convection Model. This figure is a modified version of the computational outline proposed by Vasyliunas [1970]. The five rectangular boxes represent quantities that are calculated directly in the model. Arrows connecting these boxes indicate the flow of information in the computation. Boxes with rounded corners correspond to assumptions and models that provide necessary input information. Dashed lines indicate physical effects that are presently neglected in the model.

In each time step, the computation works its way once around the logic loop of Figure 1. Starting at the top of the diagram:
1. The initial magnetospheric plasma distribution at the outer boundary of the calculation and the magnetic field model are used to compute the density of gradient/curvature drift current in the inner magnetosphere.
2. Current continuity requires that the divergence of the drift current density perpendicular to $\underline{B}$ be balanced by field-aligned currents. This allows us to compute the Birkeland current distribution into the ionosphere.
3. From the distribution of Birkeland current, a model for the ionospheric electrical conductance (height-integrated conductivity), and suitable boundary conditions, we calculate the ionospheric potential distribution V by solving the ionospheric Ohm's law equation.
4. The computed potential distribution is mapped along the magnetic field into the magnetosphere to give the magnetospheric electric field distribution.
5. We then re-evaluate the plasma distribution in the magnetosphere by following the total drift velocity ($\underline{E} \times \underline{B}$ plus gradient/curvature drift) of the magnetospheric plasma for a time step Δt, bringing us back to the top of the diagram. The system evolves in time by stepping repeatedly around this logic loop.

Model Inputs

Table 1 gives an outline of the input quantities required by the model. These quantities provide either initial or boundary conditions for the convection computation. By changing the input quantities with time through a specific event, we model the dynamic response of the plasma in the inner magnetosphere to changing external (e.g., solar wind) conditions. In general, we use the best data or independent models available to specify these input quantities. Sometimes, however, we investigate the effects of the various inputs by performing 'computer experiments,' varying one of the input parameters in a controlled way, while holding the other inputs constant.

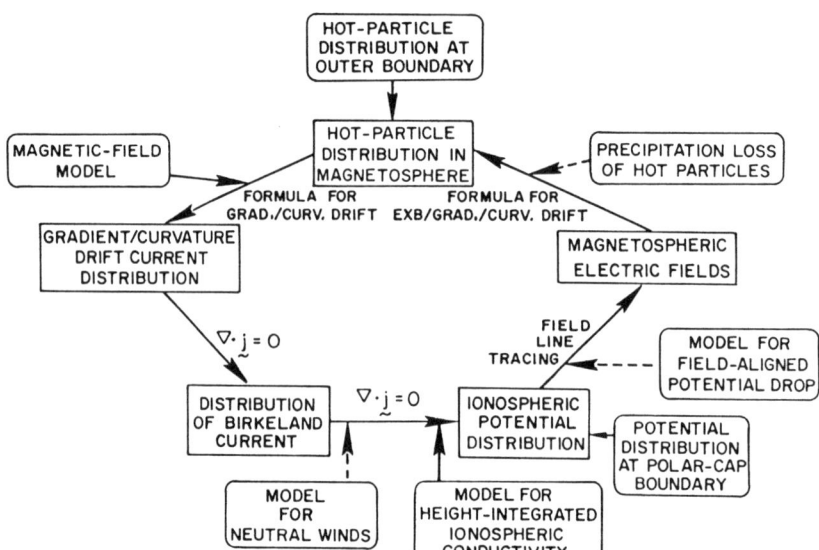

Fig. 1. Logic diagram of the Rice Convection Model. The program cycles through the entire central pentagon every time step Δt.

TABLE 1. Inputs to the Model

Quantity	Comments
1. Plasma-sheet particle distribution	• 21 species • Event specific or average values
2. Magnetic field model	• Time dependent standoff distance based on solar wind data • Ring current contribution varies with Dst
3. Ionospheric conductivity model	• Time independent part due to solar photoionization • Time dependent part due to auroral electron precipitation

4. Potential distribution at polar cap	• Assume classic convection through polar cap boundary
	• Cross polar cap potential drop:
	  a. Directly measured from polar orbiters
	  or
	  b. Inferred from IMF parameters.

Plasma Distribution at Outer Boundary

At present, we represent the plasma distribution at the boundary of and within our modeling region by an isotropic distribution of 21 species. Each species k is characterized by a charge q_k, an energy invariant λ_k ($\equiv E_k (\int ds/B)^{2/3}$), and a number invariant η_k ($\equiv n_k \int ds/B$) where E_k is the particle kinetic energy, n_k is the number density for species k, and $\int ds/B$ is the volume of a flux tube containing one unit of magnetic flux. The parameters λ_k and η_k can be shown to be invariant along a drift path (see Harel et al. [1981a] for a more detailed discussion). The spatial distribution of each species is characterized by the locus of points that define the inner edge to which the particles have penetrated. The number invariant η_k is taken to be constant between the outer boundary of the model and the computed inner edge for species k. In general, we assume a Maxwellian energy distribution at the outer boundary with temperature based on event-specific plasma sheet measurements (when available) or on average values if necessary.

Magnetic Field Model

In order to couple the magnetospheric and ionospheric portions of our calculation, we have used modified versions of standard models of the Earth's magnetic field, either the empirical model of Olson and Pfitzer [1974] or the theoretical model of Voigt [1981]. The magnetic field model that we are presently using is based on a series of equilibrium Voigt models. We introduce time dependent changes in the magnetic configuration by considering the variations in solar wind and Dst conditions through a modeled event. The solar wind ram pressure affects the standoff distance and, through this, the Chapman-Ferraro currents on the surface of the magnetopause. Measured variations in Dst are used to modify the strength of the ring current in the magnetic field models.

Global Ionospheric Conductance

The height-integrated ionospheric Pedersen and Hall conductivities used in the model consist of universal time independent base terms due to solar UV insolation and UT dependent terms due to auroral electron precipitation. The auroral enhancement terms have been inferred from event-specific measurements made by low-altitude satellites which are then extrapolated in local time and UT [Harel et al., 1981a], or have been based on an empirical model of ionospheric conductance [Spiro et al., 1982]. As a first effort at achieving self-consistency between the auroral conductivity distribution and the computed magnetospheric plasma distribution, the conductivity model is slid in latitude so that the low-latitude edge of enhanced auroral conductivity coincides with the computed plasma sheet electron inner edge.

A promising method for inferring the global distribution of Pedersen and Hall conductances within the auroral zone involves the use of auroral images from polar satellites. Photometric data from the auroral imager on Dynamics Explorer have already been used to infer auroral zone conductivities [Kamide et al., 1982]. By observing at a pair of wavelengths in the UV, it should be possible to infer both Pedersen and Hall conductances, even in the sunlit auroral zone. Routine use of auroral imagers on operational satellites would go a long way toward providing adequate coverage of this important parameter of magnetosphere/ionosphere coupling.

Electric Potential Distribution at 'Polar Cap' Boundary

The solution of the ionospheric potential equation requires that the potential V be specified at the high latitude boundary of our model. We choose a simple local time dependence for V that corresponds roughly to the results of Heppner [1977] and Heelis et al. [1976]. The maximum polar-boundary potential difference is inferred from solar wind magnetic field measurements and empirical relationships between IMF parameters and the cross polar cap potential drop [Reiff et al., 1981]. If available, one could also use direct event-specific measurements of the cross polar cap potential drop as measured by a polar orbiting satellite in a dawn-dusk orbit.

Model Outputs

The convection model self-consistently calculates five primary outputs which are intrinsic to the computational scheme of Figure 1: the density of gradient/curvature drift current in the magnetosphere, the distribution of field-aligned currents that couple the magnetosphere and ionosphere, the ionospheric potential distribution, the magnetospheric electric field, and the plasma distribution in the inner magnetosphere. These outputs are integral parts of the computational scheme.

In addition, other useful quantities can also be calculated which are of considerable geophysical interest, but which are not essential links in the computational chain. These include:
1. The distribution of horizontal Pedersen and Hall currents in the ionosphere.
2. The calculation of magnetic perturbations (theoretical magnetograms) due to the currents included in the model.
3. The calculation of the ionospheric Joule heating rate.
4. The determination of total ring current strength within the modeling region.

Thus, the convection model is able to make multiple predictions of a variety of geophysical parameters which can then be compared with available data from specific events. Large discrepancies between model predictions and measurements indicate that the input assumptions are not consistent with the actual event or that the model is neglecting relevant physical processes or both, while relative agreement of multiple predictions with data is taken to indicate that the model is adequately representing the physics of the system.

Event Simulations

We have used the computational scheme outlined in the previous sections to simulate specific geophysical events by varying the input parameters (magnetic field, cross-polar-cap potential drop, ionospheric conductance) and computing the response of the inner magnetosphere. To date, we have computer simulated three events as summarized in Table 2. While the convection model self-consistently calculates many quantities that can be compared with data for a given event, the actual comparisons are limited by the available data base which varies from event to event.

The first event simulation that we performed was for an extended substorm-like event that occurred on September 19, 1976. Model results were compared with ionospheric electric field and magnetic perturbation data from the S3-2 spacecraft [Harel et al., 1981a,b; Karty et al., 1982], with typical mid-latitude electric field data [Spiro et al., 1981], and with magnetograms from a variety of ground stations [Chen et al., 1982].

We have also modeled two magnetic storms: the storm of July 29, 1977 in which the magnetosphere

TABLE 2. Simulated Events

Event	Type	Data Comparisons
September 19, 1976[a]	Substorm	Ionospheric $\underline{E}$, ΔB, theoretical magnetograms
July 29, 1977[b]	Magnetic Storm	Dst, low-latitude edge of plasma-sheet, ionospheric $\underline{E}$, ΔB
March 22, 1979[c]	Magnetic storm	Plasma sheet/ring current data, low-latitude edge of plasma sheet, ionospheric $\underline{E}$

[a]Harel et al.[1981a,b], Spiro et al. [1981], Karty et al. [1982], Chen et al. [1982].
[b]Wolf et al. [1982].
[c]CDAW-6, work in progress.

underwent a dramatic compression, with the daytime magnetopause pushed in past geosynchronous orbit, and the storm of March 22, 1979, characterized by a more modest compression. Both of these events were subjects of Coordinated Data Analysis Workshops, CDAW-2 and CDAW-6, respectively. Since the present version of the convection model neglects local plasma production and loss processes, we have been limited to modeling only the storm main phase. Accurate modeling of storm recovery and ring current decay will require the realistic inclusion of loss processes. The results of our computer simulation of the July 29, 1977 storm have been compared with measurements of Dst, with measurements of the location of the plasma sheet inner edge, and with S3-3 electric field and magnetic perturbation data [Wolf et al., 1982].

The simulation of the March 22, 1979 storm has been used to conduct a detailed analysis of ring current injection. For this event, extensive plasma data is available to help specify the plasma distribution both in the tail plasma sheet and in the afternoon ring current region. In addition, we have made preliminary comparisons between our computed electric field patterns and measurements made in the auroral zone (STARE radar) and at mid-latitude (Saint-Santin radar).

Physical Insights Gained from the Convection Model

By computer modeling the electrodynamics of convection, we have quantitatively examined the close coupling between the magnetosphere and ionosphere. The model has provided explanations for many observed phenomena within the context of a consistent treatment of convection and has made predictions as to other effects. In this section, we present a brief summary of what we have learned from the convection model.

Electric Fields

1. <u>Verification of Shielding of Inner Magnetosphere from Convection Electric Field</u>. The convection models have consistently shown the phenomenon of inner magnetosphere shielding, i.e., the tendency for polarization electric fields produced near the plasma sheet inner edge to neutralize the convection electric field earthward of the inner edge. This effect is well known and has been studied analytically for simple cases [Karlson, 1963, 1971; Block, 1966; Vasyliunas, 1970; Swift, 1971]. The computer simulations have given added insight into the dependence of shielding time scales and shielding effectiveness on magnetospheric and ionospheric parameters [e.g., Jaggi and Wolf, 1973; Harel et al., 1981a]. In addition, the simulation of the large storm of July 29, 1977 indicated that a strong magnetospheric compression destroys shielding by

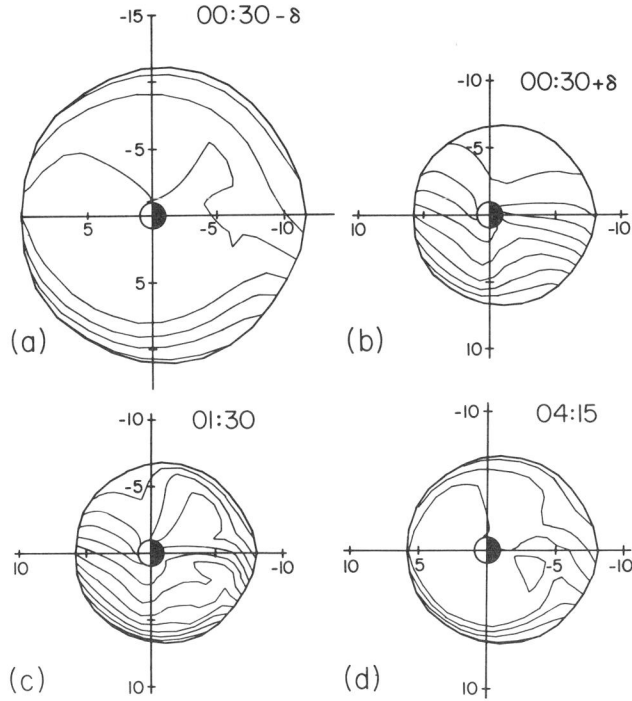

Fig. 2. Model electric-equipotential patterns for four universal times on July 29, 1977. The magnetospheric equatorial plane is shown, with the Sun to the left. Displayed equipotentials are 6 kV apart. Corotation electric fields are not included in the display. The model sudden commencement occurred at 00:30 UT. The symbol δ represents a positive infinitesimal, so the top two diagrams pertain to just before (diagram a) and just after the sudden commencement (diagram b). Note that the sudden commencement disrupted shielding, which was then gradually reestablished (diagrams c and d) (from Wolf <u>et al</u>. [1982]).

weakening or reversing the region 2 currents, leading to substantial penetration of convection electric fields to low L [Wolf et al., 1982]. This effect is shown schematically in Figure 2.

2. <u>Explanation of Rapid Subauroral Plasma Flow</u>. Satellite observations have shown the frequent occurrence of latitudinally restricted regions of intense electric fields just equatorward of the evening auroral zone in the region of the mid-latitude ionization trough [Spiro et al., 1974; Smiddy et al., 1977; Maynard, 1978; Spiro et al., 1978]. The computer simulation of the September 19, 1976 substorm event predicted the occurrence of intense electric fields in the evening trough region, and, indeed, such fields were observed during this event by the S3-2 satellite [Harel et al., 1981b]. Southwood and Wolf [1978] explained the existence of such latitudinally narrow elec-

tric field enhancements as being due to the penetration of the plasma sheet ion inner edge inside the precipitation eroded electron inner edge, conditions that were met in the September 19, 1976 simulation.

3. *Explanation of Dawn-Dusk Asymmetry in Sunward Convection Potential.* Wolf [1970] demonstrated that the sharp change in ionospheric conductivity at the dawn and dusk terminators has a dramatic effect on the electric potential distribution. Current continuity across the terminator requires electric equipotentials to be compressed near the dusk polar cap boundary and spread apart near the dawn polar cap. More sophisticated numerical calculations for the September 19, 1976 substorm bear out this conclusion [Spiro et al., 1981].

Current Systems

1. *Agreement with Standard Patterns of Ionospheric and Field-Aligned Currents (Including Current Overlap Region Near Midnight).* The typical ionospheric and field-aligned current patterns computed by the convection model agree quite well with patterns inferred from ground and satellite measurements. The region of applicability of the computational scheme used in the Rice Convection Model is limited to the inner magnetosphere and underlying ionosphere. Unfortunately, this limitation has the awkward effect of excluding the highest latitude part of the auroral zone where most of the auroral electrojet current flows. We have stepped around this problem (at least as regards the computation of ground magnetic effects and global Joule heating for the September 19, 1976 substorm event) by adding a narrow conducting band at the poleward edge of the modeling region [Karty et al., 1982]. Within this band, ionospheric conductivity and Birkeland current density are assumed independent of latitude; poleward of the band (in the polar cap) the ionosphere is assumed to have zero conductivity. The electrodynamics of the band are computed analytically from boundary conditions furnished by the computer model.

Figure 3 shows a comparison between an observational summary of field aligned currents presented by Iijima and Potemra [1978] and a typical result of the computer simulation. Of special interest here is the current overlap region apparent just before midnight where there is a central region of upward field aligned current bracketed by downward current regions. The most poleward downward current sheet corresponds to the region 1 sheet, while the two lower latitude sheets both map to the inner magnetosphere (region 2). The overlap of upward and downward currents near midnight is explained by the computer model as being due to the different configurations assumed by the inner edges of the higher energy ion species compared to lower energy ions or electrons. This effect is shown schematically

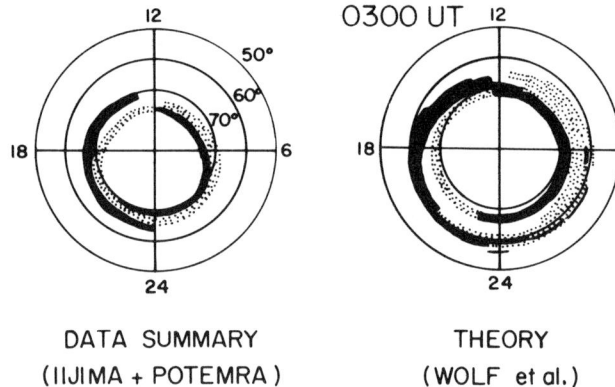

Fig. 3. Comparison of observed and theoretical Birkeland current patterns. The theoretical pattern was computed for 0300 UT on July 29, 1977, a time between two large substorms. The data summary diagram was redrawn from the paper of Iijima and Potemra [1978] and pertains to AE > 100 γ.

in Figure 4. Magnetospheric drift currents flow along contours of constant flux tube volume. As a result of the shape of the high-energy ion inner edge, current flows out of the inner edge near midnight. This current must be balanced by upward Birkeland currents from the ionosphere. Similarly, magnetospheric current flows into the more earthward, low energy ion and electron inner edges near midnight, leading to downward Birkeland currents. Mapped into the ionosphere, these are the overlap region field-aligned currents.

2. *Ground Magnetic Effects of Computed Currents.* By performing a Biot-Savart integration over the self-consistent current system calculated in our model, we can construct theoretical magnetograms corresponding to magnetic perturbations at desired points on the Earth's surface [Chen et al., 1982]. We have compared our computed theoretical magnetograms with actual magnetic data from a global network of observatories for the September 19, 1976 event. The decomposition of theoretical magnetograms into component parts (e.g., perturbations arising from auroral electrojets, field aligned currents, ring currents, etc.) has demonstrated that the magnetic perturbations due to different but related currents often nearly cancel each other, a practical extension of Fukushima's theorem [Fukushima, 1969] to conditions of inhomogeneous conductivity and curved magnetic field lines. The ability to construct synthetic magnetograms has been put to practical use in a test of the KRM method of magnetogram inversion. In this test, synthetic magnetograms computed for 96 locations and the glo-

bal conductivity distribution were provided as input to the magnetogram inversion algorithm of Kamide et al. [1981]. The electric fields and currents obtained from the magnetogram inversion technique were compared with the parent fields of the convection model [Wolf and Kamide, 1983]. The test verified the accuracy of the inversion technique for auroral zone currents and fields, given an accurate conductivity model.

3. Explanation of Dawn-Dusk Asymmetry in Low-Latitude ΔB. Substorm associated perturbations in mid- and low-latitude magnetograms have been associated with the current distribution shown schematically in Figure 5a [e.g., Kamide and Fukushima, 1972; or Crooker and McPherron, 1980]. One current loop represents the diversion of westward tail current through the westward electrojet during a substorm [Atkinson, 1967; McPherron et al., 1973], while the other current loop represents a westward partial ring current centered near dusk [Cummings and Dessler, 1967]. The latter current loop has been invoked to explain the observed dawn-dusk asymmetry in low-latitude magntic perturbations. It is not immediately clear how the two-loop current pattern of Figure 5b is related to the observed Birkeland current distribution shown in Figure 3.

Computer simulation of the September 19, 1976 event has led to an alternative explanation of the low-latitude asymmetry in ΔB and to the modified current picture shown in Figure 5b. The computer results indicate that the ΔB asymmetry is due mainly to net downward (region 1 plus region 2) Birkeland current on the dayside and net upward Birkeland current on the nightside. The partial ring current (shown centered near

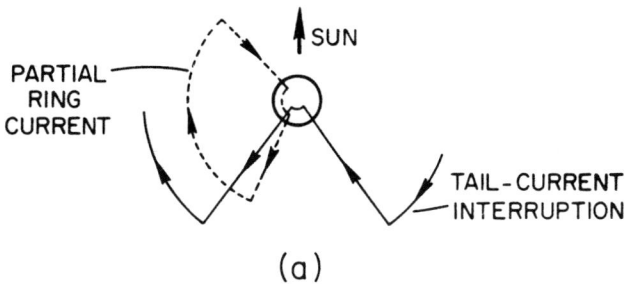

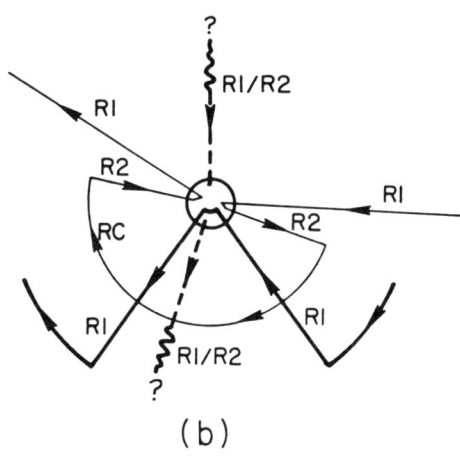

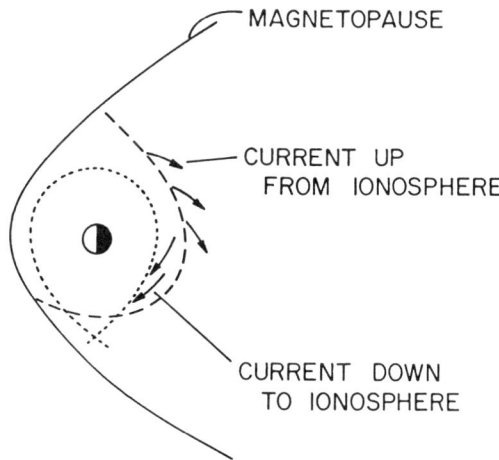

Fig. 4. Characteristic inner edge configurations of high energy ions (dashed line) and lower energy ions or electrons (dotted line). Current flows out of the high energy ion inner edge near midnight, into the lower energy ion inner edge.

Fig. 5. (a) A double current system to account for the substorm disturbance field at the Earth's surface [Crooker and McPherron, 1972]. (b) A substorm current system based on our computer simulation of magnetic disturbance. Dashed curves indicate that net Birkeland currents flow down from the magnetosphere on the day side, up to the magnetosphere on the night side. The current system denoted by the thin curve, which connects R1 Birkeland current, ionospheric north-south current, R2 Birkeland current and partial ring current, causes most of the magnetic perturbation observed from polar orbiting satellites, but causes a modest ground magnetic disturbance (from Chen et al. [1982]).

dusk in Fig. 5a) is rotated and extended in local time in Figure 5b, connecting the dawnside region 2 currents to the duskside region 2 currents. The computer results summarized schematically in Figure 5b are in good agreement with both the ground magnetic observations that motivated Figure 5a and with the observed pattern of region 2 Birkeland currents (Fig. 3), up on the dawnside and down on the duskside. The Birkeland current imbalance explanation of the asymmetry in low-latitude ΔB has been further developed by Crooker and Siscoe [1981] and Siscoe [1982] using a simple, but elegant, analytic model of the region 1/region 2 current systems.

4. <u>Generation of Region 1 Birkeland Currents on Sunward Convecting Flux Tubes</u>. Satellite observations have indicated that frequently a substantial fraction of the region 1 field aligned current occurs equatorward of the electric field reversal in the region of sunward convection [Smiddy et al., 1980; Mozer et al., 1980]. These observations stand in contrast to earlier theoretical ideas that imply that region 1 currents connect directly to the solar wind generator or to the antisunward flowing boundary layer. Recent analytic and computer calculations tentatively indicate that a channel of density depleted flux tubes located near the central axis of the magnetotail would be stable against interchange instability and would result in the generation of significant field aligned currents of region 1 sense on sunward convecting flux tubes [Karty, 1983; Karty et al., 1983]. These results resemble the classic tail current interruption picture of substorm expansion [e.g., Atkinson, 1972], but place this phenomenon in the context of a consistent treatment of convection.

Ring Current Energetics

1. <u>Formation of Ring Current by Earthward Convection of Plasma Sheet Plasma</u>. The computer simulation of the July 29, 1977 magnetic storm demonstrated that earthward convection of plasma sheet plasma is able to produce sufficient ring current to explain the observed main phase decrease in Dst. Figure 6 shows the comparison between the observed Dst profile and our theoretical profile. The theoretical values were calculated by integrating the Biot-Savart law over the computed current system and correcting for changes in Chapman-Ferraro currents at the surface of the magnetopause when the magnetosphere was compressed and re-expanded. The effects of the Chapman-Ferraro corrections can be seen in Figure 6 as jumps at the sudden compression (0030 UT) and at the partial re-expansion (0430 UT). The computed time-dependent decrease in Dst agrees reasonably well with the observed main phase decrease, even considering the neglect of charge exchange and other loss processes which would lead to ring current decay, and the neglect of direct injection of ionospheric ions into the ring current which would further decrease Dst.

2. <u>Preliminary Computer Experiments Testing Ring Current Cause-and-Effect Relationships</u>. We have used the simulation of the magnetic storm of March 22, 1979 as the basis for a detailed study of ring current injection. To isolate the effects on the ring current of different model assumptions and to test the predictions of analytic models of high-latitude electrodynamics [Siscoe, 1982], we have performed many computer runs through the event. These runs include:

a. Nominal run — This run used our best-esti-

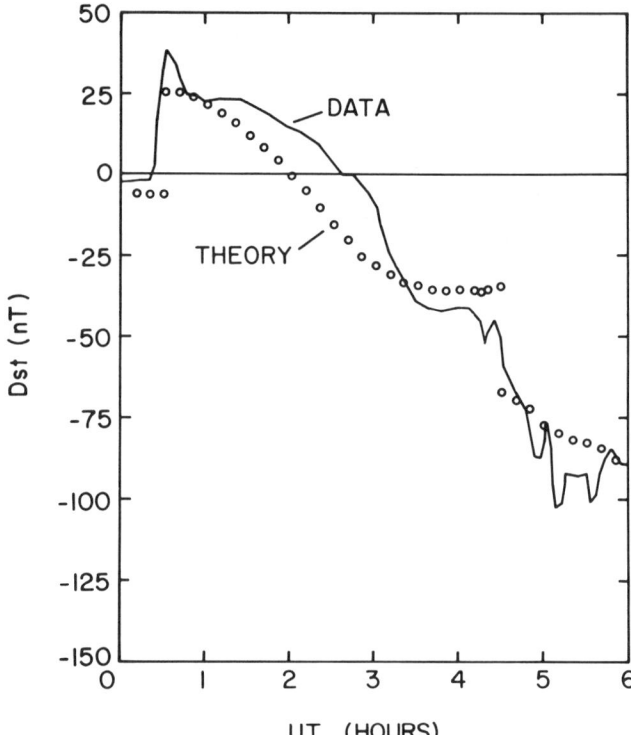

Fig. 6. Comparison of observed and predicted Dst index for magnetic storm of July 29, 1977 (from Wolf et al. [1982]).

mate conductivity model, a nominal series of magnetic field models that were varied in step with observed changes in solar wind ram pressure and Dst, and boundary plasma sheet density and temperature values chosen on the basis of ISEE-2 observations of the pre-storm plasma sheet [provided by G. Paschmann, private communication, 1982].

b. Half-pressure run — The input for this run was the same as for the nominal run except the boundary plasma sheet pressure was decreased by a factor of two, keeping the same plasma sheet temperature.

c. Increased temperature run — The conductivity and magnetic field inputs for this run were the same as for the nominal run. The plasma sheet plasma distribution was modified such that the plasma temperature was increased by a factor of approximately 2.4 while keeping the same total particle pressure.

d. Constant B run — Inputs for this run were the same as for the nominal run, except the magnetic field configuration was held constant, i.e., no compression at sudden commencement, no expansions or compressions due to changes in solar wind dynamic pressure, and no inner magnetosphere inflation due to ring current development.

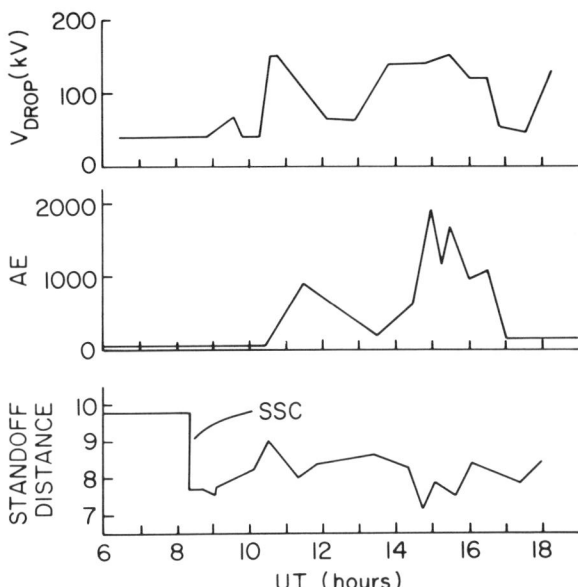

Fig. 7. Key input parameters for simulation of the March 22, 1979 magnetic storm. The three panels show polar-boundary potential drop (from IMF data), AE index (used in conductivity model), and magnetopause standoff distances (from solar-wind ρV^2).

e. Double conductivity run — This run had the same inputs as the constant magnetic field run except the nominal ionospheric conductivity was doubled everywhere.

Figure 7 shows plots of the primary input parameters used for the nominal run of the March 22, 1979 event. The AE index is the primary input to the auroral conductivity algorithm. The magnetopause standoff distance, calculated from solar wind parameters, is a primary input for calculating Chapman-Ferraro current strengths in the magnetic field model. The polar boundary potential drop is an essential input for calculating the ionospheric potential distribution.

Figure 8 shows plots of time integrated Joule heating and ring current energy for the nominal run and for the constant uncompressed magnetic field run. For the nominal run, the total Joule heating is about 1.5 times the ring current energy. Eliminating the compressions and expansions of the magnetic field model changes the total Joule heating only slightly, but significantly reduces the ring current energy by a factor of 0.6. These results are roughly consistent with our earlier simulations. For the September 19, 1976 substorm simulation, which involved no compression, Joule heating was about 2-3 times ring current energy. For the storm of July 29, 1977, which involved a dramatic compression of the magnetosphere, Joule heating was 0.5 of the ring current energy. These results clearly indicate that the storm-time compression of the magnetosphere plays an important role in ring current injection.

Figure 9 illustrates the effects of changing the input plasma distribution at the outer (plasma sheet) boundary of our calculation. Halving the plasma pressure, but keeping the temperature the same, decreases the final ring current energy by approximately a factor of 0.75. Similarly, increasing the plasma sheet temperature decreases the ring current energy,

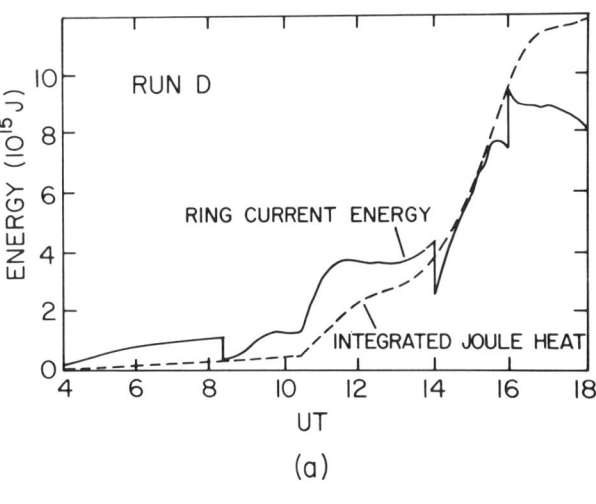

(a)

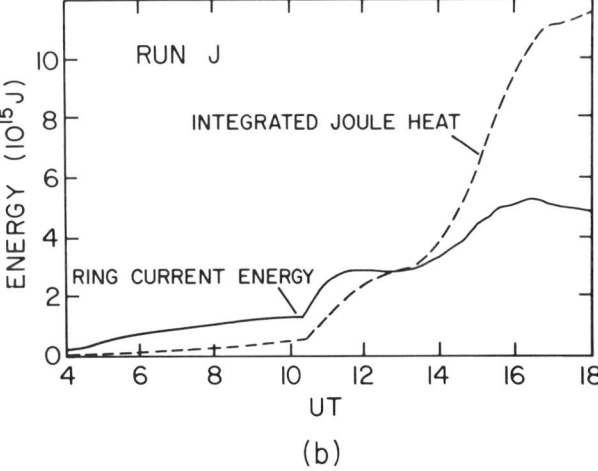

(b)

Fig. 8. Effect of magnetic-field compression on ring-current energy and Joule heating. Diagram (a) shows ring current energy and Joule heating for our nominal run, which included a "realistic" magnetic field model, including a compression at the time of the sudden commencement. Diagram (b) shows the same parameters for no magnetic field compression. (The pre-storm, uncompressed magnetic-field model was used throughout the event.)

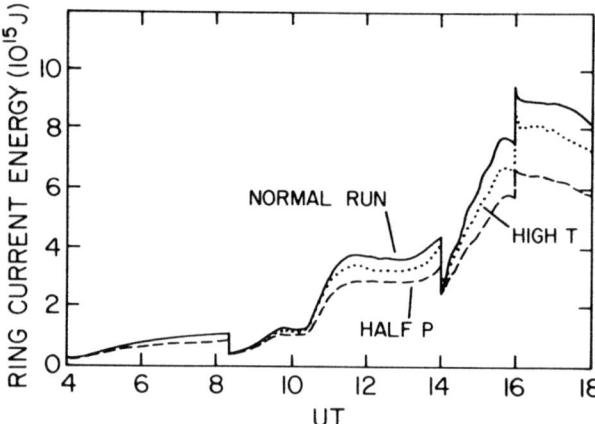

Fig. 9. Effect of the plasma-sheet boundary condition on ring-current injection.

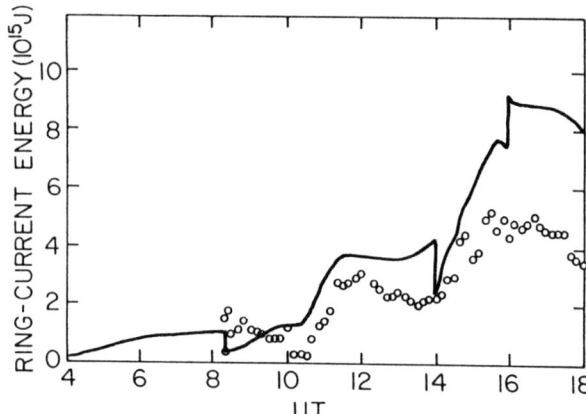

Fig. 11. Model ring current energy (solid line) and ring current energy from Dst (circles).

but only by 12%. Raising the temperature decreases the lower-energy part of the plasma distribution, and it is the lower energy portion of the plasma sheet distribution that is injected deep into the inner magnetosphere to form the storm-time ring current.

Figure 10 shows the effect of ionospheric conductivity on ring current injection. Doubling the conductivity for constant magnetic field model increases ring current energy, but only by a factor of 1.4. Increased conductivity leads to decreased shielding which allows particles to be injected deeper into the magnetosphere.

Finally, Figure 11 shows a comparison between the ring current energy estimated from the observed Dst variations and our computed ring current energies. The ring current energy is inferred from Dst by using the relation [Cummings and Dessler, 1967]:

$$\Delta B = -\frac{2}{3} B_0 (U_{ring}/U_m) \qquad (2)$$

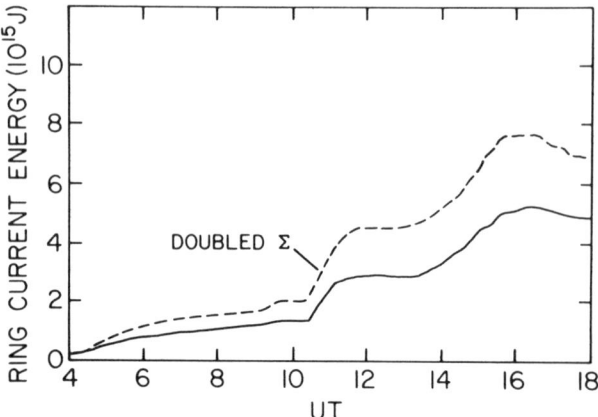

Fig. 10. Effect of conductivity on ring-current energy.

where ΔB = low latitude northward magnetic field perturbation at the Earth's surface, B_0 = surface magnetic field at the equator $\approx$ 31,000 nT, U_{ring} = ring current energy, and U_m = magnetic energy in the Earth's dipole field external to the Earth $\approx 8 \times 10$ J. The Dst values were corrected for effects due to Chapman-Ferraro currents on the magnetopause [W. P. Olson, private communication]. Figure 11 shows that the Dst inferred ring current energy is less than the theoretically predicted nominal run value. We tentatively attribute this discrepancy to the model's neglect of plasma loss processes such as charge exchange and precipitation.

3. <u>Preliminary Computer Experiments Testing Particle Arrival Times at SCATHA</u>. Measurements made by the Lockheed ion mass spectrometer on the SCATHA spacecraft show a classic ion dispersion pattern with higher energy ring current ions detected before lower energy ions [Strangeway and Johnson, 1983]. For this event, 10 keV equatorially mirroring ions were detected at SCATHA just after 1300 UT when the spacecraft was near 5.5 R_E geocentric distance in the afternoon local time sector.

Figure 12 shows the temporal development, for our nominal run, of the inner edge of equatorially mirroring particles with magnetic moment $\mu = 65$ eV/γ (corresponding to an energy of ~10 keV at SCATHA). The particles that make up the inner edge are assumed to $\underline{E} \times \underline{B}$ and gradient drift from an initial location at 10 R_E geocentric distance at 0400 UT. For this run, the inner edge near dusk does not cross over the location of SCATHA until just after 1500 UT. Similar inner-edge-development calculations were also done for the half pressure, increased temperature, constant B and doubled conductivity experimental runs. On the basis of these calculations, we tentatively conclude:

1. Particles arrive at SCATHA ~2 hours later

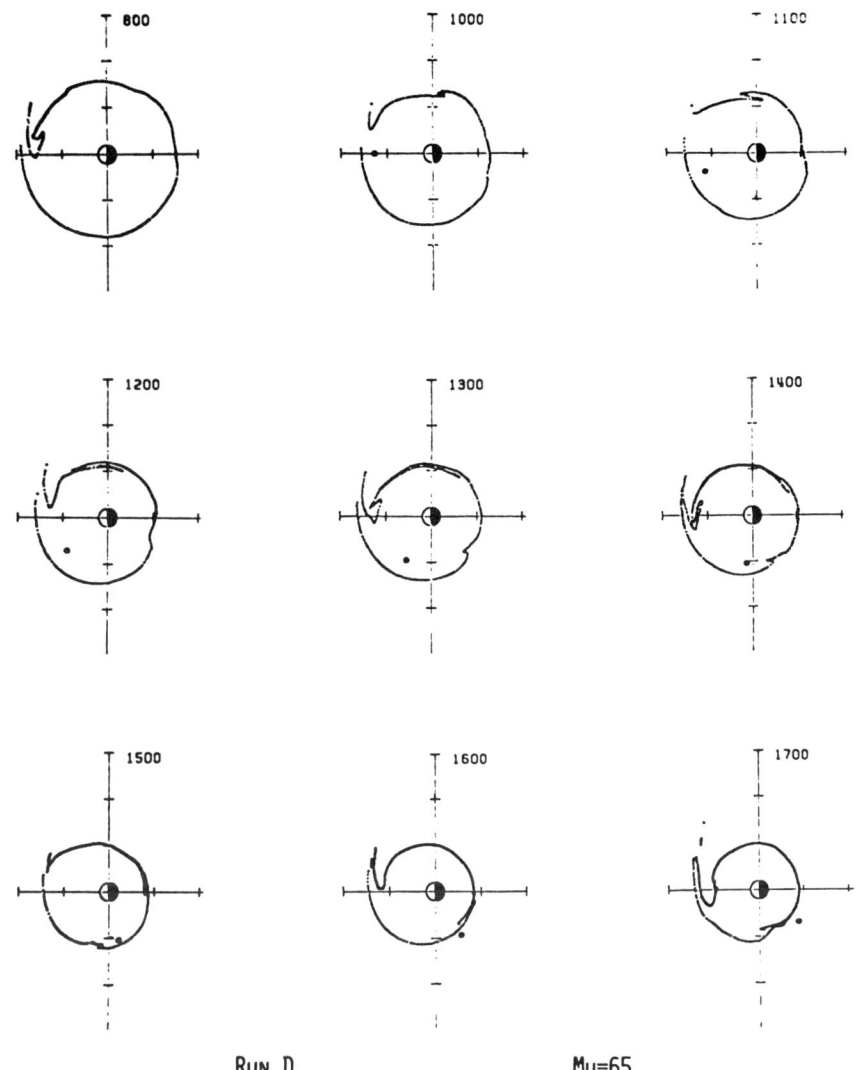

Fig. 12. Computed plasma-sheet inner edge for equatorially mirroring particles ($\mu = 65$ eV/γ, 10 keV at SCATHA) as a function of time for our nominal run. The particles started on a 10 R_E circle at 0400 UT (4.33 hr before the SSC). The small circle represents the position of the SCATHA spacecraft.

than the actual measurements for the nominal run.
2. Reducing the plasma sheet pressure decreases shielding efficiency and allows ring current ions to penetrate closer to Earth. This results in better agreement between predicted and observed particle arrival times and suggests that the inclusion of particle loss processes in the computer code would also lead to better agreement.
3. Increased plasma sheet temperature reduces low energy ion density and allows for deeper penetration of ring current ions.
4. Similarly, increased ionospheric conductivity reduces shielding efficiency and allows for deeper (and earlier) penetration of particles.

Conclusion

Over the past decade, the Rice Convection Model has provided quantitative insights into the electrodynamics of the coupled inner magnetosphere/ionosphere system. From this work has come a theoretical understanding of the relationship between the observed pattern of region 1/region 2 Birkeland currents and the older tail-current interruption picture of substorm dynamics. Using event-specific input parameters, the model has computer simulated specific geophysical events and made multiple comparisons of model results and measured quantities.

Future development of the model will concentrate on including additional physical processes that are currently neglected such as the local production and loss of plasma sheet/ring current particles, the effects of neutral winds on the distribution of electric potential, and the presence of field aligned potential drops along auroral field lines. These additional processes will be parameterized and included, one at a time, to the existing model so that the interactions between the different physical processes can be determined and complex cause and effect relationships sorted out.

Acknowledgments. The authors express sincere appreciation to G. L. Siscoe whose thought (and action) provoking questions led to several of the computer experiments described in the text. We also acknowledge helpful conversations with G. M. Erickson, M. Harel, J. L. Karty, G. Paschmann, P. H. Reiff, and G. -H. Voigt. This research was supported by the National Science Foundation under grant ATM-82-06026, by the Air Force Geophysics Laboratory under contracts F19628-80-C-0009 and F19628-83-K-0016, and by NASA grant NGR-44-006-137.

References

Alfvén, H., Theory of magnetic storms, 1, Kungl. Sv. Vet — Akademiens Handl., 111, 18(3), 1939.

Atkinson, G., The current system of geomagnetic bays, J. Geophys. Res., 72, 6063, 1967.

Atkinson, G., Magnetospheric flows and substorms, in Magnetosphere-Ionosphere Interactions, edited by K. Folkestad, p. 203, Universitetsforlaget, Oslo, 1972.

Block, L. P., On the distribution of electric fields in the magnetosphere, J. Geophys. Res., 71, 855, 1966.

Chen, C.-K., R. A. Wolf, M. Harel, and J. L. Karty, Theoretical magnetograms based on quantitative simulation of magnetospheric substorm, J. Geophys. Res., 87, 6137, 1982.

Crooker, N. U., and R. L. McPherron, On the distinction between the auroral electrojet and partial ring current systems, J. Geophys. Res., 77, 6886, 1972.

Crooker, N. U., and G. L. Siscoe, Birkeland currents as the cause of the low-latitude asymmetric disturbance field, J. Geophys. Res., 86, 11201, 1981.

Cummings, W. D., and A. J. Dessler, Field-aligned currents in the magnetosphere, J. Geophys. Res., 72, 1007, 1967.

Fukushima, N., Equivalence in ground geomagnetic effect of Chapman-Vestine's and Birkeland-Alfvén's electric current systems for polar magnetic storms, Rep. Ionos. Space Res. Jpn., 23, 219, 1969.

Harel, M., R. A. Wolf, P. H. Reiff, R. W. Spiro, W. J. Burke, F. J. Rich, and M. Smiddy, Quantitative simulation of a magnetopheric substorm, 1, Model logic and overview, J. Geophys. Res., 86, 2217, 1981a.

Harel, M., R. A. Wolf, R. W. Spiro, P. H. Reiff, C.-K. Chen, W. J. Burke, F. J. Rich, and M. Smiddy, Quantitative simulation of a magnetospheric substorm, 2, Comparison with observations, J. Geophys. Res., 86, 2242, 1981b.

Heelis, R. A., W. B. Hanson, and J. L. Burch, Ion convection velocity reversals in the dayside cleft, J. Geophys. Res., 81, 3803, 1976.

Heppner, J. P., Empirical models of high-latitude electric fields, J. Geophys. Res., 82, 115, 1977.

Iijima, T., and T. A. Potemra, Large-scale characteristics of field-aligned currents associated with substorms, J. Geophys. Res., 83, 599, 1978.

Jaggi, R. K., and R. A. Wolf, Self-consistent calculation of the motion of a sheet of ions in the magnetosphere, J. Geophys. Res., 78, 2852, 1973.

Kamide, Y., and N. Fukushima, Positive geomagnetic bays in evening high latitudes and their possible connection with partial ring current, Rep. Ionos. Space Res. Jpn., 26, 79, 1972.

Kamide, Y., A. D. Richmond, and S. Matsushita, Estimation of ionospheric electric fields, ionospheric currents, and field-aligned currents from ground magnetic records, J. Geophys. Res., 86, 801, 1981.

Kamide, Y., J. D. Craven, L. A. Frank, and S.-I. Akasofu, Simultaneously determined distributions of aurora and polar currents (abstract), EOS Trans. AGU, 63, 1059, 1982.

Karlson, E. T., Streaming of a plasma through a magnetic dipole field, Phys. Fluids, 6, 708, 1963.

Karlson, E. T., Plasma flow in the magnetosphere, 1, A two-dimensional model of stationary flow, Cosmic Electrodyn., 1, 4744, 1971.

Karty, J. L., High latitude field aligned currents, Ph.D. thesis, Rice University, Houston, TX, April, 1983.

Karty, J. L., C.-K. Chen, R. A. Wolf, M. Harel, and R. W. Spiro, Modeling of high-latitude currents in a substorm, J. Geophys. Res., 87, 777, 1982.

Karty, J. L., R. A. Wolf, and R. W. Spiro, Region one Birkeland currents connecting to sunward convecting flux tubes, paper presented at Chapman Conference on Magnetospheric Currents, Irvington, VA, 1983.

Maynard, N. C., On large poleward-directed electric fields at sub-auroral latitudes, Geophys. Res. Lett., 5, 617, 1978.

McPherron, R. L., C. T. Russell, and M. P. Aubry, Satellite studies of magnetospheric substorm on August 15, 1968, 9, Phenomenological model for substorms, J. Geophys. Res., 78, 3131, 1973.

Mozer, F. S., C. A. Cattell, M. K. Hudson, R. L. Lysak, M. Temerin, and R. B. Torbert, Satellite measurements and theories of low altitude auroral particle acceleration, Space Sci. Rev., 27, 155, 1980.

Olson, W. P., and K. A. Pfitzer, A quantitative model of the magnetospheric magnetic field, J. Geophys. Res., 79, 3839, 1974.

Reiff, P. H., R. W. Spiro, and T. W. Hill, Dependence of polar-cap potential drop on interplanetary parameters, J. Geophys. Res., 86, 7639, 1981.

Schield, M. A., J. W. Freeman, and A. J. Dessler, A source for field-aligned currents at auroral latitudes, J. Geophys. Res., 74, 247, 1969.

Senior, C., and M. Blanc, On the control of magnetospheric convection by the spatial distribution of ionospheric conductivities, submitted to J. Geophys. Res., 1983.

Siscoe, G. L., Energy coupling between regions 1 and 2 Birkeland current systems, J. Geophys. Res., 87, 5124, 1982.

Smiddy, M., M. Kelley, W. Burke, F. Rich, R. Sagalyn, B. Shuman, R. Hays, and S. Lai, Intense poleward-directed electric field near the ionospheric projection of the plasmapause, Geophys. Res. Lett., 4, 543, 1977.

Smiddy, M., W. J. Burke, M. C. Kelley, N. A. Saflekos, M. S. Gussenhoven, D. A. Hardy, and F. J. Rich, Effects of high-latitude conductivity on observed convection electric fields and Birkeland currents, J. Geophys. Res., 85, 6811, 1980.

Southwood, D. J., and R. A. Wolf, An assessment of the role of precipitation in magnetospheric convection, J. Geophys. Res., 83, 5227, 1978.

Spiro, R. W., W. B. Hanson, D. L. Sterling, and R. A. Hoffman, Midlatitude trough characteristics as observed by Atmosphere Explorer (abstract), EOS Trans. AGU, 58, 1159, 1974.

Spiro, R. W., R. A. Heelis, and W. B. Hanson, Ion convection and the formation of the midlatitude F region ionization trough, J. Geophys. Res., 83, 4255, 1978.

Spiro, R. W., M. Harel, R. A. Wolf, and P. H. Reiff, Quantitative simulation of a magnetospheric substorm, 3, Plasmaspheric electric fields and evolution of the plasmapause, J. Geophys. Res., 86, 2261, 1981.

Spiro, R. W., P. H. Reiff, and L. J. Maher, Jr., Precipitating electron energy flux and auroral zone conductances: An empirical model, J. Geophys. Res., 87, 8215, 1982.

Strangeway, B. J., and R. J. Johnson, Mass composition of substorm-related energetic ion dispersion events, J. Geophys. Res., 88, 2057, 1983.

Swift, D. W., Possible mechanisms for formation of the ring current belt, J. Geophys. Res., 76, 2276, 1971.

Vasyliunas, V. M., Mathematical models of magnetospheric convection and its coupling to the ionosphere, in Particles and Fields in the Magnetosphere, edited by B. McCormac, p. 60, D. Reidel, Hingham, MA, 1970.

Vasyliunas, V. M., The interrelationship of magnetospheric processes, in Earth's Magnetospheric Processes, edited by B. McCormac, p. 29, D. Reidel, Hingham, MA, 1972.

Voigt, G.-H., A mathematical magnetospheric field model with independent physical parameters, Planet. Space Sci., 29, 1, 1981.

Wolf, R. A., Effects of ionospheric conductivity on convective flow of plasma in the magnetosphere, J. Geophys. Res., 75, 4677, 1970.

Wolf, R. A., and Y. Kamide, Inferring electric fields and currents from ground magnetometer data: A test with theoretically derived inputs, submitted to J. Geophys. Res., 1983.

Wolf, R. A., M. Harel, R. W. Spiro, G.-H. Voigt, P. H. Reiff, and C.-K. Chen, Computer simulation of inner magnetospheric dynamics for the magnetic storm of July 29, 1977, J. Geophys. Res., 87, 5949, 1982.

COUPLING OF BIRKELAND CURRENT RINGS

G. L. Siscoe and N. U. Crooker

Department of Atmospheric Sciences, University of California, Los Angeles, CA 90024

Abstract. The Regions 1 and 2 Birkeland current patterns in the ionosphere can be idealized as two nearly concentric, bipolar circles or rings. The Region 1 ring is electrically connected to the outer (or poleward) boundary of the plasma sheet and the Region 2 ring to its inner (or equatorward) boundary. The physics governing the radii of these two rings are therefore different. In particular, in the absence of substorms the rate of change of the radius of the Region 1 ring is proportional to the polar cap potential, but the rate of change of the radius of the Region 2 ring is proportional to the rate of change of the polar cap potential. Thus, the rates of growth and decay of the radii of these rings will in general be different. Calculation shows that during dayside merging intervals, Ring 1 expands about 16 times faster than Ring 2. Under typical merging potentials, Ring 1 will cross the initial ring separation in about 1 hour. We identify situations in which the circles approach very near to each other (or actually attempt to cross each other) as necessitating the onset of a substorm to cause a sudden reduction in the radius of the inner ring. Thus, the behavior of the rings, which would otherwise be nearly independent of each other, are instead strongly coupled by the substorm process. The theory predicts the common radius of the coupled ring system (~ 19° to 22°) and its dependence on the cross-polar-cap potential (~ $0.2).

Introduction

This paper concerns the interaction between the two Birkeland current rings. The rings to which we refer are defined by an idealization of the Iijima-Potemra current pattern as shown in Figure 1. The purpose of the paper is to demonstrate, through a consideration of the equations that govern the radii of the two rings, that they are strongly united to each other. We will show that they form a coupled ring pair, expanding and contracting together. Each ring exhibits a distinctive type of behavior within the union. Ring 2 moves slowly and sedately while Ring 1 moves relatively much more quickly and in a jerky fashion in response to substorms, which play a major role in the coupling process.

The coupling is inherently dynamical in nature. The energy involved flows from the source of the convection potential through the Region 1 Birkeland current to the ionosphere and from there through the Region 2 Birkeland current to the ring current. The fact that the Birkeland currents are elements of major magnetospheric current systems is important in accounting for the interdependence of the rings. It is relevant therefore in a discussion of ring coupling to begin with a comparison of the strengths of the major magnetospheric current systems.

Comparison of Magnetospheric Current Systems

Table 1 is a list of representative values for the strengths of the known major magnetospheric current systems. The entry for the Region 1 Birkeland current system is the largest of the numbers shown, although the variation each current exhibits relative to its representative value is so large that the differences between the numbers in the table are not very significant.

The Birkeland current systems and the tail current system are capable of transferring energy from the solar wind to the magnetosphere (Atkinson, 1978). However, because the Region 2 current system connects to a radially fairly narrow portion of the inner (i.e., earthward) surface of the ring current, energy transfer to the inner magnetosphere can be inferred to be transported mainly through this system.

Since the argument that demonstrates the existence of ring coupling depends importantly on geometrical assumptions, it is useful to review these briefly here.

Geometrical Assumptions and Idealizations

We are concerned with the ionospheric expressions of the Regions 1 and 2 Birkeland current systems and with their relation to the tail, the plasma sheet and the ring current. The poleward and equatorward borders of the plasma sheet are identified with the surfaces generated by projecting Ring 1 and Ring 2, respectively, outward along magnetic field lines. Under the assumption that the poleward border of the plasma sheet separates regions of closed and open magnetic flux, Ring 1 encloses an area of open flux. The open flux area in the ionosphere is the polar cap. It maps along magnetic field lines into the tail lobes. The inter-ring annulus is the ionospheric image of the plasma sheet, formed by projecting the plasma sheet into the ionosphere along magnetic field lines. In this description the ring current is an inherent part of the plasma sheet. It is the part in which drift currents close in the magnetosphere, rather than on the magneto-

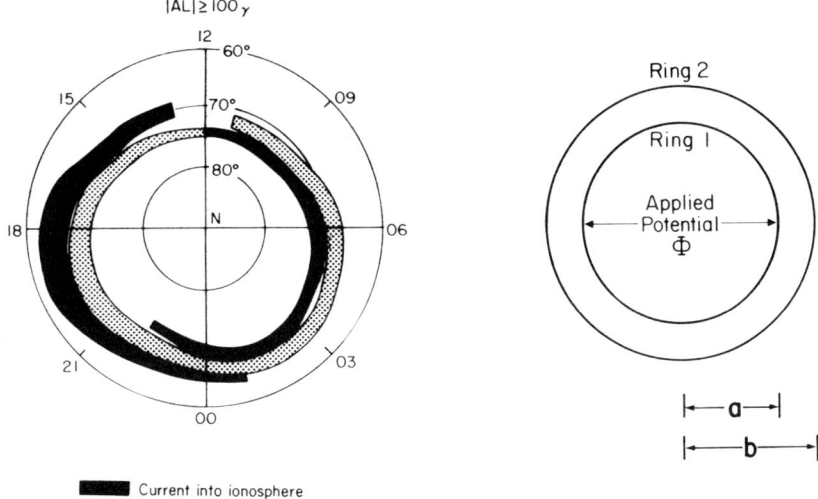

Figure 1. The Iijima-Potemra (1976) current pattern (left) and its idealization in terms of two concentric rings (right).

pause. The inner (earthward) border of the ring current is the same as the inner border of the plasma sheet. It maps along magnetic field lines into Ring 2 in the ionosphere. Field lines are assumed to be equipotential lines, although it is recognized that voltage differences can exist between different points on the same field line which are of the order of ten percent of the convection potential.

The discovery in the 1960's that the ring current tends to shield that portion of the magnetosphere lying under it from the convection electric field was a key concept leading to the recognition of ring coupling as a distinct magnetospheric process. In the context of ring phenomenology, the shielding tendency of the ring current is equivalent to a tendency of Ring 2 to be an electric equipotential.

The Equipotential Nature of Ring 2

The convection potential is impressed on Ring 1. To determine the response of the inner magnetosphere to the convection potential, its value and distribution on Ring 1 are prescribed and the resulting distribution of the potential and the current throughout the ionosphere are then calculated (e.g., Wolf and Harel, 1980). Vasyliunas (1972) showed that the ring current plasma must be taken into account in the calculation. It provides an effective Hall current in parallel with the ionospheric Hall current. The effective ring current Hall conductance Σ^* is estimated to be of the order of 10 times larger than the ionospheric Hall conductance Σ^H, a value which is approximately 20 times greater than the ionospheric Pedersen conductance Σ^P.

In steady state, the large effective Hall conductance causes the electric field originating on Ring 1 to be suppressed equatorward of Ring 2. the field there is reduced by a factor of the order of Σ^H/Σ^* compared to its value in the absence of the ring current. The property of the inner surface of the ring current to shield the volume under it from the influence of the convection potential was recognized earlier on the basis of a consideration of single particle drift orbits (Fejer, 1961; Swift, 1968; Schield et al., 1969). In the limit $\Sigma^* \to \infty$, the shielding becomes perfect, and the volume under the ring current becomes field-free, which implies that the ring current inner surface is an equipotential. Since the inner surface maps down to Ring 2 along equipotential magnetic field lines, Ring 2 is also an equipotential in this limit. The

Table 1

Current System	Representative Total Current (10^6 A)
Region 1 Birkeland	2.7[1]
Region 2 Birkeland	2.2[1]
Chapman-Ferraro	2.5[2]
Tail	2.0[3] per 10 R_e
Ring	2.0[4] per 50 nT Dst

[1] Iijima and Potemra (1978) values for low to moderately disturbed conditions ($|AL| \geq 100\gamma$, $2- \leq K_p \leq 4+$).

[2] Based on a geocentric subsolar distance of 10 R_e.

[3] Based on lobe field strength of 20 nT.

[4] Based on an equatorial line current of 4 R_e radius.

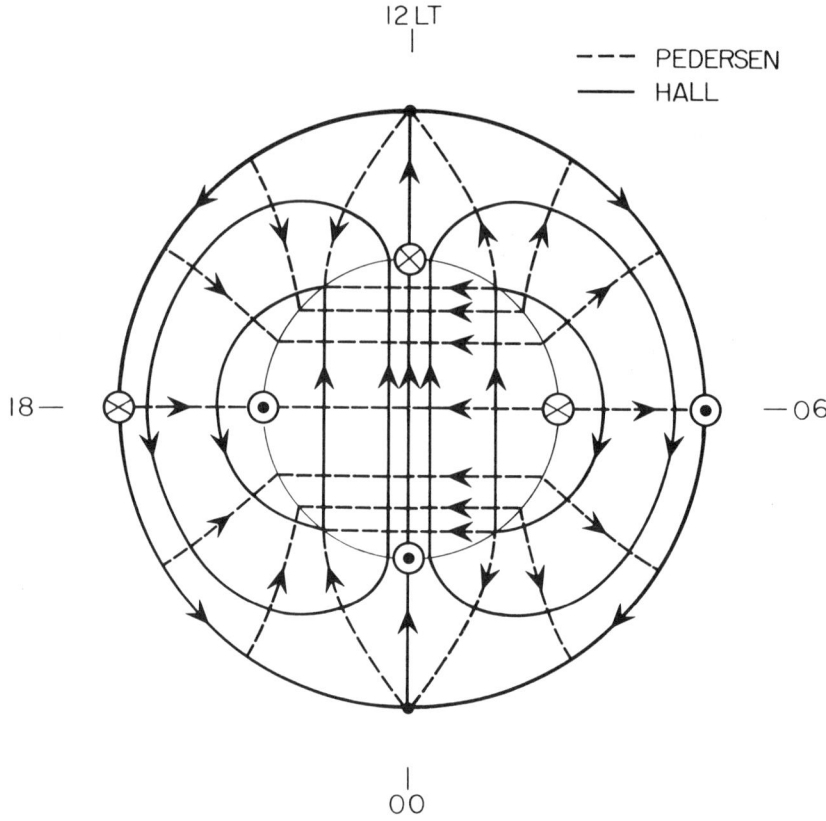

Figure 2. The patterns of Pedersen, Hall and Birkeland currents in the polar cap and inter-ring annulus (corresponding to the auroral oval) in the case where Ring 2 is an equipotential (from Crooker and Siscoe, 1981).

actual magnetospheric situation approximates the $\Sigma^* \to \infty$ limit. Ring 2 therefore approximates an equipotential.

Figure 2 shows the patterns of Hall and Pedersen currents produced by a sinusoidally distributed convection potential impressed on Ring 1 in the presence of an equipotential Ring 2 (Crooker and Siscoe, 1981). The Birkeland currents associated with Rings 1 and 2 in this case are related to the amplitude of the applied convection potential Φ, the ring radii a and b and the ionospheric conductances Σ_A^P and Σ_C^P by

$$I_1^P = \phi \left(\frac{b^2+a^2}{b^2-a^2} \Sigma_A^P + \Sigma_C^P \right) \qquad (1)$$

$$I_2^P = \frac{2ab}{b^2-a^2} \Sigma_A^P \phi \qquad (2)$$

in which subscripts A and C denote values of the conductance in the inter-ring annulus and in the polar cap, respectively. Conductances in the two regions are assumed to be uniform, but they are allowed to change across Ring 1. Superscripts P on the current symbols indicate that these are the Pedersen current contributions to the Birkeland currents. The Pedersen contributions are oriented parallel to the dawn-dusk meridian. There is no Hall current contribution to I_2 in the case of an equipotential Ring 2. There is a smaller contribution to I_1 from the Hall current than from the Pedersen current. It is oriented parallel to the noon-midnight meridian, and causes a rotation of the I_1 distribution relative to the I_2 distribution (Harel et al., 1981). Equations (1) and (2) are derived in the planar approximation, that is, under the assumption that the ring radii a and b are much smaller than the radius of the earth R_e.

To see the effect of the ring current plasma (or equivalently the effect of an equipotential Ring 2) equations (1) and (2) should be contrasted with the corresponding equations for the Birkeland currents in the absence of a ring current. These can be obtained from (1) and (2) by setting $b = \infty$:

$$I_1^P(b = \infty) = (\Sigma_A^P + \Sigma_C^P) \phi \qquad (3)$$

$$I_2^P(b = \infty) = 0 \qquad (4)$$

Equation (4) merely expresses the fact that the ring current is responsible for the existence of I. Comparison of equations (1) and (3) shows that the ring current amplifies I_1 by the factor $(b^2 + a^2)/(b^2 - a^2)$ multiplying Σ_A^P. Since Σ_A^P is generally appreciably larger than Σ_C^P, the amplification factor is nearly a direct multiplicative factor for I_1^P.

The statistical Birkeland current pattern presented by Iijima and Potemra (1976) gives a value close to five as an estimate for the amplification factor. That is, the ionosphere should draw approximately five times as much current from the source of the convection potential than it would in the absence of the ring current. If Σ_C^P is ignored in (3), one finds that in the absence of a ring current a conductance of $\Sigma_A^P = 40$ Mho is required to draw a Birkeland current of $I_1 = 2 \times 10^6$ A from a convection potential of $\Phi = 50$ kV. The values used here for the current and the potential are representative of typical conditions, but the observed conductance is closer to $\Sigma_A^P = 10$ Mho. Thus an amplification of approximately a factor of four again is seen to be needed to account for the observations.

The precise form of the amplification factor given in (1) depends on several idealizations, perhaps the most serious of which is the representation of the Birkeland current rings by mathematical lines, even though they must have finite latitudinal extent. The exercise serves to demonstrate the existence of such a factor. In the following the amplification factor and the corresponding coefficient in equation (2) will be treated as semi-empirical constants.

Equations Governing the Radii of the Birkeland Current Rings

The size of Ring 1 is fixed by the amount of magnetic flux in the tail lobes. The size of Ring 2 is determined by the location of the inner surface of the ring current. The sizes of the two rings are therefore governed by different physical processes. Consider the case of Ring 2 first.

Virtually identical formulas for the dependence of the size of Ring 2 on the convection potential and the size of Ring 1 have been derived on the basis of three different approaches (Jaggi and Wolf, 1973; Southwood, 1977; Siscoe, 1982a). Because of its use later in this paper, the approach of the last reference will be summarized here.

The ring current gains energy in an inward movement and loses energy in an outward movement. The energy is transferred to and from the ring current via the Region 2 Birkeland current during intervals when Ring 2 is not an equipotential. Such intervals occur when the applied convection potential is suddenly increased or decreased. A finite time is required for Ring 2 to reestablish its equipotential condition after a change in the convection potential. During this time the Region 2 Birkeland current crosses a finite net potential in the ionosphere, which causes energy to be transferred to or from the ring current, depending on the sense of the change in the convection potential. An increase in the convection potential transfers energy to the ring current, which causes the ring current to move inward. The inward movement has the effect of increasing the degree of shielding. The inward movement continues until the shielding has increased to the point where Ring 2 is again an equipotential. The Birkeland 2 current then crosses a net zero potential in the ionosphere, and energy transfer ceases. The process is seen to be self-limiting. A decrease in the applied convection potential reverses the steps described for the case of an increase in the potential.

Equating the energy transferred during non-equipotential Ring 2 intervals with the energy required to move the ring current radially yields an equation for the ring current position. The location of the inner surface of the ring current can be expressed in terms of the radius of Ring 2 under the assumption of dipole field geometry. The result is

$$\left(\frac{b}{R_e}\right)^{16/3} = A \frac{a}{b} \Phi \qquad (5)$$

where

$$A = \frac{\pi}{4} \frac{\eta \Sigma_A^P}{\Sigma^*} \frac{e}{K_o L_o^{8/3}} \qquad (6)$$

The factor η is the semi-empirical replacement factor for the coefficient in equation (2). It has a numerical value near 3. K_o is the kinetic energy of the ions in the ring current at the fiducial distance L_o and e is the electronic charge.

There are two limiting cases to consider, one corresponding to convection being driven by a viscous-like interaction with the solar wind (i.e., momentum transfer from the solar wind to plasma on closed field lines) and the other to convection driven by magnetic merging. The two cases will be treated separately even though they may occur simultaneously in nature.

Solution for the Viscous-Like Interaction

We wish to treat here the case in which magnetospheric convection is driven by a viscous-like interaction only. Neither dayside nor nightside magnetic merging occurs. Open magnetic flux therefore does not exist. If there is a tail, it is filled with its plasma sheet. There are no tail lobes. The definition of Ring 1 as the circumference of the open polar cap is not suitable here. We need to adopt for the moment the more general definition that Ring 1 encloses the region of anti-sunward convection, however generated. The flux enclosed by Ring 1 in this case is thus equal to the closed flux in the viscous boundary layer.

To estimate the size of Ring 1 we adopt a simple model in which the boundary layer is characterized by a thickness δ, a flow speed V_{BL} and a magnetic field strength B_{BL}. The convection potential is then

$$\phi_V = \delta \, V_{BL} \, B_{BL} \qquad (7)$$

As a further idealization we assume the magnetic field lines lie in meridian planes. Then by equating the magnetic flux inside Ring 1 between two meridian planes with the magnetic flux in the boundary layer between the same two planes, one finds

$$\frac{a}{R_e} = \alpha \, \phi^{\frac{1}{2}} \qquad (8)$$

where

$$\alpha = \left(\frac{2R}{R_e^2 V_{BL} B_P}\right)^{\frac{1}{2}} \qquad (9)$$

In (9) R is the geocentric distance to the boundary layer and B_P is the strength of the dipole field at the pole.

Combining equations (5) and (8) gives

$$\frac{b}{R_e} = (\alpha A)^{\frac{3}{19}} \phi^{\frac{9}{38}} \qquad (10)$$

To obtain numerical estimates for the ring radii in this purely viscous-like interaction case, we take $\eta = 3$, $\Sigma_A^P/\Sigma^* = 1/20$, $K_0 = 5$ keV, $L_0 = 10$, $V_{BL} = 200$ km/sec^{-1}, and $R = 15\, R_e$. Values of ϕ_V as large as 30 kV have been estimated (Crooker, 1980; Reiff et al., 1981). Although others find smaller values (e.g., Mozer and Temerin, 1982), we use $\phi_V = 30$ kV to obtain an upper limit on a and b for this case. If the ratios a/R_e and b/R_e are expressed in units of degrees, these being the polar angles of the rings in the small angle approximation, we find the angular radius of Ring 1 to be $a/R_e(\phi_V = 30$ kV$) = 6.2°$ and $b/R_e(\phi_V = 30$ kV$) = 11.5°$.

The sizes of the rings are much less than observed, and the relative inter-ring distance is much greater than observed. The result confirms what we knew beforehand, namely a pure viscous-like interaction can not account for the main part of convective phenomena. The nature of the disagreement is different than has been discussed previously. Furthermore the theory produces ring parameters that, as nearly as can be judged, are reasonable for the pure viscous-like interaction limit.

Solution for the Magnetic Merging Interaction

By comparison with the preceding discussions, the equation for the Ring 1 radius in the case of a purely magnetic merging interaction is relatively obvious. Let the magnetic flux in the tail lobe be F_L. By equating F_L to the magnetic flux enclosed by Ring 1, one finds

$$\pi a^2 B_p = F_L \qquad (11)$$

The flux in the tail lobes increases as a result of magnetic merging at the dayside magnetopause and decreases as a result of magnetic merging in the tail. If we denote the potential generated by the former as ϕ_M and by the latter as ϕ_T, then the equation for a is simply

$$\frac{d}{dt}(\pi a^2 B_p) = \phi_M - \phi_T \qquad (12)$$

It is instructive to consider the case in which only dayside merging occurs. Magnetic flux is added to the tail lobes but there is no return by means of tail merging. This situation simulates the growth phase of magnetospheric substorms. Then (12) has the solution

$$\frac{a}{R_e} = \beta \, t^{\frac{1}{2}} \qquad (13)$$

with

$$\beta = \left(\frac{\phi_M}{\pi R_e^2 B_p}\right)^{\frac{1}{2}} \qquad (14)$$

The solution for the radius of Ring 2 from (5) is then

$$\frac{b}{R_e} = (\beta A \, \phi_M)^{\frac{3}{19}} t^{\frac{3}{38}} \qquad (15)$$

One sees that the radii of the two rings expand at different rates under the convection potential that results from pure dayside merging. It is of interest to compare their relative rates of expansion,

$$\frac{db/dt}{da/dt} = \frac{1}{19}\frac{b}{a} \qquad (16)$$

Observationally one finds $b/a \sim 1.2$. Equation (16) thus reveals that Ring 2 expands about 6 percent as fast as Ring 1.

Based on a typical ring separation of 4°, Ring 1 will expand through the inter-ring distance in 2 hours under a 50 kV convection potential and in 1 hour under a convection potential at 100 kV, which is more representative of merging potentials during active times. In symbols this is

$$\frac{b-a}{da/dt} \sim 1 \text{ hour} \qquad (17)$$

There is a close correspondence between this time scale which represents a maximum inter-substorm time under a 100 kV potential and the typical

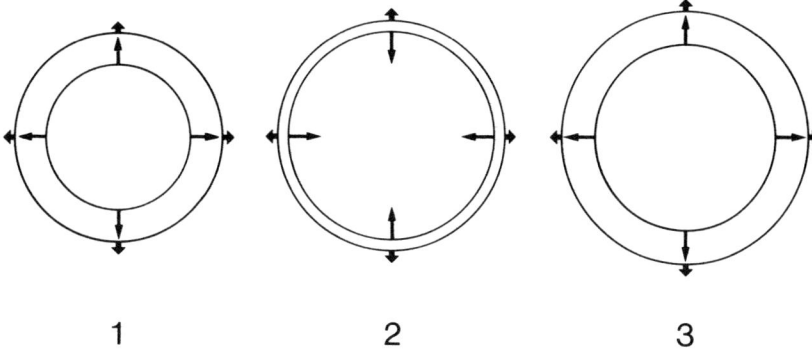

Figure 3. A sketch showing three successive stages in the expansion of Rings 1 and 2 in response to a potential applied across Ring 1. Substorms allow Ring 1 to maintain an average expansion speed equal to that of the more sluggish Ring 2 by letting it execute a series of virtually in place expansions and contractions (from Siscoe, 1982b).

duration of substorm repeat intervals within active periods.

Ring Coupling Via Substorm

Siscoe (1982b) noted that the faster expansion of the inner ring under growth phase conditions leads to a physical impossibility when the inner ring overtakes the outer ring. The correspondence between the predicted behavior of the ionospheric rings and the inferred behavior of the tail and plasma sheet during a substorm cycle was also noted. The encroachment of Ring 1 on Ring 2 under growth phase conditions corresponds morphologically to the thinning of the plasma sheet during the growth phase. The expansion phase that ensues converts open magnetic flux rapidly into closed magnetic flux. In the notation of equation (12), this process corresponds to $\Phi_T \gg \Phi_M$. The expansion of Ring 1 is thereby terminated and is followed by a contraction. In this way substorms prevent the occurrence of the physically impossible situation noted above. After the expansion phase, if dayside merging persists, the cycle is repeated. Ring 2 continues to expand slowly while Ring 1 undergoes a series of expansions and contractions associated with the growth and expansion phases of the substorm cycle (see Figure 3).

A series of substorms of diminishing amplitude will also occur when dayside merging ceases (Crooker and Siscoe, 1983), reminiscent of the model substorm sequence described by Akasofu (1977). In this case, between substorms both Φ_M and Φ_T are zero. The radius of Ring 1 is therefore constant, whereas equation (5) shows that the radius of Ring 2 will begin to contract. Thus Ring 2 encroaches on Ring 1, and the sequence of plasma sheet thinning followed by a substorm recurs.

Solution for Substorm Coupling

We have seen that the inter-ring annulus narrows regardless of whether the convection potential increases or decreases. Substorms prevent actual contact of the rings. It is therefore meaningful to speak of the two rings as a single coupled unit with a common radius. The common radius is not precisely defined because the theory does not at present predict the inter-ring separation (b-a). Also the substorm process introduces a vacillation in the value of $\underline{a}$. Observationally one finds $(b-a)/\frac{1}{2}(b+a) \stackrel{\sim}{\sim} 0.18$. Thus the qualitative behavior of the coupled ring system can be assessed from equation (5) by setting $\underline{a} = \underline{b}$. This gives

$$\frac{b}{R_e} = A^{\frac{3}{16}} \Phi^{\frac{3}{16}} \qquad (18)$$

Using the same values as before for the parameters that determine A but taking Φ in the range 50 kV to 100 kV to represent dayside merging potentials, we find the corresponding range of the polar angle of the common radius to be 19° to 22°, in satisfactory agreement with observations. The predicted weak dependence of the ring radii on Φ was shown to be consistent with data obtained from the polar orbiting S3-2 spacecraft (Crooker et al., 1983).

Directly and Indirectly Driven Processes

A hypothetical example illustrating the coupling of ring radii by the substorm process is shown in Figure 4 (Crooker and Siscoe, 1983). The convection potential is represented by a steady, approximately 20 kV potential generated by a viscous-like interaction onto which is superimposed a potential generated by dayside merging which forms an irregular sequence of intervals of increased potential. The various intervals of dayside merging exemplify different degrees of intensity and durations of the merging condition.

The lower data field shows schematically how the radii of the two rings vary in response to the changing convection potential. The vertical

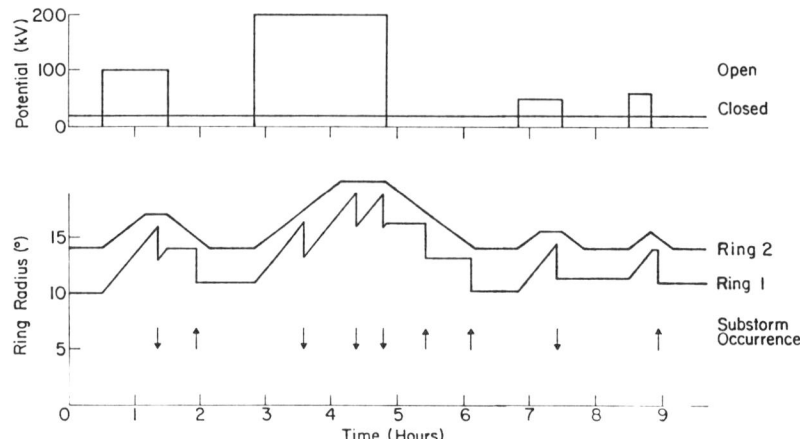

Figure 4. Schematic time variations of the cross-polar-cap potential and the radii of Rings 1 and 2. The contribution to the potential labeled OPEN is generated by dayside magnetic merging. The contribution labeled CLOSED is generated by a viscous-like (i.e., non-merging) process. The ring radii vary in response to the potential. When the ring separation shrinks to some critical value, a substorm occurs. Each substorm occurrence is marked by an arrow which indicates the north-south direction of the IMF at the time of substorm onset (from Crooker and Siscoe, 1983).

arrows mark occurrences of substorms. The directions of the arrows indicate the north-south orientation of the IMF at the time of the substorms.

At the beginning of the data interval, the angular radii of Rings 1 and 2 are set equal to 10° and 14°, respectively. During the first interval of merging potential, both rings begin to expand, Ring 1 faster than Ring 2. Before the interval ends, Ring 1 approaches close enough to Ring 2 that the criterion for a substorm is met. Ring 1 correspondingly suddenly contracts. Because the potential generated by merging continues through the substorm, Ring 1 begins to expand again. Before the substorm criterion is met a second time, the merging interval ends. The radius of Ring 1 now remains constant. Ring 2 begins to contract in order to reach its new equilibrium position under the reduced potential. Its contraction brings it close enough to Ring 1 once again to satisfy the criterion for a substorm. Accordingly Ring 1 abruptly contracts. Both rings are now back to where they started, in equilibrium with the potential generated by the viscous-like interaction.

The second merging interval persists longer. The substorm cycle is completed three times before it ends. The rings reach their greatest distention during this interval. In the succeeding quiescent period, the contraction of Ring 2 toward its quiescent equilibrium position triggers two substorms which cause Ring 1 to contract out of its way.

The final two minor merging sequences illustrate cases in which only one substorm occurs, the first during the merging interval, the second after it.

One recognizes in this textbook example a number of known varieties of substorm behavior. There are isolated substorms associated with short merging intervals and which occur either during or shortly after the interval. There is a longer merging interval during which a series of substorms rapidly follow one another, resembling the situation referred to as the convection bay (Pytte et al., 1978).

Perhaps the most significant aspect of magnetospheric dynamics illustrated by this figure is the distinction between directly and indirectly driven processes. The Dst index is a measure of ring current energy. It is therefore a function of the location of the inner surface of the ring current, that is, of the radius of Ring 2. The radius of Ring 2 is a function of Φ through equation (17). Since Φ may be regarded as a solar wind parameter, it follows that Dst is a directly driven quantity, in the sense that a directly driven quantity is determined fully by a specification of the solar wind variables (Akasofu, 1981).

On the other hand the AE index, which is a measure of substorms, is determined by the accidental (in the sense of virtually unpredictable) near contacts of the two rings as they perform their distinctive radial oscillations under the influence of changes in Φ. Thus the AE index and the substorms it measures are indirectly driven, in the sense that they depend on Φ, but they are not predictable on the basis of knowing Φ alone.

It should be cautioned that the above argument does not preclude the type of direct driving of substorms which is often referred to as external triggering. Substorms appear to be triggered by positive and negative sudden impulses and by a sudden northward turning of the IMF (see the review by McPherron, 1979). The triggering process or processes probably involve the generation of large scale inductive electric fields in the tail and would therefore neither be evident in nor require changes in the inter-ring distance.

Summary

The following is a list of the main points discussed in this paper.
1. The ring current plasma acts to make Ring 2 a nearly equipotential ring.
2. The quasi-equipotential nature of Ring 2 causes the ionosphere to draw three to five times more current from the source of the convection potential than it otherwise would.
3. There is a simple algebraic relation between the radius of Ring 2 and the convection potential (equation 5).
4. When the radius of Ring 1 is determined on the basis of a pure viscous-like interaction, one finds that the rings are smaller and relatively more separated than observed.
5. When the radius of Ring 1 is governed by dayside magnetic merging, one finds that Ring 1 expands on the order of 16 times faster than Ring 2. Substorms are needed to retain Ring 1 within Ring 2.
6. The inter-ring annulus narrows under conditions of increasing and decreasing Φ. Substorms prevent actual contact of the rings. The rings therefore form a coupled unit under the counteracting processes of annulus thinning and substorms.
7. The equation for the characteristic radius of the coupled unit predicts the observed size of the ring pair and gives the observed weak dependence of b on Φ (equation 18).
8. Dst is a directly driven quantity and AE is an indirectly driven quantity except for the phenomenon of external substorm triggering.

Acknowledgment. This research was supported by the National Science Foundation under grant number ATM81-20455.

References

Akasofu, S.-I., Physics of Magnetospheric Substorms, D. Reidel, Hingham, Mass., 1977.
Akasofu, S.-I., Energy coupling between the solar wind and the magnetosphere, Space Sci. Rev., 28, 121, 1981.
Atkinson, G., Energy flow and closure of current systems in the magnetosphere, J. Geophys. Res., 83, 1089-1103, 1978.
Crooker, N.U., The configuration of dayside merging, in Dynamics of the Magnetosphere, edited by S.-I. Akasofu, pp. 101-119, D. Reidel, Hingham, Mass., 1980.
Crooker, N. U., and G. L. Siscoe, Birkeland currents as the cause of the low-latitude asymmetric disturbance field, J. Geophys. Res., 86, 11,201-11,210, 1981.
Crooker, N. U., G. L. Siscoe, M. A. Doyle, and W. I. Burke, Ring coupling model: Implications for the ground state of the magnetosphere, submitted to J. Geophys. Res., 1983.
Crooker, N. U., and G. L. Siscoe, Ring coupling model: Implications for substorm onsets, Geophys. Res. Lett., 10, 761-764, 1983.
Fejer, J.A., The effects of energetic trapped particles on magnetospheric motions and ionospheric currents, Can. J. Phys., 39, 1409-1417, 1961.
Harel, M., R. A. Wolf, R. W. Spiro, P. H. Reiff, and C.-K. Chen, Quantitative simulation of a magnetospheric substorm, 2. Comparison with observations, J. Geophys. Res., 86, 2242-2260, 1981.
Iijima, T., and T. A. Potemra, The amplitude distribution of field-aligned currents at northern high latitudes observed by Triad, J. Geophys. Res., 81, 2165-2174, 1976.
Iijima, T., and T. A. Potemra, Large-scale characteristics of field-aligned currents associated with substorms, J. Geophys. Res., 83, 599-615, 1978.
Jaggi, R. K., and R. A. Wolf, Self-consistent calculation of the motion of a sheet of ions in the magnetosphere, J. Geophys. Res., 78, 2852-2866, 1973.
McPherron, R. L., Magnetospheric substorms, Rev. Geophys. Space Phys., 17, 657-681, 1979.
Mozer, F. S., and M. A. Temerin, The origin of magnetospheric electric fields, in book of abstracts for the 1982 Yosemite meeting on the Origin of Plasmas and Electric Fields in the Magnetosphere, distributed by J. C. Foster and F. T. Berkey, Center for Space and Space Sciences, Utah State University, Logan, Utah.
Pytte, T., R. L. McPherron, E. W. Hones, Jr., and H. I. West, Jr., Multiple-satellite studies of magnetospheric substorms: Distinction between polar magnetic substorms and convection-driven negative bays, J. Geophys. Res., 83, 663-679, 1978.
Reiff, P. H., R. W. Spiro, and T. W. Hill, Dependence of polar cap potential drop on interplanetary parameters, J. Geophys. Res., 86, 7639-7648, 1981.
Schield, M. A., J. W. Freeman, and A. J. Dessler, A source for field-aligned currents at auroral latitudes, J. Geophys. Res., 74, 247-256, 1969.
Siscoe, G. L., Energy coupling between Regions 1 and 2 Birkeland current systems, J. Geophys. Res., 87, 5124-5130, 1982a.
Siscoe, G. L., Polar cap size and potential: A predicted relationship, Geophys. Res. Lett., 9, 672-675, 1982b.
Southwood, D. J., The role of hot plasma in magnetospheric convection, J. Geophys. Res., 82, 5512-5520, 1977.
Swift, D. W., Further possible consequences of the asymmetric development of the ring current belt - Effect of variations in ionospheric con-

ductivity, Planet. Space Sci., 16, 329-342, 1968.

Vasyliunas, V. M., The interrelationship of magnetospheric processes, in Earth's Magnetospheric Processes, edited by B. M. McCormac, pp. 29-38, D. Reidel, Hingham, Mass., 1972.

Wolf, R. A., and M. Harel, Dynamics of the magnetospheric plasma, in Dynamics of the Magnetosphere, edited by S.-I. Akasofu, pp. 143-163, D. Reidel, Hingham, Mass., 1980.

REGION ONE BIRKELAND CURRENTS CONNECTING TO SUNWARD CONVECTING FLUX TUBES

J. L. Karty, R. A. Wolf, and R. W. Spiro

Department of Space Physics and Astronomy, Rice University, Houston, TX 77251

Abstract. Early theories of magnetospheric convection [e.g., Dungey, 1961; Axford and Hines, 1961] suggest that region-1 Birkeland currents flow primarily at the boundary between sunward and antisunward flowing plasma, or in an antisunward-flowing boundary layer. However, observations from S3-2 and S3-3 spacecraft indicate that region-1 currents very frequently flow on sunward-moving flux tubes, often considerably equatorward of the plasma flow reversal. Apparently, region-1 currents often flow on field lines in the interior of the plasma sheet, where the flow is subsonic and there is approximate balance between magnetic forces and pressure gradients.

Analytic stability arguments, based on current conservation, demonstrate that there may be a sector, within the plasma sheet in the night side magnetosphere, where plasma pressures are reduced relative to the surrounding regions. General physical arguments along with the stability analysis imply that a depleted region near the center of the magnetotail would be stable against the interchange instability and would generate currents of region-1 sense.

In order to simulate such a depleted region, several 'computer experiments' have been performed with the Rice Convection Model, enforcing gradients in plasma content at the tailward boundary of the calculation. (Earlier runs with the Rice Convection Model assumed, for simplicity, uniform plasma content in flux tubes drifting in from the tail.) Results of the new computer simulations indicate that gradients in flux tube content across the tailward boundary can cause region-1 currents to flow on sunward convecting closed flux tubes, in general agreement with the satellite data.

Introduction

In the last decade, Birkeland currents in the auroral regions have been extensively studied. Iijima and Potemra [1978] identified two primary sets of Birkeland currents, region-1 being the higher latitude set and region-2 being the lower latitude set (Fig. 1). Region-2 currents are usually interpreted as connecting to partial rings of gradient/curvature drift current flowing near the inner edge of the plasma sheet [Schield et al., 1969], although other interpretations have been presented [Rostoker and Boström, 1976; Hasegawa and Sato, 1979; Akasofu, 1982]. However, it is less clear to what region-1 currents connect in the magnetosphere. Recent efforts at explaining these currents have followed two major lines: a) understanding the day side currents as relating to a boundary layer phenomenon [e.g., Sonnerup, 1980] or b) as a combination of day side currents directly linked to open magnetic field lines, and night side currents connected to the plasma sheet [Stern, 1983]. It is now becoming clear that there are possibly several different sources of region-1 Birkeland currents, differentiated by generation mechanisms and local time.

Here we discuss the generation of region-1 currents on sunward convecting flux tubes in the sector extending eastward from late afternoon to post dawn. In traditional convection models of the magnetosphere, region-1 currents connect to the boundary layer, which flows mainly antisunward, or to the interface between sunward and antisunward convection. However, recent observations from the S3-3 and S3-2 satellites (e.g., see Fig. 2) indicate that most of the region-1 current frequently occurs equatorward of the electric field reversal on regions of sunward convecting flux tubes in the magnetosphere [e.g., Cattell et al., 1979; Mozer et al., 1980; Smiddy et al., 1980; Harel et al., 1981b]. Figure 2 shows an example of a dawn-dusk satellite crossing in which dusk side region-1 currents are deep within the sunward convecting region.

In this paper, we investigate whether substantial region-1 currents could possibly connect to gradient/curvature-drift currents in the subsonic-flow region of the plasma sheet. In previous work with the Rice Convection Model, in which the plasma content of plasma-sheet flux tubes was assumed uniform across the tail (i.e., independent of y), substantial region-1-sense current only appeared for very brief periods following major disruptions of the magnetosphere. In this paper we examine the implications of relaxing the assumption of uniform plasma content across the plasma sheet.

Analytic Stability Analysis

Before examining the consequences of substantial plasma content gradients in the magnetotail, we first consider what gradients are physically reasonable.

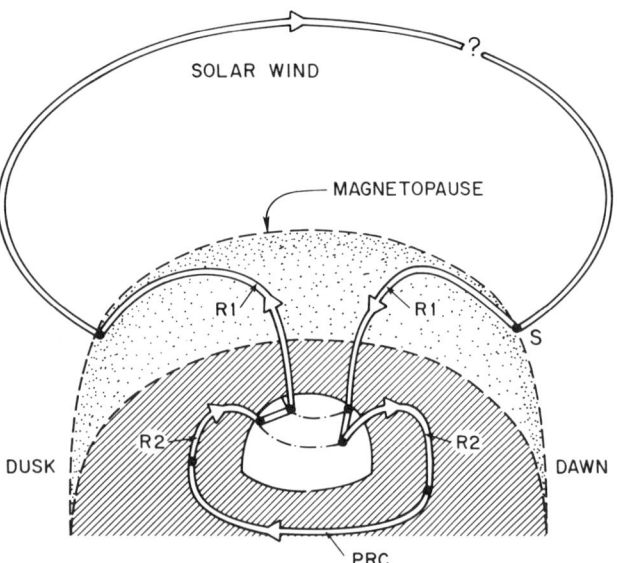

Fig. 1. Schematic diagram of the basic magnetospheric-convection current system. The view is from behind (antisunward of) the Earth, and above the equatorial plane. The lined region is the equatorial plane; the dotted region, the dayside magnetopause. The notations 'R1,' 'R2,' and 'PRC' mean 'Region-1 Birkeland Current,' 'Region-2 Birkeland Current,' and 'Partial Ring Current,' respectively [from Harel et al., 1981b].

For particles that gradient, curvature and $\underline{E} \times \underline{B}$ drift adiabatically and without loss on closed magnetic field lines, it can be shown that η, the number of particles per unit magnetic flux with given values of the first two adiabatic invariants, is an invariant along the particle drift path [Harel et al., 1981a]. Thus, the boundaries between different flux-tube population levels tend to align themselves with drift paths. Furthermore, according to conventional convection theory, particles in the central part of the plasma sheet (i.e., not close to the magnetopause) drift Earthward. If there is no loss, contours of constant η can extend from the inner plasma sheet to the distant magnetotail.

Consider the idealized stability problem shown in Figure 3. We assume a sharp interface between flux tubes with two different density levels. Specifically, n_1 and n_2 give the numbers of particles per unit magnetic flux below and above the interface, respectively. For simplicity we further assume an isotropic pitch angle distribution. The particle kinetic energy is given by

$$E = S\lambda \qquad (1a)$$

where λ is the energy invariant,

$$S \equiv [\int ds/B]^{-2/3} \qquad (1b)$$

and $\int ds/B$ is the volume of a magnetic flux tube of unit magnetic flux [Harel et al., 1981a]. For simplicity, assume that the particles all have the same charge q and the same value of the energy invariant λ. The bounce-averaged gradient/curvature-drift velocity is given by

$$\underline{v} = \frac{\lambda}{qB} \hat{z} \times \nabla S \qquad (2)$$

where B is the local magnetic field strength and q is the charge of the particle.

We assume that the ionospheric current can be represented as a height-integrated conduction current and that the magnetospheric current is due to gradient drift, curvature drift and magnetization current. As in standard MHD theory, we assume that $\underline{E} \times \underline{B}$ drift is much faster than gradient or curvature drift. Several more assumptions serve to make the problem analytically simple: the height-integrated ionospheric conductivity is assumed to be spatially uniform (which makes the Hall current divergenceless), and the gradient of S is assumed to be constant,

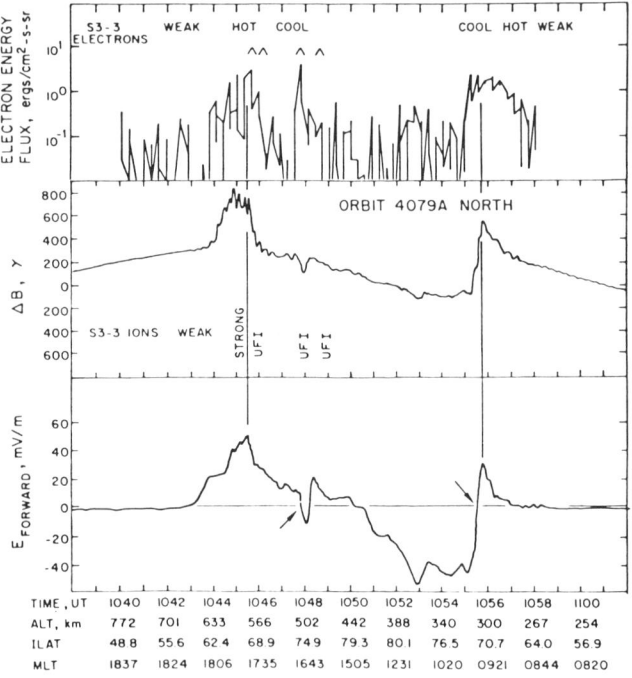

Fig. 2. Precipitating electron energy flux, east-west magnetic perturbations, ΔB, and forward component of electric field for orbit 4079 A, Northern hemisphere (S3-2 satellite data). Arrows in bottom panel indicate the most equatorward electric field reversal; vertical lines mark the equatorward edges of region-1 currents [from Harel et al., 1981a].

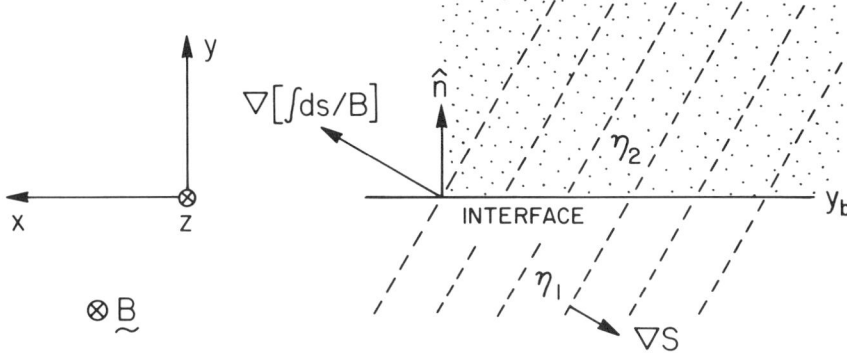

Fig. 3. Geometry for the analytic stability calculation. We are looking down on a small section of the northern ionosphere. The interface separates flux tubes with higher content (η_2) from those of lower content (η_1). The dotted lines represent contours of constant $S = [\int ds/B]^{-2/3}$.

though still in an arbitrary direction. We ignore the jump in flux-tube volume that occurs at the interface, which maps to a tangential discontinuity in the magnetosphere; the boundary currents associated with this jump do not directly affect the analysis.

Requiring that the divergence of height-integrated ionospheric current balances the divergence of flux-tube-integrated drift current yields the basic equation

$$\Sigma_p \nabla^2 V = (\eta_2 - \eta_1)\lambda \, \delta[y - y_b(x,t)]\{\partial S/\partial x + (\partial S/\partial y)(\partial y_b/\partial x)\} \quad (3)$$

where Σ_p = height-integrated Pedersen conductivity, V = electrostatic potential, and y_b = y-coordinate of the interface.

For the zero-order, steady-state situation, we further require that the flat interface shown in Figure 3 be an equipotential. We choose the reference frame in which the zero-order potential V_0 is antisymmetric about the interface. The zero-order solution is

$$V_0 = \frac{(\eta_2 - \eta_1)\lambda}{2\Sigma_p} \frac{\partial S}{\partial x} |y - y_b| \quad (4)$$

Referring to Figure 3, suppose that $\eta_2 > \eta_1$, i.e., the larger-content flux tubes are on the top, and $\partial S/\partial x < 0$, i.e., the larger-volume flux tubes are to the left. Then the electric field is directed away from the interface. (The electric field is in the +y direction above the boundary, in the -y direction below.) $\underset{\sim}{E} \times \underset{\sim}{B}$ drift is toward the left above the boundary and toward the right below. In other words, lower-content flux tubes drift toward the region of smaller flux-tube volume; higher-content flux tubes drift toward larger flux-tube volume. This is qualitatively similar to the standard interchange instability: flux tubes that are heavily loaded with plasma tend to drift away from regions of high magnetic field.

The stability of the system shown in Figure 3 can be studied by solving the problem assuming a small ripple in the interface. Namely, we assume that the interface lies at

$$y_b(x,t) = y_0 \exp[i(kx - \omega t)] \quad (5)$$

The condition that particles drifting just below the interface remain always just below the interface can be written

$$\frac{\partial y_b}{\partial t} = \frac{1}{B}\left[\frac{\partial V_0}{\partial y}(x, y_b - \epsilon, t)\frac{\partial y_b}{\partial x} + \frac{\partial V}{\partial x}(x, y_b - \epsilon, t)\right] \quad (6)$$

A similar condition can be written for particles just above the interface. The frequency ω in equation (5), which gives the time evolution of the ripple, can be obtained as follows: solve equation (3) for the electrostatic potential and substitute the result in equation (6). The resulting expression for ω is

$$\omega = \frac{i(\eta_2 - \eta_1)\lambda k}{2\Sigma_p B} \frac{\partial S}{\partial y} \quad (7)$$

In the reference frame chosen, the ripple does not propagate, but either grows or decays. The stability criterion can be summarized as follows: let $\hat{n}$ be a unit vector normal to the interface, in the direction of larger flux-tube content. If

$$\hat{n} \cdot \nabla [\int ds/B] > 0$$

then the ripple decays in time. If

$$\hat{n} \cdot \nabla [\int ds/B] < 0$$

then the ripple grows in time.

Figure 4 shows examples of stable interface configurations for the plasma sheet, viewed both in the ionosphere (as in Fig. 3) and in the magnetospheric equatorial plane. It is assumed that flux-tube volume, viewed in the equatorial plane, increases with increasing distance from the Earth. Viewed in the ionosphere, flux-tube volume decreases with increasing distance from the pole. In drawing the contours in the upper diagrams, we assumed, for stability, that $\hat{n}$ had to have a positive component in the direction of increasing flux-tube volume. These plasma-sheet interfaces cannot reach to low latitudes, and thus must become parallel (or almost parallel) to latitude lines at their low-latitude extremes.

Note that the lower right diagram in Figure 4 corresponds to a depleted region near the center of the magnetotail. A similarly shaped plasma-enhanced region in the central magnetotail would be unstable.

Figure 5 sketches the current configurations corresponding to the interface configurations shown in Figure 4. Note that these interfaces, in most cases, carry region-1-sense currents (down to the ionosphere on the dawn side, up from the ionosphere on the dusk side). However, in diagram a (b), there is little dusk side (dawn side) current. The depleted-channel configuration (c) carries region-1 current on both sides. So far, computer simulations have been performed only for the situation schematically depicted in Figures 4(c) and 5(c). The exact positioning of the region of current generation is still being investigated.

The results of the stability calculations imply

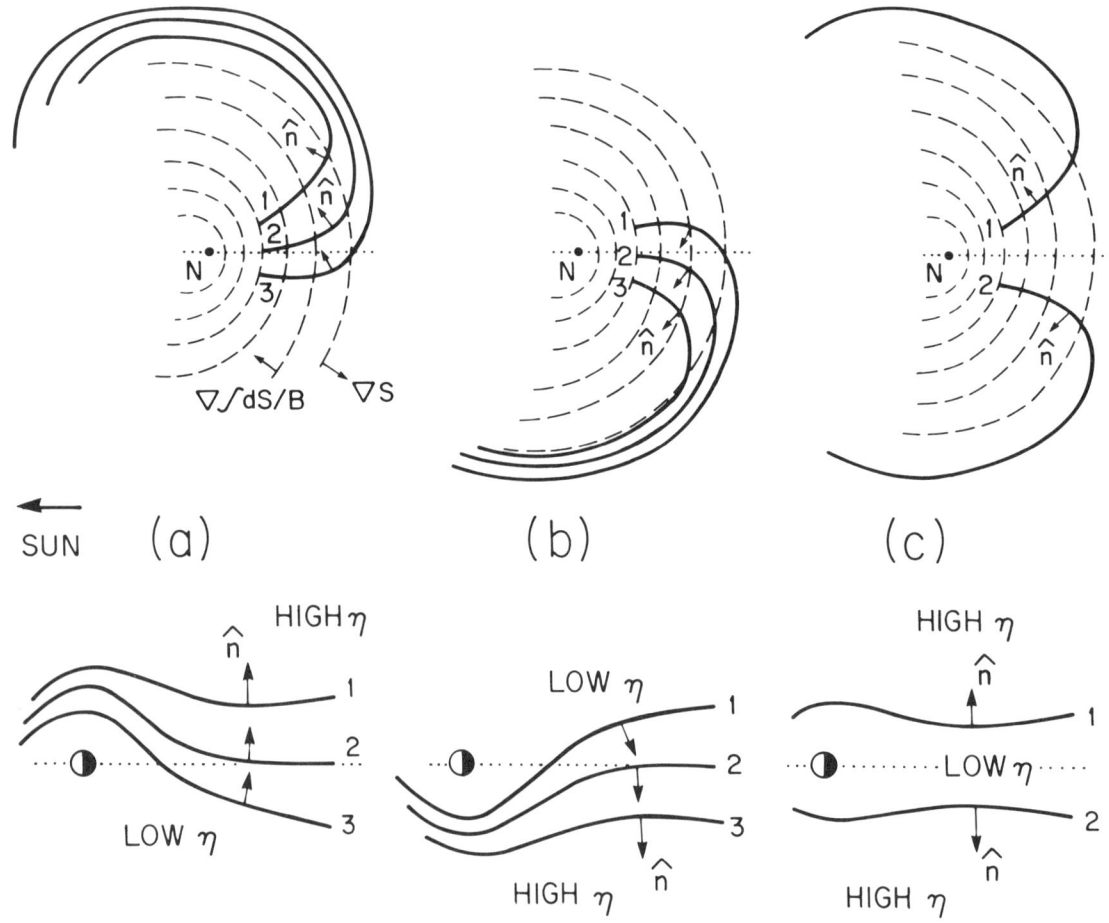

Fig. 4. Some possible stable configurations for the plasma sheet. The top diagrams show the northern ionosphere, with Sun to the left. 'N' designates the north magnetic pole. Heavy curves show possible stable interfaces. The arrows based on the heavy curves point in the direction of higher flux-tube content. The dashed curves are contours of constant flux tube volume. The lower diagrams show an equatorial view of the same interfaces.

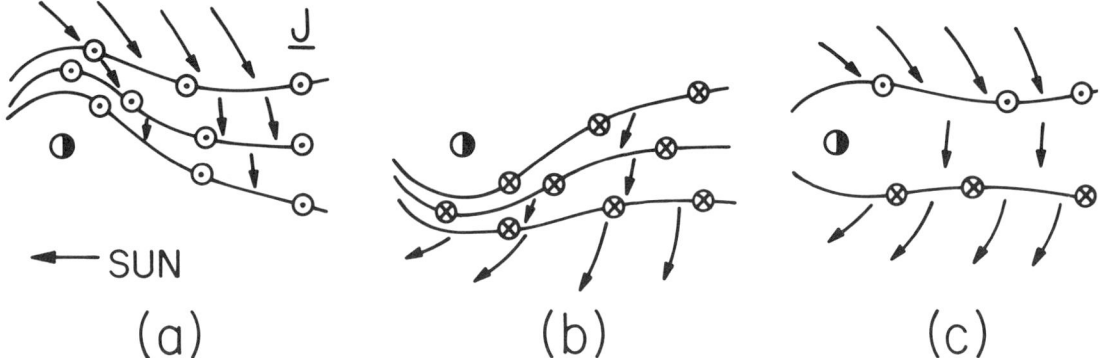

Fig. 5. Currents flowing on and near the stable interfaces shown in Figure 4. The arrows show gradient/curvature-drift currents in the equatorial plane. The symbol ⊙ represents Birkeland current from the magnetosphere down into the ionosphere. The ⊗ symbol represents current up from the ionosphere to the magnetosphere.

that a magnetotail channel-like configuration, with higher content flux tubes toward the flanks, would be stable once formed. Observations of Hones et al. [1981] show greatest sunward flows near the magnetotail axis, consistent with the depleted channel configuration, which requires larger dawn-dusk electric field in the depleted region.

Figure 6 shows how region-1 and region-2 currents fit together in this picture, including a depleted channel. The inner edge of the plasma sheet is represented as another interface, but one that does not connect to the far tail. This figure illustrates the connection of region-2 current to the partial ring current as suggested by previous Rice computer simulations (see Fig. 1) and other workers [e.g., Schield et al., 1969]. In addition, however, this figure sug-

gests a tail with a central region of plasma depletion. Gradient and curvature drift currents are shown near the flanks through the high plasma content region.

Computer Model and Results

To complement the analytic calculations, we have modified the Rice Convection Model (RCM) to include the effects of non-uniform plasma distribution in the tail plasma sheet. The RCM simulates the electrodynamics of convection by self-consistently calculating electric fields, currents, and plasma flows in the coupled magnetosphere/ionosphere system.

Variation in plasma sheet flux tube content across the tail is simulated by 'holding back' appropriate species of particles. (The Rice Con-

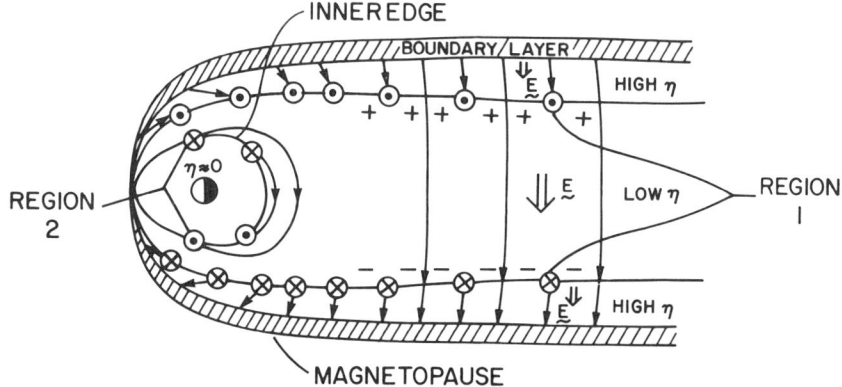

Fig. 6. Diagram combining a standard view of region-2 currents with region-1 currents flowing at the edges of a depleted channel. Black lines with arrowheads represent gradient/curvature-drift currents. Hollow arrows represent the electric fields. Plus and minus signs qualitatively indicate the charge distribution required to keep $\nabla \cdot \underline{J} = 0$ (by drawing horizontal ionospheric currents away from regions where Birkeland currents flow into the ionosphere).

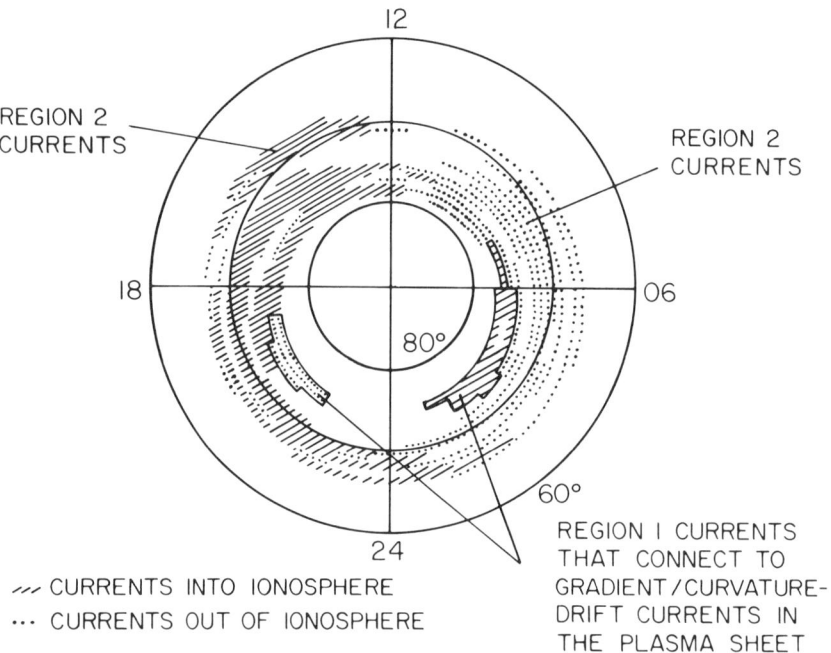

Fig. 7. Global pattern of Birkeland currents, UT = 10:40.

vection Model depicts the plasma distribution in terms of 21 'species' [e.g., Harel et al., 1981a].) Several different computer experiments, with different tailward boundary conditions, were performed for the September 19, 1976 substorm event. In one of the runs, the flux tube content in the center of the tail was 83% of the content in the flanks. In this run, this same 'holding back' procedure was used throughout the event. Particles were held by pinning the inner edge of the outermost species at the back boundary in the MLT ≈ 22 to MLT ≈ 2 sector. For the other runs more species were held back, so that the content in the center of the tail was slightly less than half the content in the flanks. In all these runs, the modeling region previously used with the RCM was approximately doubled in the equatorial area by moving the tailward boundary to ~ 20 R_E.

Preliminary results of the computer simulations indicate that the presence of a depleted channel of lower-density plasma near the axis of the magnetotail results in the generation of region 1 Birkeland currents on sunward convecting flux tubes. These currents have so far only been generated in the dawn-dusk portions of the night time sector of the magnetosphere. Figure 7 gives the global pattern of computed field-aligned currents for a time near the peak of the modeled substorm. The night time upward currents shown at the highest modeled latitudes in the post-dusk sector are of region 1 sense, and are located in the region of westward (i.e., sunward) drift; similarly, the downward currents at the highest latitudes on the dawn side are also located equatorward of the electric field reversal in the sunward flow region. Note that the computed region 1 current density is comparable to the region 2 current density across a large portion of the night time sector.

It should be emphasized that these currents of region 1 sense are located equatorward of the electric field reversal and are generated as a result of enforcing a low density region of plasma depletion near the axis of the magnetotail. As a result of the depleted plasma region, dawn to dusk cross-tail currents are diverted into the ionosphere as downward currents on the dawn side and return as upward currents on the dusk side. The depleted channel scenario resembles the tail current interruption picture of substorm expansion [e.g., Atkinson, 1972]. We have modified this tail-current interruption idea by viewing the depleted magnetotail channel as existing most of the time, not just during substorm expansions, since the phenomenon of region-1 currents on sunward-convecting flux tubes has also been observed in periods other than expansion phases. We have included in the model, however, that the depletion may be largest during the expansion phase. Our tentative conclusion is that cross-tail gradients in plasma-sheet flux-tube content tend to produce region-1 currents that connect to sunward-moving plasma-sheet flux tubes.

Conclusion

A magnetotail depleted channel, with flowing plasma inside, is stable against the interchange instability. Installing a depleted channel of low-density flux tubes near the center of the magnetotail as a boundary condition for the RCM results in the generation of region-1 currents on sunward-convecting plasma-sheet flux tubes.

Acknowledgments. The authors are pleased to acknowledge helpful conversations with G. M. Erickson, G.-H. Voigt, and T. W. Hill. This research was supported by the National Science Foundation under grants ATM79-20157 and ATM82-06026, by the Air Force Geophysics Laboratory under contracts F19628-80-C-0009 and F19628-83-K-0016, and by NASA grants NGR-44-006-137 and NGL-44-006-012.

References

Akasofu, S.-I., Hall current as a source of the cross-tail current interruption, the asymmetric main phase field, and the poleward expanding auroral bulge, Planet. Space Sci., 30, 389, 1982.

Atkinson, G., Magnetospheric flows and substorms, in Magnetosphere-Ionosphere Interactions, edited by K. Folkestad, p. 203, Universitetsforlaget, Oslo, 1972.

Axford, W. I., and C. O. Hines, A unifying theory of high-latitude geophysical phenomena, and geomagnetic storms, Can. J. Phys., 39, 1433, 1961.

Cattell, C., R. L. Lysak, R. B. Torbert, and F. S. Mozer, Observations of differences between regions of current flowing into and out of the ionosphere, Geophys. Res. Lett., 6, 621, 1979.

Dungey, J. W., Interplanetary magnetic field and the auroral zones, Phys. Rev. Lett., 6, 47, 1961.

Harel, M., R. A. Wolf, P. H. Reiff, R. W. Spiro, W. J. Burke, F. J. Rich, M. Smiddy, Quantitative simulation of a magnetospheric substorm, 1, Model logic and overview, J. Geophys. Res., 86, 2217, 1981a.

Harel, M., R. A. Wolf, R. W. Spiro, P. H. Reiff, C.-K. Chen, W. J. Burke, and F. J. Rich, Quantitative simulation of a magnetospheric substorm, 2, Comparison with observations, J. Geophys. Res., 86, 2242, 1981b.

Hasegawa, A., and T. Sato, Generation of field aligned current during substorms, in Dynamics of the Magnetosphere, edited by S.-I. Akasofu, p. 529, D. Reidel, Dordrecht, Holland, 1979.

Hones, E. W., Jr., J. Birn, S. J. Bame, J. R. Asbridge, G. Paschmann, N. Sckopke, and G. Harendel, Further determination of the characteristics of magnetospheric plasma vortices with ISEE 1 and 2, J. Geophys. Res., 86, 814, 1981.

Iijima, T., and T. A. Potemra, Large-scale characteristics of field-aligned currents associated with substorms, J. Geophys. Res., 83, 599, 1978.

Mozer, F. S., C. A. Cattell, M. K. Hudson, R. L. Lysak, M. Temerin, and R. B. Torbert, Satellite measurements and theories of low altitude auroral particle acceleration, Space Sci. Rev., 27, 155, 1980.

Rostoker, G., and R. Boström, A mechanism for driving the gross Birkeland current configuration in the auroral oval, J. Geophys. Res., 81, 235, 1976.

Schield, M. A., J. W. Freeman, and A. J. Dessler, A source for field-aligned currents at auroral latitudes, J. Geophys. Res., 74, 247, 1969.

Smiddy, M., W. J. Burke, M. C. Kelley, N. A. Saflekos, M. S. Gussenhoven, D. A. Hardy, and F. J. Rich, Effects of high latitude conductivity on observed convection electric fields and Birkeland currents, J. Geophys. Res., 85, 6811, 1980.

Sonnerup, B. U. Ö., Theory of the low-latitude boundary layer, J. Geophys. Res., 85, 2017, 1980.

Stern, D. P., The origins of Birkeland currents, Rev. Geophys. Space Phys., 21, 125, 1983.

CORRECTED GEOMAGNETIC COORDINATES FOR EPOCH 1980

Georg Gustafsson

Kiruna Geophysical Institute, S-981 27 Kiruna, Sweden

Abstract. A new set of corrected geomagnetic coordinates have been calculated. They are based on the field model IGRF-1980 for Epoch 1980. They differ from the earlier coordinates based on IGRF-1965 model by about one degree in latitude in the auroral oval region and a few degrees in longitude in certain areas.

Introduction

The corrected geomagnetic coordinates is a magnetic coordinate system that has been found useful for quantitative analysis of geophysical phenomena, and it has been widely used. Over the last years the magnetic field models have been improved, in particular the MAGSAT data have contributed. The long term variations in the internal field of the earth cause a change in magnetic coordinates corresponding to 10 km/year in certain areas of the earth. Therefore, it has been considered valuable to revise the corrected geomagnetic coordinates calculated by Gustafsson (1970), which were based on a field model for 1965.

The corrected geomagnetic coordinates have been presented in tables of geocentric coordinates on a sphere of radius 6371.2 km.

Corrected Geomagnetic Coordinates

The corrected geomagnetic coordinates, latitude and longitude in degrees, have been calculated from the field model IGRF-1980 for Epoch 1980, see Table 1. The tables are given for each 2 degrees in geocentric latitude and 5 degrees in longitude and at a distance of 6371.2 km, corresponding to the mean radius of the earth.

The physical meaning of the ovals of constant corrected geomagnetic latitude is simply a projection of circles in the geomagnetic (dipole) equatorial plane at different heights down to the earth's surface along the field lines represented by the model.

The location of the dipole pole at a latitude of 78.80 degrees, longitude -70.76 degrees in geocentric coordinates, was calculated from the field model.

When evaluating the coordinates the integration has been carried out both up and down the field line and if the deviation has been larger than 0.01 degrees in any direction at the starting point the space in the tables has been left open. This occurred near the magnetic poles.

Discussion

The coordinates presented here are based on 8 terms of the spherical harmonic development of the main field of the earth. The higher order terms decrease very rapidly with height. The contribution to the geomagnetic field of the harmonic terms of n = 2, 3, 4 and so on decrease outward proportional to r^{-4}, r^{-5}, r^{-6}, respectively (cf. Gustafsson, 1970). As an example, at 300 km the term n = 2 will be reduced by 20% and the n = 7 term by 53%. Therefore, the field becomes more and more dipole-like with height. That is also the reason why the corrected geomagnetic coordinates cannot be calculated near the equator.

The tables list the coordinates at a distance of 6371.2 km, if the coordinates should be

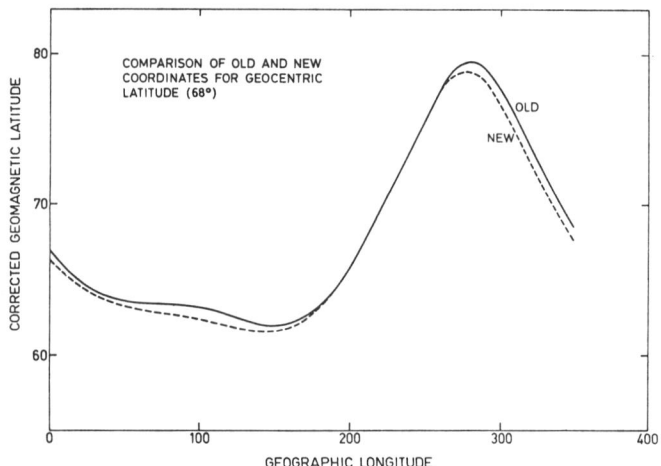

Fig. 1. A comparison between old and new corrected geomagnetic coordinates for the geocentric latitude 68°.

TABLE 1. Corrected geomagnetic coordinates (latitude and longitude) for each 2° in geocentric latitude and 5° in geocentric longitude.

NORTHERN HEMISPHERE

LATI-TUDE	LONGITUDE EAST					
	0.0	10.0	20.0	30.0	40.0	50.0
88.0	82.1 154.9	81.8 155.3	81.5 156.1	81.2 157.3	81.0 158.7	80.8 160.4
86.0	81.6 140.0	81.0 141.6	80.5 143.8	80.1 146.5	79.7 149.4	79.4 152.7
84.0	80.6 127.5	79.9 130.6	79.3 134.2	78.7 138.1	78.2 142.3	77.8 146.8
82.0	79.3 117.8	78.5 122.1	77.7 126.7	77.1 131.7	76.6 136.9	76.2 142.3
80.0	77.7 110.3	76.8 115.5	76.1 121.0	75.4 126.7	74.9 132.6	74.5 138.8
78.0	76.0 104.5	75.1 110.4	74.3 116.5	73.6 122.8	73.1 129.2	72.7 135.9
76.0	74.2 100.0	73.2 106.4	72.4 112.9	71.8 119.6	71.3 126.5	70.9 133.6
74.0	72.4 96.4	71.4 103.2	70.5 110.0	69.9 117.0	69.4 124.2	69.0 131.7
72.0	70.4 93.5	69.4 100.5	68.6 107.6	68.0 114.9	67.5 122.4	67.1 130.1
70.0	68.4 91.1	67.4 98.3	66.6 105.6	66.0 113.1	65.5 120.8	65.2 128.8
68.0	66.4 89.1	65.4 96.5	64.6 103.9	64.0 111.6	63.6 119.5	63.3 127.6
66.0	64.3 87.4	63.3 94.8	62.6 102.4	62.0 110.2	61.6 118.3	61.3 126.7
64.0	62.3 85.9	61.2 93.4	60.5 101.2	60.0 109.1	59.6 117.3	59.4 125.8
62.0	60.1 84.5	59.1 92.2	58.4 100.0	57.9 108.1	57.6 116.5	57.4 125.1
60.0	57.9 83.4	57.0 91.1	56.3 99.0	55.9 107.2	55.6 115.7	55.4 124.5
58.0	55.7 82.4	54.8 90.2	54.2 98.2	53.8 106.4	53.5 115.1	53.4 124.0
56.0	53.5 81.4	52.6 89.3	52.0 97.4	51.6 105.8	51.5 114.5	51.4 123.5
54.0	51.2 80.6	50.3 88.5	49.8 96.7	49.5 105.1	49.4 114.0	49.3 123.2
52.0	48.8 79.8	48.0 87.8	47.6 96.0	47.4 104.6	47.3 113.5	47.3 122.8
50.0	46.5 79.1	45.7 87.1	45.3 95.4	45.2 104.1	45.2 113.2	45.2 122.6
48.0	44.0 78.5	43.3 86.5	43.0 94.9	43.0 103.7	43.1 112.8	43.2 122.3
46.0	41.5 77.9	40.9 86.0	40.7 94.4	40.8 103.3	40.9 112.5	41.1 122.1
44.0	38.9 77.3	38.4 85.5	38.3 94.0	38.5 102.9	38.8 112.3	39.0 122.0
42.0	36.3 76.8	35.8 85.0	35.9 93.6	36.2 102.6	36.6 112.0	36.9 121.8
40.0	33.6 76.4	33.2 84.6	33.4 93.2	33.9 102.3	34.4 111.8	34.8 121.7
38.0	30.8 75.9	30.6 84.2	30.9 92.9	31.5 102.1	32.2 111.7	32.6 121.6
36.0	27.9 75.6	27.8 83.9	28.4 92.6	29.2 101.9	30.0 111.5	30.5 121.5
34.0	24.9 75.2	25.0 83.6	25.7 92.4	26.7 101.7	27.7 111.4	28.3 121.5
32.0	21.8 74.9	22.1 83.3	23.1 92.2	24.3 101.5	25.4 111.3	26.2 121.4
30.0	18.4 74.6	19.0 83.1	20.3 92.0	21.7 101.4	23.1 111.2	24.0 121.4
28.0	14.7 74.3	15.7 82.8	17.4 91.9	19.2 101.3	20.7 111.1	21.8 121.3
26.0	10.4 74.0	12.0 82.7	14.3 91.7	16.5 101.2	18.4 111.1	19.5 121.3
24.0	3.3 73.8	7.3 82.5	10.8 91.6	13.7 101.2	15.9 111.1	17.3 121.3

TABLE 1. (continued)

NORTHERN HEMISPHERE

LATI-TUDE	LONGITUDE EAST					
	60.0	70.0	80.0	90.0	100.0	110.0
88.0	80.6 162.3	80.5 164.4	80.4 166.6	80.3 168.8	80.3 171.0	80.3 173.3
86.0	79.1 156.2	78.9 159.8	78.7 163.6	78.6 167.4	78.6 171.3	78.6 175.1
84.0	77.5 151.5	77.3 156.4	77.1 161.4	76.9 166.5	76.8 171.5	76.8 176.5
82.0	75.8 148.0	75.6 153.8	75.4 159.7	75.2 165.7	75.1 171.7	75.1 177.6
80.0	74.1 145.1	73.8 151.7	73.6 158.3	73.5 165.1	73.4 171.8	73.3 178.5
78.0	72.4 142.8	72.1 150.0	71.9 157.2	71.7 164.5	71.6 171.9	71.5 179.2
76.0	70.6 141.0	70.3 148.6	70.1 156.3	69.9 164.1	69.8 172.0	69.7 179.7
74.0	68.7 139.4	68.5 147.4	68.3 155.5	68.1 163.8	68.0 172.0	67.8 180.2
72.0	66.9 138.1	66.7 146.4	66.5 154.9	66.3 163.5	66.1 172.1	66.0 180.6
70.0	65.0 137.1	64.8 145.6	64.6 154.3	64.5 163.2	64.3 172.1	64.1 180.9
68.0	63.1 136.1	62.9 144.9	62.8 153.9	62.6 163.0	62.4 172.1	62.2 181.2
66.0	61.2 135.3	61.0 144.3	60.9 153.5	60.7 162.8	60.5 172.1	60.3 181.4
64.0	59.2 134.7	59.1 143.8	59.0 153.1	58.8 162.6	58.6 172.2	58.3 181.6
62.0	57.3 134.1	57.2 143.4	57.0 152.8	56.9 162.5	56.6 172.1	56.4 181.8
60.0	55.3 133.6	55.2 143.0	55.1 152.6	54.9 162.3	54.7 172.1	54.4 181.9
58.0	53.3 133.2	53.2 142.7	53.1 152.4	52.9 162.2	52.7 172.1	52.4 181.9
56.0	51.3 132.9	51.2 142.5	51.1 152.2	50.9 162.1	50.7 172.1	50.4 182.0
54.0	49.3 132.6	49.2 142.3	49.1 152.1	48.9 162.0	48.7 172.0	48.3 182.0
52.0	47.3 132.4	47.2 142.1	47.0 152.0	46.8 162.0	46.6 172.0	46.3 182.0
50.0	45.2 132.2	45.1 142.0	45.0 151.9	44.8 161.9	44.5 171.9	44.2 182.0
48.0	43.2 132.1	43.1 141.9	42.9 151.9	42.7 161.8	42.5 171.9	42.2 181.9
46.0	41.1 131.9	41.0 141.9	40.8 151.8	40.6 161.8	40.4 171.8	40.1 181.9
44.0	39.0 131.9	38.9 141.8	38.7 151.8	38.5 161.7	38.2 171.8	38.0 181.8
42.0	37.0 131.8	36.8 141.8	36.6 151.8	36.3 161.7	36.1 171.7	35.9 181.7
40.0	34.9 131.8	34.7 141.8	34.4 151.8	34.2 161.7	34.0 171.6	33.8 181.6
38.0	32.8 131.7	32.6 141.8	32.3 151.8	32.0 161.6	31.8 171.5	31.7 181.5
36.0	30.6 131.7	30.4 141.8	30.1 151.8	29.8 161.6	29.7 171.5	29.5 181.4
34.0	28.5 131.7	28.3 141.8	27.9 151.8	27.6 161.6	27.5 171.4	27.4 181.3
32.0	26.4 131.7	26.1 141.8	25.7 151.8	25.4 161.5	25.3 171.3	25.2 181.2
30.0	24.2 131.7	24.0 141.8	23.5 151.7	23.2 161.5	23.1 171.3	23.1 181.1
28.0	22.1 131.7	21.8 141.8	21.3 151.7	21.0 161.5	20.9 171.2	20.9 181.1
26.0	19.9 131.6	19.6 141.8	19.1 151.7	18.7 161.4	18.7 171.1	18.7 181.0
24.0	17.7 131.6	17.4 141.8	16.9 151.7	16.5 161.4	16.4 171.1	16.6 180.9

TABLE 1. (continued)

NORTHERN HEMISPHERE

LATI- TUDE	LONGITUDE EAST					
	120.0	130.0	140.0	150.0	160.0	170.0
88.0	80.3 175.5	80.4 177.6	80.5 179.5	80.7 181.3	80.9 182.9	81.1 184.2
86.0	78.6 178.9	78.7 182.5	78.9 186.0	79.1 189.2	79.5 192.2	79.8 194.9
84.0	76.9 181.5	77.0 186.2	77.2 190.8	77.4 195.2	77.8 199.3	78.3 203.1
82.0	75.1 183.4	75.2 189.1	75.4 194.5	75.6 199.7	76.0 204.7	76.5 209.4
80.0	73.3 185.0	73.3 191.3	73.5 197.4	73.7 203.3	74.1 208.9	74.7 214.2
78.0	71.4 186.3	71.5 193.2	71.6 199.8	71.8 206.1	72.2 212.2	72.7 218.0
76.0	69.6 187.3	69.5 194.6	69.6 201.7	69.8 208.4	70.2 214.8	70.7 221.0
74.0	67.7 188.2	67.6 195.9	67.6 203.3	67.8 210.3	68.1 217.0	68.7 223.4
72.0	65.8 188.9	65.7 196.9	65.6 204.6	65.7 211.9	66.0 218.8	66.6 225.5
70.0	63.9 189.5	63.7 197.8	63.7 205.7	63.7 213.2	63.9 220.3	64.5 227.2
68.0	61.9 190.0	61.7 198.5	61.6 206.6	61.6 214.3	61.8 221.6	62.3 228.6
66.0	60.0 190.5	59.7 199.2	59.5 207.4	59.5 215.2	59.7 222.7	60.2 229.8
64.0	58.0 190.8	57.7 199.7	57.5 208.1	57.4 216.0	57.6 223.6	58.1 230.8
62.0	56.0 191.1	55.7 200.1	55.4 208.7	55.3 216.7	55.5 224.3	56.0 231.7
60.0	54.0 191.4	53.6 200.5	53.3 209.1	53.2 217.3	53.3 225.0	53.8 232.4
58.0	52.0 191.6	51.6 200.8	51.2 209.5	51.1 217.8	51.2 225.6	51.7 233.1
56.0	50.0 191.7	49.5 201.0	49.2 209.9	49.0 218.2	49.1 226.0	49.6 233.6
54.0	47.9 191.8	47.5 201.2	47.1 210.1	46.9 218.5	47.0 226.4	47.6 234.1
52.0	45.9 191.9	45.4 201.3	45.0 210.3	44.8 218.8	44.9 226.8	45.5 234.5
50.0	43.8 191.9	43.3 201.4	42.9 210.5	42.7 219.0	42.9 227.1	43.5 234.9
48.0	41.8 191.9	41.3 201.5	40.8 210.6	40.6 219.2	40.8 227.3	41.5 235.2
46.0	39.7 191.8	39.2 201.5	38.8 210.6	38.6 219.3	38.8 227.5	39.5 235.5
44.0	37.6 191.8	37.2 201.5	36.7 210.7	36.5 219.4	36.8 227.6	37.5 235.7
42.0	35.5 191.7	35.1 201.4	34.7 210.7	34.5 219.4	34.7 227.8	35.5 236.0
40.0	33.5 191.6	33.0 201.4	32.6 210.7	32.4 219.5	32.8 227.9	33.6 236.2
38.0	31.4 191.5	31.0 201.3	30.6 210.6	30.4 219.5	30.8 228.0	31.7 236.3
36.0	29.3 191.4	28.9 201.2	28.5 210.6	28.4 219.5	28.8 228.1	29.8 236.5
34.0	27.2 191.3	26.8 201.1	26.5 210.5	26.4 219.5	26.8 228.2	27.9 236.7
32.0	25.1 191.2	24.8 201.0	24.4 210.5	24.4 219.5	24.9 228.2	26.1 236.9
30.0	23.0 191.1	22.7 200.9	22.4 210.4	22.4 219.5	23.0 228.3	24.2 237.1
28.0	20.9 191.0	20.6 200.8	20.4 210.4	20.4 219.5	21.0 228.4	22.3 237.3
26.0	18.7 190.9	18.6 200.7	18.3 210.3	18.4 219.5	19.1 228.5	20.5 237.5
24.0	16.6 190.8	16.5 200.6	16.3 210.2	16.4 219.6	17.1 228.6	18.6 237.7

TABLE 1. (continued)

NORTHERN HEMISPHERE

LATI- TUDE	LONGITUDE EAST					
	180.0	190.0	200.0	210.0	220.0	230.0
88.0	81.4 185.2	81.7 185.8	82.0 186.1	82.3 185.9	82.6 185.1	82.9 183.8
86.0	80.3 197.2	80.8 199.1	81.4 200.4	82.0 201.1	82.6 200.9	83.3 199.6
84.0	78.9 206.6	79.5 209.7	80.3 212.3	81.2 214.4	82.1 215.8	83.1 216.1
82.0	77.2 213.8	78.0 217.9	78.9 221.7	80.0 225.2	81.2 228.3	82.5 230.8
80.0	75.4 219.3	76.3 224.1	77.3 228.9	78.5 233.5	79.9 238.1	81.5 242.7
78.0	73.5 223.6	74.4 229.0	75.6 234.4	76.9 239.9	78.4 245.6	80.1 251.7
76.0	71.5 227.0	72.5 232.8	73.7 238.7	75.2 244.8	76.8 251.3	78.6 258.6
74.0	69.4 229.7	70.5 235.9	71.8 242.2	73.3 248.8	75.1 255.8	77.0 263.8
72.0	67.4 232.0	68.5 238.4	69.8 245.0	71.4 251.9	73.2 259.4	75.2 267.9
70.0	65.3 233.8	66.4 240.5	67.8 247.3	69.5 254.5	71.4 262.3	73.4 271.2
68.0	63.2 235.4	64.3 242.3	65.8 249.3	67.5 256.7	69.5 264.8	71.6 273.9
66.0	61.0 236.7	62.2 243.7	63.7 250.9	65.5 258.5	67.5 266.8	69.7 276.1
64.0	58.9 237.9	60.1 245.0	61.7 252.4	63.5 260.1	65.6 268.6	67.8 278.0
62.0	56.8 238.9	58.1 246.1	59.6 253.6	61.5 261.5	63.6 270.1	65.9 279.6
60.0	54.7 239.7	56.0 247.1	57.6 254.7	59.5 262.8	61.7 271.5	63.9 281.1
58.0	52.6 240.5	53.9 248.0	55.6 255.7	57.5 263.9	59.7 272.7	62.0 282.4
56.0	50.6 241.1	51.9 248.7	53.6 256.6	55.5 264.9	57.7 273.8	60.0 283.6
54.0	48.5 241.7	49.9 249.4	51.6 257.4	53.6 265.9	55.7 274.9	58.0 284.6
52.0	46.5 242.2	47.9 250.1	49.6 258.2	51.6 266.7	53.8 275.8	56.0 285.6
50.0	44.5 242.7	46.0 250.7	47.7 258.9	49.6 267.6	51.8 276.7	54.1 286.6
48.0	42.6 243.1	44.0 251.2	45.8 259.6	47.7 268.3	49.8 277.6	52.1 287.4
46.0	40.6 243.5	42.1 251.7	43.9 260.2	45.8 269.1	47.9 278.4	50.1 288.3
44.0	38.7 243.9	40.2 252.2	42.0 260.8	43.9 269.8	45.9 279.2	48.1 289.1
42.0	36.8 244.2	38.4 252.7	40.1 261.4	42.0 270.5	44.0 280.0	46.1 289.9
40.0	34.9 244.5	36.5 253.1	38.3 262.0	40.1 271.2	42.1 280.7	44.1 290.6
38.0	33.1 244.8	34.7 253.5	36.5 262.5	38.3 271.8	40.1 281.4	42.1 291.3
36.0	31.3 245.1	32.9 254.0	34.7 263.1	36.4 272.5	38.2 282.1	40.1 292.0
34.0	29.4 245.4	31.2 254.4	32.9 263.6	34.6 273.1	36.3 282.8	38.1 292.7
32.0	27.6 245.7	29.4 254.8	31.1 264.1	32.8 273.7	34.4 283.4	36.1 293.3
30.0	25.9 246.0	27.6 255.2	29.3 264.7	30.9 274.3	32.5 284.0	34.1 293.9
28.0	24.1 246.3	25.9 255.6	27.6 265.2	29.1 274.9	30.6 284.6	32.1 294.5
26.0	22.3 246.6	24.1 256.0	25.8 265.6	27.2 275.4	28.6 285.2	30.1 295.1
24.0	20.5 246.9	22.4 256.4	24.0 266.1	25.4 275.9	26.7 285.8	28.1 295.6

TABLE 1. (continued)

NORTHERN HEMISPHERE

LATI-	LONGITUDE EAST					
TUDE	240.0	250.0	260.0	270.0	280.0	290.0
88.0	83.1 182.0	83.4 179.6	83.6 176.6	83.7 173.5	83.7 170.0	83.7 166.6
86.0	83.9 197.0	84.5 192.7	85.0 186.4	85.9 178.1	85.4 169.0	85.3 159.8
84.0	84.2 214.8	85.2 210.9	86.2 202.8	86.9 187.8	87.1 166.2	86.8 145.6
82.0	83.9 232.6	85.4 232.8	86.8 229.6	88.2 214.6	88.8 157.4	87.9 114.5
80.0	83.1 247.6	84.8 253.0	86.5 260.1	88.4 275.1		88.0 67.1
78.0	81.9 258.9	83.8 268.1	85.6 282.0	87.1 308.9	87.6 259.8	86.5 40.0
76.0	80.5 267.2	82.5 278.5	84.3 294.9	85.6 321.0	86.0 357.2	85.1 28.8
74.0	79.0 273.4	81.0 285.7	82.7 302.6	84.0 326.5	84.2 355.8	83.4 22.5
72.0	77.3 278.0	79.3 290.8	81.1 307.6	82.2 329.7	82.4 355.1	81.7 18.7
70.0	75.6 281.7	77.6 294.6	79.4 311.1	80.5 331.7	80.7 354.6	79.9 16.2
68.0	73.8 284.6	75.9 297.5	77.6 313.6	78.7 333.1	78.9 354.3	78.1 14.5
66.0	71.9 286.9	74.0 299.9	75.8 315.6	76.9 334.1	77.1 354.1	76.3 13.1
64.0	70.1 288.9	72.2 301.8	73.9 317.1	75.0 334.8	75.2 353.8	74.5 12.1
62.0	68.2 290.6	70.3 303.3	72.1 318.3	73.2 335.4	73.4 353.7	72.6 11.2
60.0	66.3 292.0	68.4 304.6	70.2 319.3	71.3 335.9	71.5 353.5	70.8 10.5
58.0	64.3 293.3	66.5 305.8	68.3 320.2	69.4 336.3	69.6 353.3	68.9 9.9
56.0	62.3 294.4	64.5 306.8	66.3 320.9	67.5 336.6	67.7 353.2	67.0 9.4
54.0	60.4 295.5	62.5 307.7	64.4 321.5	65.5 336.8	65.8 353.1	65.1 9.0
52.0	58.4 296.4	60.6 308.5	62.4 322.0	63.6 337.1	63.9 352.9	63.2 8.5
50.0	56.3 297.3	58.5 309.2	60.4 322.5	61.6 337.2	62.0 352.8	61.3 8.2
48.0	54.3 298.1	56.5 309.9	58.4 323.0	59.6 337.4	60.0 352.6	59.3 7.8
46.0	52.3 298.9	54.5 310.5	56.3 323.4	57.6 337.5	58.1 352.5	57.4 7.5
44.0	50.3 299.7	52.4 311.1	54.3 323.7	55.6 337.6	56.1 352.3	55.5 7.1
42.0	48.2 300.4	50.3 311.7	52.2 324.1	53.6 337.7	54.1 352.2	53.5 6.8
40.0	46.2 301.0	48.2 312.2	50.2 324.4	51.6 337.7	52.2 352.0	51.6 6.5
38.0	44.1 301.7	46.1 312.7	48.1 324.7	49.5 337.8	50.2 351.8	49.6 6.2
36.0	42.0 302.3	44.0 313.2	46.0 325.0	47.5 337.8	48.2 351.7	47.7 5.9
34.0	40.0 302.9	41.9 313.7	43.9 325.3	45.4 337.9	46.2 351.5	45.7 5.6
32.0	37.9 303.5	39.8 314.2	41.7 325.5	43.4 337.9	44.2 351.3	43.8 5.3
30.0	35.8 304.1	37.7 314.6	39.6 325.8	41.3 338.0	42.2 351.2	41.9 5.0
28.0	33.7 304.6	35.5 315.0	37.5 326.1	39.2 338.0	40.2 351.0	40.0 4.7
26.0	31.6 305.1	33.4 315.5	35.3 326.3	37.1 338.1	38.2 350.9	38.0 4.4
24.0	29.6 305.6	31.3 315.9	33.2 326.6	35.0 338.1	36.2 350.7	36.1 4.1

TABLE 1. (continued)

NORTHERN HEMISPHERE

LATI-	LONGITUDE EAST					
TUDE	300.0	310.0	320.0	330.0	340.0	350.0
88.0	83.6 163.4	83.4 160.6	83.2 158.3	82.9 156.6	82.7 155.5	82.4 154.9
86.0	85.0 152.0	84.5 146.0	84.0 142.1	83.4 139.8	82.8 138.9	82.1 139.0
84.0	86.1 133.1	85.1 126.1	84.2 123.1	83.2 122.4	82.3 123.2	81.4 125.0
82.0	86.5 105.8	85.1 103.9	83.7 104.9	82.5 107.2	81.3 110.2	80.2 113.8
80.0	86.0 78.6	84.4 85.0	82.8 90.3	81.3 95.3	80.0 100.2	78.8 105.2
78.0	84.9 60.2	83.2 71.4	81.5 79.6	79.9 86.5	78.4 92.7	77.2 98.6
76.0	83.5 49.0	81.7 62.2	79.9 72.0	78.3 79.9	76.8 87.0	75.4 93.6
74.0	81.9 42.0	80.1 55.8	78.3 66.3	76.6 75.0	75.0 82.6	73.6 89.6
72.0	80.2 37.3	78.4 51.2	76.5 62.1	74.7 71.1	73.1 79.1	71.7 86.4
70.0	78.5 33.9	76.7 47.8	74.7 58.8	72.9 68.1	71.2 76.3	69.7 83.8
68.0	76.7 31.4	74.8 45.1	72.9 56.2	71.0 65.6	69.2 73.9	67.7 81.7
66.0	74.9 29.5	73.0 43.0	71.0 54.1	69.0 63.6	67.2 72.0	65.6 79.8
64.0	73.0 28.0	71.1 41.3	69.0 52.4	67.0 61.9	65.1 70.4	63.5 78.2
62.0	71.2 26.8	69.2 39.9	67.1 51.0	65.0 60.5	63.0 68.9	61.4 76.9
60.0	69.3 25.8	67.3 38.7	65.0 49.7	62.9 59.2	60.9 67.7	59.3 75.7
58.0	67.4 24.9	65.3 37.7	63.0 48.7	60.8 58.1	58.7 66.6	57.0 74.6
56.0	65.4 24.1	63.3 36.9	61.0 47.8	58.6 57.2	56.5 65.7	54.8 73.6
54.0	63.5 23.5	61.3 36.1	58.9 46.9	56.4 56.4	54.3 64.8	52.5 72.8
52.0	61.6 22.9	59.3 35.5	56.7 46.2	54.2 55.6	52.0 64.0	50.1 72.0
50.0	59.6 22.4	57.3 34.9	54.6 45.6	51.9 54.9	49.6 63.3	47.7 71.3
48.0	57.6 21.9	55.2 34.3	52.4 45.0	49.6 54.3	47.2 62.7	45.3 70.6
46.0	55.7 21.5	53.2 33.8	50.2 44.5	47.3 53.8	44.7 62.1	42.8 70.0
44.0	53.7 21.0	51.1 33.4	48.0 44.0	44.9 53.3	42.2 61.6	40.2 69.5
42.0	51.7 20.6	49.0 33.0	45.8 43.6	42.5 52.8	39.6 61.1	37.5 69.0
40.0	49.8 20.3	47.0 32.6	43.5 43.2	40.0 52.4	37.0 60.6	34.8 68.5
38.0	47.8 19.9	44.9 32.2	41.3 42.8	37.5 52.0	34.3 60.2	32.0 68.0
36.0	45.8 19.5	42.8 31.8	39.0 42.5	35.0 51.6	31.5 59.8	29.1 67.6
34.0	43.9 19.2	40.7 31.5	36.7 42.1	32.3 51.3	28.6 59.4	26.0 67.2
32.0	41.9 18.8	38.6 31.1	34.3 41.8	29.7 50.9	25.5 59.1	22.8 66.9
30.0	40.0 18.5	36.6 30.8	32.0 41.5	26.9 50.6	22.4 58.7	19.4 66.6
28.0	38.1 18.1	34.5 30.4	29.6 41.1	24.1 50.3	19.0 58.4	15.6 66.2
26.0	36.1 17.8	32.5 30.1	27.3 40.8	21.2 49.9	15.3 58.1	11.2 65.9
24.0	34.2 17.4	30.4 29.7	24.9 40.4	18.2 49.6	11.0 57.8	4.5 65.6

TABLE 1. (continued)

SOUTHERN HEMISPHERE

LATI-TUDE	LONGITUDE EAST										
	0.0		10.0		20.0		30.0		40.0		50.0	
88.0	72.6	23.8	72.9	24.7	73.2	25.3	73.5	25.7	73.8	25.7	74.1	25.4
86.0	71.2	27.3	71.7	28.8	72.3	30.1	72.9	31.0	73.5	31.6	74.2	31.8
84.0	69.8	30.7	70.5	32.9	71.3	34.9	72.2	36.5	73.1	37.8	74.1	38.6
82.0	68.4	33.7	69.3	36.6	70.3	39.2	71.4	41.6	72.5	43.7	73.8	45.3
80.0	67.0	36.2	68.0	39.8	69.2	43.1	70.4	46.3	71.8	49.1	73.3	51.5
78.0	65.5	38.5	66.7	42.6	68.0	46.6	69.4	50.5	71.0	54.1	72.6	57.4
76.0	64.1	40.4	65.4	45.1	66.8	49.7	68.4	54.2	70.0	58.6	71.8	62.7
74.0	62.8	42.1	64.2	47.3	65.6	52.5	67.2	57.6	69.0	62.6	70.9	67.5
72.0	61.4	43.7	62.9	49.3	64.4	55.0	66.1	60.7	67.9	66.3	69.8	71.9
70.0	60.1	45.0	61.6	51.1	63.2	57.3	64.9	63.5	66.8	69.6	68.7	75.8
68.0	58.8	46.3	60.4	52.8	62.0	59.4	63.8	66.0	65.6	72.7	67.6	79.4
66.0	57.5	47.5	59.2	54.4	60.9	61.4	62.6	68.4	64.4	75.4	66.4	82.6
64.0	56.3	48.5	58.0	55.8	59.7	63.2	61.4	70.6	63.2	78.0	65.1	85.6
62.0	55.1	49.5	56.8	57.1	58.5	64.9	60.2	72.6	61.9	80.3	63.8	88.3
60.0	53.9	50.5	55.7	58.4	57.3	66.4	59.0	74.5	60.7	82.5	62.5	90.8
58.0	52.7	51.3	54.5	59.6	56.2	67.9	57.8	76.3	59.4	84.6	61.2	93.0
56.0	51.6	52.2	53.4	60.7	55.1	69.4	56.6	78.0	58.1	86.5	59.8	95.2
54.0	50.5	53.0	52.3	61.8	53.9	70.7	55.4	79.6	56.8	88.3	58.4	97.2
52.0	49.3	53.8	51.2	62.9	52.8	72.0	54.2	81.1	55.5	90.0	57.0	99.0
50.0	48.2	54.6	50.1	63.9	51.6	73.3	53.0	82.6	54.2	91.7	55.6	100.8
48.0	47.1	55.3	49.0	64.9	50.5	74.5	51.8	84.0	52.9	93.2	54.1	102.5
46.0	45.9	56.1	47.9	65.9	49.3	75.7	50.5	85.3	51.6	94.7	52.7	104.0
44.0	44.8	56.8	46.7	66.8	48.2	76.8	49.3	86.6	50.2	96.1	51.2	105.5
42.0	43.6	57.6	45.6	67.7	47.0	77.9	48.0	87.9	48.8	97.5	49.7	106.9
40.0	42.4	58.3	44.4	68.6	45.7	79.0	46.7	89.1	47.4	98.8	48.1	108.3
38.0	41.2	59.1	43.2	69.5	44.5	80.0	45.3	90.2	45.9	100.0	46.5	109.6
36.0	40.0	59.8	41.9	70.4	43.2	81.0	44.0	91.3	44.4	101.2	44.9	110.8
34.0	38.7	60.6	40.6	71.2	41.8	81.9	42.5	92.4	42.9	102.3	43.2	111.9
32.0	37.3	61.3	39.2	72.1	40.4	82.9	41.1	93.4	41.3	103.4	41.5	113.0
30.0	35.9	62.1	37.8	72.9	39.0	83.7	39.5	94.3	39.7	104.4	39.8	114.0
28.0	34.5	62.8	36.3	73.7	37.5	84.6	37.9	95.2	38.0	105.3	38.0	114.9
26.0	32.9	63.6	34.8	74.4	35.9	85.4	36.3	96.0	36.3	106.2	36.2	115.8
24.0	31.3	64.3	33.1	75.2	34.2	86.1	34.6	96.8	34.5	106.9	34.3	116.6
22.0	29.6	65.1	31.4	75.9	32.4	86.8	32.8	97.5	32.7	107.7	32.4	117.3
20.0	27.7	65.8	29.5	76.6	30.6	87.4	30.9	98.1	30.8	108.3	30.4	118.0
18.0	25.8	66.5	27.6	77.2	28.7	88.0	29.0	98.6	28.8	108.8	28.4	118.6
16.0	23.7	67.2	25.5	77.8	26.7	88.5	27.0	99.1	26.8	109.3	26.4	119.1
14.0	21.5	67.8	23.4	78.3	24.5	88.9	25.0	99.5	24.8	109.8	24.3	119.5
12.0	19.0	68.4	21.0	78.8	22.3	89.3	22.8	99.9	22.6	110.1	22.2	119.9

TABLE 1. (continued)

SOUTHERN HEMISPHERE

LATI-TUDE	LONGITUDE EAST										
	60.0		70.0		80.0		90.0		100.0		110.0	
88.0	74.3	24.7	74.5	23.8	74.7	22.5	74.9	21.1	75.0	19.5	75.2	17.9
86.0	74.8	31.6	75.5	30.9	76.1	29.7	76.6	28.1	77.1	25.9	77.5	23.3
84.0	75.1	38.9	76.1	38.5	77.0	37.2	77.9	35.0	78.7	31.7	79.3	27.4
82.0	75.1	46.2	76.4	46.4	77.7	45.5	79.0	43.3	80.1	39.2	81.0	33.0
80.0	74.8	53.4	76.4	54.5	78.1	54.5	79.8	52.9	81.3	48.8	82.7	41.0
78.0	74.4	60.2	76.2	62.5	78.2	63.8	80.2	63.7	82.2	61.0	84.1	53.2
76.0	73.7	66.5	75.8	70.0	78.0	73.0	80.3	75.1	82.7	75.5	85.1	71.6
74.0	72.9	72.3	75.1	77.0	77.5	81.7	80.0	86.4	82.7	91.0	85.4	95.5
72.0	72.0	77.5	74.3	83.4	76.8	89.6	79.4	96.6	82.2	105.4	84.9	118.7
70.0	70.9	82.2	73.3	89.0	75.8	96.6	78.5	105.6	81.2	117.4	83.8	135.6
68.0	69.8	86.4	72.2	94.0	74.7	102.6	77.4	113.1	80.0	126.9	82.2	154.1
66.0	68.6	90.2	70.9	98.4	73.5	107.9	76.1	119.4	78.5	134.2	80.5	154.1
64.0	67.3	93.6	69.6	102.3	72.1	112.4	74.6	124.6	76.9	139.8	78.7	159.3
62.0	65.9	96.6	68.2	105.8	70.7	116.3	73.1	128.9	75.2	144.3	76.8	163.0
60.0	64.6	99.4	66.8	108.9	69.1	119.8	71.4	132.6	73.4	147.9	74.8	165.8
58.0	63.1	101.9	65.3	111.7	67.6	122.8	69.7	135.7	71.5	150.8	72.8	168.0
56.0	61.7	104.3	63.8	114.2	65.9	125.4	68.0	138.4	69.6	153.2	70.7	169.8
54.0	60.2	106.4	62.2	116.5	64.2	127.8	66.2	140.7	67.7	155.3	68.6	171.2
52.0	58.7	108.4	60.6	118.6	62.5	129.9	64.3	142.8	65.7	157.0	66.5	172.4
50.0	57.2	110.3	58.9	120.5	60.8	131.9	62.4	144.6	63.7	158.6	64.4	173.5
48.0	55.6	112.0	57.3	122.3	59.0	133.6	60.5	146.2	61.7	159.9	62.3	174.4
46.0	54.0	113.6	55.6	123.9	57.2	135.2	58.6	147.7	59.6	161.1	60.1	175.2
44.0	52.4	115.1	53.8	125.4	55.3	136.7	56.6	149.0	57.5	162.2	58.0	175.9
42.0	50.8	116.6	52.1	126.8	53.5	138.0	54.6	150.2	55.5	163.2	55.8	176.5
40.0	49.1	117.9	50.3	128.1	51.6	139.3	52.6	151.3	53.3	164.0	53.6	177.1
38.0	47.4	119.2	48.5	129.4	49.6	140.4	50.6	152.3	51.2	164.8	51.4	177.6
36.0	45.6	120.4	46.6	130.5	47.7	141.5	48.6	153.2	49.1	165.5	49.2	178.0
34.0	43.9	121.5	44.7	131.6	45.7	142.5	46.5	154.1	47.0	166.2	47.0	178.5
32.0	42.0	122.6	42.8	132.6	43.8	143.4	44.5	154.9	44.8	166.8	44.8	178.8
30.0	40.2	123.5	40.9	133.5	41.7	144.2	42.4	155.6	42.7	167.4	42.6	179.2
28.0	38.3	124.5	39.0	134.4	39.7	145.0	40.3	156.3	40.5	167.9	40.3	179.5
26.0	36.4	125.3	37.0	135.2	37.7	145.7	38.2	156.9	38.3	168.3	38.1	179.7
24.0	34.4	126.1	35.0	135.9	35.6	146.4	36.1	157.4	36.1	168.7	35.9	180.0
22.0	32.5	126.8	32.9	136.6	33.6	147.0	34.0	157.9	34.0	169.1	33.6	180.2
20.0	30.4	127.4	30.9	137.2	31.5	147.5	31.8	158.4	31.8	169.4	31.4	180.3
18.0	28.4	128.0	28.8	137.7	29.4	148.0	29.7	158.8	29.6	169.7	29.2	180.5
16.0	26.3	128.5	26.7	138.2	27.2	148.4	27.5	159.1	27.4	169.9	26.9	180.6
14.0	24.2	129.0	24.6	138.7	25.1	148.8	25.4	159.4	25.2	170.2	24.7	180.7
12.0	22.1	129.4	22.4	139.1	23.0	149.2	23.2	159.7	23.0	170.3	22.4	180.8

TABLE 1. (continued)

SOUTHERN HEMISPHERE

LATI-	LONGITUDE EAST					
TUDE	120.0	130.0	140.0	150.0	160.0	170.0
88.0	75.4 16.3	75.9 15.0	79.4 7.5		84.0 5.8	75.7 15.2
86.0	77.8 20.4	77.8 17.6		79.8 1.7	76.9 8.2	76.5 6.4
84.0	79.7 22.2	79.8 16.6	84.4 1.3	79.1 5.8	78.4 2.0	77.6 359.2
82.0	81.6 24.7	81.8 15.3	82.7 3.8	80.7 358.7	79.7 353.4	78.5 350.2
80.0	83.6 28.8	83.8 13.2	83.3 358.7	82.1 348.4	80.7 342.6	79.0 339.9
78.0	85.5 36.1	85.8 8.9	84.9 345.9	83.2 334.2	81.2 329.6	79.2 328.7
76.0	87.1 55.2	87.8 358.2	86.0 323.6	83.7 316.0	81.3 315.5	78.9 317.5
74.0	88.1 100.6		86.3 291.9	83.5 296.7	80.9 301.8	78.3 307.1
72.0	87.3 149.3	87.6 226.1	85.4 264.9	82.7 280.1	80.0 290.0	77.3 297.9
70.0	85.6 168.3	85.6 215.2	83.9 248.7	81.4 267.7	78.7 280.3	76.0 290.2
68.0	83.7 176.1	83.6 211.0	82.0 239.3	79.8 258.7	77.2 272.7	74.6 283.8
66.0	81.6 180.2	81.4 208.8	80.1 233.5	78.0 252.3	75.5 266.7	72.9 278.5
64.0	79.6 182.8	79.3 207.5	78.0 229.5	76.1 247.5	73.7 262.0	71.2 274.1
62.0	77.5 184.5	77.2 206.5	75.9 226.7	74.1 243.8	71.8 258.2	69.4 270.4
60.0	75.4 185.7	75.0 205.9	73.8 224.5	72.0 240.9	69.9 255.0	67.5 267.3
58.0	73.2 186.6	72.8 205.3	71.7 222.9	70.0 238.6	67.9 252.4	65.6 264.7
56.0	71.1 187.3	70.6 204.9	69.5 221.5	67.9 236.7	65.8 250.2	63.6 262.4
54.0	68.9 187.9	68.5 204.6	67.4 220.4	65.8 235.0	63.8 248.3	61.6 260.4
52.0	66.7 188.4	66.2 204.3	65.2 219.5	63.6 233.7	61.7 246.7	59.6 258.7
50.0	64.5 188.8	64.0 204.0	63.0 218.7	61.5 232.5	59.6 245.3	57.5 257.1
48.0	62.3 189.2	61.8 203.8	60.8 218.0	59.3 231.4	57.5 244.0	55.5 255.7
46.0	60.1 189.5	59.6 203.6	58.6 217.3	57.2 230.5	55.4 242.9	53.4 254.5
44.0	57.9 189.7	57.4 203.4	56.4 216.8	55.0 229.6	53.3 241.9	51.3 253.4
42.0	55.7 190.0	55.1 203.3	54.2 216.3	52.9 228.9	51.2 241.0	49.2 252.4
40.0	53.4 190.2	52.9 203.1	52.0 215.8	50.7 228.2	49.1 240.1	47.1 251.4
38.0	51.2 190.3	50.7 203.0	49.8 215.4	48.5 227.6	46.9 239.4	45.0 250.6
36.0	49.0 190.5	48.4 202.8	47.6 215.0	46.4 227.0	44.8 238.7	42.9 249.8
34.0	46.7 190.6	46.2 202.7	45.3 214.7	44.2 226.5	42.7 238.0	40.8 249.1
32.0	44.5 190.7	43.9 202.6	43.1 214.3	42.0 226.0	40.5 237.4	38.7 248.4
30.0	42.2 190.8	41.7 202.4	40.9 214.0	39.8 225.5	38.4 236.8	36.5 247.7
28.0	40.0 190.9	39.4 202.3	38.7 213.7	37.7 225.1	36.2 236.3	34.4 247.2
26.0	37.7 191.0	37.2 202.2	36.5 213.4	35.5 224.7	34.1 235.8	32.2 246.6
24.0	35.4 191.0	34.9 202.0	34.3 213.1	33.4 224.3	31.9 235.4	30.1 246.1
22.0	33.2 191.1	32.7 201.9	32.1 212.9	31.2 223.9	29.8 234.9	27.9 245.6
20.0	30.9 191.1	30.5 201.8	29.9 212.6	29.0 223.6	27.6 234.5	25.7 245.1
18.0	28.7 191.1	28.2 201.7	27.7 212.4	26.9 223.3	25.5 234.1	23.6 244.7
16.0	26.4 191.1	26.0 201.5	25.5 212.2	24.8 223.0	23.4 233.8	21.4 244.3
14.0	24.1 191.0	23.8 201.4	23.4 211.9	22.6 222.7	21.2 233.4	19.3 243.9
12.0	21.9 191.0	21.5 201.3	21.2 211.7	20.5 222.4	19.1 233.1	17.1 243.5

TABLE 1. (continued)

SOUTHERN HEMISPHERE

LATI-	LONGITUDE EAST					
TUDE	180.0	190.0	200.0	210.0	220.0	230.0
88.0	75.2 15.4	74.7 14.9	74.3 14.2	73.9 13.5	73.5 12.9	73.1 12.4
86.0	76.0 5.2	75.3 4.5	74.5 4.3	73.8 4.5	73.1 4.9	72.4 5.7
84.0	76.7 357.5	75.6 356.7	74.6 356.7	73.5 357.3	72.5 358.4	71.5 359.9
82.0	77.1 348.7	75.7 348.4	74.4 349.2	73.0 350.6	71.7 352.5	70.5 354.8
80.0	77.3 339.3	75.6 340.1	73.9 341.9	72.3 344.3	70.7 347.1	69.3 350.2
78.0	77.1 329.8	75.1 332.1	73.2 335.1	71.3 338.5	69.6 342.2	68.0 346.1
76.0	76.6 320.7	74.4 324.6	72.3 328.8	70.2 333.2	68.3 337.8	66.6 342.4
74.0	75.8 312.4	73.4 317.7	71.2 323.1	69.0 328.5	67.0 333.9	65.1 339.2
72.0	74.7 305.0	72.2 311.6	69.9 318.0	67.6 324.2	65.5 330.3	63.5 336.2
70.0	73.4 298.6	70.9 306.3	68.5 313.5	66.2 320.4	64.0 327.1	61.9 333.6
68.0	72.0 293.2	69.4 301.7	67.0 309.6	64.6 317.0	62.4 324.2	60.3 331.1
66.0	70.4 288.6	67.8 297.6	65.4 306.0	63.0 314.0	60.8 321.6	58.6 328.9
64.0	68.6 284.6	66.1 294.1	63.7 302.9	61.3 311.2	59.1 319.2	56.9 326.9
62.0	66.9 281.2	64.4 291.0	62.0 300.1	59.6 308.7	57.4 317.0	55.2 325.1
60.0	65.0 278.3	62.6 288.3	60.2 297.6	57.8 306.5	55.6 315.0	53.4 323.3
58.0	63.1 275.7	60.7 285.9	58.4 295.4	56.1 304.4	53.8 313.2	51.7 321.7
56.0	61.2 273.5	58.8 283.7	56.5 293.3	54.2 302.6	52.0 311.5	49.9 320.2
54.0	59.3 271.5	56.9 281.8	54.6 291.5	52.4 300.8	50.2 309.9	48.1 318.8
52.0	57.3 269.7	55.0 280.0	52.7 289.8	50.5 299.2	48.4 308.4	46.3 317.5
50.0	55.3 268.1	53.0 278.4	50.8 288.3	48.6 297.8	46.5 307.1	44.5 316.2
48.0	53.3 266.7	51.0 277.0	48.8 286.9	46.7 296.4	44.7 305.8	42.7 315.0
46.0	51.2 265.4	49.0 275.7	46.9 285.6	44.8 295.1	42.8 304.6	40.9 313.9
44.0	49.2 264.2	47.0 274.5	44.9 284.3	42.9 294.0	40.9 303.4	39.0 312.8
42.0	47.1 263.1	45.0 273.4	42.9 283.2	40.9 292.9	39.0 302.4	37.2 311.8
40.0	45.1 262.1	42.9 272.3	40.9 282.2	38.9 291.8	37.1 301.4	35.3 310.9
38.0	43.0 261.2	40.9 271.4	38.9 281.2	37.0 290.9	35.2 300.4	33.5 310.0
36.0	40.9 260.4	38.8 270.5	36.8 280.3	35.0 290.0	33.2 299.6	31.6 309.1
34.0	38.7 259.6	36.7 269.7	34.8 279.5	33.0 289.1	31.3 298.8	29.7 308.3
32.0	36.6 258.8	34.6 268.9	32.7 278.7	31.0 288.4	29.3 298.0	27.8 307.6
30.0	34.5 258.2	32.5 268.2	30.6 277.9	28.9 287.6	27.4 297.3	25.9 306.9
28.0	32.3 257.5	30.3 267.5	28.5 277.2	26.9 286.9	25.4 296.6	23.9 306.2
26.0	30.2 256.9	28.2 266.8	26.4 276.6	24.8 286.3	23.4 296.0	22.0 305.6
24.0	28.0 256.3	26.0 266.2	24.3 275.9	22.8 285.7	21.4 295.4	20.0 305.1
22.0	25.8 255.8	23.8 265.6	22.2 275.4	20.7 285.1	19.4 294.8	18.1 304.5
20.0	23.6 255.3	21.7 265.1	20.0 274.8	18.7 284.5	17.4 294.3	16.1 304.0
18.0	21.4 254.8	19.5 264.6	17.9 274.3	16.6 284.0	15.3 293.8	14.1 303.6
16.0	19.3 254.3	17.3 264.1	15.8 273.8	14.5 283.6	13.3 293.4	12.1 303.1
14.0	17.1 253.9	15.1 263.7	13.6 273.4	12.4 283.1	11.3 292.9	10.1 302.7
12.0	14.9 253.5	12.9 263.2	11.5 272.9	10.4 282.7	9.2 292.5	8.0 302.3

TABLE 1. (continued)

SOUTHERN HEMISPHERE

LATI-TUDE	LONGITUDE EAST											
	240.0		250.0		260.0		270.0		280.0		290.0	
88.0	72.8	12.2	72.5	12.2	72.2	12.5	72.0	13.1	71.8	13.9	71.7	14.9
86.0	71.8	6.7	71.3	7.8	70.8	9.2	70.4	10.7	70.1	12.4	69.9	14.1
84.0	70.7	1.7	69.9	3.7	69.2	5.8	68.6	8.1	68.2	10.5	67.9	13.0
82.0	69.4	357.3	68.4	0.0	67.5	2.9	66.8	5.8	66.3	8.8	66.0	11.9
80.0	68.0	353.5	66.8	356.9	65.8	0.3	65.0	3.8	64.4	7.4	64.0	10.9
78.0	66.5	350.1	65.2	354.1	64.0	358.1	63.1	2.1	62.5	6.1	62.0	10.0
76.0	64.9	347.1	63.5	351.7	62.2	356.2	61.2	0.6	60.5	4.9	60.0	9.2
74.0	63.3	344.4	61.8	349.5	60.4	354.4	59.3	359.2	58.5	3.9	58.1	8.5
72.0	61.7	342.0	60.0	347.5	58.6	352.9	57.4	358.0	56.6	3.0	56.1	7.8
70.0	60.0	339.8	58.3	345.8	56.7	351.5	55.5	357.0	54.6	2.2	54.1	7.2
68.0	58.3	337.8	56.5	344.2	54.9	350.2	53.5	356.0	52.6	1.4	52.1	6.7
66.0	56.6	336.0	54.7	342.7	53.0	349.1	51.6	355.1	50.6	0.7	50.1	6.1
64.0	54.8	334.3	52.9	341.4	51.2	348.0	49.7	354.3	48.6	0.1	48.1	5.7
62.0	53.1	332.8	51.1	340.1	49.3	347.0	47.8	353.5	46.7	359.5	46.1	5.2
60.0	51.3	331.3	49.3	338.9	47.5	346.1	45.9	352.8	44.7	359.0	44.1	4.8
58.0	49.6	329.9	47.6	337.8	45.6	345.2	44.0	352.1	42.8	358.5	42.2	4.4
56.0	47.8	328.7	45.8	336.8	43.8	344.4	42.1	351.5	40.8	358.0	40.2	4.1
54.0	46.1	327.4	44.0	335.8	42.0	343.6	40.3	350.9	38.9	357.6	38.3	3.7
52.0	44.3	326.3	42.3	334.8	40.3	342.9	38.4	350.3	37.0	357.1	36.4	3.4
50.0	42.5	325.2	40.5	333.9	38.5	342.1	36.6	349.8	35.2	356.7	34.5	3.1
48.0	40.7	324.1	38.8	333.0	36.7	341.4	34.8	349.2	33.3	356.3	32.6	2.8
46.0	39.0	323.1	37.0	332.1	35.0	340.7	33.0	348.7	31.5	356.0	30.7	2.6
44.0	37.2	322.1	35.3	331.3	33.2	340.0	31.3	348.2	29.6	355.6	28.8	2.3
42.0	35.4	321.2	33.5	330.4	31.5	339.4	29.5	347.7	27.8	355.2	27.0	2.1
40.0	33.6	320.3	31.8	329.7	29.8	338.7	27.7	347.2	26.0	354.9	25.1	1.9
38.0	31.8	319.5	30.0	328.9	28.1	338.1	26.0	346.7	24.2	354.5	23.2	1.7
36.0	30.0	318.7	28.3	328.2	26.3	337.4	24.2	346.2	22.3	354.2	21.4	1.5
34.0	28.1	317.9	26.5	327.5	24.6	336.8	22.5	345.7	20.5	353.9	19.5	1.4
32.0	26.3	317.2	24.7	326.8	22.8	336.3	20.7	345.3	18.7	353.6	17.6	1.2
30.0	24.4	316.5	22.9	326.2	21.1	335.7	18.9	344.8	16.9	353.3	15.7	1.1
28.0	22.6	315.9	21.1	325.6	19.3	335.2	17.2	344.4	15.0	353.0	13.9	1.0
26.0	20.7	315.3	19.2	325.0	17.5	334.6	15.4	343.9	13.2	352.7	11.9	0.9
24.0	18.8	314.8	17.4	324.5	15.7	334.1	13.5	343.5	11.3	352.4	10.0	0.8
22.0	16.8	314.2	15.5	324.0	13.8	333.7	11.7	343.1	9.4	352.2	8.1	0.8
20.0	14.9	313.7	13.6	323.5	12.0	333.2	9.9	342.7	7.5	351.9	6.0	0.7
18.0	12.9	313.3	11.6	323.1	10.0	332.8	8.0	342.3	5.6	351.7	4.0	0.7
16.0	10.9	312.9	9.7	322.6	8.1	332.4	6.1	342.0	3.7	351.4	1.4	0.7
14.0	8.9	312.5	7.6	322.2	6.1	332.0	4.1	341.6	1.6	351.2	0.0	0.7
12.0	6.8	312.1	5.5	321.9	3.9	331.6						

TABLE 1. (continued)

SOUTHERN HEMISPHERE

LATI-TUDE	LONGITUDE EAST											
	300.0		310.0		320.0		330.0		340.0		350.0	
88.0	71.7	16.0	71.7	17.3	71.8	18.7	71.9	20.0	72.1	21.4	72.3	22.6
86.0	69.8	16.0	69.8	17.9	69.9	19.9	70.1	21.9	70.4	23.8	70.7	25.6
84.0	67.8	15.5	67.8	18.1	67.9	20.7	68.2	23.3	68.6	25.9	69.2	28.3
82.0	65.8	15.0	65.8	18.1	66.0	21.2	66.4	24.4	66.9	27.6	67.6	30.7
80.0	63.9	14.5	63.9	18.0	64.2	21.6	64.6	25.3	65.2	29.0	66.0	32.6
78.0	61.9	14.0	62.0	17.9	62.3	21.9	62.8	26.0	63.6	30.1	64.5	34.3
76.0	59.9	13.5	60.0	17.8	60.4	22.1	61.1	26.6	62.0	31.1	63.0	35.7
74.0	57.9	13.0	58.1	17.6	58.6	22.3	59.4	27.0	60.4	31.9	61.5	37.0
72.0	55.9	12.6	56.2	17.4	56.8	22.3	57.7	27.4	58.8	32.6	60.0	38.1
70.0	54.0	12.2	54.3	17.2	55.0	22.3	56.0	27.7	57.2	33.3	58.6	39.1
68.0	52.0	11.8	52.4	17.0	53.2	22.3	54.4	27.9	55.7	33.8	57.2	39.9
66.0	50.0	11.4	50.5	16.8	51.5	22.3	52.8	28.1	54.3	34.2	55.9	40.7
64.0	48.1	11.1	48.7	16.5	49.7	22.2	51.2	28.2	52.8	34.7	54.6	41.4
62.0	46.1	10.7	46.8	16.3	48.0	22.2	49.6	28.4	51.4	35.0	53.3	42.1
60.0	44.2	10.4	45.0	16.1	46.3	22.1	48.1	28.5	50.0	35.4	52.0	42.7
58.0	42.3	10.1	43.2	15.9	44.7	22.0	46.6	28.6	48.7	35.7	50.8	43.3
56.0	40.4	9.9	41.4	15.8	43.0	22.0	45.1	28.7	47.3	36.0	49.5	43.9
54.0	38.5	9.6	39.6	15.6	41.4	21.9	43.6	28.8	46.0	36.3	48.3	44.4
52.0	36.6	9.4	37.8	15.5	39.8	21.9	42.2	28.9	44.7	36.7	47.1	45.0
50.0	34.8	9.2	36.1	15.3	38.2	21.9	40.7	29.1	43.4	37.0	46.0	45.5
48.0	32.9	9.0	34.3	15.3	36.6	21.9	39.3	29.3	42.1	37.3	44.8	46.1
46.0	31.1	8.9	32.6	15.2	35.0	22.0	37.9	29.5	40.8	37.7	43.6	46.6
44.0	29.2	8.7	30.9	15.2	33.5	22.1	36.5	29.7	39.6	38.1	42.4	47.2
42.0	27.4	8.6	29.2	15.2	31.9	22.2	35.1	30.0	38.3	38.5	41.2	47.8
40.0	25.6	8.6	27.5	15.2	30.3	22.4	33.7	30.3	37.0	38.9	40.0	48.4
38.0	23.7	8.5	25.8	15.3	28.8	22.6	32.3	30.6	35.7	39.4	38.7	49.0
36.0	21.9	8.5	24.0	15.5	27.2	22.9	30.8	31.0	34.3	39.9	37.4	49.6
34.0	20.1	8.5	22.3	15.6	25.7	23.2	29.4	31.4	33.0	40.5	36.1	50.2
32.0	18.3	8.5	20.6	15.8	24.1	23.5	27.9	31.9	31.6	41.1	34.8	50.9
30.0	16.4	8.6	18.9	16.1	22.5	23.9	26.4	32.4	30.2	41.7	33.4	51.6
28.0	14.6	8.7	17.2	16.4	20.9	24.4	24.9	33.0	28.7	42.3	31.9	52.3
26.0	12.7	8.8	15.5	16.7	19.3	24.9	23.3	33.6	27.1	43.0	30.4	53.1
24.0	10.8	9.0	13.7	17.1	17.6	25.4	21.7	34.3	25.5	43.8	28.7	53.8
22.0	8.9	9.1	12.0	17.5	15.9	26.0	20.0	35.0	23.8	44.5	27.0	54.6
20.0	7.0	9.3	10.2	17.9	14.2	26.6	18.2	35.7	22.0	45.3	25.2	55.3
18.0	5.1	9.6	8.4	18.4	12.3	27.3	16.4	36.4	20.1	46.0	23.3	56.1
16.0	3.1	9.8	6.6	18.9	10.4	28.0	14.3	37.2	18.0	46.8	21.2	56.8
14.0	0.7	10.1	4.7	19.4	8.3	28.7	12.1	38.0	15.7	47.6	18.9	57.5

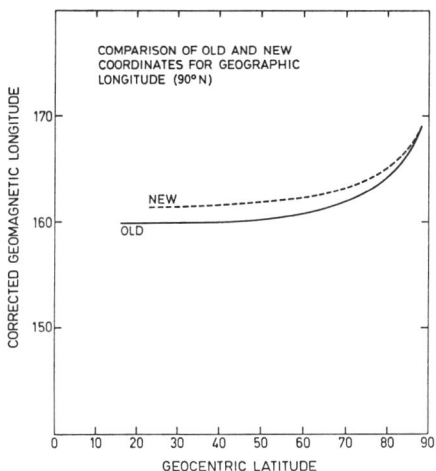

Fig. 2. A comparison between old and new corrected geomagnetic coordinates for the geocentric longitude 90°, northern hemisphere.

transformed to another distance, say to the surface of the geoid, the dipole approximation can be used. The corrected geomagnetic latitude L_{ch} at height h can be approximated from the latitude L_{ca} at the a = 6371.2 km height by the dipole relation

$$\cos^2(L_{ch}) = \frac{a+h}{a} \cos^2(L_{ca})$$

A comparison between the old (Gustafsson, 1970) and the new coordinates presented here shows a difference of about 1° in latitude, see Figure 1 and a difference of 2° for the longitude shown in Figure 2. However, in certain areas the longitudes may differ up to 5° which means a difference in magnetic time based on these coordinates of 20 minutes.

References

Gustafsson, G., A revised corrected geomagnetic coordinate system, Arkiv för Geofysik 5, 595-616, 1970.

HIGH-LATITUDE FIELD-ALIGNED CURRENTS PATTERNS IN CONNECTION WITH MAGNETOSPHERIC STRUCTURE

Y. I. Feldstein, R. G. Afonina, B. A. Belov, A. E. Levitin,

D. S. Faermark

Institute of Terrestrial Magnetism, Ionosphere and Radio Wave Propagation of the Academy of Sciences of the USSR, Troitsk, Moscow Region 142092, USSR

V. Y. Gaidukov

Institute of Applied Geophysics, Goscomhydromet, Moscow, USSR

Abstract. Distributions of field-aligned currents $J_\parallel$ determined from surface geomagnetic field variations are projected to the magnetospheric tail and the equatorial cross section of the magnetosphere. The field-aligned currents are calculated under various conditions in interplanetary space. The effects of various ionospheric conductivity distributions on the $J_\parallel$ pattern is also considered. The high-latitude system of field-aligned current (Region 1) projects onto the plasma sheet of the magnetospheric tail and to the magnetospheric regions in the immediate proximity to the magnetopause for $B_z = B_y = 0$. The field-aligned currents of opposite direction in the lower-latitude region (Region 2) are located completely in the closed magnetosphere and can only be projected onto its equatorial cross section. The boundary between Regions 1 and 2 of the field-aligned currents projects to the area near the inner boundary of the plasma sheet. For the condition with $B_z = -4$nT and $B_y = 0$, the inflow and outflow of the field-aligned currents at ionosphere altitudes are associated with the same magnetospheric regions. For $B_z = 4$nT and $B_y = 0$ an additional system of high latitude field-aligned currents appears, which flows into the afternoon sector and away from the prenoon sector. These currents project to the high-latitude region of the mantle near the magnetopause. When $B_z = 0$ and $B_y = \pm 6$nT the field-aligned currents with the same direction cover the near-noon sector from the dusk to dawn hours in both the mantle and the equatorial cross-section of the magnetosphere. The field-aligned current directions in these regions are opposite and systematically reverse with changing B_y polarity. The three-dimensional current systems coupling the magnetosphere and the ionosphere under various conditions in interplanetary space are discussed. These systems are consistant with the distribution of the ionospheric currents which are shown by ground-based observations to close the large-scale field-aligned currents. Field-aligned current systems are proposed according to the regions of their generation in the magnetosphere and depending on the interplanetary conditions.

Introduction

The methods for discriminating the magnetic field variations in high latitudes of the northern hemisphere which are associated with the interplanetary magnetic field (IMF) components and with the solar wind density and velocity were described in (Belov et al., 1977; Levitin et al., 1982). Also presented were the equivalent current systems associated with item variations and references to other studies devoted to the problem.

The magnetic field variations were used by Levitin et al., (1982) to calculate the distribution of large-scale field-aligned currents, $J_\parallel$, in summer assuming a uniform integral conductivity of the ionosphere. For the case of a non-uniform conductivity the appropriate algorithms to calculate $J_\parallel$ were proposed by Faermark (1980) and Kamide et al. (1981).

Field-aligned current was determined from the relation

$$J_\parallel = \text{Div } \bar{J}$$

where $\bar{J}$ is the altitude-integrated horizontal ionospheric current

$$J = \hat{\Sigma} \, \bar{E}$$

where E is the electric field in the ionosphere and $\hat{\Sigma}$ is the tensor of integral conductivity. The electric field is related to "phi" the potential ψ which is described by a second-order partial differential equation with the boundary condition; $\psi = 0$ on the geomagnetic parallel $\Lambda = 58°$.

The field-aligned currents are closed through the ionosphere by the currents, J_p, which coincide, in the case of a uniform conductivity of the

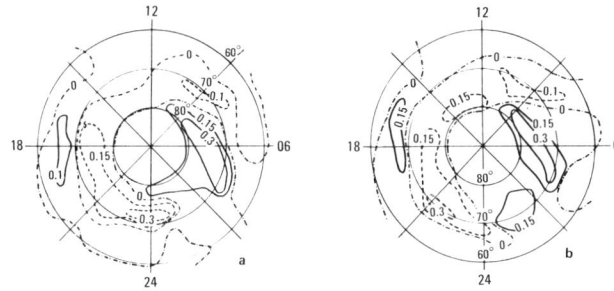

Fig. 1. Distribution of field-aligned currents during equinox for $B_z = B_y = 0$ and various ionospheric conductivities Σ. Numbers on contours for $J_\parallel$ are expressed in $\mu A/m^2$. The solid lines denote currents flowing into the ionosphere; the dashed lines correspond to outflowing currents. The current distribution corresponds to the following mean values of solar wind plasma; $n = 4$ cm^{-3}, and $v = 500$ km/s. The coordinate system is in corrected geomagnetic latitude and LMT. The contours shown in this, and subsequent figures enclose regions with field-aligned currents of equal intensity.
 (a) magnetically quiet (Kp < 3) conditions, UT$_{mean}$;
 (b) magnetically disturbed (Kp > 3) conditions, UT$_{mean}$.

ionosphere, with the Pedersen currents. In the case of non-uniform conductivity, these currents are determined from the relations

$$J_p = \nabla F; \quad \text{Div}(-\Sigma \nabla \psi) = \Delta F$$

Distributions of Field-Aligned Currents

Figures 1a and 1b present the distributions of $J_\parallel$ during equinox when the conditions of interplanetary space are fixed with the IMF components $B_y = B_z = 0$, $n = 4$ cm^{-3}, $v = 500$ km/s, and with the integrated ionospheric conductivity varying with UT and particle precipitation intensity. The component of Σ controlled by particle precipitation ($\hat{\Sigma}_c$) was taken from Wallis and Budzinski (1981), and the part controlled by solar radiation ($\hat{\Sigma}_w$) from Mehte (1979). The Sun's declination, $\chi = 0°$. The conductivity $\hat{\Sigma}_c$ varies from quiet (Kp < 3) to disturbed (Kp > 3) conditions. The conductivity $\hat{\Sigma}_w$ varies with season and with UT, because the illumination of the ionosphere above 60° latitude is a function of the tilt of the geomagnetic dipole with respect to the plane of ecliptic.

The contours in Figures 1a and 1b show the regions of inflowing (solid lines) and outflowing (dashed lines) $J_\parallel$ in units in $\mu A/m^2$ in a corrected geomagnetic latitude-local geomagnetic time solar plot. Although the ionospheric conductivity varies substantially, the general distribution of the large-scale $J_\parallel$ is preserved. Namely, the dusk and dawn sectors contain a pair of $J_\parallel$. In the dawn sector the inflowing current is located at higher latitudes compared with the outflowing current, and in the dusk sector the direction of the currents systematically reverses. The largest variations of the $J_\parallel$ distributions associated with conductivity, solar illumination or particle precipitation are observed in the night sector at auroral latitudes. The distribution of the type presented in Figures 1a and 1b is characteristic of magnetically quiet conditions with $B_y = B_z = 0$ and is consistent with the distribution inferred from satellite measurements (Zumda and Armstrong, 1974; Iijima and Potemra, 1976a).

The $J_\parallel$ distribution varies with changing IMF orientation and intensity. Figures 2 a-c show $J_\parallel$ at high latitudes for $B_z = -4$nT, $B_y = 0$ (Figure 2a); $B_z = 0$, $B_y = -6$nT (Figure 2b); and $B_z = 4$nT, $B_y = 0$ (Figure 2c).

The general pattern of the $J_\parallel$ distribution shown in Figure 1 is preserved during the period with $B_z = -4$, $B_y = 0$, but, the $J_\parallel$ intensities are larger in its latter case (both current density and the integral values).

For large values of the IMF azimuthal component

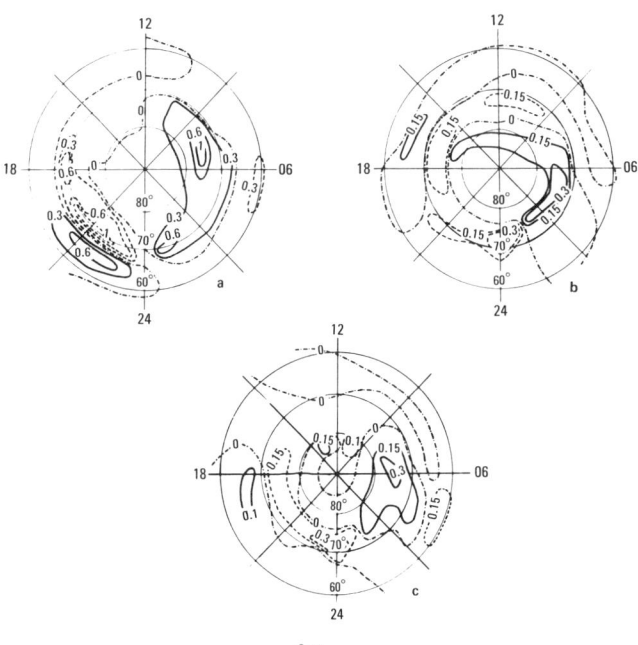

Fig. 2. Distributions of $J_\parallel$ during equinox for various IMF conditions: (a) $B_z = -4$nT, $B_y = 0$; (b) $B_z = 0$, $B_y = -6$nT; (c) $B_z = 4$nT, $B_y = 0$. The conductivities correspond to the conditions UT$_{mean}$, Kp > 3 (Fig. 2a) and Kp < 3 (Fig. 2, b, c). The other designations are the same as in Fig. 1.

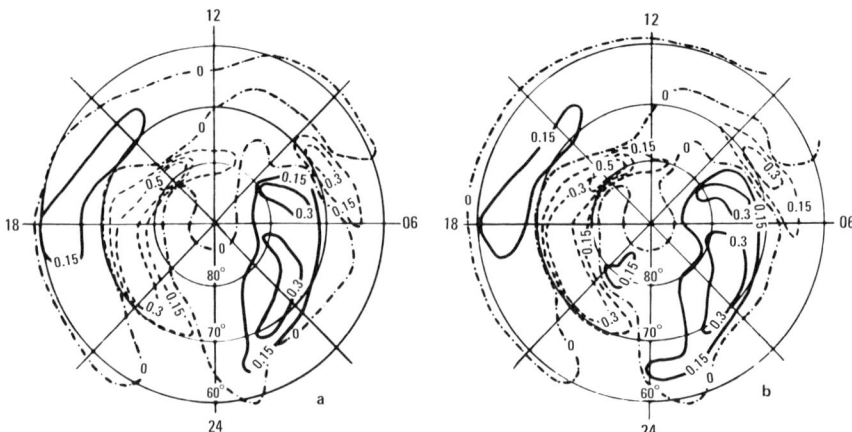

Fig. 3. Distribution of $J_{\parallel}$ during summer with $B_z = B_y = 0$ for various distributions of ionospheric conductivities: (a) Kp < 3, 1700 UT; (b) Kp < 3, 0500 UT.

B_y, the most significant variations are observed in the day-side sector. For $B_y = -6nT$ in the northern hemisphere, $J_{\parallel}$ flows into the ionosphere above 80° latitude and flows away from the ionosphere below 78°. The change of the B_y sign is accompmanied by the change of current direction in high-latitude day-side sector.

When the IMF component is northward a pair of $J_{\parallel}$ occurs near noon above 80° with currents flowing into the ionosphere at afternoon hours and away from the ionosphere at prenoon hours. Control of the $J_{\parallel}$ distributions by the interplanetary medium parameters has also been deduced from satellite observations (McDiarmid et al., 1978 a, b; Iijima and Potemra, 1982). Detailed discussion of these relationships may be found in Levitin et al. (1982) and Troshichev, (1982).

The nonuniformity of conductivity in summer has even a smaller effect on the $J_{\parallel}$ distribution. Figures 3a and b show the $J_{\parallel}$ distributions under the situation with conductivity $B_z = B_y = 0$ at 0500 and 1700 UT, the extreme levels of illumination of the polar ionosphere. The $J_{\parallel}$ distributions differ within the limits of the determination accuracy and the smoothing involved in drawing the isolines.

Figures 4a and 4b show the $J_{\parallel}$ distribution during summer for various levels of particle precipitation. Despite the extreme conditions of ionospheric conductivity, the $J_{\parallel}$ distributions differ little because of the high values of the conductivity associated with solar radiation. The comparison of these results with the results of calculating $J_{\parallel}$ from a uniform ionospheric conductivity (Feldstein et al., 1982) reveals that the uniform conductivity of the ionosphere in summer is a good approximation. The absolute values of $J_{\parallel}$ were somewhat overestimated by Feldstein et al. (1982) because their calculations were made assuming that $\Sigma_H = \Sigma_P$.

Figures 5a and b show the $J_{\parallel}$ distributions with $B_z = 0$ and $B_y = -6nT$ and $B_z = 4nT$ and $B_y = 0$. The most significant feature in Figure 5a is the $J_{\parallel}$ distribution between 75° and 85° latitude in the dayside sector, which is of the form of two longitude-aligned curtain currents. The situation during $B_z > 0$ on the day-side sector is characterized by a pair of field-aligned currents of different directions in the pre- and after-noon hours. All the $J_{\parallel}$ distributions were obtained by making allowance for nonuniform conductivity of the ionosphere in summer, and are quite consistent with the results of the calculations of Feldstein et al. (1982) who assumed a uniform conductivity in summer.

The distributions of $J_{\parallel}$ for $B_z = B_y = 0$ are similar to those obtained by Matsushita and Xu (1982b) for solar diurnal variations in the polar regions. For $B_y = -6nT$ and $B_z = 0$, the $J_{\parallel}$ distributions are similar (but with opposite flow directions) to the field-aligned currents presented by Matsushita and Xu (1982a) for the away sector of IMF.

The comparison between the $J_{\parallel}$ distributions during equinox and summer has revealed the following features:

(1) Enhancement of $J_{\parallel}$ in the near-midnight sector at invariant latitude $\Lambda \sim 70°$ during

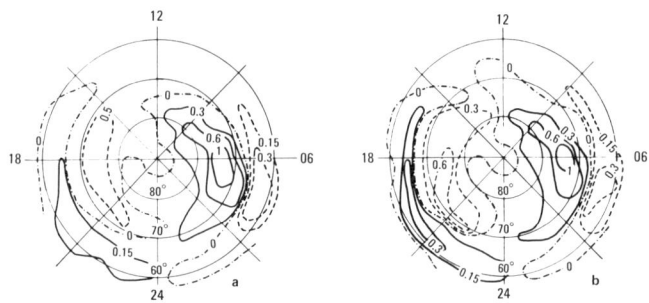

Fig. 4. Distributions of $J_{\parallel}$ during summer $B_z = -4nT$ and $B_y = 0$ for various distributions of ionospheric conductivities: (a) Kp < 3, UT_{mean}; (b) Kp > 3, UT_{mean}.

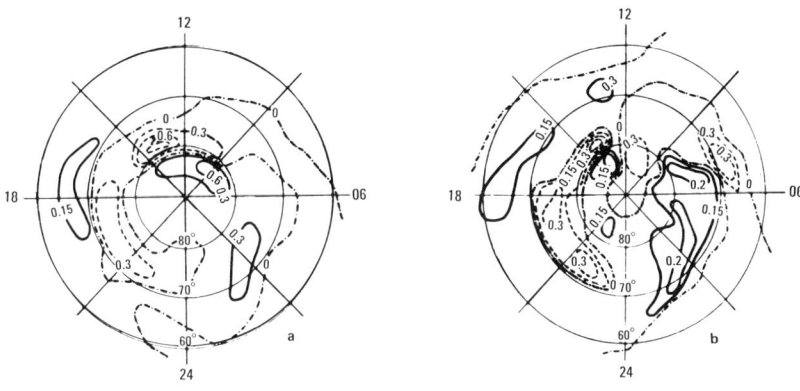

Fig. 5. Distribution of $J_\parallel$ during summer with various IMF conditions: (a) $B_z = 0$, $B_y = -6nT$; (b) $B_z = 4nT$, $B_y = 0$. The conductivities correspond to the conditions at 1700 UT, Kp < 3.

equinox with $B_z = B_y = 0$ and $B_z = -4nT$, $B_y = 0$. Such enhancements are in variance with the data presented by Fujii et al., (1981).

(2) Weakening of $J_\parallel$ in the day-side sector at $\Lambda \sim 80°$ during equinox with $B_z = 0$, $B_y = 6nT$ and $B_z = 4nT$, $B_y = 0$; and at $\Lambda \sim 78°$ at early dusk hours for $B_z = B_y = 0$. This decrease of $J_\parallel$ intensity agrees with the similar variations of $J_\parallel$ described by Fujii et al. (1981) using the Triad data in comparing between summer and winter.

Table 1 presents the values of the total field-aligned current $I = \int J_\parallel dS$ flowing into (+) and away from (−) the ionosphere during three seasons and for the IMF components $B_z > 0$ and < 0, $B_y = \pm 1nT$ and nearly zero (denoted by I_+ and I_-). The results presented in Table 1 show that the inflowing and outflowing currents equal each other to within 10% for all six systems of field-aligned currents, irrespective of season. In equinox, I is 10-20% weaker compared with winter and summer for $I(B_y)$ and $I(B_z)$. This trend is violated fo I_+ and I_-, especially under the situation for $B_z < 0$ in summer when the field aligned current is almost 1.5 times weaker compared with winter and equinox.

TABLE 1. Total values (in units 10^3A) of the field-aligned currents flowing into (I_+) and away from (I_-) the ionosphere for various values of IMF components.

	$B_z > 0$, quiet conditions						$B_z < 0$, disturbed conditions					
	$B_y=1nT$		$B_z=1nT$				$B_y=1nT$		$B_z=-1nT$			
	I_+	I_-	I_+	I_-	I_+	I_-	I_+	I_-	I_+	I_-	I_+	I_-
Summer												
0500 UT	309	348	393	408	2121	2340	450	468	1068	1065	2499	2571
UT_m	288	324	372	384	1977	2202	429	447	993	987	2313	2388
1700 UT	282	318	366	378	1899	2124	423	444	975	969	2295	2370
Equinox												
0500 UT	234	249	372	369	2274	2271	402	408	972	1098	3732	3783
UT_m	207	222	348	345	2037	2040	372	375	903	1023	3417	3471
1700 UT	186	201	336	333	1893	1899	342	348	822	939	3162	3219
Winter												
0500 UT	270	261	438	447	2034	2016	462	408	999	1146	3531	3267
UT_m	270	261	435	441	2046	2037	462	408	1014	1152	3534	3276
1700 UT	258	249	405	408	1965	1953	444	393	981	1107	3339	3108

Field-Aligned Currents and the Structure of the Magnetosphere

To elucidate the nature of field-aligned currents, one has to know the magnetospheric regions where they are generated. The observations above the ionosphere have shown that the large-scale $J_\parallel$ are generated in remote regions of the magnetosphere. This conclusion was drawn from the coincidence of the electric field direction at ionospheric altitudes with the direction of the closure currents J_p through the ionosphere. The ionosphere is a load in the magnetosphere-ionosphere circuit, and the $J_\parallel$ source is located in the magnetosphere. Mapping of the regions of field-aligned current inflow and outflow from ionospheric altitudes to the magnetosphere is still the only possibility to find out, though approximately, where the generation regions of these currents are located.

Potemra (1977) mapped the distribution of $J_\parallel$ from altitudes of ~ 800 km above the Earth's surface to the equatorial plane of the magnetosphere. The Fairfield-Mead magnetic field model (Fairfield and Mead, 1975) was used, and the $J_\parallel$ distribution was taken according to Iijima and Potemra (1976b, 1978). In the dayside sector, the $J_\parallel$ are closed through the equatorial plane without being mapped onto the magnetopause or magnetospheric tail. In the dusk sector, the currents flowing away from the ionosphere are projected onto the plasma sheet, and the projection of the boundary between the inflowing and outflowing $J_\parallel$ coincides with the inner boundary of the plasma sheet according to Vasyluinas (1968). It may be assumed, therefore, that the formation of an Alfvén layer in this region can be one of the mechanisms of current generation. The formation of such a layer may result from the spatial separation of the high-energy protons and electrons drifting around the Earth (Schield et al., 1969) or from the drift of relatively cold electrons through the boundary of the forbidden region formed by high-energy protons (Lyatsky, 1978).

We will consider the mapping of the $J_\parallel$ distributions shown in Figures 1 and 2 to the magnetosphere. For this purpose, use may be made of various models of the Earth's magnetic field including the semiempirical model proposed by Fairfield and Mead (1975) and the quantitative model proposed by Alekseev and Shabansky (1972, 1978) and by Tsyganenko (1979). Application of the three models has shown that they yield the qualitatively identical results in the $J_\parallel$ location relative to the characteristic structural regions of the magnetosphere. We shall use the Fairfield-Mead model (1978) in the discussion below because all the sources of the magnetospheric field were automatically included when it was constructed.

Figure 6 shows the projection of $J_\parallel$ to the equatorial cross section of the magnetosphere (below) and to the magnetotail cross section at a $-10R_E$ geocentric distance (above) during equinox. The positions of the regions of the maximum

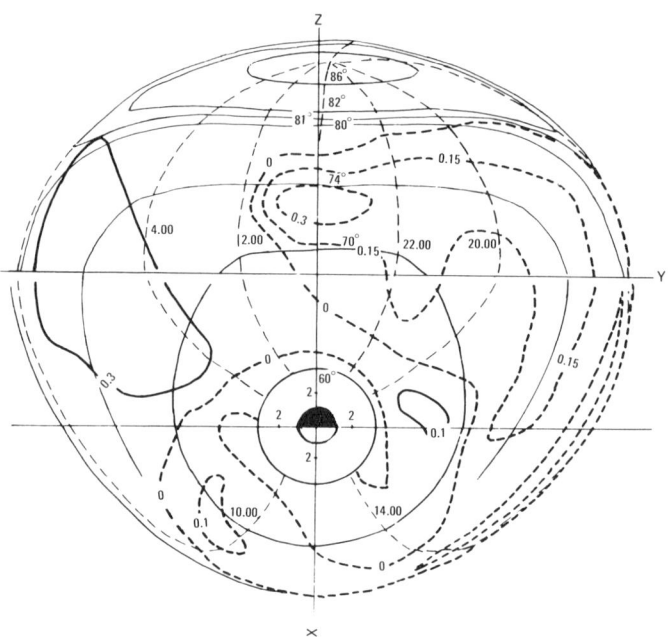

Fig. 6. The projection of the distribution of $J_\parallel$ for $B_z = B_y = 0$ during equinox to the equatorial cross section (below) and to the tail cross-section at $X = -10\ R_E$ (above). The solid and dashed contours show, respectively, the inflowing and outflowing field-aligned currents. The values of $J_\parallel$ are in units $\mu A/m^2$ and are indicated on the figure. The magnetospheric model of Fairfield and Mead (1975) for $Kp < 2$ was used; the dipole tilt angle is zero. The thin solid lines show the geomagnetic latitudes, the dashed lines are the local time meridians.

and zero values of $J_\parallel$ are also presented. The values of $J_\parallel$ in $\mu A/m^2$ are indicated. The thin continuous lines show the geomagnetic latitudes, and the dashed lines the meridians of local geomagnetic time. This magnetospheric model corresponds to a zero tilt angle of the dipole axis. The model of quiet magnetosphere was used for the IMF condition with $B_y = B_z = 0$.

The high-latitude region of field-aligned currents flowing into the ionosphere in the dawn sector and outflowing in the dusk sector (Region 1 according to Iijima and Potemra 1976a, b) is projected onto the magnetotail plasma sheet and onto the magnetospheric regions adjoining the magnetopause. The maximum $J_\parallel$ contours cover the plasma sheet at a fairly great distances along the Z-axis from the neutral sheet. However, $J_\parallel$ of rather high intensities are projected onto the regions adjoining the magnetopause on the dusk and dawn sides of the magnetosphere. In these regions, the field lines are closed (Doyle et al., 1981; Klumpar, 1979). The closure takes place

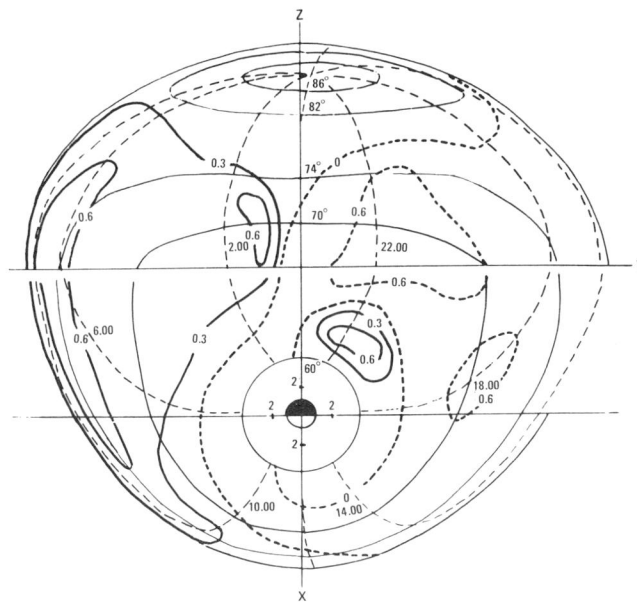

Fig. 7. The projection of the distributions of $J_\parallel$ with $B_z = -4nT$, $B_y = 0$ during equinox. The designations are the same as in Figure 6. The magnetospheric model for Kp > 2 was used.

partly through the equatorial plane on the day side and partly through the magnetotail plasma sheet. The field-aligned currents adjoining the magnetopause may result from the plasma motion in the low-latitude boundary layer which was recorded by satellites as they traversed the magnetopause (Eastman et al., 1976; Haerendel and Paschman, 1982).

The lower-latitude zone of the field-aligned current region (Region 2 according to Iijima and Potemra 1976a, b) with direction opposite to that in Region 1, is projected entirely to the equatorial plane. Therefore, $J_\parallel$ are all localized in the closed magnetosphere. The boundary between Region 1 and 2 of field-aligned currents is located in the equatorial plane on the night-side near $-7R_E$, and on the day side near the magnetopause, (i.e. near the inner boundary of the plasma sheet according to Vasyliunas, 1968). Thus, with $B_z = B_y = 0$:

(1) The Region 2 field-aligned currents are projected onto the magnetospheric regions located nearer to the Earth from the inner boundary of the plasma sheet.

(2) The Region 1 field-aligned currents are projected onto the magnetotail plasma sheet and onto the boundary layer near the magnetopause in the dawn and dusk sectors.

Figure 7 shows the contours of $J_\parallel$ in the equatorial cross section of the magnetosphere, and in the magnetotail cross section (the designations are the same as in Figue 6) for $B_z = -4nT$ and $B_y = 0$. In the case of the southward IMF component (B_z < 0), geomagnetic disturbances are observed on the Earth's surface. Therefore, use is made of the Fairfield-Mead model of the Earth's magnetic field for Kp > 2. From the data presented, it follows that the location of the inflowing and outflowing $J_\parallel$, relative to the magnetospheric regions, remains the same as in Figure 6, the only difference being the increased intensity of $J_\parallel$. However, the boundary between the poleward and equatorward region of field-aligned currents in the equatorial cross section of the magnetosphere is shifted toward the Earth.

It may be concluded from the above considerations that during $B_z < 0$ the currents flow in and out at ionospheric altitudes from the same magnetospheric regions as during $B_z = B_y = 0$. The equatorial part of the $J_\parallel$ system is projected entirely to the closed magnetosphere, and its poleward portion projects to the magnetotail plasma sheet and to the magnetopause. In this case, the most intense $J_\parallel$ are projected onto the plasma sheet. As B_z increases in the negative direction, the $J_\parallel$ intensity also increases.

Figure 8 presents the results of the projection of the complex $J_\parallel$ distribution to the magnetosphere for $B_z = 4nT$ and $B_y = 0$. An additional feature evident here is the system of the poleward currents flowing into the afternoon sector, and away from the prenoon sector. They are all projected to the magnetospheric tail in the high-latitude part of the mantle in the vicinity of the magnetopause. Thus, the northward IMF orientation in the magnetosphere gives rise to additional

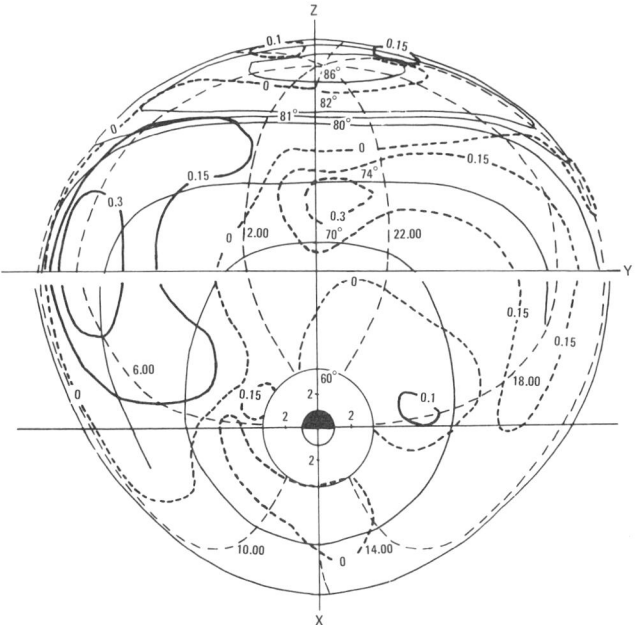

Fig. 8. The projections of distributions of $J_\parallel$ with $B_z = 4nT$, $B_y = 0$ during equinox. The designations are the same as in Figure 6. The magnetospheric model for Kp < 2 was used.

field-aligned currents which originate from processes in the high-latitude region of the plasma mantle and the magnetopause. The interface between the equatorward and poleward parts of the $J_\parallel$ system is located in the equatorial plane of the magnetosphere in the dusk sector $\geqslant 10\ R_E$, which should be expected under such extremely magnetically quiet conditons.

The distribution of the IMF B_y component-controlled $J_\parallel$ projected to the magnetosphere is peculiar (Figure 9, for $B_y = -6nT$, $B_z = 0$). The field-aligned currents of the same direction cover the near-noon sector from the dusk to dawn hours in both the mantle and the equatorial plane. The $J_\parallel$ directions in these regions are opposite and reverse with changing B_y-component polarity. Such variation in $J_\parallel$ follow from the three-dimensional model of currents proposed by Lyatsky (1978).

Three-Dimensional Current Systems

The field-aligned currents from the magnetosphere are closed by ionospheric currents J_p. The distribution of such currents was presented by Feldstein et al. (1983). For the By < 0 situation in the day sector near $\Lambda \sim 80°$ the currents J_p flow to the equator, in agreement with the field-aligned current system presented in Figure 9.

The plasma motions in the boundary regions of the magnetosphere give rise to a three-dimensional current system on the day side in which the field-aligned currents flow into the ionosphere in the dawn sector and away from the dusk sector (Sonnerup, 1980). An enhancement of this current system with increasing solar wind velocity, or intensifying the magnetic field merging processes in the dayside magnetosphere during the periods with Bz < 0 will, result in a decrease of the geocentric distance of the magnetopause and a shift of the dayside cusp to lower latitudes (Maezawa, 1974; Lyatsky and Maltsev, 1975; Akasofu et al., 1981). The effects of this current system can also contribute to the Earth's magnetic field during geomagnetic storms (Ivanov and Mikerina, 1975). Such a field-aligned current system is shown in Figures 6 and 7, and should be closed by the ionospheric currents J_p flowing in the poleward region from dawn to dusk. Such currents, calculated from surface magnetic fields, were determined by Feldstein et al. (1983). Use was made of the IMF-independent variations of the geomagnetic field. In particular, an intensive eastward current J_p exists in the dayside sector between $\Lambda = 80°$ and $84°$. An ionospheric current of such direction must exist permanently in the polar cap. The $J_\parallel$ associated with this current will intensify when $B_z < 0$. Thus, the three-dimensional current system with the configuration discussed by Maezawa, (1974); Lyatsky and Maltsev, (1975); Akasofu et al., (1981); and Ivanov and Mikerina, (1975) exists in the magnetosphere permanently, rather than only during the periods with $B_z < 0$ or during magnetospheric substorm

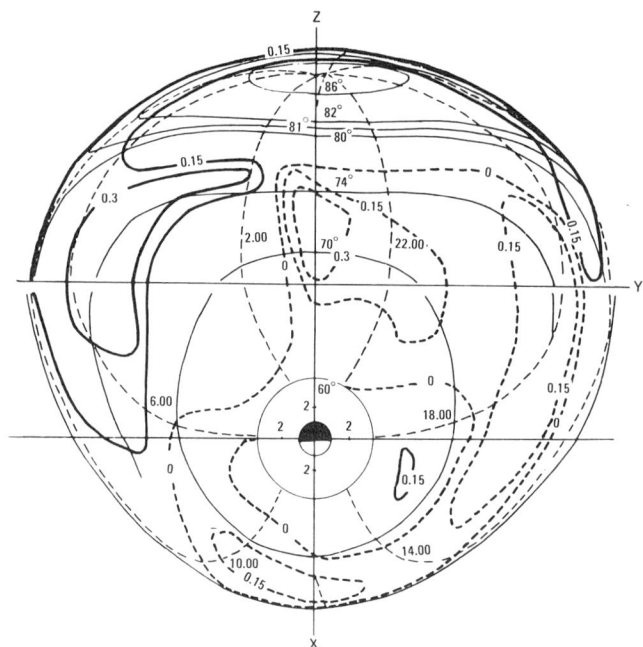

Fig. 9. The projection of the distribution of $J_\parallel$ with $B_z = 0$ and $B_y = -6nT$ during equinox. The magnetospheric model for Kp < 2 was used.

expansion phases. This permanent system is only enhanced during disturbances. However, such an enhancement was interpreted in the papers cited above as a new three-dimensional current system in the dayside magnetosphere.

In the night sector, the three-dimensional current system is coupled to magnetospheric convection which is responsible for this system. During the periods with $B_z < 0$, a portion of the magnetotail currents flowing in the plasma sheet may be deflected to the ionosphere (Akasofu, 1977).

During the intervals with $B_z > 0$, a new field-aligned current system appears which projects to the high-latitude part of the mantle near the magnetopause (Figure 8). The direction and location of these currents in the remote part of the magnetosphere suggest that they may arise from the deflection to the ionosphere of a portion of the currents flowing along the magnetotail periphery.

Figure 10 shows a possible version of such a large-scale current system which can explain the observed effects. The discontinuity of the currents in the high-latitude part of the tail gives rise to a $J_\parallel$ flowing into the ionosphere in the dusk sector and away from the dawn sector. The closure of field-aligned currents in the dayside sector of the polar cap should be accompanied by the appearance of westward currents, J_p, at ionospheric altitudes. In fact, the calculations of J_p from the observed variations of the geomagnetic field in the intervals with $B_z > 0$ give an

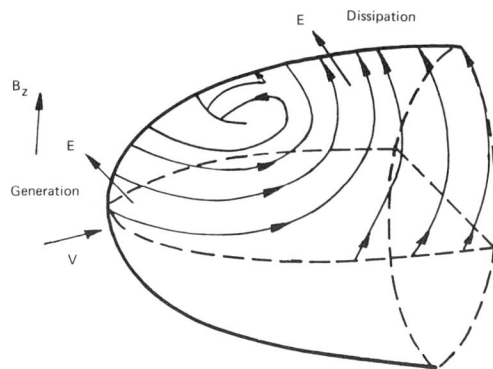

Fig. 10. A schematic diagram of currents (denoted by the arrows) on the magnetopause. A fraction of the magnetopause currents are deflected into the magnetosphere and flow into the ionosphere during noon hours. The regions of generation and dissipation of electromagnetic energy are indicated.

intense westward current in the dayside sector of polar cap. This result indicates the existence of a three-dimensional current system of the assumed configuration. During the intervals with $B_z > 0$, such system may arise from the processes occuring on high latitudes in the tail which result in plasma mantle thinning near the midnight meridian (Sckopke and Paschman, 1978; Sckopke et al., 1976) and the processes of merging of the interplanetary and geomagnetic fields in the remote region of the cusp (Ivanov and Evdokimova, 1975).

Field-aligned currents were recorded (Hones et al., 1982) in many north hemisphere crossing of the ISEEE-1 and 2 satellites in the magnetopause. These currents flow along field lines antiparallel to the geomagnetic field. Such a direction of $J_\parallel$ follows from the above proposed three-dimensional current system during $B_z > 0$. Apparently, B_z was positive during the ISEE measurements (King 1979). This situation is drastically different from the case with $B_z < 0$, when the merging process occurs in the vicinity of the equatorial plane (Nishida, 1982; Sonnerup, et al., 1981).

Discussion

The projection of the distribution of $J_\parallel$ to the magnetosphere permits the currents to be related to characteristic magnetospheric regions. Depending on particular situation in interplanetary space, the field-aligned currents are generated in one or another magnetospheric region. Such regions are:

(1) The plasma sheet and the dayside magnetosphere (the low-latitude boundary layer) adjoining the magnetopause, where the poleward $J_\parallel$ are generated. This occurs with $B_z = B_y = 0$ and $B_z < 0$, $B_y = 0$.

(2) The region between the inner boundary of the plasma sheet and the plasmapause, where diffusive precipitation of soft auroral particles are observed. In this region with $B_z < 0$, $B_y = 0$, and $B_z = B_y = 0$, the equatorward system of currents are generated.

(3) The low-latitude boundary layer and the entry layer in the dayside magnetosphere, the nightside plasma mantle adjoining the magnetopause. The field-aligned currents are generated in there regions under the situation with $B_y > 0$ or $B_y < 0$ and $B_z = 0$. The current flow directions are opposite in the entry layer and in the mantle. The direction of the currents also varies depending on B_y polarity. The dayside cusp region $J_\parallel$ of the same direction maps to a fairly extended local time interval on either side of noon.

(4) The region of plasma mantle adjoining the magnetopause, where $J_\parallel$ projects to $\Lambda > 80°$ in the day sector, are generated with $B_z > 0$, $B_y = 0$. The $J_\parallel$ flow into the ionosphere in the afternoon sector and away from the ionosphere in the prenoon sector.

In conclusion, the large-scale field-aligned currents are generated in various remote regions of the magnetosphere. In some of the regions, the currents are generated continuously and they are probably due to the existence of large-scale convection (in the plasma sheet, above the low-altitude boundary layer, the region between the inner boundary of plasma sheet and the plasmapause). In other regions, the generation of currents and their directions are due to the interactions of solar wind and IMF with the geomagnetic field (in the low latitude boundary layer, the entry layer, the part of the mantle adjoining the magnetopause).

The classification of $J_\parallel$ according to the region of their generation and condition in the interplanetary medium seems to be preferable, compared with the division of the space-time distribution of $J_\parallel$ into Regions 1, 2, 3 (Region 3 refers to the cusp system). With such a classification, the same system will include the field-aligned currents generated by the same processes, but belongs to different zones. For example, $J_\parallel$ generated during $B_y > 0$, or < 0 and $B_z = 0$ cover Regions 1 and 3. The day sector of Regions 1 and 3 may comprise $J_\parallel$ of quite different origins. Namely the field-aligned currents controlled by the IMF components $B_z > 0$ and $B_y < 0$ or $By > 0$ are located at the latitudes of Region 3, and the field-aligned currents under the situations with $B_z = B_y = 0$ or $B_z < 0$ and $B_y > 0$ or $B_y < 0$ are located at the latitudes of Region 1. It is appropriate to take the next step and replace the classification of the field-aligned currents according to their location in Regions 1, 2, or 3 by a new classification. In the new classification proposed above, $J_\parallel$ are grouped according to their generation region, and depending on the conditions in interplanetary medium. The grouping of $J_\parallel$ according to their generation region was also proposed earlier (Saflekos et al., 1982); and the possible sources of the large-scale field aligned currents were reviewed by Potemra (1982).

References

Akasofu, S.-I., Physics of magnetospheric substorms, D. Reidel, Dordrecht, 1977.

Akasofu, S.-I., M. Roederer, G.K. Corrick and D.N. Covey, Equatorward shift of the cusp during magnetospheric substorms, Planet. Space Sci., 29, 317, 1981.

Alekseev, I.I. and V.P. Shabansky, A model of a magnetic field in the geomagnetosphere, Planet. Space Sci., 20, 117, 1972.

Alekseev, I.I., Regular magnetic field in the Earth magnetosphere, Geomagnetism and Aeronomie, 18, 656, 1978.

Belov, B.A., R.G. Afonina, A.E. Levitin and Y.I. Feldstein Influence of interplanetary magnetic field (IMF) vector components on north polar cap geomagnetic field in Magnetic field variations and aurorae, ed. Y.I. Feldstein, IZMIRAN, Moscow, 15, 1977.

Doyle, M.A., F.J. Rich, W.J. Burke and M. Smiddy, Field-aligned current and electric fields observed in the region of dayside cusp, J. Geophys. Res., 86, 5656, 1981.

Eastman, T.E., E.W. Hones, J.S. Bame and J.R. Asbridge. The magnetospheric boundary layer: site of plasma, momentum and energy transfer from the magnetosheath into the magnetosphere, Geophys. Res. Letters, 3, 685, 1976.

Faermark, D. S., Method of electric and magnetic field calculation in high-latitude ionosphere, in solar wind and magnetospheric investigations, ed. A.E. Levitin, IZMIRAN, Moscow, 122, 1980.

Fairfield, D.N. and G.D. Mead, Magnetospheric mapping with a quantitative geomagnetic field model, J. Geophys. Res., 80, 535, 1975.

Feldstein, Y.I., R.G. Afonina, B.A. Belov and A.E. Levitin, Magnetic field and field-aligned current variations in the polar cusp controlled by interplanetary medium parameters, Planet. Space Sci., 30, 635, 1982.

Feldstein Y.I., A.E. Levitin and N.G. Vorfolomeeva, Field-aligned currents and magnetosphere structire, Geomagnetism and Aeronomia, 23, 290, 1983.

Fujii, R.T., Iijima, T.A. Potemra and M. Sugiura, Seasonal dependence of large-scale Birkeland currents, Geophys. Res., Lett., 8, 1103, 1981

Haerendel, G. and G. Paschman, Interaction of the solar wind with the dayside magnetosphere, in Magnetospheric plasma physics, ed. A. Nishida, D. Reidel, 59, 1982.

Hones, E.W., B.U.O. Sonnerup, S.J. Bame, G. Paschman and C.T. Russell, Reverse drapping of magnetic field lines in the boundary layer, Geophys. Res. Lett., 9, 523, 1982.

Iijima, T. and T.A. Potemra, the amplitude distribution of field-aligned currents at northern high latitudes observed by Triad, J. Geophys. Res., 81, 2165, 1976a.

Iijima, T. and Potemra T. A., Field-aligned currents in the dayside cusp observed by Triad, J. Geophys. Res., 81, 5971, 1976b.

Iijima, T. and T.A. Potemra, Large-scale characteristics of field-aligned currents associated with substorms, J. Geophys. Res., 83, 599, 1978.

Iijima, T. and T.A. Potemra, The relationship between interplanetary quantites and Birkeland current densities, Geophys. Res. Lett., 9, 442, 1982.

Ivanov, K.G. and N.V. Mikerina, "Magnetic" and "helium" regions of interplanetary plasma stream and the structure of the main phase of magnetic storm. Geomagnetism and Aeronomia, 15, 881, 1975.

Ivanov K.G. and L.V. Evdokimova, Electric field of polar caps presuming a magnetopause to be an anizotropic rotating discontinuity, Geomagnetism and Aeronomia, 15, 303, 1975.

Kamide, Y., A.D. Richmond and S. Matsushita, Estimation of ionospheric electric fields, ionospheric currents and field-aligned currents from ground magnetic records, J. Geophys. Res., 86, 801, 1981.

King, J.H., Interplanetary medium data book, 1975-1978, NSSDC/WDC-A-79-08, 1979.

Klumper, D.M., Relationships between auroral particle distributions and magnetic field perturbations associated with field-aligned currents, J. Geophys. Res., 84, 6524, 1979.

Lyatsky, V.B., The current system of magnetosphere-ionosphere distrubances, Leningard, Publ. House Nauka, 1978.

Levitin, A.E., R.G. Afonina, B.A. Belov and Y.I. Feldstein, Geomagnetic variations and field-aligned currents at northern high-latitudes and their relations to the solar wind parameters, Phil. Trans. R. Soc., London, A304, 253, 1982.

Maezawa, K., Dependence of the magnetopause position of the southward interplanetary magnetic field, Planet. Space Sci., 22, 1443, 1974.

Maltsev Yu.P. and W.B. Lyatsky, Field-aligned currents and erosion of the dayside magnetosphere, Planet, Space Sci., 23, 1257, 1975.

Matsushita, S. and W.Y. Xu, Equivalent ionospheric current systems representing IMF sector effects on the polar geomagnetic field, Planet. Space Sci., 30, 641, 1982a.

Matsushita, S. and W.Y. Xu, Equivalent ionospheric current systems representing solar daily variations of the polar geomagnetic field, J. Geophys. Res., 87, 8241, 1982b.

McDiarmid, I.B., J.R. Burrows and M.D. Wilson, Comparison of magnetic field perturbation at high latitudes with charged particles and IMF measurements, J. Geophys. Res., 83, 681, 1978a.

McDiarmid, I.B., J.R. Burrows and M.D. Wilson, Magnetic field perturbations in the dayside cleft and their relationship to the IMF, J. Geophys. Res., 83, 5753, 1978b.

McDiarmid, I.B., and J.R. Burrows and M.D. Wilson, Large-scale magnetic field perturbations and particle measurements at 1400 km on the dayside, J. Geophys. Res., 84, 1431, 1979.

Mehte, N.G., Ionospheric electrodymanics and its coupling to the magnetosphere, Ph.D. Thesis, Univ. of Calif., San. Diego, 1979.

Nishida A., IMF control of the Earth's magnetosphere, STP Synposium, Ottawa, 1982.

Potemra, T.A., Large-scale characteristics of field-aligned currents determined from the Triad magnetometer experiment, in Dynamical and chemical coupling, eds. B. Grandel and J.A. Holtet, D. Reidel, Dordrecht, 337, 1977.

Potemra, T.A. Birkeland currents: present understanding and some remaining questions, Nobel Symposium N54, Kiruna, Sweden, 1982.

Saflekos, N.A., R.E. Sheehan and R.L. Carovillano, Global nature of field-aligned currents and their relation to auroral phenomena, Rev. Geophys. Space Phys., 20, 709, 1982.

Schield, M.A., J.W. Freeman and A.J. Dessler, A source for field-aligned currents at auroral latitudes, J. Geophys. Res., 74, 247, 1969.

Sckopke, N., G. Paschmann, H. Rosenbauer and D.H. Fairfield, Influence of the interplanetary magnetic field on the occurrence and thickness of the plasma mantle, J. Geophys. Res., 81, 2687, 1976.

Sckopke, N. and G. Paschmann, The plasma mantle. A survey of magnetotail boundary layer observations, J. Atmosph. Terr. Phys., 40, 261, 1978.

Sonnerup, B.U.O., Theory of the low-latitude boundary layer J. Geophys. Res., 85, 2017, 1980.

Sonnerup, B.U.O., G. Paschmann, I. Papamastorakis. N. Sckopke, G. Haerendel. S.J. Bame, J.R. Asbridge, J.T. Gosling and C.T. Russell, Evidence for magnetic field reconnection at the Earth's magnetosphere, J. Geophts. Res., 86, 10049, 1981.

Troshichev, O.A., Polar magnetic disturbances and field-aligned currents, Space Sci. Reviews. 32, 275, 1982.

Tsyganenko, N.A., Subroutines and tables for geomagnetic field, computations, materials of the WDCB, Sov. Geophys. Comm., Acad. Sci., USSR, Moscow, 1979.

Vasyliunas, V.M., A survey of low-energy electrons in the evening sector of the magnetosphere with OGO-1 and OGO-3, J. Geophys. Res., 73, 2839, 1968.

Vasyliunas, V.M., A survey of low-energy electrons in the evening sector of the magnetosphere with OGO-1 and OGO-3, J. Geophys. Res., 73, 2839, 1968.

Wallis, D.D. and E.E. Budzinski, Empirical models of height integrated conductivities, J. Geophys. Res., 86, 125, 1981.

Zmuda A.J. and J.C. Armstrong, The diurnal flow pattern of field aligned currents, J. Geophys. Res., 79, 4611, 1974.

ELECTRIC FIELDS AND CURRENTS OBSERVED BY S3-2 IN THE VICINITY OF DISCRETE ARCS

William J. Burke

USAF Geophysics Laboratory, Hanscom AFB, MA 01731

Abstract. The high time resolution of the electric and magnetic field detectors on the polar orbiting satellite S3-2 made it possible to examine the details of auroral events down to discrete-arc scales. Depending on the instantaneous look direction of an electron detector, information about field-aligned accelerations above the satellite could also be obtained. Case studies of four arc events, three in the auroral oval and one in the polar cap, have been completed.

Field-aligned currents associated with arcs in the auroral oval appeared as matched pairs of oppositely directed current sheets. Magnetic deflections, almost exclusively in the east-west direction departed from and returned to baselines established by the large-scale Region 1/Region 2 currents. The upward currents had intensities of up to 145 $\mu A/m^2$ and were carried by electrons that were accelerated through field aligned potential drops. The relationship between $j_\parallel$ and $\phi_\parallel$ is not inconsistent with predictions of a laminar flow model. The most intense return (downward) currents were in the 10 to 15 $\mu A/m^2$ range. At satellite altitudes near 1000 km, these currents approximate the critical limit for current driven, ion cyclotron instabilities.

The arc in the polar cap was sun-aligned and was found in a region of intense convective shear, with the electric field pointing toward the center of the arc. The field-aligned currents consisted of three sheets; two with currents flowing into and one out of the ionosphere. The upward current was carried by polar-rain electrons that had undergone a field-aligned acceleration of approximately 1 kV.

Introduction

Since the launching of S3-2 and S3-3 in the mid-seventies it has been possible to study auroral processes on scales down to those of discrete arcs using satellite borne instrumentation. Measurements of electric fields and charged particle distributions by S3-3 at altitudes between 2000 and 8000 km has revolutionized our understanding of auroral acceleration processes. For it is partially in this altitude range that field-aligned potential drops frequently occur. Complementary information on the electrodynamic structure of arcs has come from the similarly equipped S3-2 satellite flying at altitudes below that of field-aligned potential drops. One significant difference between the two satellites was the presence of a high-resolution magnetometer on S3-2 which allows an examination of field-aligned currents associated with discrete arcs.

The purpose of this paper is to review briefly the results of three detailed studies of discrete arc characteristics (Burke et al., 1980; 1982; 1983) based on observations by S3-2 in the auroral oval and polar cap. The main conclusions that we hope to demonstrate are that: (1) In the auroral oval the field-aligned currents associated with discrete arcs appear as matched pairs of upward and downward currents superimposed on the large-scale Region 1/Region 2 system. (2) Whereas upward field-aligned currents can have intensities greater than 100 $\mu A/m^2$, the downward currents are limited to < 15 $\mu A/m^2$. This suggests quite different physical processes occurs when energetic and cold electrons are the main current carriers. (3) In the winter polar cap the field-aligned currents associated with polar cap arcs can appear as triple sheets; an upward current surrounded by two downward sheets. The upward currents have intensities of a few $\mu A/m^2$ and are carried by polar rain electrons that have been accelerated through potential drops of approximately 1 kV. (4) For the polar cap arc the relationship between the field-aligned current density ($j_\parallel$) and potential drop ($\phi_\parallel$) described by Lyons (1981) provides a reasonable guide for data interpretation. In the case of the very intense auroral oval arcs the relationship is not as clear. This may be due to errors derived from the slower time resolution of particle detectors and magnetometers.

The Air Force scientific satellite S3-2 was launched into polar orbit in December, 1975, with an initial apogee, perigee and inclination of 1557 km, 240 km and 96.3°, respectively. It was spin-stabilized with a nominal spin period of 20 sec, and a spin axis that is nearly perpendicular to the orbital plane in a cart-wheel sense. Figure 1 shows that the scientific package carried by S3-2 includes: (1) a dc electric field experiment consisting of two dipoles, (2) a triaxial fluxgate magnetometer, (3) an energetic electron spectrometer, (4) an ion drift meter, and (5) a thermal electron probe. The scientific package is described in detailed by Burke et al. (1980) and is not repeated here. We do note, however, that the electron spectrometer looked perpendicular to the spin axis and required a full

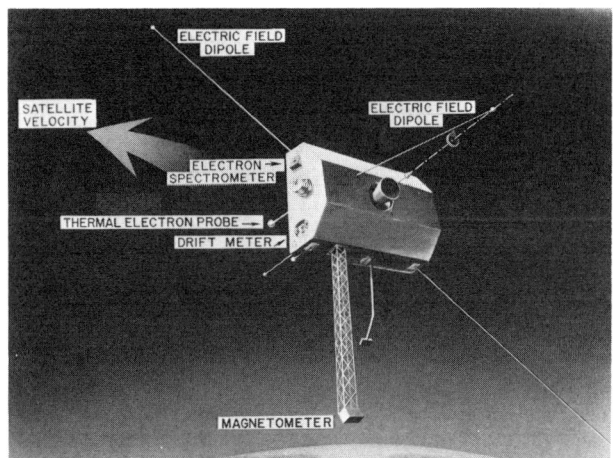

Fig. 1 Scientific complement of S3-2.

second to compile a 32 point spectrum. Thus, useful electron information about arc-scale events was obtained occasionally rather than systematically.

Data are presented here in a satellite coordinate system with X positive along the satellite velocity, Z positive toward local nadir and Y completes the right hand system. With the S3-2 orbit near the dawn-dusk meridian Y is positive in the antisunward direction (Smiddy et al., 1981). The symbol E_x is used to represent the meridional component of the electric field and ΔB_y represents the east-west magnetic deflection due to longitudinally extended, field-aligned current sheets. In our satellite coordinate system plots of ΔB_y versus time, regions of positive (negative) slope correspond to regions of field-aligned currents into (out of) the ionosphere.

Observations

We now turn to S3-2 measurements taken in the vicinity of discrete arc-like structures in the auroral oval and polar cap. Data from the magnetometer are considered first then from the electron spectrometer.

The first case is of an arc in the late evening sector of the auroral oval that was studied by Burke et al. (1980) to show that data retrieved from S3-2 could indeed be used to obtain reliable information on spatial scales previously accessible only to rocket experiments. Defense Meteorological Satellite Program (DMSP) imagery taken at northern high latitudes on January 11, 1976, is shown in Figure 2. The superimposed grid gives the corrected geomagnetic longitude and latitude projected along magnetic field lines to an altitude of 100 km. The trajectory of S3-2 Rev. 517 projected to the same altitude is represented by the lighter dashed line. The temporal separation between the DMSP and S3-2 trajectory crossover was 14 minutes,

with S3-2 measurements taken after those of DMSP. During quiet times, auroral forms tend to be stationary in magnetic latitude and local time. In the 14 minutes between the DMSP and S3-2 cross over the Earth rotated $\sim 3.5°$ in longitude. Thus, relative to the arc, the S3-2 orbit should be moved 3.5° to the east. That is, away from the folded arc and well into the straight line arc segment. As represented by the heavier dashed line the S3-2 trajectory crossed the arc at an angle of $\sim 30°$ away from the normal incidence. Simultaneous magnetometer mesurements from Dixon Island (marked by the X in Figure 2) indicate that auroral currents were very stable during the 14 mintues separating the DMSP and S3-2 overpasses.

The field-aligned currents associated with this arc may be derived from the magnetometer trace shown in Figure 3. We note that the arc appears as a spike-like feature superimposed on the large-scale Region 1/Region 2 current system. The upward current was mostly carried by precipitating, auroral electrons with energies of a few keV. The downward or return current, located equatorward of the arc reached an intensity of ~ 10 $\mu A/m^2$. Here the current carriers are upward moving cold electrons. Measurements from the thermal plasma probe showed that the ionospheric electron density and temperature were $3 \times 10^9 m^{-3}$ and 3150°K, respectively. To carry a current of 10 $\mu A/m^2$ these electrons must have an upward drift speed that is 7% of their thermal speed. This is approximately equal to the critical drift for the onset of O^+ cyclotron turbulence with $T_e/T_i \sim 3$ (Kindel and Kennel, 1971). Turbulent waves act as scattering centers that anomalously resist the upward motion of current carrying electrons. To maintain the required field-aligned current ($j_\parallel$) a field-aligned electric field ($E_\parallel$) component is required. As the satellite passed through the return current the electric field signature measured by the spinning dipole became irregular. Combining precise satellite attitude information with simultaneous measurements from the drift-meter Burke et al. (1980) showed that the local value of $E_\parallel$ was ~ 10 mV/m.

Two more examples of arc-like structures that were detected on opposite sides of the auroral oval during the substorm are given in Figure 4. In this figure we have plotted ΔB_y as measured on the dawn and dusk sides of the southern hemisphere auroral oval as functions of UT, invariant latitude, magnetic local time and altitude. The first event begins at 0953:06 UT ($\Lambda = 60°$ MLT = 05.4) and appears as a 100 nT excursion of ΔB_y that lasts for a total of 4 seconds. The second event is a 300 nT excursion that begins at 1148:16 UT ($\Lambda = 67°$, MLT = 19.3) and lasts 7 seconds. As in the previous example, these excursions are from baselines established by the large-scale Region 1/Region 2 current system. In both cases the excursions are located near the equatorward edge of the region 1 current

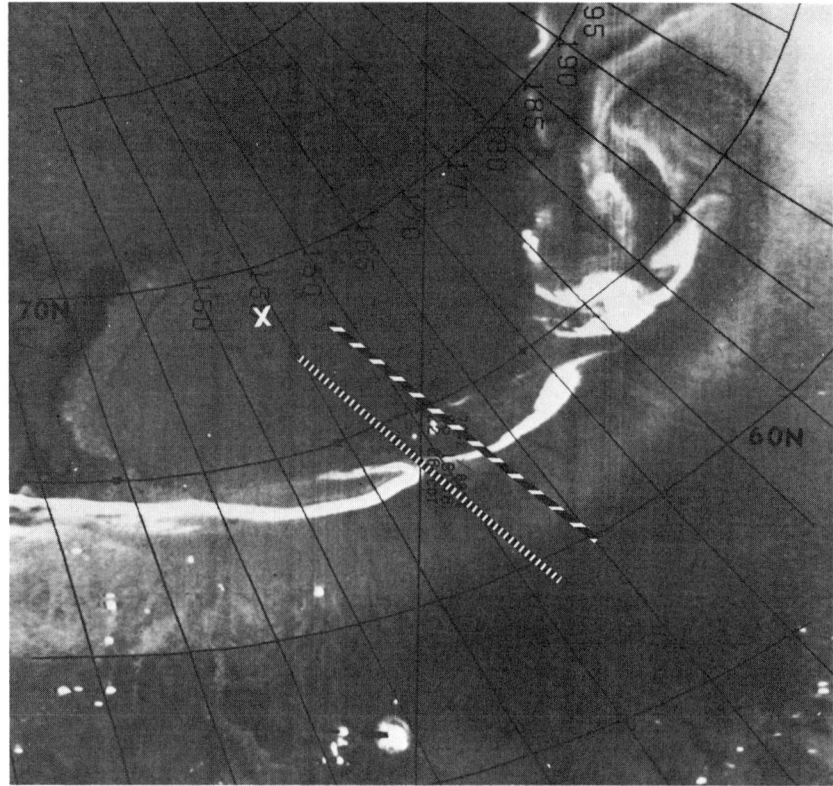

Fig. 2 DMSP imagery from the midnight sector 11 January 1976. The heavy (light) dashed lines give the corrected (uncorrected) trajectory of S3-2 projected to 100 km altitude for Rev 517. The symbol X marks the location of the Dixon Island magnetometer station.

sheet. The upward current sheets associated with these events were traversed in 0.25 seconds. Thus, the upward current sheets had latitudinal thicknesses of < 2 km and average current densities of 45 $\mu A/m^2$ and 135 $\mu A/m^2$. The return current sheets were latitudinally more extended with $j_\parallel$ in the 10 to 15 $\mu A/m^2$ range.

Since the electron spectrometer on S3-2 requires a full second to compile a 32 point spectrum care must be exercised interpreting measurements in the vicinity of the spatially narrow, upward current sheets. Figure 5 provides sequential measurements of directional differential flux taken near the dusk side event. Dashed lines indicate the one count per sampling period flux level. To the right of the figure are the UT's at which the spectrum compilation began and the pitch angle (α) sampled half way through the second. In compiling spectra, energy channels were alternately sampled in decreasing (odd cases) then increasing energy. Poleward of the event (spectrum 1), the electron flux level was close to or below instrumental sensitivity. This was also true of the highest energy channels sampled in spectrum 2. Spectrum 2 is marked by a burst of electrons with a peak flux of 8 x 10^9 electrons/(cm^2-sec-ster) at 3.5 keV and $\alpha = 151°$. The slopes of the low energy portions of spectra 2 and 3 are relatively flat. A peak flux of 5 x 10^9 electrons/(cm^2-sec-ster-eV) at 1.7 keV was measured while the detector was looking close to the field line (spectrum 3). Following the broad thermal spectrum 4 the measured flux rapidly dropped to background. Obviously, there is a high degree of spatial-temporal aliasing in spectra 2 and 3. However, two purely phemonenological remarks can be made. First, a detailed examination of magnetometer and particle detector data records shows that the 3.5 keV electron burst of spectrum 2 was encountered 1/4 seconds prior to (2 km poleward of) the 135 $\mu A/m^2$ current sheet. Second, a flux of 1.2 x 10^{10} electrons/(cm^2-sec-ster-eV) for electrons with energies of 80 eV, seen in spectrum 2, is the highest level ever observed in S3-2's lowest energy channel. In the case of the dawn-side event a similar burst of electrons was also observed with a peak flux of 8 x 10^8 electrons/(cm^2-sec-ster-eV) at ~3.0 keV and $\alpha = 90°$. In this case, however, the burst exactly coincided with passage through the upward current sheet as detected by the magnetometer.

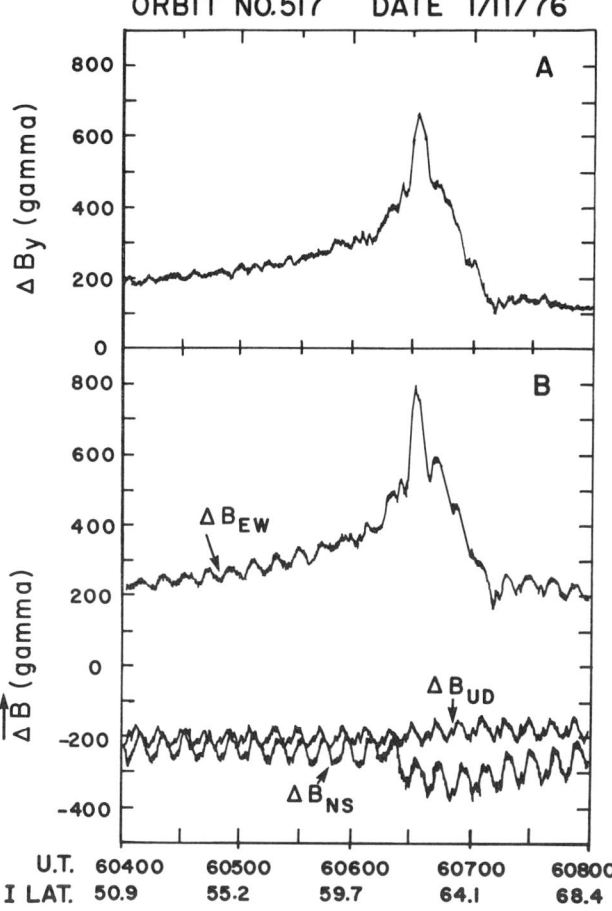

Fig. 3 S3-2 magnetometer measurements along trajectory shown in Figure 2. The upper panel gives the deviation of the satellite coordinate component from the IGRF model. The bottom three traces give the total field deflections in geomagnetic coordinates.

Before giving further consideration to these electron measurements and their relationship to the field-aligned current intensities let us consider the structure of discrete arcs in the polar cap. Figure 6 is a plot of E_x, ΔB_y the directional flux of electrons $(cm^2\text{-sec-ster})^{-1}$ and electron pitch angles measured during an S3-2 crossing of the winter polar cap. During the orbit the satellite passed with 1° of the magnetic pole along the MLT dawn-dusk meridian. Average values of E_x at 5 second intervals are represented by a fine, solid line in the top panel. Positive values of E_x in the northern hemisphere correspond to regions of sunward convection. Values of ΔB_y are also given in the top panel by a line whose thickness corresponds to a 1 bit (5 nT) uncertainly in the data. Eight regions of negative slope in the ΔB_y trace marked in Figure 6 are regions with current directed out of the ionosphere.

At the of time this high-latitude pass the hourly + 7.4 nT (King, 1979). Large scale variations in E_x and ΔB_y are poor indicators of auroral oval boundaries. Clear signatures of Region 1/Region 2 currents are not easily discerned. On the evening side there are reversals of E_x at 1956:06 UT (Λ =76°) and 1957:48 UT (Λ =82°). Both reversals (events 1 and 2) are associated with currents out of the ionosphere and locally enhanced fluxes of electrons. Event 2 marks a transition in both E_x and the electron flux. Equatorward of event 2 electron

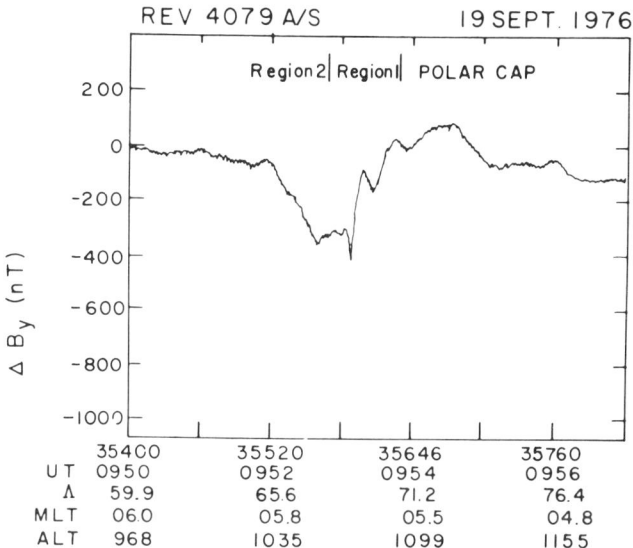

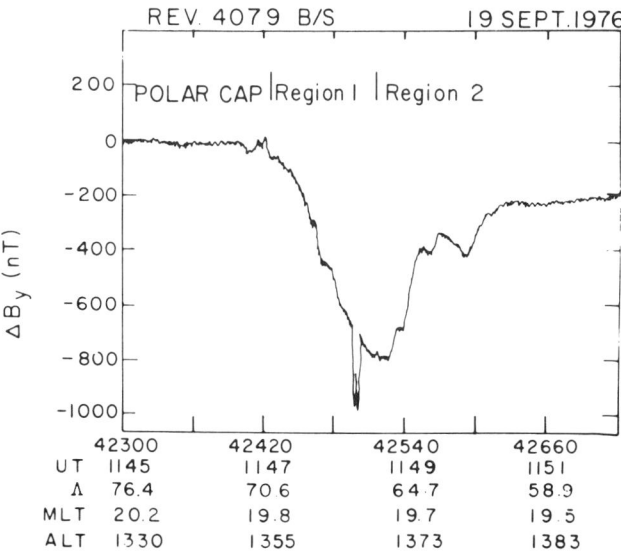

Fig. 4 S3-2 magnetometer measurements taken at southern high latitudes during a substorm period. The data are presented in satellite coordinates, Y component.

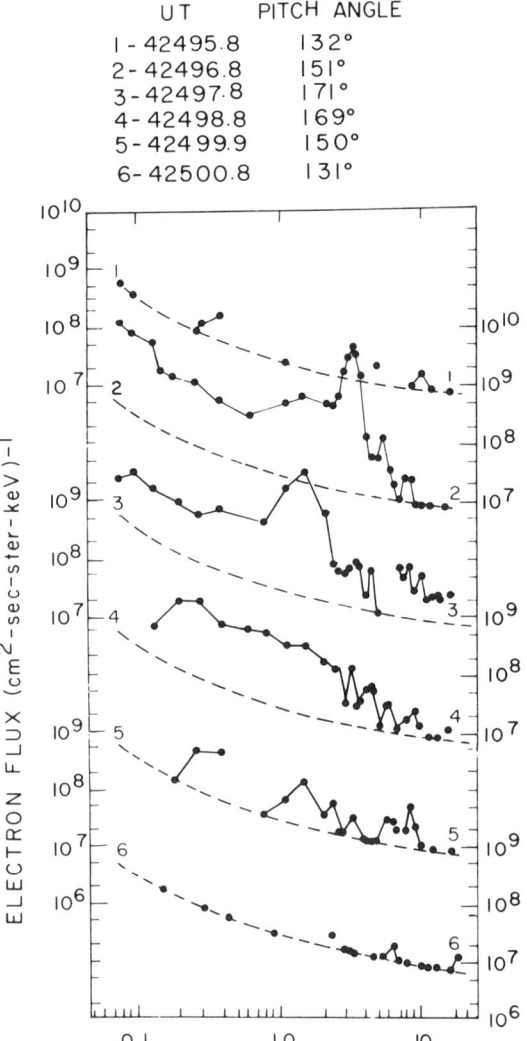

Fig. 5 Directional differential fluxes measured in the vicinity of the intense dusk side, field-aligned current structure.

fluxes vary between 5×10^7 and 3×10^8 (cm^2-sec-ster)$^{-1}$ and are fairly isotropic over the downcoming hemisphere. Poleward of the event the electron flux decreases abruptly to values near the lowest sensitivity of the detector, a level typical of polar rain (Winningham and Heikkila, 1974). On the morning side the last, clear observation of polar rain occurs at 2003:30 UT ($\Lambda = 81.6°$). Based on the occurrence of polar rain we set the evening and morning side boundaries of the polar cap at 82° and 81.6°, respectively. Events 3 through 7 lie embedded in polar rain within the polar cap. For reasons of economy only event 6 is considered in detail.

Figure 7 is an expansion of the data in the vicinity of events 6 and 7. In this plot E_x values are 1 second averages. Between 2002:52 and 2002:57 UT (event 6) the electric field reversed direction and ΔB_y decreased by 127 nT, indicating a $j_\parallel$ of 2.8 $\mu A/m^2$. This value is consistent with the electron flux of 1.2×10^9 (cm^2-sec-ster)$^{-1}$. We note that the magnetic trace does not indicate a matched pair of current sheets. Rather the ΔB_y trace departs from a value of -100 nT at 2002:30 UT and returns to it at 2003:17 UT. Figure 8 is a schematic of event 6 showing that the return currents is divided into two segments one on each side of the upward current sheet.

Several important aspects of polar cap arcs can be deduced from consideration of the distribution function of electrons measured in the upward current sheet. In Figure 9 we have plotted the electron distribution functions versus energy measured by S3-2 at pitch angles of $\sim 90°$ in event 6 (triangles) and measured by a zenith looking detector on a DMSP satellite (dots) as it passed over a visible polar cap arc (Hardy et al., 1982). In this semi-logarithmic format a Maxwellian distribution would appear as a straight line with a negative slope.

For electrons with energies greater than 500 eV both distributions have similar shapes with maxima near 1 keV. The flux maximum measured during event 6 exceeds that of the DMSP event by a factor of 5 and is sufficiently energetic to produce a visible arc. The fact that the distributions have maxima suggests that they are both made up of two populations. Those with energies below the maxima are either secondary or degraded primary electrons. Electrons with energies at or greater than the maxima are primaries that have undergone field-aligned accelerations. A least squares fit of the four data points at and beyond the maximum of the event 6 distribution function is represented by the dashed line in Figure 9. The data are consistent with a primary population that has a temperature of 220 eV and has been accelerated through a potential 1 kV. If we assume that the parent population in the magnetosphere was isotropic, then we can calculate its density from the intercept of the unaccelerated distribution function. In convenient units (1)

$$f_0 \text{ (cm}^{-6}\text{sec}^{-3}) = 8.61 \times 10^{-25} n(\text{cm}^{-3})/T^{1.5}(\text{eV})$$

We estimate the density of the parent population to be 2.2 cm^{-3}. Anticipating the discussion section, it is also useful to calculate the quantity.

$$j_0 = nq \sqrt{kT/2\pi m} \qquad (2)$$

where j_0 is the field-aligned current density from an isotropic magnetosphere using this technique we may also glean further information about the two intense upward current sheets. In Figure 10 we have plotted the distribution functions measured at the time of the morning (upper) and dusk (lower) side electron bursts. The burst electrons come from magnetospheric populations

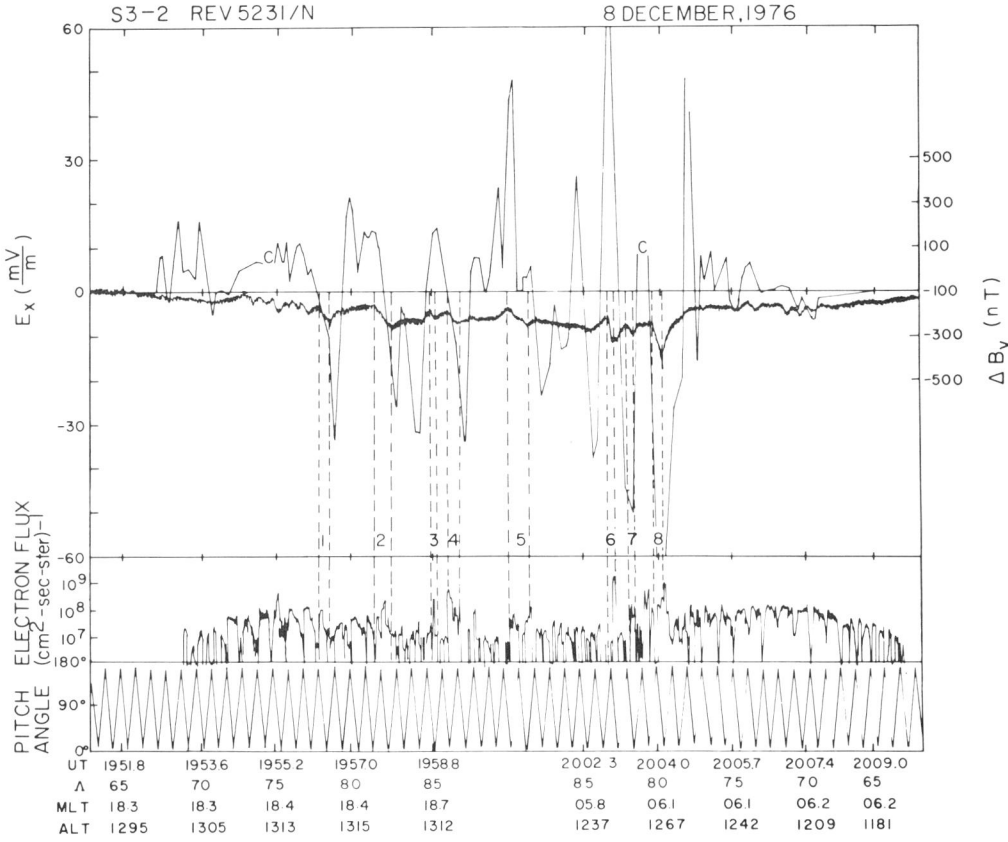

Fig. 6 Electric field, magnetic field, electron flux and pitch angle measurements taken at northern (winter) high latitudes during a period of northward IMF.

that were accelerated through potential drops of 3.0 and 3.5 kV in the morning and evening event, respectively. In the case of the morning side event we calculate a temperature of 200 eV and a density of 0.2 cm^{-3}. The evening side burst electrons come from two parent populations; one with a temperature of 200 eV and a density of 0.92 cm^{-3} and the other with a temperature of 4 keV and a density of 1.1 cm^{-3}. The characteristics of the polar cap arc, morning side event and evening side event are summarized in Table 1. Note that the values of j_0 are a factor of ten less than those presented in a similar table by Burke et al.(1983).

Discussion

Two essentially different approaches to the relationship between field-aligned currents and potential drops have appeared in the literature. The first approach concerns the transmission of electrical information between magnetospheric generators and ionospheric loads via kinetic Alfven waves (Goertz and Boswell, 1979; Lysak and Carlson, 1981). The second approach assumes a potential drop along magnetic field lines and calculates the field-aligned current reaching the ionosphere (Knight, 1973; Lemaire and Scherer, 1973; 1974; Whipple, 1977; Fridman and Lemaire, 1980; Lyons, 1981). The relationship was derived by considering collisionless, single-particle trajectories in a dipole magnetic field. Assuming an isotropic Maxwellian parent population Lyons (1981) has shown that the field-aligned current density carried out of the ionosphere by precipitating electrons is

$$j_\| = j_o(B_i/B_v) \left[1-(1-B_v/B_i) \exp\left(\frac{-q\phi_\|/kT}{B_i/B_v - 1}\right)\right] \quad (3)$$

where j_o is defined in equation (2); B_i and B_v are the magnetic field strengths in the ionosphere and at the top of the potential drop, respectively. The altitude of the potential drop determines how much it is able to open the atmospheric loss cone. Consider the limit $B_i = B_v$ in which a potential drop exists in a narrow region just above the ionosphere. No matter how large the potential drop $j_\| = j_o$. In the limit $q\phi_\|/kT \to \infty$, $j_\| = j_o B_i/B_v$. Figure 11 is a plot of $j_\|/j_o$ as a function of $q\phi_\|/kT$ for six values of B_i/B_v, as given in equation 3. This model was verified by Lyons (1981) in his analysis

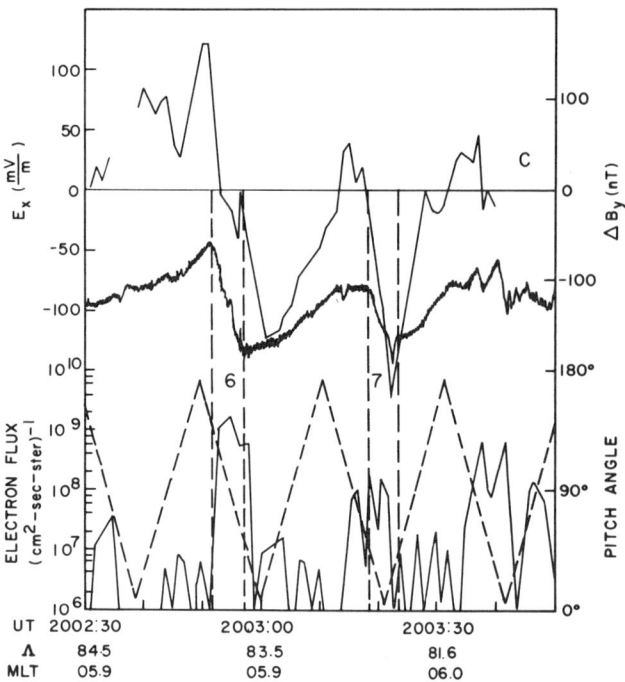

Fig. 7 Expanded plot of particle and field measurements in vicinity of Event 6.

of electron measurements by rocket borne instruments over an auroral arc.

In the case of the polar cap arc the value of $j_\parallel/j_0$ was 4.9 and $q\phi_\parallel/kT$ was 4.5. This point is plotted on Figure 11 by the symbol P. With no adjustment we see that there is agreement between this single particle model and the S3-2 measurements.

Crossing the morning event the S3-2 magnetometer measured an upward current density of 45 $\mu A/m^2$. The values of j_0 and $q\phi_\parallel/kT$ inferred from the particle detector measurements are .12 $\mu A/m^2$ and 15, respectively. This event designated by the letter M in Figure 11, lies well above all of the current-voltage characteristic curves.

TABLE 1

Event	n(cm^{-3})	kT(keV)	$\phi_\parallel$ (kV)	j=($\mu A/m^{-2}$)
Polar Cap	1.45	.22	1.0	0.57
Morning (Cold)	.30	.20	3.0	0.12
Evening (Cold)	.92	.20	3.5	0.35
Evening (Hot)	1.10	4.0	3.5	1.9

In the evening sector event we measured two populations, one hot (kT = 4 keV) and one cold (kT = 200 eV). We now attempt to evaluate their relative contributions to $j_\parallel$. Magnetometer data presented in Figure 4 shows that $j_\parallel$ = 135 $\mu A/m^2$. The values of j_0 for the hot and cold electrons are 1.9 and .35 $\mu A/m^2$, respectively; $\phi_\parallel$ = 3.5 kV. Using the symbols H and C in Figure 11 we have marked the current-voltage relationship one gets by assuming that either the hot (H) or the cold (C) electrons is solely responsible for $j_\parallel$. Again, both points are well above the theoretical curves. Thus, if the theoretical relationships between $j_\parallel$ and $\phi_\parallel$ are valid then the current cannot be carried by just one of the populations.

In order to reconcile S3-2 measurements with

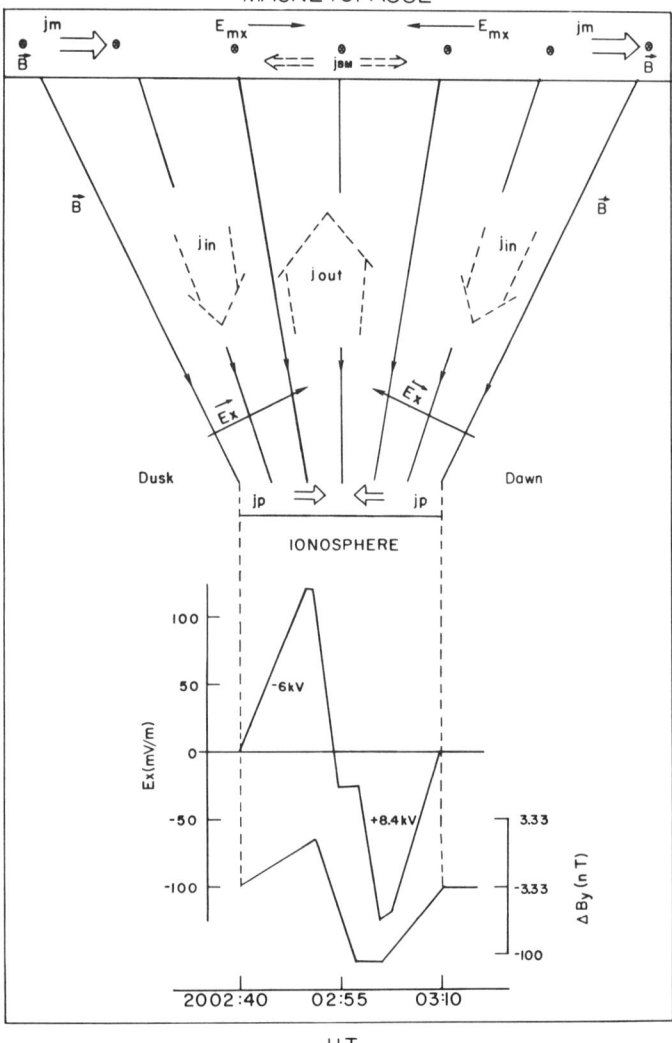

Fig. 8 Two dimensional projection of currents associated with Event 6.

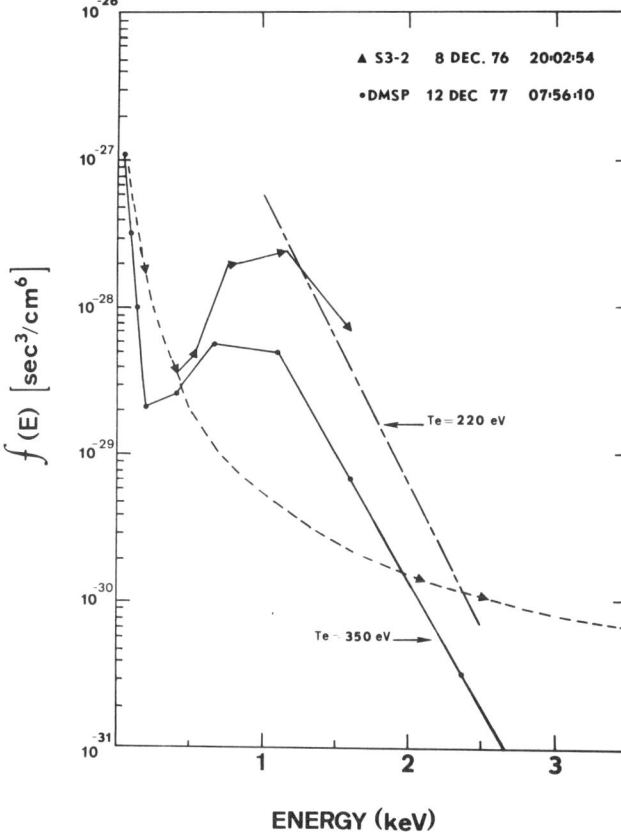

Fig. 9 Distribution functions of electrons above Event 6 (triangles) and a polar cap arc observed by DMSP (circles).

bursts electrons were sampled at angles of 30° and 90° away from the direction of the field lines. At these angles the use of the distribution functions in Figure 11 to estimate f_0 of the parent population inevitably leads to underestimates of magnetospheric densities.

As the resolution of scientific magnetometers on satellites increases the intense field-aligned currents reported here will become more commonplace events. At this conference Sugiura has reported MAGSAT measurements of $j_\parallel$ approaching 800 $\mu A/m^2$. In themselves the S3-2 measurements cannot be used to verify the model presented in Equation 3. However, in conjunction with the model they can be used to understand the magnetospheric conditions that give rise to such intense field-aligned currents. Because the model is based on Liouville's theorem it should have a high degree of applicability. Any breakdown of the Liouville theorem by wave-particle interactions would lead to significant heating of current carrying electrons. In both the evening and morning events, S3-2 measured electron temperatures were ~200 eV, a number typical of the distant plasma sheet.

theoretical predictions it is necessary to assume that we have underestimated $\Phi_\parallel$, n or both in analyzing the electron measurements given in Figure 10. For example, if $\Phi_\parallel$ has a latitudinal gradient, a common feature of inverted-V precipitation, then our estimate of $\Phi_\parallel$ would only be a lower bound. This may provide a partial explanation of the evening side event. Recall that the evening side electron burst was detected 0.25 sec prior to (2 km poleward of) the intense current sheet. The morning side electron burst exactly coincided with the 45 $\mu A/m^2$ upward current sheet. Thus, a gradient in $\Phi_\parallel$ is not a likely explanation of discrepancies between observations and theory in this case. Underestimates of the densities of magnetospheric populations can occur when accelerated electron beams are sampled at pitch angles away from the directions of magnetic field lines. Sampling electrons simultaneously at several pitch angles, Burch et al. (1976; 1979) showed that electrons in the peak channel have maximum fluxes along the magnetic field lines. Electrons with energies above that of the peak are not so field-aligned. In the cases of the evening and morning side

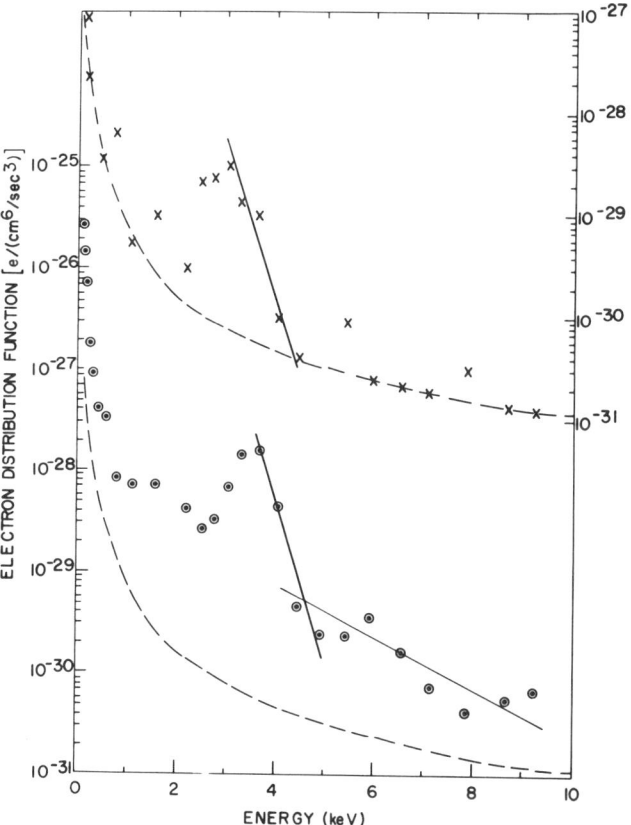

Fig. 10 Distribution functions associated with electron burst events in the vicinity of the two field-aligned current events shown in Figure 4.

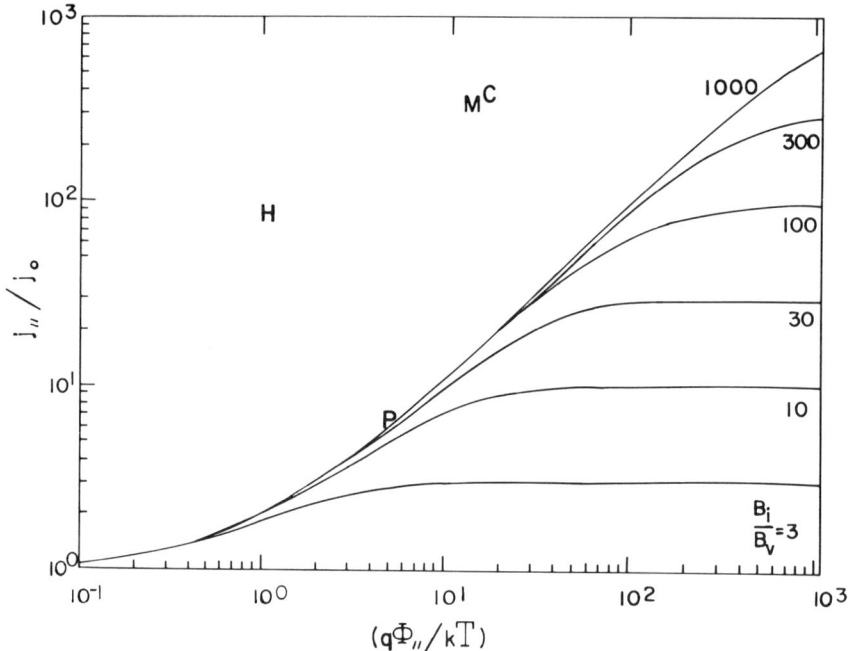

Fig. 11 Plots of $j_\parallel$ as a function of $\phi_\parallel$ based on equation (3) for six values of B_i/B_v. Values of $j_\parallel/J_0$ and $q\phi_\parallel/kT$ from the magnetometer and electron spectrometer measurements in the polar cap and the two intense current sheets are plotted for reference.

In the case of the morning side event, the model predicts that, for $q\phi_\parallel/kT = 15$, the value of $j_\parallel/j_0$ should be 12, or $j_0 = 3.75$ μA/M^2. From equation (2) we calculate a parent electron density of 9.4 cm^{-3}.

In the case of the evening side event if we assume that $\phi_\parallel = 10$ kV in the current sheet the model predicts that the hot electron population should carry a current of ~ 8 μA/m^2. Thus, the cold electrons must carry the remaining 137 μA/m^2. With this potential drop $q\phi_\parallel/kT = 50$ or the cold electrons and the model predicts that $j_\parallel/j_0 = 40$ or $j_0 = 3.4$ μA/m^2. The cold electron density in the source region is 9.2 cm^{-3}. If we assume $\phi_\parallel = 40$ kV, the hot electrons carry 19 μA/m^2 and the cold electron density required by the model is 2.2 cm^{-3}.

References

Burch, J.L., S.A. Fields, W.B. Hanson, R.A. Heelis, R.A. Hoffman and R.W. Janetzke, Characteristics of auroral electron acceleration regions observed by Atmosphere Explorer C, J. Geophys. Res., 81: 2223, 1976.

Burch, J.L., S.A. Fields, R.A. Heelis, Polar cap electron acceleration regions, J. Geophys. Res., 84: 5863, 1979.

Burke, W.J., Electric Fields, Birkeland currents and electron precipitation in the vicinity of discrete auroral arcs, in Physics of Auroral Arc Formation, ed. by S.-I. Akasofu, and J.R. Kan, American Geophysical Union, Geophysical Monograph 25, Washington, D.C., p. 164, 1981.

Burke, W.J., D.A. Hardy, F.J. Rich, M.C. Kelley, M. Smiddy, B. Shuman, R.C. Sagalyn, R.P. Vancour, P.J.L. Wildman and S.T. Lai, Electrodynamic structure of the late evening sector of the auroral zone, J. Geophys. Res., 85: 1179, 1980.

Burke, W.J., M.S. Gussenhoven, M.C. Kelley, D.A. Hardy and F.J. Rich, Electric and magnetic field characteristics of discrete arcs in the polar cap, J. Geophys. Res., 87: 2431, 1982.

Burke, W.J., M. Silevitch and D.A. Hardy, Observations of small-scale auroral vortices by the S3-2 satellite, J. Geophys. Res., 88: 3127, 1983.

Friden, M., and J. Lemaire, Relationship between auroral electron fluxes and field-aligned electric potential difference, J. Geophys. Res., 85: 664, 1980.

Goertz, C.K. and R.W. Boswell, Magnetosphere-ionosphere coupling, J. Geophys. Res., 84: 7239, 1979.

Hardy, D.A., W.J. Burke and M.S. Gussenhoven, DMSP optical and electron measurements in the vicinity of polar cap arcs, J. Geophys. Res., 87: 2413, 1982.

Kindel, J.M., and C.F. Kennell, Topside current instabilities, J. Geophys. Res., 76: 3065, 1971

King, J.H., Interplanetary medium data book supplement I, 1975-1978, NSSDC/WDC-A-R, G79-08,

Goddard Space Flight Center, Greenbelt, MD, 1979.

Knight, L., Parallel electric fields, Planet. Space Sci., 21: 741, 1973.

Lemaire, J. and M. Scherer, Plasma sheet particle precipitation: a kinetic model, Planet. Space Sci., 21: 281, 1973.

Lemaire, J. and M. Scherer, Ionosphere-plasma-sheet parallel electric fields, Planet. Space Sci., 22: 1485, 1974.

Lyons, L.R., The field-aligned current versus electric potential relation and auroral electrodynamics, in Physics of Auroral Arc Formation, ed. by S.-I. Akasofu and J.R. Kan, American Geophysical Union, Geophysical Monograph 25, Washington D.C., p. 252, 1981.

Lysak, R.L. and C.W. Carlson, The effect of microscopic turbulence on magnetosphere ionosphere coupling, Geophys. Res. Lett., 8: 269, 1981.

Smiddy, M., W.J. Burke, M.C. Kelley, N.A. Saflekos, M.S. Gussenhoven, D.A. Hardy and F.J. Rich, Effects of high latitude conductivity on observed convection electric fields and Birkeland currents, J. Geophys. Res., 85: 6811, 1980.

Whipple, E.C., The signature of parallel electric fields in a collisionless plasma, J. Geophys. Res., 82: 1525, 1977.

Winningham, J.D. and W.J. Heikkila, Polar cap auroral electron fluxes observed with ISIS-1, J. Geophys. Res., 79: 949, 1974.

ASSOCIATION OF FIELD-ALIGNED CURRENTS WITH SMALL-SCALE AURORAL PHENOMENA

Cynthia Cattell

Space Sciences Laboratory, University of California, Berkeley, California 94720

Abstract. Comparisons of theoretically deduced and experimentally measured properties of the field-aligned current regions to the occurrences of electrostatic shocks, electrostatic ion cyclotron (EIC) waves, ion beams, ion conics, and double layers observed within them can yield useful new information and constraints on which plasma processes may be important. Using statistical studies of data from the S3-3 satellite, it is found that the differences between upward and downward current regions and between Region I and Region II currents can explain many of the observed asymmetries between phenomena observed in the evening and morning auroral zone. Statistical studies also imply that EIC waves are destabilized by upward currents; that EIC waves cannot account for the bulk of the energization which produces conics; that connection of Region I currents to the plasmasheet boundary is consistent with many of the observed properties of electrostatic shocks; and that double layers provide the parallel potential to produce inverted-V precipitation in the evening auroral zone. The parallel electric field distribution associated with electrostatic shocks (which probably produce discrete arcs) is still not fully understood, particularly in the morning auroral zone.

Introduction

In this paper, two related aspects of field-aligned currents will be discussed: (1) field-aligned currents as a source of free energy for auroral acceleration processes, and (2) constraints put on plasma processes by their observed association with the currents. Both experimental evidence and theoretical considerations suggest that the current carriers in upward and downward current regions have different distribution functions and source regions, leading to differences in the plasma properties.

Early kinetic models by *Lemaire and Scherer* [1973, 1974], *Knight* [1973], and more recent studies by *Fridman and Lemaire* [1979], and *Chiu and Schulz* [1978] show that a larger potential difference between the plasmasheet and the ionosphere is required for currents out of the ionosphere than for currents into the ionosphere. The large parallel potential which is required to accelerate plasmasheet electrons to carry the current in the upward current regions keeps the low energy ionospheric electrons out of the region of parallel potential. Cold ionospheric ions are not suppressed, and, therefore, the electron temperature is higher than the ion temperature. Downward currents can be carried by the large reservoir of cold ionospheric electrons with only a small potential drop. The average ion and electron temperatures in this plasma are probably comparable. In addition, the particle densities should be higher in downward current regions than in upward ones. Because the important parameter for determining the instability of the plasma is the drift velocity (which is proportional to the current density divided by the particle density), plasma in upward current regions should be more unstable. The drift velocity for a given current in a flux tube also depends on altitude, as described by *Lysak and Hudson* [1979]. For a density profile fit to S3-3 data, the drift velocity has a broad maximum at altitudes from ~5000 to 8000 km. These differences put constraints on the plasma instabilities that can occur.

There are also likely to be more subtle differences between the Region I (poleward) and the Region II (equatorward) current sheets defined by *Iijima and Potemra* [1976], since they map to different regions in the distant magnetosphere. For example, one might expect to see differences between the phenomena observed in the morning Region II sheet and the evening Region I sheet, even though both are upward, since the distribution of source particles is different.

The above arguments are partially supported by experimental evidence. The work on current carriers was reviewed by *Klumpar* [1983] at this conference. Comparisons between currents carried by measured energetic particles and the current measured by magnetometers made both on satellites [*Cattell*, 1981; *Klumpar et al.*, 1976] and on rockets [*Cloutier and Anderson*, 1975; *Potemra*, 1979, and references therein] show that energetic particles often carry at least half the current in upward currents, but can never account for the downward current. Most of these studies suffered from lack of high time resolution. Recent studies from the DE satellites presented at this conference confirm that the observed upward current can usually be accounted for by the observed energetic electrons [*Sugiura et al.*, 1983a, 1983b], and that downward current is (at least sometimes) carried by beams of cold electrons accelerated upward from the ionosphere by a small potential (< 100 V) [*Burch et al.*, 1983], in agreement with ISIS observations [*Klumpar and Heikkila*, 1982]. Density data from several passes through the auroral zone by the S3-3 satellite show that the density is higher in the downward current region (see Figure 4, *Mozer et al.*, 1979). Figure 6 from *Mozer and Temerin* [1982] also shows that the density is lower in the region of upward current. Regions of low density in the auroral zone have also been observed on DE-1 [*Gurnett et al.*, 1982]. It should be noted that, although the average current direction in a given region is upward (downward), there are often many smaller scale length currents in the opposite direction.

To understand the effects of these differences, the occurrence of the following small-scale (compared to the ~5° scale size of the Region I or II currents) phenomena in the Region I and Region II currents and in upward and downward currents will be described: (1) electrostatic shocks [*Mozer et al.*, 1977, 1980], which are latitudinally narrow (a few energetic ion gyroradii), longitudinally extensive regions of strong electric fields oriented primarily perpendicular to the magnetic field; (2) ion beams [*Ghielmetti et al.*, 1978; *Gorney et al.*, 1981], which have distribution functions peaked parallel (south pole) or anti-parallel (north pole) to the magnetic field and are evidence of parallel electric fields below the satellite; (3) ion conics [*Gorney et al.*, 1981; *Sharp et al.*, 1977], which have maxima in the distribution func-

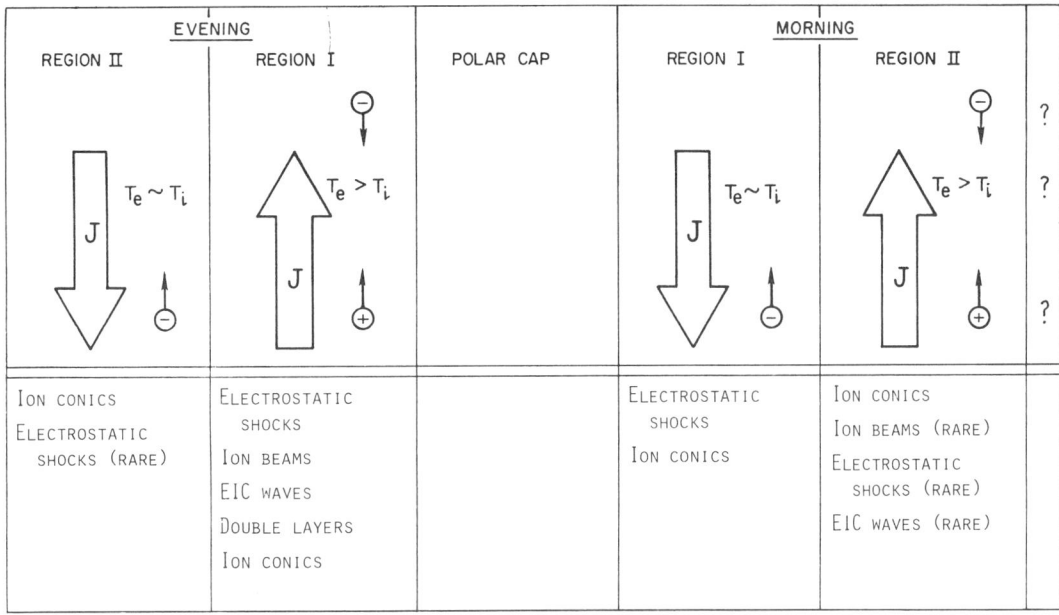

Fig. 1. Schematic drawing of the occurrence of small-scale phenomena with respect to the morning and evening Region I/II current system.

tion at a pitch angle between 90° and 180° (north pole) and are evidence for acceleration perpendicular to the magnetic field, with subsequent motion along the field line due to the magnetic mirror force; (4) electrostatic hydrogen cyclotron waves [*Kintner et al.*, 1978; *Temerin et al.*, 1979]; (5) and double layers [*Temerin et al.*, 1982], which are regions of electric field parallel to the magnetic field confined to a short distance (few Debye lengths) along it [*Block*, 1975].

The data were obtained by instruments on board the polar-orbiting S3-3 satellite [*Cattell*, 1982]. The electric field, density, and wave data were obtained by the UCB double probe [*Mozer et al.*, 1979]. The current measurements were inferred from the magnetometer data, which had to be averaged over the 18-s spin period to obtain the necessary amplitude resolution [*Mozer et al.*, 1979]. The particle fluxes were measured by the Aerospace Corporation detectors [*Mizera and Fennell*, 1977] and by the Lockheed Palo Alto Research Laboratories detectors [*Shelley et al.*, 1976; *Sharp et al.*, 1977]. The morphology of convection electric fields, field-aligned currents, electrostatic shocks, waves, upflowing ions, and precipitating electrons in the auroral zone determined from the S3-3 data at altitudes of ~1 R_E has been described for individual events in many papers [*Mozer et al.*, 1980 and references therein; *Mizera et al.*, 1982 and references therein]. In this paper, therefore, primarily the statistical relationships are described.

The next section presents a schematic picture of the phenomena occurring in the evening and morning auroral zones and discusses the questions to be addressed. The third section describes the statistics for each individual phenomenon. The results are summarized, and both answered and unanswered questions are discussed in the final section.

Overview of the Association of Field-Aligned Currents and Auroral Processes

In Figure 1, a schematic view of the relationship between the processes of interest and the field-aligned current regions is outlined. Some of the questions raised by these observations are listed in Table 1 and should be kept in mind while reading the detailed statistics of each phenomenon presented in the next section.

In the evening, the equatorward current sheet (Region II), which is downward, is carried by ionospheric electrons, and T_e, the electron temperature, is probably approximately equal to T_i, the ion temperature, since the bulk of the particles are of ionospheric origin. This current generally occurs well poleward of the plasmapause corotation boundary [*Cattell et al.*, 1979a,b; *Mozer et al.*, 1980; *Torbert et al.*, 1981]. The evening poleward (Region I) current is upward, carried by plasmasheet electrons, and therefore, T_e is larger than T_i. This current often occurs close to the polar cap convection boundary and is frequently equatorward of that boundary. In contrast, the morning Region I current almost always occurs coincident with the polar cap convection boundary [*Cattell et al.*, 1979a,b; *Torbert et al.*, 1981; *Mozer et al.*, 1980; *Smiddy et al.*, 1980]. This downward current is carried by ionospheric electrons. The Region II current in the morning is upward, which suggests that it is carried by plasmasheet electrons. However, no actual comparisons of the currents and particles have been presented for this region.

The first notable fact about Figure 1 is that, although they are occasionally observed in all regions, electrostatic shocks occur predominantly in the Region I currents. Therefore, electrostatic shocks in the evening occur in large-scale upward currents while those in the morning occur in large-scale downward currents. This leads to question 4 in Table 1: "Why do electrostatic shocks occur primarily in the Region I currents rather than in association with a particular direction of current?" Next, it can be seen that evidence for parallel electric fields (ion beams and/or direct measurement of $E_\parallel$ in double layers) has only been observed in upward current regions (evening RI and morning RII) in the data set presently available. This leads to question 5: "Why does evidence for upward parallel electric fields occur only in regions of large-scale upward current?" and its corollary, No. 6: "Is there evidence for downward $E_\parallel$ associated with downward currents and is this expected?"

TABLE 1. Questions

General:

1. What are the differences between Region I and Region II currents?

2. What are the differences between upward (out of the ionosphere) and downward currents?

3. What are the differences between large-scale (i.e., scale size of RI or RII) and small-scale currents?

Specific:

4. Why do electrostatic shocks occur primarily in the Region I currents rather than in association with a particular direction of current?

5. Why does evidence for upward parallel electric fields occur only in regions of large-scale upward current?

6. Is there evidence for downward $E_\parallel$ associated with downward currents and is this expected?

7. Do ion beams or currents drive EIC waves?

8. Do EIC waves explain the observations of conics?

Figure 1 also shows that electrostatic ion cyclotron (EIC) waves have not been observed in all the current regions. (Note that only hydrogen cyclotron waves will be discussed herein). What information does this yield about the free energy source for the waves? Question 7 is: "Do ion beams or currents drive EIC waves?"

The process that energizes ion conics is not yet fully understood. Does the fact that ion conics occur in all the current regions, while, for example, EIC waves occur only in upward current regions, put any constraints on the acceleration mechanism (Question 8)?

Finally, all of the above observations suggest that there are systematic differences in the current regions that are not yet fully understood and that investigating these differences may lead to new insights about auroral acceleration. This is the reason for questions 1 - 3.

Description of Data

A. Electrostatic Shocks

One of the early dramatic discoveries by the S3-3 satellite was the existence of latitudinally narrow, large ($\lesssim$ 1 V/m) electric fields oriented primarily perpendicular to the magnetic field [*Mozer et al.*, 1977] which were called 'electrostatic shocks' [*Swift*, 1975, 1979; *Hudson and Mozer*, 1978]. These structures have been associated with the parallel electric field acceleration that produces discrete auroral arcs [*Torbert and Mozer*, 1978; *Mozer et al.*, 1980; *Kletzing et al.*, 1983]. They are seen at auroral latitudes at all local times [*Mozer et al.*, 1980]. Figure 2 [*Bennett et al.*, 1983] shows that the probability of observing an electrostatic shock is approximately constant for all magnetic local times (although multiple shocks are more common in the afternoon and evening sectors).

Figure 3 plots the magnitude of the shock electric field versus the magnitude, direction, and location of the current sheet in which it occurred. Morning events (2 - 10 MLT) are plotted with open circles, evening events (14 - 22 MLT), with solid dots. (The details of the study from which Figures 3, 5, and 8 are taken can be found in *Cattell et al.*, 1979). Most shocks are observed in the poleward (RI) current, so that shocks in the evening (morning) occur primarily in regions of large-scale upward (downward) current. The largest electrostatic shocks are observed in the evening, upward RI current sheet. (This MLT-dependence of the magnitude is confirmed by a recent study with

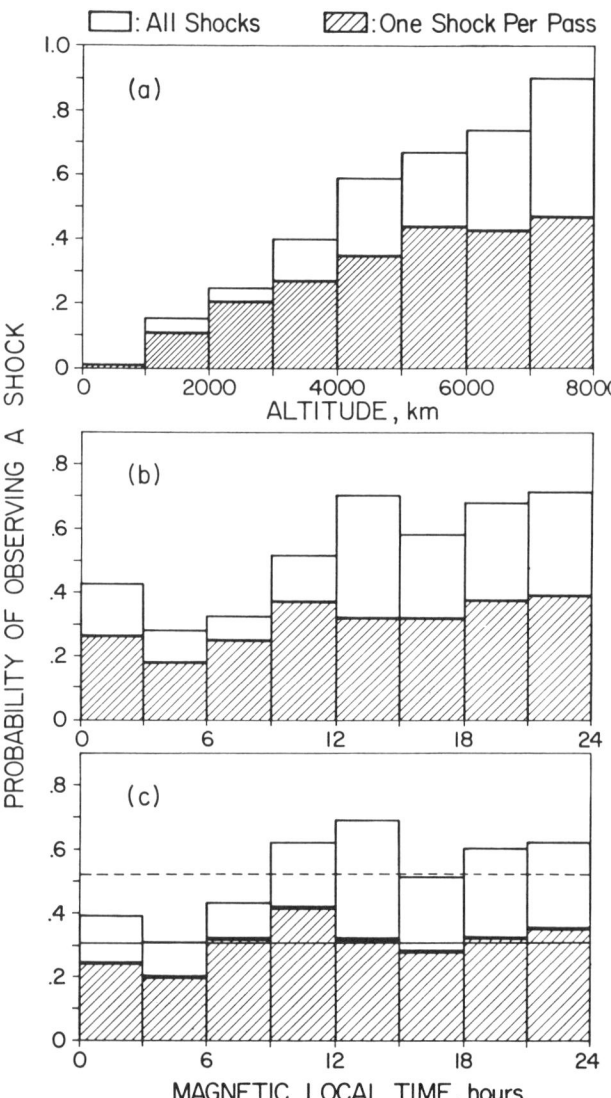

Fig. 2. The probabilities of observing an electrostatic shock for (a) bins of every 1000 km altitude; (b) bins of 3 hours magnetic local time; and (c) bins of 3 hours magnetic local time, normalized for altitude. The normalization is made using the known altitude distribution, since the satellite did not sample uniformly in altitude at all magnetic local times. The shaded regions were created using only one shock per auroral zone pass, while the unshaded regions included all shocks from each pass (from *Bennett et al.*, 1983).

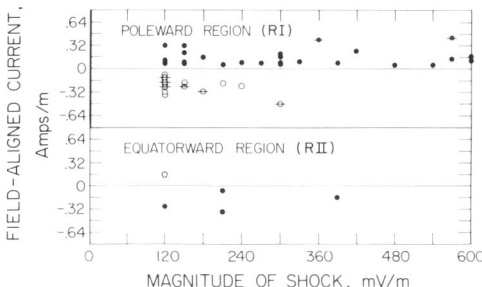

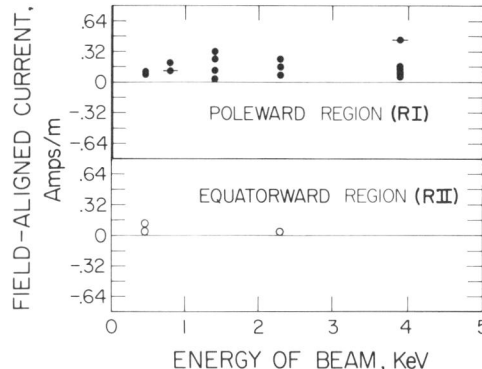

Fig. 3. Plot of the magnitude of electrostatic shocks versus field-aligned current. Positive numbers are currents out of the ionosphere; negative numbers are currents into the ionosphere. Open circles refer to events between 2 and 10 MLT; dots refer to events between 14 and 22 MLT. Circles with horizontal bars indicate that the event occurred in an auroral pass with 3 current sheets. [*Cattell et al., 1979a*].

Fig. 5. Plot of maximum beam energy versus field-aligned current (same format as Figure 3) [*Cattell et al., 1979*].

a larger statistical base [*Bennett et al., 1983*]).

The explanation for the occurrence of electrostatic shocks in Region I may be found in the connection of that region to the plasmasheet boundary. Recent studies of electric field and particle data from the ISEE-1 satellite [*Cattell et al., 1982*] have revealed that large, highly variable electric fields similar to electrostatic shocks (since the electric field instrument on ISEE has only one pair of probes, the direction of the field with respect to the magnetic field cannot be determined) are commonly measured when the satellite crosses the plasmasheet boundary. An example is presented in Figure 4. The distribution of the large field events is consistent with equipotential mapping along magnetic field lines of the low altitude S3-3 events [*Levin et al., 1983*]. This, in turn, is consistent with the suggestion of *Lysak and Hudson* [1979] (further described by *Lysak and Dum* [1983] and *Cattell et al.* [1981]) that electrostatic shocks are generated by propagation of a kinetic Alfvén wave [*Hasegawa*, 1976] on the plasmasheet boundary into the ionosphere. Associated with the wave is a field-aligned current which becomes unstable to electrostatic waves at altitudes near 1 R_E where, as described in the Introduction, the drift velocity maximizes. If one assumes that the distribution of magnitudes of the perturbation on the plasmasheet boundary is not dependent on the local time (which is not inconsistent with the data), so that the associated distribution of currents is the same, the difference in the magnitude of shocks in the morning and evening might be explained by the fact that the drift velocity associated with a given current is larger in upward currents than in downward currents. Therefore, the plasma in the evening Region I would be more unstable, leading to larger electrostatic shocks.

B. Ion Beams

Many studies of specific events [*Mozer et al.*, 1980; *Mizera et al.*, 1982; *Temerin et al.*, 1981a and b] have provided evidence that ion beams are accelerated in the parallel electric field associated with electrostatic shocks. A recent study [*Bennett et al.*, 1983] suggests that most energetic ion beams are collocated with shocks, whereas less energetic beams are not. In addition, ion beams occur much more frequently in the evening than in the morning [*Ghielmetti et al.*, 1978; *Gorney et al.*, 1981]. In this section, the relationship of currents, ion beams, and electrostatic shocks is described to provide information on parallel electric field acceleration.

In Figure 5, the maximum energy channel in which the ion beam was observed is plotted versus the magnitude, direction, and region of the field-aligned current in which it occurred (open circles refer to morning events; solid dots, to evening events). The ion beams all occurred in regions of large scale upward current (i.e., Region I in the evening and Region II in the morning). This is certainly not surprising, because ion acceleration out of the ionosphere parallel to the magnetic field requires an upward electric field. Since electromagnetic energy is being converted to particle energy, $\mathbf{j} \cdot \mathbf{E} > 0$, and the current must also be upward. Most ion beams occurred in Region I, in agreement

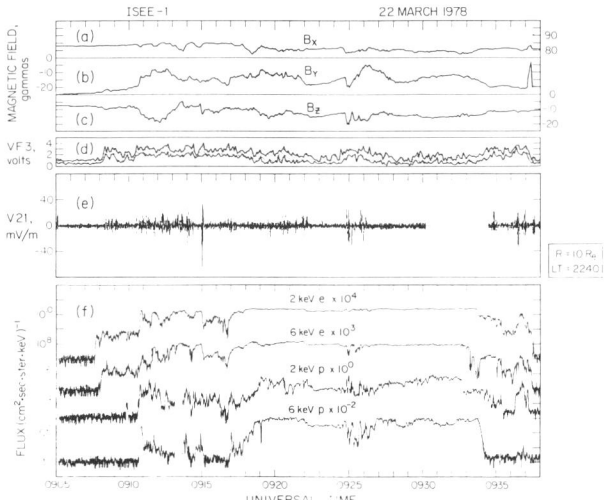

Fig. 4. Electric and magnetic field and particle data from the ISEE-1 satellite: (a) the spacecraft potential, which is indicative of the plasma density (decreases correspond to increases in density); (b) the raw electric field data; (c) the angle between the electric field boom and the sun-satellite line; (d) fluxes of energetic electrons and protons; (e)-(h) the three components of the magnetic field in GSE coordinates and the magnitude of the magnetic field. The plasmasheet boundary can be seen in the decrease in the spacecraft potential, increase in the particle fluxes, and decrease in the magnetic field magnitude. Large fluctuating electric fields are coincident with the boundary [*Cattell et al.*, 1982].

TABLE 2. 2 - 10 MLT

Current Sheet	"Polar Cap"*	Poleward (RI)	Equatorward (RII)	Multiple Pairs**
All Shocks, 42 events	3 (7%)	29 (69%)	8 (19%)	2 (5%)
Shocks with beams, 9 events	2 (22%)	1 (11%)	5 (56%)	1 (11%)

*A third sheet, poleward of RI, is sometimes observed in the morning (see *Cattell et al.,* 1979a).
**Cases where more than one pair of current sheets were observed, so that Regions I and II could not be identified.

with the observation of *Ghielmetti et al.* [1978] and *Gorney et al.* [1981] that most beams were in the evening.

The observations in the evening sector conform to expectations in that the Region I current has been associated with discrete aurora (see *Akasofu,* 1977; *Potemra et al.,* 1979; and references therein), so that evidence for parallel electric field acceleration is expected there. The association of electrostatic shocks, ion beams, Region I currents, and aurora is consistent.

The picture in the morning sector is not as clear. A preliminary study has been made of a larger set of morning events (an event is one electrostatic shock) than those included in Figure 5 in an attempt to clarify the situation. The results (summarized in Table 2) are that: (1) in agreement with the more limited event set in Figure 5, most (~70%) electrostatic shocks occur in Region I (downward current); and (2) most of the shocks which have ion beams (~60%) occur in Region II (upward current). The one case of an electrostatic shock with an ion beam in Region I occurred in a small-scale region of upward current. (Note that 'small-scale current' means scale sizes of ~18 seconds or ~70 km (the resolution limit for the S3-3 magnetometer), whereas the duration of a shock is often less than 1 second.) However, there are many other cases of shocks located within small-scale upward currents in the large scale downward Region I sheet which have no ion beams. It is surprising that the small-scale size upward currents in a region of large-scale downward current only rarely are associated with upward parallel electric fields. Question 3 in Table 1 was raised by these observations.

C. Electrostatic Ion Cyclotron Waves

Both ion beams [*Kintner et al.,* 1979; *Kaufmann and Kintner,* 1982] and field-aligned currents [*Cattell,* 1981] have been correlated with electrostatic ion cyclotron waves observed by S3-3. Although many theoretical, numerical, and laboratory studies [*Kindel and Kennel,* 1971; *Lysak et al.,* 1980; *Böhmer et al.,* 1976; *Hauck et al.,* 1978; *Pritchett et al.,* 1981; *Okuda and Ashour-Abdalla,* 1983; and others] have been made of the instability, the question of whether beams or currents drive the waves in the auroral zone is still not resolved.

Figure 6 [*Cattell,* 1981] is a plot of the spectral density in the 128 Hz filter (for events identified as ion cyclotron waves in the broadband data) versus the magnitude of the field-aligned current density. More than 90% of the currents were upward. Figure 7 [*Kintner et al.,* 1979] presents the correlation between maximum ion energy and spectral density for the same events. Over 90% of the wave events were associated with upflowing ions; approximately 20% of these events were conics.

The correlation between the field-aligned current density and the wave spectral density may be degraded by the low time resolution of the current measurement. Data from the S3-2 [*Burke,* 1981] and DE-2 [*Sugiura et al.,* 1983] show that there is usually fine structure in the currents and that the upward current density may be very large (100's of $\mu A/m^2$) in a narrow region. In addition, the drift velocity, not the current density, is the important quantity, and it correlates with the growth rate, rather than the wave amplitude. Therefore, a linear correlation between current and spectral density would not be expected.

A recent study of the linear dispersion relation for electrostatic waves [*Bergmann,* 1982], has shown that, for current and particle parameters based on S3-3 measurements, the observed properties of the EIC waves are reproduced only in the case where $T_e > T_i$ and the instability is driven by field-aligned current carried by the hot electrons. These properties are observed only in the upward current regions. The fact that more than 90% of the EIC events occurred in upward current regions is, therefore, very suggestive that the waves are current-driven.

An additional piece of evidence in favor of the current driven mechanism is the fact that EIC waves are observed above ~5000 km [*Kintner et al.,* 1979]. This is the altitude range where the drift velocity maximizes for a given current density [*Lysak and Hudson,* 1979]. In this scenario, the good correlation between ion beams and the wave spectral density is a result of the fact that upward current regions in general require large parallel potential drops.

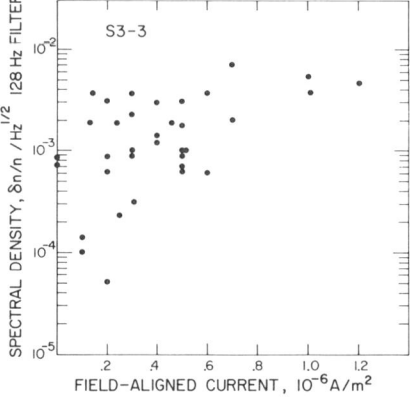

Fig. 6. The spectral density in the 128 Hz filter versus the field-aligned current measured within one spin period of the wave event [*Cattell,* 1981].

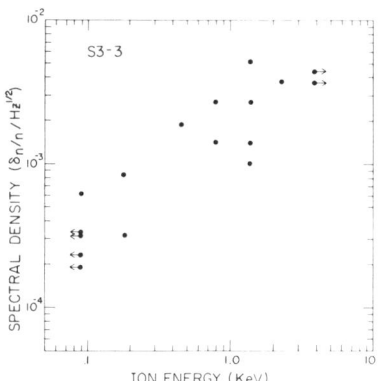

Fig. 7. The spectral density in the 128 Hz filter versus maximum ion energy for upstreaming ion event [*Kintner et al.*, 1979].

D. Ion conics

Conic ions, which provide evidence for energization perpendicular to the magnetic field, are commonly observed on the S3-3 satellite at altitudes of ~2000 - 8000 km [*Sharp et al.*, 1977; *Gorney et al.*, 1981] and on the ISIS satellites at altitudes below ~2000 km [*Klumpar*, 1979]. The acceleration mechanism which is most favored in the literature is heating by electrostatic ion cyclotron waves [*Lysak et al.*, 1980; *Ungstrup et al.*, 1979; *Okuda and Ashour-Abdalla*, 1983; *Dusenbery and Lyons*, 1981]. Comparisons of the location of ion conics with that of EIC waves suggest that EIC waves cannot account for the bulk of the ion energization.

The distribution of the ion conics in the various current sheets (Figure 8) shows that ion conics are observed in both morning and evening Region I and Region II current sheets with approximately equal probability (although conics are less common in the morning Region II sheet than elsewhere). The distribution of the ion conics and that of electrostatic ion cyclotron waves are, therefore, different — ion conics occur in all current regions, whereas EIC waves occur primarily in the evening Region I current. (Note that both EIC waves and conics are observed in the cusp, which is not included in this study). This is the first hint that some and possibly most conics are heated by some other mechanism. However, it could possibly be dismissed by noting that many of the observed conics have pitch angle maxima which are consistent with heating at altitudes below the satellite, and so correlation between the ions and waves might not be expected at the location of S3-3.

It is instructive, therefore, to compare the altitude distribution of conics to that of EIC waves. *Gorney et al.* [1981] (see their Figure 6) stated that the probability of observing a conic was approximately constant with altitude for $K_p \leq 3$ and increased with altitude above ~4000 km for $K_p > 3$. Most 90° conics occurred at altitudes below 3000 km, consistent with the pitch angle distributions observed by *Klumpar* [1979]. In addition, most of the cases that Gorney et al. mapped back to the mirror point to determine the source altitude were energized at altitudes $\leq$ 3000 km. It is therefore likely that the primary energization of conics (at least for quiet times) occurs below ~3000 km. *Kintner et al.* [1979] showed that EIC waves are observed almost exclusively above ~4500 km. Therefore, statistical arguments suggest that it is very unlikely that EIC waves energize most conic ions.

Other mechanisms for energizing conics have been suggested in the literature, including: (1) large gradients in the perpendicular electric field [*Mozer et al.*, 1980; *Lysak*, 1981; *Greenspan and Whipple*, 1982; *Yang and Kan*, 1983]; (2) waves at the lower hybrid frequency, and other VLF waves [*Mozer et al.*, 1980; *Lysak*, 1981; *Chang and Coppi*, 1981; *Gorney et al.*, 1982; *Kintner and Gorney*, 1983; *Roth and Hudson*, 1983]; and (3) oxygen cyclotron waves [*Lysak et al.*, 1980; *Ashour-Abdalla et al.*, 1981]. Mechanism (1) may provide some energization, although statistical arguments similar to those used against EIC waves show that it cannot account for all the conics (for example, the magnitude of shocks is much smaller at altitudes < 4500 km and they are less common; shocks occur primarily in Region I, whereas conics are in both Regions I and II). Mechanism (3) cannot be specifically ruled out by the above statistical arguments, since oxygen cyclotron waves have not been observed on S3-3, probably due to their low frequency and large Doppler shifts. Mechanism (2), therefore, appears to be most viable, since VLF waves are very common in the auroral zone [*Temerin*, 1981]. Theoretical work, however, implies that preheating of the ions is required before the ions can be resonant with the waves.

E. Double layers

Although evidence for parallel electric fields (ion beams) and theoretical arguments requiring a parallel potential in upward current regions have been discussed, we have not yet addressed the nature of the parallel electric field. The S3-3 data provide evidence for parallel electric fields on two scale sizes — $\leq$0.1° (electrostatic shocks) and ~1° ("inverted V") [*Mozer et al.*, 1980]. Three possible distributions of the parallel potential drop for the same perpendicular signature are pictured in Figure 9. (Note that all three could be either 'V'-shaped or 'S'-shaped. The S-shaped configuration is more common in the auroral zone on both scale sizes [*Mozer et al.*, 1980; *Temerin et al.*, 1981]). In 9a, all of the parallel potential drop occurs in a very narrow region. This configuration, called a "strong double layer," has not been observed in the auroral zone by S3-3 on the ~1° scale size [*Boehm and Mozer*, 1981]. The opposite case, in which the potential is distributed over a broad altitude range, is shown in Figure 9c. This could be supported by, for example, anomalous resistivity [*Hudson et al.*, 1978]. In Figure 9b, the potential drop occurs in a series of small potential drops. These "weak double layers" (in association with solitary waves) have been observed

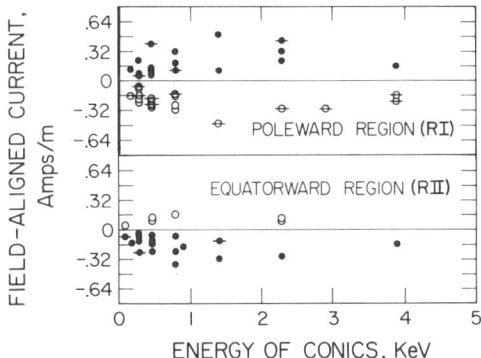

Fig. 8. Plot of the maximum conic energy versus field-aligned current (same format as Figure 3) [*Cattell et al.*, 1979].

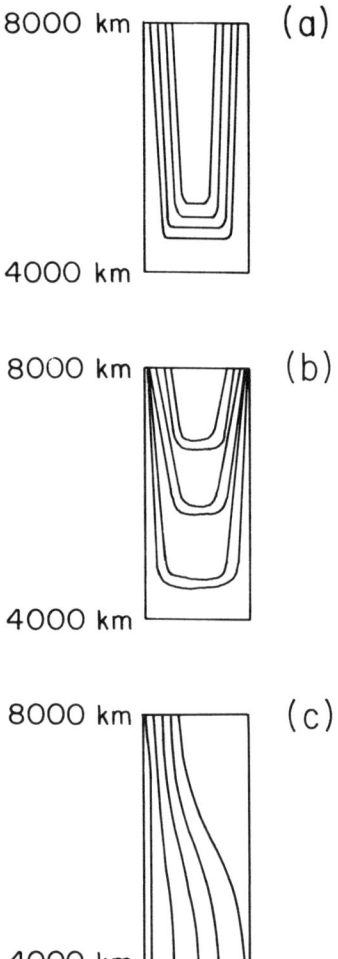

Fig. 9. Cartoon of 3 different parallel potential distributions for a given perpendicular electric field distribution (not drawn to scale): (a) strong double layer; (b) weak double layers; (c) distribution potential.

in the auroral zone for the first time by the S3-3 satellite [*Temerin et al.*, 1982].

An example of both is presented in Figure 10. The top two traces, which are the components of the electric field perpendicular to the magnetic field, show EIC waves modulated by low frequency turbulence. In the parallel component, symmetric pulses (solitary waves with no net potential drop) and asymmetric pulses (double layers with a net potential drop) are indicated.

Figure 11 shows how double layers are related to field-aligned currents and the other phenomena discussed above. The first two bar graphs indicate when EIC waves and double layers (DL) were observed. In the cross-hatched region the detector was close to saturation, making positive identification difficult. The field-aligned current density (positive out of the ionosphere) is plotted in the first panel. In the second panel, which shows E_x, the equatorward, perpendicular component of the electric field, the signature of a paired electrostatic shock can be seen (a northward spike at ~85413 s, followed by a southward spike at ~85448). The magnetic field B along one spinning boom (indicative of the pitch angle) is plotted in the third panel. The next four panels show the fluxes of electrons measured by the Lockheed detectors [*Shelley et al.*, 1976], which look down the magnetic field (at particles moving up the magnetic field) at the zero crossing with positive slope of B. The fluxes of ions from the Aerospace detectors [*Mizera and Fennell*, 1977], which look down the field at the minimum of B, are plotted in the bottom three panels. The EIC waves, double layers, upflowing ions, downflowing electrons, and the electrostatic shock are all collocated in the region of the largest upward field-aligned current.

Another example which contained two regions of double layers is shown in Figure 12 [*Mozer and Temerin*, 1982]. Both were coincident with EIC waves, upward field-aligned currents, ion beams, and downward electron fluxes. In addition, plasma density measurements were available and showed that the density was reduced by ~75% in the upward current regions. There was a small electrostatic shock at the edge of one of the two double layer regions. In all of the events studied, double layers were collocated with upward currents, density depletions, upgoing ion beams, downgoing electrons and EIC waves [*Mozer and Temerin*, 1982]. Double layers were usually not associated with electrostatic shocks.

As can be seen in Figure 10, double layers are observed approximately 5% of the time (in the regions where they are observed) and have typical electric fields of ~10 mV/m (pointing out of the ionosphere). This corresponds to an average field of ~0.5 mV/m, producing a potential drop of a few kilovolts over an altitude of several thousand kilometers. This is large enough to account for the observed parallel electric field acceleration, although it should be emphasized that the altitude distribution of double layers has not yet been determined. Since the double layer regions usually occur on spatial scales of 0.5 - 1° invariant latitude, they provide the potential drop on inverted-V precipitation scales, rather than on the very narrow scale associated with single discrete arcs [*Mozer and Temerin*, 1982]. This is

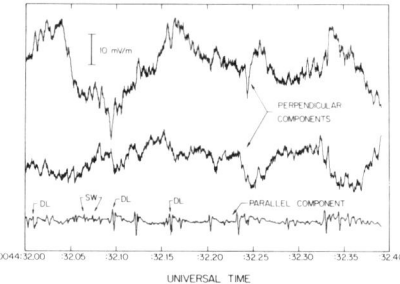
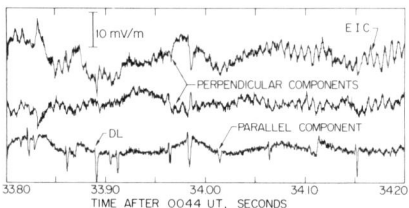

Fig. 10. The two perpendicular and one parallel electric field components. Examples of double layers (DL), solitary waves (SW), and electrostatic ion cyclotron waves (EIC) are indicated [*Temerin et al.*, 1982].

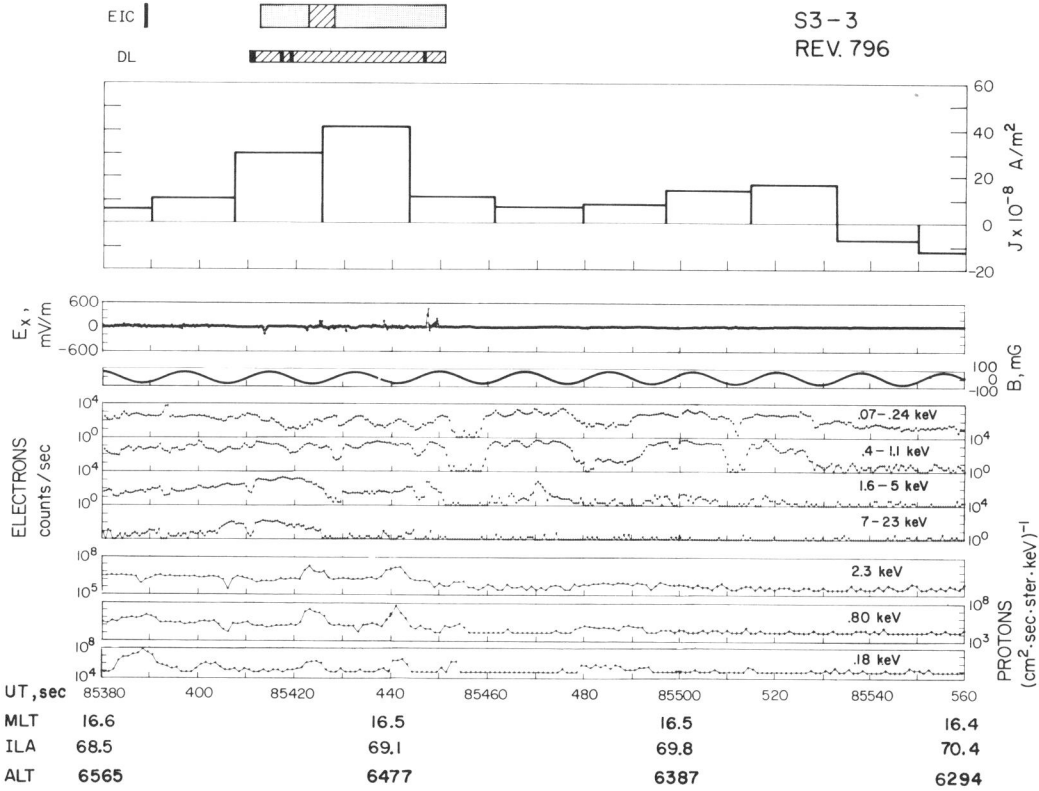

Fig. 11. An example of the association of EIC waves, double layers (DL), field-aligned currents, electrostatic shocks, electron fluxes, and ion fluxes in the evening auroral zone.

consistent with the fact that double layers are rarely collocated with or adjacent to electrostatic shocks. The distribution of the parallel potential drop within electrostatic shocks, which are associated with discrete aurora [*Kletzing et al.*, 1983 *Torbert and Mozer*, 1978] is not yet fully understood.

The features of the observed weak ($e\phi/T_e \lesssim 1$) double layers are consistent with simulations of current-driven systems [*Sato and Okuda*, 1981; *Kindel et al.*, 1981; *Hudson et al.*, 1983], rather than with systems with a fixed potential drop, which produce strong double layers ($e\phi/T_e \gg 1$). In the simulations, many weak double layers form in association with ion phase space holes. Formation of these holes requires $T_e/T_i \gg 1$, which probably occurs only in upward current regions, as discussed in the Introduction. The double layers are produced by a dynamic interaction between the current and localized regions of negative potential [*Hasagawa and Sato*, 1982; *Lotko and Kennel*, 1983; [*Hudson et al.*, 1983]. The field-aligned current plays a critical role, first, in exciting the turbulence in which negative potential perturbations arise, and second, by intensifying such perturbation to form double layers.

Discussion and Conclusions

In the previous section, correlations of electrostatic shocks, EIC waves, ion beams, ion conics, and field-aligned currents have been presented. These correlations have been combined with known or theoretically deduced properties of the various current regions to provide information on auroral processes and, in particular, to answer the questions listed in Table 1.

Statistical evidence on the altitude distribution and location of EIC waves and ion conics implies that the observed EIC waves are not the primary mechanism for heating ions (Question 8). Similar evidence suggests that acceleration via gradients in the electric field also cannot account for most of the observed perpendicular ion energization. Both of these mechanisms undoubtably provide some of the energization. In particular, the gradient mechanism may be important during active times, when *Gorney et al.* [1981] suggest that there is a second, higher altitude source of the conics. Since the plasma is less unstable to current-driven instabilities at the low conic source altitudes (because the drift velocity maximizes at ~ 1 R_E altitude), it seems unlikely that a current-driven mode can explain the low altitude conics. Alternatively, VLF waves may provide the necessary heating in the the low altitude source. These waves are driven by electron beams. Upflowing electron beams are observed in downward current regions and are very narrow at low altitudes [*Klumpar and Heikkila*, 1982; *Burch et al.*, 1983]. Since downgoing electron beams (which occur in upward current regions) become more unstable at lower altitudes due to magnetic focussing, the maximum wave amplitudes should occur at altitudes of several thousand kilometers [*Maggs and Lotko*, 1981], in agreement with the source altitude of conics. Questions still remain in the theory of heating via VLF waves. Recent studies of the S3-3 wave data have revealed the existence of strong waves just below the ion cyclotron frequency which may also provide some heating [*Lysak and Temerin*, 1983; *Temerin and Lysak*, 1983].

The question of whether ion beams or field-aligned currents provide the free energy to destabilize electrostatic ion cyclotron

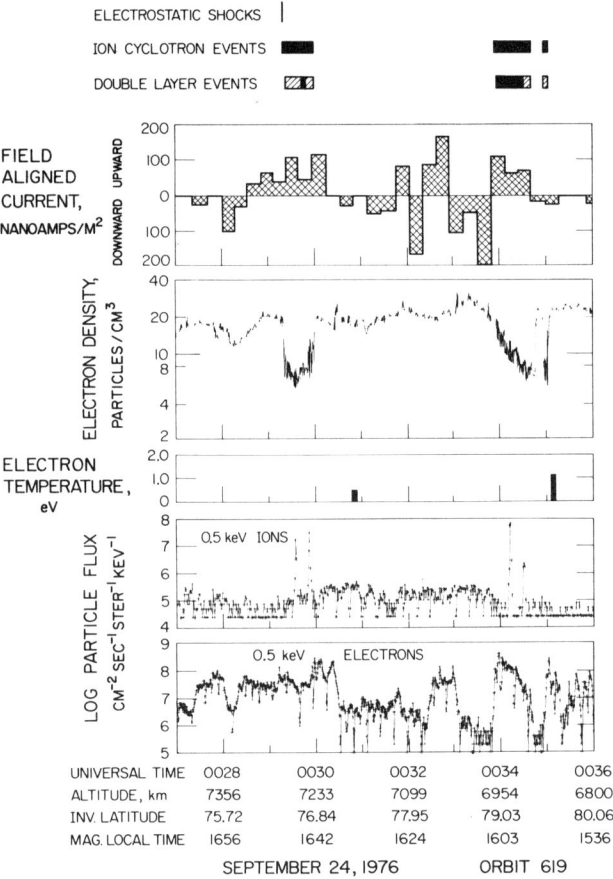

Fig. 12. Double layers, electrostatic shocks, ion cyclotron waves, field-aligned currents, plasma density and temperatures, and particle fluxes for a nine minute interval on Sept. 24, 1976.

waves (Question 7) remains undecided. The available evidence, combined with theoretical studies, favors the hypothesis that the waves are driven by upward currents carried by energetic electrons. Ion beams are observed with the EIC waves because upward currents have large parallel potential drops. Why the potential (and, therefore, the ion beam energy) correlates with the wave spectral density is not explained by this model.

Many of the observed features of electrostatic shocks are consistent with the idea that they are produced by the current associated with perturbations on the plasmasheet boundary. For example, if Region I currents map to the plasmasheet boundary, the fact that shocks occur almost exclusively in Region I is easily explained (Question 4). The morning/evening asymmetry and the altitude dependence of the magnitude of electrostatic shocks are then a result, respectively, of the larger drift velocity and temperature ratio in upward currents and the drift velocity maximum at altitudes of $\sim 1 R_E$.

Electric fields pointing out of the ionosphere occur almost exclusively in the large-scale regions of upward current, and are more common in the evening Region I than in the morning Region II upward current. The latter may be due to the fact that Region II maps to the ring current [Schield et al., 1969], which has a higher density than the plasmasheet boundary, and may, therefore, not require as large a potential drop to support a given current (an aspect of Question 2). The reason that upward parallel electric fields are not commonly seen (only one example thus far) in small-scale upward currents within the large-scale downward current is not yet understood. In light of recent observations by DE [Sugiura et al., 1983a,b] of very small-scale currents with high current densities, the question of the importance of the scale size of the current for supporting parallel electric fields is extremely interesting and important (Questions 3 and 5). Some theoretical work on the importance of the scale size of the currents for EIC waves has been done [Bakshi et al., 1983].

Although the fact that downward currents do not require large potential drops was discussed in the Introduction, evidence (or lack thereof) for downward parallel fields in the S3-3 data has not yet been described (Question 6). A study of downward ion beams by Ghielmetti et al. [1979] concluded that the observations were consistent with mechanisms other than parallel electric fields. Only one example of a downward ion beam which could be best explained by a downward parallel electric field [Fennell et al., 1979] has been found. This event (which occurred during a large substorm) was associated with one of the largest currents ever observed on S3-3. This large current might, therefore, have been associated with a larger potential drop than usually occurs in downward currents. It is unlikely that downward parallel electric fields occur very often, since most mechanisms for supporting parallel electric fields either seem most consistent with theories that require $T_e > T_i$ (e.g., double layers) and/or some critical drift velocity, (e.g., anomalous resistivity) or produce only upward fields (e.g., mirror-force supported $E_\parallel$).

The association of field-aligned currents (evening RI), double layers, ion beams, and downward electron fluxes can explain the 'inverted-V' scale parallel electric field acceleration in the evening auroral zone. EIC waves are also always observed with these events. Both the double layers and the EIC waves observations are consistent with current-driven mechanisms.

The distribution of the parallel electric field in electrostatic shocks is not, in general, understood (although there are some cases (for example, Figure 11) where the observations are consistent with the possibility that the associated parallel potential occurs in double layers). Several papers have presented evidence from the S3-3 satellite that electrostatic shocks provide the acceleration mechanism for discrete arcs [Torbert and Mozer, 1978; Mozer et al., 1980; Kletzing et al., 1983]. In addition, the parallel field associated with the morning shock events must be different from that associated with the evening events. There is much experimental and theoretical work remaining in this area to provide a complete understanding of auroral phenomena in both morning and evening. Comparisons of properties of currents and auroral processes can yield useful new information for these studies.

Acknowledgments. The author thanks R. D. Sharp of Lockheed Palo Alto Laboratories and J. F. Fennell, D. Gorney, and P. Mizera, of the Aerospace Corporation for use of their particle data and discussions of it. J. Wygant listened to early versions of the talk and provided useful criticisms. Discussions of the data and theory with M. Hudson, W. Lotko, F. S. Mozer, and M. Temerin are gratefully acknowledged. Finally, the author would like to thank C. Overhoff for her assistance in producing the manuscript and S. Primbsch for programming. This work was supported by ONR contract N00014-81-C-0006.

References

Akasofu, S.-I., *Physics of Magnetospheric Substorms*, D. Reidel, Hingham, Mass., 1977.

Ashour-Abdalla, M., H. Okuda, and L. Z. Cheng, Acceleration of heavy ions on auroral field lines, *Geophys. Res. Lett., 8*, 795, 1981.

Bakshi, P., G. Ganguli, and P. Palmadesso, Finite width currents, magnetic shear and the current driven ion cyclotron instability, preprint, 1983.

Bennett, E. L., M. Temerin, and F. S. Mozer, The distribution of auroral electrostatic shocks below 8000 km altitude, *J. Geophys. Res.*, in press 1983.

Bergmann, Rachelle, Electrostatic ion (hydrogen) cyclotron and acoustic wave instabilities in regions of upward field-aligned current and upward ion beams, submitted to *J. Geophys. Res.*, 1982.

Block, L. P., Double layers, in *Physics of Hot Plasma in the Magnetosphere*, edited by B. Hultqvist and H. Stenflo, Plenum Pub. Co., New York, 1975.

Boehm, M. H., and F. S. Mozer, An S3-3 search for confined regions of large parallel electric fields, *Geophys. Res. Lett., 8*, 607, 1981.

Böhmer, H., J. D. Hauck, and N. Rynn, Ion-beam excitation of electrostatic ion-cyclotron waves, *Phys. Fluids, 19*, 450, 1976.

Burch, J. L., P. H. Reiff, and M. Sugiura, Upward electron beams measured by DE-1: A primary source of dayside Region-1 Birkeland currents, this volume, 1983.

Burke, W. J., Electric fields, Birkeland currents, and electron precipitation in the vicinity of discrete auroral arcs, in *Physics of Auroral Arc Formation, AGU Geophysical Monograph 25*, edited by S.-I. Akasofu and J. R. Kan, p. 164, American Geophysical Union, Washington, D.C., 1981.

Cattell, C. A., The relationship of field-aligned currents to electrostatic ion cyclotron waves, *J. Geophys. Res., 86*, 3641, 1981.

Cattell, C. A., S3-3 satellite instrumentation and data, in *The IMS Source Book*, edited by C. T. Russell and D. J. Southwood, p. 91, American Geophysical Union, Washington, D.C., 1982.

Cattell, Cynthia, Robert Lysak, R. B. Torbert, and F. S. Mozer, Observations of differences between regions of current flowing into and out of the ionosphere, *Geophys. Res. Lett., 6*, 621, 1979a.

Cattell, C. A., R. B. Torbert, and F. S. Mozer, The relationship of field-aligned currents to plasma convection boundaries (abstract), *Eos Trans. AGU, 60*, 348, 1979b.

Cattell, C. A., M. K. Hudson, R. L. Lysak, D. W. Potter, M. Temerin, R. B. Torbert, and F. S. Mozer, Observations of electrostatic shocks and associated plasma instabilities by the S3-3 satellite, in *Relation between Laboratory and Space Plasmas*, edited by H. Kikuchi, p. 115, D. Reidel, Hingham, Mass., 1981.

Cattell, C. A., M. Kim, R. P. Lin, and F. S. Mozer, Observations of large electric fields near the plasmasheet boundary by ISEE-1, *Geophys. Res. Lett., 9*, 539, 1982.

Chang, T., and B. Coppi, Lower hybrid acceleration and ion evolution in the suprauroral region, *Geophys. Res. Lett., 8*, 1253, 1981.

Chiu, Y. T., and J. M. Cornwall, Electrostatic model of a quiet auroral arc, *J. Geophys. Res., 85*, 543, 1980.

Chiu, Y. T., and M. Schulz, Self-consistent particle and parallel electrostatic field distributions in the magnetospheric-ionospheric auroral region, *J. Geophys. Res., 83*, 629, 1978.

Chiu, Y. T., M. Schulz, and J. M. Cornwall, Effects of auroral-particle anisotropies and mirror forces on high-latitude electric fields, in *Physics of Auroral Arc Formation, AGU Geophysical Monograph 25*, edited by S.-I. Akasofu and J. R. Kan, p. 234, American Geophysical Union, Washington, D.C., 1981.

Cloutier, P. A., and H. R. Anderson, Observations of Birkeland currents, *Space Sci. Rev., 17*, 563, 1975.

Correll, D. L., N. Rynn, and H. Böhmer, Onset, growth and saturation of the current-driven ion cyclotron instability, *Phys. Fluids, 18*, 1800, 1975.

Dusenbery, P. B., and L. R. Lyons, Generation of ion conic distribution by upgoing ionospheric electrons, *J. Geophys. Res., 86*, 7627, 1981.

Fennell, J. F., P. F. Mizera, and D. R. Croley, Jr., Observations of ion and electron distributions during the July 29 and 30, 1977 storm period, *Proc. Magnetospheric Boundary Layers Conf., Alpbach, ESA SP-148*, p. 97, 1979.

Fridman, M., and J. Lemaire, Relationship between auroral electron fluxes and field aligned potential difference, *J. Geophys. Res., 85*, 664, 1980.

Ghielmetti, A., R. G. Johnson, R. D. Sharp, and E. G. Shelley, The latitudinal, diurnal, and altitudinal distribution of upflowing ions of ionospheric origin, *Geophys. Res. Lett., 5*, 59, 1978.

Ghielmetti, A., R. D. Sharp, E. G. Shelley, and R. G. Johnson, Downward flowing ions and evidence for injection of ionospheric ions into the plasma sheet, *J. Geophys. Res., 84*, 5781, 1979.

Gorney, D. J., A. Clarke, D. Croley, J. Fennell, J. Luhman, and P. Mizera, The distribution of ion beams and conics below 8000 km, *J. Geophys. Res., 86*, 83, 1981.

Gorney, D. J., S. R. Church, and P. F. Mizera, On the ion harmonic structure in auroral zone waves: The effect of ion conic damping of auroral hiss, *J. Geophys. Res., 87*, 10479, 1982.

Greenspan, M. E., and E. C. Whipple, The effect of oblique double layers on particle magnetic moment and gyrophase, in *Proc. Yosemite Conf. on Origins of Plasmas and Electric Fields*, Yosemite Ca., 1982.

Gurnett, D. A., A. M. Persoon, S. D. Shawhan, Auroral zone electron densities from DE-1 plasma wave observations (abstract), *Eos Trans. AGU, 63*, 1056, 1982.

Hasegawa, A., Particle acceleration by MHD surface wave and formation of aurora, *J. Geophys. Res., 81*, 5083, 1976.

Hasegawa, A., and T. Sato, Existence of a negative potential solitary wave structure and formation of a double layer, *Phys. Fluids, 25*, 632, 1982.

Hauck, J. D., Böhmer, H., N. Rynn, and G. Benford, Ion beam excitation of ion-cyclotron waves and ion heating in plasmas with drifting electrons, *J. Plasma Phys., 19*, 253, 1976.

Hudson, M. K., R. L. Lysak, and F. S. Mozer, Magnetic field-aligned potential drops due to electrostatic ion cyclotron turbulence, *Geophys. Res. Lett., 5*, 143, 1978.

Hudson, M. K., W. Lotko, I. Roth, and E. Witt, Solitary waves and double layers on auroral field lines, *J. Geophys. Res., 88*, 916, 1983.

Hudson, M. K., and F. S. Mozer, Electrostatic shocks, double layers and anomalous resistivity in the magnetosphere, *Geophys. Res. Lett., 5*, 131, 1978.

Hudson, M. K., and D. W. Potter, Electrostatic shocks in the auroral magnetosphere, in *Physics of Auroral Arc Formation, AGU Geophysical Monograph 25*, edited by S.-I. Akasofu and J. R. Kan, p. 260, American Geophysical Union, Washington, D.C., 1981.

Iijima, T., and T. A. Potemra, The amplitude distribution of field-aligned currents at northern high latitudes observed by Triad, *J. Geophys. Res., 81*, 2165, 1976.

Kaufmann, R. L., and P. M. Kintner, Upgoing ion beams, 1. Microscopic analysis, *J. Geophys. Res., 87*, 10487, 1982.

Kindel, J. M., and C. F. Kennel, Topside current instabilities, *J. Geophys. Res., 76*, 3055, 1971.

Kindel, J. M., C. Barnes, and D. W. Forslund, Anomalous dc resistivity and double layers in the auroral ionosphere, in *Physics of Auroral Arc Formation, AGU Geophysical Monograph 25*, edited by S.-I. Akasofu and J. R. Kan, p. 296, American Geophysical Union, Washington, D.C., 1981.

Kintner, P. M., M. C. Kelley, R. D. Sharp, A. G. Ghielmetti, M. Temerin, C. A. Cattell, and P. Mizera, Simultaneous observations of energetic (keV) upstreaming ions EHC waves, *J. Geophys. Res., 84*, 7201, 1979.

Kintner, P. M., M. C. Kelley, and F. S. Mozer, Electrostatic hydrogen cyclotron waves near one earth radius altitude in the polar magnetosphere, *Geophys. Res. Lett., 5*, 139, 1978.

Kletzing, C., C. Cattell, F. S. Mozer, S.-I. Akasofu, and K. Makita, Evidence for electrostatic shocks as the source of discrete auroral arcs, *J. Geophys. Res., 88*, 4105, 1983.

Klumpar, D. M., Transversely accelerated ions: An ionospheric source of hot magnetospheric ions, *J. Geophys. Res., 84*, 4229, 1979.

Klumpar, D. M., What are the carriers of the Birkeland currents? A review of the ISIS contributions, this volume, 1983.

Klumpar, D. M., and W. J. Heikkila, Electrons in the ionospheric source cone: Evidence for runaway electrons as carriers of downward Birkeland current, *Geophys. Res. Lett., 9*, 873, 1982.

Knight, S., Parallel electric fields, *Planet. Space Sci., 21*, 741, 1973.

Lemaire, J., and M. Scherer, Plasma sheet particle precipitation: A kinetic model, *Planet. Space Sci., 21*, 281, 1973.

Lemaire, J., and M. Scherer, Field-aligned currents and parallel electric fields, *Planet. Space Sci., 22*, 1485, 1974.

Levin, S., K. Whitley, and F. S. Mozer, A statistical study of large electric field events in the earth's magnetotail, submitted to *J. Geophys. Res.*, 1983.

Lotko, W., and C. F. Kennel, Spiky ion acoustic waves in collisionless auroral plasma, *J. Geophys. Res., 88,* 381, 1983.

Lysak, R. L., Electron and ion acceleration by strong electrostatic turbulence, in *Physics of Auroral Arc Formation, AGU Geophysical Monograph 25,* edited by S.-I. Akasofu and J. R. Kan, p. 444, American Geophysical Union, Washington, D.C., 1981.

Lysak, R. L., and C. T. Dum, Dynamics of magnetosphere-ionosphere coupling including turbulent transport, *J. Geophys. Res., 88,* 356, 1983.

Lysak, R. L., and M. K. Hudson, Coherent anomalous resistivity in the region of electrostatic shocks, *Geophys. Res. Lett., 6,* 661, 1979.

Lysak, R. L., M. K. Hudson, and M. Temerin, Ion heating by strong electrostatic ion cyclotron turbulence, *J. Geophys. Res., 85,* 678, 1980.

Lysak, R., W. Lotko, M. Hudson, and E. Witt, Formation of double layers on auroral field lines, *Proc. of Symp. on Plasma Double Layers,* Risø, Denmark, in press, 1982.

Lysak, R. L., and M. Temerin, Generation of Alfvén-ion cyclotron waves on auroral field lines in the presence of heavy ions, *Geophys. Res. Lett.,* in press 1983.

Maggs, J. E., and W. Lotko, Altitude dependent model of the auroral beam and beam-generated electrostatic noise, *J. Geophys. Res., 86,* 3439, 1981.

Mizera, P. F., and J. F. Fennell, Signatures of electric fields from high and low altitude particle distributions, *Geophys. Res. Lett., 4,* 311, 1977.

Mizera, P. F., J. F. Fennell, D. R. Croley, Jr., A. L. Vampola, F. S. Mozer, R. B. Torbert, M. Temerin, R. L. Lysak, M. K. Hudson, C. A. Cattell, R. G. Johnson, R. D. Sharp, P. M. Kintner, and M. C. Kelley, The aurora inferred from S3-3 particles and fields, *J. Geophys. Res., 86,* 2329, 1981.

Mozer, F. S., C. A. Cattell, M. K. Hudson, R. L. Lysak, M. Temerin, and R. B. Torbert, Satellite measurements and theories of low altitude auroral particle acceleration, *Space Sci. Rev., 27,* 155, 1980.

Mozer, F. S., C. A. Cattell, M. Temerin, R. B. Torbert, S. von Glinski, M. Woldorff, and J. Wygant, The dc and ac electric field, plasma density, plasma temperature, and field-aligned current experiments on the S3-3 satellite, *J. Geophys. Res., 84,* 5875, 1979.

Mozer, F. S., C. W. Carlson, M. K. Hudson, R. B. Torbert, B. Parady, J. Yatteau, and M. C. Kelley, Observations of paired electrostatic shocks in the polar magnetosphere, *Phys. Rev. Lett., 38,* 292, 1977.

Mozer, F. S., and M. Temerin, Solitary waves and double layers as the source of parallel electric fields in the auroral acceleration region, in *Proceedings of Nobel Symposium No. 54 at Kiruna, Sweden, March, 1982,* edited by B. Hultqvist and T. Hagfors, Plenum Pub. Co., New York, in press, 1982.

Okuda H., and M. Ashour-Abdalla, Acceleration of hydrogen ions and conic formation along auroral field lines, *J. Geophys. Res., 88,* 899, 1983.

Potemra, T. A., Current systems in the earth's magnetosphere, *Rev. Geophys. Space Phys., 17,* 641, 1979.

Potemra, T. A., T. Iijima, and N. A. Suflekos, Large-scale characteristics of Birkeland currents, in *Dynamics of the Magnetosphere,* edited by S.-I. Akasofu, p. 165, D. Reidel, Hingham, Mass., 1979.

Pritchett, P. L., M. Ashour-Abdalla, and J. M. Dawson, Simulation of the current-driven electrostatic ion cyclotron instability, *Geophys. Res. Lett., 8,* 611, 1981.

Sato, T., and H. Okuda, Numerical simulations on ion acoustic double layers, *J. Geophys. Res., 86,* 3357, 1981.

Schield, M. A., J. W. Freeman, and A. J. Desler, A source of field-aligned currents at auroral latitudes, *J. Geophys. Res., 74,* 247, 1969.

Sharp, R. D., R. G. Johnson, and E. G. Shelley, Observation of an ionospheric acceleration mechanism producing energetic (keV) ions primarily normal to the geomagnetic field direction, *J. Geophys. Res., 82,* 3224, 1977.

Sharp, R. D., R. G. Johnson, and E. G. Shelley, Energetic particle measurements from within ionospheric structures responsible for auroral acceleration, *J. Geophys. Res., 84,* 480, 1979.

Shelley, E. G., R. D. Sharp, and R. G. Johnson, Satellite observations of an ionospheric acceleration mechanism, *Geophys. Res. Lett., 3,* 654, 1976.

Smiddy, M., W. J. Burke, M. C. Kelley, N. A. Saflekos, M. S. Gussenhoven, D. A. Hardy, and F. J. Rich, Effects of high-latitude conductivity on observed convection electric fields and Birkeland currents, *J. Geophys. Res., 85,* 6811, 1980.

Sugiura, M., N. C. Maynard, J. L. Burch, and J. D. Winningham, Initial results of DE-1 and -2 observations of field-aligned currents, this volume, 1983*a.*

Sugiura, M., T. Iyemori, R. A. Hoffman, N. C. Maynard, L. H. Brace, J. D. Winningham, and J. L. Burch, Relationships between field-aligned currents, electron fluxes precipitation, and electric fields as observed by the DE-2 satellite, this volume, 1983*b.*

Swift, D. W., On the formation of auroral arcs and acceleration of auroral electrons, *J. Geophys. Res., 80,* 2096, 1975.

Swift, D. W., An equipotential model for auroral arcs: The theory of two-dimensional laminar electrostatic shocks, *J. Geophys. Res., 84,* 6427, 1979.

Temerin, M. A., The polarization, frequency, and wavelengths of high latitude turbulence, *J. Geophys. Res., 83,* 2609, 1978.

Temerin, M. A., Plasma waves on auroral field lines, in *Physics of Auroral Arc Formation, AGU Geophysical Monograph 25,* edited by S.-I. Akasofu and J. R. Kan, p. 351, American Geophysical Union, Washington, D.C., 1981.

Temerin, M., M. H. Boehm, and F. S. Mozer, Paired electrostatic shocks, *Geophys. Res. Lett., 8,* 799, 1981*a.*

Temerin, M., C. Cattell, R. Lysak, M. K. Hudson, R. B. Torbert, F. S. Mozer, R. D. Sharp, and P. M. Kintner, The small-scale structure of electrostatic shocks, *J. Geophys. Res., 86,* 11278, 1981*b.*

Temerin, M., M. Woldorff, and F. S. Mozer, Nonlinear steepening of the electrostatic ion cyclotron wave, *Phys. Rev. Lett., 43,* 1941, 1979.

Temerin, M., and R. Lysak, Electromagnetic ion cyclotron (ELF) waves generated by auroral electron precipitation, submitted to *J. Geophys. Res.,* 1983.

Torbert, R. B., and F. S. Mozer, Electrostatic shocks as the source of discrete auroral arcs, *Geophys. Res. Lett., 5,* 135, 1978.

Torbert, R. B., C. A. Cattell, F. S. Mozer, and C.-I. Meng, The boundary of the polar cap and its relation to electric fields, field-aligned currents, and auroral particle precipitation, in *Physics of Auroral Arc Formation, AGU Geophysical Monograph 25,* edited by S.-I. Akasofu and J. R. Kan, p. 143, American Geophysical Union, Washington, D.C., 1981.

Ungstrup, E., D. M. Klumpar, and W. J. Heikkila, Heating of ions to superthermal energies in the topside ionosphere with electrostatic ion cyclotron waves, *J. Geophys. Res., 84,* 4289, 1979.

Yang, W. H., and J. R. Kan, Generation of conic ions by auroral electric fields, *J. Geophys. Res., 88,* 465, 1983.

THREE-DIMENSIONAL POTENTIAL STRUCTURE ASSOCIATED WITH BIRKELAND CURRENTS

Lars Block

Department of Plasma Physics, Royal Institute of Technology,
S-100 44 Stockholm, Sweden

Abstract. The U-shaped equipotential structures believed to exist above auroral arcs with upward Birkeland currents pose certain problems in terms of particle, momentum and current conservation. Supply of electrons must cover losses due to precipitation. The supply mechanism must be consistent with the observed energy gain and precipitation field-alignment.

Three-dimensional models of the potential structure which may account for the above requirements are discussed. Particles can be injected into the structure along equipotential surfaces, both from the ends of the Birkeland current sheet and sideways by means of an electric field component tangential to the sheet. The generator can inject particles sideways through equipotential surfaces against the electric field. These particles carry the perpendicular current, but they can only to a limited extent carry the Birkeland currents, which are mostly carried by ambient particles pushed away by the space charge generated parallel electric field. A consequence of this is that especially a thermo-electric generator replaces cold ambient electrons by hot gradient-drifting populations.

The nature of the parallel electric field is also discussed. The observed current sheet (or arc) thickness is inconsistent with a uniformly distributed parallel electric field. A field concentrated in a single strong double layer is inconsistent with the observed precipitation spectrum with substantial fluxes of electrons with energies below the total parallel potential drop. The multitude of weak double layers and solitons seen on the S3-3 satellite seems to agree with known facts.

Introduction

Since it was realized that electric fields parallel to the magnetic field contribute to the acceleration of auroral electrons, qualitative two-dimensional models of the potential distribution above auroral arcs have been proposed (Block, 1969, 1972; Carlqvist and Boström, 1970; Gurnett 1972; Swift 1975). These include both U-, V-, and S-shaped equipotential lines. Quantitative models based on measurements by backscatter radars (de la Beaujardière and Vondrak, 1982) and rocket-borne instruments (Marklund, 1983) have also been constructed. In these models the ionospheric latitude potential profile is integrated from the measured electric fields. The corresponding high altitude, magnetospheric potential profile is then found by subtracting the voltage equivalent of the energy at the peak in the precipitation spectrum. The potential distribution at intermediate altitudes can of course only be qualitatively sketched subject to the high and low altitude boundary conditions. These models of individual arcs are consistent with several satellite observations, e.g. INJUN 5 (Frank and Gurnett 1971; Gurnett, 1972), S3-3 (Mozer et al. 1977; 1979; Mozer, 1981; Temerin, 1981) and AE-C (Heelis et al., 1981).

Theoretical models supporting these ideas have been developed by Lennartsson (1973, 1977, 1980), Swift (1976), Kan and Akasofu (1979) and Lyons (1980, 1981). They are based on current continuity, assuming a hot, thermal magnetospheric plasma and taking geomagnetic mirror effects into account. Lyons (1980) tested his model against measurements from a rocket experiment (Maynard et al., 1977; Evans et al., 1977) with remarkably good consistency between the inferred high altitude potential and the measured ionospheric potential. However, the theory gave no information on the transition between the two potentials.

A striking feature of the above mentioned quantitative models based on measurements in individual arcs is that in all cases so far analyzed the high altitude electric field reverses, i.e. the equipotential lines are always U- or V-shaped, never S-shaped (Marklund, 1983). The reason for this is that the maximum field aligned potential drop always dominates over the total horizontal potential drop across the arc at ionospheric altitudes. This is surprising since it seems to be in conflict with the results from S3-3 where single electrostatic shocks are often observed with no nearby electric field reversal (Mozer et al., 1980). A possible explanation to this is given by Atkinson (1982).

The Need For Three-Dimensional Models

Although the two-dimensional models are probably correct, they do not suffice for explaining among other things the cause of their formation, the total plasma flow taking into account the finite

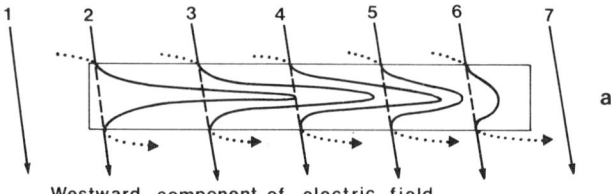

Westward component of electric field

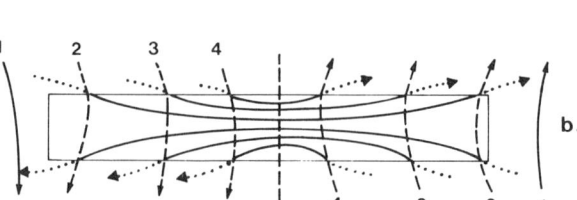

Westward electric field | Eastward electric field

Zero tangential electric field

Fig. 1. Three-dimensional equipotential models proposed by Atkinson (1982). They are all consistent with the usual U- or V-shaped structures in vertical north-south oriented cross-sections. The view is downwards along magnetic field lines. the solid lines are equipotentials at high altitude and the dashed lines are the same at ionospheric altitude. The rectangles enclose regions with parallel electric field. The length scale is compressed. With proper length scale the high altitude equipotentials outside the rectangles would be more east-west directed, consistent with the large scale magnetospheric convection (indicated by the dotted lines which were not shown in Atkinson's original figure).

arc length, how the precipitating particles are injected into the acceleration region, i.e. the generator problem, and the observed energy and pitch-angle spectra inside and outside the arcs. In order to understand these processes both the three-dimensional potential structure, the dynamics and the detailed distribution of the parallel electric field must be considered.

Three different three-dimensional models, shown in Figure 1, were proposed by Atkinson (1978, 1981) and a fourth one by Block (1981), see Figure 2. These were designed for understanding the eastward or westward plasma convection caused by the strong electric field at high altitude at the arms of the equipotential U:s. Atkinson (1982) also considered those parts of the current closure that are due to polarization current in two of his models.

It is the purpose of this paper to consider some elementary requirements that must be put on any three-dimensional model, mainly from the particle point of view.

Particle Conservation

The rate of electron losses due to precipitation in a strong arc from a geomagnetic flux tube with one square meter cross section in the E-layer in the presence of a $10 \mu A/m^2$ Birkeland current is of the order of 10^{14} electrons/sm^2. The total electron content of such a flux tube above the region where the plasma is too dense for formation of parallel electric fields, can be estimated to be of the order of 10^{15} - 10^{17} electrons/m^2 at L = 6, so if there is no external supply the flux tube would be drained within 10 - 1000 seconds. As seen from the ground, the typical duration of an arc lies in that range. Then it either fades out or moves away.

Charge conservation in the flux tube demands either an equally efficient ion sink, an electron source, or any combination thereof, which in macroscopic terms is simply current continuity. That will be considered more in detail in the next section. In this section only the supply of plasma is considered. If the loss of negative charge due to electron precipitation is partly or entirely balanced by ion losses, e.g. through curvature or

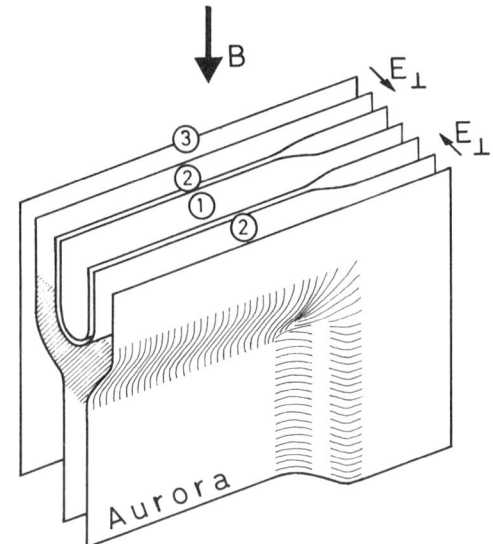

Fig. 2. Perspective view of equipotential surfaces above an auroral arc at a perpendicular electric field reversal. Electrons injected in region (1) will fall through the entire DL potential drop. Electrons drifting along surfaces in region (2) at shaded altitudes will fall through part of the drop. Electrons on surface (3) will not see any parallel E-field.

gradient drift away from the flux tube, a corresponding supply of plasma must exist in steady state (which does not necessarily prevail). The purpose of this section is to estimate roughly to what extent that can occur in different three-dimensional potential structures.

The most obvious way to supply plasma is through ExB-drift, which can be either along the arc as in Block's model (Figure 2) or from the sides due to a tangential electric field as in Atkinson's models a and b (Figure 1).

In case of supply only along the arc associated with a Birkeland current sheet the influx of electrons at one end is of the order of

$$\Phi = \int_{z_o}^{z_1} n_e (E_\perp/B) w dz \approx <n_e> (E_{\perp o}/B_o) R_E w_o L^4/4$$

where z is a coordinate along the magnetic field line with z_o being the minimum altitude at which there is a significant parallel electric field. R_E is the earth's radius, w is the thickness of the current sheet (arc) at the altitude z, L is the McIlwain parameter, and index o indicates values at $z = z_o$. If we assume an average electron density $<n_e> = 10^7$ m^{-3}, L = 6 and $E_{\perp o}$ = 100 mV/m we get $\Phi \approx 5 \cdot 10^{19}$ electrons per second per meter sheet thickness, which is barely enough for a 500 km long arc, if the outflux at the other end is negligible.

A similar calculation for supply from one side (north or south of the current sheet) shows that a tangential electric field of the order of one mV/m in the arc frame of reference would suffice for a 10 km thick sheet (arc). Atkinson's model a (Figure 1) is equivalent to Block's model (Figure 2) with a superimposed tangential electric field of this kind. It is important to note, however, that the average field component tangential to the arc must be very much smaller than the component across the arc, i.e. the solid flow lines 1-7 in Figure 1a should be bent more along the Birkeland current sheet as shown by the dotted lines, since arcs and Birkeland current sheets are generally aligned with the plasma flow. On the other hand the equipotential surfaces (flow lines) must converge very much at the inflow into the region with very strong perpendicular field (the electrostatic "shock", as observed on S3-3, cf. Mozer et al., 1979; Mozer, 1981) and diverge at the outflow to accomodate the total shock potential of some kilovolts, also in the upstream and downstream non-shocked regions. The anti-sunward convection in the polar cap has, for example, been seen to converge already in the dayside cleft region, since the general polar cap convection flow is usually concentrated in relatively thin regions near the convection reversal at a few thousand kilometers altitude, as originally found by the INJUN 5 satellite (see e.g. Gurnett, 1972) where only one electric field component perpendicular to the magnetic field was measured. Heelis et al. (1976) reported measurements on the AE-C satellite of complete vector ion drift velocities, which clearly show some cases of converging flow near the cleft (cf. especially their Figure 4). A similar convergence must also exist near midnight for the sunward convection equatorward of the reversal.

If all auroral arcs are associated with convection reversals, i.e. U- or V-shaped equipotential structures above the arc (which is true for all arcs so far quantitatively analyzed, cf. Marklund, 1983), then as many reversals must exist at high altitudes as there are multiple parallel arcs in the E-region. This is consistent with the fact that most arcs are not situated at the large scale convection reversal (the polar cap boundary). Burke et al. (1979) report a four-lobed pattern consistent with the cross-polar-cap arc recently seen by the UV imaging experiment on Dynamics Explorer (Frank et al., 1982). However, many more lobes are required to explain multiple parallel arcs. They should be relatively small scaled (a few hundred kilometers long or less) within either the anti-sunward convection region in the polar cap or the sunward convection equatorward of the main reversal. These closed flow patterns can best be described by Atkinson's model c in Figure 1. Evidence of such small scale convection lobes is given by Burke (1983), who reports on electric field measurements on S3-2. Several electric field reversals can be seen in his Figure 6, both in the polar cap and at lower latitudes in the pre- and post-midnight sectors.

The supply of low energy particles to such closed convection loops is possible neither from the ends, nor sideways by tangential fields. Some supply from the ionosphere may be possible if the structure contains some parts with downward electric field, which need not be strong, to drive a sufficient number of electrons upward. That means that one arm of the U must be slightly tilted towards the other arm as shown in Figure 3. Hence, ionospheric electrons are pulled up in one region and precipitated again in another one. The same is of course also happening for many of the ions in the upward ion beams and conics. These ideas were suggested by Block and Fälthammar (1969). It seems also likely, however, that the precipitation causes a depletion of the ambient cold plasma as proposed by Block and Fälthammar (1968). After extended plasma depletion the arc structure may break down.

In summary, four ways to account for the losses of electrons due to precipitation have been discussed, namely
(i) depletion of the ambient electron population in the flux tube,
(ii) plasma drift along the arc due to a northward and southward electric field,
(iii) plasma drift perpendicular to the arc due to a weak tangential electric field in the arc frame of reference,
(iv) upward ionospheric electron flow within the same convection cell, which requires a downward electric field somewhere, as shown in Figure 3.

Quantitatively, each of the first three can

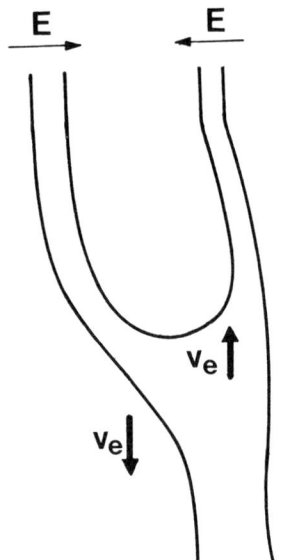

Fig. 3. Suggested deformation of the U-shaped equipotential lines associated with the model in Fig. 1c. The supply of electrons in closed convection loops on closed field lines must be from the ionosphere. Hence, the reversed inclination of the righ arm of the U, which causes a downward electric field pulling ionospheric electrons upwards.

barely explain the duration, length and width, respectively, of typical arcs. The fourth one can go on virtually for ever. It seems very likely that all or most of these processes play some role in maintaining arcs and Birkeland current sheets, in varying proportions, as reflected by auroral dynamics. In particular it should be stressed that the tangential electric field of mechanism (iii) is _in the arc frame of reference_, regardless of how the arc moves. It does not necessarily exist in the E-layer, but always at and above the altitudes with parallel electric fields. A striking example of that is substorm breakup when the aurora quickly moves poleward without a corresponding ionospheric longitudinal electric field. It is tempting to speculate that the flux tube has been drained of plasma and the Birkeland current, maintained by circuit inductance, desperately needs to move to new flux tubes with fresh plasma.

The Magnetospheric Generator And Charge Conservation

Current continuity and charge conservation cannot be understood without considering the nature of the generator. There are essentially two kinds of generators in the magnetosphere, described by two terms in the momentum equation of the plasma

$$\underline{J} \times \underline{B} = \rho d\underline{v}/dt + \nabla p \quad (1)$$

with obvious symbols. The current $\underline{J}$ that enters is only the perpendicular current $\underline{J}_\perp$. The first term on the RHS accounts for an MHD dynamo generator, provided $d\underline{v}/dt$ represents retardation. In a stationary state $d\underline{v}/dt = (\underline{v}\cdot\underline{\nabla})\underline{v}$. The corresponding current polarizes (depolarizes) the plasma as it is accelerated (retarded) by the $\underline{J} \times \underline{B}$ force. The gain (loss) of particle kinetic drift energy is exactly balanced by the loss (gain) of potential energy due to the polarization displacement (cf. Equation (4) below). The second term ∇p represents a thermoelectric generator, since it converts heat into electric energy. The particles within the generator are gradient and curvature drifting against the electric field. They loose thermal energy since their total energy due to gyration (μB), oscillation along the field lines, and electrostatic potential is constant (the variation of kinetic energy due to perpendicular drift is here assumed to be small).

Solving (1) for $\underline{J}$ and taking the divergence we obtain after some vector manipulation

$$\underline{\nabla}\underline{J}_\perp = -\underline{\nabla}\underline{J}_\| = 2(\underline{B}/B^3)\times\underline{\nabla}B\cdot(\rho\underline{\dot{v}}+\underline{\nabla}p)-(\underline{B}/B^2)\underline{\nabla}\times(\rho\underline{\dot{v}}) \quad (2)$$

where $\underline{\dot{v}} = d\underline{v}/dt$. This MHD-equation accounts for the coupling between the internal generator current $\underline{J}_\perp$ and the Birkeland currents $\underline{J}_\|$. However, it does not explain the coupling in terms of particle flow. Are $\underline{J}_\perp$ and $\underline{J}_\|$ carried by the same or different particles, and what kind of particles?

The MHD dynamo. Rostoker and Boström (1976) have described this in detail. The nature of the dynamo is illustrated in Figure 4. A channel flow in the $-z$-direction in the presence of the magnetic field B_y is retarded by the current

$$J_x = \rho \dot{v}_z/B_y \quad (3)$$

The two terms containing $\underline{\dot{v}}$ in equation (2) represent the divergence of $\underline{J}_\perp$ due to variation along $\underline{J}_\perp$ of B and $\rho\dot{v}$, respectively. In terms of particle drifts the current J_x in (3) is of course due to the inertia term in the particle drift equation

$$\underline{u}_\perp = -(\underline{B}/qB^2)\times(q\underline{E} - \mu\underline{\nabla}B - md\underline{u}/dt) \quad (4)$$

It should be noted that this retarding effect must, for energy reasons, be associated with divergent flow, as in the outflow regions of the potential models illustrated in Figures 1 and 2. In the inflow regions the plasma is accelerated, implying a load (motor) on a generator situated somewhere else. The current J_x of equation (3) is identical with the polarization current discussed by Atkinson (1982).

Note that all retarded particles contribute to J_x in proportion to their mass but independent of

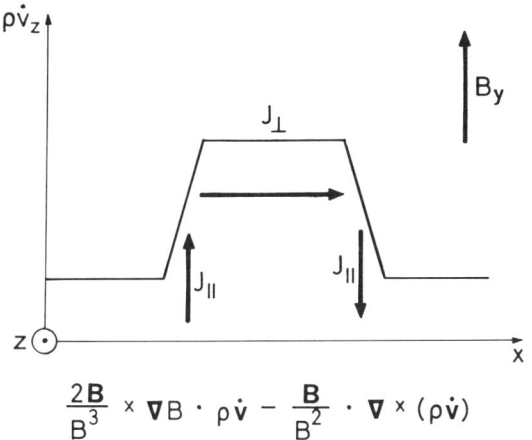

Fig. 4. An MHD-dynamo couples to Birkeland currents in regions with a finite curl of the mass flow (upper part). For energy reasons, the flow must diverge (lower part), requiring a tangential electric field that varies through the Birkeland current sheet.

gyration energy. Note also that both in the inflow and in the outflow regions there must be a tangential electric field that varies in magnitude across the arc. Mathematically that is required when the electric field is curl-free.

The thermo-electric generator. The last term in equation (2) and the term containing ∇p in equation (3) are associated with a thermoelectric generator, since the pressure p is proportional to the temperature. There are three important differences between the two kinds of generators, namely

(i) the thermo-electric generator converts its internal heat, not bulk kinetic energy, into electric energy,
(ii) the current supplied by a thermo-electric generator is carried mainly by energetic (hot) particles, i.e. the particles contribute to $J_\perp$ in proportion to their gyration and field-aligned kinetic energy, independent of their mass,
(iii) the pressure gradient ∇p must have a component perpendicular to ∇B in order to couple to the Birkeland currents, i.e. constant-p-lines must be misaligned with constant-B-lines (cf. Figure 5), unlike the dynamo which works for all directions of $\dot{v}$.

We may note that the magnetospheric convection is sub-sonic, i.e. the heat energy dominates over the kinetic energy. Thus, (i) suggests that thermo-electric generators probably play an important role in driving the magnetospheric currents.

According to (ii) $J_\perp$ is due to hot particles, since they have the largest gradient and curvature drifts. As is well known from equation (4) the former is proportional to the gyration energy and the latter to the parallel particle energy. Both drifts are along constant-B-lines when projected onto the equatorial plane. A consequence of this, expressed by (iii), is that no divergence of $J_\perp$ produced by a thermo-electric generator can result if constant-p and constant-B-lines are colinear. It can be seen that the misalignment shown in Figure 5 is consistent both with the Birkeland current systems 1 and 2 and with the shape of the auroral oval, being most equatorward near midnight. This is a further argument for the importance of thermo-electric generators in the magnetosphere.

Figure 5 implies that the generators consist of high pressure ridges around the earth. These can be identified with the hot plasma sheet.

It is of course possible that the plasma flow along a high pressure ridge is retarded, so that both types of generators operate simultaneously in the same place.

A feature common to the two kinds of generators is that they produce Birkeland currents at boundaries between plasmas with different properties, e.g. different temperatures in the case of the thermo-electric generator. The curl $\underline{v}$, associated with the MHD-dynamo, implies that the fast convecting plasma is injected from a different region of the magnetosphere, presumably from the tail. Many rocket observations have indicated that intense precipitation occurs near boundaries where plasma properties change (see e.g. Edwards et al., 1976; Maynard et al., 1977, and Evans et al., 1977).

Charge conservation. The generators inject charges of one sign into the flux tubes with Birkeland currents, and remove charges of the opposite sign. The parallel potential drop adjusts itself so as to drive exactly that Birkeland current which is required to maintain current continuity.

There are two ways in which the charges injected into a flux tube can be discharged by Birkeland currents, either by pitch-angle scattering of the injected particles into the loss cone, or by setting up a parallel electric field that pushes

THERMO-ELECTRIC GENERATOR

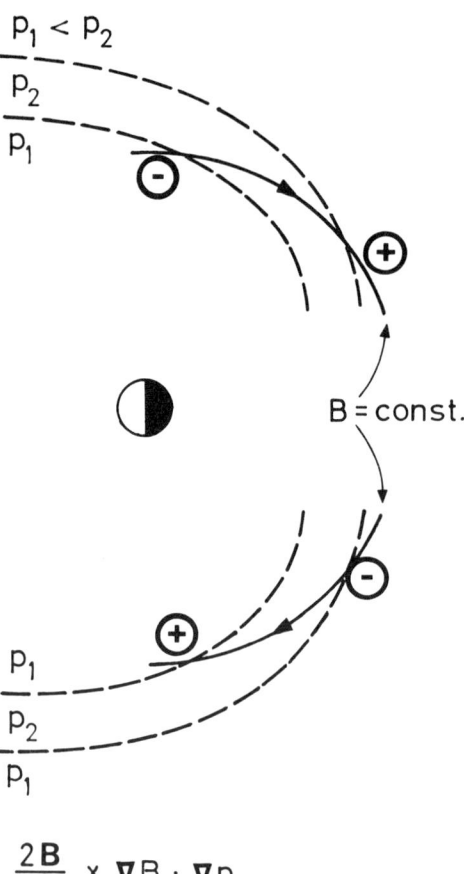

Fig. 5. A thermo-electric magnetospheric generator due to high pressure ridges, as shown, is consistent with the shape of the auroral oval and with the gross Birkeland current configuration. Dashed lines indicate constant pressure, solid lines constant magnetic field, along which particles gradient-drift.

both injected, primarily low energy particles, and ambient particles of the same charge sign down to the ionosphere, or pulls up oppositely charged particles from the ionosphere. Clearly, both processes occur in various proportions. A clear example of the former process is given by Iversen et al. (1983) who on the GEOS-2 satellite found strong correlation between injection of >22 keV electrons, VLF waves with frequencies near the electron gyro frequency, and X-rays detected on balloons magnetically conjugate with the satellite. However, only the latter process will be discussed here.

When the generator is a dynamo, many injected low energy particles may be pushed down by the parallel electric field to the ionosphere. The thermo-electric generator is different, in that it chiefly injects hot particles which are more difficult to precipitate. It is, thus, most likely that a thermo-electric generator will replace ambient low energy particles by hot injected particles. That effect will be less important for an MHD dynamo. A striking example of this effect is provided by Shepherd et al. (1980), who on GEOS-2 found several sudden injections of >30 keV electrons and simultaneous disappearance of 50-500 eV electrons, well correlated with 6300 A emissions observed form the ground at the magnetically conjugate point.

The emissions lasted of the order of an hour which indicates that the <500 eV electrons where stored at some intermediate altitude range between the magnetic mirror and a parallel potential drop, which may have gradually moved downwards, towards the ionosphere. Convection at intermediate altitudes into the region may also have contributed to the long-lasting emissions.

Observations on GEOS-2 have shown several cases of injection of gradient-drifting high energy particles with significant densities (at least approaching one tenth per cm^3), which is impossible unless ambient electrons are pushed down or pulled up along the field lines (see e.g. Kremser et al., 1982, where the injected charge density can be estimated from their Figure 2).

The Distribution Of The Parallel Electric Field

Several mechanisms for production of parallel electric fields have been discussed in the literature. (For a review see Fälthammar, 1978). These will not be discussed here. Only some simple consequences of three different models will be considered and related to observations.

Evans (1974) found good agreement between an observed auroral precipitation spectrum and that of a model in which a magnetic field-aligned potential drop accelerated electrons downwards and reflected the backscattered and secondary population. Crucial features of this model were that all particles from the magnetosphere had gained an energy equal to the potential drop, and that the spectra of the upward and downward fluxes below that energy would be identical. This has been questioned by e.g. Whalen and Daly (1979) who examined spectra from five sounding rocket experiments, where the downward fluxes exceeded the upward fluxes by a factor 2 or more. This fact contradicts the assumption that all magnetospheric particles injected into the acceleration region had traversed the total potential drop. There are two possible ways out of that dilemma, if indeed such a potential drop exists, either that intense wave-particle interaction within or below the acceleration region absorbs some of the precipitation energy, or that the potential distribution is such that some particles can be

Distribution of $E_{\parallel}$

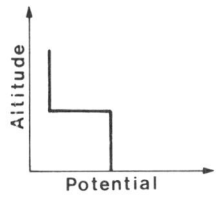

a. One single double layer does not allow for electrons accelerated through part of the potential drop

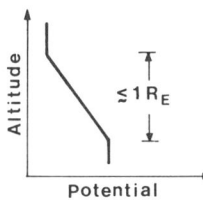

b. Uniform $E_{\parallel}$ over $\leq 1 R_E$ accelerates low energy electrons too quickly to account for arc length or even width

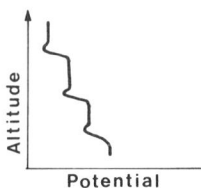

c. Many double layers with soliton like overshoots as seen by S3-3 is consistent with known facts

Low energy electrons injected within the potential drop must be allowed to drift all the way along or into the arc region

Fig. 6. Some parallel electric field models are inconsistent with observations.

injected within the acceleration region. The former process is very unlikely, since it implies an enormous heat input into the ambient particles, which would be heated at a rate of several eV/s, even if the wave-particle interaction were spread out over an altitude range of the order of an earth radius (Block, 1975).

It must be concluded that some electrons must be injected into the flux tube within the acceleration region in such a way that they fall through only part of the potential drop. Hence, the parallel electric field cannot be concentrated in one single double layer, as in Figure 6a, but it must be extended over a large altitude interval. The first guess may then be a uniform or at least monotonic field spread out over, say one earth radius (see Figure 6b). Electrons injected into such a field will very quickly be precipitated and will not have a chance to drift very far into the Birkeland current sheet region. An elementary calculation, assuming that the parallel electric field is proportional to the gradient of the magnetic field, extending over five earth radii, with a total potential drop of 10 kV, deposits an electron in the ionosphere in about four seconds if falling through the entire potential drop, and even faster if it falls through part of it. During this time it can only $\underline{ExB}$-drift about one kilometer if $E = 10$ mV/m in the ionosphere.

It may be concluded that low energy particles injected into such a field and falling through only part of the potential drop cannot reach the central parts of the arc and Birkeland current sheet. This contradicts the observations of Whalen and Daly (1979).

If, however, the acceleration of newly injected cold electrons can be delayed for some time, during which they can drift sufficiently far into the arc region, the model need not be in conflict with observations. The multitude of weak double layers with soliton-like overshoots, recently observed on the S3-3 satellite (Temerin et al., 1982; Hudson et al., 1983), may provide this delay, since cold electrons can be trapped for a while above the potential minima (cf. Figure 6c). A model with multiple weak double layers in series above each other along auroral field lines but without overshoots was proposed by Block (1972).

An alternative explanation of persistent low energy precipitation could be provided by a dynamic model in which an extended uniform field region like that in Figure 6b moves downward and then depletes the plasma at progressively lower altitudes.

Burch et al. (1979) report observations in the polar cap of field-aligned low energy electron beams which they interpret in terms of thermal electrons originating from within the acceleration region.

A spectral hardening of electron precipitation towards the center of the current sheet or arc is often reported (e.g. Edwards et al., 1976). That is consistent with injection sideways due to a tangential electric field with some delay mechanism to account for the still finite low energy precipitation in the center.

Summary

The present discussion may be summarized with the aid of Figure 7. The following main points have been made:

1. The magnetospheric generator can move mainly high energy particles well outside the loss cone against the electric field into the U- or V-shaped potential region associated with Birkeland currents. These particles (energetic 90° particles in the upper part of Figure 7) provide the space charge required for maintaining the parallel electric field and also the closure of the Birkeland current system.

2. The Birkeland currents themselves are mainly carried by colder particles drifting above and within the acceleration altitude interval (middle and lower part of Figure 7) along equipotential surfaces into the precipitation region, either strictly along the arc or with a transverse component implying a weak tangential electric

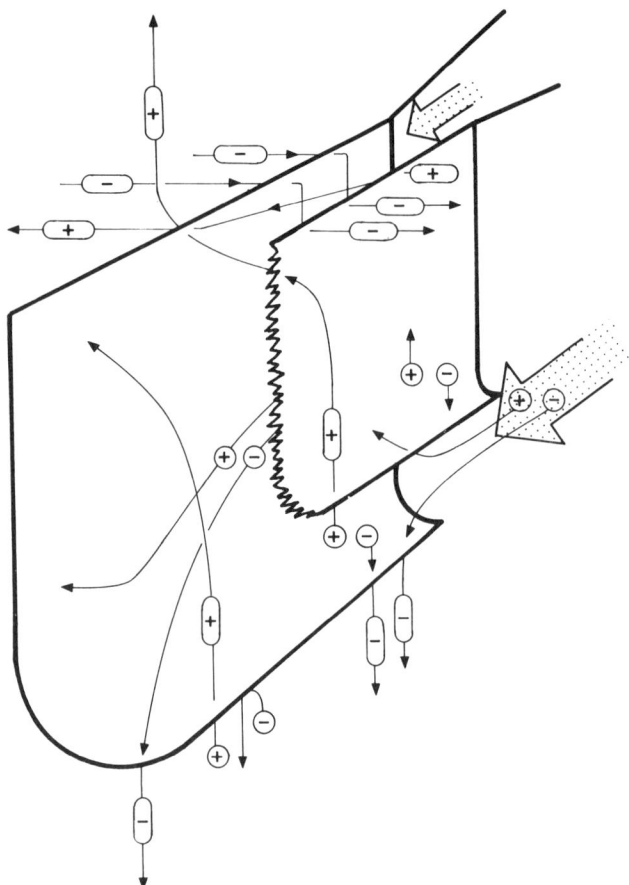

Fig. 7. Outline of particle orbits in one half of the potential U-structure (cf. Fig.2) of a Birkeland current sheet driven by a thermo-electric generator. The dot-filled arrows show flow of cold plasma. Circles with + and - represent low energy particles with corresponding charge. Horizontally and vertically elongated closed loops with + and - indicate particles with high perpendicular and parallel energy, respectively. Perpendicular currents are mainly carried by hot, gradient and curvature drifting particles, and Birkeland currents by originally cold particles. Hence, ambient cold plasma is gradually replaced by hot drifting populations, unless inflow of cold plasma is balancing the loss.

field. A certain depletion of the auroral flux tubes in the topside ionosphere is also taking place (cold particles in lower part of Figure 7) as predicted by Block and Fälthammar (1968), since ambient electrons are pushed down and ions pulled up.

3. In especially multiple parallel arcs the high altitude plasma convection must be closed within a loop of relatively limited extent (< 1000 km). The supply of low energy electrons must then be from the ionosphere in parts of the loop adjacent to the upward Birkeland current region, as suggested by Block and Fälthammar (1969).

4. The total parallel potential drop can neither be concentrated in a single strong double layer, nor can it be monotonically distributed over a large altitude range of say about one earth-radius. Both of these models are inconsistent with observed precipitation spectra. A delay mechanism for the acceleration of cold electrons must exist, enabling some of them to drift into the center of the precipitation region. The solitons and multiple double layers observed on S3-3 (Mozer et al., 1982) seem to provide such a delay.

5. There are two kinds of generators in the magnetosphere, namely MHD dynamos and thermo-electric generators. Both of these couple to Birkeland currents at boundaries between plasmas of different properties, e.g. temperatures and/or densities. Many rocket observations have indicated such boundaries associated with intense precipitation.

References

Atkinson, G., Review of auroral currents and auroral arcs, J. Geomagn. Geoelectr., 30, 435-447, 1978.

Atkinson, G., Duality of the magnetic-flux-tube and electric-current descriptions of magnetospheric plasma and energy flow, Rev. Geophys. Space Phys., 19, 617, 1981.

Atkinson, G., Inverted V's and/or discrete arcs: A three-dimensional phenomenon at boundaries between magnetic flux tubes, J. Geophys. Res., 87, 1528-1534, 1982.

Block, L.P., Some Phenomena in the Polar Ionosphere, 9th Int. Conf. on Phenomena in Ionized Gases, Bucharest, Rumania, Report 69-30, Dept of Electron and Plasma Physics, Royal Institute of Technology, Stockholm, 1969.

Block, L.P., Potential Double Layers in the Ionosphere, Cosmic Electrodynamics, 3, 349-376.

Block, L.P., Double Layers in Hultqvist and Stenflo (eds), Physics of the Hot Plasma in the Magnetosphere, Plenum Pub. Co., New York, p. 229-249, 1975.

Block, L.P., Double Layers in the Laboratory and Above the Aurora, in Physics of Auroral Arc Formation, Eds. S.-I. Akasofu and J.R. Kan, AGU Geophysical Monograph 25, 218-225, 1981.

Block, L.P. and C.-G. Fälthammar, Effects of field-aligned current on the structure of the ionosphere, J. Geophys. Res., 73, 4807-4812, 1968.

Block, L.P. and C.-G. Fälthammar, Field-aligned currents and auroral precipitation, in Atmospheric Emissions, eds. B.M. McCormac and A. Omholt, Reinhold Book Corp, 285-292, 1969.

Burch, J.L., S.A. Fields and R.A. Heelis, Polar cap electron acceleration regions, J. Geophys. Res, 84, 5863-5874, 1979.

Burke, W.J., Electric fields and currents observed by S3-2 in the vicinity of discrete arcs, This volume, 1983.

Burke, W.J., M.C. Kelley, R.C. Sagalyn, M. Smiddy and S.T. Lai, Polar cap electric field structures with a northward interplanetary magnetic field, Geophys. Res. Letters, 6, 21-24, 1979.

Carlqvist, P., and R. Boström, Space charge regions above the aurora, J. Geophys. Res., 75, 7140-7146, 1970.

de la Beaujardière, O. and R. Vondrak, Chatanika radar observations of the electrostatic potential distribution of an auroral arc, J. Geophys. Res., 87, 797-810, 1982.

Edwards, T., D.A. Bryant, M.J. Smith, U. Fahleson, C.-G. Fälthammar and A. Pedersen, Electric fields and energetic particle precipitation in an auroral arc, in Magnetospheric Particles and Fields, ed. B.M. McCormac, Reidel Publ. Co., 285-289, 1976.

Evans, D.S., Precipitating electron fluxes formed by a magnetic field aligned potential difference, J. Geophys. Res., 79, 2853-2858, 1974.

Evans, D.S., N.C. Maynard, J. Troim, T. Jacobsen, and A. Egeland, Auroral vector electric field and particle comparisons, 2. Electrodynamics of an arc, J. Geophys. Res., 82, 2235-2249, 1977.

Fälthammar, C.-G., Generation mechanisms for magnetic field-aligned electric fields in the magnetosphere, J. Geomag. Geoelectr., 30, 419-434, 1978.

Frank, L.A., and D.A. Gurnett, Distributions of plasmas and electric fields over the auroral zones and polar caps, J. Geophys. Res., 76, 6829-6846, 1971.

Frank, L.A., J.D. Craven, J.L. Burch and J.D. Winningham, Polar views of the earth's aurora with Dynamic Explorer, Geophys. Res. Letters, 9, 1001-1004, 1982.

Gurnett, D.A., Electric field and plasma observations in the magnetosphere, in Critical Problems of Magnetospheric Physics, ed. E.R. Dyer, IUCSTP Secretariat, c/o National Academy of Sciences, Washinton, D.C. 123-138, 1972.

Heelis, R.A., W.B. Hanson and J.L. Burch, Ion convection velocity reversals in the dayside cleft, J. Geophys. Res., 81, 3803-3809, 1976.

Heelis, R.A., W.B. Hansen and J.L. Burch, AE-C observations of electric fields around auroral arcs, in Physics of Auroral Arc Formation, Eds. S.-I. Akasofu and J.R. Kan, AGU Geophysical Monograph 25, 154-163, 1981.

Hudson, M.K., W. Lotko, I. Roth and E. Witt, Solitary waves and double layers on auroral field lines, J. Geophys. Res., 88, 916-926, 1983.

Iversen, I.B., L.P. Block, K. Brönstad, U. Fahleson, R. Grard, G. Haerendel, H. Junginger, A. Korth, G. Kremser, M.M. Madsen, J. Niskanen, W. Riedler, P. Tanskanen, K. Torkar, and S. Ullaland, Simultaneous observations of a pulsation event with balloons and a geostationary satellite on August 12, 1978, submitted for publication in J. Geophys. Res., 1983.

Kan, J.R. and S.-I. Akasofu, A model of the auroral electric field, J. Geophys. Res., 84, 507-512, 1979.

Kremser, G., J. Bjordal, L.P. Block, K. Brönstad, M. Håvåg, I.B. Iversen, J. Kangas, A. Korth, M.M. Madesn, J. Niskanen, W. Riedler, J. Stadsnes, P. Tanskanen, K. Torkar, and S. Ullaland, Co-ordinated balloon-satellite observations of energetic particles at the onset of a magnetospheric substorm, J. Geophys. Res., 87, 4445-4453, 1982.

Lennartsson, O.W., Ionospheric electric field and current distribution associated with high altitude electric field inhomogeneities, Planetary Space Sci., 21, 2089-2112, 1973.

Lennartsson, O.W., On the role of magnetic mirroring in auroral phenomena, Astrophys. Space Sci., 51, 461-495, 1977.

Lennartsson, W., on the consequences of the interaction between the auroral plasma and the geomagnetic field, Planetary Space Sci., 28, 135-147, 1980.

Lyons, L.R. Generation of large-scale regions of auroral currents, electric, potentials, and precipitation by the divergence of the convection electric field, J. Geophys. Res., 85, 17-24, 1980.

Lyons, L.R., Discrete aurora as the direct result of an inferred, high-altitude generating potential distribution, J. Geophys. Res., 86, 1-8, 1981.

Marklund, G., Auroral arc classification scheme based on the observed arc associated electric field pattern, accepted for publication in Planetary and Space Sci., 1983.

Maynard, N.C., D.S. Evans, B. Maehlum, and A. Egeland, Auroral vector electric field and particle comparisons, 1. Premidnight convection topology, J. Geophys. Res., 82, 2227-2234, 1977.

Mozer, F.S., The low altitude electric field structure of discrete auroral arcs, in Physics of Auroral Arc Formation, eds. S.I. Akasofu and J.R. Kan, AGU Geophysical Monograph 25, 136-142, 1981.

Mozer, F.S., C.W. Carlson, M.K. Hudson, R.B. Torber, B. Parady, J. Yatteu, and M.C. Kelley, Observations of paired electrostatic shocks in the polar magnetosphere, Phys. Rev. Lett., 38, 292-295, 1977.

Mozer, F.S., C.A. Cattell, M. Temerin, R.B. Torbert, S. von Glinski, M. Woldorff, and J. Wygant, The DC and AC electric field, plasma density, plasma temperature, and field-aligned current experiments on the S3-3 satellite, J. Geophys. Res., 84, 5875-5884, 1979.

Mozer, F.S., C.A. Cattell, M.K. Hudson, R.L. Lysak, M. Temerin, and R.B. Torbert, Satellite measurements and theories of low altitude auroral particle acceleration, Space Sci. Rev., 27, 155-213, 1980.

Rostoker, G. and R. Boström, A mechanism for

driving the gross Birkeland current configuration in the auroral oval, J. Geophys. Res., 81, 235-244, 1976.

Shepherd, G.G., R. Boström, H. Derblom, C.-G. Fälthammar, R. Gendrin, K. Kaila, A. Korth, A. Pedersen, R. Pellinen, G. Wrenn, Plasma and field signatures of poleward propagating auroral precipitation observed at the foot of the GEOS-2 field line, J. Geophys. Res., 85, 4587-4601, 1980.

Swift, D.W., On the formation of auroral arcs and acceleration of auroral electrons, J. Geophys. Res., 80, 2096-2108, 1975.

Swift, D.W., An equipotential model for auroral arcs:2. Numerical Solutions, J. Geophys. Res., 81, 3935, 1976.

Temerin, M.A., Plasma waves on auroral field lines, in Physics of Auroral Arc Formation, eds. S.-I. Akasofu and J.R. Kan, AGU Geophysical Monograph 25, 351-358, 1981.

Temerin, M., K. Cerny, W. Lotko, and F.S. Mozer, Observations of double layers and solitary waves in the auroral plasma, Phys. Rev. Letters, 48, 1175-1179, 1982.

Whalen, B.A. and P.W. Daly, Do field-aligned auroral particle distributions imply acceleration by quasi-static parallel electric fields?, J. Geophys. Res., 84, 4175-4182, 1979.

THE ROLE OF CURRENTS IN PLASMA REDISTRIBUTION

G. Atkinson

Herzberg Institute of Astrophysics, National Research Council of Canada, 100 Sussex Drive, Ottawa, Ontario K1A 0R6, Canada

Abstract. Cross-field currents in the outer magnetosphere, both polarization and field-gradient drift are carried by protons and positive ions because of their greater mass and energy. Field-aligned currents are believed to be carried by electrons because of their greater mobility. If a given current-system contains both, then there is a plasma density build-up on flux tubes containing downward field-aligned current and a plasma decrease on flux tubes containing upward current. Thus field-aligned currents are the signature of a redistribution of plasma in the magnetosphere. Three particular examples are studied: (1) the familiar example of a drifting plasma blob, (2) Region II currents and the inner edge of plasma penetration, (3) the dusk Region I and the electrojet closure currents. These last are sufficiently intense to indicate significant plasma redistribution both in and in between substorms. The result is a revision of models of convection in from the tail and of the magnetospheric properties. The upward currents form a single continuous current sheet separating strongly inflated flux tubes from more dipolar flux tubes. The substorm fits naturally into the model and plasma depletion on flux tubes carrying upward field aligned currents is responsible for triggering, expansion and westward development of the substorm as well as between substorm flows.

Introduction

There are many circumstances in space plasmas in which the current is carried by one charge carrier in part of the current circuit and a different charge carrier in another part of the circuit. In general this leads to enhancements and depletions of the species at locations where the change in carrier occurs. In this paper we will be concerned in particular with the case where currents along magnetic field lines are carried by electrons and currents across field lines by positive ions (usually protons).

These effects have been studied for the ionospheric feet of field lines where horizontal currents due to ion drifts in the ionosphere close by field-aligned electron drifts (Block and Falthammer, 1968). On field lines where downward field aligned currents enter the ionosphere a plasma depletion in the ionosphere is expected since both the upward drifting electrons and the horizontally drifting ions are moving away from the ionospheric foot of the field line. This effect has been used to explain observations of depletions in the topside ionosphere by Klumpar (1970) and by Rich et al. (1980).

A similar effect is expected in the outer magnetosphere where cross-field currents are believed to be usually carried by positive ions and protons rather than by electrons. This applies both to polarization currents (inertial effects) since the mass of the proton is much greater than the electrons, and to currents associated with plasma pressure since the bulk of the energy of the plasma in the outer magnetosphere resides with the protons rather than the electrons. Thus if field-aligned currents are carried by electrons, a plasma density enhancement would be expected on flux tubes containing a downward field-aligned current since current continuity requires that both electrons and protons flow onto the flux tubes. Conversely a plasma density depletion is expected on flux tubes containing upward field-aligned current. Thus near-earth observations of field-aligned currents gives information on outer magnetospheric mass motions. The density changes will be seen as temporal ($\partial\rho/\partial t$) in a reference frame moving with the convecting flux tube, or as temporal plus spatial ($\partial\rho/\partial t + \underline{v}\cdot\nabla\rho$) in a stationary reference frame in which the convection velocity is $\underline{v}$.

The above effect is not new, Atkinson (1967b) and Unti and Atkinson (1968) suggested it as an explanation of the westward travelling surge of substorms. In particular it was proposed that the plasma reduction on flux tubes associated with the upward field-aligned currents allowed reconnection of tail field lines and/or collapse of inflated field lines. Alfven (1981) discussed the plasma transport-current relationship for astrophysical applications in general.

In the next section of this paper, the implications of magnetospheric field-aligned currents on mass transfer in the magnetosphere will be explored for three cases: (i) a plasma blob consisting of energetic protons and cold electrons subject to magnetic field gradient drifts. This very familiar case is studied as an introduction to the ideas; (ii) the region II currents. This is a familiar model but brings out some new aspects; (iii) the dusk region I currents and the Harang Discontinuity. This requires some rethinking of nightside processes.

Drifting Plasma Blob

This is a text book situation that has been around for many years, and is often discussed in the form of localized enhancements and instabilities of the ring current (e.g. Fejer, 1961). It is illustrated in figure 1. The plasma blob consists of energetic westward drifting protons and cold electrons. At the eastern end of the blob, plasma is depleted on a particular flux tube as electrons drift to the ionosphere (opposite to the current direction) and protons drift westward out of the flux tube. At the western end electrons drift up from the ionosphere and protons westward onto flux tubes thus increasing the plasma density. Thus the field aligned currents at ionospheric heights are the signature of plasma depletions and enhancements in the outer magnetosphere. The current is of course upwards at the eastern end, westward in the blob, downward at the western end and closes in the ionosphere. This example is included because it is well understood and simple. The physics is the same in the more complicated situations to follow.

Region II Currents

The model used in this discussion is one in which region II currents are due to Alfven-layer shielding, consistent with the ideas of Karlson (1963) and Block (1969). This process was later included in self-consistent models of the magnetospheric flow by Vasyliunas (1970, 1972) computer simulations thereof; Jaggi and Wolf, 1973; Wolf, 1974; Hamel and Wolf, 1976; and the narrow-channel approximation (Atkinson, 1982).

In all these self-consistent models, the bulk of the plasma is on magnetic flux tubes with feet in auroral latitudes; that is more or less in the auroral oval. As the plasma convects and field-gradient drifts in local time its L-value must vary so that it stays on flux tubes within the

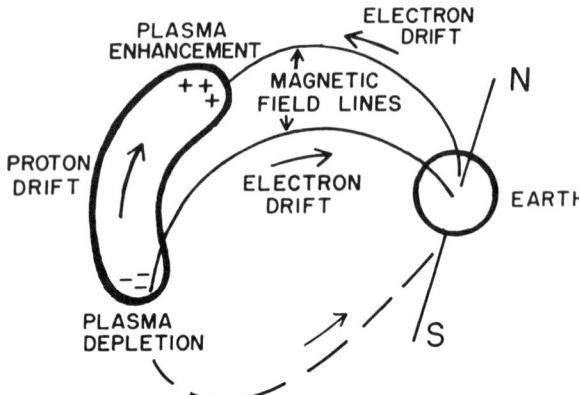

Fig. 1. Current continuity and plasma depletion and enhancement for a drifting plasma blob of energetic protons and cold electrons.

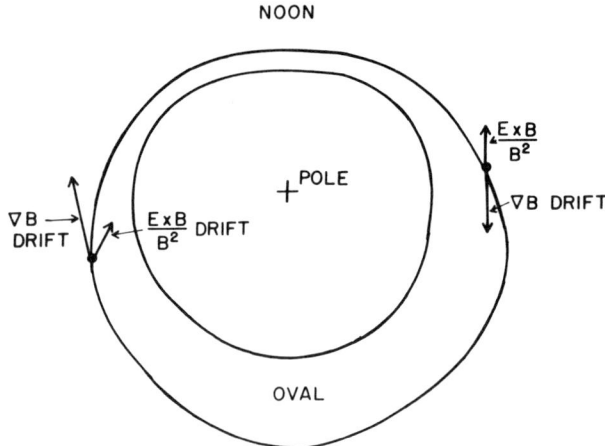

Fig. 2. View of the oval showing the mapping along field lines of magnetic field gradient drift and electric field convection.

auroral oval. The magnetic-field gradient drift maps along field lines to the ionosphere as circular paths of nearly constant geomagnetic latitude, whereas the oval is at considerably higher geomagnetic latitudes, on the dayside. Hence if we consider the lower latitude boundary of the oval, energetic protons subject to magnetic field gradient drift have a drift component (as mapped to the ionosphere) normal to and out of the dusk side oval boundary and poleward into the oval on the dawnside (see figure 2). This normal component of field-gradient drift must be opposed by an equal and opposite component of electric field convection if the plasma is to stay within the oval.

The normal component of the electric field convective velocity at the southern oval boundary implies that there must be a mid-latitude convection associated with this model in the steady state as was pointed out in Atkinson (1982). It is suggested that flows observed in the nightside mid-latitude trough (Evans et al, 1982; Rich et al, 1980) may result from this effect.

The above model is totally consistent with the conclusion that region II field-aligned currents indicate plasma depletion or enhancement in the frame of the convection. On the dusk side, as a flux tube convects poleward into the oval, protons drift across field lines onto it and electrons are injected from the ionosphere consistent with downward currents and density enhancement. The converse occurs at the dawnside southern oval boundary.

Dusk Region I and Harang Discontinuity Currents

On the nightside where there are three field-aligned current sheets, the middle one will be referred to as the Harang discontinuity current for the purposes of this paper. Region I currents probably are part of a circuit involving magnetopause and magnetotail plasma sheet currents. They would not be expected in a non-convecting magnetosphere since the electric fields and ionospheric currents would be zero. Hence, as pointed out in Atkinson (1978), the current system is associated with distortions in the shape of the magnetosphere from the shape of an equilibrium non-convecting magnetosphere. These distortions are the lowering of the cusp on the dayside and the existence of more closed flux on the nightside. (An identical description is that the foot of the polar cap magnetic flux is more toward the sun than in a non-convecting magnetosphere). The excess dusk to dawn currents corresponding to these shape changes flow via region I currents to the ionosphere.

Figure 3 is a sketch of the type of model usually visualized and is for illustrative purposes only. We shall shortly replace it by one which includes the effects of plasma depletion by field-aligned currents. The shaded area represents a boundary between dipolar field lines and fully tail-like field lines. The along-tail thickness of this boundary is totally unknown and should not be judged from the diagram. In a model with no convection, tail currents flow right across the tail as shown in figure 3a. In a model with convection, the increase in the amount of closed flux on the nightside corresponds to either the shaded region becoming thicker in the along-tail direction, or to it moving down the tail. In any case, with more closed flux tubes on the nightside less current flows across the tail and the excess is diverted to the ionosphere as illustrated in figure 3b. Again it is emphasized that the figure is illustrative and that the discussion to follow applies quite generally.

If region I currents connect to tail currents, as indicated in figure 3b, then the same processes of mass depletion and enhancement should be at work. In particular in association with the dusk side region I current, energetic protons are drifting to the magnetopause and electrons to the ionosphere thereby depleting the plasma on flux tubes. The author hypothesizes that <u>this plasma depletion is the major process by which plasma is removed from inflated tail-like flux tubes allowing them to collapse to a dipole-like form.</u> The remainder of this paper will be spent in showing how this process produces the observed properties of magnetospheric flow and also how the substorm fits into the picture.

(a) <u>Cross-oval potential.</u> The potential can be predicted in order of magnitude from the size of the currents. For northern plus southern hemispheres the current, I, associated with region I currents is typically 2 to 6 million amperes (Potemra et al, 1979). Thus the rate of removal of electron-proton pairs by the process is $I/e = 1$ to 4×10^{25} pairs per second. If the length of flux tube involved is 10^8 m and the electron den-

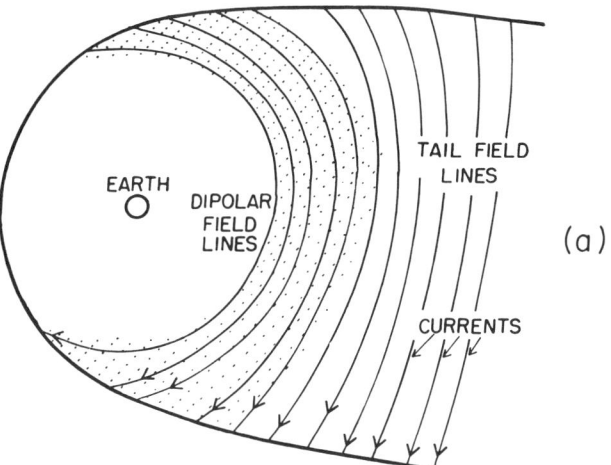

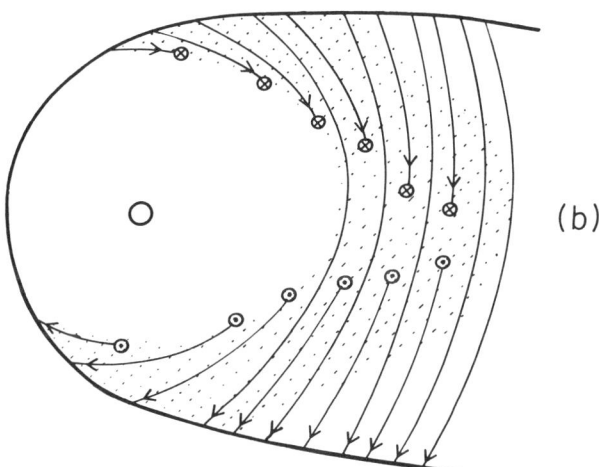

Fig. 3. Equatorial section of a reasonable model for the magnetosphere showing currents for (a) a non-convecting magnetosphere, (b) a convecting magnetosphere. The dotted region indicates transition between tail-like and dipolar field lines. In b, current flow is field-aligned to the ionosphere (arrow tails), away from the ionosphere (arrow heads).

sity is $.3$ cm^{-3}, this represents a flux tube with a cross sectional area of $.3$ to 1.3×10^{12} m^2. For a field strength of 30 nT, the rate at which flux tubes are emptied is 10 to 40 kV. This is the rate at which flux would collapse from tail-like to dipolar and is typical of the cross-oval potential.

(b) <u>Flow model.</u> Injection of flux on the evening side is not consistent with the usual flow

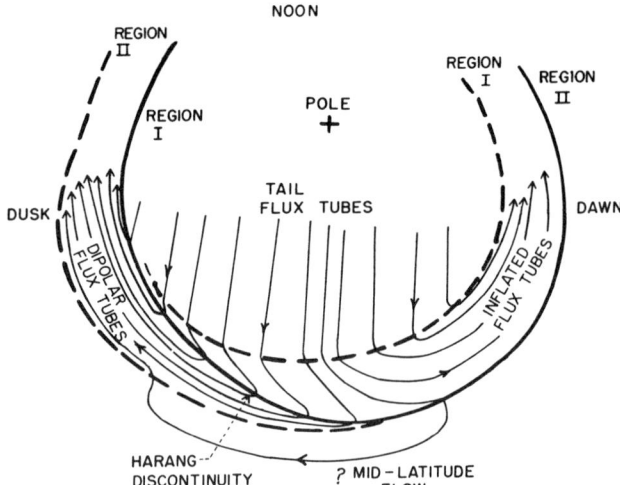

Fig. 4. Polar cap view showing the relationship of currents to the magnetospheric flux-tube topology and flow.

model. Hence we must look for a flow model for the nightside which is consistent with both this injection and with other magnetospheric limitations. A model is illustrated in figure 4. It is surprisingly similar to previous models. The major difference is in the nature of the flux tubes in the various regions and the interpretation of the boundary current sheets. We note the following points:

1. The upward currents (Region I dusk, Harang discontinuity and region II dawn) form a continuous sheet separating more or less dipolar and greatly inflated plasma sheet flux tubes.

2. The dawn-midnight region I downward current separates inflated flux tubes from tail flux tubes, defined in turn as sunward and antisunward convecting.

3. The dusk-midnight region II downward current is the Alfven layer limit of penetration of the energetic plasma that remains after passage through the upward current sheet.

4. Flux tubes between region I and II currents are dipolar at dusk, but inflated plasma sheet at dawn.

5. A correct convective flow must satisfy both outer magnetospheric and ionospheric current continuity. We have already described closure in the outer magnetosphere. The flow reversal at the Harang Discontinuity allows Pedersen current to feed the upward current. The direction change of the flow at the region I downward current is consistent with the existence of a steep ionospheric conductivity gradient. The westward-electrojet Hall current closes with the southward flowing Pedersen current, which in turn feeds the Harang Discontinuity current.

6. The current closure in the outer magnetosphere is still somewhat speculative. A reasonable model is shown in the figure 5 equatorial section of the magnetosphere. Dots and crosses in circles indicate field aligned current from and to the ionosphere respectively. The bifurcating current flow from the Harang Discontinuity is consistent with proton drift in dipolar fieldlines on one side and inflated tail-like field lines on the other. The current on the dipolar side is carried by those protons which are convected through the current sheet. All field-aligned current sheets should be thicker than indicated in the diagrams with dimensions of one to ten earth radii.

Features which support this model are:

1. The current system is very similar to the empirical one of Potemra et al (1979).

2. The convection pattern is very similar to the empirical ones of Heppner (1977).

3. The Harang Discontinuity separates regions of diffuse and discrete arcs (Nielsen and Greenwald, 1979) consistent with it separating dipolar and inflated field lines.

4. The latitude of the downward currents is very similar at dawn and dusk (Potemra et al, 1979) as might be expected if it is a boundary of inflated flux tubes.

5. The model agrees with the Crooker and Siscoe (1981) conclusion that a substantial part of the electrojet currents close via region I type currents and not by partial ring currents.

6. The model is consistent with the wide band of discrete auroras stretching around the morning side in active times, but only a single arc or very few arcs on the evening side (Akasofu, 1976).

7. The model indicates a spiral boundary for the inner edge of the plasma sheet consistent with observation (see, for example, figure 1b in Atkinson, 1972, which is a composite of diagrams

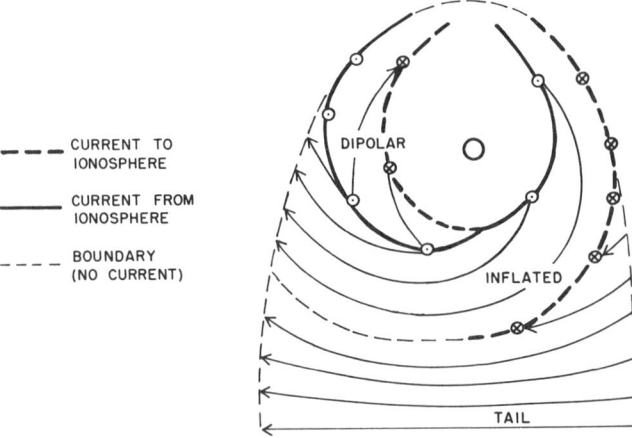

Fig. 5. Equatorial section of the magnetosphere showing current closure. All the boundaries and current sheets are much thicker and more diffuse than indicated.

by Vasyliunas, 1968, and Frank, 1971). The same boundary should also be the spiral injection boundary of Mauk and McIlwain, 1974.

8. Magnetospheric substorms fit naturally into the model as will now be discussed at some length.

(c) <u>The Substorm.</u> Several features of the substorm are reproduced by the model on the single assumption that the upward currents are increased locally in the Harang discontinuity section of the upward currents. These include:

1. The initial brightening occurs in the southernmost arcs (Akasofu, 1964).
2. The initial brightening usually occurs on an arc within the Harang discontinuity (Nielson and Greenwald, 1979).
3. The northward motion of this arc in the expansive phase is accounted for (Akasofu, 1964). In this model the enhanced current increases the plasma depletion rate in the outer magnetosphere and flux tubes collapse more rapidly than they can convect away. As a result the boundary moves northward to form the bulge.
4. The west edge of the bulge moves at the speed of the drifting protons producing the westward travelling surge.

The above is consistent with many of the features of the classical substorm model including collapse of flux tubes, current diversion to the ionosphere, convective flow and the westward travelling surge as initially proposed in Atkinson (1966, 1967a, 1967b, 1967c) and Unti and Atkinson (1968). Features of the classical model that are not intrinsically part of this model, but are not inconsistent with it, are the reconnection in the tail and the existence of a growth phase. Much further work has been done on this classical model (see the review by McPherron, 1979).

There has been considerable discussion in the last few years of the "triggering mechanism" of magnetospheric substorms; that is the instability responsible for their onset, or what causes the initiation and growth of current diversion? The instability that best fits the flow model developed in this paper is a coupled ionosphere-magnetosphere instability. The feedback loop in this instability is illustrated in figure 6. An enhancement of ionospheric conductivity increases the currents. The increase in field aligned currents and consequent collapse of flux tubes enhances the electron precipitation in both discrete and diffuse events. The precipitation increases the ionospheric conductivity thereby closing the feedback loop. This instability was proposed in Atkinson (1972, 1979). It is totally consistent with the magnetospheric model discussed in this paper and is particularly appealing because a single mechanism (plasma depletion by currents) explains (i) the triggering mechanism, (ii) the northward expansion, (iii) the westward travelling surge and (iv) the quiet time flow.

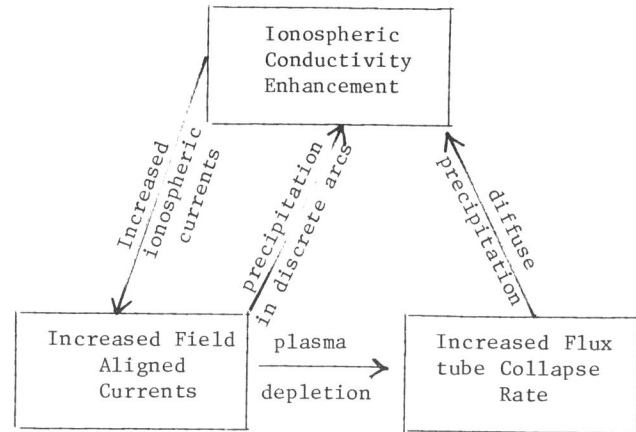

Fig. 6. The feedback system of a coupled ionosphere-magnetosphere instability.

Summary

1. If field-aligned currents are carried by electrons and cross-field currents by protons, plasma depletions occur in the magnetosphere on flux tubes carrying upward current and enhancements on flux tubes carrying downward current.
2. A drifting plasma blob undergoes plasma enhancement at the western end and depletion at the eastern end by this mechanism.
3. Applying this to steady-state region II currents leads to the conclusion that there must be a component of the convective flow normal to the current sheets. As flux tubes flow into or out of the auroral oval through the region II currents the plasma is enhanced or depleted by the mechanism in I above thereby maintaining a stationary plasma distribution. This produces a flow at sub-auroral latitudes.
4. The process is applied to upward region I and Harang discontinuity currents. It is proposed that the resulting plasma depletion is the major process in the magnetosphere by which plasma-sheet flux tubes (inflated into a tail-like shape) are able to collapse to a dipolar form. Appropriate cross-oval potentials are predicted from typical currents.
5. The resulting model requires some changes in models of the magnetosphere. The upward currents (region I evening, Harang, region II morning) form a continuous system separating inflated flux tubes from dipolar. Morning region I downward current separates tail from plasma sheet and evening region II downward current is an energetic plasma boundary.
6. The usual nightside convective flow pattern satisfies current continuity in the ionosphere (allowing for the high conductivity of the oval and the electrojets) and feeds current to the upward current system which closes with both

cross-tail currents and the partial-ring-current type of current.

7. The model explains a number of observations that do not fit existing models.

8. The substorm fits naturally into the model, particularly if it is assumed to be due to a coupled ionosphere-magnetosphere instability. The triggering, expansion and westward development of the substorm and the between substorm flow are all dominated by the single mechanism in 1 above, namely plasma depletion in flux tubes carrying upward current.

References

Akasofu, S.-I., The development of the auroral substorm, Planetary Space Sci., 12, 273, 1964.

Akasofu, S.-I., Recent progress in studies of DMSP auroral photographs, Space Sci. Rev., 19, 169-215, 1976.

Alfven, H., Cosmic Plasma, Reidel Publ. Co., Dordrecht, Holland, 1981.

Atkinson, G., A theory of polar substorms, J. Geophys. Res., 71, 5157-5164, 1966.

Atkinson, G., Polar magnetic substorms, J. Geophys. Res. 72, 1491-1494, 1967a.

Atkinson, G., An approximate flow equation for geomagnetic flux tubes and its application to polar substorms, J. Geophys. Res. 72, 5373-5382, 1967b.

Atkinson, G., The current system of geomagnetic bays, J. Geophys. Res., 6063-6067, 1967c.

Atkinson, G., Magnetospheric flows and substorms, in Magnetosphere-Ionosphere Interactions, edited by K. Folkestad, Universitetsferlaget, Oslo, pp. 203-216, 1972.

Atkinson, G., Energy flow and closure of current systems in the magnetosphere, J. Geophys. Res. 83, 1089-1103, 1978.

Atkinson, G., The expansive phase of the magnetospheric substorm, in Dynamics of the magnetosphere, edited by S.-I. Akasofu, D. Reidel Publishing Company, 1979.

Atkinson, G., The narrow-channel approximation for magnetospheric flows, J. Geophys. Res., 87, 7489-7503, 1982.

Block, L.P., On the distribution of electric fields in the magnetosphere, J. Geophys. Res., 71, 855, 1966.

Block, L.P., and C.G. Falthammer, Effects of field-aligned currents on the structure of the ionosphere, J. Geophys. Res., 73, 4807-4812, 1968.

Crooker, N.U., and G.L. Siscoe, Birkeland currents as the cause of the low-latitude assymetric disturbance field, J. Geophys. Res., 86, 11201-11210, 1981.

Evans, J.V., J.M. Holt, W.L. Oliver, R.H. Wand, Studies of the mid-latitude and high latitude ionosphere from Millstone Hill during solar maximum conditions, MIT, Haystack Observatory, Westford, Massachusetts, 1982.

Fejer, J.A., The effects of energetic trapped particles on magnetospheric motions and ionospheric currents, Can. J. Phys., 39, 1409, 1961.

Feldstein, Y.I., and G.V. Starkov, Dynamics of auroral belt and polar geomagnetic disturbances, Planet. Space Sci., 15, 209-229, 1967.

Frank, L.A., Relationship of the plasma sheet, Ring current, trapping boundary, and plasma pause near the magnetic equator and local midnight, J. Geophys. Res., 76, 2265, 1971.

Harel, M., and R.A. Wolf, Convection, in Physics of Solar-Planetary Environments, Vol. II, ed. D.J. Williams, Amer. Geophys. Un., Washington, D.C. 1976.

Heppner, J.P., Empirical models of high-latitude electric fields, J. Geophys. Res., 82, 1115, 1977.

Jaggi, R.K. and R.A. Wolf, Self consistent calculation of the motion of sheets of ions in the magnetosphere, J. Geophys. Res., 78, 2842, 1973.

Karlson, E.T., Streaming of plasma through a magnetic dipole field, Phys. Fluids, 6, 798, 1963.

Klumpar, D.M., Relationships between particle distributions and magnetic field perturbations associated with field-aligned currents, J. Geophys. Res., 84, 6524-6532, 1979.

Mauk, B.H. and C.E. McIlwain, Correlation of Kp with the substorm-injected plasma boundary, J. Geophys. Res., 79, 3193, 1974.

McPherron, Magnetospheric substorms, Rev. Geophys. and Space Phys., 17, 1979.

Nielson, E. and R.A. Greenwald, Electron flow and visual aurora at the Harang discontinuity, J. Geophys. Res., 84, 4189-4200, 1979.

Potemra, T.A., T. Iijima, and N. Saflekos, Large-scale characteristics of Birkeland currents, in Dynamics of the Magnetosphere, edited by S.-I. Akasofu, D. Reidel Publ. Co., Dordrecht, Holland, 1979.

Rich, F.J., W.J. Burke, M.C. Kelley, and M. Smiddy, Observations of field-aligned currents in association with strong convection electric fields at subauroral latitudes, J. Geophys. Res., 85, 2335, 1980.

Unti, T. and G. Atkinson, Two-Dimensional Chapman-Ferraro problem with neutral sheet, J. Geophys. Res., 73, 7319-7327, 1968.

Vasyliunas, V.M., A survey of low-energy electrons in the evening sector of the magnetosphere with OGO1 and OGO3, J. Geophys. Res., 73, 2839, 1968.

Vasyliunas, V.M., Mathematical models of magnetospheric convection and its coupling to the ionosphere, in Particles and Fields in the Magnetosphere, ed. B. McCormac, D. Reidel, Dordrecht, Holland, 1970.

Vasyliunas, V.M., The interrelationship of magnetospheric processes, in Earth's Magnetospheric Processes, ed. B. McCormac, D. Reidel, Dordrecht, Holland, 1972.

Wolf, R.A., Calculations of magnetospheric electric fields, in Magnetospheric Physics, ed. B. McCormac, D. Reidel, Dordrecht, Holland, 1974.

THE ROLE OF FIELD-ALIGNED CURRENT FILAMENTS IN
GENERATING MORNING SECTOR Pi 1 PULSATIONS AT SUB-AURORAL LATITUDES

M. J. Engebretson and S. J. Solberg

Department of Physics, Augsburg College, Minneapolis, MN 55454

L. J. Cahill, Jr.

Department of Physics, University of Minnesota, Minneapolis, MN 55455

R. L. Arnoldy

Space Science Center, University of New Hampshire, Durham, NH 03824

Abstract. Recent studies of digitally sampled two-axis search coil magnetometer data from Siple Station, Antarctica have revealed an unusual class of asymmetric pulsations at the equatorward edge of the morning sector auroral zone. We have reviewed three-axis data taken with this same instrumentation in 1974 to further characterize the vector properties of these one-sided pulsations and to place further constraints on possible source mechanisms. Data will be presented showing that ionospheric and filamentary field-aligned currents, as modeled by Arnoldy et al. [1982] and Engebretson et al. [1983a] are likely to be major contributors to these Pi 1 signatures. If these models are correct, filamentary structures in morning sector Region II Birkeland currents are produced by the ionospheric response to localized impulsive precipitation of energetic electrons from trapped orbits in the magnetospheric plasma sheet (pulsating aurora).

Introduction

Observations of irregular variations of the earth's magnetic field in the ULF frequency range ($1 \leq T \leq 600$ s) have been recorded and analyzed for at least several decades. Modern classification schemes have denoted these irregular variations as Pi 1 (1-40 s period) and Pi 2 ($T \geq 40$ s), although subtypes of these classifications, most notably the impulsive or burstlike Pi(b) and the more continuous Pi(c) categories, appear to be triggered by substantially different mechanisms [Jacobs, 1970; Heacock and Hunsucker, 1981]. Because most earlier observations used analog instrumentation with limited bandwidth to record these pulsations, it was not always straightforward to distinguish these subtypes and classify individual observations properly.

Several recent studies have indicated that the ionosphere may play an important role in the generation of certain Pi pulsations, typically in response to pulsations in precipitation of energetic electrons [Reid and Phillips, 1971; Reid, 1976; Heacock and Hunsucker, 1977; Kokubun et al., 1981; Ward et al., 1982]. Arnoldy et al. [1982], using digitally recorded magnetic field observations along with simultaneous photometer and riometer data from Siple, Antarctica, argued that the small time delays between optical and magnetometer/riometer pulses could only be satisfied by an ionospheric source. The high time resolution of the data also made it possible to notice, as Reid [1976] did earlier, that the Pi 1 pulsations thus generated were nonsinusoidal. The fact that these pulsations resembled transients more than broadband waves led both Arnoldy et al. [1982] and Lanzerotti and Rosenberg [1983] to suggest that their generation involved an ionospheric response to the impulsive and localized precipitation of magnetospheric electrons (the source of pulsating aurora). These studies thus could separate the question of the origin of the quasi-periodic precipitation, presumably due to magnetospheric processes [Davidson, 1979, Chiu et al., 1983] from the task of characterizing the Pi pulsations themselves.

Arnoldy et al. [1982] explained their observations of Pi 1 by including the effects of ionospheric polarization, with resulting Cowling currents. They were able to reproduce the predominantly unidirectional nature of the magnetic pulsations and account for the time delays between peaks in optical and magnetic signatures. Engebretson et al. [1983a] found similar directions and phase relationships using data similar to that of Arnoldy et al. [1982] but including only horizontal magnetic field observations, but also discovered in a few events a rotation of the direction of the pulsations. The presence of a large, unbalanced field-aligned current filament could easily explain the rotations, but Engebretson et al. [1983a] showed that such a filament, if it existed, could not simply be identified with the energetic electron burst responsible for the original ionization and brightening of the pulsating aurora, both because of the 1-s time delay and because of the limited number of current carriers in the energetic electron burst. It was calculated that a field-aligned current filament of order of magnitude 1/10 of the amplitude of the transient ionospheric current would be sufficient to produce the observed rotation of the successive Pi pulses as the source region (auroral patch) passed overhead. They noted that the current carried by the energetic electrons responsible for the original ionization of the auroral patch was at least two orders of magnitude lower still. They instead suggested that transient filaments of upward field-aligned current, presumably carried by thermal or low-energy electrons, could be caused by the ionospheric response to the original energetic electron precipitation. These transient upward current filaments and associated return currents would result in a structuring of the region of upward field-aligned current in the morning sector, such as we have recently observed with magnetometers on board the Dynamics Explorer satellites [Cahill et al., 1982] and MAGSAT [Engebretson et al., 1983b].

This interpretation reverses the temporal sequence proposed by Bösinger et al. [1981] and Pashin et al. [1982], who explained certain substorm-related Pi 1-2 pulsations as a consequence of fluctuations of localized upward-going Birkeland currents, but shares with them the idea of an intimate association between Pi pulsations and localized field-aligned currents (FAC). The data to be presented here, as well as other recent work which will be reviewed, deal with morning sector Pi associated with pulsating aurora. The mechanism we will propose later in this paper is not meant to apply to other local times or substorm phases.

Two recent studies have given strong evidence of an intimate connection between morning sector FAC, ionospheric structure, and the occurrence of Pi 1. Senior et al. [1982], using radar observations of morning sector auroral features correlated with TRIAD satellite magnetometer data, noted that localized small enhancements in Region II Birkeland currents, as inferred from TRIAD data, were related to localized enhancements in E region ionization observed by the radar. Although a one to one correspondence could not be made in most cases, correlation was best whenever the radar field of view and the satellite trajectory coincided. Calculated levels of precipitating energetic electron fluxes again were far below levels necessary to account for the field-aligned current perturbations. Most recently Engebretson et al. [1983b] compared Siple magnetometer, photometer, and riometer data with MAGSAT satellite magnetometer data obtained during three active days in early 1980. During the periods each day when MAGSAT flew over Siple (with local time separation of less than three hours) it was found that asymmetric Pi 1 waves were observed on the ground if and only if the morning sector Region II Birkeland currents extended equatorward at least to Siple's latitude. Fine structure was observed in the Birkeland currents at times when asymmetric pulsations were observed on the ground, although no firm association could be made between individual current structures seen on the satellite and pulsations seen on the ground.

It is our purpose in this paper to present new data which provide additional independent but indirect evidence for the association of Pi 1 with both ionospheric and subsequent Birkeland currents and to compare these data with an extension of the model of Engebretson et al. [1983a]. We will present data from a study of asymmetric pulsation observations obtained at Siple, Antartica (60 degree invariant latitude) in 1974, one of the earliest years of operation of Siple Station. During this year full three-axis search coil magnetometer data were digitally recorded at a rate of 20 vector samples per second. The availability of vertical (Z axis) component data as well as data in the horizontal directions provides an additional test of the mechanisms described above. The 1974 data set is limited in other ways, however, because of the lack of simultaneous photometer or riometer data. Although the lack of such corroborative data prevents us from making positive identification of some asymmetric pulsations, we will show evidence that amplitudes, directions, and phases of the Z component of these pulsations are all consistent with the model to be presented.

Observations

The Siple micropulsation detector consists of three orthogonal permeable core search coils and associated analog and digital electronics. The search coils and preamplifiers are deployed with the Z sensor up along the geomagnetic field line, the Y sensor directed magnetically east perpendicular to the geomagnetic field line and the X sensor directed south along the geomagnetic meridian, completing an orthogonal system. The sensitivity of the search coil, dependent upon frequency as documented in Taylor et al., [1975], is 0.4 pico Tesla per digitization unit for sinusoidal variations at 1 Hz and scales roughly as 0.4 pT * Hz=const from .001 Hz to 10 Hz. The signal from each axis is sampled 20 times per second using a 12-bit analog-to-digital converter, for a Nyquist frequency of 10 Hz.

Data from three days in the austral winter of 1974, June 27, August 5, and September 1, were chosen because of the high level of Pi 1 and Pi 2 pulsation activity evident on summary filter plots (the plot format is described in Taylor et al., 1975). Medium resolution waveform plots were made of all active periods during these days by averaging the full 20 points per second data for various periods as indicated below. We will show a representative sample of this three dimensional magnetic field data set, including typical examples of asymmetric pulsations in two or all three components of the magnetic field. During several periods pulsation activity in all three

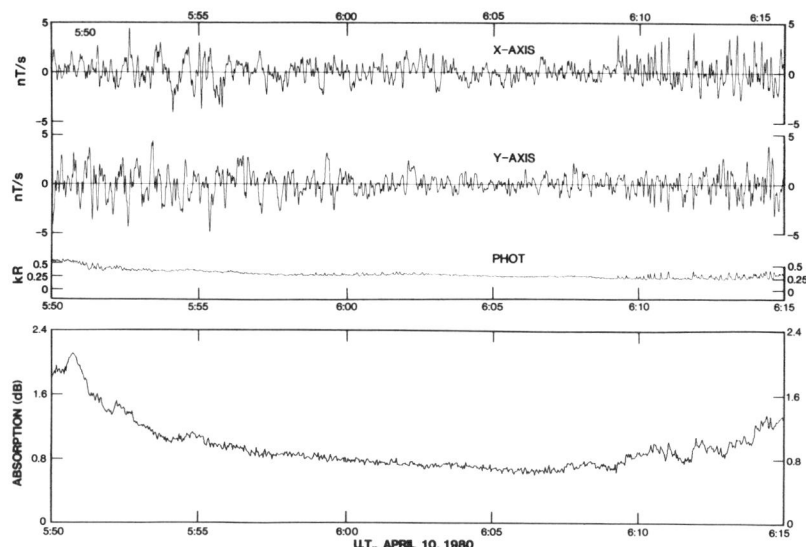

Fig. 1. Correlated search coil magnetometer (dB/dt), photometer, and riometer data from Siple, Antarctica from 0550 to 0615 UT April 10, 1980. In this and subsequent plots positive X points south and positive Y points east.

vector directions reached or exceeded the full scale range of the digitizing electronics, indicating unusually intense activity at these times.

We show first data from April 10, 1980, during the period of MAGSAT-Siple correlations [Engebretson et al., 1983b], in order to display the ground signatures of the observed asymmetric pulsations from several devices. The top two traces of Figure 1 are 0.5 second averages of the dB/dt signal from the X (positive south) and Y (positive east) search coil magnetometer sensors. The third trace is the 4278 Å signal from the Siple photometer (also using 0.5 second averaged data). The bottom trace indicates the level of ionospheric absorption in dB measured by the 30 MHz riometer at 2-s intervals. Pulsations are clearly shown simultaneously in all detectors near the end of the interval shown, beginning near 0610 UT. Sharp increases in optical brightness are followed within approximately 1 second (as seen in expanded plots) by sharp increases in ionospheric absorption and one-sided excursions in the magnetic field traces.

Note that in the middle of the interval shown there are no asymmetric pulses in magnetic field, no pulses in auroral light, and only smaller increases in riometer absorption. When MAGSAT crossed the invariant latitude of Siple at 0602 UT the equatorward edge of the Region II Birkeland current extended to $\Lambda = 60.5$, nearly to the invariant latitude of Siple.

Shown in Figure 2 is a plot of 1-s average values of dB/dt from 0600 to 0625 UT June 27, 1974. Local time at Siple is UT - 5.5 hours, so this event occurred shortly after local midnight.

Asymmetric pulsations are seen to occur in all three components near the peak of activity, from 0608 to 0615. Most pulses are oriented toward the north, with variable large Y components indicating variation from NE through NW, and with consistently upward Z components. Lines in the figure denote a well-correlated set of four successive pulses at 0608:20, 50, 0609:10 and 25. In each case the X component is negative (north) and the Z component is positive (up). In three of the four cases there is a clear and large negative Y (westward) component. Although pulsations in Z are not always correlated one-to-one with pulses in both X and Y during this 25 minute interval, the major pulses in Z occur at the same time as large amplitude pulses in the X and/or Y (horizontal) components.

The gradual increase and decrease in amplitude over a 20 minute interval are suggestive of the motion of an auroral patch through the field of view of the Siple magnetometer. Given typical observed values of the drift speed of 0.1 to 1 km/s [Oguti and Watanabe, 1976], a patch could travel between 120 and 1200 km during this time interval, and could easily move into and out of the magnetometer's range of sensitivity.

Notable in the next figure, Figure 3, is the low amplitude in the Z component observed during a series of pulsations between 0726 and 0740 UT August 5, 1974. Both the background and the peak amplitude of the Z pulsations were typically much lower than that of the X or Y components in most of the events we observed. Pulsations are again often clearly synchronized between all three components. The correlation is especially clear between 0736 and 0740, when nearly every westward

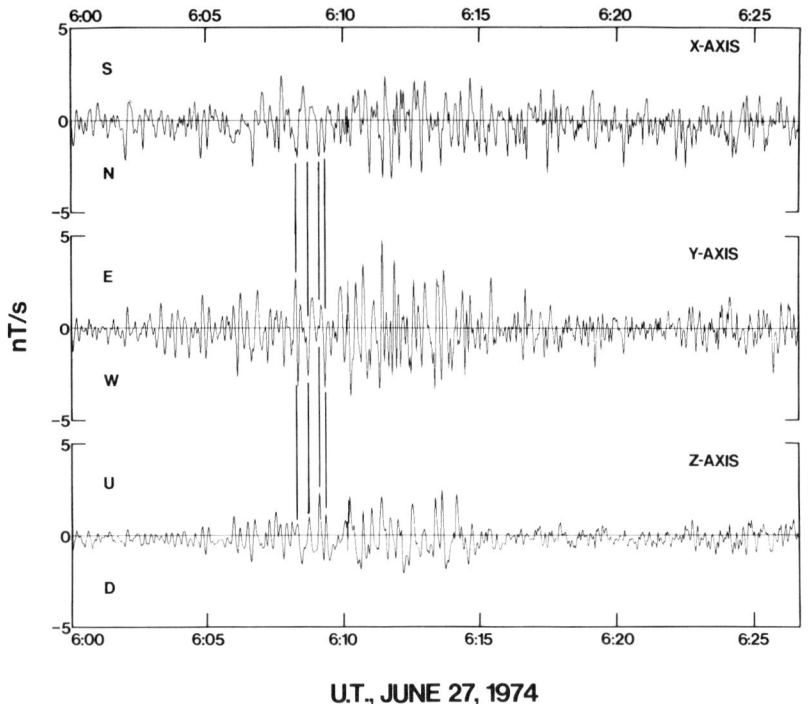

Fig. 2. Three-axis search coil magnetometer data from Siple, Antarctica from 0600 to 0625 UT June 27, 1974. The original 20 vector samples per second data has been averaged to provide 1 point per second as shown in this figure.

(Y) pulse is accompanied by a downward (Z) pulse. Pulses at 0737:05 and 0737:20 are exceptions; here the Z pulse is upward and the strongest horizontal component is toward the north (X > 0). The most straightforward interpretation of these changes in direction is that the two sets of pulses originate in different auroral patches. Two earlier pulses, at 0728 and 0730:30, are also synchronized with similar pulses in the Y component. These may represent distant signals from one or the other of the patches that is drifting into the Siple magnetometer's field of view.

The portion of Figure 3 between 0736:25 and 0737:43 is shown in more detail in Figure 4 using three point averages of the original 20 points per second data. During pulsations 1-4 the vertical

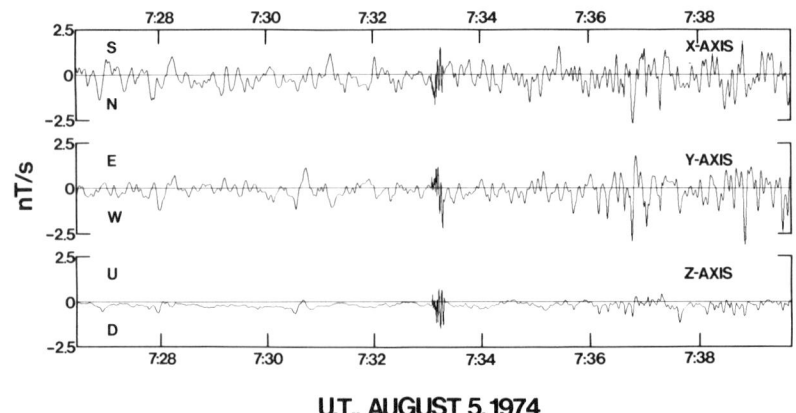

Fig. 3. Three-axis search coil magnetometer data from Siple, Antarctica from 0726 to 0740 UT August 5, 1974. Data shown are 0.5 second averages of the original 20 vector sample per second data. The rapid variation observed in all three components near the center of the interval is a calibration mark.

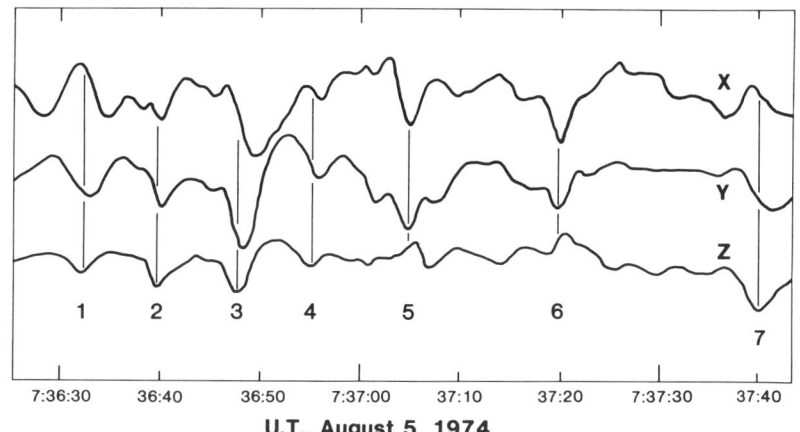

Fig. 4. High resolution plot of dB/dt from 0736:25 to 0737:43 UT August 5, 1974. Data shown are .15 s averages of the original 20 vector sample per second data.

component is downward and clearly leads the horizontal components; Z leads the Y component by 0.5, 0.2, 0.3, and 0.3 s respectively (measured on full resolution plots), and in each case the associated X component peaks even later. A similar but more extreme phase difference is noted in pulsation 7, where the Z component leads the Y component by 1.0 s and the X by an even greater amount. Pulsations 5 and 6, however, are characterized by an upward Z component, and it is clear that the Y and X components here lead the Z component. In pulsation 5 Y leads X by 0.2 s and Z by 0.7 s; in pulsation 6 Y leads X by 0.1 s and Z by 0.8 s. As noted in Figure 5, these pulsations may come from a separate patch of pulsating aurora. It may also be worth noting that these pulsations appear to be part of a more closely spaced set than are the others noted in Figure 4.

Figure 5 shows another period later in the same day during which clear, low amplitude asymmetric pulses appeared. Although both upward and downward pulsations are evident in the Z component, we will emphasize the set of upward pulses in Z observed from 0838 to 0842 U.T. In each case these pulses are accompanied by larger horizontal pulses which shift from southeast to south to southwest. Pulses after 0843 are accompanied by little or no vertical perturbation. Although we have no riometer and photometer data to help resolve spatial movement, it is possible that the rotation of pulsation direction observed here is similar to that shown in Figure 4 of Engebretson et al. [1983a], in which a pulsating auroral patch was assumed to have drifted over Siple from the northwest to the southeast. We will show later that all the components observed here are consistent with the qualitative model developed by Engebretson et al. [1983a].

A tabulation of the 162 visually identified asymmetric pulses occurring during the three days surveyed disclosed that 71% had a negative Z (downward) component, 20% a positive Z (upward) component, and 9% nearly zero Z component amplitude (less than 10% of the horizontal amplitude).

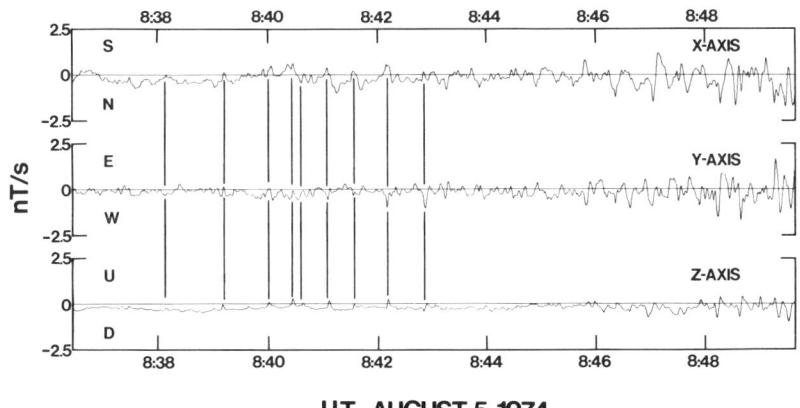

Fig. 5. Three axis search coil magnetometer data from Siple, Antarctica from 0836 to 0850 UT August 5, 1974, as in Figure 3.

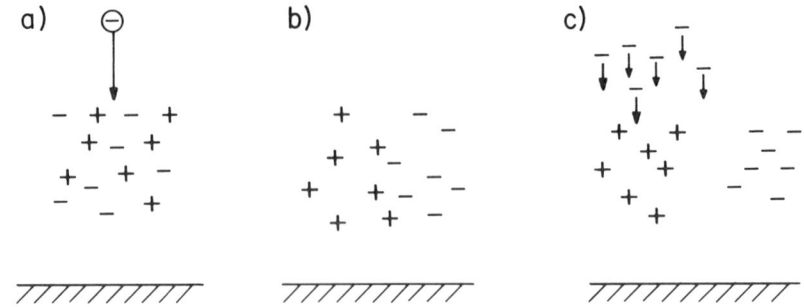

Fig. 6. Model of ionospheric generation of asymmetric Pi 1 pulsations stimulated by patchy precipitation of energetic electrons. (a) Precipitating energetic electrons greatly increase the ionization in a localized region of the ionosphere. (b) The dominant ionospheric electric field separates the electrons from the less mobile ions, which remain approximately in the location of the original precipitation. This separation sets up a polarization electric field and a Cowling current which redirects the electron velocity but maintains the charge separation. (c) The charge imbalance at the foot of the region of original energetic electron precipitation is neutralized as thermal electrons are attracted down the field lines. This forms a filament of field-aligned current with current density much larger than the original flux of energetic electrons.

Surveys of the largest amplitude events in each interval gave quite similar results, 78%, 17%, and 5% respectively. Using the model to be discussed below, this suggests that the majority of the pulses originated in regions poleward (south) of Siple. This conclusion is identical to that reached by Engebretson et al. [1983a] on the basis of the predominance of westward-directed horizontal pulses observed during active days in 1980.

Model

The model for the generation of asymmetric Pi 1 pulsations developed by Arnoldy et al. [1982] and extended by Engebretson et al. [1983a] is diagrammed in Figure 6, but without a consideration of the return current necessary to close the magnetosphere-ionosphere circuit. As such, our model should be understood as a first attempt to describe the relevant phenomena, and not as a self-consistent or quantitative explanation. A similar argument leading to filamentary Birkeland currents associated with spatial variations in ionospheric conductivity was presented by Coroniti and Kennel [1972], but in another context.

Panel (a) shows a patch of precipitating energetic electrons (>= 10 Kev) causing increased ionization in a localized area (d = 10 to 100 km) of the ionospheric E region. Under the influence of the dominant morning sector equatorward ionospheric electric field, the electrons will move as part of the westward electrojet (panel b). The much less mobile ions will be essentially stationary, and as a result the newly created charge distribution will be polarized. Because the surrounding regions of the ionosphere have much lower levels of ionization, with typical nighttime recombination time constants $\approx$ 5-10 s [Siren et al., 1980], the electrojet cannot quickly supply enough electrons to neutralize the positive ion charge left at the site of the original precipitation. This has two consequences: (1) The charge separation establishes a polarization electric field which drives an equatorward Cowling current. This current, in addition to the enhancement in the Hall current (electrojet), provides an ionospheric current source for the observed magnetic perturbations which is approximately constant in direction [Engebretson et al., 1983a; Arnoldy et al., 1982]. (2) Because the Cowling current acts to move the newly dissociated electrons from their origin in the precipitation region, however, it cannot help neutralize the ionic charges. The fact that conductivity along the field line is higher than either Hall or Pedersen conductivity suggests that thermal electrons will be drawn downward along the field lines traversed by the original energetic precipitating electrons (panel c). The filamentary current thus flowing up the field line will be much larger than that carried by the precipitating energetic electrons responsible for the optical pulsations and original ionization. This current will also contribute to the observed magnetic pulsations, but with a direction which depends on the relative position of the observer.

Figure 7 displays in qualitative form the directions and intensities of asymmetric Pi 1 pulsations expected using this model. Panel a of Figure 7 is an adaptation of Figure 16 of Kamide and Rostoker [1977] showing field-aligned and ionospheric currents (electrojets) appropriate for the southern auroral zone. Panel b of Figure 7 shows the direction of increased current flow due to the Hall effect. In this figure a representative set of locations relative to an observer at Siple is shown. Because of the finite times of precipitation of energetic electrons and ionospheric recombination, only short segments of line currents are drawn [Engebretson et al., 1983a].

ENGEBRETSON ET AL. 337

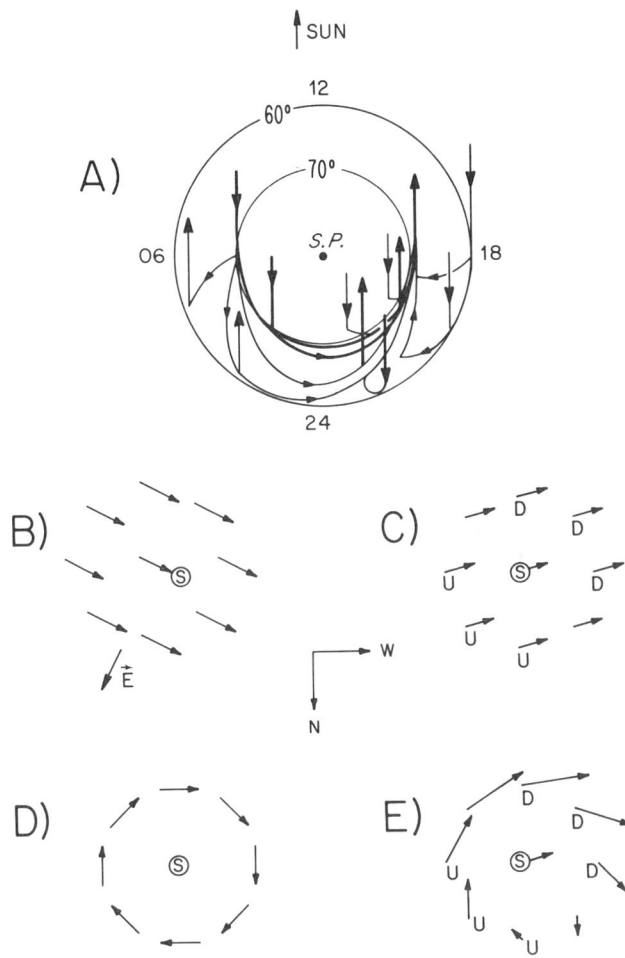

Fig. 7. Model of current systems and magnetic field perturbations associated with pulsating aurorae. (a) Antarctic auroral zone current systems, after Kamide and Rostoker [1977]. The south magnetic pole is in the center of the diagram. Vertical heavy arrows represent the Birkeland (upward and downward) currents. Smaller arrows represent auroral electrojets.
(b) Ionospheric (horizontal) current enhancements due to enhanced precipitation of energetic electrons associated with a pulsating aurora. A current is shown as a heavy line at each of eight directions from Siple, denoted by the circled S.
(c) Magnetic field perturbations at Siple due to the ionospheric current enhancements described in (b). The tail of each delta B vector begins at the respective field-aligned current location relative to Siple. Horizontal directions are indicated by the direction of the arrow, and the direction of the vertical component is indicated by U (up), D (down), or blank (near zero deflection).
(d) Magnetic field perturbation at Siple as in (c), but due to upward field-aligned currents. The driving mechanisms and charge source for these currents is discussed in the text.
(e) Magnetic field perturbations at Siple as in (c), but from the sum of ionospheric and field-aligned current contributions.

Charge polarization effects produce a Cowling current (not shown) oriented along the direction of the electric field. Panel c indicates the change in magnetic field at the observing point caused by this increased resultant horizontal current. The horizontal component of the delta B vector observed at Siple associated with each relative location is plotted at that location. The vertical component, if any, is denoted by a U (up) or D (down) near the tail of each vector.

In panel d we have plotted, in the same format as above, the magnetic perturbations expected at Siple from an upward field-aligned current filament. For example, an upward filament west of Siple would produce a northward perturbation of B at Siple. We therefore plot a northward vector west (to the right) of the circled S. In panel e we have plotted the qualitative vector sum of the two contributions to delta B. It can be seen that the dominant direction of horizontal magnetic perturbation is southwest to west, but that for pulsation patches and associated field-aligned current filaments north of Siple a variety of directions is possible. Because the field-aligned current filaments will not contribute a vertical perturbation, the net vertical variation is predicted to be as in panel c.

Discussion

Three-axis observations of magnetic pulsations can be used in three ways to test the validity of the above model. Vector orientations must be consistent with those predicted in the model, and relative amplitudes and phases of horizontal and vertical components must be consistent with the model and with known physical effects of ground induction. Although the relative amplitude test will be shown to be inconclusive, the other two tests provide additional indirect evidence for the validity of the model proposed here.

An elementary Biot-Savart calculation of the magnetic field generated by the ionospheric part of the current system proposed in the above model shows that the ratio of vertical to horizontal amplitude will equal the ratio of horizontal distance from the current (perpendicular to the current) to the height of the current above the earth. Because the range of a ground magnetometer is typically more than 100 km, it is expected that a substantial portion of pulsations observed at Siple would have ratios of the Z component to horizontal components of > 1.0 if the source were purely ionospheric. A field-aligned current filament, on the other hand, will cause no vertical perturbation regardless of its orientation relative to the observer, as long as the field line can be considered sufficiently vertical. Thus, if pulsations are frequently larger in the Z than in the X or Y components, the pulsation source must be primarily or only an ionospheric current. If pulsations have only small components in the Z direction, field-aligned currents must be significant contributors to the total magnetic perturbation. The average ratio observed for clearly visible pulses during the three days studied was 0.372, and in no case was the vertical component larger than 0.9 times the horizontal.

The presence of ground induction effects, however, can produce a similar reduction in vertical components of the perturbation magnetic field [Chapman and Bartels, 1940; Park, 1974; and Nopper and Hermance, 1974] so that a comparison of vertical amplitude ratios cannot be used to infer the presence of field-aligned filaments. The equivalent skin depth in the earth's crust for 5-20 s period magnetic variations can be estimated to lie between 5 and 100 km, depending on properties of the crust in the region below the observing site. Even with the reasonable assumption of a poorly conducting crust below Siple, this skin depth would be small enough to produce significant diminution of the Z component signal relative to horizontal components. Park [1974] predicts a maximum vertical/horizontal ratio of 0.5 for any orientation and position of an ionospheric current under these conditions. We occasionally observed pulses with vertical/ horizontal amplitude ratio of up to 0.9, which appear to violate the induction models of Park [1974]. Although it is possible that in such cases we are observing the near cancellation of oppositely directed horizontal components from ionospheric and field-aligned currents (Figure 7e), there is enough uncertainty in both data and models that we cannot claim that this test constitutes a clear case for the presence of field-aligned current filaments.

However, both Park [1974] and Nopper and Hermance [1974] also predict a phase lag of over 20 degrees for the vertical component under these conditions, and we have noted considerable variation in phases, including lags of up to this amount (Figure 4). The lags in pulses 5 and 6 of that figure are just what would be expected if earth induction effects were to modify the signal of an ionospheric current. In the other pulses in Figure 4, however, the Z component leads the horizontal. These data are not consistent with a purely ionospheric source, but are consistent with our model of a transient ionospheric current followed quickly by a field-aligned current filament. A delayed magnetic perturbation from a field-aligned current filament will serve to extend the horizontal components, but cannot influence the Z component. Hence the Z component will peak earlier than the horizontal, even though it may have been delayed relative to the original (ionospheric) horizontal components.

An additional argument for the validity of our model and indirectly for the presence of field-aligned currents can be made on the basis of the directions of the three-dimensional perturbations. In nearly every case downward pulses had westward components, and upward pulses had predominantly southward components, directions consistent with the morning sector current geometry used in Figure 7. Pulsations in Figure 2 are exceptions, but the time of observation in this figure is near 0030 LT, during which period the electrojet currents are oriented quite differently at subauroral latitudes.

Summary

Three dimensional high time resolution observations of asymmetric magnetic pulsations in dB/dt appear to support a model of ionospheric generation of Pi 1 pulsations as a response to localized precipitation of energetic electrons. Changes in direction of the pulsations and observed phase shifts suggest that field-aligned current filaments may be set up by the patchy electron precipitation and may in turn contribute to the observed magnetic signature. Further support for this model must await a coordinated study using observations at several sites, ideally in conjunction with observations of field-aligned currents from low-altitude satellites.

Acknowledgements. We thank the Office of Polar Programs of the National Science Foundation for the establishment and support of Siple Station, S. B. Mende and T. J. Rosenberg for permission to use unpublished data from the Siple photometer and riometer, respectively, and W. Baumjohann and the referee of this paper for helpful criticism. This research was supported by NSF grant DPP-81-20957.

References

Arnoldy, R. L., K. Dragoon, L. J. Cahill, Jr., S. B. Mende, and T. J. Rosenberg, Detailed correlations of magnetic field and riometer observations at L=4.2 with pulsating aurora, J. Geophys. Res., 87, 10449, 1982.

Bösinger T., K. Alanko, J. Kangas, H. Opgenoorth, and W. Baumjohann, Correlations between Pi B type magnetic micropulsations, auroras and equivalent current structures during two isolated substorms, J. Atmos. Terr. Phys., 43, 933, 1981.

Cahill, L. J., Jr., M. J. Engebretson, and M. Sugiura, Magnetic fluctuations in and near field-aligned current sheets (abstract), EOS Trans. AGU, 63, 388, 1982.

Chapman, S., and J. Bartels, Geomagnetism, London: Oxford University Press, 1940.

Chiu, Y. T., M. Schulz, J. F. Fennell, and A. M. Kishi, Mirror instability and the origin of morningside auroral structure, J. Geophys. Res, 88, 4041, 1983.

Coroniti, F. V., and C. F. Kennel, Polarization of the auroral electrojet, J. Geophys. Res., 77, 2835, 1972.

Davidson, G. T., Self-modulated VLF wave-electron interactions in the magnetosphere: a cause of auroral pulsations, J. Geophys. Res., 84, 6517, 1979.

Engebretson, M. J., L. J. Cahill, Jr., R. L. Arnoldy, S. B. Mende, and T. J. Rosenberg, Correlated irregular magnetic pulsations and optical emissions observed at Siple Station, Antarctica, J. Geophys. Res., 88, 4841, 1983a.

Engebretson, M. J., L. J. Cahill, Jr., T. A. Potemra, L. J. Zanetti, R. L. Arnoldy, S. B. Mende, and T. J. Rosenberg, The relationship between irregular magnetic pulsations and field-aligned currents, submitted to the Journal of Geophysical Research, 1983b.

Heacock, R. R., and R. D. Hunsucker, A study of concurrent magnetic field and particle precipitation pulsations, 0.005 to 0.5 Hz, recorded near College, Alaska, J. Atmos. Terr. Phys., 39, 487, 1977.

Heacock, R. R., and R. D. Hunsucker, Type Pi 1-2 magnetic field pulsations, Space Sci. Rev., 28, 191, 1981.

Jacobs, J. A., Geomagnetic micropulsations, New York: Springer-Verlag, 1970.

Kamide, Y., and G. Rostoker, The spatial relationship of field-aligned currents and auroral electrojets to the distribution of nightside auroras, J. Geophys. Res., 82, 5589, 1977.

Kokubun, S., K. Hayashi, T. Oguti, K. Tsuruda, S. Machida, T. Kitamura, O. Saka, and T. Watanabe, Correlations between very low frequency chorus bursts and impulsive magnetic variations at L=4.5, Can. J. Phys., 59, 1034, 1981.

Lanzerotti, L. J., and T. J. Rosenberg, Impulsive particle precipitation and concurrent magnetic field changes observed in conjugate areas near L=4, J. Geophys. Res., in press, 1983.

Nopper, R. W. Jr., and J. F. Hermance, Phase relations between polar magnetic substorm fields at the surface of a finitely conducting earth, J. Geophys. Res., 79, 4799, 1974.

Oguti, T. and T. Watanabe, Quasi-periodic poleward propagation of on-off switching aurora and associated geomagnetic pulsations in the dawn, J. Atmos. Terr. Phys., 38, 543, 1976.

Park, D., Magnetic field at the earth's surface produced by a horizontal line current, J. Geophys. Res., 79, 4802, 1974.

Pashin, A. B., K. H. Glassmeier, W. Baumjohann, O. M. Raspopov, A. G. Yahnin, H. J. Opgenoorth, and R. J. Pellinen, Pi 2 magnetic pulsations, auroral break-ups, and the substorm current wedge: a case study, J. Geophys., 51, 223, 1982.

Reid, J. S., An ionospheric origin for Pi 1 micropulsations, Planet. Space Sci., 24, 705, 1976.

Reid, J. S., and J. Phillips, Time lags in the auroral zone ionosphere, Planet. Space Sci., 19, 959, 1971.

Senior, C., R. M. Robinson, and T. A. Potemra, Relationship between field-aligned currents, diffuse auroral precipitation and the westward electrojet in the early morning sector, J. Geophys. Res., 87, 10469, 1982.

Siren, J. C., T. J. Rosenberg, D. Detrick, and L. J. Lanzerotti, Conjugate observations of electron microburst groups by bremsstrahlung x-rays and riometer techniques, J. Geophys. Res., 85, 6760, 1980.

Taylor, W. W. L., B. K. Parady, P. B. Lewis, R. L. Arnoldy, and L. J. Cahill, Jr., Initial results from the search coil magnetometer at Siple, Antarctica, J. Geophys. Res., 80, 4762, 1975.

Ward, I. A., M. Lester, and R. W. Thomas, Pulsing hiss, pulsating aurora, and micropulsations, J. Atmos. Terr. Phys., 44, 938, 1982.

ROTATIONALLY-INDUCED BIRKELAND CURRENT SYSTEMS

T. W. Hill

Space Physics and Astronomy Department, Rice University, Houston, TX 77251

Abstract. Rotational effects which are negligible in Earth's magnetosphere become dominant in the magnetospheres of the outer planets where they give rise to Birkeland current circuits coupling the ionospheric and magnetospheric motions. The centrifugal force of corotation produces an azimuthal ring current in the magnetosphere while the Coriolis force produces a current parallel to the plasma flow. The acceleration current also becomes significant when deviations from strict corotation are appreciable. In general, none of these currents are divergence-free, and closure is provided by ionospheric conduction currents via connecting Birkeland currents. Thus the ionospheric conductivity regulates magnetospheric motions much as it does in the terrestrial case. The effects of such currents have been clearly observed in the magnetospheres of Jupiter and Saturn.

Birkeland current circuits also transmit planetary angular momentum to external sinks such as conducting satellites (e.g., Io), plasma production sites (e.g., the Io torus), and/or the surrounding solar wind (as proposed for Uranus).

Introduction

The global-scale Birkeland current system(s) in the terrestrial magnetosphere can be attributed, either directly or indirectly, to the solar-wind/magnetosphere interaction. By contrast, the principal Birkeland current systems of the giant planets (certainly Jupiter, probably Saturn, and possibly also Uranus and Neptune) are generated internally by the rotation of the planet and the (partial) corotation of its magnetosphere. The importance of rotation vis-a-vis solar wind interaction as a magnetospheric energy source (and hence current generator) was first elucidated for Jupiter's magnetosphere by Brice and Ioannidis [1970] and has been firmly established by many subsequent theoretical and observational studies of the Jovian magnetosphere (see Hill et al. [1983a] for a review and references). The same argument can be applied plausibly, if not conclusively, to the magnetosphere of Saturn and, if they exist, also to the magnetospheres of Uranus and Neptune [see, e.g., Siscoe, 1979].

The basic argument can be summarized as follows. In the terrestrial magnetosphere, bulk plasma motion is dominated by Earth's rotation within the plasmasphere, extending to 5 (±2) Earth radii. Beyond the plasmapause, plasma motion is instead dominated by solar-wind influences; aurora, for example, and all the attendant effects, are entirely attributable to the solar-wind interaction.

At the giant planets, however, the plasmasphere, i.e., the sphere of influence of planetary rotation, essentially fills the magnetosphere, and the magnetospheric dynamics, if any, necessarily involve rotation. This argument is based on scaling laws concerning the strength of the solar-wind/magnetosphere interaction which are plausible if not compelling [e.g., Brice and Ioannidis, 1970; Kennel, 1973; Hill and Michel, 1975; Siscoe, 1979]; the argument has been verified dramatically in the case of Jupiter [e.g., Hill et al., 1983a, and references therein], and verified at least partially in the case of Saturn, where both rotational and solar-wind effects are evident [e.g., Schardt, 1983]; it remains at this time only a plausible conjecture with respect to Uranus and Neptune, where even the existence of a magnetosphere has not been established. The rotational effects are important at the giant planets chiefly because of their faster spin rates and larger magnetic moments compared to terrestrial values (the latter being measured in the case of Jupiter and Saturn and 'anticipated' in the case of Uranus and Neptune); the dilution of the solar wind with increasing heliocentric distance is also weakly involved in the comparison (see Siscoe [1979] for the relevant scaling laws).

The centrifugal force of corotation exceeds that of planetary gravity beyond the synchronous orbit which occurs at 6.6 R_E, 2.3 R_J, 1.9 R_S, 3.4 R_U, and ~4 R_N in the magnetospheres of Earth, Jupiter, Saturn, Uranus, and Neptune, respectively. Typical plasmapause radii, by contrast, are 4-6 R_E, 50-100 R_J, and 15-20 R_S (they are unknown for Uranus and Neptune). Thus, unlike Earth, the giant planetary magnetospheres have a significant 'middle ground,' outside the synchronous orbit but inside the plasmapause, wherein the effects of rotation should dominate those of both planetary gravity and the solar wind (see Siscoe [1979] for a more complete discussion).

Rotational Currents

Rotational effects are manifested in the form of magnetospheric currents which, in general, are coupled to ionospheric conduction currents via connecting Birkeland currents (the latter being generally assumed, as in the terrestrial case, to flow with negligible resistance). The magnetospheric currents are conveniently derived from the MHD equation of motion

$$\rho d\underline{v}/dt = \underline{j} \times \underline{B} - \nabla P + \rho\Omega^2\underline{r} + 2\rho\underline{v} \times \underline{\Omega} \quad (1)$$

where ρ and P are plasma mass density and pressure, $\underline{j}$ is current density, $\underline{B}$ is magnetic field, $\underline{\Omega}$ is the planetary spin vector, $\underline{r}$ is the axial radius vector of a cylindrical coordinate system aligned with the spin axis, and $\underline{v}$ is the plasma bulk velocity relative to the corotating reference frame. If the equation were written in the inertial frame of reference, the centrifugal and Coriolis forces (last two terms) would be incorporated in the acceleration term (left-hand side), which is often neglected altogether in terrestrial studies; one thus derives the pressure-gradient current

$$\underline{j}_\perp = \underline{B} \times \nabla P/B^2 \quad (2)$$

which is the dominant contributor to internal terrestrial magnetospheric currents. (The subscript "$\perp$" means perpendicular to $\underline{B}$.) For a rotation-dominated magnetosphere one can adopt the opposite approximation, neglecting the pressure-gradient term to obtain

$$\underline{j}_\perp = (\rho/B^2)\, \underline{B} \times (d\underline{v}/dt - \Omega^2\underline{r} - 2\underline{v} \times \underline{\Omega}) \quad (3)$$

Although the pressure-gradient term is not everywhere negligible in the Jovian and Saturnian magnetospheres, we shall ignore it here in order to elucidate the novel effects of the rotation-induced currents.

If the pressure gradient force is assumed negligible, it follows [Hill and Michel, 1976; Siscoe, 1977] that the plasma is confined to a thin equatorial disc or 'plasma sheet,' and (3) can be integrated across the sheet to obtain

$$\underline{J}_\perp = \eta\,[\Omega^2 r\hat{\phi} + 2\Omega\underline{v} - \hat{z} \times (d\underline{v}/dt)] \quad (4)$$

where (r,ϕ,z) is a spin-aligned cylindrical coordinate system, and it is assumed that the magnetic moment is aligned with $\underline{\Omega}$ so that $\underline{B} = -B\hat{z}$ in the equatorial plane (the polarity appropriate to Jupiter and Saturn, opposite Earth's). The sheet-current density is defined by

$$\underline{J}_\perp = \int \underline{j}_\perp\, dz \quad (5)$$

and the flux-tube content (plasma mass contained per unit magnetic flux) by

$$\eta = \int \rho dz/B \quad (6)$$

(The integrals in (5) and (6) are across the thickness of the equatorial current sheet.) The flux-tube content η is conserved along convection streamlines in the absence of plasma sources and sinks [e.g., Hill et al., 1981].

A detailed discussion of the cold-plasma approximation (neglect of ∇P) and the related thin-sheet approximation is given by Vasyliunas [1983].

The divergence of the equatorial magnetospheric current (4) must, in a steady state, be compensated by Birkeland currents whose closure in the ionosphere imposes a relationship between $\underline{J}$ and the electric field $\underline{E}$. This relationship, as in studies of terrestrial magnetosphere/ionosphere coupling [e.g., Vasyliunas, 1970; Wolf, 1983], is obtained by applying the current-continuity equation and Ohm's law; the result for a dipole field geometry is

$$\nabla_\perp \cdot \underline{J}_\perp + \frac{1}{r}\,[4\frac{\partial}{\partial r}(\Sigma r E_r) + \frac{\partial}{\partial \phi}(\Sigma E_\phi)] = 0 \quad (7)$$

[Hill et al., 1981], where Σ is the height-integrated Pedersen conductivity of the ionosphere at the location that maps magnetically to the equatorial point (r,ϕ). (Equation (7) involves two approximations: (1) $r \gg R_p$, the planetary radius, and (2) $\underline{E} \cdot \nabla \Sigma_H \ll \nabla \cdot (\Sigma_p \underline{E})$ where Σ_H and Σ_p denote Hall and Pedersen conductivities, respectively.)

If $\underline{v}$ and $\underline{E}$ are related in the magnetosphere by the MHD approximation

$$\underline{E} + \underline{v} \times \underline{B} = 0 \quad (8)$$

and if $\underline{E}$ is further assumed to be derivable from a potential (the steady-state assumption), then a closed set of equations is obtained (4, 7, and 8) from which the convection pattern and associated electric field and current pattern are, in principle, derivable from a given set of appropriate boundary conditions. The resulting differential equations are, however, grotesquely non-linear [Hill et al., 1981], and their solution would be a formidable problem even if the appropriate boundary conditions were known, which they are not. In the meantime, certain general conclusions can be deduced from the structure of the governing equations.

The three terms on the right-hand side of (4) will be referred to respectively as the centrifugal, Coriolis, and acceleration currents, after the forces from which they derive. The first two are obviously of zero order and first order, respectively, in the ratio $v/\Omega r$; the third term is of second order, at least in a steady state where $d\underline{v}/dt = (\underline{v} \cdot \nabla)\,\underline{v}$. It is therefore legitimate (as well as instructive) to consider the three terms separately and consecutively.

The Centrifugal Current

Consider first the centrifugal current

$$\underline{J}_c = \eta\Omega^2 r\hat{\phi} \quad (9)$$

which is strictly azimuthal. Its divergence is simply

$$\nabla \cdot \underline{J}_c = \Omega^2 \, \partial \eta / \partial \phi \quad (10)$$

Any such divergence will, in a steady state, be compensated by a predominantly zonal (azimuthal) Pedersen current in the ionosphere connected by Birkeland currents as illustrated in Figure 1 [Dessler, 1980a]. The topology of this current system is equivalent to that of a 'partial ring current' in the terrestrial magnetosphere [e.g., Boström, 1964].

Note from (10) that the Birkeland current is toward the planet where $\partial \eta / \partial \phi < 0$ and vice-versa; this implies that the magnetospheric $\underline{E} \times \underline{B}$ drift tends to be outward at local maxima of η (as a function of longitude) and inward at local minima. In effect, the more massive flux tubes at a given radius 'fall' outward in the centrifugal force field, being replaced by less massive flux tubes 'lifted' inward against the same force field. The rate of this convection or flux-tube interchange at a given radius is evidently proportional to the local azimuthal variation of flux-tube content (10) and inversely proportional to the height-integrated ionospheric Pedersen conductivity. A quantitative evaluation of this rate depends on details of the convection/interchange geometry, for which no generally accepted model exists.

The greatest uncertainty concerns the scale length of the azimuthal variations associated with convective interchange. On the basis of the magnetic-anomaly model [Dessler and Hill, 1975], it was proposed initially by Vasyliunas [1978] that a global two-cell convection system results from the global-scale longitudinal asymmetry of flux-tube content in the Io plasma torus. Such an asymmetry, fixed in the corotating reference frame, has been inferred from remote optical, radio, and energetic-electron observations and forms the basis of the magnetic-anomaly model, although its effects, and even its existence, remain controversial (see Hill et al. [1983a] for a review and references). The resulting two-cell convection system [Vasyliunas, 1978] presumably would resemble the standard terrestrial convection system except that it would be fixed in the corotating reference frame. The current system of Figure 1 has been discussed in detail by Dessler [1980a] in this context. The equations governing this corotating convection system have been derived independently by Chen [1977] and Hill et al. [1981]; solutions for illustrative but oversimplified cases have been presented by Hill et al. [1981] and by Summers and Siscoe [1982], but a solution for realistic boundary conditions has not yet appeared.

At the other end of the spectrum we have flux-tube interchange models in which the relevant azimuthal variations of flux-tube content are assumed to be of microscopic scale (compared to the radial distance) and to be randomly distributed in longitude and time. Models of this type are generally formulated in terms of a radial diffusion equation describing a convective random walk of flux tubes in response to externally-imposed electromagnetic perturbations [e.g., Fälthammar, 1968]. The small-scale azimuthal gradients and the corresponding small-scale Birkeland current systems of the topology shown in Figure 1 are seldom considered explicitly in such models, and the applicability of a diffusion formalism to such internally-driven interchange motions is open to question. Siscoe and Summers [1981], however, have explicitly treated the small-scale Birkeland current system associated with small-scale interchange motions and have thereby derived an effective diffusion coefficient that depends explicitly on the scale size of the convection eddy cells. The diffusion equation derived by Siscoe and Summers differs from that appropriate to externally-driven interchange diffusion (it is non-linear while the latter is linear), but they find that the two types of diffusion equation have qualitatively similar solutions.

To the extent that the flux-tube interchange process is diffusive in character, the results of Siscoe and Summers [1981] show how the diffusion rate depends on the scale size of the convection cells, which must be specified independently in their analysis. The relevant scale size has not been derived theoretically; Siscoe and Summers obtained a satisfactory fit to Voyager 1 observations of plasma density near the orbit of Io ($r = 6\, R_J$) by assuming a scale size $\Delta r = 0.1\, R_J$ at Io's orbit, increasing linearly with radius. Unfortunately, the Voyager 1 encounter with the Io plasma torus was severely time aliased so that the observed variations can equally well be interpreted in terms of a radial gradient [e.g., Siscoe and Summers, 1981] or a longitudinal gradient [e.g., Hill et al., 1981]. Further observations (or perhaps innovative analysis of existing observations) are required to settle this question.

The internally driven interchange process described above is analogous to the Rayleigh-Taylor gravitational instability with the following modifications: (a) gravity is replaced by the centrifugal force, (b) planar geometry is replaced by dipole geometry, and (c) fluid viscosity is replaced by ionospheric drag (as manifested in Pedersen conductivity). By analogy with the Rayleigh-Taylor instability, one may plausibly infer that the smallest scale perturbations are the most unstable, down to scale sizes comparable to that of the radial gradient that drives the instability. Such a scale size would be consistent with the value $0.1\, R_J$ adopted by Siscoe and Summers. (This argument would, of course, presuppose the absence of any significant large-scale azimuthal asymmetry of the mass-loading rate, i.e., it would preclude 'magnetic-anomaly' effects by assumption.)

In summary, it is fairly certain that outward transport of plasma from the Io torus is accomplished by means of convection cells accompanied by Birkeland current systems of the type shown in Figure 1, and that the ionospheric conductivity thereby regulates the rate of outward transport. (Another factor, perhaps indeed the dominant factor, that regulates the transport rate is the pressure of the less dense but hotter plasma occupying magnetospheric flux tubes that must move inward to replace the more massive flux tubes falling outward from the Io torus under the centrifugal force [Siscoe et al., 1981]; this regulatory effect lies outside the scope of our discussion in that it involves the ∇P term in equation (1).) On the other hand it is not at all certain whether these convection cells are systematic and global in scale (as in the corotating convection model) or stochastic and microscopic in scale (as in the interchange diffusion model), and on this question hinges the choice of the appropriate theoretical formulation.

Similar considerations presumably also apply to the magnetospheres of Saturn and, if they exist, of Uranus and Neptune. In the case of Saturn, theoretical work has been limited in part by the lack of unambiguous information on the nature and location(s) of the internal plasma source(s), although it is clear that such source(s) exists. In the cases of Uranus and Neptune, we have no in situ data and there is some question [Voigt et al., 1983] as to whether these magnetospheres could sustain the internal sources of plasma necessary to support such phenomena.

The centrifugal current (9) converts mechanical to electromagnetic energy at a rate dP/dA per unit equatorial area (independent of frame of reference) given by

$$\left(\frac{dP}{dA}\right)_c = -\underline{J}_c \cdot \underline{E} = -\eta \Omega^2 r E_\phi = \eta B \Omega^2 r v_r$$

$$= -\sigma v_r \partial \Phi_c / \partial r \quad (11)$$

where

$$\sigma \equiv \eta B \quad (12)$$

is the mass density per unit equatorial area and

$$\Phi_c \equiv -\Omega^2 r^2 / 2 \quad (13)$$

is the centrifugal potential energy per unit mass. Thus, outflowing plasma transfers energy from the centrifugal potential field to the stressed magnetic field ($\underline{J}_c \cdot \underline{E} < 0$ when $v_r > 0$), and inflowing plasma does the opposite. The two rates do not exactly cancel; instead, their difference is proportional to the net outward mass transport rate, and represents a net extraction of centrifugal potential energy to power dissipa-

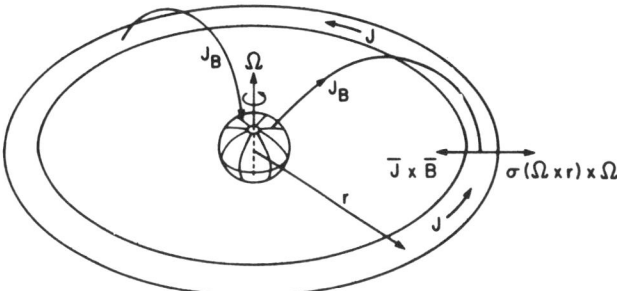

Fig. 1. The centrifugal force is associated with an azimuthal ring current. Any azimuthal gradient of flux-tube plasma content produces a divergence of this ring current which drives a closure current in the ionosphere. [Reproduced from Dessler, 1980a.]

tive processes in the system including, but not necessarily limited to, auroral emissions, atmospheric Joule heating, and radio emissions. This is demonstrated by integrating (11) over the equatorial area contained between the two circles $r = r_1, r_2$:

$$P(r_1 \leqslant r \leqslant r_2) = \dot{M} \Omega^2 (r_2^2 - r_1^2)/2 \quad (14)$$

where

$$\dot{M} \equiv \int_0^{2\pi} \sigma v_r r d\phi \quad (15)$$

is the net rate of outward mass transport, assumed to be independent of r between r_1 and r_2 (i.e., local plasma sources and sinks are neglected). The result (14) is due to Dessler [1980b], who identified outward mass transport as the principal means by which rotational energy is extracted from the Jovian magnetosphere to power various dissipative processes, the most energetic of which is apparently the aurora. A more detailed discussion is given by Eviatar and Siscoe [1980].

The centrifugal potential (13) is merely a representation, in the corotating frame of reference, of the rotational energy (with respect to the inertial frame of reference) that is extracted from the planet to maintain the magnetospheric plasma in a state of (near) corotation. The extraction of planetary rotational energy requires, for its description, consideration of the first-order (Coriolis) current.

The Coriolis Current

Any plasma motion $\underline{v}$ relative to the corotating reference frame gives rise to a Coriolis current given by the second term of (4)

$$\underline{J}_C = 2\eta \underline{\Omega} \underline{v} \quad (16)$$

Equation (16) permits the following simple interpretation (although a second order effect in $(v/\Omega r)$ modifies this interpretation somewhat as will be discussed presently). When $\underline{v}$ has an outward component ($v_r > 0$), the $\underline{J}_C \times \underline{B}$ force has a prograde azimuthal component $[(\underline{J}_C \times \underline{B})_\phi > 0]$, and vice-versa; the magnetospheric $\underline{J} \times \underline{B}$ force tends to spin up any outward-moving plasma and to spin down any inward-moving plasma. The spin-up or spin-down torque must ultimately be provided by the atmosphere through a meridional Birkeland current system having the topology shown in Figure 2. The direction of current flow in Figure 2 is appropriate to a spin-up torque (i.e., outward-moving plasma) for the assumed magnetic polarity (that of Jupiter and Saturn, opposite Earth's; for a spin-down torque (i.e., inward-moving plasma), the polarity is reversed. (Here, as elsewhere in this paper, a reversal of planetary magnetic-field polarity would produce a reversal of all electromagnetic quantities $(\underline{J},\underline{E},\underline{B})$ but not of the mechanical quantities $(\underline{v}, d\underline{v}/dt)$.) The $\underline{J} \times \underline{B}$ forces in Figure 2 have the net effect of transmitting angular momentum from the planet to the magnetospheric plasma, as is appropriate for outward-moving plasma.

It is of interest to note that the Coriolis current is effectively an image Hall current in the following sense. The Coriolis current flows everywhere parallel to $\underline{v}$ (16) and hence perpendicular to $\underline{E}$; it therefore involves no mechanical/electromagnetic energy transfer in the corotating frame of reference. In particular, the Coriolis current does not 'tap' magnetospheric rotational energy to drive aurora and other dissipative processes in the atmosphere, as does the centrifugal current.

The Coriolis current is, however, essential for

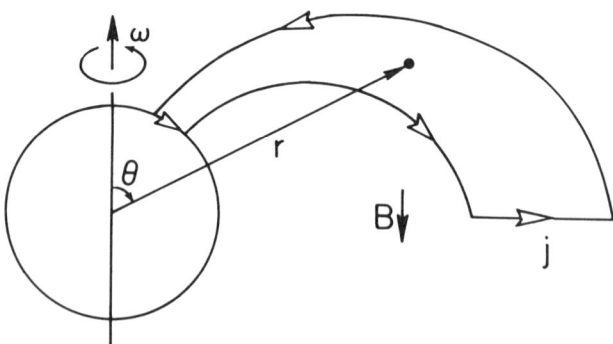

Fig. 2. The meridional current system that exchanges angular momentum between the planetary atmosphere and the magnetospheric plasma. The polarity shown is appropriate to outward transport of plasma (and angular momentum); the reverse polarity applies to inward transport. The same current-system topology occurs whenever angular momentum is extracted from the planet, e.g., by a conducting satellite or by the solar wind. [Reproduced from Hill, 1979.]

maintaining the (near) corotation of the magnetosphere in the first place. The electric fields $\underline{E}'$ in the inertial frame and $\underline{E}$ in the corotating frame are related by

$$\underline{E}' = \underline{E} - \Omega\, r\hat{\phi} \times \underline{B} \qquad (17)$$

and the Coriolis current draws energy from the corotation electric field at the rate

$$\underline{J}_C \cdot \underline{E}' = 2\sigma \Omega^2 r v_r = \sigma \frac{d}{dt}(\Omega^2 r^2) \qquad (18)$$

This is the rate at which planetary rotational energy is deposited in regions of magnetospheric outflow (and extracted from regions of inflow). Exactly one-half of this energy is accounted for by the kinetic energy of rotation of the magnetospheric plasma; the other half resides in the radially stretched magnetic field that is implied by the centrifugal current (cf., equations (11) and (18)), and it is this latter half that is available, at least in principle, to power the aurora and other dissipative phenomena [Dessler, 1980b; Eviatar and Siscoe, 1980].

What is missing from the above energy budget is the kinetic energy of the convection itself; to include this we must consider the second-order acceleration current.

The Acceleration Current

The acceleration of the plasma flow relative to the corotating frame is accomplished by the third term of (4), the acceleration current

$$\underline{J}_A = (\eta/B)\, \underline{B} \times (d\underline{v}/dt) \qquad (19)$$

Recall that this current is derived from the definition

$$\underline{J}_A \times \underline{B} = \sigma\, d\underline{v}/dt \qquad (20)$$

and that it therefore satisfies

$$\underline{J}_A \cdot \underline{E} = \underline{v} \cdot (\underline{J}_A \times \underline{B}) = \sigma \underline{v} \cdot (d\underline{v}/dt) = \sigma \frac{d}{dt}\left(\frac{v^2}{2}\right) \qquad (21)$$

Thus the acceleration current, together with its ionospheric closure currents, forms the circuit that energizes the convection system as viewed in the corotating reference frame.

The total rate at which mechanical energy (kinetic and potential) is converted to electromagnetic energy, as evaluated in the corotating frame of reference, is, from (11) and (21),

$$-\underline{E} \cdot \underline{J} = -\underline{E} \cdot \underline{J}_C - \underline{E} \cdot \underline{J}_A = \frac{\sigma}{2}\frac{d}{dt}(\Omega^2 r^2 - v^2) \qquad (22)$$

Outflowing plasma draws energy from the centrifugal potential field:

$$\frac{\sigma}{2} \frac{d}{dt} (\Omega^2 r^2) = \sigma \Omega^2 r v_r > 0 \quad (23)$$

and some portion of this energy is expended (simultaneously) in acceleration of that plasma flow, the remainder ($-\underline{J} \cdot \underline{E}$) being made available to power dissipative processes elsewhere in the system (e.g., ionospheric Joule heating, auroral phenomena, and the required compensating inflow of plasma elsewhere in the magnetosphere). The internally-generated convection is thus a net source of electromagnetic energy as viewed in the corotating frame of reference.

The transformation to the inertial reference frame (as already discussed) moves the energy source from the magnetosphere to the planetary atmosphere. If we reevaluate (22) in the inertial frame (denoted by primes), we obtain

$$\underline{J} \cdot \underline{E}' = \underline{J} \cdot (\underline{E} - \Omega r \hat{\phi} \times \underline{B}) = \underline{J} \cdot \underline{E} + \underline{\Omega} \cdot \underline{T} \quad (24)$$

where

$$\underline{T} = \hat{z} r J_r B \quad (25)$$

is the $\underline{J} \times \underline{B}$ torque (per unit equatorial area) on the plasma. This torque is balanced by an equal but opposite torque on the atmosphere, whereby planetary spin energy is extracted to maintain, to a first approximation, the corotation of the magnetosphere. The deviations from this first approximation are represented by the first term on the right-hand-side of (24), i.e., by (22).

The inertial observer thus recognizes the true energy source (planetary rotation), although the formulation of the problem in the rotating frame has two advantages: (1) the mathematics is thereby simplified to the extent that corotation is a good approximation, because the acceleration current, which renders the governing equation non-linear, is necessarily non-negligible in the inertial frame but is negligible for sufficiently small r in the corotating frame; and (2) many magnetospheric phenomena, especially those that are remotely observable such as atmospheric heating, auroral emissions, and the related radio emissions, depend on the electric field as measured in the corotating frame of the atmosphere.

The total radial current component is

$$J_r = 2\eta\Omega v_r + \eta (d\underline{v}/dt)_\phi$$

$$= 2\eta\Omega v_r + \eta(dv_\phi/dt) + \eta v_r v_\phi/r$$

$$= (\eta/r) \frac{d}{dt} [r(v_\phi + \Omega r)] \quad (26)$$

and the $\underline{J} \times \underline{B}$ torque (per unit equatorial area) with respect to the spin axis is thus

$$T = r J_r B = \sigma \frac{d}{dt} (rv'_\phi) \quad (27)$$

The right-hand side is the mass density (per unit equatorial area) times the rate of change of angular momentum per unit mass as measured in the inertial reference frame; thus (27) is an expression of the conservation law for angular momentum.

If we combine (27) with the magnetosphere-atmosphere coupling equation (7), and if we arbitrarily set $\partial/\partial\phi = 0$ (see below), we obtain an ordinary first-order differential equation for $v'_\phi(r)$ whose analytic solution is straightforward if slightly cumbersome [Hill, 1979]. The solution has different forms for outflow ($v_r > 0$) and inflow ($v_r < 0$), as illustrated in Figure 3 [Hill et al., 1982]. The figure shows angular momentum per unit mass (rv'_ϕ) in the inertial frame vs. axial distance r/r_0 where r_0 is defined by

$$r_0/R_P = (\pi \Sigma B_P^2 R_P^2 / \dot{M}_0)^{1/4} \quad (28)$$

R_P and B_P are the planetary radius and surface equatorial field strength, respectively, and

$$\dot{M}_0 = \eta_0 \Phi_0 \quad (29)$$

where η_0 is the mean value of η in regions of outflow, and

$$\Phi_0 = \sum_0 \int E_\phi r d\phi \quad (30)$$

is the potential drop summed across all regions of outflow. ($\dot{M}_0$ is thus the total rate of outward mass transport — see Hill et al. [1981].) The inflow solution is shown in Figure 3 for two sample values (0.1 and 1) of the ratio ($\dot{M}_i/\dot{M}_0$) of inflow to outflow mass transport rates — see Hill et al. [1982] for further discussion.

The mathematical device of setting $\partial/\partial\phi = 0$ is not expected to yield a useful 'approximation' in any local sense, but the result may be interpreted as indicating the average behavior of $v'_\phi(r)$ averaged over longitude at a given radius, because the average value of the $\partial/\partial\phi$ terms must vanish in such an average. The outflow solution (Figure 3) is, in fact, capable of fitting adequately the general trend of the available measurements of $v'_\phi(r)$ [Hill, 1980], although this 'fit' should be viewed with some caution in view of the absence of any clear evidence for the expected corresponding inflow signature [Hill et al., 1982].

Other Dynamo Current Systems

The current system of Figure 2 represents a general class of 'Faraday disc dynamo' or 'unipolar inductor' current systems that extract rotational energy from the planetary atmosphere (specifically, from its Pedersen-conducting layer) and deposit that energy (less any dissipative losses due to atmospheric Joule heating, auroral acceleration, etc.) in an external 'load'

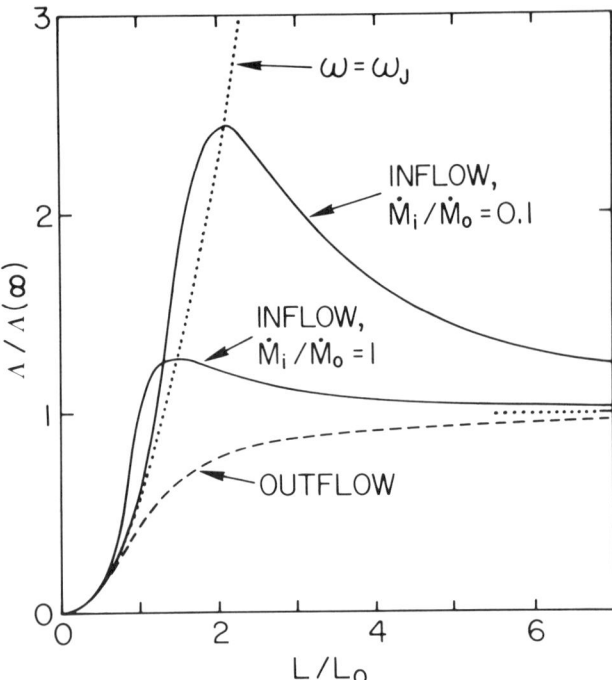

Fig. 3. Normalized angular momentum per unit mass versus normalized axial distance, for plasma transported radially in the magnetosphere. The curves are analytic solutions of the atmosphere-magnetosphere coupling equation with the idealized meridional current system of Figure 2. [Reproduced from Hill et al., 1982.]

consisting of a plasma or other conductor that is not corotating with the atmosphere. In the case considered above the external load is provided by the inertia of the plasma transported outward in the magnetosphere. Other candidates that come to mind for the load of such a disc-dynamo circuit include (1) the local production of plasma from a non-corotating source, (2) a conducting satellite or ring system within the magnetosphere, and (3) the surrounding solar wind. We conclude with a brief discussion of these.

Plasma Production

Plasma is continually produced by photoionization and electron-impact ionization of a neutral gas torus co-orbiting Jupiter with the satellite Io [e.g., Brown et al., 1983, and references therein]. The ions, when produced, are in Kepler orbits with azimuthal speed

$$v_K \approx (GM/r)^{1/2} \qquad (31)$$

where G is the gravitation constant and M is the mass of Jupiter; this is less than the local corotation speed by the ratio

$$v_K/\Omega r \approx (2.3/L)^{3/2} \approx 0.24 \quad (L = 6) \qquad (32)$$

and a radial current is required to accelerate the newly-created plasma to the local flow velocity $\underline{v}$ in the corotating frame (or $\underline{v} + \Omega r \hat{\phi}$ in the inertial frame). This is equivalent to an acceleration current, and its magnitude can be obtained from the momentum equation (1) by adding a mass-source term $S_m \underline{v}$ to the left-hand side and a momentum-source term $\underline{S}_p = S_m(v_K - \Omega r)\hat{\phi}$ to the right-hand side, where S_m is the mass-loading rate (kg/m³-s). The resulting current is

$$\underline{J}_P = S_\eta (\Omega r - v_K)\hat{r} \qquad (33)$$

where

$$S_\eta = \int S_m dz/B \qquad (34)$$

and we have neglected $v/\Omega r$. The same result can be obtained by noting that each new ion undergoes a net radial displacement of one gyroradius $[m(\Omega r - v_K)/qB]$ as it is picked up by the flow; hence the name 'pick-up' current [Goertz, 1980; Hill et al., 1983a]. Combining (33) with the coupling equation (7) under the assumption $\partial/\partial\phi = 0$ yields an expression for v_ϕ, the departure from corotation that is required to drive the closing ionospheric currents [Pontius and Hill, 1982]; the result corresponds to $v_\phi/\Omega r \approx -4\%$, in agreement with the Doppler shift of observed emission lines from torus ions [Brown, 1983].

The Birkeland current system associated with ion production in the torus is probably responsible for certain Jovian decametric radio emissions (those that are not modulated by Io's orbital phase but are otherwise similar to the Io-phase-modulated emissions) — see for example, Goldstein and Goertz [1983]. Similar currents are undoubtedly driven by plasma production in the Saturnian magnetosphere, although the rate and location(s) of such plasma production are less certain [e.g., Lazarus and McNutt, 1983] and no observational evidence of such currents has yet been reported.

The Io Current System

The concept of a unipolar dynamo Birkeland-current circuit being set up by the relative motion of Jupiter and Io was originally introduced [Piddington and Drake, 1968; Goldreich and Lynden-Bell, 1969] to explain the pronounced modulation of Jovian decametric emissions associated with the orbital phase of Io. The topology is that of Figure 2, with Io providing the equatorial connection between upward and downward Birkeland currents. The magnitude of the current in this case is determined simply by the relative motional EMF (~430 kV) and by the effective

Pedersen conductances of Jupiter and of Io (neither of which is very well known).

Like any Birkeland current system, the Io-Jupiter current system is initially established by the propagation of Alfvén waves between Io and the Jovian ionosphere. A steady-state current system is established after a few round-trip Alfvén-wave transit times. The dense Io plasma torus reduces the Alfvén speed to the point that a wave signal apparently cannot travel from Io to Jupiter and back during the time that Io takes to move its own diameter relative to the surrounding plasma; thus a steady-state current system linking Io with Jupiter is apparently not established, but is instead replaced by a series of standing Alfvén waves or 'wings' trailing Io in the corotating plasma wake [Neubauer, 1980]. The magnitude of the Birkeland current is then determined by the Alfvén conductance $(1/\mu_0 v_A)$ of the plasma torus rather than the Pedersen conductance of Io (see Hill et al. [1983a] for further discussion and references), although it presumably has many of the same effects as the earlier steady-state model, including the excitation of the Io-phase-modulated Jovian decametric radio emissions [see Carr et al., 1983, and references therein].

A magnetic signature attributable to such a current system was detected by Voyager 1 at its close encounter with the Io magnetic flux tube [Acuña et al., 1981]. Evidence of qualitatively similar current systems was seen by Voyager 1 near the Saturnian satellite Titan [Ness et al., 1982] and perhaps by Pioneer 11 near the Jovian satellite Ganymede [Kivelson and Winge, 1976]. In these two cases the interaction appears to be more Venus-like than Io-like, i.e., the Birkeland currents tend to be closed by acceleration currents in the surrounding plasma rather than by Pedersen currents in the ionosphere of the mother planet.

It is not known whether similar current systems are excited by any other outer planetary satellites or, for example, by the Saturnian rings.

Solar-Wind Coupling

As a final example, the solar wind may provide the load of the Faraday disc dynamo and thus extract rotational energy from the planet. Such a circuit will be established whenever magnetic interconnection or some other quasi-viscous interaction allows an exchange of momentum between the solar wind and the outer magnetosphere, thus causing the plasma in the outer magnetosphere to deviate from corotation. Such an interaction was proposed for Jupiter's magnetosphere by Gold [1976]; it probably occurs at Jupiter but its effects are evidently small compared to those already discussed in connection with the internal plasma source. A similar conclusion may or may not apply to Saturn, which also contains significant internal plasma

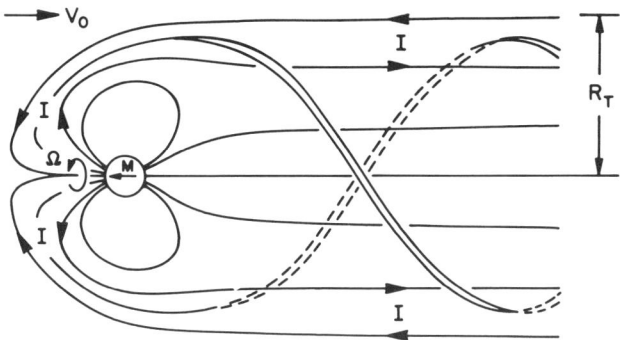

Fig. 4. Sketch of the twisted tail magnetic field and associated Birkeland current system for a 'pole-on' magnetosphere in the presence of solar-wind/magnetosphere coupling. It is anticipated that this geometry may apply to Uranus during the 1986 Voyager encounter. [Reproduced from Hill et al., 1983b.]

sources. The disc dynamo interaction undoubtedly occurs in Earth's polar cap, where its effects are, however, quite negligible compared to those of convection driven directly by the solar wind.

In the case of Uranus, however, both the solar-wind driven convection and the internal plasma source may be relatively ineffective compared to other planets studied thus far [Voigt et al., 1983], and a disc-dynamo mechanism with the solar wind as the load has been proposed [Hill et al., 1983b] to explain the production of a polar Birkeland current system and attendant aurora. (Auroral emissions from Uranus have been detected with the IUE spacecraft — see Caldwell et al. [1983] and references therein.)

The geometry of the proposed current system and the associated helical twist of the tail field is illustrated in Figure 4 [Hill et al., 1983b]. Because of the pole-on configuration that is anticipated during the 1986 Voyager encounter (with the planetary dipole moment roughly aligned with the solar wind), the geometry is unusually simple and allows the development of an analytic model for the current system. The magnitude of the current and of the associated power extracted from the rotation of the atmosphere can be calculated for given (assumed) values of the magnetic moment and the ionospheric Pedersen conductivity.

In conclusion, it may be worth noting that the total spin-down torque, associated with all the above current systems of the Figure 2 type, is insufficient to perceptibly modify any of the outer-planetary spin periods or satellite orbital periods during the age of the solar system. The torque is, however, sufficient to make planetary rotation a significant, and sometimes dominant, power source for magnetospheric phenomena.

Acknowledgments. I am grateful to V. M. Vasyliunas for his incisive comments. This work was supported by the National Science Foundation (Atmospheric Sciences Division, Grant ATM80-19425) and by the National Aeronautics and Space Administration (Planetary Division, Grant NAGW-168).

References

Acuña, M. H., F. M. Neubauer, and N. F. Ness, Standing Alfvén wave current system at Io: Voyager 1 observations, J. Geophys. Res., 86, 8513, 1981.

Boström, R., A model of the auroral electrojets, J. Geophys. Res., 69, 4983, 1964.

Brice, N. M., and G. A. Ioannidis, The magnetospheres of Jupiter and Earth, Icarus, 13, 173, 1970.

Brown, R. A., Observed departure of the Io plasma torus from rigid corotation with Jupiter, Astrophys. J., 268, L47, 1983.

Brown, R. A., C. B. Pilcher, and D. F. Strobel, Spectrophotometric studies of the Io plasma torus, in Physics of the Jovian Magnetosphere (Chapt. 6), A. J. Dessler, ed., Cambridge University Press, 1983.

Caldwell, J., R. Wagener, T. Owen, M. Combes, and Th. Encrenaz, Tentative confirmation of an aurora on Uranus, Nature, 303, 310, 1983.

Carr, T. D., M. D. Desch, and J. K. Alexander, Phenomenology of magnetospheric radio emissions, in Physics of the Jovian Magnetosphere (Chapt. 7), A. J. Dessler, ed., Cambridge University Press, 1983.

Chen, C.-K., Topics in planetary plasmaspheres, Ph.D. thesis, University of California, Los Angeles, 1977.

Dessler, A. J., Corotating Birkeland currents in Jupiter's magnetosphere: an Io plasma-torus source, Planet. Space Sci., 28, 781, 1980a.

Dessler, A. J., Mass-injection rate from Io into the Io plasma torus, Icarus, 44, 291, 1980b.

Dessler, A. J., and T. W. Hill, High-order magnetic multipoles as a source of gross asymmetry in the distant Jovian magnetosphere, Geophys. Res. Lett., 2, 567, 1975.

Eviatar, A., and G. L. Siscoe, Limit on rotational energy available to excite Jovian aurora, Geophys. Res. Lett., 7, 1085, 1980.

Fälthammar, C.-G., Radial diffusion by violation of the third adiabatic invariant, in Earth's Particles and Fields, (p. 157), B. M. McCormac, ed., Reingold, New York, 1968.

Goertz, C. K., Io's interaction with the plasma torus, J. Geophys. Res., 85, 2949, 1980.

Gold, T, The magnetosphere of Jupiter, J. Geophys. Res., 81, 3401, 1976.

Goldreich, P., and D. Lynden-Bell, Io, a Jovian unipolar inductor, Astrophys. J., 156, 59, 1969.

Goldstein, M. L., and C. K. Goertz, Theory of radio emissions and plasma waves, in Physics of the Jovian Magnetosphere (Chapt. 9), A. J. Dessler, ed., Cambridge University Press, 1983.

Hill, T. W., Inertial limit on corotation, J. Geophys. Res., 84, 6554, 1979.

Hill, T. W., Corotation lag in Jupiter's magnetosphere: A comparison of observation and theory, Science, 207, 301, 1980.

Hill, T. W., A. J. Dessler, and L. J. Maher, Corotating magnetospheric convection, J. Geophys. Res., 86, 9020, 1981.

Hill, T. W., A. J. Dessler, and C. K. Goertz, Magnetospheric models, in Physics of the Jovian Magnetosphere (Chapt. 10), A. J. Dessler, ed., Cambridge University Press, 1983a.

Hill, T. W., A. J. Dessler, and M. E. Rassbach, Aurora on Uranus: A Faraday disc dynamo mechanism, Planet. Space Sci., (in press), 1983b.

Hill, T. W., C. K. Goertz, and M. F. Thomsen, Some consequences of corotating magnetospheric convection, J. Geophys. Res., 87, 8311, 1982.

Hill, T. W., and F. C. Michel, Planetary magnetospheres, Rev. Geophys. Space Phys., 13, 967, 1975.

Hill, T. W., and F. C. Michel, Heavy ions from the Galilean satellites and the centrifugal distortion of the Jovian magnetosphere, J. Geophys. Res., 81, 4561, 1976.

Kennel, C. F., Magnetospheres of the planets, Space Sci. Rev., 14, 511, 1973.

Kivelson, M. G., and G. R. Winge, Field-aligned currents in the Jovian magnetosphere: Pioneer 10 and 11, J. Geophys. Res., 81, 5853, 1976.

Lazarus, A. J., and R. L. McNutt, Jr., Low energy plasma ion observations in Saturn's magnetosphere, J. Geophys. Res., (in press), 1983.

Ness, N. F., M. H. Acuna, and K. W. Behannon, The induced magnetosphere of Titan, J. Geophys. Res., 87, 1369, 1982.

Neubauer, F. M., Nonlinear standing Alfvén wave current system at Io: Theory, J. Geophys. Res., 85, 1171, 1980.

Piddington, J. H., and J. F. Drake, Electrodynamic effects of Jupiter's satellite Io, Nature, 217, 935, 1968.

Pontius, D. H., Jr., and T. W. Hill, Departure from corotation of the Io plasma torus: Local plasma production, Geophys. Res. Lett., 9, 1321, 1982.

Schardt, A. W., The magnetosphere of Saturn, Rev. Geophys. Space Phys., 21, 390, 1983.

Siscoe, G. L., On the equatorial confinement and velocity space distribution of satellite ions in Jupiter's magnetosphere, J. Geophys. Res., 82, 1641, 1977.

Siscoe, G. L., Towards a comparative theory of magnetospheres, in Solar System Plasma Physics, C. F. Kennel, L. J. Lanzerotti, and E. N. Parker, eds., North-Holland, Amsterdam, 1979.

Siscoe, G. L., and D. Summers, Centrifugally driven diffusion of Iogenic plasma, J. Geophys. Res., 86, 8471, 1981.

Siscoe, G. L., A. Eviatar, R. M. Thorne, J. D. Richardson, F. Bagenal, and J. D. Sullivan, Ring current impoundment of the Io plasma torus, J. Geophys. Res., 86, 8480, 1981.

Summers, D., and G. L. Siscoe, Solutions to the equations for corotating magnetospheric convection, Astrophys. J., 261, 677, 1982.

Vasyliunas, V. M., Mathematical models of magnetospheric convection and its coupling to the ionosphere, in Particles and Fields in the Magnetosphere, B. M. McCormac, ed., (pp. 60-71), D. Reidel, Dordrecht, Netherlands, 1970.

Vasyliunas, V. M., A mechanism for plasma convection in the inner Jovian magnetosphere, Cospar Program/Abstracts, p. 66, Innsbruck, Austria. 29 May-10 June, 1978.

Vasyliunas, V. M., Plasma distribution and flow, in Physics of the Jovian Magnetosphere (Chapt. 11), A. J. Dessler, ed., Cambridge University Press, 1983.

Voigt, G.-H., T. W. Hill, and A. J. Dessler, The magnetosphere of Uranus: Plasma sources, convection, and field configuration, Astrophys. J., 266, 390, 1983.

Wolf, R. A., The quasi-static (slow-flow) region of the magnetosphere, in Solar-Terrestrial Physics: Principles and Theoretical Foundations, R. L. Carovillano and J. M. Forbes, eds., D. Reidel, Boston, MA, (in press), 1983.

DYNAMICS OF FIELD-ALIGNED CURRENT SOURCES AT EARTH AND JUPITER

D. D. Barbosa

Institute of Geophysics and Planetary Physics,
University of California, Los Angeles, CA 90024

Abstract. This paper examines several mechanisms for generation of field-aligned currents in the magnetospheres of Earth and Jupiter. Implications for planetary radio emission in relation to field-aligned current systems are discussed.

Introduction

Birkeland or field-aligned currents (FAC) had been inferred to exist in the polar ionosphere quite some time ago. Recently, there has been a tremendous effort to synthesize all aspects of high-latitude observations of FAC, auroral displays, electric fields, and neutral winds into a coherent global description of the electrodynamics. The ultimate goal is to understand the nature of the solar wind interaction with the Earth's magnetosphere. This paper will examine theoretical mechanisms for the generation of FAC at both Earth and Jupiter to emphasize basic similarities in the dynamical processes that are involved.

Terrestrial Birkeland Currents

A convenient starting point is the summary in Figure 1 of field-aligned current observations made by the Triad satellite (Iijima and Potemra, 1976). The principal feature of this diagram is the unambiguous separation of the current systems into three components: the circumpolar region 1 current which lies poleward of the neighboring region 2 current, and the cusp current system which occurs in the region spanning 10-14 hours L.T. There is also a complex interleaving of the region 1 and 2 currents from 22-24 L.T. in the vicinity of the Harang discontinuity [see Kamide and Vickrey (1983) for a summary]. The essential properties of these individual current systems were predicted theoretically, but as a whole, the interrelationship and morphology of the systems are only now becoming clear.

The strongest currents are the region 1 system referred to as primary currents in the solar wind interaction with Earth. Early studies of the aurora and high-latitude magnetic variations by Birkeland and Alfven led them to allege that FAC were a causitive element. But the knowledge of a viable electric source was lacking and the proposal of Chapman, that ionospheric electric currents were a closed system with an energy source tied to atmospheric winds, prevailed instead. After the theoretical and experimental confirmation of the solar wind, however, all logical justification for high-latitude electromagnetic activity shifted in favor of the Birkeland-Alfven hypothesis. The presence of the magnetospheric convection system was adduced by Axford and Hines (1961) and its relation to high-latitude geophysical phenomena was elucidated. The link to the solar wind was made through what they invoked as a "viscous-like" but unspecified ion-ion interaction of external solar wind with internal terrestrial plasma. The proposal of a direct access to the terrestrial convection system via a transitional boundary layer was made by Parker (1967, 1969) and Coleman (1970, 1971). These papers established the foundation for a dynamic or dynamo theory of the solar wind interaction based on solar wind particle inertia coupled with a dissipative sink for momentum (the ionosphere) through what has been confirmed experimentally as the region 1 current system.

The Chapman-Ferraro current system was designed to confine and separate the Earth's magnetic field from the excluded solar wind, thus obviating any solar wind-ionosphere interaction. As such, the situation is similar to inviscid fluid flow around an object which produces no drag. The inclusion of a viscous interaction in a thin boundary layer where the velocity shear is significant allows the fluid to impart a net force on the object. The modification and extension of the Chapman-Ferraro current to allow for finite inertia of the solar wind and finite resistivity of the ionosphere results in a similar effect.

Figure 2 represents solar wind plasma impinging on a region of magnetic field lines which have an electrical connection to conducting end plates. The solar wind is assumed to contain no magnetic field. Electrons and ions will be impelled in opposite directions by the Lorentz force and an electric current along the interface will flow. Charge will build up at the edges of where the plasma flow exists ($\nabla \times V \neq 0$) setting up the convection electric field $E = -V \times B/c$. If the end plates were perfect conductors, such a charge separation could not exist and the electric field would be shorted out. The result would be to produce a surface current (ideal Chapman-Ferraro) which excludes the plasma from entry. For finite conductivity of the end plates, a limited dis-

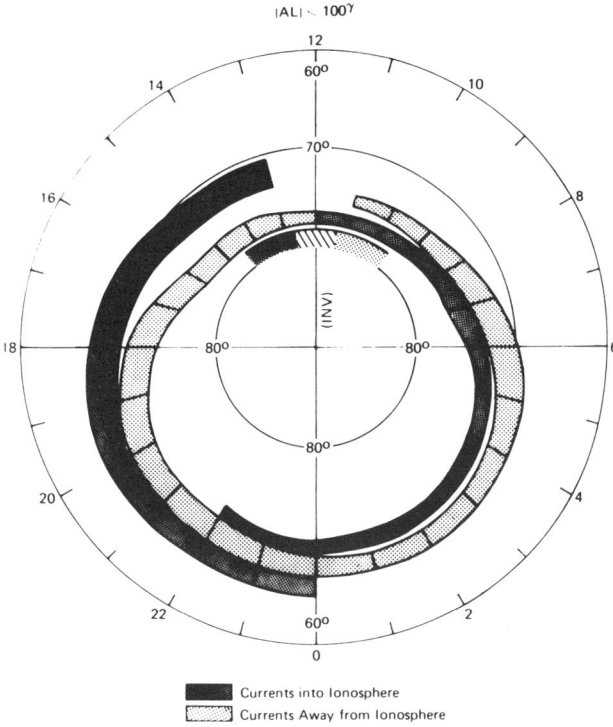

Fig. 1. A summary plot of the major field-aligned current systems (Region 1, Region 2, and the cusp system) observed by the Triad satellite (from Iijima and Potemra, 1976).

charge current flows in the medium and plasma intrudes into the boundary layer. In a steady state the power loss in the end plates is supplied by the solar wind kinetic energy and the actual current flow is sustained in a collisionless plasma by the self-generated inertial drift resulting from the solar wind deceleration

$$\vec{V}_{in} = \frac{mc}{q} \frac{\vec{g}_{in} \times \vec{B}}{B^2} , \qquad (1)$$

where g_{in} is the inertial acceleration resulting from the slowing of the solar wind. The driven flow of $J_{in} = \Sigma n_i q_i V_{in}$ through E produces an electromotive power density $J \cdot E < 0$ to maintain the global magnetospheric convection system.

In Figure 3 we have indicated the boundary layer flow (exaggerated in size) on the front and sides of the magnetosphere. The current continuity is such that the inertial current is supplied by the Chapman-Ferraro (CF) system. The discharge current at the inner edge of the boundary layer enters the ionosphere as region 1 FAC. Note also that these are the only currents that enter the ionosphere as a part of the braking mechanism.

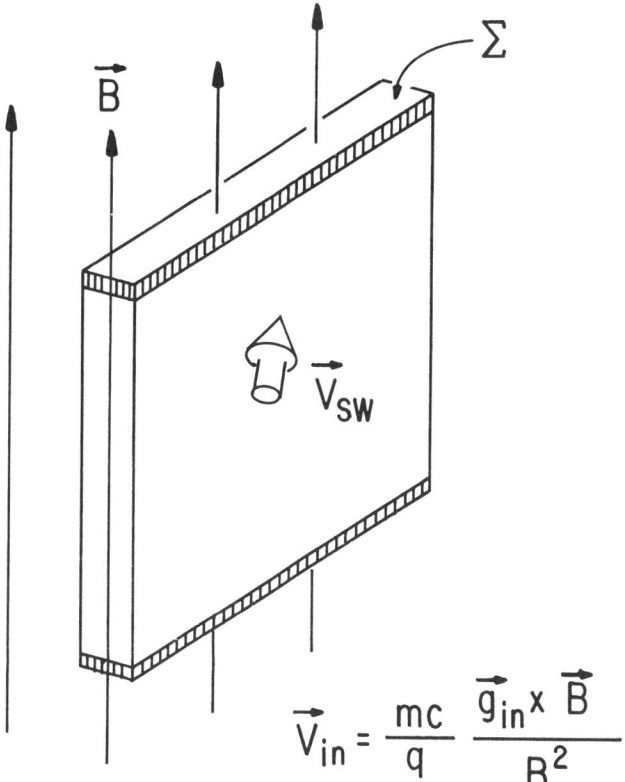

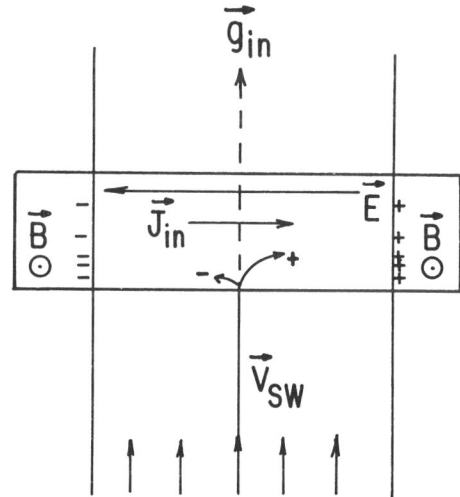

Fig. 2. A sketch illustrating the effect of solar wind plasma breaching the magnetopause. Finite conductivity of the ionospheric "end plates" restricts the flow of field-aligned discharge current allowing a convection electric field $E \simeq -V \times B/c$ to be sustained in the boundary layer by the solar wind bulk flow. The deceleration of the solar wind is accompanied by a mass and charge dependent inertial drift V_{in} of particles which produces the braking electric current such that $J_{in} \cdot E < 0$.

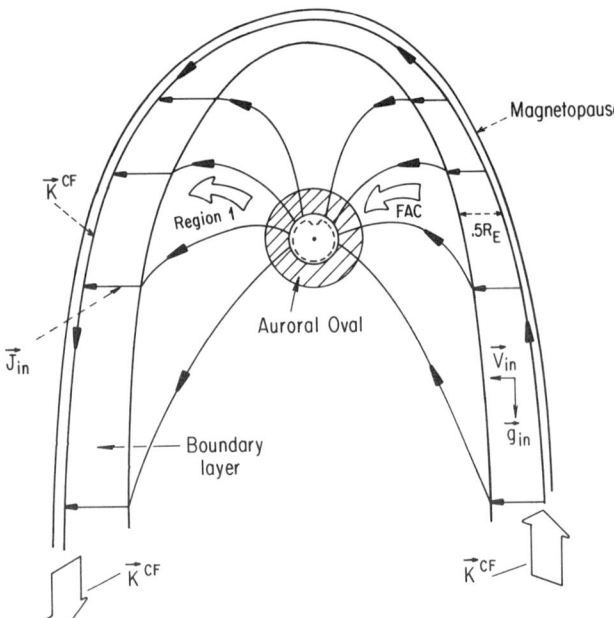

Fig. 3. Overview of the boundary layer dynamo current system. The Chapman-Ferraro magnetopause surface current K^{CF} is the source for the inertial solar wind braking current $J_{in} = \nabla_\perp \cdot K^{CF}$ which is zero only for the case of perfect Chapman-Ferraro shielding. Under actual conditions, the shielding is limited so that the region 1 field-aligned current flows completing the solar wind-magnetosphere-ionosphere circuit. Note that the deficit of solar wind momentum is absorbed by the terrestrial atmosphere through ion-neutral collisions in the ionosphere. Thus, the region 1 currents, rather than just implicit, are fundamental to this model of the solar wind-ionosphere interaction.

That is, it is not necessary to have a return FAC flow to the outer edge of the boundary layer at both dawn and dusk sides, nor must current close through the cusp system as Eastman et al. (1976) have supposed. Current continuity is maintained via the CF system which encircles over the top and bottom of the magnetosphere in the usual way (see Atkinson, 1978) and the inertial braking currents are a shunting of the CF system through the ionosphere via region 1 FAC to accomodate tangential drag. The electrical potential available across the boundary layer is $\Phi = 38$ kV for a boundary layer flow 400 km/s across a 30 nT magnetic field. If the region 1 current is 2 MA, this results in a power supply of 7.6×10^{10} W in each of the layers. The transverse current in the layer is supplied mostly by ions while the FAC from and to the ionosphere are most likely mobile electrons. For a density of $n = 10$ cm^{-3} and a deceleration rate of 1.6 km/s^2 [(200 km/s)^2/(4 R_E)], the current density is 1×10^{-9} A/m^2 and the current per unit length is 3×10^{-2} A/m for a height of 5 R_E. These numbers indicate that the inertial drift (I) is of the correct magnitude to supply the 2×10^6 A of the region 1 current system.

This is a physical description of the dynamic generation of region 1 FAC. If we now add a very small interplanetary magnetic field B^{IMF} to the solar wind, we should not expect a discontinuous change in the nature of the solar wind magnetosphere interaction but a modification of the preceding to allow for the effects of, say, the B_z^{IMF} component. This would result in the sustained generation of additional CF shielding currents to exclude the B_z^{IMF} from penetration. Of course, the CF current increases (decreases) for $B_z^{IMF} < 0$ ($B_z^{IMF} > 0$) away from the base line of $B_z^{IMF} = 0$. Finally, local ohmic dissipation J^2/σ in the boundary layer leading to the heating of electrons and ions can result from anomalous wave-particle interactions (see Gurnett et al., 1979a).

The next relevant current system is the cusp currents which display a pronounced and systematic variation with B_y^{IMF} (Potemra, 1978, 1979). We emphasize two things here: (1) the cusp current is poleward and separate from the region 1 system and (2) it is smaller in amperage than the region 1 system by at least the local time extent since the line current ratio is less than or equal to unity (Iijima and Potemra, 1976). The first point indicates that the source regions (and possibly the generation mechanisms also) are different. We have suggested previously that the region 1 system is related to the low-latitude boundary layer on the dayside and flanks of the magnetosphere (joining continuously to the nightside plasma sheet) and that the cusp current system is associated with the tail magnetopause leeward of the earth. The second point indicates that the cusp current (and therefore the solar wind-magnetotail interaction) produces only a small perturbation on the global convection system.

There is evidence indicating that the field lines emanating from the cusp trace the magnetopause surface of the magnetic tail (see Fairfield, 1977). In Figure 4 we have sketched a possible mapping of the polar field lines into a lobe of the tail. The limit of the polar cap is separate from the region 1 circle to distinguish those field lines that are completely blown back into the tail from those that are closed but stressed in the boundary layer. The mapping is such that the line C1-C2-C3 delimits the circumference of a quadrant of the tail at the magnetopause, and lines of constant magnetic longitude are distorted as shown. Close to the magnetic pole, circles of constant latitude remain quasicircular in the lobe, but are severely distorted at smaller latitudes away from the pole. The pole point is also displaced upwards toward the top of the lobe as flux is squeezed around and underneath the pole point.

This type of configuration allows a close tie between the cusp current system and the east-west component of the interplanetary magnetic field

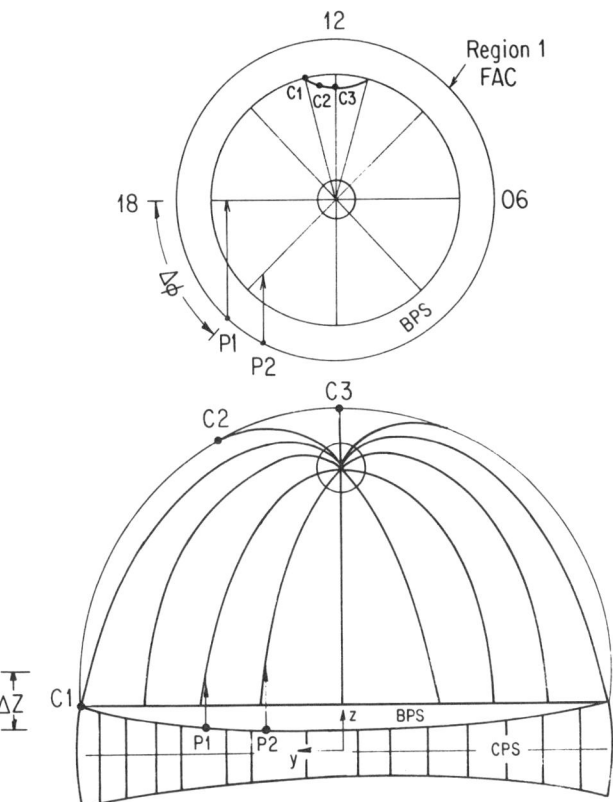

Fig. 4. A suggested mapping of polar cap field lines to the tail. The top figure is a polar view showing contours of constant magnetic longitude extending from the magnetic pole. Also indicated are points C1-C2-C3 in the polar cap associated with the cusp. The bottom figure is a cross-section of the tail viewed from the tail showing the mapped trace of the polar cap contours. The mapping is such that C1-C2-C3 define the periphery of a quadrant of the tail (see also Figure 6) which leads to the asymmetric redistribution of magnetic flux in the tail lobe. In this picture the solar wind interaction with the magnetotail occurs via the tail "mantle" just inside the magnetopause and the consequent effects upon the ionosphere are localized to the cusp region by this mapping. The central plasma sheet (CPS) and boundary plasma sheet (BPS) are also indicated.

B_y^{IMF}. Recently, Cowley (1981) has proposed such a model for the solar wind interaction with the tail that exhibits many of the requisite features. We will follow a different approach [found also by Primdahl and Spangslev (1981)] which emphasizes the generation of shielding currents related to B_y^{IMF}. We have sketched in Figure 5 the cross section of a cylindrical conductor (viewed from the tail) which is immersed in a flow plasma with a magnetic field ($B_y^{IMF} < 0$) draped over the cylinder. This simple picture immediately indicates that a shielding current will be induced in the cylinder close to the circumference with the sense that it is towards (away) from the earth in the Northern (Southern) lobes. The effect will also be to produce a torque about the axis of the cylinder which will bend or twist the tail in the manner shown. If the interplanetary field with magnitude $|B_y^{IMF}| = B_0$ is totally excluded from the tail, the distribution of shielding current is $K_s = cB_0 \sin\theta/4\pi$ along the surface of the cylindrical tail of radius a and the total current per lobe is $I_s = cB_0 a/2\pi$ which may be evaluated as

$$I_s = 500 \text{ kA } (B_0/5nT) (a/10R_E) . \qquad (2)$$

Allowance for some penetration of the IMF into the tail reduces the shielding current and thus (2) represents an upper limit to the cusp current which is compatible with the observations of Iijima and Potemra (1976).

Since the B_y^{IMF} shielding current has the same sense throughout a lobe, current continuity suggests the closure circuit in Figure 6 which shows a diagram of the current path through the magnetosphere. Upon entering the ionosphere, the current must access closed field lines and the neighboring region 1 system. It passes through boundary layer field lines close to the magnetopause and note al-

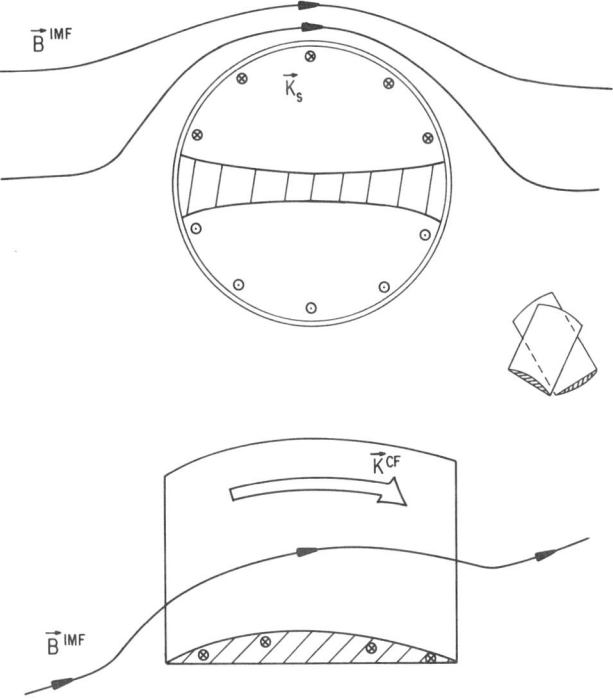

Fig. 5. A cross-section of the tail is illustrated with an external (solar wind) magnetic field draped over it. This type of magnetic configuration requires a surface current flow as shown to exclude the magnetic field from penetration.

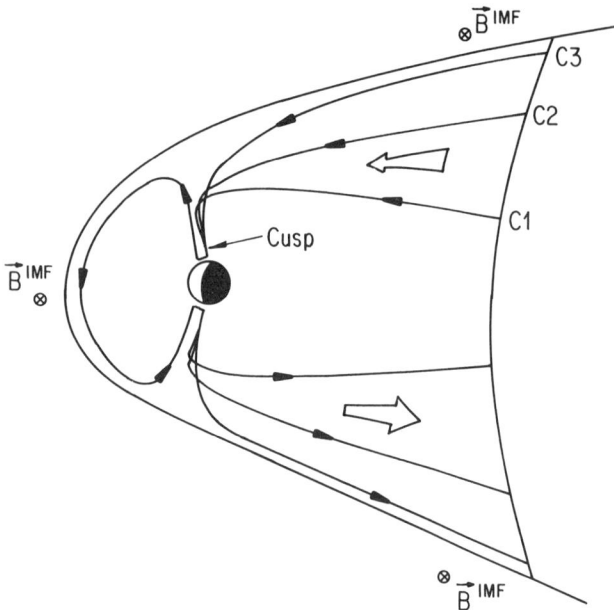

Fig. 6. A perspective view of field-aligned current from the tail magnetopause entering the ionosphere by route of the cusp. The circuit is completed by field-aligned current flow along the dayside magnetopause from hemisphere to hemisphere.

so that the contribution to the B_y^{IMF} shielding system on the dayside is also compatible with the current flow from the Northern hemisphere to the Southern hemisphere. Finally, the current exits from the Southern hemisphere cusp in the complementary manner. Note also that the flow path from the cusp to the region 1 locale in the ionosphere is afforded by a localized north-to-south electric field in the vicinity of local noon and the Hall current produced will result in a westward DPY current system. We will present shortly a theoretical study of the current and electric field patterns associated with the B_y^{IMF} effect, but for now we may report that the organization of all B_y^{IMF} and associated ionospheric effects [including the electric field phenomenology of Heppner (1972)] is reasonably well-understood by the schema of Barbosa (1979b).

Finally, we may remark on an aspect of the field line mapping in Figure 4. We have indicated two arrows directed from points P1 and P2 from the plasma sheet, over the boundary, and into the lobe. The configuration of the field lines is such that proceeding in the z direction in the tail maps to a direction essentially in longitude in the ionosphere $\Delta z \rightarrow \Delta\phi$. Also, the differential in longitude increases with increasing distance y from the axis of the tail. Although the topic of substorms is not dealt with here, it should be noted that the mapping suggests the possibility of convoluted or folded features in auroral arc structures during substorm activity. For instance, the so-called westward travelling surge [see Figure 9 of Burke (1982)] has often been attributed to an MHD instability close to the ionosphere. Alternatively, the longitudinal feature may simply be a signature of the mapping indicated in Figure 4. If field lines progressively north of the point P1 are dynamically becoming closed and laden with plasma, the fold in the ionospheric surge may be associated with this geometry. The recovery of the surge would indicate that the field lines around substorm recovery are not as severely bent suggesting that the substorm process involves a relaxation of the magnetic field configuration to a state which is not as stressed similar to the phenomenon of solar flares (e.g., Barbosa, 1979a).

Jovian Birkeland Currents

Field-aligned currents at Jupiter have been inferred from radio observations of the decametric emission (DAM) [see Carr et al., 1983]. The Io-controlled DAM has long been assumed to be associated with FAC resulting from the electrodynamic interaction of the satellite with the Jovian ionosphere. This effect, as well as all of the generation processes discussed here for Jupiter, results from the differential motion of the conducting elements (e.g., the moon's ionosphere) relative to that of the corotating atmosphere of Jupiter.

The simplest demonstration of FAC in a corotation-dominated magnetosphere deals with the source of plasma for the Io plasma torus. Ionization of neutral atoms near the orbit of Io will lead to acceleration of the ions to the corotation velocity $V_{CR} \simeq 75$ km/s there. This "pickup" process imparts a gyrospeed that is comparable to the corotation speed since the new ion is essentially at rest prior to pickup. From the equations of motion it may be shown that the acceleration results in a radial displacement outwards of the ion such that the guiding center will be a Larmor radius farther out than the initial radial position of the ion. The ionization electron will be picked up at essentially the same radial distance. Thus, an ion electrical current outwards is produced and in order to maintain charge neutrality, a field-aligned current (electron carriers) closes through the Jovian ionosphere (Goertz, 1980). The current flow particle-by-particle can be followed and understood clearly at each step and the route through the ionosphere has the effect of $J \cdot E < 0$ so that all power is derived from the rotation of the planet, as expected.

There is a similarity here to the discussion in the previous section regarding the braking of the solar wind. The difference is that the pickup ions are being accelerated and represent a load to the planetary dynamo. The fact that this is the same effect (with the sign reversed) arising from finite particle inertia can be seen by applying (1) to the problem. Immediately after ionization, the new ion will experience upon acceleration to corotation an inertial force directed opposite to that of the corotation direction so that V_{in} is directed radially outwards. The magnitude of the drift is $V_{in} = g_{in}/\Omega_{ci} \simeq v_\perp^2/r_i\Omega_{ci} \simeq v_\perp$ to displace the ion a Larmor radius outwards during the first quarter period. Thus, the inertial aspect of the pickup process is evident, and much of the physics regarding FAC generation applied to the solar wind earlier is operating here in reverse. There is also a hint here that at Jupiter variations in azimuthal velocity of ions are intimately related to variations in radial velocity. The next FAC system considered concerns the transport of Iogenic plasma to the outermost region of the magnetosphere. We write the momentum equation in the corotating frame of Jupiter as

$$\rho(\frac{\partial}{\partial t} + \vec{v} \cdot \nabla) \vec{v} = -\nabla p + \frac{1}{c} \vec{J} \times \vec{B}$$

$$-2\rho(\vec{\omega} \times \vec{v}) - \rho\vec{\omega} \times (\vec{\omega} \times \vec{r}) \quad . \quad (3)$$

Here ρ is the mass density, p is the plasma kinetic pressure, $\vec{\omega} = \Omega_J \hat{z}$ is the angular velocity vector of Jupiter, and v = 0 means rigid corotation of plasma. A very simple model for the Lorentz force that elucidates many features of plasma transport is given by

$$\frac{1}{c} \vec{J} \times \vec{B} = -\alpha \vec{v} \quad (4)$$

representing the effect of resistive drag by the ionosphere. For a dipole field, it may be shown that (e.g., Nishida and Watanabe, 1981)

$$\alpha = \frac{2\Sigma B_J^2 \cos\theta}{c^2 \ell L^6} \quad (5)$$

where Σ is the height-integrated ionospheric Pedersen conductivity, B_J is the equatorial surface magnetic field, θ is the polar angle of the field line with McIlwain parameter L, and ℓ is the half-thickness of the plasma sheet at the magnetic equator. A simple solution of (3) is obtained for small v by balancing $-\alpha v$ with only the Coriolis force for the azimuthal component (v_ϕ) and only the centrifugal force for the radial component (v_r). The result is

$$v_r \simeq R_c \Omega_J r/2 \quad (6)$$

and

$$\tan \psi = -v_\phi/v_r \simeq R_c \quad (7)$$

where R_c is a small, dimensionless parameter representing the ratio of the Coriolis force on the plasma to the resistive drag force

$$R_c = \frac{\rho\Omega_J c^2 \ell L^6}{\Sigma B_J^2 \cos\theta} \quad . \quad (8)$$

Thus, for a radial dependence of ρ which is slower than L^{-6}, the plasma will be transported out in a spiral pattern with a spiral angle that increases with radial distance. The actual structure of the plasma is believed to be in the form of blobs or "eddies" (Siscoe and Summers, 1981). This is the consequence of a centrifugal Rayleigh-Taylor instability operating at the outer edge of the Io plasma torus to produce convective cells of plasma which exude out and migrate along the spiral streamlines. This process is powered by the rotation of the planet and FAC will be present at the edges of the blobs throughout the inner and middle magnetosphere. By these same considerations, there should also exist a FAC system at the inner edge of the warm Io plasma torus due to both ion pickup and radial transport outwards (e.g., Gurnett, et al., 1979b).

The power available to magnetospheric plasma is communicated by FAC which brakes the rotational motion of the planet. The power is limited, however, by local ohmic losses in the ionosphere. That is, if one computes the mechanical power, the electrical power density, and the ohmic losses, the general result for a rotating MHD generator is derived as

$$P_M = - \vec{T} \cdot \vec{\omega} = \sigma E_{CR}^2 x \quad , \quad (9)$$

$$P_E = - \vec{J} \cdot \vec{E} = \sigma E_{CR}^2 x(1 - x) \quad , \quad (10)$$

and

$$P_O = J^2/\sigma = \sigma E_{CR}^2 x^2 \quad . \quad (11)$$

Here, $x = (\Omega_J - \Omega)/\Omega_J$ is the fractional departure of magnetospheric plasma from rigid corotation, σ is the ionospheric Pedersen conductivity, and $E_{CR} = 2\Omega_J R_J B_J \sin\theta \cos\theta/c$. Energy balance is provided by

$$P_M = P_E + P_O \quad (12)$$

Thus, the electrical power available to magnetospheric plasma P_E is limited by ohmic dissipation in the ionosphere [see Hill et al. (1983) for an application of the above to the planet Uranus].

The above relations indicate that 1) for $x < 1/2$ small dissipative losses occur and a good efficiency P_E/P_M is obtained, 2) for $x \simeq 1$, little power is being invested in magnetospheric plasma and almost all of the rotational kinetic energy is being expended in heating the ionosphere, and 3) for $x = 1/2$, maximal power out is reached with $P_E = P_O = \sigma E_{CR}^2/4$. This implies that the middle magnetosphere of Jupiter where a considerable breakdown of corotation occurs ($x \simeq 1/2$) is a prime location for intense auroral activity and magnetospheric energy deposition at a rate ~ 6×10^{13} W (Barbosa et al., 1981; Barbosa, 1981; Dessler, 1980; Eviatar and Siscoe, 1980).

Planetary Radio Emissions

Several studies have attempted to establish a correlation between auroral kilometric radiation (AKR) and auroral particle phenomena like FAC (Voots et al., 1877; Green et al., 1979, 1982; Saflekos et al., 1983). Part of the difficulty is related to the aspect that field-aligned currents do not directly generate radio noise, merely being an energy link for the auroral chain of processes that include a small component (1%) of electromagnetic radio emission. The strongest terrestrial AKR is associated with substorms and the nighttime sector 20-24 hours L.T. Although much is known about the phenomenology of both AKR and FAC during substorms, there is still a large gap in our understanding of the physical processes involved. It is hoped that future in-situ studies of the auroral region by satellites such as the Dynamics Explorer will be able to provide needed information in this area.

At Jupiter, the principal new findings regarding DAM have been obtained by the Voyager spacecraft. The most interesting aspect is the reported local time dependence of the Io-independent component (Alexander, et al., 1981). All theoretical interpretations of DAM related to FAC have tacitly assumed that the noise is emanating from the Io flux tube or the Io L shell. Yet, the aspect of a local time dependence seems difficult to reconcile with a region that is so interior to the magnetosphere. If FAC are involved in the generation of the noise, what process can sensibly produce this type of effect in a corotation-dominated region? Also relevant is the fact that the Io-dependent emissions do not exhibit local time effects (Carr et al., 1983).

It is conceivable that the assumption of covicinity of the emission L shell is not valid. As far as we know, there has never been any evidence advanced that the two DAM emissions are spatially related. For lack of knowledge of any other regulating factors or other FAC systems, the two components have been grouped together in identity. However, we have just now described several other FAC systems associated with Iogenic plasma transport and for one of them there is good theoretical justification for a FAC dissipative output of ~ 6×10^{13} W. Since the strong breakdown of corotation occurs outside of 18 R_J, this region is more susceptible to the influence of the solar wind and dawn-dusk asymmetries of the magnetosphere's configuration. Also, any variability in solar wind parameters is more likely to influence a FAC system in the middle magnetosphere than one at Io's orbit. Correlation studies like that of Terasawa et al. (1978), if validated, give support to the alternative picture that the source of Io-independent DAM is located on higher latitude L shells. The separation in magnetic latitude from L = 6 to L = 20 is not very large and it is still possible to retain the known beaming properties of the emission. However, the separation is large in the sense that the physical regions for FAC generation are completely different.

There is accumulating evidence for the existence of FAC in the region exterior to the Io plasma torus. It is possible that our later understanding of the FAC systems at Jupiter and the DAM radio emissions will progress to the point that the word Io can be eliminated from the reference to the emission known as the Io-independent DAM.

Acknowledgements. Support for this work was provided by the National Aeronautics and Space Administration and by the National Science Foundation.

References

Alexander, J.K., T.D. Carr, J.R. Thieman, J.J. Schauble, and A.C. Riddle, Synoptic observations of Jupiter's radio emissions: Average statistical properties observed by Voyager, J. Geophys. Res., 86, 8529, 1981.

Atkinson, G., Energy flow and closure of current systems in the magnetosphere, J. Geophys. Res., 83, 1039, 1978.

Axford, W.I. and C.O. Hines, A unifying theory of high-latitude geophysical phenomena and geomagnetic storms, Can. J. Phys., 39, 1433, 1961.

Barbosa, D.D., Dynamo action in a thin slab, Astrophys. J., 228, 909, 1979a.

Barbosa, D.D., High-latitude field-aligned current sources and induced electric fields, J. Geophys. Res., 84, 5175, 1979b.

Barbosa, D.D., On the injection and scattering of protons in Jupiter's magnetosphere, J. Geophys. Res., 86, 8981, 1981.

Barbosa, D.D., F.L. Scarf, W.S. Kurth, and D.A. Gurnett, Broadband electrostatic noise and field-aligned currents in Jupiter's middle magnetosphere, J. Geophys. Res., 86, 8357, 1981.

Burke, W.J., Magnetosphere-ionosphere coupling: Contributions from IMS satellite observations, Rev. Geophys. Space Phys., 20, 685, 1982.

Carr, T.D., M.D. Desch, and J.K. Alexander, Phenomenology of magnetospheric radio emissions, in Physics of the Jovian Magnetosphere, ed. by A.J. Dessler, Cambridge University Press, New York, 1983.

Coleman, P.J. Jr., Tangential drag on the magnetospheric cavity, Cosmic Electrodynamics, 1, 145, 1970.

Coleman, P.J. Jr., A model of the geomagnetic cavity, Rad. Sci., 6, 321, 1971.

Cowley, S.W., Magnetospheric asymmetries associated with the Y-component of the IMF, Planet. Space Sci., 29, 79, 1981.

Dessler, A.J., Mass-injection rate from Io into the Io plasma torus, Icarus, 44, 291, 1980.

Eastman, T.E., E.W. Hones, S.J. Bame, and J.R. Asbridge, The magnetospheric boundary layer: Site of plasma, momentum, and energy transfer from the magnetosheath into the magnetosphere, Geophys. Res. Lett. 3, 685, 1976.

Eviatar, A. and G.L. Siscoe, Limit on rotational energy available to excite the Jovian aurora, Geophys. Res. Lett., 7, 1085, 1980.

Fairfield, D.H., Electric and magnetic fields in the high-latitude magnetosphere, Rev. Geophys. Space Phys., 15, 285, 1977.

Goertz, C.K., Io's interaction with the plasma torus, J. Geophys. Res., 85, 1949, 1980.

Green, J.L., D.A. Gurnett, and R.A. Hoffman, A correlation between auroral kilometric radiation and inverted-V electron precipitation, J. Geophys. Res., 84, 5216, 1979.

Green, J.L., N.A. Saflekos, D.A. Gurnett, and T.A. Potemra, A correlation between auroral kilometric radiation and field-aligned currents, J. Geophys. Res., 87, 10463, 1982.

Gurnett, D.A., R.R. Anderson, B.T. Tsurutani, E.J. Smith, G. Paschmann, G. Haerendel, S.J. Bame, and C.T. Russell, Plasma wave turbulence at the magnetopause: Observations from ISEE 1 and 2, J. Geophys. Res., 84, 7043, 1979a.

Gurnett, D.A., W.S. Kurth, and F.L. Scarf, Auroral hiss observed near the Io plasma torus, Nature, 280, 767, 1979b.

Heppner, J.P., Electric fields in the magnetosphere, in Critical Problems of Magnetospheric Physics, ed. by E.R. Dyer, p. 107, National Academy of Science, Washington, D.C. 1972.

Hill, T.W., A.J. Dessler, and M.E. Rassbach, Aurora on Uranus: A Faraday disc dynamo mechanism, preprint, 1983.

Iijima, T. and T.A. Potemra, Field-aligned currents in the dayside cusp, J. Geophys. Res., 81, 5971, 1976.

Kamide, Y. and J.F. Vickrey, Variability of the Harang discontinuity as observed by the Chatanika radar and the IMS Alaska magnetometer chain, Geophys. Res. Lett., 10, 159, 1983.

Nishida, A. and Y. Watanabe, Joule heating of the Jovian ionosphere by corotation enforcement currents, J. Geophys. Res., 86, 9945, 1981.

Parker, E.N., Small-scale equilibrium of the magnetopause and its consequences, J. Geophys. Res. 72, 4365, 1967.

Parker, E.N., Solar wind interaction with the geomagnetic field, Rev. Geophys. Space Phys., 7, 3, 1969.

Potemra, T.A., Observations of Birkeland currents with the Triad satellite, Astrophys. Space Sci., 58, 207, 1978.

Potemra, T.A., Current systems in the Earth's magnetosphere, Rev. Geophys. Space Phys., 17, 640, 1979.

Primdahl, F. and F. Spangslev, Cusp region and auroral zone field aligned currents, Ann. Geophys., 37, 529, 1981.

Saflekos, N.A., R.E. Sheehan, and R.L. Carovillano, Global nature of field-aligned currents and their relation to auroral phenomena, Rev. Geophys. Space Phys., 20, 709, 1983.

Siscoe, G.L. and D. Summers, Centrifugally driven diffusion of Iogenic plasma, J. Geophys. Res., 86, 8471, 1981.

Terasawa, T., K. Maezawa, and S. Machida, Solar wind effect on Jupiter's non Io-related radio emission, Nature, 273, 131, 1978.

Voots, G.R., D.A. Gurnett, and S.-I. Akasofu, Auroral kilometric radiation as an indicator of auroral magnetic disturbances, J. Geophys. Res., 82, 2259, 1977.